ISBN 978-0-266-87531-4
PIBN 10902898

UNIVERSITY OF ILLINOIS BULLETIN

ISSUED WEEKLY

Vol. XIV JUNE 18, 1917 No. 42

[Entered as second-class matter Dec. 11, 1912, at the Post Office at Urbana, Ill., under the Act of Aug. 24, 1912.]

PERCENTAGE OF EXTRACTION OF BITUMINOUS COAL WITH SPECIAL REFERENCE TO ILLINOIS CONDITIONS

BY

C. M. YOUNG

ILLINOIS COAL MINING INVESTIGATIONS

COÖPERATIVE AGREEMENT

(This Report was prepared under a Coöperative Agreement between the Engineering Experiment Station of the University of Illinois, the Illinois State Geological Survey, and the U. S. Bureau of Mines)

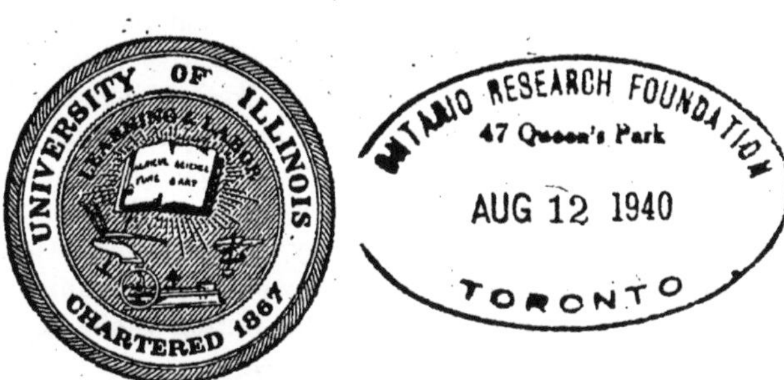

BULLETIN No. 100

ENGINEERING EXPERIMENT STATION

PUBLISHED BY THE UNIVERSITY OF ILLINOIS, URBANA

EUROPEAN AGENT
CHAPMAN & HALL, LTD., LONDON

THE Engineering Experiment Station was established by act of the Board of Trustees, December 8, 1903. It is the purpose of the Station to carry on investigations along various lines of engineering and to study problems of importance to professional engineers and to the manufacturing, railway, mining, constructional, and industrial interests of the State.

The control of the Engineering Experiment Station is vested in the heads of the several departments of the College of Engineering. These constitute the Station Staff and, with the Director, determine the character of the investigations to be undertaken. The work is carried on under the supervision of the Staff, sometimes by research fellows as graduate work, sometimes by members of the instructional staff of the College of Engineering, but more frequently by investigators belonging to the Station corps.

The volume and number at the top of the title page of the cover are merely arbitrary numbers and refer to the general publications of the University of Illinois; *either above the title or below the seal is given the number of the Engineering Experiment Station bulletin or circular which should be used in referring to these publications.*

The present bulletin is issued under a coöperative agreement between the Engineering Experiment Station of the University of Illinois, the State Geological Survey, and the United States Bureau of Mines. The reports of this coöperative investigation are issued in the form of bulletins by the Engineering Experiment Station, the State Geological Survey and the United States Bureau of Mines. For bulletins issued by the Engineering Experiment Station, address Engineering Experiment Station, Urbana, Illinois; for those issued by the State Geological Survey, address State Geological Survey, Urbana, Illinois; and for those issued by the United States Bureau of Mines, address the Director, United States Bureau of Mines, Washington, D. C.

UNIVERSITY OF ILLINOIS
ENGINEERING EXPERIMENT STATION

BULLETIN No. 100 JUNE, 1917

PERCENTAGE OF EXTRACTION OF BITUMINOUS COAL WITH SPECIAL REFERENCE TO ILLINOIS CONDITIONS

BY
C. M. YOUNG
ASSISTANT PROFESSOR OF MINING RESEARCH

ENGINEERING EXPERIMENT STATION
PUBLISHED BY THE UNIVERSITY OF ILLINOIS, URBANA

CONTENTS

LIST OF FIGURES

LIST OF TABLES

PERCENTAGE OF EXTRACTION OF BITUMINOUS COAL WITH SPECIAL REFERENCE TO ILLINOIS CONDITIONS

INTRODUCTION

1. *Preliminary Statement.*—The purpose of the discussion presented in this bulletin is to record the results now being obtained in recovering coal in the mines in Illinois and in other bituminous coal mining districts of the United States. A brief discussion is also presented with reference to recovery in the principal European countries. Where the methods employed are now producing an unusually good percentage of extraction, the conditions under which the mining is carried on are described in considerable detail with the belief that they may suggest changes in practice which will be helpful to those who are now endeavoring to recover a greater percentage of the coal in the ground.

Most of the data presented were obtained from those operating the mines, and represent, therefore, calculations or estimates based upon thorough familiarity with conditions. Some of the methods by which high extraction is attained in other districts are described with the hope that the coal producers of Illinois may find herein suggestions which will prove helpful in their efforts to attain higher recovery.

It has been impossible to include in a single publication all the material available concerning the physical conditions encountered and the methods adopted in the various coal fields, but there will be found in the bibliography a list of books and articles in which these subjects are covered in greater detail. It is the present purpose to begin at an early date a more extended investigation of the plans and dimensions of mine workings in Illinois with reference to the cost of production and the percentage of extraction.

2. *Acknowledgments.*—The writer wishes to acknowledge his indebtedness to Professor H. H. Stoek, head of the Department of Mining Engineering, University of Illinois, and to Mr. F. W. DeWolf, Director of the Illinois State Geological Survey, and also to Mr. G. S. Rice, Chief Mining Engineer, United States Bureau of Mines, under whose direction the work of the Illinois Coal Mining

Investigations is being carried on. Professor Stoek has been especially helpful in the collection of material for the present bulletin. Many operators and engineers throughout the country have contributed statements concerning the districts with which they are familiar. The state mine inspectors have assisted in the work by suggesting mines at which particularly good records of extraction have been made and also mines at which new methods are being tried with a view of increasing the percentage.

3. *Summary.*—The facts and information presented in this bulletin include:

(1) A general statement of the importance of the problem of increasing the percentage of extraction of the coal in the ground in order to utilize the coal resources to a greater extent than at present, and, if possible, to decrease the cost of producing coal; also an account of previous efforts made to compile data upon this subject.

(2) A statement with reference to the conditions which have influenced the development of American coal mining methods and which must be considered in changing these methods in order to obtain more nearly complete recovery.

(3) A record of the recovery of coal in Illinois in the past, and a discussion of the efforts now being made to increase the percentage of extraction.

(4) An account of methods adopted in other states and in certain European countries by which higher percentages of extraction are being obtained.

(5) A brief history of the development of English mining practice, upon which American practice is founded.

(6) A short bibliography with reference to the subject of coal mining methods.

4. *Conclusions.*—A summary of conclusions suggested by a study of the data and information contained herein is presented as follows:

(1) In general in America probably one ton of coal has been left in the mine for every ton brought to the surface.

(2) An effort is being made in many sections of the United States and in a number of Illinois mines to decrease this loss of coal.

(3) The low percentage of recovery in the United States is largely due to economic conditions and to efforts to produce cheap coal.

(4) Where economic conditions have been favorable, percentages of recovery have been obtained in the United States quite as high as in any of the foreign countries in which usually the economic conditions have not been such as to make the production of cheap coal the determining element in the choice of a method.

(5) The low price at which much of the coal land in the United States has been bought has not offered an inducement to save the coal.

(6) The best results in recovery are now being obtained in districts where the value of coal land is high.

(7) As a general rule better extraction is being obtained in West Virginia and Pennsylvania than in the Middle West.

(8) In view of the results being obtained in some other districts, under conditions no more favorable than those in Illinois, the percentage of extraction in Illinois should be increased.

(9) The best results are being obtained by the larger and stronger companies which can afford to plan for the future.

(10) The low value of the smaller sizes of coal in the past has been a drawback to pillar drawing, because very often pillar coal has contained more of the small sizes than room coal. With the increasing use of small sizes in mechanical stokers, the price will undoubtedly advance to nearly the same level as that of the larger sizes; thus this drawback to greater recovery will gradually disappear.

(11) One of the reasons why newer methods have not been generally tried is to be found in the prejudice, too common in coal mining practice, against innovations, and in the fact that mining methods have been based largely upon previous practice in other countries or in other states.

(12) Many of the attempts to draw pillars have been unsystematic. Upon such unsystematic work are based many of the opinions concerning the technical and commercial practicability of pillar drawing and the prejudices against it.

(13) Subsidence of the surface must be regarded as a necessary accompaniment of mining. Instead of trying to prevent

subsidence, the pillars should be removed systematically so that the surface subsidence will occur uniformly and not in isolated spots. Although there may be a temporary disturbance of the surface, after a short time its condition will be as good or nearly as good as before the mining.

(14) In Illinois, at the present time, more than fifty per cent of the coal is frequently left in the ground in an effort to prevent squeezes and subsidence; even then it is not at all certain that the desired result is accomplished.

(15) The best results may be obtained by driving room entries to their full length, then by beginning the rooms at the inby end of the entry, in order that pillar drawing may begin as soon as the inby room is finished.

(16) To be effective, pillar drawing must begin as promptly as possible after the rooms are worked out.

(17) Where pillars are left to be drawn subsequently, the coal is usually lost, because the pillars are crushed through squeezes, or because it is not found economical or convenient to take the coal out and at the same time to keep up the output of the mine with the compartively small amount of coal left. In other words, unless pillar drawing follows very closely after the first working, very little pillar coal is obtainable.

(18) In many districts poor top has prevented taking out the full thickness of the coal, and one of the great losses is that due to coal left in the roof. This loss has been overcome in some cases very successfully, and should be carefully studied.

(19) The reported percentages of extraction are usually too high because, in estimating, often only the section mined is considered and no account is taken of top or bottom coal left unmined. Also frequently only limited areas of the mine are considered instead of the mine as a whole.

(20) At different mines in the same region where physical conditions are practically the same, the mining methods vary widely with regard to length of rooms, number of rooms in a panel, thickness of barrier pillars, etc. This variation in practice suggests the advisability of a detailed study to determine, if possible, a standard method for a given set of conditions.

CHAPTER I

Mining Methods and Conditions in Relation to Extraction

5. *Introduction.*—The subject of the percentage of coal extracted from the mines in the United States has received very meager attention, except in the case of individual mines or companies. The only comprehensive official study of an extended coal mining area has been in the anthracite district of Pennsylvania, where the high value of the coal and the knowledge that the supply is limited early stimulated an interest in the subject. This interest led to the appointment of the Coal Waste Commission which reported in 1893.*

In 1905 H. H. Stoek† published a table of coal pillar data which contained percentages of extraction gathered largely by correspondence. See Table 1, page 23.

In 1914 A. W. Hesse‡ collected as much information as possible on this subject, which is summarized in Table 2, page 24.

In previous bulletins of the Coöperative Coal Mining Investigations tables of pillar data and percentages of extraction were given. These are summarized in Table 3, pages 25, 26, 27.

Doubtless many of the figures in these tables and others on the percentage of recovery are open to question, but they represented the best and most nearly complete information available when they were published. There are several reasons for questioning the accuracy of the figures on extraction. Chief among them is the fact that estimates are usually based upon areas which are too small or upon insufficient data. A single panel or a single lease is sometimes used as a unit upon which to base estimates, and often the areas thus selected are favorably located. While the estimates may represent results obtained with the given method of mining, they by no means represent the average results for the mine as a whole. The panel selected for measurement is usually one in which there has been no squeeze, while all about it there may be squeezed areas in which large amounts of coal have been lost. Many estimates are

---

* Report of Commission Appointed to Investigate the Waste of Coal Mining with the View to the Utilizing of the Waste, 1893.

† Mines and Minerals, Vol. 26, p. 107, 1905, and International Library of Technology, Vol. 150, par. 40, p. 60.

‡ "Maximum Coal Recovery," W. Va. Min. Inst., June 3, 1914; Coll. Eng., Vol. 35, p. 13; and Coal Age, Vol. 5, p. 1051.

based upon possible future recovery from pillars, which may or may not be obtained.

Estimates covering extraction frequently do not take account of top and bottom coal left in the mine, and the values reported often refer only to the section of the coal actually mined. In mining a coal bed ten feet thick, for instance, two feet of top coal may be left unmined. The maximum percentage of extraction from mining eight feet, in this case, would be eighty per cent of the total coal in the bed. If, then, fifty per cent of the eight feet mined is obtained, only forty per cent of the total coal in the bed is recovered.

The only accurate method of estimation is to divide the actual amount of coal mined, as determined by the tonnage for which the miner is paid, by the amount of coal in the ground as determined by multiplying a given area by the average thickness shown in a large number of sections of the bed.

Even where great care is exercised, results are often subject to errors. The causes for these have been outlined by Smyth* as follows: Inaccuracies in railroad weights of possibly five to ten per cent, inaccuracies in estimation of coal used at the mines frequently amounting to ten per cent, inaccuracies in estimating the mean thickness of the bed amounting, even in very uniform beds, probably to five per cent, difficulties of obtaining final figures until a mine is worked out.

The present condition of the coal mining industry in this country is a natural result of the course and character of its development. In general, only those beds and even parts of beds have been worked, the exploitation of which would result in the largest immediate profits. Those methods of mining which were cheapest and which promised the largest profit on the coal produced have been followed, often without regard to the possible injury of the mine or the resulting loss of coal. There has been, moreover, no restriction of market, and in many cases districts have been opened when there has been very little demand for coal in the surrounding territory, but when conditions of operation and transportation have been such as to make it possible for coal from these districts to enter markets already supplied. The result has been cheap coal, produced by wasteful methods.

Another result has been over-development of the industry. The opening in nearly all districts of too many mines has resulted in the

* Smyth, John O., Personal Communication.

idleness of many mines during a large part of each year with the accompanying increase in the cost of production. In dull periods coal has frequently been sold for less than the cost of production in order that mines might be kept in operation and certain fixed charges met. This subject was taken up by Bush and Moorshead in 1911 in a paper* before the American Mining Congress in which it was said that the production in this country exceeded the consumption first in 1891, and that the difference between consumption and capacity for production had steadily increased. The strike of 1910 in Illinois, Indiana, and the Southwest emphasized the over-capacity of the mines of that region. Though the mines of Illinois were idle during six months of the year, the production of 45,900,246 tons was only ten per cent less than the production of the previous year. The mines of Oklahoma, Arkansas, and Missouri were also idle during six months of 1910 because of the strike, but the production showed an average decrease of only twenty per cent. It was also said that the possible capacity of West Virginia mines was seventy-five per cent more than the total production, that the output in the Pittsburgh and the No. 8 Ohio districts was reduced to thirty per cent of the normal production during the three or four months of each year when navigation on the lakes was closed, and that few properties during the three preceding years—1909, 1910, and 1911—had been operated more than 225 working days per year.

This over-production, with its small profit or even loss in the operation of mines, results in a natural tendency to employ only those methods which will insure cheap coal. It is natural, also, that under these conditions there should exist an attitude of hesitancy with regard to the adoption of new or different methods. Neither the coal producer nor the public has as yet become aroused to the full realization of the fact that the natural resources of the country are not inexhaustible. The coal mining engineer of America accordingly, has not had as his problem the development of methods of extraction which would result in the largest percentage of ultimate recovery, but rather the development of methods which would result in the lowest cost of production. In many cases, however, as is shown by the detailed descriptions given later, where economic conditions have seemed to warrant it, methods have been developed by

---

*Bush, B. F., and Moorshead, A. J., "The Condition of the Bituminous Coal Industry," Proc. Amer. Min. Cong., p. 246, 1911.

American engineers and coal producers which have given a percentage of recovery equal to that secured in any European country.

The fact should be borne in mind, when comparisons are made between mining methods in different countries, that, while it is true that the percentage of extraction is less in this country than in most of the European countries, the cost of coal to the consumer and the profit to the producer are also less.

The subject of the comparative cost of production of coal and of the comparative profits realized in Great Britain and in the United States was taken up by Rice* substantially as follows: the average value of coal in the United States on cars at the mine in 1913 is reported as $1.18 per short ton for bituminous coal and $2.13 per short ton for anthracite. In Wales, in 1913, the average value per short ton at the mines for all kinds of coal was $2.55, and in Great Britain as a whole, $2.21. In the German Empire the average value for all kinds was $2.27, and for Westphalia it was $2.37 per short ton. Net mining profits in Great Britain and in Germany are between twenty-five and fifty cents per ton, while profits in the United States for bituminous coal are probably not more than five cents per ton.

It is a matter of course that more expensive methods of mining cannot be adopted without increasing the cost of the coal, and under the conditions which have prevailed in the coal industry for many years there could be no material increase in the cost of coal to the producer without a corresponding increase in the selling price. The prevailing opinion, however, that the percentage of recovery cannot be greatly increased without an increase in the cost of production is questionable, and certainly this increase in cost would not be as great as is generally believed. This is a matter which can be conclusively determined only by actual trial of new methods extending over a sufficient period to insure the reliability of the results. The fact that the adoption of methods which result in an increase in the percentage of extraction has been possible in some districts with little or no increase in cost at least furnishes a reason for thinking that similar changes could be made in other districts with similar results.

Careful planning of operations over long periods and steady working are necessary in order to obtain a high percentage of ex-

* Rice, G. S. "Mining Costs and Selling Prices of Coal in the United States and Europe, with Special Reference to Export Trade," Second Pan-American Scientific Congress.

traction. At present these conditions are impossible in many districts and can be attained only by centralized control of production and selling price, which will provide against alternation of idle and rush periods with the disorganization which accompanies them. Under existing conditions it is feared by operators that the necessary coöperation would be interpreted and attacked as a violation of anti-trust laws. In some of the European countries syndicates working in coöperation with the governments regulate the output of the mines and the selling price of coal with results which are said to be highly satisfactory and conducive to a high recovery.*

One of the chief commercial factors affecting the choice of a method has been the cost of coal in the ground. This has generally been very low, and the loss of coal, therefore, has not been considered a serious matter. Even at the present time the value of coal rights in the southern Illinois field, where the No. 6 bed is worked, is estimated at not more than $100 to $150 an acre, and it has been only a very short time since such coal rights could be purchased for less than $50 an acre. The thickness of this coal is somewhat variable, being in some places fourteen feet or more, but, if we assume that only about seven feet is worked, the output will amount to about 12,000 tons per acre and the cost of coal in the ground will be about one cent per ton. A great deal of the coal in the state, however, has been bought at a very much lower figure. In some cases also there is a second bed of coal which will be available later, and when this is considered, the cost of coal in the ground will be much less than one cent per ton.

This phase of the subject was discussed by Rice,† in 1909, as follows:

"The influencing conditions causing the great losses that are at present incurred are:

1. Cheapness of 'coal in place'; that is, in the seam.
2. Low market prices, resulting from extreme competition.
3. Character of the seam, roof, and floor as determining the method of mining.
4. Surface subsidence due to mining.
5. Interlaced boundary ownerships.

---

* Scholz, Carl, "The Economics of the Coal Industry," Proc. Amer. Min. Cong., p. 241, 1911.

† Rice, George S., "Mining-Wastes and Mining-Costs in Illinois," Trans. Amer. Inst. Min. Eng., Vol. 40, p. 31, 1909.

6. Carelessness in mining operations.

The first two factors, taken together, are the controlling ones in most mining operations in influencing the choice of a mining system. The majority of Illinois operators are sufficiently progressive to find ways and means to take out practically all the coal under a given area if it could be made evident that it paid to do so. That many do not do all that can be done in this direction is apparent; but if, without unusual investment, a profit of operation could be shown in taking out all the coal over the profit made by present methods, the industry could undoubtedly find men to accomplish the task. In other words, from an engineering standpoint practically all the coal under a given area can be taken out. It is a question of cost.

"*Cheapness of Coal in Place.*—This is chiefly due to the great abundance of coal. Except in the barren northern one-fourth of the State, lying north of the outcrop of the coal-basin, the development of a tract depends primarily not on the possibility of finding coal in that particular locality, but on the question whether it is a suitable place, from a market standpoint, to open a mine, the thickness of seam and the quality of the coal being considered.

"The price of coal rights varies from $10 per superficial acre in the middle part of Illinois, away from the mining centers to $100 per acre near developed mines. Or, in the case of leasing, from 2 cents per ton run-of-mine hoisted, in the southern part of the State, to 5 cents in the northern part. The cost of the fee is relatively so much cheaper per ton than leasing that the latter system is not much used. The ownership of the coal by the operator is conducive to better mining, but relative to other items that go to make up the total, the cost of the 'coal in place' is so low as to be almost negligible. In central Illinois, in some cases, at a cost of only $10 per acre, two workable seams, from 6 to 8 ft. thick, are obtained. Allowing only 50 per cent yield of the two seams, 13,000 tons would be produced per acre, the purchase cost thus being 1/13 of a cent per ton, or about 1/1000 of the total cost of production in central Illinois. In the Wilmington long-wall field the average cost of the coal rights is about $50 per acre. The seam there, although it averages a trifle less than 3 ft. in thickness, produces about 5,000 tons per acre. The cost is therefore about 1 cent per ton in place, which is 1/130 of the total cost of production. Hence, it may be seen that there is little incentive, from the standpoint of the purchase price of the coal, to save the latter in mining operation."

The cost of coal rights has very greatly increased since Rice's discussion, and there is every reason to believe that the value of coal in the ground will be much greater in the future than it has been in the past. During most of the productive period, however, the coal in the ground over a considerable part of the State has been worth not more than one-tenth of a cent per ton, and under these circumstances the loss of coal has, naturally, not been considered a serious matter. What has been important, and still is, is the extraction of coal at low cost, and the subject of high recovery is one of increasing importance at the present time.

Every ton of coal left in the ground represents the loss of a possible profit. Every ton of such coal represents a loss in increased value. An acre of coal left in the ground at any time means the extraction of another acre at some later time when the value of coal in the ground will be greater. In other words, producers are now extracting coal, worth possibly $150 an acre, which might be left until it would reach even a greater value if it were not for the fact that coal was wasted in the ground when it was worth only fifty dollars an acre.

A low percentage of extraction increases the cost of production, because, for a given output, the workings must cover a larger area. This involves longer haulage roads, and, consequently, a greater investment in rails and trolley wire, greater maintenance expense, greater consumption of power, lower output per unit of equipment, and lower output per man. With long haulage roads there is greater chance for derangements of the track or for falls of roof, which may cause the stopping of haulage until the trouble is removed or may result in wrecks if the trouble is not discovered in time. Another source of danger lies in the greater haulage speed which must be employed on long roads if the output is to be maintained.

The cost of ventilation is also higher, because these larger workings require a larger quantity of air to maintain safe conditions, and more power is required to circulate air through the longer passages. There is, moreover, a greater loss of power because of the more numerous stoppings, which are often inefficient.

Another difficulty accompanying the spreading of the workings over a large area is that of providing the intensive supervision which is highly desirable in coal mining, particularly where skilled workmen have been replaced by comparatively unskilled laborers, and the pick has been replaced by explosives and mining machines. Unless the cost of operation is to be increased by the employment of a larger number of foremen or face bosses, this intensive supervision can be obtained only by concentration of the workings.

6. *Subsidence.*—One reason for the use of methods involving low extraction is the desire to maintain the original surface of the ground. Rice* discussed surface subsidence as follows:

> "The influence of this factor upon the yield results from the high value of Illinois lands for agricultural purposes. . . . If the long-wall system were applied to the thick seams, when applicable at all, it would cause a considerable derangement of the surface, and when the latter is so nearly level as the prairie-land of Central Illinois, it makes the question of subsidence a serious one. . . . However, until the agricultural land in the United States becomes insufficient to fill the needs of the population, which would be reflected in a continual increase of price for farming land, the money-loss from temporarily destroying the surface in places is relatively small, as compared with the selling price of the coal mined from the seam. Taking the average value of the surface at $125 per acre, if 80 per cent be rendered worthless the immediate money-loss would be $100 per acre. A seam 6 ft. thick would contain per acre 11,000 tons of coal in place, yielding, at 90 per cent, 9,900 tons. The damage done by practically destroying the surface would be only 1 cent per ton. If the land-prices should rise to an amount two or three times as great as the value stated, this loss would still not prohibit mining."

As far as the long-wall district is concerned, very little if any damage has resulted from subsidence, and little attention has been given to the subject. The most noticeable effects are generally temporary, and farm operations are not hindered.

The subject of the relation of surface values to subsidence in Illinois has been considered by L. E. Young in Bulletin 17 of this series.† Fig. 1 is a map reproduced from this bulletin, on which the approximate values of farm lands are shown. Table 4, page 28, shows the relation between coal values and surface values. The land values indicated on the map and set forth in the table suggest that it might be possible, at least in many cases, to mine coal at a small profit even if the value of the surface were totally destroyed. There is no need, however, for assuming the permanent destruction of the surface or even its serious permanent injury. Generally any damage resulting from subsidence could be largely or wholly re-

* Rice, Op. cit., p. 40.

† Young, Lewis E., "Surface Subsidence in Illinois," Ill. Coal Min. Invest., Bulletin 17, p. 52, 1916.

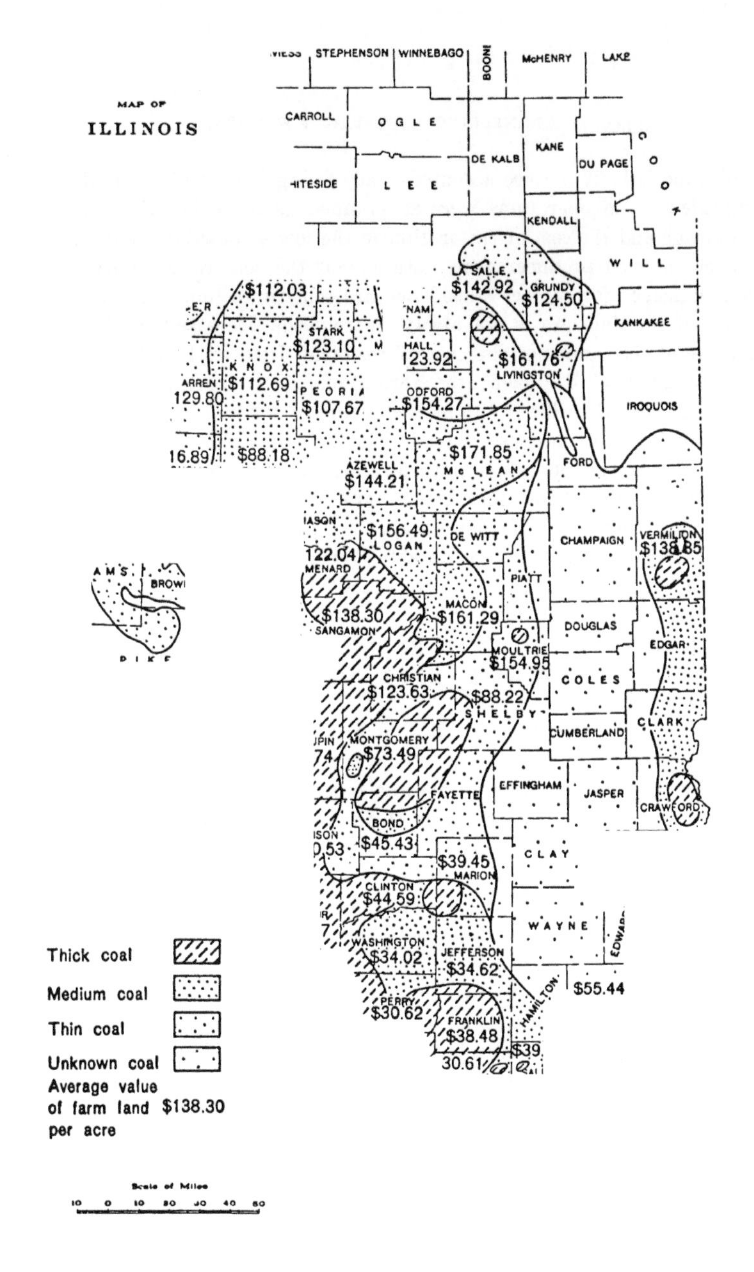

FIG. 1. MAP SHOWING THICKNESSES OF COAL AND VALUES OF FARM LANDS, AS GIVEN BY THE 1910 CENSUS REPORTS

paired, especially if it were accepted that mining is certain to result in subsidence and operations were so planned as to reduce the surface damage and the cost of restoration to the lowest possible amount.

When the coal producer owns nothing but the coal rights, unreasonable damages for surface subsidence are sometimes imposed. The measure of the damages is not always merely the decreased productive value of the land, nor the cost of restoring it to its former condition by artificial drainage; the formation of a small pond has often been claimed to lower the market value of a farm to a considerable extent, simply because it made the farm less sightly. Under these circumstances, operators naturally desire to avoid disturbing the surface for they know that an attempt will be made to recover damages and that, even if they escape the payment of exorbitant amounts, they will incur considerable expense in defending the suits. An effort, accordingly, is often made to conduct the mining operations in a manner which will not result in surface subsidence. The result is that the loss in the ground represents an important percentage of the coal, in many cases more than half that contained in the area worked.

It is very important that the allowable damages for surface subsidence be regulated by some law. This law should fix the damage payable by the coal producer, in case he is legally responsible, upon the basis of the actual damage done to the surface. Under such a law the operators would know the extent and character of their responsibility and could, without fear of excessive or unreasonable damages, proceed according to methods which would yield the highest possible percentage of extraction justifiable under such conditions.

7. *Squeezes.*—Closely related to, but not identical with, the subject of surface subsidence is that of squeezes. There may be subsidence without a squeeze, but with the conditions in Illinois a squeeze is usually followed by subsidence. The removal of a portion of a deposit throws additional weight upon the pillars left and if these pillars are not strong enough to support this additional weight, they will crack and crush, causing a movement of the overlying material. This movement is called a "squeeze," or sometimes a "creep." Large quantities of coal are often left in the mine in the form of pillars in an effort to prevent a squeeze, which may not only interfere with the operation of the mine and entail a loss of

the coal in the squeezed area, but large areas of the mine may become inaccessible for future economical working. A more nearly complete extraction of the coal properly carried out should, however, result in less damage from squeezes.

There are two ways in which a squeeze may be prevented or stopped; first, by the use of a support strong enough to prevent any movement of the overlying rock, and secondly, by a fracture of the rock above the excavated portion so that the weight on the pillars will not be sufficient to crush them.

The first method may be employed by either leaving natural supports (coal pillars) of sufficient size and strength to hold the roof without any movement, or by the use of artificial supports, such as timber or iron columns, or sand or culm filling. The cost of timber or iron would prohibit their use if a large percentage of coal was to be extracted. Filling is not generally non-compressible, but it occupies most of the space from which coal has been removed, prevents any scaling off of pillars, and eliminates any possibility of movement of pillars. The filling method, however, is hardly to be considered feasible in Illinois because of the cost. In the Upper Silesian coal field where this method is most extensively used, the cost is from twelve to eighteen cents per ton.*

Another difficulty arises from the fact that the material would have to be flushed into the mine with water, and then the water would have to be pumped out. This water would probably have an injurious effect on the clay bottom. The material, moreover, would have to be brought from a distance unless the value of the land should be so small as to permit the use of material from the neighboring surface, and this condition would rarely prevail in Illinois.

In the leaving of coal pillars of sufficient strength to prevent roof movement, the amount of coal which must be left varies with local conditions. It is difficult, if not impossible, to determine this factor in advance, and in attempting to approach as closely as possible the limit of safety, it often occurs that too much coal is removed. Even if the limit is not passed so far as immediate movement is concerned, it may be passed with reference to ultimate movement and the crushing of the pillars.

Apparently, so far as a large part of the State of Illinois is con-

---

* Gullachsen, Berent Conrad, "The Working of the Thick Coal Seams in Upper Silesia," Trans. Inst. Min. Engrs. Vol. 42, p. 209, 1911.

cerned, it is necessary to leave in the ground about one-half of the area of the coal if movement of the overlying beds is to be prevented.

The desirable dimensions of the rooms and the pillars vary widely from wide pillars between wide rooms to narrow pillars between narrow rooms. One company had squeezes when it drove 25-foot rooms on 50-foot centers, but it has had no trouble with 30-foot rooms with 60-foot centers. This is a question not simply of the crushing strength of the coal nor of the ability of the bottom to withstand pressure, but of the effect on the pillars of scaling at the sides. In other words, the strength of the pillar is not determined merely by its original size but by its effective size after the scaling action, which may follow the extraction of the room coal, has occurred. This scaling action is increased by the shattering effect of explosives.

The use of coal in the ground to prevent squeezes and subsidence, which is what abandoning of pillar coal amounts to, ought to be considered only as a last resort. It has been found that squeezes can be prevented by the removal of so little coal on the advance as to leave a solid support, and by the complete removal on the retreat so that the roof is left entirely without support. This process prevents the gradual settling which occurs when some support is left and produces a sharp bending, or localization of stress, sufficient to cause a rupture of the overlying rock and prevent the transference of weight from the mined-out area to the standing coal. This is the only certain method which has been found for the prevention of squeezes unless an absurdly large quantity of coal is abandoned. The means by which squeezes may be prevented vary under different conditions, but the essential consideration is that the roof of the mined-out area shall be left absolutely without support either from coal or from timber, so that it must fall.

TABLE 1[1]

DIMENSIONS OF ROOMS AND OF ROOM PILLARS AND PERCENTAGE OF EXTRACTION

LOCALITY AND COAL SEAM	Width of Room Feet	Width of Room Pillars Feet	Per Cent of Coal Left in First Working	Total Per Cent of Coal Recovered
Alabama:				
Newcastle seam	25	10–20	40	85
Thin seam	20	10–20	30	90
Blue Creek	25	25	35	85
Blocton	25	25	30	80
Flat Top	25	25	35	85
Arkansas:				
Sebastian County	18–30, usually 24	12	30	70
Colorado:				
Trinidad series	20–25	20–25	50	50–80
Illinois:				
Springfield	30	15–20	60	60
Staunton (machine mines)	30	30	60	60
Indian Territory	18–30, usually 24	10–12	30	70
Iowa	21–30	9–15	35	90
Maryland:				
Georges Creek	12–16	40–100	75	90
Pennsylvania:				
Connellsville	12	42	60	95
Connellsville	12	84	80	90
Pittsburgh	21–24	12–18	40	85–95
Clearfield	21–24	15	40–50	80–95
West Virginia:				
Fairmont	18–24	18–40	40–60	95
Clarksburg, Pittsburg seam	18	32	64	90
Mineral County, Pittsburg seam	12	48	80	95 (?)
Tucker County, Davis or Kittanning seam	18	22	35	90 (?)
Putnam County, Pittsburg seam	30	15	34	75–85
New River, Sewell-Nuttall seam	24	26	60	60–80
Thacker, (Thacker) Lower Kittanning seam	18–21	42	67	80
Logan, Lower Kittanning seam	21	20	57	....
Kanawha (No. 2 Gas), Lower Kittanning seam	28	12–20	40	....
Pocahontas, Double Entry, No. 3 Pocahontas seam	18	42	68	80
Pocahontas, Panel System, No. 3 Pocahontas seam	18	42	....	95[2]
Raleigh, Sewell-Nuttall seam	25	25	50	90

[1]H. H. Stoek, Mines and Minerals, Vol. 26, p. 107, 1905, and International Library of Technology, Vol. 150, par. 40, p. 60.

[2]Estimated.

The widths of rooms and room-pillars given in Table 1 show that there were few cases in which plans were made for the extraction of more coal from pillars than from the advance workings.

PRINCIPAL FACTORS GOVERNING RECOVERY OF COAL IN DIFFERENT DISTRICTS

Item	Operating District and State	Per Cent of Recovery Entire Seam	Ultimate Recovery Compared to Present	Period of Operation Years	Average Height of Seam, Feet	Roof Coal Carried	Nature of Top	Nature of Bottom	Pillar-Drawings	Clay Veins Encountered
1	Colorado, Southern	80–90	Same	5–30	8.5	24 in.	Slate	Soft slate	Yes	Dikes
2	Colorado, Other districts	60–65	Same	5–30	8.5	18–24 in.	Very soft	Soft slate	Yes	None
3	Colorado, Other districts	75–80 65	Better	10–35	3–7	Few places	Sandstone, poor shale	Same as top	Yes	None
4	Michigan, Saginaw District	80–90	Better	15	3	None	Black slate	Fire clay	Where allowed	None
5	Illinois, Central	50	Same	20	8	None	Slate, clod, and limestone	Fire clay	None	None
6	Illinois, Southern	65–70	Same	20	8	Yes	Sandy shale	Fire clay	To some extent	None
7	Illinois, Springfield District	55–75	Increase [2]	20–25	6–7	Some places	Hard shale	Fire clay	Where allowed	None
8	Illinois, Williamson and Saline Counties	55–75	Increase [2]	18	5.5	Some places	Hard shale	Fire clay	Where allowed	None
9	Illinois, Sherrard Field	90	Same	20	3.7	None	Blue rock and cap rock	Slate and sand rock	Yes	Yes
10	Pennsylvania, Extreme southwest section	72.5	Better	3	7.5	10 in.	1.5–3 ft. draw slate	Soft fire clay	Yes	None
11	Pennsylvania, Westmoreland Co.	84	Below	25–35	6.7	None	Slate	Fire clay	Yes	None
12	Pennsylvania, Somerset Co.	95	Same	8	3.9	None	Hard black slate	Limestone	Yes	Very few
13	Maryland, Georges Creek Field	97	Same	12	3	None	Sand rock	Sand rock	Yes	Yes
14	Maryland, Georges Creek Field	88	Same	94	9	18 in.	Gray slate, coal, dark shale	Hard gray shale	Yes	None
15	Ohio, Belmont County	60	Same	....	5–6	None	Slate and shale	Fire clay	None	None
16	Ohio, E. Harrison County	70–75	Same	....	3.7–5	None	10 in. firm slate	Fire clay	None	None
17	West Virginia	90	Same	50	8	Some places	Varies	Fire clay	Yes	Yes
18	Alabama, no reply									
19	Tennessee, no reply									
20	Kansas, no reply									
21	Iowa, no reply									
22	Kentucky, no reply									

[1] A. W. Hesse, "Maximum Coal Recovery," W. Va. Min. Inst., June 3, 1914; Coll. Eng., Vol. 35, p. 13; and Coal Age, Vol. 5, p. 1051.
[2] Operated by panel method, all others by room-and-pillar.

TABLE 3

DIMENSIONS OF WORKINGS AND ESTIMATED PERCENTAGES OF EXTRACTION IN ILLINOIS MINES

District	No. of Coal Bed	No. of mine (Coöp. Investigations)	Depth of hoisting shaft	Method		Entry width			Entry pillar width			Barrier pillar width		No. rooms on room entry	Room		Width of room pillar	Room neck		Dist. from entry to full room width	Dist. between room centers	Width of room stump	Width of cross-cuts	Has mine had squeezes?	Operators' Estimate of Percentage Extracted
				Unmodified Room-and-Pillar	Panel	Main	Cross	Room	Main	Cross	Room	Main	Cross		Width	Length		Width	Length						
II	2	12	125	×	—	8	8	—	12	12	—	23	—	—	24	150	18	8	10	20	42	34	8	No	44
	2	13	114	—	×	9	9	9	20	20	20	30	30	16	24	250	20	9	4	20	44	35	9	No	55
	2	14	135	×	—	8	8	—	18	18	—	30	—	—	28	250	22	18	12	22	50	42	8	No	46
	—	15	160	×	—	8	8	—	18	18	—	20	—	—	25	240	17	8	10	20	42	34	10	No	49
III	1	17	210	×	—	8	8	—	30	30	—	50	—	—	20	250	20	8	7	14	40	32	12	No	96
	1	18	69	×	—	12	12	—	42	15	—	36	—	—	26	250	19	8	7	12	45	37	12	Yes	96
	1	19	70	×	—	12	12	—	18	18	—	18	—	—	24	250	18	8	7	25	42	34	8	No	50
	2	22	60	×	—	6	6	—	..	..	—	..	—	—	15	250	15	6	15	25	30	24	15	Yes	45
	1	24	40	×	—	12	12	—	20	20	—	15	—	—	24	300	21	8	7	29	45	37	15	No	70
IV	5	26	196	×	—	8	8	—	30	21	—	50	—	—	26	190	6	8	8	30	32	24	8	No	56
		29	185	×	—	8	8	—	36	24	—	40	—	—	24	180	8	8	7	18	32	24	8	No	57
		30	60	×	—	10	8	—	15	12	—	25	—	—	22	210	10	8	7	18	32	24	13	No	57
		34	200	×	—	8	10	—	30	34	—	38	—	—	28	180	10	10	9	30	38	28	4	Yes	42
		41	365	×	—	16	16	—	20	30	—	80	—	—	25	200	10	12	9	15	35	23	8	No	56
		25	185	—	×	8	8	8	36	30	24	50	24	23	24	180	6	8	7	18	30	22	8	Yes	54
		27	170	—	×	12	12	12	20	15	12	50	50	30	30	200	8	8	9	20	38	30	8	No	53
		28	150	—	×	8	8	8	25	20	8	40	20	18	24	180	10	8	10	15	34	26	8	Yes	54
		32	68	—	×	8	8	8	30	30	24	20	20	30	20	150	12	8	10	15	32	24	8	No	57
		33	285	—	×	8	8	8	20	20	20	50	40	15	30	150	6	8	10	35	36	28	6	No	64
		35	187	—	×	8	8	8	35	35	30	60	50	14	26	200	9	8	9	27	35	27	8	Yes	56
		36	238	—	×	12	12	12	30	24	10	30	20	24	25	240	10	8	9	27	35	27	12	Yes	48
		37	235	—	×	10	10	10	30	30	20	60	40	18	30	240	12	12	15	15	42	30	10	Yes	41
		38	245	—	×	10	10	10	27	24	20	35	35	25	30	200	6	10	9	20	36	26	10	No	62
		39	204	—	×	12	12	12	30	30	20	30	30	15	30	250	12	12	11	18	42	30	10	Yes	65
		40	270	—	×	10	10	10	30	24	20	35	35	27	24	200	6	8	9	18	30	22	7	Yes	50
V	5	43	160	×	—	14	14	—	30	30	—	40	—	—	30	250	20	18	10	20	50	32	18	Yes	65
		44	270	×	—	15	18	—	25	25	—	50	—	—	22	250	24	18	12	18	46	28	18	Yes	59
		45	320	×	—	16	16	—	24	20	—	..	—	—	27	250	20	19	12	30	47	28	15	No	64
		46	450	×	—	14	14	—	20	20	—	50	—	—	33	200	7	19	12	25	40	21	19	Yes	82
		47	90	×	—	14	19	—	20	12	—	20	—	—	22	250	12	8	6	13	34	26	12	No	68
		48	76½	×	—	14	14	—	25	25	—	50	—	—	24	300	21	12	6	15	45	33	12	No	59
		49	337	×	—	14	14	—	30	21	—	35	—	—	24	250	9	12	9	25	33	21	12	Yes	73

DIMENSIONS OF WORKINGS AND ESTIMATED PERCENTAGES OF EXTRACTION IN ILLINOIS MINES

District	No. of Coal Bed	No. of mine (Coop. Investigations)	Depth of hoisting shaft	Method: Unmodified Room-and-Pillar	Method: Panel	Entry width: Main	Entry width: Cross	Entry width: Room	Entry Pillar width: Main	Entry Pillar width: Cross	Entry Pillar width: Room	Barrier Pillar width: Main	Barrier Pillar width: Cross	No. rooms on room entry	Room: Width	Room: Length	Width of room pillar	Room neck: Width	Room neck: Length	Dist. from entry to full room width	Dist. between room centers	Width of room stump	Width of crosscuts	Has mine had squeezes?	Operators' Estimate of Percentage Extracted
VI	6	50	726	X	—	12	12	—	40	100	—	35	—	—	22	250	25	12	24	34	47	35	22	Yes	58
		55	160	X	—	10	9	—	50	50	—	50	—	—	25	250	26	15	12	19	51	30	15	Yes	46
		56	580	X	—	12	12	—	30	100	—	40	—	—	26	250	26	12	23	45	52	40	11	No	50
		59	200	X	—	9	9	—	20	50	—	20	—	—	23	225	15	12	16	32	38	26	15	Yes	68
		60	220	X	—	12	12	—	18	20	—	20	—	—	21	300	12	12	6	24	33	21	12	Yes	51
		62	190	X	—	10	10	—	30	75	—	30	—	—	20	250	15	10	15	15	35	25	10	Yes	51
		63	140	X	—	10	10	—	30	40	—	30	—	—	22	500	18	10	12	18	40	30	12	Yes	..
		64	120	X	—	11	11	—	19	38	—	22	—	—	20	300	15	9	8	15	35	26	8	No	64
		65	90	X	—	12	12	—	20	10	—	20	—	—	21	250	9	10	6	24	30	20	10	Yes	58
		51	494	—	X	9	9	9	30	30	30	100	100	16	24	250	16	9	18	38	40	31	18	Yes	52
		52	640	—	X	12	12	12	50	50	38	150	100	30	25	250	25	12	18	41	50	38	12	No	58
		53	450	—	X	12	16	16	28	50	50	100	100	35	29	300	11	20	18	18	40	22	..	Yes	65
		54	380	—	X	15	7	7	50	50	30	150	150	21	20	250	25	9	11	35	45	34	7	Yes	58
		57	320	—	X	12	12	12	50	38	38	100	100	20	25	300	25	12	10	32	50	38	18	No	55
		58	516	—	X	12	12	12	25	20	20	125	125	32	22	200	12	12	15	34	34	22	12	Yes	56
VII	6	68	387	X	—	21	16	—	40	40	—	27	—	—	32	235	33	9	15	27	65	56	20	Yes	50
		70	92	X	—	14	14	—	50	50	—	50	—	—	30	300	30	21	25	50	60	39	21	No	50
		72	287	X	—	21	21	—	40	40	—	..	—	—	30	300	30	21	6	..	60	39	21	No	56
		78	160	X	—	21	21	—	55	45	—	55	—	—	32	250	41	21	24	34	73	52	21	Yes	56
		79	127	X	—	21	21	—	60	30	—	60	—	—	30	300	30	21	12	24	60	39	21	No	62
		80	145	X	—	12	14	—	18	35	—	60	—	—	30	250	30	21	18	38	60	39	21	No	59
		81	200	X	—	21	21	—	80	40	—	60	—	—	35	250	45	21	25	40	80	59	21	Yes	55
		82	192	X	—	14	21	—	42	42	—	60	—	—	40	250	40	21	18	36	80	59	30	No	63
		84	320	X	—	21	21	—	39	39	—	40	—	—	30	240	30	21	15	25	60	39	21	No	60
		87	707	X	—	12	12	—	40	40	—	150	—	—	30	300	30	12	20	50	60	48	20	No	58
		88	85	X	—	12	12	—	20	20	—	40	—	—	29	250	17	12	6	16	46	34	12	Yes	65
		89	85	X	—	8	8	—	20	15	—	40	—	—	25	200	20	8	10	24	45	37	8	Yes	48
		90	160	X	—	21	9	—	60	30	—	30	—	—	33	250	17	9	12	24	50	41	12	No	63
		66	332	—	X	21	21	21	40	40	40	110	75	30	30	300	30	21	25	50	60	39	21	Yes	56
		67	320	—	X	12	12	12	30	30	24	60	60	30	30	300	10	9	6	32	40	31	9	No	63
		69	290	—	X	12	12	12	30	30	30	60	60	27	23	200	22	12	9	30	45	33	12	No	47
		71	194	—	X	10	10	10	40	40	40	75	60	..	38	200	27	9	18	36	65	56	10	Yes	51
		73	318	—	X	21	21	21	60	40	13	60	60	13	30	300	20	21	12	30	50	29	21	No	60
		74	330	—	X	12	12	12	60	60	12	50	50	14	30	265	30	12	12	30	60	48	12	Yes	50
		75	310	—	X	12	12	12	50	40	25	50	50	25	30	250	18	12	18	30	48	36	12	Yes	52
		76	370	—	X	10	10	10	50	40	..	75	75	..	28	225	27	10	20	35	55	45	10	Yes	44
		77	462	—	X	14	14	14	50	50	40	100	100	40	30	250	25	22	22	28	55	33	21	Yes	63
		83	140	—	X	21	21	21	35	35	12	35	35	12	27	600	43	21	12	24	70	49	21	Yes	50
		85	440	—	X	12	12	12	50	50	25	50	50	25	35	250	19	9	9	25	44	35	12	Yes	55
		86	536	—	X	12	8	12	100	50	16	100	50	16	26	400	25	12	18	30	50	32	8	No	52

TABLE 3 (Concluded)

DIMENSIONS OF WORKINGS AND ESTIMATED PERCENTAGES OF EXTRACTION IN ILLINOIS MINES

District	No. of Coal Bed	No. of mine (Coöp. Investigations)	Depth of hoisting shaft	Method: Unmodified Room-and-Pillar	Method: Panel	Entry width: Main	Entry width: Cross	Entry width: Room	Entry Pillar width: Main	Entry Pillar width: Cross	Entry Pillar width: Room	Barrier Pillar width: Main	Barrier Pillar width: Cross	No. rooms on room entry	Room: Width	Room: Length	Width of room pillar	Room neck: Width	Room neck: Length	Dist. from entry to full room width	Dist. between room centers	Width of room stump	Width of cross-cuts	Has mine had squeezes?	Operators' Estimate of Percentage Extracted
VIII	6	91	217	×	—	9	9	—	25	21	—	40	—	—	43	200	4	9	19	18	47	38	5	Yes	55
	6	92	240	×	—	7	7	—	35	30	—	60	—	—	24	240	3	9	18	18	27	18	9	Yes	68
	6	93	186	×	—	8	9	—	21	21	—	50	—	—	24	200	16	9	9	9	40	31	9	Yes	75
	7	94	90	×	—	7	6	—	16	12	—	17	—	—	24	240	6	9	9	9	30	21	9	No	81
	6	95	90	×	—	6	6	—	..	..	—	..	—	—	25	500	11	9	12	12	36	27	9	No	82
	7	97	223	×	—	7	8	—	21	21	—	50	—	—	21	150	9	9	12	12	30	21	7	No	68
Averages by districts II	2		133	—	—	8	8	—	17	17	—	25	—	—	25	222	19	8	9	21	45	36	9	—	48
III	1 & 2		90	×	—	10	10	—	28	21	—	30	—	—	22	260	18	8	9	21	40	33	12	—	71
IV	5		202	—	—	10	10	—	27	25	—	44	—	—	26	197	9	9	9	21	35	26	9	—	55
V	5		243	×	—	15	16	—	25	22	—	41	—	—	26	250	16	15	10	21	42	27	15	—	67
VI	6		348	—	—	11	11	—	33	33	—	81	—	—	23	262	18	11	14	28	41	30	13	—	56
VII	6		279	—	—	16	15	—	46	39	—	60	—	—	31	277	28	16	16	32	58	42	17	—	56
VIII	6 & 7		174	×	—	7	7	—	24	21	—	43	—	—	27	255	8	9	13	13	35	26	8	—	71
Aver. of 30 panel mines			306	—	×	12	12	12	39	35	25	68	56	23	27	251	18	12	13	24	45	33	11	...	55
Aver. of 48 room-and-pillar-mines.........			208	×	—	12	12	..	31	27	..	46	..	..	26	250	19	12	12	24	45	33	13	...	54

TABLE 4[1]

VALUES[2] OF SURFACE AND OF COAL RIGHTS BY COUNTIES IN ILLINOIS

County	Value of Coal per Acre	Number of Coal Bed	Average Surface Value, Census of 1910
Bond	$ 25	6	$ 45.43
Bureau	10–100	2	114.53
Christian	10– 50	6	123.63
Franklin	35–100	6	38.48
Fulton	15–100	5	88.18
Gallatin	20– 25	5	48.60
Grundy	10– 25	2	75.52
Henry	135	6	112.03
Jackson	25– 75	2, 6	31.27
La Salle	10–100	2, 5	142.92
Livingston	10– 50	6	161.76
Logan	20– 50	5	156.49
Macoupin	15– 50	6	69.74
Madison	10– 40	6	70.53
Marion	20	6	39.45
Marshall	15	2	123.92
McLean	15	5	171.85
Menard	25– 30	6	122.04
Montgomery	25– 50	6	73.49
Morgan	20– 30	6	124.28
Peoria	20– 50	5	107.67
Perry	25	6	30.62
Putnam	15	2	104.69
Randolph	25	6	36.11
St. Clair	10–100	6	81.57
Saline	50–150	5	39.88
Sangamon	20–100	5, 6	138.30
Scott	10– 40	2	83.21
Shelby	10– 25	6, 5	88.72
Vermilion	100–150	6, 7	138.85
Warren	15	1, 2	129.80
Will	15	2	104.08
Williamson	50–150	6	30.61
Woodford	15	2	154.27

[1] Young, Lewis E., "Surface Subsidence in Illinois," Ill. Coal Min. Invest., Bulletin 17, p. 55, 1916.

[2] These prices are not offered as an authoritative basis for valuation but indicate in a general manner the prices at which coal has been sold or at which it is held in some of the important counties.

CHAPTER II

Extraction in Illinois

8. *Plan for Division of State into Districts.*—At the beginning of the work of the Coöperative Coal Mining Investigations, the State was divided into districts in order that those beds which are similar in general conditions might be studied and considered together. This subdivision into districts is shown by Fig. 2, and the districts are described in Table 5.

Table 5

Districts into which the State has been Divided for the Purposes of Investigation

Investigations, District	Coal Seam	Method of Mining	Counties	Investigations Numbers for Mines Examined
I	2	Long-wall	Bureau, Grundy, La Salle, Marshall, Putnam, Will, Woodford..........	1 to 11
II	2	Room-and-pillar	Jackson..........................	12 to 16
III	1 and 2	Room-and-pillar	Brown, Calhoun, Cass, Fulton, Greene, Hancock, Henry, Jersey, Knox, McDonough, Mercer, Morgan, Rock Island, Schuyler, Scott, Warren..........................	17 to 24
IV	5	Room-and-pillar	Cass, DeWitt, Fulton, Knox, Logan, Macon, Mason, McLean, Menard, Peoria, Sangamon, Schuyler, Tazewell, Woodford..................	25 to 42
V	5	Room-and-pillar	Gallatin, Saline..................	43 to 49
VI	6 (east of Duquoin anticline)	Room-and-pillar	Franklin, Jackson, Perry, Williamson............................	50 to 65
VII	6 (west of Duqouin anticline)	Room-and-pillar	Bond, Christian, Clinton, Macoupin, Madison, Marion, Montgomery, Moultrie, Perry, Randolph, Sangamon, Shelby, St. Clair, Washington..........................	66 to 90
VIII	6 and 7 (Danville)	Room-and-pillar	Edgar, Vermilion................	91 to 97

In the present publication the conditions prevailing and the methods followed in the various districts are described, and the extent to which these affect the percentage of recovery is discussed. Material and information has been gathered at various times, and

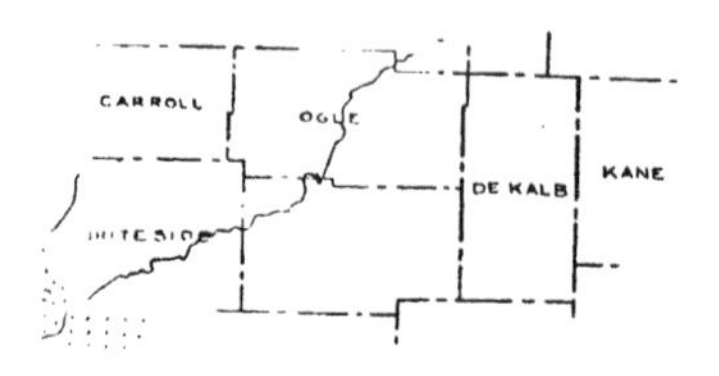

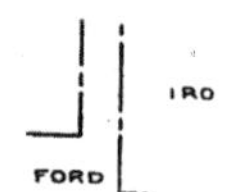

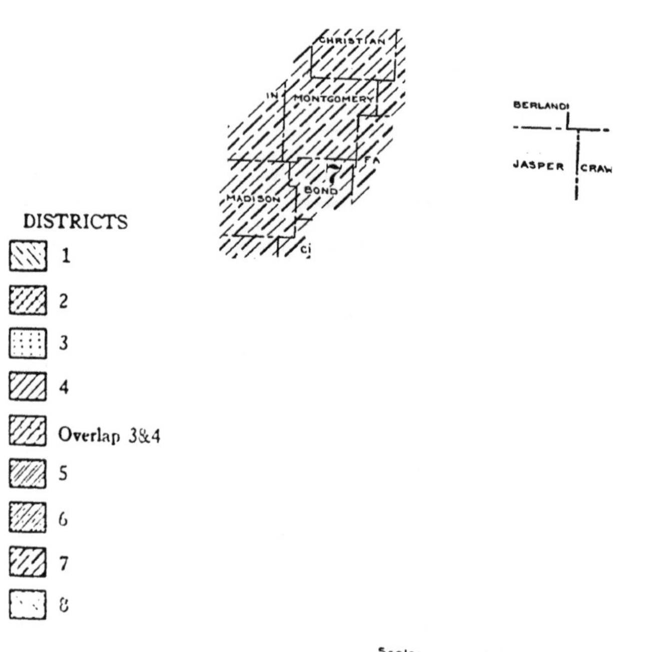

FIG. 2. MAP OF DISTRICTS OF COÖPERATIVE COAL MINING INVESTIGATIONS

some of it, especially that relating to physical conditions and usual methods of operation, has been published in previous bulletins of this series. These facts are summarized in Bulletin 13.*

9. *Conditions Affecting Extraction.*—Since there is an immense quantity of coal underlying the state and only a comparatively small portion has been extracted, it is perhaps natural that little serious attention has been given to the subject of high recovery. Those controlling production have been concerned principally with other phases of the subject, not because they have been indifferent to the highest possible utilization of resources, but because they have believed that the methods in use were giving the lowest possible cost of production; and low cost of production has been regarded as a necessity for the development of the Illinois fields in competition with other coal fields.

Table 3, rearranged from Bulletin 13, gives the dimensions of the workings and the estimates of recovery for the mines examined by the Coöperative Coal Mining Investigations.

The values for the percentage of extraction given in the last column of Table 3 are, in most cases, founded upon estimates furnished by the operators. In many instances subsequent investigation has shown that these values are not correct. There are only a few mines in the state from which it has been possible to obtain accurate data on recovery because of the lack of information on which such data could be based. Generally, it has been found that persons estimating the percentage of recovery have been inclined to use values too high and have failed to take into account some of the sources of loss. Later figures on extraction, the most trustworthy it has been possible to obtain, will be found in the descriptions of the districts.

10. *District I.*—The No. 2 bed varies in thickness from two feet, eight inches to four feet, the average thickness being about three feet, two inches. On the east side of the LaSalle anticline the thickness of cover ranges from 40 to 200 feet; on the west side the bed lies at a depth of 350 to 550 feet. In the eastern, or Wilmington, section the roof is a smooth gray shale, though sandstone is found in some places. In the western or LaSalle field the roof is a gray

* Andros, S. O., "Coal Mining in Illinois," Ill. Coal Min. Invest., Bul. 13, 1915.

shale. In the Wilmington field the floor is a dark gray fire clay varying in thickness from a few inches to several feet. When this clay is wet, it heaves badly under pressure. In the LaSalle field the floor is fire clay, but a hard sandstone is sometimes found immediately beneath the coal.*

Nearly all the coal produced in this district is mined by the long-wall method, and this method, of course, gives the highest possible percentage of recovery. G. S. Rice says that at one mine in which a record was kept for six years the loss of coal from all causes was five per cent.†

11. *District II.*—The No. 2 seam is found under shallow cover ranging from 25 to 160 feet. In most places the floor is sandstone, but shale or clay is occasionally found. In places a wet and fluid sand is found about thirty feet below the surface, and it has a marked effect upon surface subsidence, causing the formation of rather deep pits instead of gentle sags. The bed is divided into two benches by a shale parting, varying in thickness from one-eighth inch to thirty-six feet. The bottom bench varies in thickness from 3½ to 4 feet, and the top bench has an average thickness of two feet. Where the parting between the benches is less than four inches thick, the two benches of the seam are worked as one and the working faces in rooms and entries are from six to seven feet high. Where the parting is more than four inches thick, only the lower bench is mined and the parting becomes the mine roof. When both benches are worked and the bed is more than six feet thick, only the lower six feet of coal are mined, eight to twelve inches of top coal being left; but if the coal is not more than six feet thick the full thickness of the bed is mined, and the gray shale overlying the coal becomes the roof.

With one exception, the mines examined are operated by the unmodified room-and-pillar method. Operations are carried on without close adherence to the projected sizes of rooms and pillars. The result of this practice is a rather high percentage of extraction, as pillars are gouged to a considerable extent.‡ At one mine in this

* Andros, S. O., "Coal Mining Practice in District I," Ill. Coal Min. Invest., Bul. 5, p. 10, 1914.

† Rice, George S., "Mining-Wastes and Mining-Costs in Illinois," Trans. Amer. Inst. Min. Engrs., Vol. 40, p. 31, 1909.

‡ Andros, S. O., "Coal Mining Practice in District II," Ill. Coal Min. Invest., Bul. 7, p. 9, 1914.

district which is operated on the panel system and in which a serious attempt is made to remove pillars as far as possible, the percentage of extraction is probably higher than at any other mine in southern Illinois. At this mine the shaft is 115 feet deep. There are triple main and cross entries, each ten feet wide, with 20-foot entry pillars. Barrier pillars on main and cross entries are twenty feet wide. Rooms are twenty feet wide with 10-foot pillars. All cross-cuts are eight feet wide. Although there are no exact figures on the percentage of recovery, it is evident from the dimensions of the workings that about two-thirds of the coal is extracted in the first working. Since by slabbing pillars, forty to fifty per cent of the pillar coal is also obtained, the final recovery probably amounts to about eighty per cent. The rooms are widened about thirty feet before the end is reached, little or no pillar coal is left beyond this point, and as much of the remainder of the pillars as possible is taken out by slabbing.

The possibility of extracting a large amount of pillar coal depends upon the character of the top which may be allowed to fall without serious consequences, because the shale and sand overlying the coal seal the opening so that the influx of water is not seriously increased by a break. When the top falls, the necks of the rooms are boarded up and the water is handled by a pump.

12. *District III.*—The No. 1 and No. 2 beds are worked. The cover overlying the coal is thin. The topography of the surface in many places is rolling, with hills about 150 feet high near Matherville. Bed No. 2 lies at depths of seven feet to one hundred feet with an average cover of fifty-five feet. Bed No. 1 averages four feet in thickness and is broken in places by small faults, slips, clay veins, and rolls. A poorly developed parting divides the bed into two benches, the upper of which is in most places about two feet thick.

The immediate roof in the northwestern part of the district is of hard black shale which is easy to support. In the southern part a bituminous calcareous shale, two to five inches thick, lies in places immediately over the coal. This shale, called clod, is hard when first exposed to the air but after exposure softens and falls. Throughout the district the cap rock is limestone. In limited areas where the shale is missing, this limestone forms the immediate roof. Above

2

the cap rock occurs a dense, fine-grained, non-crystalline limestone locally called "blue rock."

Below bed No. 1 there occurs in places an irregular band of hard bone, three to six inches thick. The floor proper is of light gray micaceous fire clay which contains plant stems and roots. This clay heaves badly when wet and sometimes swells enough to fill the entry. In parts of some mines a carbonaceous shale lies between the fire clay floor and the coal; sometimes this shale is supplanted by sandstone. These casual deposits are called "false bottoms."

Bed No. 2 varies in thickness from 1 foot, 10 inches to 4 feet, and averages 2 feet, 6 inches. The bed has a slight dip to the east. A band of mother coal and iron pyrites persists throughout the bed. This occurs about fourteen inches from the roof. The immediate roof is of smooth and regular calcareous shale, known locally as soapstone. The floor is of soft gray fire clay which contains nodular concretions of iron pyrites called sulphur balls. The coal in this district lies near the surface, but at no point is the overburden stripped.

Except at two mines, the mining system is the simplest form of double-entry room-and-pillar. Table 3 shows the dimensions of workings in the mines examined. The coal is gained during the first working with a waste of pillar coal amounting to about 45 per cent of the bed. At the two exceptions 75 per cent of the pillar coal is recovered on the retreat, a large percentage for Illinois room-and-pillar mines.

A main entry and a parallel air-course, each six feet high and eight feet wide, are driven from each side of the shaft toward the boundaries. At right angles to these main entries, pairs of cross entries are driven every 500 feet. On the cross entries, after leaving a barrier pillar of 50 feet, rooms are turned on 45-foot centers. Room necks are 7 feet long and 8 feet wide, and are widened to the left at angles of about 45 degrees; thus they reach the full room width of 26 feet at distance of 14 feet from the beginning of the widening. After the first room on each entry has been holed through, the room-pillar cross-cuts are closed by gob stoppings, and the line of No. 1 rooms is kept open; thus two additional air-courses are provided.

After the entry has been driven to the limit and the rooms on it have been worked out, the last pillar on the entry is drawn; then the other room pillars are drawn until the pillar between rooms

3 and 4 is reached. The room pillars between the main entry and room 4 are left to protect the main entry and air-course. The method of drawing pillars is illustrated in Fig. 3. When the room is driven

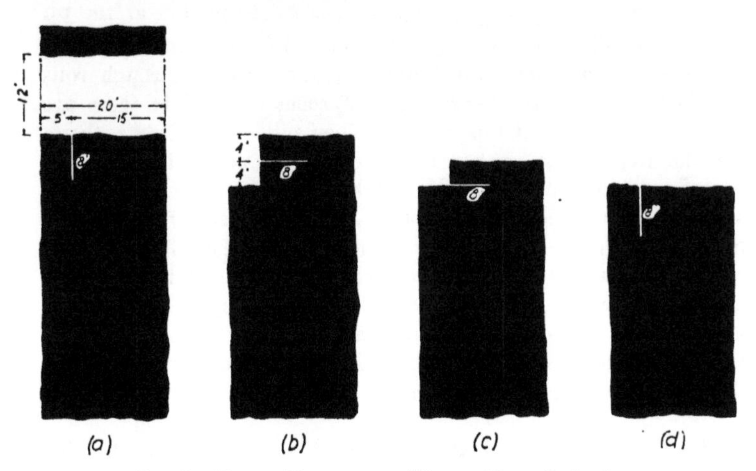

FIG. 3. PILLAR DRAWING AT MATHERVILLE, ILLINOIS

up to its full length, a 12-foot cut is made across the end of the pillar (a), a 5-foot slab about 8 feet long is shot from the side of the pillar, a 4-foot slab is shot from the end (b), and the end of the pillar is squared up by shooting off another 4-foot slab (c). Beginning again at (d), the process is repeated.

The hard roof is easy to support and often stands while 25 to 200 feet of pillars are being drawn. When the weight of the roof becomes too heavy, the roof breaks at the pillar ends. The cracking of the props gives ample warning of the break, and work is discontinued until the roof falls. The interval between the first heavy cracking of props and the breaking of the roof is usually not more than twelve hours.

A break line of about twenty-five degrees with the face of the rooms is roughly maintained. When roof falls prevent access to the squared-up pillar ends, a 12-foot cut is again made completely through the pillar, as at the face of the room when drawing began, and with this new pillar end the procedure continues; consequently, very little pillar coal is lost. Carl Scholz, President of the Coal

Valley Mining Company, states that at mine No. 3 at Matherville the loss of pillar coal does not exceed four per cent.

At the No. 3 mine of the Coal Valley Mining Company, the cost of producing coal is much less on pillars than on advance work in rooms. Room coal costs on the average $1.25 per ton at the pit mouth, and pillar coal costs $1.015. This difference in cost exists because track, yardage, bottom digging, and driving through rolls and slips are properly charged against room coal, while there are no such charges against pillar coal. When pillars are drawn, therefore, the average cost per ton for the total production is materially reduced. At this mine rooms are worked with one man at the face, but two men are placed at each pillar and at the face of each entry. Only one man has been injured in connection with the pillar drawing.

With the extraction of such a large percentage of the bed surface subsidence is to be expected. The topography of the surface is rolling, and subsidence is usually indicated by cracks in the hillsides. The largest single area affected was reported to be one acre which subsided from 6 to 12 inches.*

13. *District IV.*—In District IV the No. 5 coal is mined. The average thickness of this coal is 4 feet, 8 inches according to data taken at 240 mines and given in the Thirty-first Annual Coal Report of Illinois. The No. 5 bed outcrops in Peoria, Fulton, and Knox Counties, but is found at greater depths toward the east. It lies from 300 to 600 feet below the surface in Macon County, 400 feet in McLean, and from 260 to 300 feet in Logan.

The roof is of black sheety shale varying in thickness from a few inches to 35 feet and containing occasionally "niggerheads" of pyrite. In many mines there is, in places, a layer of pyrite two or three inches thick between the coal and the shale. Where this layer is present, the shale is protected from the air and stays up; where it is not present, the shale falls badly and sometimes caves to a height of 35 feet. A limestone occurs above the shale in most mines, though in a few places a fine grained micaceous sandstone is found. In some cases the shale is absent, and the cap rock becomes the roof.

A great many clay veins extend through the coal and the roof shale; there are also small faults, slips, and rolls, and places where

*Andros, S. O., "Coal Mining Practice in District III," Ill. Coal Min. Invest., Bul. 9, pp. 11 et seq., 1915; "Coal Mining in Illinois," Ill. Coal Min. Invest., Bul. 13, 1915.

the coal has been eroded and the space has been filled with drift. It is difficult, therefore, to calculate the total tonnage and to project any plan of operation. In many places the coal adheres to the roof and separates from it with difficulty. In one mine about an inch of coal is left to protect the roof shale from the air. In most mines the floor consists of a dark gray clay which heaves badly when wet.

Operations are conducted on the unmodified room-and-pillar system or on the so-called panel system. Dimensions of workings are given in Table 3. There are also four mines in the district which are operated on the long-wall system. Mining methods have not been given very careful attention, and the variations in the coal bed tend to minimize the effect of such attention as has been given. The method of mining generally practiced in the district involves the running of parallel main entries from the shaft toward the boundaries, and the turning of cross entries from the main entries at intervals of 350 to 400 feet. Rooms are turned off these cross entries on 30-foot to 42-foot centers, and are driven 20 to 30 feet wide. Room pillars average 9 feet in width and rooms 26 feet, but pillars are gouged as the miner pleases. This haphazard method is productive of so many squeezes that in some mines a modification of the system has been employed in which stub or room entries are turned off the cross entries. This method approaches the panel system and is called, locally, "block-room-and-pillar." Sometimes a sufficiently large cross barrier pillar is left to confine a squeeze to the block in which it originates, but generally the barrier pillar is gouged and squeezes ride over it unchecked until they reach a horseback or some ungouged pillar which is large enough to stop them. In several mines squeezes originating in rooms have traveled to the main barrier pillar and to the solid coal at the entry face. In one mine an entry was saved from a threatened squeeze by very heavy timbering ahead of the squeeze.

Eleven of the sixteen mines examined are at present operated on this semi-panel system, but the relative dimensions of room and room pillar have not been changed from previous operations. These dimensions are not safe under the roof found in the district. Room width is not uniform, but rooms are narrowed to avoid horsebacks and widened again where the coal resumes its normal thickness. There is a temptation to get all the coal possible on the advance, because the numerous rolls make uncertain the total tonnage which can be extracted from any area, and the rolls interfere seriously with any projected plan since cutting through them is expensive.

Pillars are drawn in only a few mines, and in these drawing is not done systematically but is confined to shooting slabs off the thickest parts of the pillars. Room pillars are tapered to cross-cuts in nearly all mines. In one case an attempt was made to draw pillars, and a track was laid along the rib, but objections were raised by the miners to this position of the track, and the attempt was abandoned. Principally because of the insufficient pillar-width, the floor of fire clay heaves badly even when dry.*

Nineteen mines were examined in this district, and the estimates of the percentage of recovery furnished at seventeen ranged from 55 to 75 per cent. averaging 67.26 per cent. It is probable that most of the estimates are too high for, although the gouging of pillars tends toward high percentage of extraction, careless methods always result in the loss of much larger quantities of coal than is supposed. One company, which has given careful attention to the forms and dimensions of its workings, is extracting about 70 per cent of the coal. It is doubtful however, if the extraction throughout the district as a whole amounts to 60 per cent.

14. *District V.*—Bed No. 5 in Saline and Gallatin Counties lies at a depth of 25 to 450 feet, being nearest the surface along the southern portion of the district. The bed varies in thickness from 4 to 8 feet, and averages 5⅓ feet in Saline County and 4 feet in Gallatin County.

The roof of the No. 5 coal in this district is of shale which is sometimes laminated and interbedded locally with bone and stringers of coal for a distance of 3 feet above the seam. The roof usually contains many concretions of iron pyrites called "niggerheads." It breaks quickly when wide spans are left supported, and it is drawn when it shows a plainly marked parting not more than 4 inches above the coal; but such a parting rarely occurs and the coal bed is so thin that the top coal cannot profitably be left in place. There are numerous falls which can be avoided only by making entries narrower than at present.

---

* Andros, S. O. "Coal Mining Practice in District IV," Ill. Coal Min. Invest., Bul. 12, pp. 15 and 19, 1915.

The floor is of fire clay which in places contains much sand and heaves badly when wet. The bed contains many hills and rolls causing grades as high as 15 per cent in the entries of some mines. The coal is not pinched out at these hills, but follows the contours with undiminished thickness. In some mines about 9 inches of bottom coal is left below a "blue band," but as this bottom coal is not of good quality, increased facility in shooting compensates for the loss of coal.

The room-and-pillar system of mining is used exclusively, a main haulage entry and a parallel air-course being noted in every mine examined except one,— in which triple main entries were driven, two for intake air and one for return air and haulage. In the smaller mines and in many of the larger ones, the dimensions of workings are not suited to the roof conditions. The main entries vary in width from 14 to 16 feet. A few shaft pillars have been gouged. The room stumps, which are left when rooms are turned off the cross entries, are generally small. The closing of entries by roof falls may often be attributed to local squeezes which ride over the room stumps. Table 3 gives dimensions of workings for each mine examined.

The custom of driving wide rooms and entries, of leaving narrow pillars throughout the mine, and of obtaining all the coal possible on the advance without attempting to draw pillars has resulted in a high percentage of extraction for Illinois mines. The percentages given in Table 3 were calculated from the most nearly exact data obtainable at the time of their publication but are unquestionably too high. This reported extraction, averaging 67.1 per cent for the seven mines examined, was accomplished only with greatly increased expense for cleaning up.* One of the large operators of this district reports an average recovery at ten mines of 60.5 per cent over a 5-year period with a maximum of 72 per cent and a minimum of 52 per cent where the cover varies from 60 to 414 feet. Pillar drawing is not practiced, and it would be impossible to gain the percentages of coal given in the table if the dimensions given were adhered to, but pillars are gouged to such an extent that there should be a higher percentage of extraction than is calculated from the dimensions of rooms and pillars in the table.

---

*Andros, S. O., "Coal Mining Practice in District V," Ill. Coal Min. Invest., Bul. 6, pp. 9 and 12, 1914.

15. *District VI.*—This district has experienced a rapid development, because the No. 6 coal commands a ready market; consequently mining on a large scale is possible. Bed No. 6 lies close to the surface along the Duquoin anticline* but dips sharply to the east, reaching a depth of 726 feet at Sesser. A general uplift has brought it to the surface along an east-west line extending through Carterville to Marion and along a southeast line from Marion to the boundary of the district. East of the area affected by the Duquoin anticline, the bed has a pronounced dip to the north. Along the outcrop line there are a few slopes and strippings, but the steep dip of the bed leaves only a small acreage with thin cover, and the remaining openings are shafts. The seam itself is thick, ranging from 7½ to 14 feet and averaging, as shown by 130 borings, 9 feet, 5 inches. A clean persistent parting of mother coal lies 14 to 24 inches below the top of the bed, and a second parting generally appears 5 to 8 inches lower down. Above the upper parting the coal occurs in layers 3 to 6 inches thick, with partings of mother coal between them.

The immediate roof consists of a gray shale 15 to 110 feet thick. This shale does not stand well when the coal is removed, and the top coal is generally left as a roof, at least until the rooms are finished. The bottom is generally of clay, four inches to eight feet thick, below which is limestone. There is only one persistent band of impurity in the bed. This, which is known as the blue band, generally consists of bone or shaly coal and is found uniformly at a height of 18 to 30 inches from the bottom. Its thickness varies from ½-inch to 2½ inches.†

The large number of squeezes which have occurred in mines of District VI would seem to indicate the presence of one or more thick beds of strong rock among the overlying strata. A study of the logs of numerous wells does not, however, show the presence of any continuous strong bed which would be a serious obstacle to the introduction of methods allowing a larger percentage of extraction. The State Geological Survey makes the following statement concerning the overlying limestone: "Over a large part of the area within 25 feet of the coal is a limestone cap rock which in places rests upon the coal, except for the draw slate that lies between. Where the lime-

*Andros, S. O., "Coal Mining Practice in District VI," Ill. Coal Min. Invest., Bul. 8, p. 11, 1914.

†Shaw, E. W., and Savage, T. E., U. S. Geol. Sur., Folio No. 185.

stone cap rock is not present within 25 feet of the coal it may be entirely absent, or lie at a considerably greater distance above the coal, amounting in some places possibly to as much as 100 feet."* The limestone cap rock is of variable thickness up to about 11 feet, the average thickness being 4 to 5 feet.† In some places sandstones are found at various distances above the coal, but none of these seems to be close enough to the coal to affect the choice of a mining method. In other words, it seems that there is no layer of rock, sufficiently near the coal to require serious consideration, which cannot be broken by careful attention to the proper methods. An examination of bore hole records of the Connellsville district of Pennsylvania, where the percentage of coal extracted is very high, indicates that there is more difficulty in breaking the overlying layers of rock in that district than would be experienced in most cases in District VI of Illinois. At a few mines an unusually wide room pillar is left in the middle of a panel for the purpose of limiting the area affected by a squeeze.

According to Table 3, all mines in the district, except strippings, are worked by the room-and-pillar method or by the panel method. Where the latter is employed, frequently no attention is paid to panel pillars so that the advantage of this method in the stopping of squeezes is largely lost. Practice is not uniform in regard to the number of rooms, which may be as low as 14 or as high as 30, turned from a room entry. The description of mining practice in this district‡ given in Bulletin 8 of the Coöperative Investigations says, "The immediate roof overlying the coal falls in slabs after short exposure to the air and top coal is usually left to protect it, but the cap rock is a tough coherent shale which does not break easily. The first mines opened in the district had widths of rooms and pillars unsuitable for this tough cap rock. New mines as they were opened adopted the dimensions of the older mines and a great waste has resulted through the loss of pillar coal. It will never be possible in this district to draw any considerable portion of the pillars where rooms 20 to 29 feet wide are driven with narrow room

---

*Cady, Gilbert H., "Coal Resources of District VI," Ill. Coal Min. Invest., Coöperative Agreement, Bul. 15, p. 88, 1916.

†Ibid, p. 32.

‡Andros, S. O., "Coal Mining Practice in District VI," Ill. Coal Min. Invest., Bul. 8, p. 12, 1914.

pillars. Fear of yardage charges has been an important factor in maintaining the present improper dimensions. . . . With present dimensions when rooms have been driven 200 to 300 feet there is a large area of unsupported cap rock. If an attempt is made to draw pillars under such conditions a squeeze is usually started which often rides over room and entry pillars and sometimes affects a large acreage. In one mine 85 acres were squeezed; in another, 80."

Early operations were carried on without regard to the possible production of squeezes. Pillars were gouged out or entirely removed whenever the demand for coal seemed to excuse this procedure; a natural consequence was the occurrence of squeezes. At one mine there have been five squeezes of which two involved about 80 acres each, one about 40, one about 20, and one possibly 10. The present plan for the future operation of this mine contemplates leaving barrier pillars 150 feet wide along the important entries, and removing this pillar coal later. It is believed that this plan will confine roof movement to the worked out areas and that the entry pillars and barriers can be extracted later. This plan is much the same as that shown by Fig. 32, page 102.

At another mine a large squeeze approached within 125 feet of the air shaft and caused a depression on the surface which necessitated the regrading of a considerable amount of track, including the track scales. Practice at this mine represents one extreme, since no attempt is made at room-pillar drawing beyond driving cross-cuts about 30 feet wide at the ends of rooms, and as much coal as possible is taken on the advance. An attempt will be made to take out barrier and entry pillars on the retreat. Rooms are driven 25 feet wide on 45-foot centers, and cross-cuts are 25 feet wide. The excavated area is about 50 per cent and the top coal, which is only 18 inches thick, is left up.

The figures for recovery of coal given in Table 3 are unquestionably too high although they were based on the best information available at the time. The average percentage of extraction in District VI is not more than 50 per cent, and it is probably nearer 45 per cent. The maps of mines may show an excavated area of 50 per cent or even more, but they do not take into account the unmined top coal. The thickness of coal taken out is generally about 7 feet and top coal ranging from a few inches to 4 or 5 feet in thickness is left. Even if the top coal is ignored, the extraction is not so high as the estimates generally indicate because of losses in squeezed areas and boundary barriers.

Special investigations on the subject of recovery made at several mines in Franklin County gave results which are summarized as follows:

> At one mine, the recovery in worked out areas where pillars are not drawn is about 65 per cent; where the pillars are taken, it is about 75 per cent.
>
> At another mine, close observations were made in connection with a study of subsidence. In a panel where the extraction was considered good and possibly above the average for the mine 40 per cent of the coal is left as pillars. Two feet or 20 per cent of the thickness of the bed is also left as top coal, and the loss from this cause would be 20 per cent of the remaining 60 per cent or 12 per cent of the total. The total loss is then at least 52 per cent. No attempt has been made to extract room pillars, but some entry pillars are taken, and top coal is taken over the area in which these pillars are drawn.
>
> At one of the mines where the thickness of the coal is greater than the average, little pillar work has been done. The coal varies from 9½ to nearly 16 feet in thickness,* and about 9 feet of it is taken out. Generally about one foot of coal is left on the bottom to avoid the possibility of taking up a bed of "black jack" which is not easily distinguishable from coal. This black jack is probably a coal of very high ash content. The leaving of bottom coal results in the elevation of the working place in the bed so that the top coal left is only 3 feet or even less in thickness. In a few places in this mine some pillar coal has been taken out. Where this was done, break-throughs about 24 feet wide were driven at the ends of the rooms. Then work was commenced at the ends of the pillars, and coal was taken out by pick work. This work seems to have been successful, but it has not been followed systematically. The largest number of pillars which have been taken together was six, and no attempt was made to obtain a break in the roof.

* Cady, G. H., "Coal Resources of District VI," Ill. Coal Min. Invest., Bul. 15, p. 58, 1916.

The leaving of coal on the bottom is a practice followed at only a few other mines. In some cases where the blue band is thick the mining is done above it, and a portion of the upper part of the bed, ordinarily included in the top coal, is taken down. At two mines where this method is followed in part, the blue band and the coal below it are left in where the blue band is thick and the top coal is taken to within about ten inches of the roof, at which point there is a parting. Greater care is required to prevent the breaking of the top coal where this is done.

At one of the mines in the southern part of Franklin County a little pillar coal is drawn, though pillar coal is not depended upon for an important part of the output. Rooms are 25 feet wide with 20-foot pillars. Rooms are holed through into those of adjoining panels. When the rooms have reached their full lengths, cross-cuts 24 feet wide are driven across the ends of the pillars. In addition to these cross-cuts the pillars are probably slabbed to some extent. The coal is about 9 feet thick, and about 1½ feet of top coal is left up. No bottom coal is left. The barrier pillars are about 100 feet thick. Break-throughs are 21 feet wide. It seems hardly proper to speak of this kind of work as the extraction of pillar coal, but it represents a practice which is common in this district.

At one of the mines, pillars are drawn, beginning in the middle of a panel, in six rooms at a time; then another group of six pillars is attacked, one pillar being left untouched between the groups. This is simply another method of attempting to get as much coal as possible before being driven out by a fall of the top, and is not an attempt to break the cap rock.

Systematic work in the recovery of pillar coal as done at one of the mines is illustrated in Fig. 4. The mine is operated on the block system commonly called the panel system, though the panels are not kept sufficiently isolated to warrant the use of the term. Cross entries are driven at intervals of 1,370 feet, and panel entries are driven through from cross entry to cross entry. On each pair of panel entries twenty-eight rooms are turned to each side on 40-foot centers. A barrier pillar 125 feet wide is left along the cross entry. After the room entry is driven the rooms are necked, but only the first fourteen rooms on each side of the room entry are worked, and these are finished before the rooms at the other end of the panel are driven. Break-throughs are normally 11 feet wide; those at the ends of the rooms are 24 feet wide. Pillar drawing is commenced when the rooms at one end of the panel are finished, and the coal is taken out through the

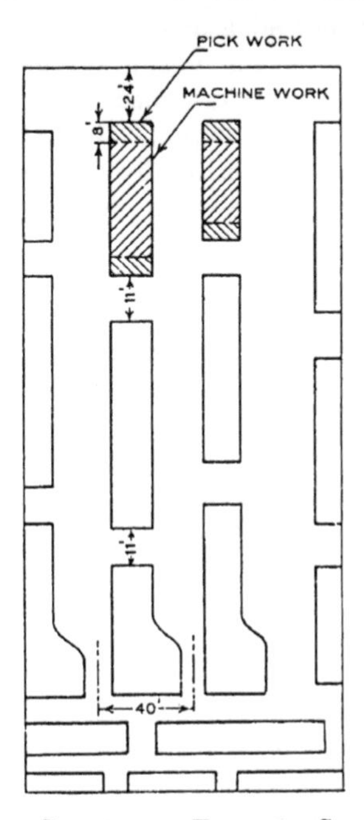

FIG. 4. PILLAR DRAWING IN FRANKLIN COUNTY, ILLINOIS

first cross entry, that is, the one next to room No. 1. The coal from the remaining portion of the panel is taken out through the next cross entry, that is, the one next to room No. 28. The advantage of this method lies in the fact that the extraction of the coal from the second half of the panel is not interfered with, as far as haulage and ventilation are concerned, by movements caused by pillar drawing in the first half. The pillar coal is attacked first by the driving of 24-foot cross-cuts at the ends of the rooms; then other cuts are made in the pillar with the breast machine in such manner as to leave stumps between the cuts and the break-throughs. These stumps are removed as far as possible by pick work, but the miners are not always able to finish as much work with the machines as is desired since this work is frequently interrupted by movements of the roof. As much of the coal as is possible is then taken out with picks. While the work at this mine is as systematic as that at any mine in southern Illinois, the company has no exact record of the amount of pillar coal extracted, but it is known that the pick-mined coal amounts to approximately 10 per cent of the output. When the pillars are not drawn, the recovery is estimated by the operators to be about 65 per cent; when they are taken, the estimates run about 75 per cent.

Through the courtesy of the Franklin County Coal Operators' Association (Illinois), data have been made available regarding the extraction of coal in that county as presented in the following paragraphs:

The coal mined is the No. 6 bed of the State Geological Survey classification. Measurement of 113 sections taken in twelve of the largest mines in the county gave an average thickness of 9.2 feet of coal, the average minimum thickness for the same twelve mines being 8 feet, and the average maximum thickness 10.64 feet. The blue band, which is characteristic of the No. 6 coal bed, varied from ¼ to 2 inches in thickness, and its average distance from the floor was 21.5 inches. Owing to the difficulty of keeping up the shaly material above the coal bed, the top coal is almost universally left as roof protection, and, up to the present time, very little of this top coal has been recovered, although some operators are expecting to recover it at a later date in connection with pillar drawing. In one of the twelve mines from which the data were obtained, top coal was not left in the rooms. This, however, is exceptional practice, the average thickness of the top coal left in the twelve mines being 1½ feet. The average thickness of coal mined was 7.46 feet, and the average tonnage per acre to January 1, 1916 was 6,627 tons. This is equivalent to 40.7 per cent extraction, if it is assumed that all the 9.2-foot bed is available for shipment, or to 41.6 per cent if it is assumed that the blue band and refuse discarded in the loading, or 0.2 foot, is deducted from the thickness of the bed. A very careful estimate for each of the twelve mines noted, made by dividing the total amount of coal in the area mined up to January 1, 1916 into the actual shipments since the mine began operating, gave percentages of extraction varying from 37.7 to 49.5, or an average of 41.4 per cent.*

For six of the twelve mines, data were available for the average percentage of extraction in the portion of the bed actually mined; that is, the total thickness less the top coal left up to protect the roof. This average is 48.65. These mines are all comparatively new mines, and in only a few cases has any portion of the workings reached the boundary so as to permit drawing the pillars in return workings. At many of the mines it is hoped to increase the percentages of extraction through subsequent pillar drawings, but the amount of such increase is, of course, problematical. In many instances squeezes have already occurred, but as a general thing only the room pillars have been affected.

The twelve mines under discussion are representative of the practice in Franklin County and to a great extent of that of southern Illinois. In a number of these mines experiments are now being conducted to determine in what respect present methods of working may be modified to yield a larger percentage of extraction. Although these mines are operating under practically the same physical conditions and all on the panel system, the variation in the detailed operations, such as the number of rooms per panel, or the width of barrier pillars, indicate the necessity for a critical comparative study of details to determine the best method for the given conditions.

Investigations in Williamson County supplied the following facts:

At one mine rooms are 21 feet wide on 40-foot centers. From 2 to 2½ feet of top coal is generally left up on the advance. In one part of the mine the coal is 11 feet thick and only 7½ feet of it is taken out. It is estimated that 40 to 50 per cent of the pillar coal is won. The top coal, which is the

* The Peabody Coal Company reports that the percentages of extraction at its four mines in this district are 67, 63, 55, and 55.

best part of the bed, is taken out when the pillars are drawn. There are no definite records on recovery, but it is probable that 55 or 60 per cent is gained. If 7½ of the 11 feet are removed, and 40 per cent of the pillars and all the top coal over the area in which the pillars are drawn are removed, the extraction is about 60 per cent.

In another mine a rather high percentage of extraction is attained because of favorable conditions which permit the leaving of small pillars. The coal is 9 feet thick and the depth only 100 feet. Rooms are 24 feet wide on 35-foot centers. In some cases top coal, about 20 inches thick, is left if the machine runners think the top is insecure. Probably from 65 to 70 per cent of the coal is taken out. No pillar work could be done with rooms and pillars of these dimensions, but on one side of the mine 15-foot pillars are now being left with the intention of taking them out on the retreat.

At some mines in the western part of Williamson County considerable trouble has been experienced, because large quantities of water enter when the top is broken. The cover here is only about 100 feet thick and there are only 3 to 4 feet of solid rock. The rooms of one mine are 20 feet wide and are driven on 40-foot centers, although they are sometimes crowded. Some rooms were driven on 32-foot centers, but the pillars were not sufficient to prevent squeezes. Entries are 12 feet wide and entry pillars 20 and 25 feet wide. The coal is 5 to 11 feet thick with an average thickness of 8 feet. Top coal averages 20 inches in thickness. Above the bed is a shale; above this is a so-called soapstone, ranging from 2 to 8 feet in thickness and averaging about 4 feet; and above this is a black shale. In some places a draw slate from 1 to 2 feet thick occurs above the coal, and above this is limestone 1 to 3½ feet thick. Where is no draw slate, there is no limestone. An unconsolidated sand is found in some places above the coal. The pillars are sometimes slabbed a little to compensate for the coal left in entry and barrier pillars. When this slabbing is done, the extraction amounts to about 50 per cent. In some cases extra cross-cuts are taken, and the extraction is thereby increased to about 75 per cent. On the whole, the extraction is estimated to be about 60 per cent,— an estimate which is probably reliable since the work is more carefully done here than at many other mines of the district. Where more than 60 per cent of the coal is taken out in this mine, the roof breaks and water enters in large quantities. Although the amount of water is influenced by precipitation, the flow is continuous. In one case where the rock had been broken and a large quantity of water had entered the mine, it was thought that the strata was drained to some extent and that it would be possible to allow the top to break at a slightly higher elevation. It was found, however, that the new break allowed a large amount of water to enter the mine, and it has been impossible for the company to do any pillar work. It is planned, as some of the workings reach the boundaries, to draw pillar coal. In these cases pumps will already have been installed, and it will be possible to conduct the water to these. The water, moreover, will be entering in abandoned places and not between the workers and the shafts. Some of the workings are now not far from the boundaries, and the plan can be put into operation in the near future.

At another mine the coal is 9 feet thick, the top is of white shale, and the bottom of fire clay. The top coal is about 2 feet thick. The mine is operated on the panel system. No pillars are drawn until the rooms on an entry have been finished; then a cross-cut three machine-cuts wide, or about 20 feet, is driven at the ends of the pillars in about half the rooms on the inside end of the stub. Following this, another cross-cut is made farther back in the pillar leaving a stump about 10 feet wide. The distance between the first and second cuts varies according to conditions, or according to the judgment or inclination of the machine men. If it is made farther back, some of the pillar coal is lost. This operation is repeated until the first break-through in each pillar is reached, the remainder of the pillar being left standing until all the rooms on the stub are finished; then the room stump and the entry pillars are drawn. No effort is made to obtain a break in the roof, and the leaving of stumps of pillars is likely, by partially sustaining the roof, to bring on a squeeze. At present the driving of rooms without necks is being tried, the purpose being to avoid payment for narrow work; and it is believed there will not be sufficient difference between the support left under that system and that left under the present system to endanger the entries. These rooms without necks are turned six machine-cuts wide and are widened to seven cuts beyond the first cross-cut.

At another mine the average thickness of the coal is 9 feet, 4 inches. The top coal is about 2 feet thick and is left up until the pillars have been partly drawn back, being taken down just before the track is removed. There is generally a good parting between the main bed and the top coal, which is said to be poorer than the main bed. Above the bed is shale of unknown thickness, which has never broken high enough to expose any other rock above it. This shale slacks when exposed to the air. It does not form a very good top and most of the entries are timbered. The bottom is generally of clay, but in some places limestone appears next to the coal. Rooms are 20 feet wide on 30- to 35-foot centers and are 185 to 190 feet long. Stub entries are turned on 400-foot centers, and 16 to 18 rooms are turned from a stub. The room pillars are gouged to a considerable extent. It is planned that all the rooms in a panel shall be driven to their full length before pillar drawing is commenced, but this plan is not always followed, and squeezing sometimes commences before all the pillars can be attacked. This, of course, is promoted by the gouging of room pillars. When the rooms are finished, cross-cuts 20 feet wide are driven at the face with breast machines. The rest of the pillar work is generally done with picks though machines are used when possible. Movement of the roof, however, generally interferes with machine work after the first cross-cut. The pick work generally consists of slabbing along the sides of pillars, but machines are sometimes used. Squeezes have always been confined to the panels, and no entries have been lost until the entry pillars have been drawn. A careful computation, based upon a comparison between the actual area worked and the number of tons hoisted, shows that the extraction at this mine has been 48.89 per cent. This is one of the most thoroughly worked mines in the No. 6 bed, and the estimation of percentage of extraction is undoubtedly as close as any that has been made. The results found furnish one of the reasons for the statement that extraction in most mines of the district is less than the operators of the mines believe it to be.

At a Perry County mine the general system is the same as in Williamson County. Room entries are driven through from cross entry to cross entry. A somewhat closer adherence to the panel system is to be noted, however, in that 25-foot pillars are left at the ends of rooms. Rooms are 24 feet wide on 60-foot centers, and they are driven 250 feet long. Break-throughs are staggered. When the room is completed, an 18-foot cross-cut is driven through the pillar at the end. Top coal, about 3 feet thick (the best of the bed in quality* as it is at the mine last mentioned), is left up until pillar drawing commences. Pillar drawing is commenced in the middle of the panel. After the completion of the cross-cut at the end of the first pillar attacked, the top coal is loosened by a light shot near each rib. Work on the pillars is then prosecuted by making a cut through the pillar, if the condition of the top will permit, wide enough to leave an 8-foot stump at the end of the pillar and another of the same dimension next to the nearest break-through. These two 8-foot stumps and whatever is left by the machines are taken out by hand work. Two men are used on solid work and two on the machine. This method has been found successful and a considerable amount of pillar coal has been recovered, but pillar drawing is not a necessary part of the system and is not always carried out. Where the pillar coal and the top coal are taken, the recovery is said to be from 75 to 80 per cent of the coal in the area actually worked.

Several plans are now being tried or considered for the more nearly complete extraction of the coal in this district. One of these, which has so far been given only an incomplete test, is a panel long-wall method. Double entries, which would have been the room entries of a panel under the ordinary methods of operation, were driven 340 feet long. At the end, two rooms were driven on each side separated by 25-foot pillars. The rooms at the extreme end of the block on each side were 9 feet wide and the ones further back were 18 feet wide; each was 200 feet long. They were connected at the ends so that ventilation was obtained, the course of the air current being as shown

---

*For a discussion of the differences between the top coal and the remainder of the bed see "Chemical Study of Illinois Coals," by S. W. Parr, Ill. Coal Min. Invest., Bul. 3, p. 49, 1916.

in Fig. 5. Then the outby ribs of the 18-foot rooms were worked as long-wall faces by continuous-cutting chain machines making a 6-foot cut. The top behind the working face was propped. It was the intention to support the immediate roof until the face had advanced some distance and then to make an attempt to break the overlying rock by the withdrawal of the props. This plan was found to be impossible,

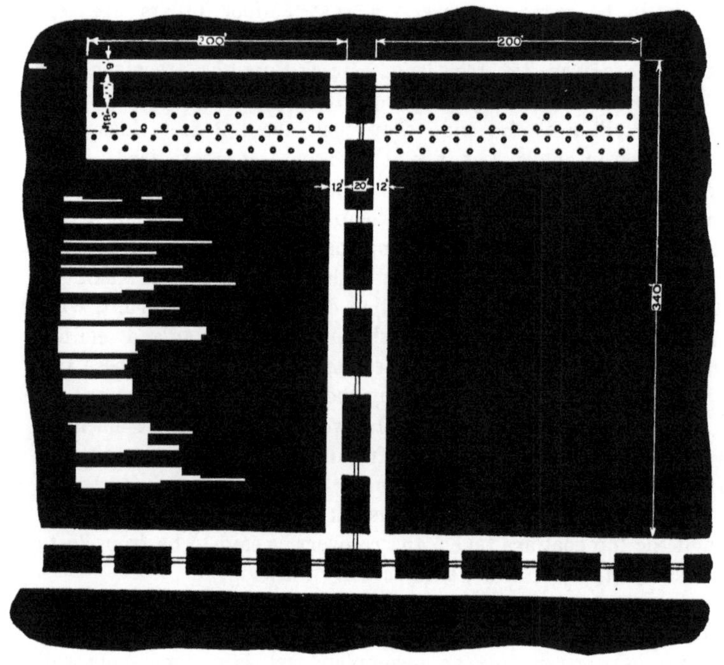

FIG. 5. PANEL LONG-WALL

however, as the top fell when the face had advanced only about 40 feet. Other conditions made it necessary to discontinue the experiment temporarily. In operating by this method, sprags were placed in the cutting behind the machine to prevent the premature fall of the coal. No trouble was experienced in getting the coal down; it was produced very rapidly and was easily handled. At present it is not known whether the top can be broken along the desired line, but it will be seen that this line is only 400 feet long and that it is interrupted in the center by the entry pillar. Even if it is not possible to break the top and to work the coal back continuously on two longwall faces, it seems that the attack can be repeated farther back in the block and that coal can be produced as cheaply as by the ordinary method; also that a much higher percentage of extraction can be attained. If it should be necessary to follow the method by repeated attacks on the block, there would be some resemblance to the "single room" method successfully worked in West Virginia.

Various other plans for higher extraction have been suggested and some have been partly applied, but the great demand for coal, which has been stimulated by the European War, has caused coal producers to concentrate all their attention upon the immediate production of a large tonnage. Anything in the nature of experimental work will be postponed until the return of more nearly normal conditions, but there is reason to believe that successful efforts will be made to increase the percentage of extraction and that the present large loss will be greatly decreased.

16. *District VII.*—The coal worked in District VII is the No. 6 bed on the west side of the Duquoin anticline. The thickness varies from 2½ feet to 14 feet and averages about 7 feet. There is a well defined parting plane in the coal about 18 inches from the roof. Where the roof is of black shale and where the coal is 7 feet or more in thickness, the upper bench or "top coal" is left. The roof is a non-calcareous black shale, a calcareous gray shale called locally "white top" or "soapstone," an unconsolidated dark gray or black shale called "clod" and made up of fragments of varying size and hardness extremely difficult to support, or a hard gray limestone called "rock top." A poorly defined cleat or cleavage in the coal may be seen in some places. The floor throughout the district is of fire clay which generally heaves when wet.

The thickness of the coal is almost ideal for easy working and for large production; some of the mines have obtained daily capacities which rank among the highest in the world. The older mines have been worked without much regard to system, but the newer ones are more carefully planned. The planning, however, is directed toward large daily production rather than toward a high percentage of extraction.

Varying roof conditions often make different entry and room

widths necessary in different sections of a mine. In many mines the entries and rooms under rock top are too wide and the pillars too narrow,—a condition responsible for squeezes which sometimes have endangered even the shaft. Squeezes have occurred in thirteen of the twenty-five mines examined in this district; they have generally begun in sections in which the roof was of limestone. In mines in which the rooms are not frequently surveyed there is no definite knowledge of room-pillar width except at cross-cuts. Table 3 gives dimensions of workings at each of the mines examined.

In ten of the mines examined where the immediate roof was of thick black shale, top coal was left to prevent variations of temperature and humidity from affecting the shale of the roof proper, which spalls badly when exposed to the air. Where no top coal is left, this black shale usually falls with the coal or is drawn. Where there is less than four inches of shale between the coal and the limestone, the shale is drawn. In some mines where the latter is more than four inches thick it is propped; in others it is drawn, unless it is more than two feet in thickness.* The Peabody Coal Company reports extractions at its mines in this district of 65, 62, 60 and 50 per cent. At the twenty-five mines examined, the average estimated percentage of recovery furnished by the operators was 55.5 per cent. So far as is known, no efforts have been made to extract a higher percentage of coal. The reason for this attitude is to be found partly in the condition of the surface, which in many places is so nearly level that any noticeable subsidence disturbs the drainage and effects the value of the surface for agricultural purposes.

At present the abandonment of a large percentage of the coal is a result of the difficulties experienced in attempting to secure satisfactory agreements with the owners of the surface. In some cases the owners of coal rights are not the owners of the surface and are not free from responsibility for surface damage. Since the operators have found that the estimates of damage caused by subsidence are likely to be very high, it has become the custom to operate the mines under methods which will avoid subsidence. Unfortunately it has not been possible to estimate the exact amount of coal which must be left in the ground, and squeezes and subsidences have sometimes occurred when it was thought that sufficient coal had been left.

---

*Andros, S. O., "Coal Mining Practice in District VII." Ill. Coal. Min. Invest., Bul. 4, p. 11, 1914.

Even where low value of the land or good drainage reduces the cost of possible injury to the surface by subsidence, no effort is made to secure a higher extraction. The occurrence of squeezes is feared, and experience shows that the only way to prevent them without radically changing the system of mining is to leave large amounts of coal in the form of pillars. One company which was formerly getting 50 to 60

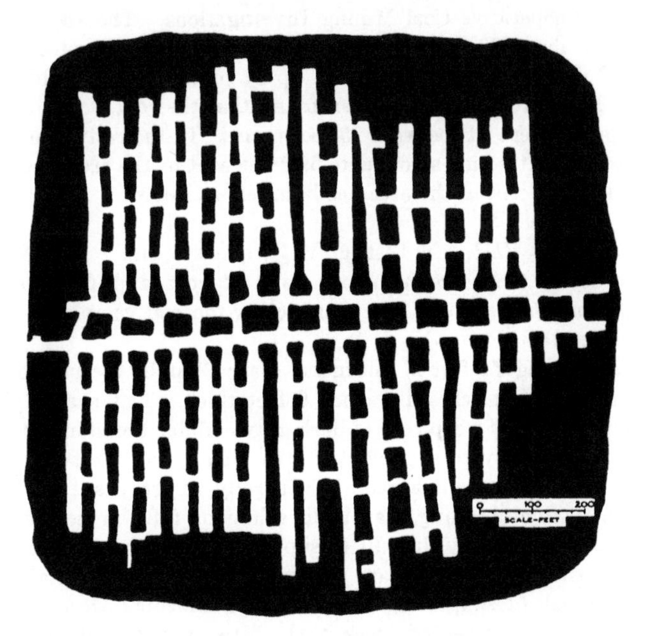

FIG. 6. PLAN OF AN OPERATION IN MACOUPIN COUNTY, ILLINOIS, SHOWING EXTRACTION IN A LIMITED AREA

per cent of the coal with frequent squeezes and subsidence has changed the dimensions of rooms and pillars so that now only 40 to 50 per cent is obtained. Thus far, with the new dimensions, squeezes have not occurred. No effort has been made to extract pillars systematically with the purpose of breaking the cap rock and thus preventing a squeeze by relieving the stress on the pillars, but there is nothing to indicate that this plan could not be carried out.

The plan of a portion of one of the mines is shown as Fig. 6. This may be taken as a fairly typical projection of large mines in this district. In many of the mines, including parts of the one illustrated, the workings are on a panel system, but probably not enough attention is paid to the matter of leaving pillars sufficiently wide to prevent the spread of squeezes beyond the boundaries of the panels. This illustration is presented, because the mine was surveyed with unusual care in connection with an investigation of subsidence which is being carried on by the Coöperative Coal Mining Investigations. The rooms and pillars are about 30 feet wide; this dimension was adopted with the belief that the top would be held up by pillars of this width left between 30-foot rooms. It had been found that the roof would fall if 25-foot pillars were left between 25-foot rooms. In the restricted area measured, the extraction amounts to 59.2 per cent of the area worked; that is, 40.8 per cent of the area has been left as pillars.

17. *District VIII.*—In District VIII, seams 6 and 7 are mined. In both seams there are numerous rolls of roof and floor called "faults," or "horsebacks." In many cases the roll completely displaces the coal.

Seam 6 averages 6 feet in thickness. Near Danville the immediate roof is of grayish black shale about 6 feet thick. This shale, lying between the coal and a cap rock of dark gray nodular limestone, makes a roof which is easy to support. In the vicinity of Westville and Georgetown, the immediate roof is generally of gray shale which shows no distinct bedding, has little cohesion, falls in conchoidal masses, and is extremely difficult to support. Stringers of coal, furthermore, extend from the seam proper into the roof material and render the task of supporting the roof more difficult. Occasionally there are 3 to 4 inches of black shale between the coal and the gray shale which forms the cap rock. Wherever this black shale is broken, air and moisture disintegrate the gray shale cap rock, and the roof becomes unsupportable. In all parts of the Danville district the floor is of soft fire clay.

Seam 7 varies in thickness from 2½ to 5½ feet, the average being 5 feet. The coal has two benches separated by a clay band one inch thick, which persists throughout the bed from 6 to 8 inches above the floor. This bed also has numerous rolls.

While the stripping operations, which are important in this district, are conducted in the No. 7 bed, the largest underground operations are in the No. 6 bed. The mines are operated on the room-and-pillar method, or a modification of it, but the numerous rolls in the roof prevent close adherence to the system. The frequent occurrence

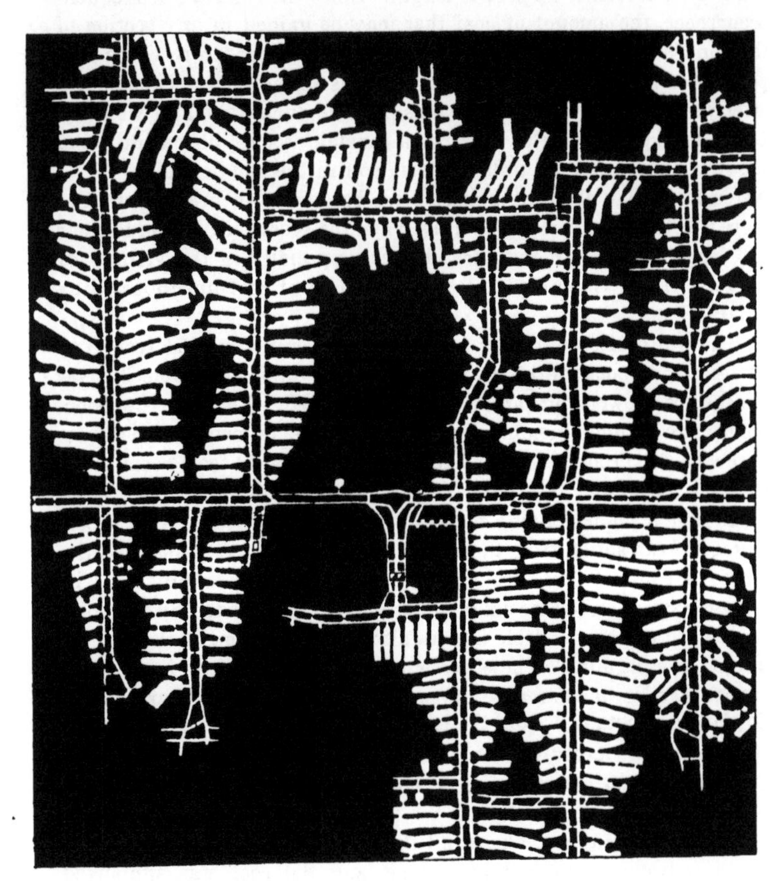

FIG. 7. PLAN OF MINE IN VERMILION COUNTY

of rolls has a marked effect upon the manner of driving rooms. In a roll area it is difficult to support the roof, and the expense of driving through the hard rock of the roll is great; consequently, when a roll is encountered in driving rooms it is customary to change the direction of the room and to drive it parallel with the roll until the coal resumes its normal condition, as shown in Fig. 7, which is a map of a mine typical of the district. Often it is necessary to abandon a room before it has been driven to its proper length. Since the rolls are of frequent occurrence, the amount of coal that may be gained in any section of the mine is problematical; consequently, the operator, on reaching that portion of the coal where the seam regains its normal thickness, will attempt to get as much of the coal as possible during the first working. Little attempt is made to preserve a constant room-pillar width, and the practice of gouging pillars is common in the smaller mines.* No systematic pillar drawing is attempted, because with present practice there is little pillar coal left to draw when the rooms are driven to their full length. The roof is so treacherous, especially in the vicinity of the rolls, that it is not safe to leave wide spans of roof unsupported by pillars.

The width of room pillars at the mines examined varied from 4 to 16 feet, and room widths varied from 21 to 43 feet. Table 3 gives dimensions of workings at each mine examined. Very narrow room pillars were found in mine No. 91, where the following dimensions were recorded; room centers, 47 feet; room widths, 43 feet; room pillar width, 4 feet.

Although pillar gouging in the district has resulted in a high percentage of extraction from the bed in the first working, it has caused a subsequent loss of coal through squeezes due to narrow pillars. The average extraction for the six mines examined, as reported by the operators, is 70 per cent. Table 3 gives also the percentage of the bed extracted at each mine. These percentages were calculated from measurements made in the mines and were checked by records of production per acre obtained from the books of each operating company and by planimeter measurements of mine maps. The Peabody Coal Company reports an extraction of 66 per cent at its mine in this district.

At one of the mines, almost all the pillar coal was extracted after all the advance work had been done, and the roof was supported largely by the rolls which occurred at intervals of 60 to 100 feet. At another mine pillars are being extracted, and it is estimated, that the total recovery will amount to about 85 per cent.

---

* Andros, S. O., "Coal Mining Practice in District VIII," Ill. Coal Min. Invest., Bul. 2, p. 16, 1914.

18. *Conclusion.*—It will be seen that nearly all the work in Illinois described as pillar drawing is unsystematic. It is merely incidental to the mining of room coal, and preparation for it is rarely made in laying out the mines. There are no apparent reasons, so far as physical conditions are concerned, except in a few instances, why plans could not be made for leaving pillars large enough to support the top during the advance work and for recovering the pillars on the retreat. Squeezes could thus be avoided, and the percentage of extraction could be increased materially. The commercial conditions which seem to make such a course difficult could probably be overcome, except in those cases in which subsidence of the surface subjects the operators to claims for damages in excess of amounts which would seem to be reasonable compensation for the injury done. The law covering payments for damages due to subsidence ought to be made so clear that there could be no doubt concerning the amount to be paid, and this amount should be limited to a fair compensation for the injury actually done.

At present there is promise of a considerable improvement with regard to the percentage of coal extracted from Illinois mines. The subject is receiving more and more attention on the part of coal producers and careful planning of the work with a view to high extraction as well as to low cost will follow as the natural result of greater interest on the part of the operators.

CHAPTER III

Methods and Recovery in the United States

19. *Early Methods in the United States.*—This chapter presents a discussion of early methods of mining coal in the United States, information regarding the percentage of coal recovered in different districts, and descriptions of the most advanced methods employed for obtaining high extraction, especially those which are applicable to conditions in Illinois. In collecting this material all available sources of information have been utilized. The descriptions of methods have been taken largely from the technical literature of coal mining, but in instances in which the correctness of the description seemed in doubt, or in which statements concerning the percentage of extraction seemed to need verification, the subjects have been reviewed by persons familiar with the local conditions.

In response to the large number of inquiries sent out, many persons have furnished the desired information in as nearly complete form as possible, but in many cases there has been available no authentic information on the subject of recovery. The estimates are necessarily more or less approximate because the conditions are such that it is practically impossible to obtain correct values, or the subject has not been considered of sufficient importance by the operators to warrant the expenditure of the time and money necessary for obtaining the values. It is believed that the values for percentage of extraction given in the following pages represent the most reliable information obtainable on the subject, but they are not presented as being absolutely correct.

At the time mining was begun here, this country was a colonial possession of Great Britain; the methods of mining to which immigrants were accustomed were those of Great Britain, and the application of these methods to mining problems in America was a matter of course. The development of the early English methods is discussed in the appendix. The coal miners of this country, furthermore, have been for the most part men who received their training in the work in England, Scotland, or Wales, or children of such men, and not until a comparatively recent date did these miners loss their dominance in the American coal fields. The conditions under which coal was found in this country were also not very different from those in Great Britain. It was natural, therefore, that bituminous coal mining practice in this country should correspond to that of Great Britain at the time the industry began here.

Mining in this country was begun in the Richmond (Virginia) basin about the middle of the eighteenth century. There seems to be no clear record of the methods followed, but it is known that a pillar system was employed, and that, as the coal was reached in some places at a depth of several hundred feet, a considerable amount of the coal was left in the ground. It is said that the pillars were to be extracted on the retreat, but no definite record is found to indicate that this was done.

Western Pennsylvania was the next district to take up coal mining on an important scale. Maryland and West Virginia followed, basing their early methods for the most part on what had been done in Pennsylvania.

20. *Pennsylvania.*—The early history of coal mining in the western Pennsylvania district is typical of that in other sections of the country, having a similar hilly topography. When coal mining was commenced, an abundant supply of coal was found outcropping on the hills in the neighborhood of Pittsburgh, and these seams were attacked by numerous small mines on the outcrop. As the workings were extended under cover, the single entry system was followed, and as it was impossible to obtain good ventilation with this system, the rooms were driven to only a short distance, and the entry itself was not long. Later the double entry method was employed, in which two parallel entries were used, respectively, for intake and return air. Since the distance to which rooms could be driven was limited, it was impossible to work any large territory by this method; hence, as the size of the mines increased, the cross entry system was introduced. The underground developments were the same whether the coal was reached by drifts, by slopes, or by shafts.

Among the many experiments tried in the Pittsburgh bed was that involving the use of double rooms with double necks, or of double rooms with single necks, but the amount of timber required for posts made these methods too expensive and by 1906 they were in use in a very few mines. The long-wall system also was tried and abandoned.*

No record has been found of the time at which the drawing of pillars was commenced, and it is probable that this method was followed more or less from the beginning in such mines as were systematically developed.

Toward the end of the last century, the double entry system had been further developed by the turning of room or butt entries from the cross or face entries. In some mines a few of the entries were driven to the boundary, and then all the rooms were opened at once, but some of the center rooms would sometimes reach their limits before

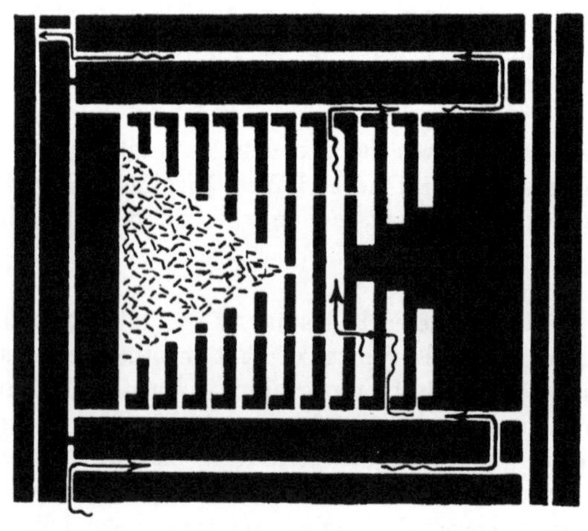

FIG. 8. OLD METHOD OF ROOM-AND-PILLAR IN PITTSBURGH, PA., DISTRICT

those whose pillars should have been drawn first. In other cases only the inby half of the rooms on each entry was turned first, while in still others one entry was completely exhausted before any side work was done on its parallel entry. In most cases, a room-and-pillar method was used with double entries, each about 9 feet wide. Main entries were separated by a pillar 51 feet wide with cut-throughs for ventilation. The main entries were driven on the butt of the coal, and face entries were turned from them about 1,000 feet apart. From these face entries, secondary butt entries or room entries were driven

*Dixon, Charlton, "A New Method of Coal Mining," Mines and Minerals, Vol. 27, p. 32, 1906.

about 400 feet apart. Rooms about 20 feet wide and 200 feet long were turned on the face of the coal. The room necks were 21 feet long and 9 feet wide. Room pillars were 15 or 20 feet wide, according to the cover above the coal. The rooms were turned from the butt entries as fast as these were driven, room pillars being drawn as mining pro-

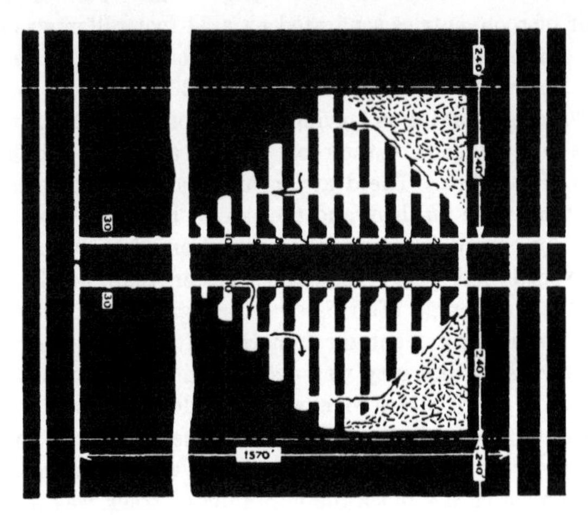

FIG. 9. IMPROVED METHOD OF ROOM-AND-PILLAR IN PITTSBURGH, PA., DISTRICT

gressed.* The objectionable features of this method are:—poor ventilation, dangerous gob, entries filled with fallen dirt requiring expense for cleaning up, maximum extent of track for the minimum quantity of coal, thus greatly increasing the cost of animal haulage, loss of thousands of tons of coal, compulsory driving of narrow work in room turning, and squeezes, which damage the coal and greatly increase the hazard of mining.

The difficulty of ventilation becomes most serious when the rooms from one entry are holed through into those approaching from a neighboring entry (see Fig. 8). Often this will occur two-thirds of the distance up each of these entries; thus all the pillars below the short circuit are deprived of proper ventilation at a point where it is constantly needed. After the pillars are drawn and the roof falls, there is no appreciable movement of air through the gob, and it often fills with explosive gas.

*Auchmuty, H. L., in Coal and Metal Miners' Pocket Book, 9th ed., p. 295.

A method described by Dixon, in the article previously referred to, was soon adopted with various modifications, although it is possible that it was already in use in one or more places at that time. According to this method the territory was laid off into blocks (see Fig. 9) 1,570 feet long, allowing for a barrier pillar 200 feet wide along the main entries and for another 200 feet along the next pair of face entries. A pair of room entries separated by a 54-foot pillar was driven through the center of this block, and thirty rooms were turned from each entry. Rooms were 240 feet long, about 26 feet wide, and were driven on 39-foot centers, thus leaving 13-foot pillars. Room turning was begun at the inby ends of the room entries, a reversal of the common practice of the time. The drawing of pillars was commenced as soon as the rooms were finished, and the line of break was kept at the proper angle by carefully timing the extraction of pillars. In this method the ventilation was considerably better than in the earlier method; but the air current, after passing through the district of pillar work on one room entry, went through the advancing rooms turned from the other. This difficulty was avoided in later methods by exhausting one room entry before room work was done on the other. The roof in the entries was easily maintained because the entries, with the exception of those on which rooms were being worked, were in solid coal. At the finishing of a block the minimum of track was in use for the minimum of coal passing over it. Track was not left in place awaiting the withdrawal of entry pillars; therefore it was not exposed to the corrosive action of mine water. Since the room pillars were attacked immediately, there was little danger of deterioration of coal or of trouble from falls. Most of the props could be recovered as they had not been subjected to any great pressure. Under the old system 50 per cent of the wood rails in rooms were lost while awaiting the attack on the ribs, and about 75 per cent of the posts were lost.

Referring to the conditions and the methods employed in mining, F. W. Cunningham* said in 1910:— "The operator in the Pittsburgh coal field, with the price of coal where it is to-day, must get the largest percentage of lump with the least amount of fine coal, and this by machine mining, in order that he may compete with coal operators in other fields." The rooms, therefore, are made as wide as possible to obtain the greatest percentage of lump coal, and the pillars are left

* Cunningham, F. W., "The Best Methods of Removing Coal Pillars," Proc. Coal Min. Inst. Amer., p. 275, 1910; p. 35, 1911.

as narrow as possible, because the greatest percentage of crushed coal comes from them. This fact explains why the use of narrow rooms and wide pillars, common in the Connellsville district, does not appeal to operators in the Pittsburgh district. It also explains the loss of much pillar coal, because a period of dull market results in the stopping of pillar work and only large coal from the rooms is marketed. A large number of rooms, accordingly, may be driven up to their limits

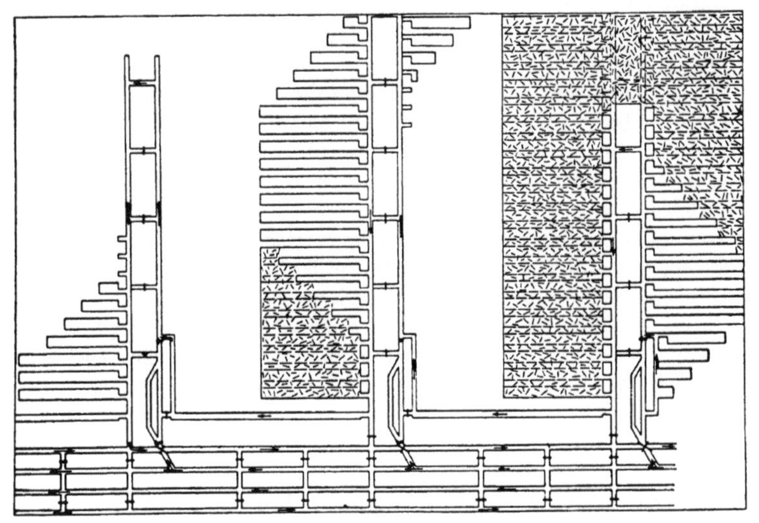

FIG. 10. MODERN METHOD IN PITTSBURGH DISTRICT

without the extraction of room pillars, and the recovery of these pillars is unprofitable after the rooms have stood for a number of years.

Fig. 10 illustrates the series of operations incident to one method of extraction of stump and chain pillars. In this method, rooms are turned and worked out progressively along one of a pair of room entries, probably the last, the pillars being drawn back as soon as the rooms are finished. There is thus a diagonal line of rooms advancing and another diagonal line, practically at right angles to this, retreating. On the other room entry of the pair, the driving of rooms is commenced at the inby end and proceeds outward. In some instances the entry pillars have been drawn on the retreat as illustrated in Fig. 10, and in others they have been left until all the rooms and room pillars have been finished. In the latter case it has sometimes been possible to obtain the coal from these entry pillars, but frequently all or part of it has been lost. The method illustrated in Fig. 10 was in use in the Pittsburgh district proper, that is in the high coal along the Monongahela River. Cunningham says that the extraction by this method would average about 80 per cent. Some companies, however, claim an extraction of 90 per cent. Some differences in percentages of extraction may be accounted for by the different methods followed in estimating: the whole bed from the limestone to the top of the seam may be taken into consideration, or the thickness of the slate partings may be subtracted.

One of the principal reasons for taking the rooms turned from one of a pair of butt entries on the advance and those turned from the parallel butt on the retreat was that this procedure made it possible to have the air current always blowing from the room work to the pillar work. This constantly moving current of air prevented gases set free by the pillar work from being carried to men working in advancing places. The miners in the pillar workings used locked safety lamps, while those in the room workings used open lamps.

Until about 1910, mining machines were used in the Pittsburgh district only in room and entry work, while pillar coal was undercut with picks. A method designed to permit the mining of pillars by machines is illustrated by Fig. 11. Cunningham gave the following facts concerning this method:—24-foot rooms are turned on 39-foot centers. After the room is worked out with a machine, a cut about 25 feet wide is made across the end of the pillar; then another cut of the same width is made far enough back on the pillar to leave a stump 5 to 8 feet wide, and the stump is removed by pick work after the machine work is finished. The stumps serve to protect the machine runners and the machine by supporting the top. It was said that at one mine where this system was used 70 per cent of the pillar coal was extracted with machines, and 30 per cent was pick mined. Good falls were obtained, and no ribs were lost. The recovery of timber was not so good as in the Connellsville region or in mines where there is no refuse gobbed along the roadways.*

An old and common method of working is illustrated by Fig.

---

* For another description of the method in use about 1910 see Schellenberg, F. C., "Systematic Exploitation in the Pittsburgh Coal Seam," Trans. Amer. Inst. Min. Engrs., Vol. 41, p. 225. 1910.

12. Track is laid in the middle of the room, and the room pillars are made as narrow as possible after the room has advanced 100 feet, or to the first cut-through. Commonly no attempt is made to recover the pillar coal beyond this point, though it is often recovered nearer the

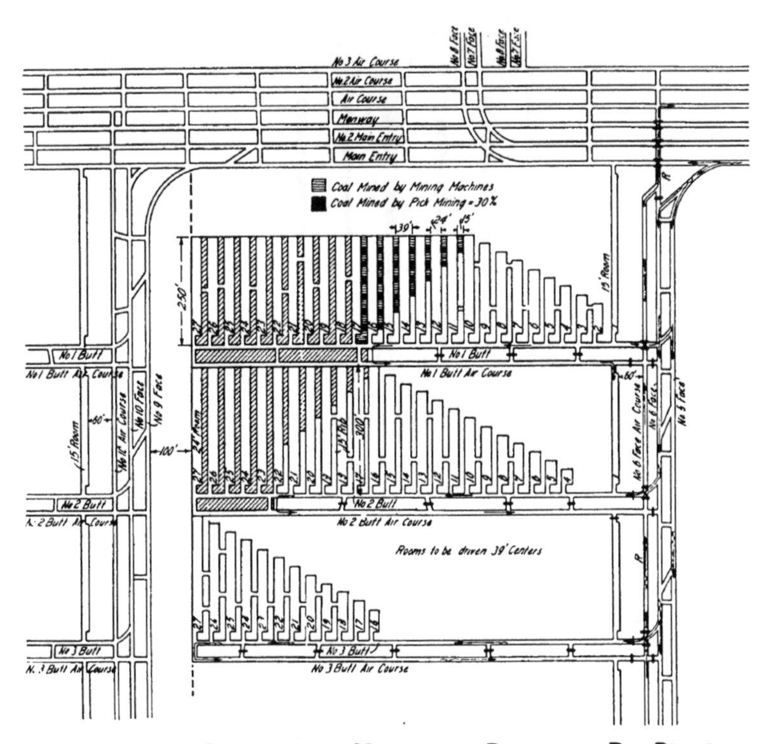

FIG. 11. PILLAR DRAWING WITH MACHINES IN PITTSBURGH, PA., DISTRICT

entries by working the pillar along the side of the fall. One of the chief operators in the Pittsburgh district claimed a recovery of 90 per cent of marketable coal by this method, but it seems that such a recovery could be made only over an area of a few acres and that the recovery over the entire area of the mine would be much lower. Schellenberg expressed the opinion that the recovery would not be more than 60 per cent if an area of 10 acres were considered.

In the discussion of Cunningham's article, G. S. Baton said that entry pillars were not often recovered in the Pittsburgh district except under remarkably favorable conditions. In his opinion not more than 40 per cent of the entry pillars were recovered where there was much

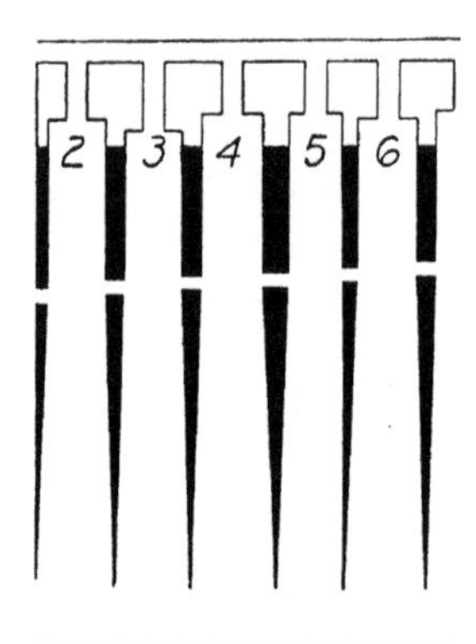

FIG. 12. TAPERED PILLARS

overburden. It is probable that a larger percentage than this is being recovered now in the more carefully operated mines. Another practice in pillar drawing which had been used in the Pittsburgh district and in other districts is illustrated by Fig. 13. A curtain of coal is left to keep out the gob when drawing room pillars, and the loss of coal amounts to about as much as it does where the pillars are narrowed and not drawn.*

It has been seen that in most of the methods employed in the Pittsburgh district, the pillar coal is taken out by pick work, and it has been within only a very recent period that even the room coal has been undercut by machines. Machine mining of pillars is cheaper than pick work and operators have recently introduced this more advanced method wherever it seemed possible. The immediate reason for this action has been the increased cost of production, largely due to high wages and expenses caused by changes in the laws affecting mining, without a corresponding advance in selling price. Since the demand

* Cunningham, F. W., Op. Cit., p. 275, 1910.

is still greater for the lump coal than for the smaller sizes, it has been necessary to increase the size of the pillars to prevent the objectionable crushing of the pillar coal. Another reason for increasing the width of the pillars is to be found in the practical difficulty of using machines on very narrow pillars. The thickness of room pillars varies, but the most common distances between the centers are 33, 36, 39, and 42 feet. With 33-foot room-centers, the room pillars are lost entirely; with 36-foot room-centers, about 55 per cent of the pillars are recovered; and

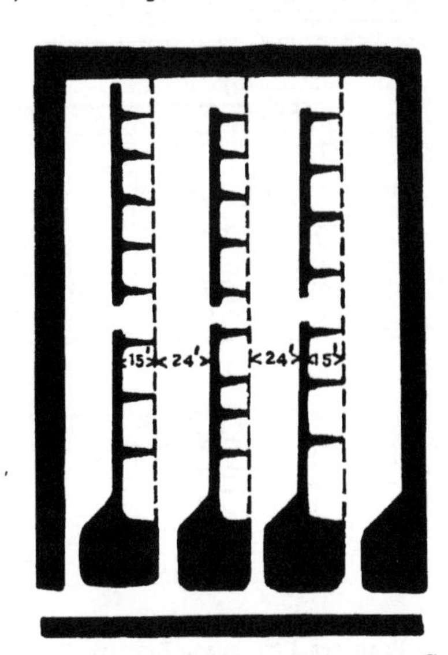

FIG. 13. PILLAR DRAWING, CURTAIN OF COAL

with 39- to 42-foot centers, from 60 to 70 per cent of the pillars are recovered.* Many miners at the present time have not the skill to do the best pillar work, even if the cost were not too high, and for this reason, if a higher percentage of pillar coal is to be won, cutting with machines, which can be operated successfully only on wider pillars, must be employed.

A method adopted for future working at the Marianna and the Hazel mines of the Pittsburgh-Buffalo Company in Pennsylvania and

* Edwards, J. C., and Gibb, H. M., "An Ideal Method of Mining," Mines and Minerals, Vol. 33, p. 665; and Edwards, J. C., "Machine Mining in Room Pillars," Mines and Minerals, Vol. 34, p. 591.

the Annabelle mine of the Four States Coal Company in West Virginia is illustrated by Fig. 14.* The plan as outlined is intended for coal under a cover of 300 to 500 feet. There are two sets of triple main

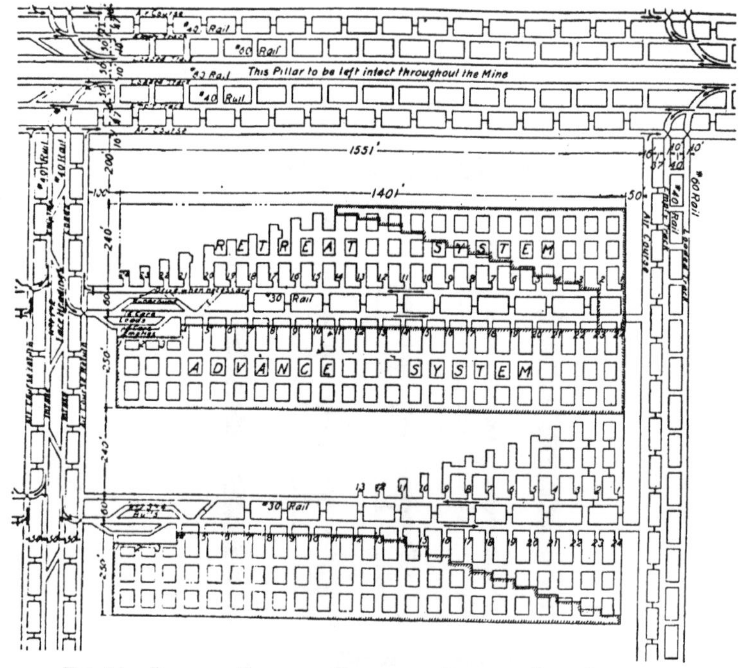

FIG. 14. PROPOSED PLAN FOR PITTSBURGH-BUFFALO COAL COMPANY

entries separated by an unbroken pillar 50 feet thick. The operation really includes two distinct mines except that the coal goes over the same tipple. The loaded and empty haulage roads on each side are driven 10 feet wide, and the airway is driven 16 feet wide. These widths are necessary in order to avoid the expense of a fourth 10-foot entry.

The panel system planned is known as the "half advancing and half retreating" system. The panels are divided into blocks 500 feet wide by entries driven "end on" in pairs, and from these butt entries

*Ibid.

the rooms are turned on the face of the coal. The butt entries on the side toward which the development of the panel is progressing, that is on the inby side, are termed "advance headings"; the exterior or outby entries are "retreat headings." A chain breast machine is used

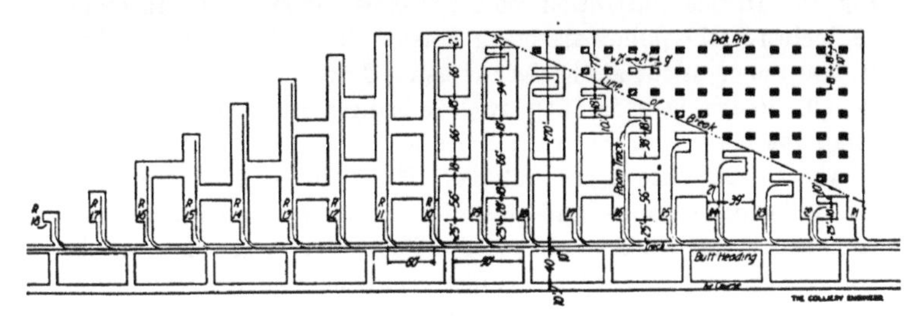

FIG. 15. EXTRACTION OF PILLARS UNDER DRAW SLATE

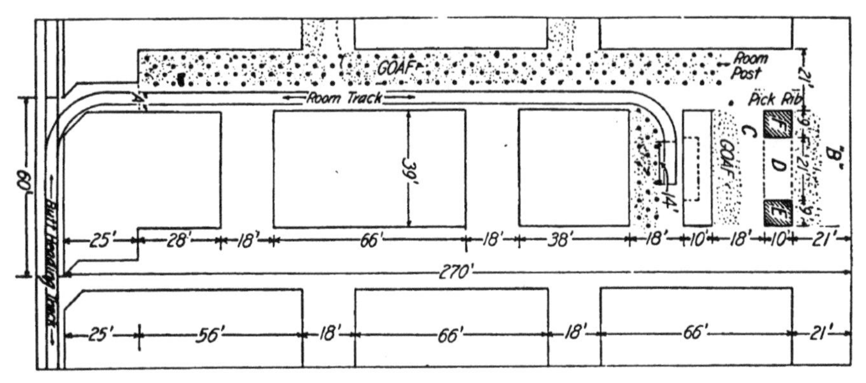

FIG. 16. DETAIL OF PILLAR WORK UNDER DRAW SLATE

in rooms and entries and a short-wall machine on pillars. Both machines continue in use until the last room on the retreat entry is completed; then the short-wall machine is left to finish the pillars, and the breast machine is transferred to another pair of butt entries under development. By the time room 14 is turned, room 2 has been finished, and work can be begun on the pillar between rooms 1 and 2. Rooms on the advance entry are 255 feet long and those on the retreat entry 246 feet long from the entry centers; this difference in length is made, because the chain pillar and entry stumps are brought back with the room pillars on the retreat entries. The method has been worked out for two general conditions; first, where a draw slate is encounted, and secondly, where there is no draw slate.

The method to be used where draw slate is encountered is illustrated by Fig. 15. In this illustration room 1 is shown as finished. In rooms 2 to 9, inclusive, the pillars are being drawn. Room 10 has reached its limit, and the cross-cut at the face is being driven through the pillar to room 9. Rooms 11 to 18, inclusive, are being driven. Fig. 16 shows in detail the method of recovering the pillars by the short-wall machine where draw slate is encountered. From the point *A*, the track is laid in 14-foot sections, and steel ties are used; consequently, the track is easily assembled or detached. Curved rails are used in the same way so as to give easy access to the cross-cuts. After the cross-cut *B* is finished, the curves and two 14-foot sections of the track are detached, and an 18-foot cut is made in the pillar at *C* by working on the butt of the coal and leaving a stump *D*, 10 by 39 feet. The draw slate from the first cut is gobbed in the room proper. The remainder of the draw slate from this cross-cut is gobbed in the outby part of the cross-cut, and the track is laid in the inby part. A cut is next made through the stump *D*, into the gob above, and small blocks or stumps *E* and *F* are left on each side of the cut to be taken out with the pick. After the block *E* has been removed, all the tracks in the cross-cut, except the two curve rails, are removed; when the block *F* has been removed, the curve rails and the two 14-foot sections of straight track are taken up, and the operation of driving through the pillar is repeated as the illustration shows. In this way the room pillar is extracted back to the point *A*.

The method of operation in the second case, where the draw slate is not encountered, is the same as in the first case, except for the manner of attacking the pillars which is illustrated in detail by Fig. 17. After the room has been completed and the cross-cut *B* driven, the two curve rails and seven sections of track are detached; thus the track is left in position to be assembled quickly for easy access to the cross-cut *H*, which is next driven. The 39- by 94-foot pillar is then split from *H* to *B*, and a 9- by 94-foot pillar is left on each side; then an 18-foot cross-cut *D* is driven through the 9- by 94-foot pillar on the rib nearest the gob, and a stump is left to be removed by the pick. The other 9- by 94-foot block is removed in the same way, and the operation is continued until the whole pillar has been removed. The entry stumps and chain pillars of the butt headings are won in the same manner as the room pillars.

Both of these methods are in use at several mines and are meeting with success. In both methods 90 per cent of the coal won is cut by machines. This percentage can be increased considerably, since it has been demonstrated that under favorable conditions part of the pick blocks can be recovered by the machine. One-third of the coal is mined from the rooms and two-thirds from room pillars. When draw slate is encountered, the 39- by 10-foot stump pillar will always afford ample

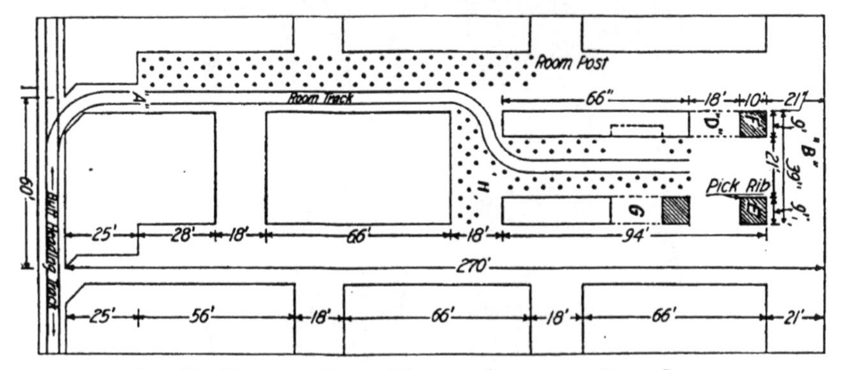

FIG. 17. DETAIL OF PILLAR WORK IN ABSENCE OF DRAW SLATE

protection to the miners; and where no draw slate is found, the two 9- by 94-foot stump pillars together with the timbering will give adequate protection.

Among the plans tried in the Pittsburgh district with the object of reducing the cost of mining by substituting more machine work for pick work is one (Fig. 18) which promised to be successful, but failed, because the miners demanded room-turning prices for the short rooms.* This added cost would have defeated any other advantage of the system. The method, however, seems to be based upon principles which may find application under other circumstances. The method was adopted, because a provision of the mining scale of the Pittsburgh district prohibited the drawing of ribs by machines unless short-wall machines were used. The plan was to continue the employment of the breast machines already in use and thus to increase the percentage

* Affelder, W. L., "Rib Drawing by Machinery," Coal Min. Inst. Amer., p. 232, 1912; and Personal Communication.

of room coal. It seemed profitable to increase the percentage of room coal since the cost for machine-cut run-of-mine coal was 45.28 cents per ton, including a differential of 2/3 of a cent on account of rolls, while the price for pick mining was 64.64 cents per ton; the average price for cutting, loading, and pick work was 51.92 cents. This average was based on the assumption that the working was regular with rooms 25 feet wide on 40-foot centers. Machine work was done with the common breast machines. The mining system was the ordinary method

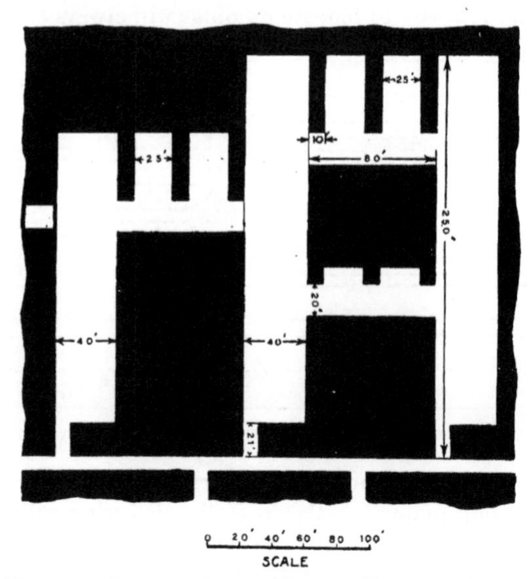

FIG. 18. METHOD OF REDUCING PILLAR WORK IN PITTSBURGH, PA., DISTRICT

of machine mining with 25-foot rooms on 40-foot centers with cross-cuts three runs wide and room necks 21 feet long. Rooms were 250 feet long, but as the neck was 21 feet long, the length of the actual room was 229 feet. All estimations of the percentage of extraction were based on a block 229 feet long and 120 feet wide. In this old system of mining, 65.7 per cent of the coal was produced as machine coal and 34.3 per cent as pillar coal drawn with the pick. The mine was operated on a run-of-mine basis. The new system was started with 36-foot rooms on 117-foot centers; but it was expected that if the system should prove successful, these dimensions would be changed to 40-foot rooms on 120-foot centers. The illustration shows the form and dimension of rooms. Two 25-foot rooms were driven from the cross-cut with 10-foot pillars on each side. These pillars were somewhat weak, but the rooms were not long, and the pillars were drawn quickly. It was admitted that the system would reduce tonnage for a time if the mines developed were not sufficiently advanced, but it was claimed that when the main rooms had been driven to within 50 feet of their intended length, the production from them would be much greater than from the ribs of three rooms of the older system. As soon as the pillars were drawn, the recovery of the so-called "rooster" coal would rapidly increase the output. In the Pittsburgh bed, the rooster coal lies above the draw slate and a laminated coal 12 inches to 2 feet thick. Although in most mines this coal is not taken out, in the Pan-handle district it seems to be better fuel than the bottom coal and is being mined in several places. The cover in the section in which this method was employed was from 75 to 125 feet thick. The small pillars were extracted by pick work, and the rooster coal also was obtained at pick prices.

21. *Connellsville District.*—The methods of production in the Connellsville field have become more intensive than those in the other districts of western Pennsylvania. The excellent quality of the Con-nellsville coke and the fact that the coal from which it is produced is found in only a limited area have made the coal so valuable that it has been found advisable to pay particular attention to high percent-age of extraction. In this district, there has been, moreover, no objection to the crushed coal from pillars as the product goes to the coke ovens where fine coal is desirable. The same fact influenced the relative sizes of rooms and pillars. While in the gas coal district it was thought desirable to take as great a quantity as possible from rooms, in the Connellsville district the practice of getting a large part of the coal from the pillars has developed in order that the percentage of extraction may be as high as possible. It is of interest to trace the more recent developments in mining practice here, because the extrac-tion in the better planned mines of this district is unusually high.

Conditions and methods in this district are discussed by F. C. Keighley* as follows:

---

* Keighley, F. C., "Mining Coal with Friable Roof and Soft Floor," W. Va. Coal Min. Inst., Dec. 10, 1914; Coal Age, Vol. 7, p. 1008.

"The coking coal known variously as the No. 8, Pittsburgh, or Connellsville seam has in places an extremely bad roof. This difficulty is strongly marked in the Connellsville basin, but can be found in other troughs. . . . The thickness of the coal and its softness might lead one to anticipate that it could be mined cheaply, but the friable roof creates a difficulty which its other qualities cannot overbalance. . . .

. . . "I have projected workings at depths ranging from 200 to 700 feet, using a dozen or more different schemes, and have never found any marked difficulty in protecting the main headings and air-courses of any mine. But I have experienced some trouble occasionally in protecting the branch, or butt, headings, and I have always had more or less trouble with the rib coal in the rooms.

"It is true that in many cases the roof will fall in headings and air-courses in spite of all that can be done, but such falls are gradual and are removed as part of the regular mine operations. On the other hand, when falls occur in the rooms they often come suddenly and cover a large area, and the break extends so far above the coal that they often reduce and may entirely cut off the production from a certain section of the mine, not only for a day, but perhaps for weeks at a time.

"It is clear, then, that any improved method of mining must provide for the protection of the rooms rather than for the care of the headings. . . .

"Panels have been projected 1,000 feet wide and 2,000 to 4,000 feet long. . . . Such a panel is subdivided into a number of smaller panels that are themselves served by two parallel entries driven at right angles or at some other angle to the flat or face heading depending on the pitch of the coal. These are known as butt headings. These sub-panels are generally 1,000 feet long and 300 to 600 feet wide. . . .

"Various widths have been chosen for rooms with 8 to 30 feet as limits, but the general belief is that rooms and headings should be driven 10 feet wide in the Connellsville region. There seems to be no opportunity to improve on this width. The best work has been obtained with large room ribs—50, 70 and 90 feet thick; but the success has not been as great as might be expected, though, with room ribs of the two larger dimensions, and with any ordinary care in mining, a general creep or squeeze cannot occur.

"This is not true when room ribs are made of smaller dimensions, such as 30, 40 or even 50 feet. In the initial stage of rib drawing with such light ribs great success is secured, but when trouble occurs it is usually in the form of a general squeeze or creep that almost paralyzes the output. It has often seemed for a time that the small ribs in the rooms resulted in cheaper mining; but when a squeeze or creep took place the small rib did not permit of the driving of a new road with safety and profit, and consequently the coal remaining in the rib could not be taken out.

"With 70-, 80-, or 90-foot ribs there is always sufficient coal left to permit driving a new road with safety through the pillar no matter

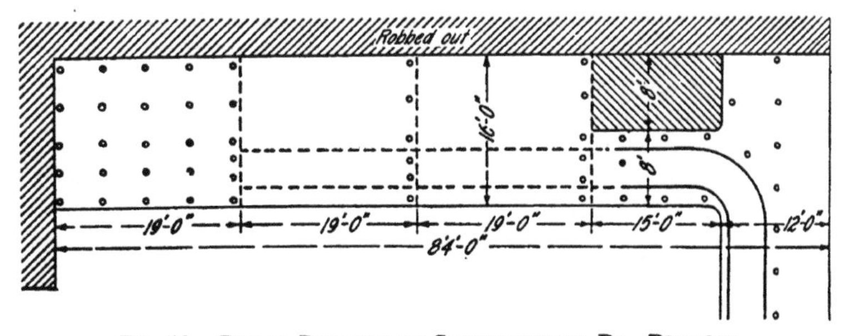

FIG. 19. PILLAR DRAWING IN CONNELLSVILLE, PA., DISTRICT

how badly the roof may have fallen or the coal be shattered on the edges of the pillars.

"Nearly all experienced miners concede that with narrow ribs only 50 per cent of the coal is recovered. The best results claimed is 65 per cent, while 90 and 95 per cent has often been recovered with the larger-rib system. The problem is whether the heavy cost of timber and the still greater cost of labor will counterbalance the loss of from 35 to 50 per cent of coal. I am disposed to believe that the larger rib, making a larger yield possible, will assure a handsome margin."

Fig. 19 illustrates a modern method of pillar drawing.* In this method a cut is made across the pillar, and an 8-foot stump is left; then this stump is taken in a retreating direction in from two to four sections, according to the width of the pillar. In the example given the rooms are 12 feet wide and are on 84-foot centers. After

* Cunningham, F. W., "The Best Method of Removing Coal Pillars," Coal Min. Inst. Amer., p. 275, 1910; p. 35, 1911.

the removal of the coal in each section, the props are drawn and the roof is allowed to fall, the break being controlled by a row of props across the end of the cut. After the last section of the stump has been taken out, the last of the track drawn, and the roof dropped across the room on the line of the end of the remaining pillar, another cut is made across the pillar and the process continues. While the falls represented in the sketches appear to be large, a better break is obtained with these than with short falls. This method has proved to be safer for the miners and to give a greater recovery of posts and coal than the methods which preceded it.

A highly developed and systematized room-and-pillar method is the so-called concentration method used in some of the mines of the H. C. Frick Coke Company,* and developed largely by Patrick Mullen, one of the company's inspectors. This method was designed to satisfy certain requirements, among which were safety of operation, completeness of extraction, reduction of cost through the greater use of machines, and an increase of daily output per man. A patent covering this method has been applied for.

It is well understood that liability to accidents is decreased by close supervision. Under the older system in the Connellsville region, it was possible for the face boss to visit each working place only once in two or three days. In order to increase the amount of supervision without increasing the number of officials, the plan of getting the working faces closer together was tried; this plan, however, necessitated a decrease in the number of working places and in the number of workmen, and it was realized that only by increasing the production of each miner could the output of the mine be kept up.

The only possible way of increasing the output per man was by replacing pick work with machine work. This substitution was made in room work; but it was found that, on account of narrow headings and narrow rooms with large room centers, machines in the narrow work alone would not accomplish the desired results, since the bulk of the coal comes from the pillars. The problem of the use of machines for pillar extraction, which was an entirely new one in the Connellsville district, has been worked out very successfully (see Fig. 20).

* Mullen, Patrick, "New Mining Methods as Practiced by the H. C. Frick Coke Company," Proc. Engrs. Soc. W. Pa., Vol. 32, p. 714, 1916; Coal Age, Vol. 10, p. 700, 1916; Howarth, W. H., "Mining by Concentration Method," Coal Min. Inst. Amer., Dec. 22, 1916; Coal Age, Vol. 9, p. 125, 1916.

The mine is blocked by driving at *A* double butt entries, 10 feet wide on 50-foot centers and 1,200 feet long, across the panel with break-throughs every 100 feet; the pairs of butt entries being driven 350 feet apart, the panel is divided into blocks about 350 by 1,200 feet. These blocks are then subdivided into blocks about 90 by 100 feet, by 12-foot face rooms at *B* 350 feet long on 112-foot centers, driven

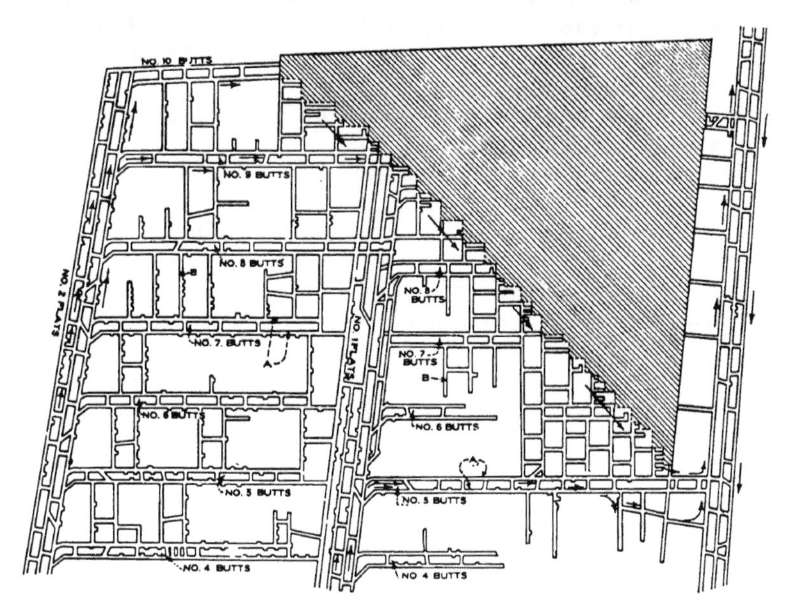

Fig. 20. Concentration Method in Connellsville, Pa., District

at right angles to the butt entries and connected by 10-foot break-throughs on 100-foot centers. A pillar of this size is considered ample to support any thickness of cover under any conditions of floor or cover to be found in the Connellsville region. In this manner a whole panel can be prepared for the intensive part of the work in which butt rooms are driven from the face rooms 10 feet wide on 25-foot centers.

As the main face room advances, the necks of the butt rooms to be driven are excavated to a depth of three machine cuts. After a main face room has been advanced 50 feet, there are available for the machine to cut two places which allow a production of forty tons; and when the room has advanced to a point where the first cross-cut is turned off, there are three places to cut in each main face room yielding sixty tons. This main room may continue to the end of the section or to the end of the coal field, butts or producing entries being turned off at projected distances. It is necessary that the operation be carefully planned and that the proper order of the work be closely adhered to. Fig. 21 shows the general schedule of operations with the position of the line of roof fracture at different dates. It is claimed that the general plan can be easily modified to suit all conditions such as depth of cover, presence or absence of draw slate, and nature of coal, bottom, and roof.

The projection takes three forms known as (a) maximum, (b) medium, and (c) minimum, according to the rate at which coal is produced (see Fig. 22.) The maximum plan is applicable where the thickness of cover does not exceed 125 feet, where the coal is hard, and where the general physical conditions of roof and bottom are good. The medium plan is applicable where the cover does not exceed 250 feet with the same physical conditions of the coal, bottom, and roof as for the maximum plan. The minimum plan may be applied to coal underlying any thickness of cover; the coal may be hard or soft, and the physical conditions of roof and bottom may be good or bad, provided, of course, that mining machines in any form can be used.

With the minimum plan, the butt rooms are driven in succession so that each room is 50 feet beyond the one succeeding. Two butt rooms advancing furnish 40 tons and one butt rib retreating furnishes 40 tons, or a total of 80 tons on the retreat; the main face room advancing yields 60 tons, or a total of 140 tons from one main face room. These quantities apply, of course, to coal of the thickness of that mined in the Connellsville basin — about 7 feet.

The medium plan will yield the same tonnage from the advancing main rooms, but the retreating work is so arranged that the face of each butt room is 30 feet behind that of the preceding room. This arrangement allows three butt rooms to be advanced at a time with a production of 60 tons, while two butt ribs are being extracted with a production of 80 tons; thus 140 tons are taken from the butt rooms and ribs and 60 tons from the advancing main rooms, a total of 200 tons for each main room.

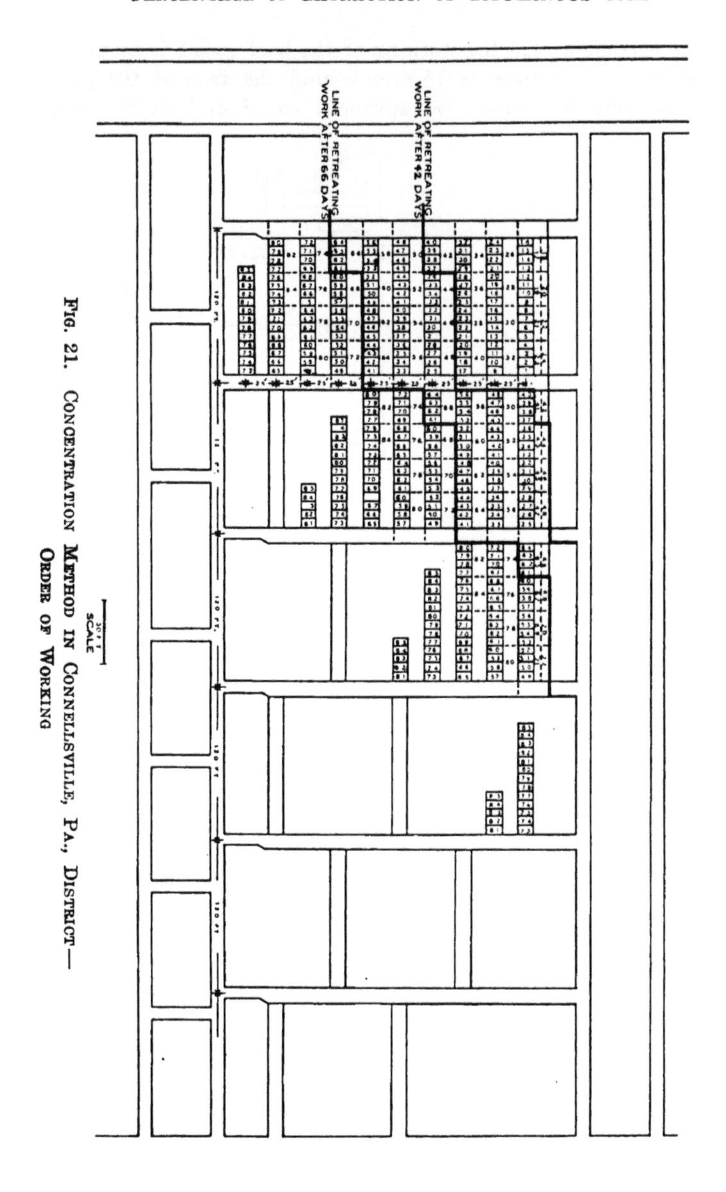

Fig. 21. Concentration Method in Connellsville, Pa., District—Order of Working

In the maximum plan the working of the butt rooms is so timed that the face of one room is 15 feet behind the face of the preceding one; four butt rooms are advanced and four butt ribs are

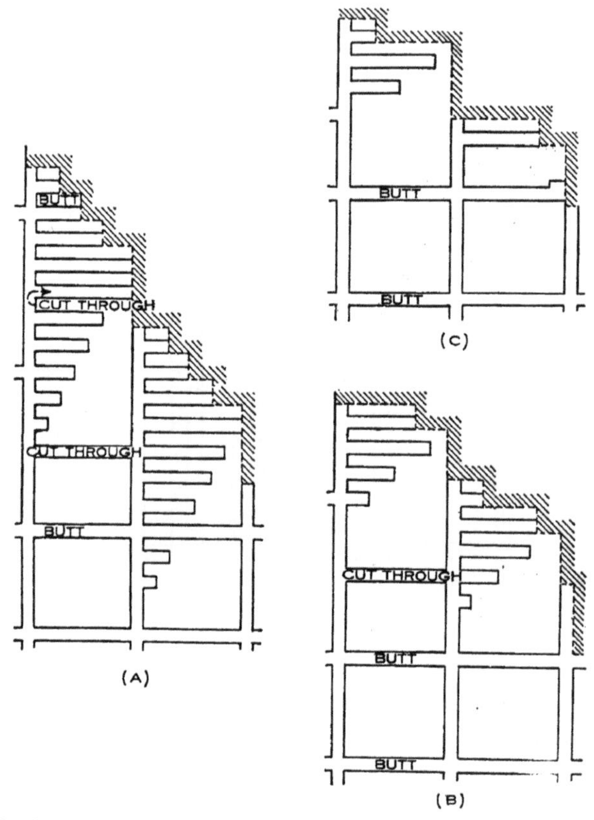

FIG. 22. CONCENTRATION METHOD—MAXIMUM, MEDIUM, AND MINIMUM PLANS

simultaneously withdrawn. The four advancing butt rooms will produce 80 tons and the four retreating butt ribs will produce 160 tons. With the 60 tons produced from the advanced main room, there is thus produced 300 tons for each main room.

The work is thoroughly systematized and proceeds with great regularity. After the miner has cleaned up his place and the day's run is completed, the machine crew enters and cuts the place to a depth of approximately 7 feet. Following the machine crew, the timber men reset any posts which the machine men have removed, post up any cross bars which have been notched in the coal over the machine cut, and put the place in good condition according to a prescribed system of timbering. The timber men are followed by the driller who bores the holes with an electrically driven drill. The driller is followed by the shot firer who charges and tamps the hole, and after an examination of the conditions, fires the charge. After the coal has been shot down, empty cars are placed by the gathering locomotives so that when the loader arrives at his working place in the morning he finds it in safe condition, the coal ready to load, and the empties in place. Miners loading under these conditions regularly obtain 18 to 20 tons per shift. The average of the loaders for short-wall mining machines in all mines of the company for the month of August, 1916, was approximately nineteen tons per shift. At mines where there is a full equipment of mining machines, the machine coal runs from 80 to 90 per cent of the total output. The recovery under the concentration system is from 90 to 92 per cent, while under the ordinary methods it is 80 to 85 per cent.* In the values given, the top or bottom coal left in place is not considered. The average thickness of top coal left is about 6 inches and the values for extraction, based on the entire thickness of the seam would be somewhat lower than the values given. Coal is left for two reasons. In the entries, from 6 to 8 inches of top coal is left as a protection. In the room work, such top or bottom coal is left in place as is necessary to keep the sulphur content of the coke made from the coal down to the required amount. It is found that the highest sulphur content of the bed occurs at the top or at the bottom and, by frequent analyses, it is determined how much of this top and bottom coal may be left.

22. *Central Pennsylvania.*—A method known locally as the "Big Pillar System" has been developed to meet conditions incident to the soft bottom in the Lower Kittanning, "B," or Miller, bed in the southern and eastern parts of Cambria County, and in the adjoining territory.†

The physical conditions for which this system was developed include a hard roof, very difficult to break, and a soft fire clay bot-

*Dawson, T. W., Personal Communication.

†Silliman, W. A., "Big Pillar System of Mining," Proc. Coal Min. Inst. Amer., p. 76, 1911.

tom. A sand rock from 10 to 40 feet thick occurs above the coal; but between this sand rock and the coal there is usually from 1 to 6 feet of slate or sandy shale, which is more or less affected by the air and which breaks away from the sand rock, especially in the roadways. The falling of this top tends to relieve the pressure, but not enough to prevent squeezing. The worst squeezes are encountered where the sandstone is only 10 feet thick.

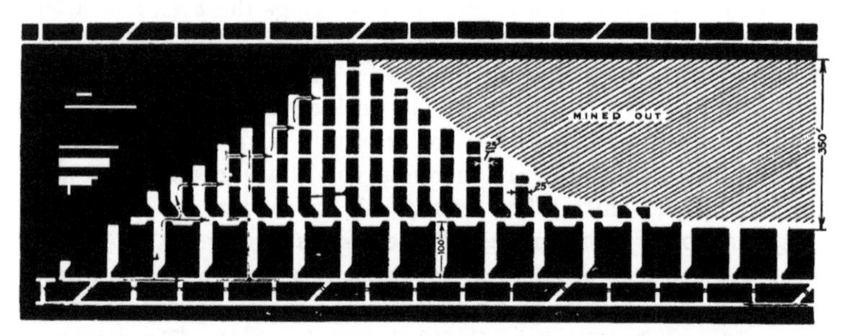

FIG. 23. "BIG PILLAR" METHOD USED IN CAMBRIA COUNTY, PA.

Under these conditions, the ordinary system of turning rooms with 40-, 50-, or 60-foot centers does not work satisfactorily. When the room pillars are drawn back to the stump, under the ordinary system the pressure is so great that a stump, even 30 or 40 feet square, will not protect the entry. Instead of breaks occurring in the roof along the line of the stumps the bottom breaks and heaves, and squeezes occur. Since the coal is soft and has a columnar fracture, the stump is badly crushed, and no amount of timbering is sufficient to prevent the closing of the entry. Thousands of feet of entry and much coal have been lost as the result of squeezes in a bed of this kind. The bed is only about 3½ feet thick so that it is necessary to take up the bottom, and the provision of space for storing bottom is one of the considerations involved in planning this system. The average dip is about eight per cent.

In the "Big Pillar" method (Fig. 23), haulage entries are driven on the strike and rooms are turned up the pitch. Entries are 10 or 12 feet wide. Rooms are turned on 100-foot centers from the entries, and, at a distance of from 100 to 125 feet from the entry, rooms called "crooked" rooms are turned at right angles, that is, parallel with the entry. There is thus left along the side of the entry a series of blocks 75 by 75 feet or 75 by 100 feet, according to the length of the rooms driven from the entry. As soon as a crooked room has intersected the straight room toward which it is being driven, an intermediate room is turned up the pitch from the crooked room; thus the rooms above the crooked rooms have 50-foot centers. The straight rooms are driven to such distances that the roof will break at the edge of the big pillar or by settling will relieve the strain. Sometimes they are only 250 feet long, and sometimes, under heavy cover, as much as 400 feet.

When the straight rooms are started, they are widened on the outby side so that the cross, or crooked, room can be turned off the straight rib, a matter of importance because of the necessity of storing bottom which is taken up in the roadway. Beyond the crooked rooms, the straight rooms are widened on the inby side; thus the men who drive a room are able to start the drawing of the pillar as soon as the room is finished. The room pillars are drawn back to the crooked rooms, and the irregular little blocks caused by the necks of these rooms are removed as completely as possible. The big block is then left standing to serve as a barrier to protect the entry, and if the space mined out is sufficiently broad, the roof will usually break. Even if the roof does not break, the strain seems to be relieved before reaching the entry. The upper edge of the big pillar may be badly crushed, and the roadway of the room may be heaved almost down to the entry, but the entry itself will be practically unaffected.

When the entry is finished and the stumps are being drawn, the system presents a special advantage in that a better output can be obtained than with the smaller stumps. Where stumps are small, the output is limited to the work of two gangs, but with the big pillars eight or ten places may be worked at all times on the retreat. The big pillars are split by a room driven up from the entry at the same time that a skip is taken along the rib of the old room. These two working places are cut through to the old falls about the same time, and the intervening portions of the pillars are brought back. This method leads to large recovery of coal, although there are no statements available concerning the exact percentage. There is some loss in the extraction of the pillars, and the coal at the edge of the big pillars is badly crushed. The method is not used in other beds in the same district, because the conditions are better.

In the Somerset County district, there has been developed a panel system which permits a high degree of concentration of work and a large percentage of extraction. The coal is low, and the miner is obliged to push his cars to the face of the room and to drop them down to the entry. The seam often dips from two to five per cent, and butt entries are driven off the main haulage slope on a grade of one per cent in favor of the load. From these, entries are driven to the rise at convenient distances, from which rooms are turned on the strike of the bed. Not only does the method result in a high percentage of extraction and facilitate the handling of the cars by hand in the rooms, but it also concentrates the work of mining. Two men working together will produce from ten to twelve tons of pick-mined coal per day, but when the men work singly it has been found that a good miner can load from seven to eight tons per day.* Under this system, the total extraction is reported to be about 93 per cent. About 50 per cent of the coal comes from rooms and entries, and the remainder from pillars.† This method is similar to that illustrated in Fig. 32, page 102, which shows a plan of operation of the Carbon Coal Company in West Virginia.

Because of the recognized objections to the room-and-pillar system, much attention has been given to the possibility of employing the long-wall system in the Pittsburgh bed and in other beds of western Pennsylvania. So far as can be learned, there is only one mine at which a long-wall system is being used in these districts, although there is an approach to it in some so-called "panel long-wall" or "block long-wall" methods. In these methods, dependence, however, is not placed on the weight to break down the coal; in fact, weight at the face is prevented so far as possible by causing the roof to break near the face.

It is worth while to review some of the experiments in the introduction of long-wall methods, because it is only by these methods that complete extraction is attained, although there is a close approach to it in the best applications of some forms of pillar working.

One of the attempts‡ was started about 1899 in the Lower Kittanning, "B," or Miller, bed where the coal was from 3 feet, 6 inches

---

* Majer, John, "Mining by Concentration Methods," Coal Age, Vol. 9, p. 345, 1916.
† Coxe, Edward H., Personal Communication.
‡ Claghorn, Clarence R., "A Modified Long-wall System," Mines and Minerals, Vol. 22, p. 16; Thomas, J. I., "Mechanical Conveyors as Applied to Long-wall Mining," Proc. Coal Min. Inst. Amer., p. 55, 1907; Delano, Warren, Personal Communication.

to 3 feet, 10 inches in thickness, and had an average dip of eight per cent. The roof was of blue slate and the floor of hard fire clay. Most of the coal lay under a fairly level surface about 174 feet thick, near the top of which was a moderately hard sandstone. The plan adopted involved blocking out the mine with entries and taking out the coal in each entry on the retreat. The faces were 250 to 300 feet

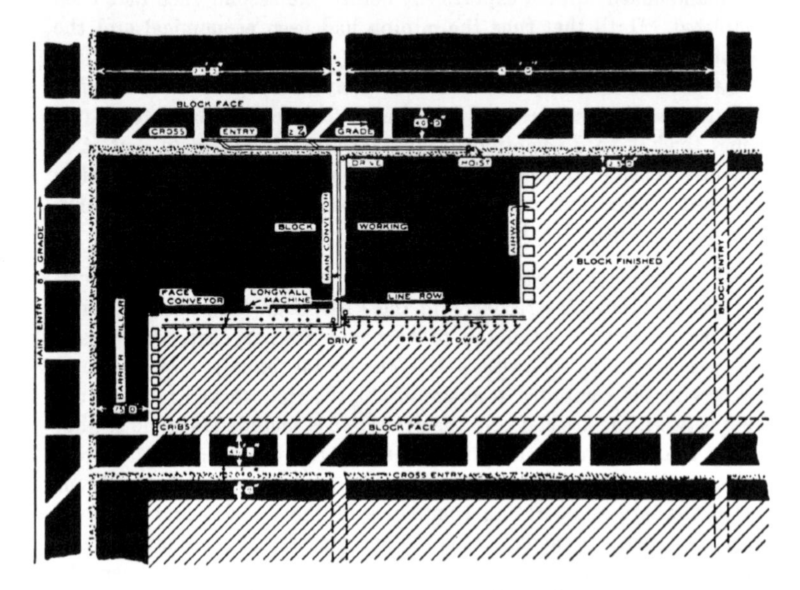

FIG. 24. BLOCK LONG-WALL WITH FACE CONVEYORS

long. One machine-cut produced from 125 to 150 tons of coal. Break-rows were made of stout, round, hardwood posts, 6 to 8 inches in diameter. These were capped with a 2-inch lid of soft wood set on a little slack to facilitate drawing. It was found that the track along the face took up too much space, and a conveyor, which was put in made it possible to set the props closer to the face. At first there were two faces, one slightly in advance of the other, each served by a conveyor which delivered coal to cars let down the block entry. Later, as shown by Fig. 24, a conveyor, by which the coal was lowered to the cars on the level, was installed in the block. Because of the unfavorable trade conditions in 1907, operations could not be carried on with the regularity essential to the success of the system as then used. The attempt, accordingly, was temporarily abandoned, but a revival of the plan is being seriously considered.

Another attempt at long-wall mining was made with very similar mechanical arrangements in the Cement seam near Johnstown. In this instance two faces were cut, but a shortage of power compelled the abandonment of the experiment before the second face had been completed. Until that time the mining had been economical and the recovery was almost perfect.*

At present the Maryland Coal Company of Pennsylvania is using eight long-wall conveyors at St. Michael. The coal is about forty-two inches thick. No description of the operation is available, but it is evidently considered successful, as the number of conveyors is being increased. Two other companies in Pennsylvania and one in Maryland have recently decided to employ the same method.†

23. *Summary of Facts Relating to the Percentage of Recovery in Pennsylvania.*—The recovery in the Pittsburgh bed, not including the coke district, is estimated to be about 80 per cent. This estimate is made on the following basis: The actual tonnage mined is compared with the computed tonnage of the district worked out; everything between the fire clay and the drawslate is included, and no deduction is made for impurities in the bed or for the average thickness of four inches left on the bottom. This computation is obtained from one of the largest operating companies of the district and is based upon actual measurements. It is the opinion of this company, assuming that this method of calculation is used, that 85 per cent is probably the best possible recovery in this district. Some other companies claim an extraction of 90 to 95 per cent, but this is calculated after deducting the coal left in the bottom, the bearing-in bands, and any other impurities in the bed which are taken out and not weighed. Such high recoveries, of course, imply careful planning for the extraction of pillars and for the execution of this work without delay.‡

Another company the workings of which lie along the Monongahela River south of Pittsburgh, estimates the recovery as 86.7 to 90.6 per cent.

* Moore, M. G., Personal Communication.
† Link-Belt Company, Personal Communication.
‡ Schluederberg, G. W., Personal Communication.

In the Connellsville district the best practice gives from 80 to 85 per cent with the methods ordinarily used there. With the new "concentration" method of the H. C. Frick Coke Company a recovery of from 90 to 92 per cent is obtained.*

In the Johnstown district it seems impossible to obtain estimates of the percentage of extraction, because all the seams in that district vary in thickness within short distances, and are somewhat cut up by rolls. In some seams, for example, the Miller seam in the vicinity of South Fork, the recovery is almost perfect. The conditions are favorable, the roof being well adapted to extraction of pillar stumps in retreating.†

One of the companies operating in Jefferson County claims an extraction of 90 per cent. The operations are in the Lower Freeport bed, and conditions are somewhat peculiar because of bad roof, lack of uniformity of the seam, and faults. Each district requires individual development before an estimate can be made of the proportion of faults to the whole area, and it is impossible to make an accurate estimate of recovery until a district has been completely worked out. The value given represents the proportion of coal extracted from the area mined in which coal existed, and does not apply to the area of faults.‡

One operator in Clearfield County estimates an extraction of 95 per cent, based upon the amount of coal mined up to December, 1916.¶

One of the companies operating in Somerset County estimates that, where mines are operated in an area of less than 300 acres and under a cover not exceeding 200 feet, the recovery should be, and in a number of instances is, in excess of 90 per cent. In the case of a property of 1,000 or more acres, where the coal extends underneath a hill giving cover of 300 to 700 feet, the recovery is from 85 to 90 per cent. Low coal, faults, and adverse grades still further reduce this percentage.§

The attainment of the higher percentages in Pennsylvania has been reached only within very recent years, and is not yet by any means universal. There are still in operation a large number of old mines,

* Dawson, T. W., Personal Communication.
† Moore, H. G., Personal Communication.
‡ Van Horn, H. M., Personal Communication.
¶ Personal Communication.
§ Delaney, E. A., Personal Communication.

mostly small, in which high percentages of extraction are not obtained. The more recent operations are planned for, and give, probably as high a yield of coal as can be expected from the area worked.

24. *Maryland.*—In the Georges Creek region of Maryland, the Big-vein coal has been mined for about one hundred years, and the methods used there furnish an illustration of progress in coal mining engineering which is especially interesting in view of the increasing attention given to the percentage of recovery.

In a paper on maximum recovery of coal,* H. V. Hesse discussed the wasteful early methods of mining in the Georges Creek region and suggested economic methods, as follows:

"A region of uniform and unusually severe conditions in the bituminous fields has been selected to illustrate the results obtained over a long period. The Georges Creek region of Maryland, with remarkable deposit of semi-bituminous 'Big-vein' coal, has operated in this seam and shipped to the market for nigh unto a hundred years. . . . More than one miner still lives who 'dug coal' before the war with the South, and . . . he tells of the detail method of extracting the coal, on account of which thousands of tons lie buried to-day, much beyond recovery. . . .

"The 'Big-vein' seam occupies the geologic horizon of the Pittsburgh bed, but differs considerably in structure and quality from the coal of Pittsburgh, Connellsville, and Fairmont. . . . The top coal averaging 2 feet thick is left up for a roof. Where this comes down the strata immediately above promptly follows. Very little of this top coal is therefore recovered. Both roof and breast of the seam contain slips known among the miners as 'horsebacks,' which frequently fall out without any warning. The coal is soft and the 'butts' and 'faces' entirely absent.

"The methods of extraction in vogue at different periods in the history of this field have established the fact that it is impossible to maintain wide working places for any length of time. Headings are driven 8 feet wide and rooms from 12 to 15 feet. In the earlier days there was practically no definite system of extraction, headings and rooms being driven at random and no pillars recovered. Fig. 25 shows such a method in use about 1850. This is reproduced from an actual survey made under the most trying circumstances. . . . It is estimated that fully 55 per cent of the original coal, not counting the top coal, remains and it is expected to recover at least one-half of this, or 27 per cent of the original, by careful operation and the use of about double the amount of timber necessary under a good system of mining. The maximum cover over

* Hesse, H. V., "Maximum Recovery of Coal," Proc. W. Va. Coal Min. Inst., p. 75, 1908; and Mines and Minerals, Vol. 29, p. 373.

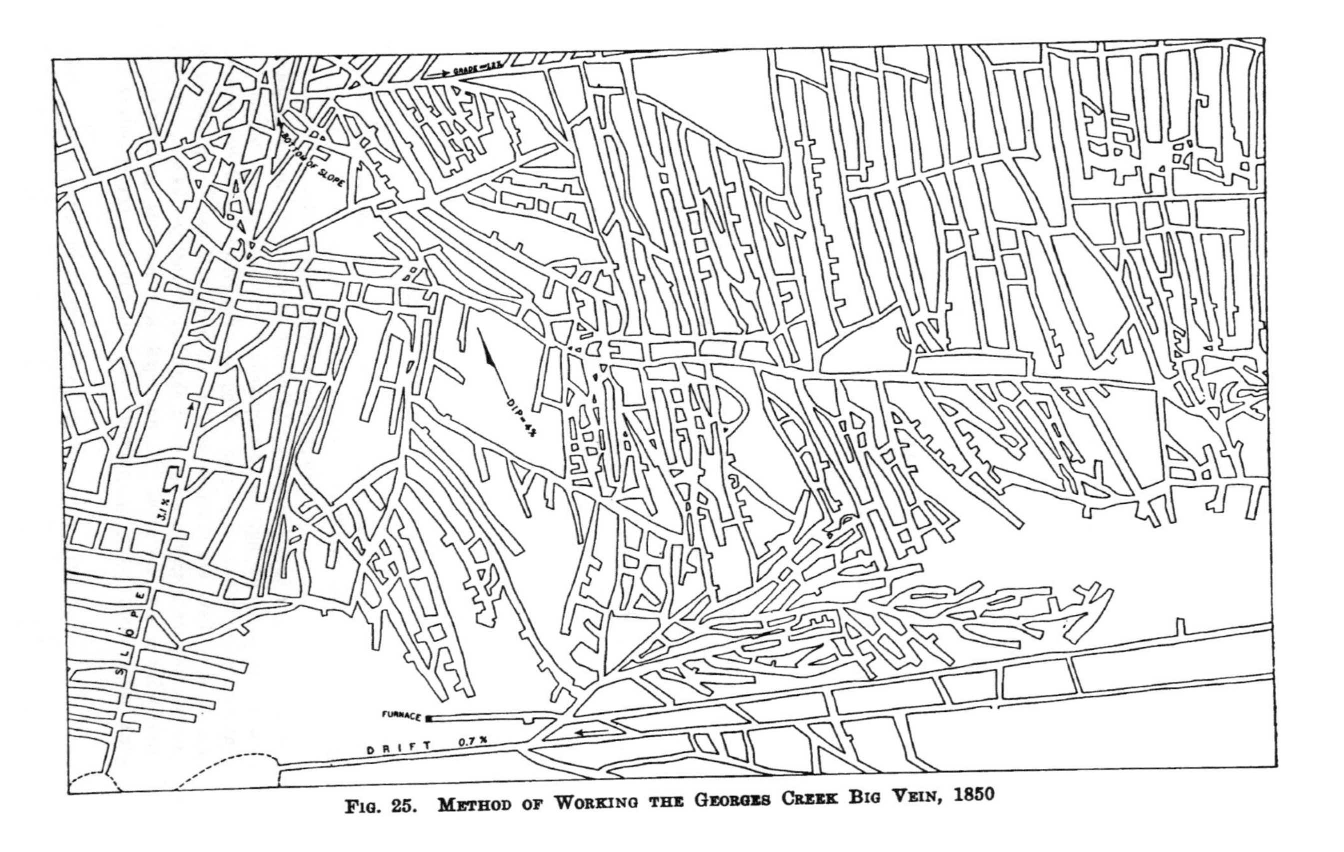

FIG. 25. METHOD OF WORKING THE GEORGES CREEK BIG VEIN, 1850

this district is 300 feet and the few comparatively large pillars, which were inadvertently left standing at irregular intervals, saved the balance from being crushed. In other sections of the same mine where a similar method was followed, but these large pillars not left in, the workings are entirely closed and the remaining pillars, containing over 50 per cent of the original coal, probably lost forever. Fortunately mining operations during this period were not conducted on a large scale and, consequently, the territory thus affected is limited to a very small portion of the company's holdings.

Fig. 26 "illustrates two methods followed during the years between 1870 and 1880. These workings are inaccessible to surveys at the present time owing to the creeps and squeezes induced by the irregular method of robbing the small pillars. . . . In the first method . . . the rooms were 14 feet wide and pillars 26 feet. These pillars were found to be totally inadequate and extracting them impossible. Cross-cutting the pillars at frequent intervals was then attempted after completion of the rooms, but this was generally accompanied by creeps closing a whole district at a time. The maximum height of the superincumbent strata in this territory is 200 feet.

"The second method shown on Fig. 26 was adopted later. . . . By this method headings were driven from the main entry on the rise of the seam at intervals of 1,000 feet to the level above, and two pairs of cross-headings turned to the right. The rooms were driven from these cross-headings at 50-foot intervals and 14 feet wide, leaving a pillar of 36 feet. The length of the rooms varied from 300 feet to 550 feet. These pillars were also of insufficient size, robbing was conducted spasmodically and although more coal was recovered than in the adjoining districts a great deal was lost. In addition to the small pillars, the method of robbing them was calculated to promote squeezes. It appears to have been the method to hold the strata with props until sufficient coal had been removed to enable the weight to break the props. As a general rule, however, before this was attained the weight had induced a creep which is well known to have no limits within a territory of small pillars.

Fig. 27 "represents a method in use in 1890. . . . Rooms were turned as shown from all headings on 100-foot centers and pillars split by half rooms. The length of rooms varied from 300 feet to 600 feet and they were 14 feet wide, leaving pillars 42½ feet wide. These pillars were not strong enough to support the overlying strata of 500 feet and the usual creep resulted when pillar drawing commenced. . . .

Fig. 28 "shows a method adopted in 1900. The maximum dip is 15 per cent and the greatest thickness of superincumbent strata 425 feet. The slope, together with parallel air-course and manway, are sunk on the heaviest dip of the coal and double entries turned off to right and left at intervals of 1,000 feet on grades of 1¼ per cent to 2¼ per cent in favor of the loads. From these haulways, cross headings are deflected at intervals of 240 feet at an angle of about 25 degrees and driven on a grade of 4 per cent to 7 per cent. Rooms varying in length from 100 to 800 feet are turned on the rise of the coal from

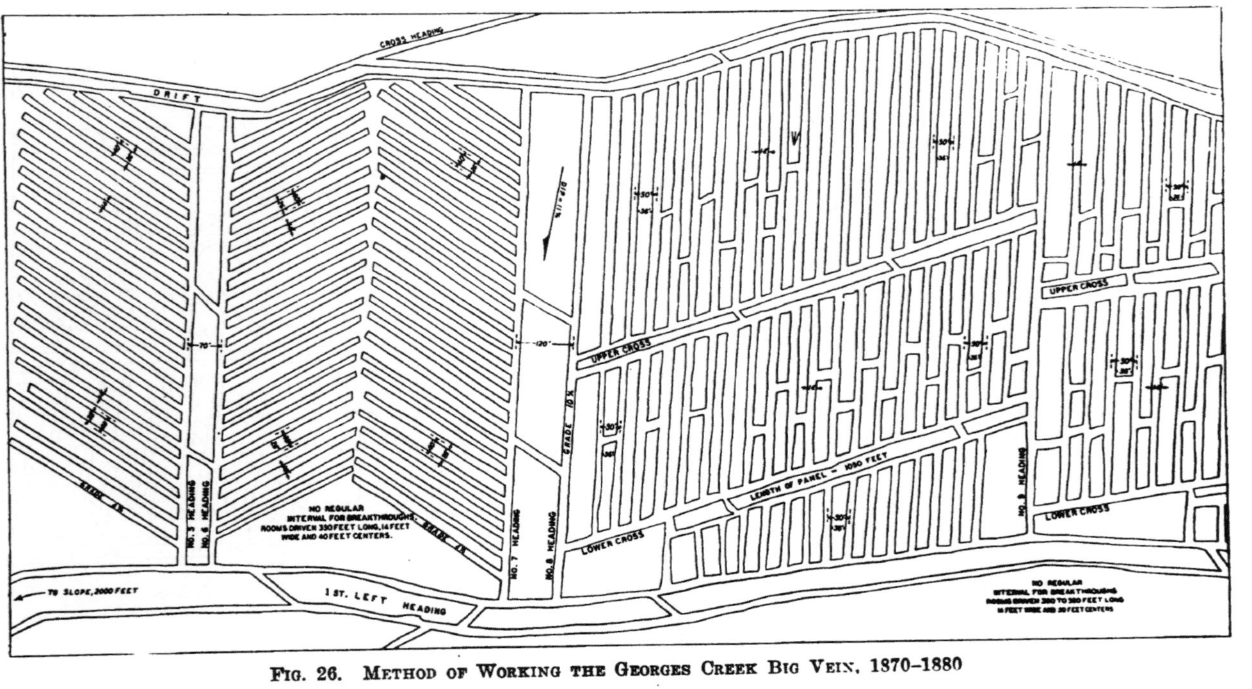

FIG. 26. METHOD OF WORKING THE GEORGES CREEK BIG VEIN, 1870–1880

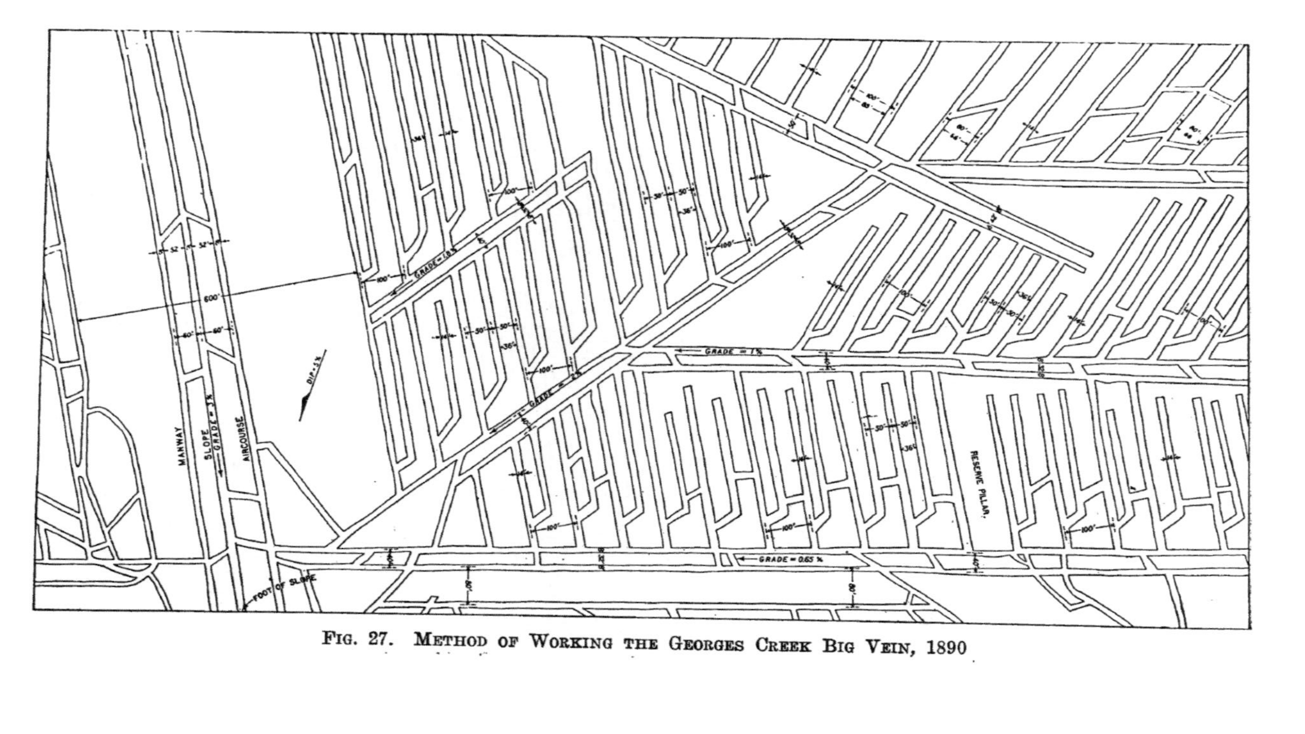

FIG. 27. METHOD OF WORKING THE GEORGES CREEK BIG VEIN, 1890

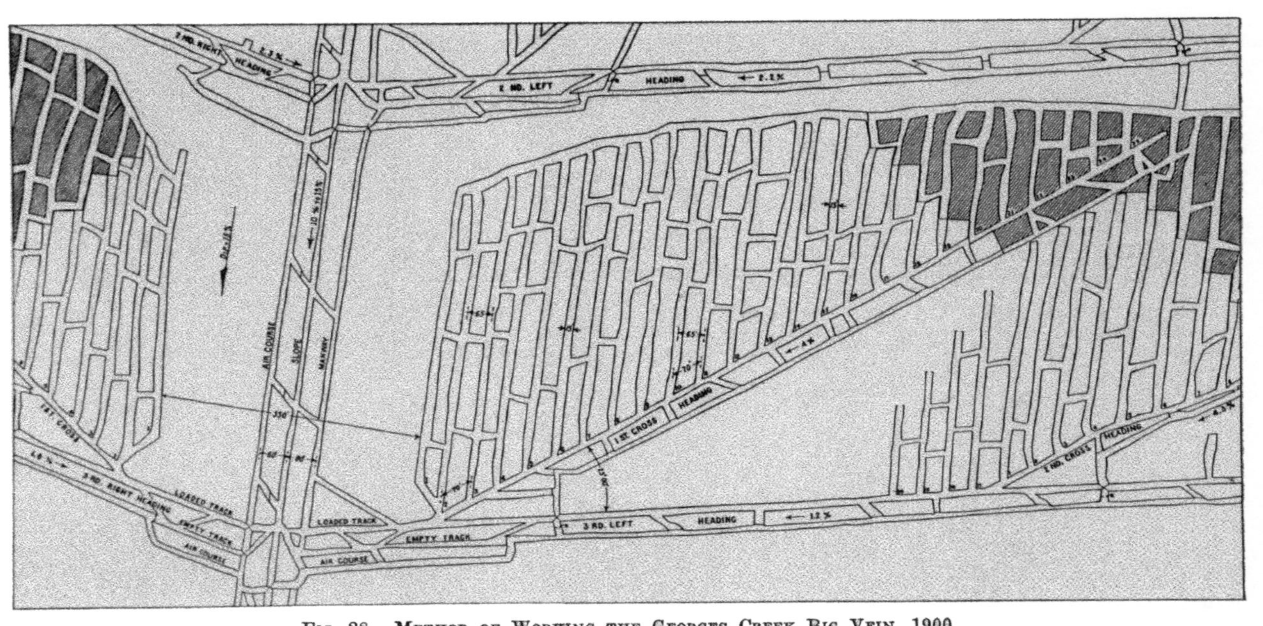

Fig. 28. Method of Working the Georges Creek Big Vein, 1900

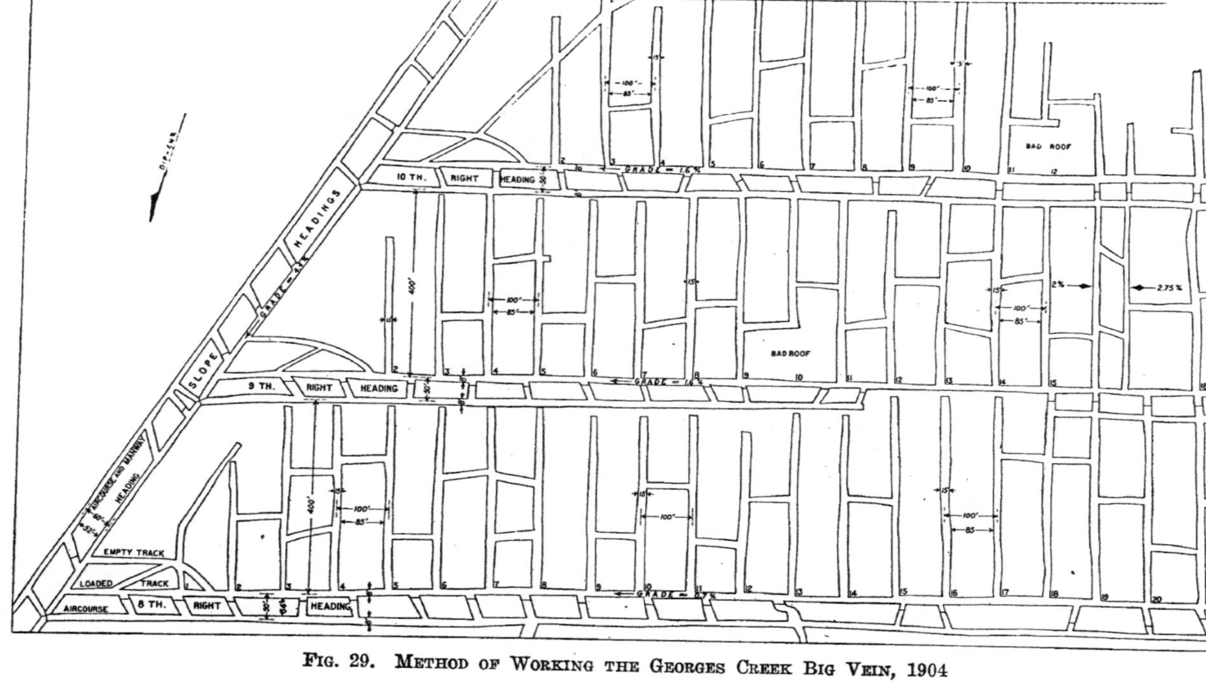

FIG. 29. METHOD OF WORKING THE GEORGES CREEK BIG VEIN, 1904

these cross-headings. The rooms are driven 15 feet wide on 65-foot centers, leaving pillars 50 feet wide. Twenty-five rooms are driven in each of these diagonal panels. Unusually large protecting pillars are left along the main haulage roads. This system has been found to be especially adapted to rapid gathering of cars thus ensuring a large tonnage. It has been found, however, that a very large recovery from the pillars is impossible, owing to the many sharp angles, which, in a thick seam of soft coal, are always difficult and ofttimes impossible to extract. This sharp-angle method was even resorted to formerly in cross-cutting the pillars preparatory to drawing them, but this has been changed to a rectangular method, thereby increasing the actual percentage of pillar coal recovered from 80 per cent to 83 per cent. The distance of rooms apart has also been increased in the last few years to 100-foot centers giving pillars 85 feet thick. It is expected that the extraction of these will show a further increase in the percentage of yield from pillars. The present yield from headings, rooms, and pillars under this system is about 90 per cent, considering the recovery from headings and rooms as 100 per cent.

Fig. 29 "illustrates a method instituted in the latter part of 1904. The main haulway is an extension of the slope from the opposite side of the basin. Double entries are turned off from this entry, on 1½-per cent grade, 400 feet apart, from which rooms are driven directly on the rise of the coal. Rooms are from 13 feet to 15 feet wide and . . . they are driven at 100-foot intervals, leaving a pillar 85 feet wide. The length of a panel is about 2,500 feet, containing 22 rooms. There are five such panels in this district and when completed it is proposed to draw the pillars in a retreating fashion with the line of pillar work on an angle of 45 degrees across the whole district. A similar method in another district . . . is yielding 88½ per cent from the pillars with a total recovery of 94 per cent from headings, rooms, and pillars . . . the greatest height of the overlying strata is 250 feet."

George S. Brackett states* that in 1898 he made some careful estimates of the percentage of recovery over a period of a year in the Georges Creek region of Maryland. The data for the computations were obtained from two mines which were worked under the following general conditions:

> The thickness of coal was 7 feet, 3 inches; the inclination was 5 to 18 degrees; the system of mining was the room-and-pillar retreating method. All the entries were driven to the boundary before any rooms were opened, and a good line was maintained on the drawing of pillars. No. 1 mine had moderate grades, and a better roof than No. 2. No. 2 had grades

* Personal Communication.

as steep as 18 per cent, and the roof was decidedly heavy on pillar workings.

The following results were obtained:

	Total Per Cent of Extraction	Per Cent of Pillars Obtained, Including Chain and Barrier
No. 1 Mine	97.6	97.0
No. 2 Mine	82.1	71.3

The average total recovery of the two mines was nearly 90 per cent.

25. *West Virginia.*—The modern methods used in the large mines of West Virginia are among the most advanced found in this country. In many parts of the state, coal mining is carried on by large corporations which are financially able to conduct operations with a view to ultimate economy. In most cases this ability has resulted in the development of methods of operation which lead to very high percentages of extraction.

In the Fairmont district in the northern part of the state, the more recently opened mines are planned for large production and a high percentage of extraction. According to the West Virginia Geological Survey, the Pittsburgh bed, which is the one mined in this district, contains more than 7 feet of clean coal* and the average total thickness of the bed is about 8 feet. The newer mines are projected on the panel system; the last rooms on each room entry are turned first, and the pillars are drawn immediately, a line of break being maintained at an angle of about 45 degrees with the entries. A plan of operation used in this district is illustrated by Fig. 30. The method of attacking pillars is shown in Fig. 31. One of the principal operators in the Fairmont district estimates the recovery, where mines are laid out systematically on the panel system, to be from 85 to 90 per cent of the entire seam.†

A company operating to the south of Fairmont estimates that 85 per cent is a good recovery. In the workings of this company, the rooms represent 27 per cent of the total area; the pillars down to the heading stump, 39 per cent; and the chain and barrier pillars, 34 per cent. The recovery in these three classes of working would be respectively, 100, 90, and 70 per cent. This would give a total yield

* W. Va. Geol. Sur., Vol. II., p. 180, 1903.
†Smyth, J. G., Personal Communication.

of 86 per cent. This company has no very accurate figures on recovery.*

In the mines in the Freeport coal bed in the Piedmont, or Elk Garden, district, there is ordinarily a good shale roof, although there

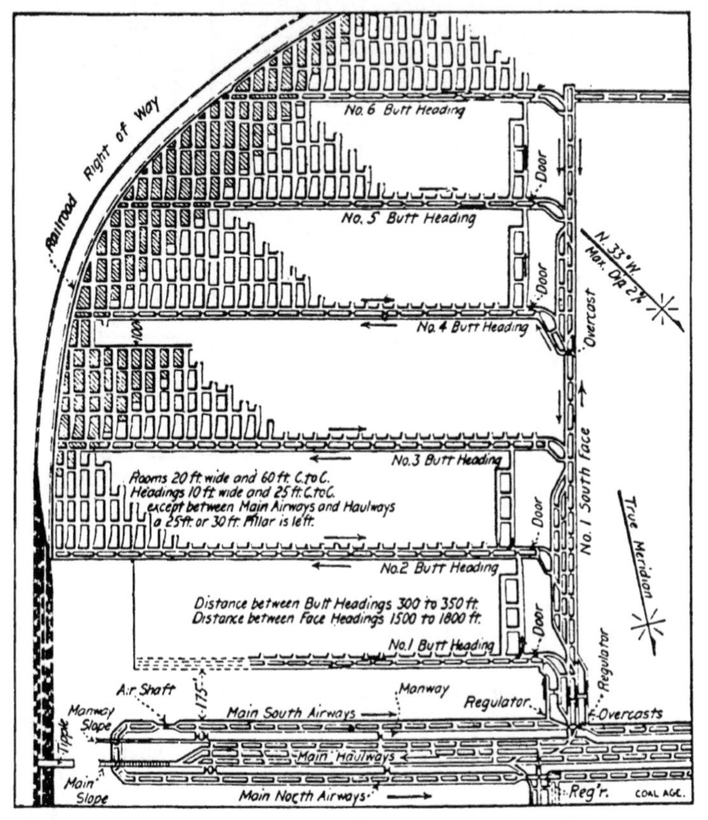

FIG. 30. PLAN OF WORKING—FAIRMONT, WEST VIRGINIA, DISTRICT

are places in which the shale is replaced by sandstone. In the mines in the Kittanning bed there are from 3 to 14 feet of shale above the coal, and above this are 40 feet of sandstone, although there are places where the sandstone forms the roof for short dis-

*Brackett, George S., Personal Communication.

tances. The percentage of extraction in this region is about 90 per cent; in most of the new mines, however, an extraction of 97 or

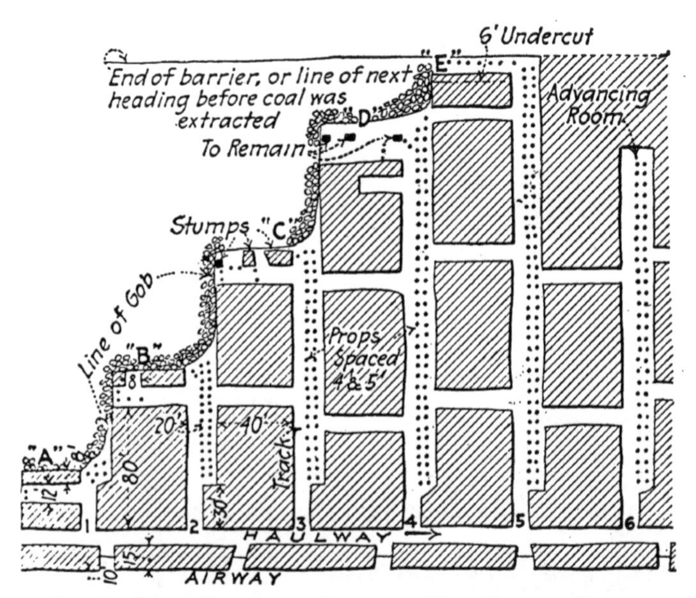

FIG. 31. PILLAR DRAWING IN FAIRMONT, WEST VIRGINIA, DISTRICT

98 per cent is being reached. This rate of recovery is higher than it formerly was, because changes have been made in the order of driving rooms and drawing pillars. The earlier custom was to drive long entries, from which rooms were turned on the advance. The room pillars could not be drawn until the entry was finished because of the danger of squeezes, and as this process sometimes occupied four or five years, falls occurred which made it impossible to recover a high percentage of the pillars. Under the present system, room entries are driven long enough for 20 rooms, and the inside room is turned first. Work on room pillars is commenced as soon as rooms 19 and 20 are finished, and room and entry pillars are taken out rapidly. Nearly all the pillar work is done with picks, and there is little machine work carried on in the district. The pillars are attacked by cross-cuts, and a stump about four feet wide

is left next to the end. This stump is removed as soon as the cut through the pillar is finished.*

In the Central West Virginia district the operations are in the Lower Kittanning bed which averages 6½ feet in thickness.† The conditions are almost ideal for a high rate of recovery. The bottom consists of hard shale. The immediate roof is of bone coal 3 to 10 inches thick, above which are shales of varying hardness, 10 feet to 15 feet thick. Above this layer occurs sandstone of an average thickness of 10 feet, and above this, shale and overlying earth. Nowhere is the overburden greater than 90 feet in thickness. The extremely favorable nature of the roof is shown by the fact that a complete break is easily obtained with as few as 3 or 4 rooms. In more recent workings 20-foot rooms are turned on 50-foot centers. The rooms are driven 300 feet long, and a 50-foot pillar is left between the heads of the rooms and the adjoining air-course. This pillar is never pierced except in case of extreme necessity. The 30-foot room pillars are taken out by driving cross-cuts through them every 30 feet retreating. The entry pillars are taken out with the room pillars. If the entry stumps and the barrier pillars at the ends of the rooms are considered, it is estimated that 75 per cent of the coal obtained is taken out as pillar coal and 25 per cent as room coal. The cost of room and of pillar coal is about the same. Bischoff estimated that the recovery is at least 90 per cent, and possibly 95 per cent, although no accurate records have been kept. In view of the unusually favorable conditions, it seems probable that this estimate is correct as there is no apparent reason for loss, except that represented by the small amount of coal which the loaders fail to shovel up. This statement, of course, refers to only the area of actual mining operations.

The Pittsburgh bed is being worked also in Braxton and Gilmer Counties about 75 miles south of the workings just mentioned. While the same method of working is followed, the physical conditions are different, and the extraction is not more than 75 per cent. The coal is 6 to 8 feet thick. The bottom is of fire clay 8 to 15 feet thick, and the immediate roof is of fire clay 2 to 6 feet thick. Above this occurs a sand rock thicker than that found in the neighborhood of Elkins, with a heavier overburden. In this southern district the

* Personal Communication.
† Bischoff, J. W., Personal Communication.

coast of pillar coal is about three cents greater per ton that the cost of room coal.*

The Kanawha region is unlike the fields farther south in that

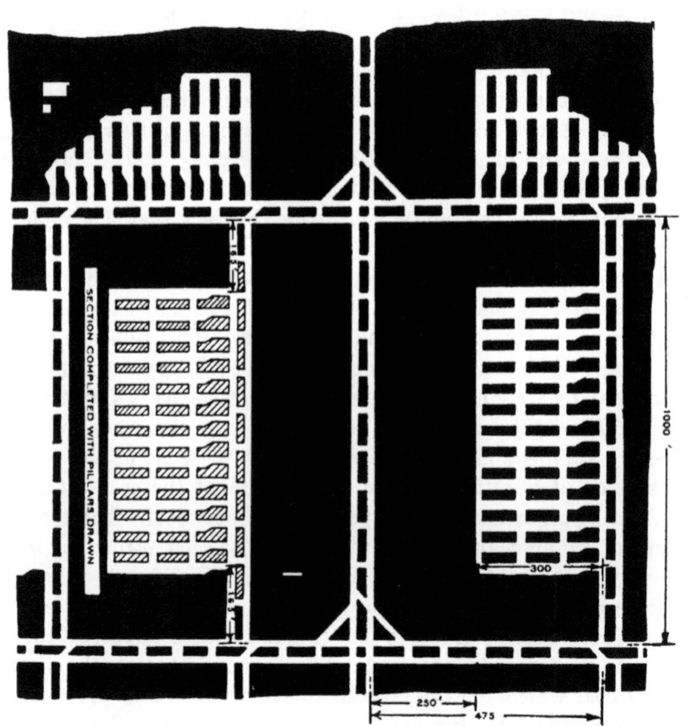

FIG. 32. WIDE BARRIER PILLARS AND ROOM STUMPS, KANAWHA DISTRICT, WEST VIRGINIA

there is a larger number of operating companies with a corresponding lesser concentration of ownership. Because of the number of individual operations it is impossible to give any general or standard method, but the room-and-pillar method is universally used. At least one operator is leaving a large barrier pillar, Fig. 32, and a large room stump for entry protection. The first break-through is driven about 80 feet from the entry, and break-throughs are kept

* Bischoff, J. W., Personal Communication.

perfectly lined up. Rooms are driven in order, and room pillars are drawn back to the first break-through as soon as the adjoining rooms are completed. The recovery under this method is said to be about 90 per cent and where the roof is extremely good as much as 95 to 97 per cent.*

In a paper on the removal of coal from the No. 2 Gas Seam in the Kanawha district, J. J. Marshall reported a very high percentage of recovery and gave the following facts concerning the seam:† The coal bed is made up of two benches separated by a solid parting the thickness of which varies from 10 inches to 40 feet. It has been found that it is not economical to remove this slate when its thickness is more than 24 inches. The aggregate thickness of the two benches averages about 9 feet, the upper bench ranging from 4 feet, 6 inches to 5 feet, 6 inches of clean coal and the lower bench from 3 feet, 6 inches to 4 feet of clean coal. Where it is impossible to mine both benches together, only the upper bench has been taken. The thickness of cover varies from a few feet to 100 feet. After the ordinary method of driving rooms and of drawing pillars on the advance, the mine described has been developed until it is now in position for the butt entries to be worked on the retreating system. On June 30, 1911, the percentage of recovery was said to be as shown by Table 6, nearly all the coal being mined by pick work:

TABLE 6

PERCENTAGE OF EXTRACTION IN KANAWHA DISTRICT

	Total Acres	Percentage of Recovery
High Coal, Both Benches	84.61	91.8
Upper Coal, Upper Bench Only	67.87	98.7
	152.48	94.9

Computations of areas are made from the mine map, and the method of computation does not insure the accuracy of the percentages given. It seems that the values are too high. If there is a loss of 4 feet or more in thickness across the working face, it is recorded, but if the loss is less than this, it is too small to show on the map which is drawn to the scale of 100 feet to the inch, and the recovery

* Cabell, C. A., Personal Communication.

† Marshall, J. J., "The Removal of Coal from the No. 2 Gas Seam in the Kanawha District," Proc. W. Va. Coal Min. Inst., p. 303, 1911.

is regarded as practically complete. It was said to be seldom necessary to record a loss, especially in the upper coal.

In the Cabin Creek portion of the Kanawha district little attention has been paid to the extraction of pillars until recently, and the extraction has amounted to only about 50 per cent. For the past 3 or 4 years, however, the extraction has been about 85 per cent. Since the proper sizes of pillars are now known and the men have a better understanding of pillar work, it is expected that the percentage of extraction will show some further increase.*

Two mines in the New River field, one in the Fire Creek bed and the other in the Sewell bed, have a recovery which is considered practically complete.† At those two particular mines the roof conditions are very favorable; in other sections of the field where they are not so good and where less attention is paid to recovery, it is thought that a fair average extraction is about 90 per cent.‡

The Pocahontas district, in the southern part of West Virginia, is one of the most important coal producing regions in the country, largely because of the high quality of the coal. Pocahontas coal is low in volatile matter and therefore is nearly smokeless; it contains little ash and little sulphur, and it makes an excellent coke. Because of these characteristics there is large demand for it. It is extensively used in coke production, in power plants, in the navy, and in domestic heaters. For coking purposes, however, Pocahontas coal is not used so extensively at present as it was a few years ago. Several beds are being operated, but the principal mines are in the Pocahontas No. 3 bed. The seam varies in thickness from about 4 feet on the west to about 10 feet on the east, but the change is gradual and the thickness is quite uniform within the area of a single mine. This seam has a fire-clay or slate bottom, and a draw-slate roof.¶ It always has one streak of bone about 2 inches thick to which the coal adheres on both sides; consequently when a piece of bone is thrown out, about twice as much coal is lost.

The No. 4 bed, which has two streaks of similar bone, is found 75 to 80 feet above the No. 3. Above this bed occurs a seam of interstratified coal and slate locally termed a "black rash." This rash contains on the average about 25 per cent of ash, and it is considered

* Keely, Josiah, Personal Communication.
† Personal Communication.
‡ Cunningham, J. S., Personal Communication.
¶ Eavenson, H. N., Personal Communication.

worthless, but sometimes over rather large areas there occurs in it a streak of clean coal from 6 to 8 inches in thickness. Miners are supposed to leave all this rash in their working places, and most of it is left, but sometimes some of this clean coal is loaded out. This fact will explain the higher yield from the No. 4 bed. In other words, the coal actually mined and loaded sometimes has a thickness greater than that considered in calculating the contents of the bed. The fact that a considerable amount of good coal is lost with the bone explains the smaller yield in mines in the No. 3 bed. It is believed that if it were not for this loss the extraction would average about 95 per cent.

As a rule, throughout southern West Virginia, the coal lands are held by land-holding companies, which lease to operating companies. The royalty is generally 10 cents per ton of coal and 15 cents per ton of coke, with a yearly minimum.

The beds are nearly flat and quite regular, of an almost ideal mining height, generally with good roof and bottom, and with little gas and water in the drift mines. It is possible, therefore, to lay out a definite plan of mining in advance, and to follow such a plan more closely than in sections where natural conditions are less favorable. In many cases the landholders specify that the coal shall be mined in accordance with certain plans, and prescribe a minimum of extraction. Certain departures from the standard methods, however, are permitted where it seems advisable.

Fig. 33 illustrates the plan of development formulated by the Pocahontas Coal and Coke Company,* which may be carried out by one of three possible procedures as follows:

Panel No. 1.—Drive rooms on 3rd cross entry as soon as come to, begin robbing as soon as second room is completed and rob advancing on 2nd and 3rd cross entries to within 100 feet of 2nd cross entry, on 1st cross entry drive last room first and rob retreating as shown, taking out the barrier pillar left on 2nd cross entry.

Panel No. 2.—Drive entries to the limit before turning rooms except as shown, turn last room on 3rd cross entry first, begin robbing at inside corner of panel, develop rooms only fast enough to keep in advance of robbing and bring robbing back with uniform breakline until completed to barrier pillars.

* Stoek, H. H., "Pocahontas Region Mining Methods," Mines and Minerals, Vol. 29, p. 395. Stow, Audley H., "Mining in the Pocahontas Field," Coal Age, Vol. 3, p. 594, 1913.

Panel No. 3.—Continuous panel, drive entries to the limit before turning rooms except as shown, turn last room on 1st cross entry first, begin robbing as soon as second room is completed, develop

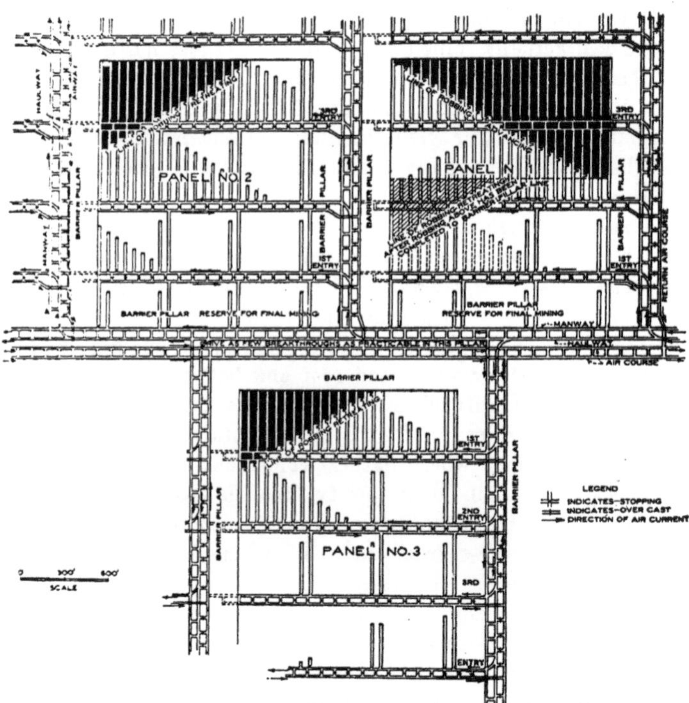

Fig. 33. Plan of Working of Pocahontas Coal and Coke Company

rooms only fast enough to keep in advance of robbing, and bring robbing back with uniform breakline until limit of mining is reached.

According to W. H. Grady, chief mine inspector of the Pocahontas Coal and Coke Company,* the essential advantages of this plan of mining include: provision for tonnage during the development period, provision for meeting the market demand, large barrier pillars insuring against squeezes and rendering impossible the de-

* Grady, W. H., "Some Details of Mining Methods with Special Reference to the Maximum of Recovery," W. Va. Coal Min. Inst., Dec. 1913; Coal Age, Vol. 5, p. 156, 1913.

struction of coal over an extended area, 4-entry system for all extensive main entries with two as intakes and two as returns with breakthroughs intervening only at the points where the cross entries turn off, rendering unnecessary the building of expensive masonry brattices every 80 feet, and insuring the maximum quantity of air for ventilation at a minimum cost for brattices and ventilating power, and cross entries with narrow chain pillars which permit the rapid advance of the entry.

The success of this method, with regard to high output, is shown by the values given in Table 7, taken from Grady's article. The percentage of recovery is based on the thickness of the total seam, including the portion rejected. The lower percentages of extraction shown in Table 7 were reached where pillars were being robbed after standing for many years. Operations had not proceeded far enough at the date of the paper quoted to permit a definite statement as to how great the final recovery would be. A later statement* has been received to the effect that the mines of the lessees of the Pocahontas Coal and Coke Company will probably show an average recovery of 90 per cent. Since some of these operations have extended over many years and since they were not so well managed formerly as at present, it is probable that the recovery now is more than 90 per cent.

In the Pocahontas district, much attention has been given to the subject of recovery, and the values which have been supplied by operating companies are as accurate as such values can be made. H.

TABLE 7

RECOVERY OF COAL IN MINES OF POCAHONTAS COAL AND COKE COMPANY

Plant	Thickness of Seam in Feet	Acres of Entry Mined	Acres of Rooms Mined	Acres of Pillars Mined	Total Acres Mined	Total Tonnage Mined	Tons Mined per Acre	Theoretical Tons per Acre	Percentage of Recovery	Proportion of Seam Rejected
1	6.15	3.06	4.57	11.03	18.66	165,254	8,856	9,922	89.3	0.24
2	5.65	4.40	4.80	14.80	24.00	188,391	8,185	9,115	89.79	0.22
3	5.16	2.68	6.52	15.80	25.00	180,386	7,215	8,325	86.6	0.22
4	4.42	5.88	8.65	13.09	27.62	192,437	6,960	7,131	97.6	0.23
5	5.94	7.00	10.09	19.20	36.29	334,005	9,203	9,582	96.0	0.22
6	4.32	2.11	3.64	9.20	15.04	94,427	6,278	6,969	90.0	0.31
7	5.34	3.31	6.34	0.00	9.65	83,000	8,601	8,614	99.8	0.20
8	5.42	3.72	6.06	9.72	19.50	144,769	8,181	8,777	93.2	0.20
9	4.65	8.10	16.80	2.34	27.24	201,044	7,380	7,534	98.0	0.18
10	8.03	5.20	8.47	10.09	23.76	262,975	11,068	12,923	85.6	0.23

* Eavenson, Howard N., Personal Communication.

N. Eavenson, whose communication has already been referred to with regard to the character of the beds mined, says that the measurements of areas worked out are as close as it is possible to get them on a large scale. The thicknesses given are those of the clean coal, and do not include any bone or black rash. In many instances a record has been kept of coal left in small areas, and the values shown by these tests agree very closely with those given in Table 8. It will be seen from the table that the amount of extraction for mines 9, 10, and 11 is given as more than 100 per cent. This record is explained by the statement previously made concerning the loading out of coal supposed to be left in the mines. At No. 9' the rash is much cleaner than at the other mines of the company, and while the seam is thicker, it carries only a very small amount of dirt; thus a higher percentage of clean coal is given.

TABLE 8

STATEMENT OF THICKNESSES AND RECOVERIES, ALL MINES, UNITED STATES COAL AND COKE COMPANY 1902 TO 1916, INCLUSIVE

Mine No.	Area Worked Out per Cent		Average Thickness Clean Coal	Net Tons Recovered Per Acre Foot	Percentage of Recovery	No. of Seam Worked	Date of First Shipments
	Rooms and Entries	Pillars					
1	45.5	54.5	5.52	1746	97.0	3 & 4	1903
2	59.7	40.3	5.67	1790	99.4	4	1902
3	61.2	38.8	4.56	1429	79.4	4	1903
4	57.3	42.6	6.19	1581	87.8	3	1904
5	63.5	36.5	6.88	1606	89.2	3	1904
6	63.6	36.4	6.09	1769	98.3	4	1903
7	66.9	33.1	6.27	1770	98.3	4	1905
8	70.8	29.2	5.98	1728	96.0	4	1905
9	74.2	25.8	7.24	1908	106.0	4	1908
10	85.1	14.9	5.31	1806	100.3	3 & 4	1907
11	73.3	26.7	5.26	1807	100.4	3 & 4	1907
12	69.8	30.2	8.22	1622	90.0	3	1908
	65.9	34.1	5.95	1738	96.5		

At the No. 3 mine, which shows the lowest percentage of extraction, the roof is exceedingly bad. Above the coal there occurs a layer of shale and slate, from 5 to 10 feet thick, which it is impossible to support even by close timbering. The mining practice is fully as good at this mine as at the others, but the yield is much less because of the more difficult conditions. The table shows the areas worked out at different mines and the percentages recovered. It is Eavenson's opinion that the average recovery in the larger mines throughout the Pocahontas field is fully 90 per cent, and that in many mines even a higher figure is reached.

One of the landholding associations of the Pocahontas district furnishes information* concerning some of the operations of its lessees, and submits a table (Table 9) showing the percentage of recovery. It takes account of operations up to January 1, 1917.

TABLE 9

PERCENTAGE OF RECOVERY ON LIVE WORK AND ROBBING
POCAHONTAS DISTRICT
January 1, 1917

	Area Mined		Percentage of Recovery			
No. of Mine	Live Work, Per cent	Robbing, Per cent	Live Work (Assumed)	Gross Robbing	Net Robbing, Assuming that 37½ per cent of Gross Robbing was Mined as Live Work	Total to Date
1	16.1	83.9	97	84.6	77.2[1]	86.6
2	.3	99.7	97	73.4	59.2	73.5
3	9.4	90.6	97	81.2	71.7	82.7
4	24.2	75.8	97	80.6	70 7	84.6
5	38.3	61.7	97	76.9	64.8	84.6
6	12.1	87.9	97	87.0	81.0	88.2
7	15.5	84.5	97	77.5	65.8	80.6
8	15.4	84.6	97	79.3	68.7	82.0
9	51.8	48.2	97	83.3	73.9	90.4
10	53.7	46.3	97	83.4	75.2	90.7
11	63.5	36.5	97	86.0	79.4	93.0
12	51.7	48.3	97	81.9	72.8	89.7

[1]Values obtained by assuming that 37.5 per cent of the pillar work is done under the same conditions as live work, i. e., with recovery of 97 per cent; thus $0.625X + 0.375 \times 97 = 84.6$, $X = 77.2$.

It will be noted that the extraction at mine No. 2, the first of the leases to exhaust the No. 3 Pocahontas seam, is not more than 73 or 74 per cent. In addition to the losses which will be mentioned, there was a considerable loss here of top coal. A thickness of 18 to 24 inches was left up in the first mining with the expectation that it would be recovered on the retreat, but most of this coal was ultimately lost on account of the bad roof. It is also possible that the loss in the coke yard at this plant, where the maximum number of ovens was run in proportion to the output, amounted to almost double the average tabulated amount.

An inspection of the table shows that, in most cases, the highest percentage of recovery has been reached at those mines where the

* Lincoln, J. J., Personal Communication.

pillar work has been least in proportion to the live work, that is, at those still in the earlier stages of working. Future operations in these mines may be expected to lower these values, but it would seem that 80 per cent would be a very conservative estimate of the average amount of coal that should be won in the various mines of the property up to the exhaustion of all properties under lease.

J. J. Lincoln has discussed losses in the Pocahontas field, and the facts brought out are of general interest in connection with the subject of coal recovery. It is said that losses may be considered under three headings; (a) mixing of coal with refuse, (b) loss in drawing pillars, and (c) loss in coke making.

There is loss in removing the bone and pyrite from the coal, as some coal adheres to the refuse. Under the present mining methods, this loss is from 2 to 4 per cent, and occurs in both new work and robbing.

The following losses are to be expected in pillar drawing in addition to loss of coal attached to refuse: (1) In drawing each stumps are occasionally crushed by the pressure of the top before broken rock from the adjacent gob from covering the coal. This loss will run from 3 to 10 per cent, according to conditions. (2) When the pillars are drawn by splitting, a similar loss, frequently greater, occurs. (3) As the drawing progresses small sections of stumps are occasionally crushed by the pressure of the top before they can be removed. (4) Stumps, sections of pillars, or entire pillars may be crushed by the weight before they can be removed, or may be surrounded and cut off by the broken top. In the mines where actual losses from this source are closely recorded they do not reach 1 per cent, and there is no mine in which they will reach 2 per cent. The third loss is not directly chargeable to mining, but occurs in the making of coke. Where the tonnage of coke produced is used as a measure of the amount of coal taken out, this loss becomes significant in calculating the percentage of coal won. The ratio used by the company in all calculations of tonnage has always been 1.6 tons of coal to one ton of coke. This ratio assumes an actual average yield of 62½ per cent of coke, but in practice this yield is not obtained, the average yield under existing conditions being nearer 55 per cent. This loss has always been charged, with the other losses, directly against the mining. This ratio cannot be used directly in determining the actual amount of coal mined, because only a part of the coal is coked. The loss varies in amount from 1 to 5 per cent, according to the conditions in the different operations, and the actual percentage of extraction is slightly higher than that given in the table because of this error.

Losses in mining and coke making may then be tabulated as follows:

"(a) Removing Refuse—Coal thrown into gob with bone and sulphur bands . .	2 per cent to 4 per cent
(b) Robbing:—	
1. Stumping, or Splitting pillars . .	2 per cent to 10 per cent
2. Sections or Stumps .	0 per cent to 2 per cent
3. Pillars lost	0 per cent to 2 per cent
(c) Additional coal consumed in coke making over and above the amount covered by the constant of calculation, 1.6 tons equals 1 ton coke	1 per cent to 5 per cent
Total . .	5 per cent to 23 per cent"

A peculiar system followed by the Gay Coal and Coke Company of Logan, West Virginia, and called a single-room system* has resulted from an attempt to apply the long-wall method to a seam the average thickness of which is 5 feet, 7 inches. This seam dips to the southwest about 1½ per cent, is practically free from partings, and is of the nature of splint coal, the bottom bench being rather strong, and the top bench somewhat friable. The average thickness of cover does not exceed 500 feet while the maximum is less than 1,000.

A block of coal was cut by two entries 600 feet long, Nos. 4 and 5 (Fig. 34). These were connected at their extremities, which where 300 feet apart. The purpose was to take out the block of coal thus formed in a retreating direction by commencing at the inner end and by working outward with a face about 300 feet long. One hundred beech or hickory posts were used to support the roof near the face. The top of each was covered with 1-inch poplar, and the

* Gay, H. S., "A Single-room System," Proc. Coal Min. Inst. Amer., p. 157, 1906; Mines and Minerals, Vol. 27, p. 325, 1906; Personal Communication.

bottom with 1/16-inch sheet steel. Each post was mounted on a hydraulic head weighing about 700 pounds and tested to a pressure of 3,000 pounds per square inch. These were set along the face 6 feet apart in parallel rows 3 feet apart. The cost of the equipment was approximately $5,000.

When the walls became more than 30 feet apart, rows of props on 15-foot centers were set 8 to 10 feet apart. When the distance between the walls had reached 60 feet, the portable posts were put into use, a row of 50 posts on 6-foot centers being set 10 feet from

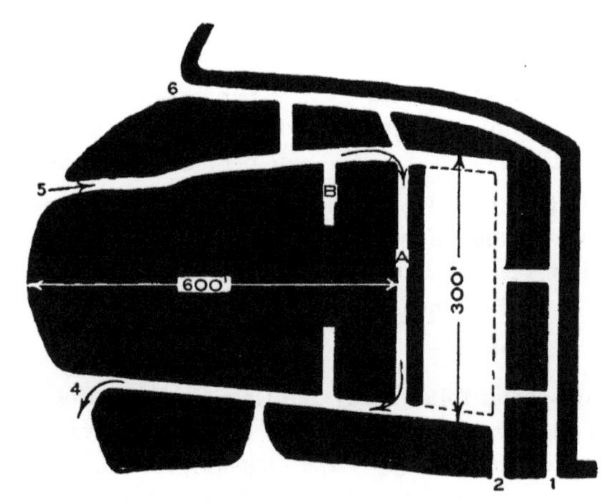

FIG. 34. SINGLE ROOM METHOD, LOGAN COUNTY, WEST VIRGINIA

the face. The heads were covered with wooden cap pieces, and the plungers were raised by a pressure of 50 pounds per square inch. When the face had advanced 6 feet farther, the other 50 posts were put in; as the work progressed, the first row was moved 6 feet ahead of the second, the posts being moved one at a time. An occasional row of posts similar to the first was also set as a precautionary measure.

When the walls were 100 feet apart, it was thought advisable to blast down the roof. The portable posts were set in a single row 6 feet from the face. Examination after the rock had fallen showed that the immediate roof consisted of a seam of strong sand slate at least 30 feet thick without any sign of a parting and that difficulty would be encountered in attempting to apply the long-wall method in this mine.

The advantages to be derived by working on a long face were so great that the company devised another plan which involved working out the remainder of the block by a system of rooms 80 feet wide, parallel with the long-wall face and separated by 30-foot pillars. Each room was to be opened by driving a sub-entry across from entry No. 4 to entry No. 5 (Fig. 34), and thereafter the manner of working was to be identical in every respect with that of the long-wall system.

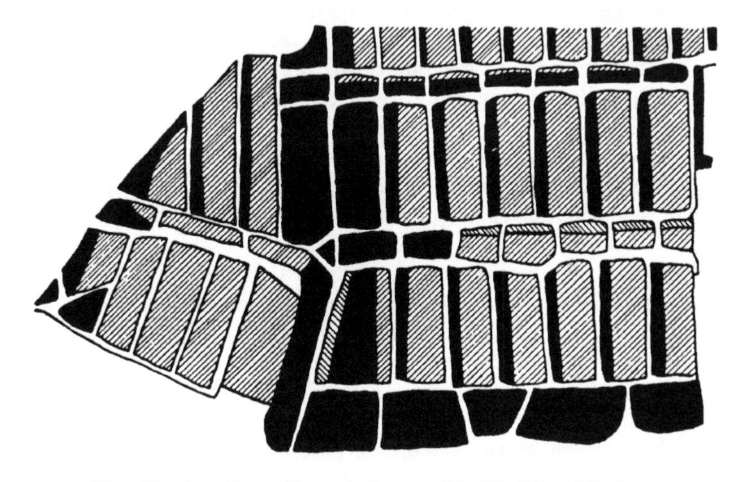

FIG. 35. BIG ROOM METHOD, LOGAN COUNTY, WEST VIRGINIA

From the experience gained it was thought that, with the roof in normal condition, rooms could be worked 90 feet wide with 30-foot pillars. Entries Nos. 4 and 5 were therefore continued eastward, and sub-entries were spaced for the rooms as shown on the map (Fig. 35). The last of the sub-entries was widened to 40 feet, and a single row of ordinary props, 8 to 10 inches in diameter, was set close to the face on 15-foot centers. When the face had moved 10 feet farther, a second row was set. When the room had reached a width of sixty feet, a row of portable posts was set on 10-foot centers, and at 70 feet another row was set. The ordinary props were used for detecting the action of the roof.

One of the advantages of this system of mining is that a large amount of coal per employee may be obtained. In fact the production per man is considerably greater than with the room-and-pillar method, and greater even than would be possible with the long-wall method. The highest rating in this seam for car distribution, exclusive of this mine, is 13 tons per loader; this mine is rated at 20 tons per loader. It is the opinion of Mr. Gay that it would be possible to produce about 11 tons per inside employee per 9-hour day for five days a week. The method also results in the recovery of a very high percentage of coal. A calculation based upon the number of tons shipped and the area excavated, according to planimeter measurement, indicates that the extraction was 85.9 per cent.

Since this description was published, the system has been modified to reduce the narrow work, but the general plan has been followed. Instead of driving a single room to form the working face, parallel rooms separated by an 18-foot pillar are driven; thus the use of brattices is unnecessary, and ventilation is improved. The hydraulic posts were soon abandoned, as posts without the hydraulic heads are cheaper, and they are easily recovered. One of the most important facts concerning the operation is that there has not been a single fatal accident in the mine since work was begun.

Another system in which an effort was made to obtain the advantages of long-wall working was tried a few years ago in West Virginia.* In developing this system (Fig. 36) triple entries are driven from the outcrop, near which double entries are turned off at right angles. From these, entries are driven parallel with the main entry; thus blocks of coal about 900 feet wide are cut off. Block entries are turned from the main entry and from these side entries, parallel with the cross entry, spaced about 500 feet apart, and driven for about 800 feet; thus the coal is blocked into areas approximately 500 by 800 feet. In working these blocks, a room is turned first at the end of the block entries to form a working face for the long-wall machine. The blocks are then worked back toward the main entry for 500 feet; thus a barrier of 300 feet protects each main entry. Track is laid along the face as near the coal as possible, and is moved as the face progresses. The roof is allowed to fall, but the line of break is kept at the correct distance from the face by three or more rows

* James, W. E., "Block System of Retreating Long-wall," Proc. W. Va. Coal Min. Inst., p. 137, 1911; Cabell, C. A., Personal Communication, and Patent Specifications.

of props, the last row being moved forward after a cut is made. The trolley wires are hung on hangers as usual, but to keep them tight they are carried on portable drums. Current for the motor and machines along the face is taken from the trolley wire on the entry

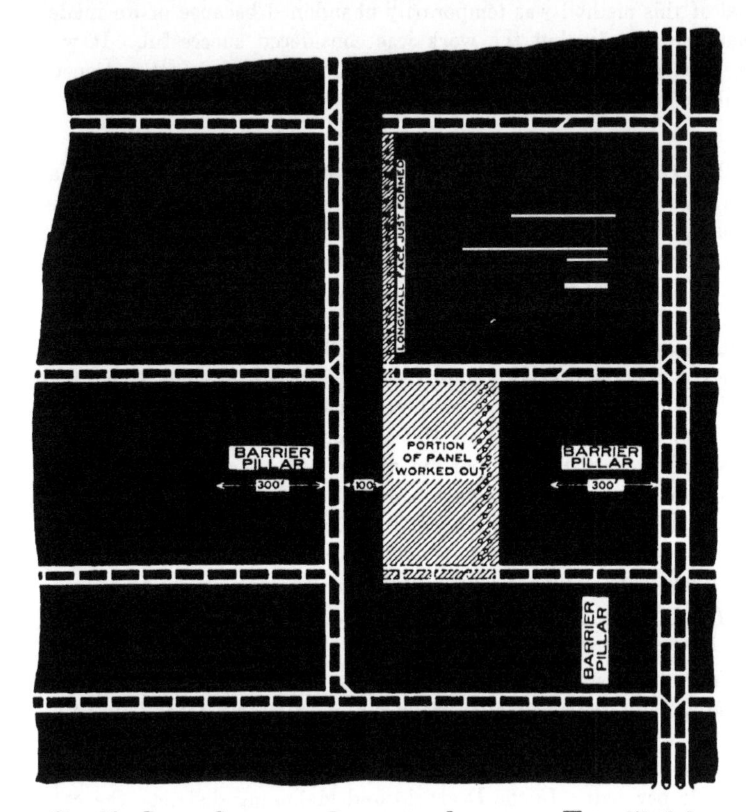

FIG. 36. BLOCK SYSTEM OF RETREATING LONG-WALL, WEST VIRGINIA

by means of a cable which is also coiled on a portable drum. Ventilation is controlled by placing a regulator in the return air course of each block. Each block or face is provided with a separate supply of fresh air by having overcasts placed at the air courses to admit the return air into the main air course entry. No doors are required at any points in the mine.

It was claimed for this method that a block will produce 440 tons per day with the use of only 500 feet of track in addition to that on the entries, that practically all the coal is obtained, and that the work of the rolling stock and cutting machines is concentrated. The trial of this method was temporarily abandoned because of an inadequate car supply, but the work was considered successful. It was not carried far enough to provide reliable data for an estimation of total extraction.

26. *Ohio.*—The literature of coal mining contains little information regarding conditions in Ohio. Some of the operations in the Hocking Valley district have been described in articles which state that large quantities of coal are left in the roof because of the poor quality of the product. An article on Hisylvania Mine No. 23 states* that the bed mined in the Hocking Valley district is the Middle Kittanning, Hocking Valley, or No. 6. The bottom consists of a few inches of fire clay overlying hard rock. The roof is of shale, 6 to 8 feet thick. The coal bed consists of three benches. The thickness of good coal is about 6 feet, and above this is about 5½ feet of a poorer coal separated from the lower portion of the bed by a distinct parting. This upper bed, with the upper bench of the lower bed, is known as top coal. In this district all coal in excess of 6 feet, and in many places in excess of 4½ feet, is to be credited to this upper bench which has a maximum thickness of 10 feet.

James Pritchard† estimates the percentages of extraction in the districts as follows:

In the Pittsburgh vein district, the Cambridge field, and the Hocking field, districts which produce approximately three-fourths of the coal of the state, the rate of extraction will range from 60 to 70 per cent. In the Massillon and Jackson fields, the rate of extraction may reach 85 per cent. In the Deerfield and Mahoning districts, the rate of extraction may reach 85 per cent. Throughout the remainder of the state, the maximum percentage of extraction will run from 60 to 70 per cent. The average rate of recovery is approximately 60 per cent, with a minimum of 55 per cent and a maximum of 75 per cent.

---

* Burroughs, W. G., "Hisylvania Mine No. 23," Coll. Eng., Vol. 34, p. 421.

† Pritchard, James, Chief Deputy and Safety Commissioner of Mines, Personal Communication.

The average value represents the most common percentage of extraction.

The conditions in the Pittsburgh vein district of eastern Ohio are described by Roby* as follows:

The extraction is limited by various physical and commercial conditions. The roof consists of a bed of weak coal of poor quality, variable in thickness, and separated from the main bed by a layer of drawslate which disintegrates on exposure. Overlying the roof coal is an unstratified "soapstone" 4 to 8 feet thick, and above this is a thick layer of hard limestone. The country is hilly, and the total cover varies from 30 to 600 feet in thickness. The character of the roof makes it necessary to leave larger pillars than would be required under a good roof. The roof coal is left up, and as it is poor in quality and not marketable, it is not considered in making estimates of extraction. The room-and-pillar method is used. There has been much discussion concerning the possibility of applying the long-wall method, but the fragility of the roof and the tendency of all rock below the limestone to shear off at the solid face have seemed to make the method impracticable.

It is desirable that rooms be worked out as quickly as possible because of the tendency of the roof and pillars to fail. Because of these conditions, the numerous interruptions which have been caused by strikes and business depressions have tended to make the rate of recovery lower than would have been the case with uninterrupted operation.†

J. C. Haring states that the recovery in the Massillon district has not exceeded 75 per cent. The highest recovery in the district is probably at the Pocock No. 4 mine where it is estimated that fully 90 per cent of the coal has been obtained.

At Steubenville, the mine of the LaBelle Iron Works has been operated on the long-wall system since 1913; prior to that time the room-and-pillar method was employed.‡ The bed is the Lower Freeport which is a little over 3 feet in thickness and has a good shale roof.

At present a considerable amount of stripping is being done in the No. 8 coal in the vicinity of Steubenville. The coal outcrops on

* Roby, J. J., Personal Communication.

† Haring, J. C., Personal Communication.

‡ Burroughs, W. G., "Long-wall Mining at Steubenville, Ohio," Coal Age, Vol. 11, p. 697.

the slopes of hills, and the rapid increase in the thickness of the overburden restricts the operations to a narrow belt, although they may extend for a considerable distance along the outcrop.

27. *Kentucky.*—The development of coal mining in Kentucky has been comparatively recent, and some of the Kentucky fields are still so young that it is impossible to obtain estimates of recovery. The state may be divided broadly into two districts, the western district being closely allied with the fields of Indiana and Illinois, and the eastern district with those of West Virginia and Virginia.

The statements presented in the following paragraphs concerning the western district are made on the authority of N. G. Alford,* who says that the fields contain about 38.3 per cent of the coal-bearing areas of this state and that in 1912 47.7 per cent of the total production came from this district. The smallness of many operations is shown by the statement that 21 per cent of the mines produced less than 10,000 tons per annum each; 51 per cent produced less than 60,000 tons each; 23 per cent produced more than 100,000 tons each; and two companies, operating 18 mines, produced 2,750,000 tons each.

Generally, the rate of recovery in the mines of western Kentucky is about 66 2/3 per cent, although in some instances it is as low as 44 per cent. Without an exception the mines of this district are developed on the room-and-pillar system with double or triple entries. With the exception of two or three isolated operations, all the coal is produced from three seams.

Most of the coal comes from the No. 9 and No. 11 beds, the former producing about three-fourths of the total output of the field. This bed is present in eight counties and approaches 5 feet in thickness. In most places it is reached by shafts of 300 feet or less in depth, although there are some local surface depressions which permit access by slopes or drifts. It has a black shale roof and a soft fire-clay bottom.

The No. 11 seam lies from 40 to 100 feet above the No. 9, and follows the latter in commercial importance. Its average thickness is 6 feet. Above the coal is a stratum of limestone of thickness varying from a few inches to 40 feet. This limestone is usually separated

---

* Alford, Newell G., "Problems Encountered in Kentucky Coal Mining," Ky. Min. Inst., 1913; Coal Age, Vol. 5, p. 674, and Personal Communication.

from the coal by a thin stratum of heavy laminated clay 6 to 24 inches thick. This top adheres uncertainly to the limestone above it, and presents a constant danger. Near the outcrop the top becomes very treacherous. The bottom of this seam consists of soft fire clay which frequently heaves in haulage entries that have been opened for some time. About half the mines in this seam are shaft mines, and the remainder are drift mines.

The third seam in commercial importance is No. 12, which is found best developed in Clay and Webster Counties. Its approximate depth below the surface is 225 feet. Its average thickness is 7 feet. The bottom is of fire clay which is high in calcium and which disintegrates rapidly when drainage water is directed through it in ditches. The roof consists of light gray disintegrated shale 10 to 15 feet thick. If all the coal is removed, this top will fall to a height of 6 or 8 feet, and heavy timber sets, thoroughly and solidly lagged, are required to support it. Because of this condition it has been found necessary to leave 16 inches of top coal as a roof. Sixty per cent of this top coal is recovered from rooms, but no attempt is made to recover it from entries. When the development of the No. 12 seam was begun, rooms were driven 21 feet wide on 33-foot centers; but this width of pillar was found to be too narrow, and it has been increased to 20 feet, with rooms 21 feet wide. Under these conditions a recovery of 44 per cent is the best which has been reached up to this time. This low percentage of recovery is in part due to physical conditions, and in part to over-development and keen competition.

Several factors contribute to limiting the recovery in the No. 9 and No. 11 beds. The most important of these is probably inadequate planning of future workings. Frequently pillars are left too small; consequently the bottom heaves, and the pillars are crushed. Partial recovery of pillar coal by taking slabs off the ribs is not general, and the total recovery of pillars has not been attempted. The operators in this district hold the opinion that pillar robbing in the No. 11 seam is particularly hazardous and impractical, because heavy limestone overlies the seam; in the Connellsville district of Pennsylvania, however, pillars are successfully drawn under a heavy limestone. Alford expresses the opinion that if the workings in the No. 11 seam were properly laid out and started, little difficulty would be found in increasing the percentage of recovery.

In this district there is an over-production with a limited market; consequently competition is keen. When business conditions are normal, the margin of profit is so small as to preclude any costly improvements, and little money has been expended in experiments. Another source of loss lies in the waste at the tipples in the summer, because the demands of consumers are more exacting in times of dull markets. Good coal attached to lumps of pyrite is often discarded in large quantities, and so far this waste has been accepted as unavoidable. This district furnishes one of the best illustrations of the effects of over-development and lack of harmony of interests. There are, however, some operations which are carried out on a considerable scale and with careful attention to the proper planning of the work.

The estimates covering production mentioned previously are confirmed by a personal communication from another operator, S. S. Lanier, who has been a close observer in the district for thirty years and who estimates that the extraction is about 65 per cent.

The eastern Kentucky district is of such recent development that estimates of production are not very reliable. H. D. Easton, operating in the southeastern part of the state, thinks it safe to say that a recovery of 90 per cent is being reached in the Straight Creek seam in Bell County, but much trouble has been experienced from squeezes due to lack of systematic working.

Mines are operated on the room-and-pillar system with rooms turned from both sides of the room entries. Rooms are generally 40 feet wide with 20-foot pillars. Cross entries are driven about 1,200 feet apart, and these usually extend to the property line or to the outcrop. It has been the practice to extract the pillar coal on the retreat, and so far as possible to keep the face lined up over a sufficient distance to get a fall of roof. Room tracks are swung across the face of the pillar and are moved as the pillar is drawn back. If the pillars are narrow, the room tracks are not moved even though the coal has to be shoveled 15 or 20 feet.

There has been no very systematic work in pillar recovery in the southeastern part of the state, and pillars or stumps have been left scattered promiscuously, with the result that many costly squeezes have occurred. One company has lost an entire mine as a result of this practice. It has been the general opinion that it would be impossible to get a clean break in the overlying strata because of the solid sandstone above the coal, but such breaks have been obtained very successfully, even in rather limited areas.

The Harland district, the Hazard district, and the Elkhorn district are all too newly developed to provide a reliable basis for estimating recovery, but it is possible that the percentage of recovery will be very high.

In eastern Kentucky the surface is of no great value, and it is almost invariably owned by the coal companies so that the necessity for sustaining it is not a factor affecting the percentage of extraction.

There are some operations in the central southern part of the state, but the mines are not yet sufficiently developed to yield adequate data for reliable estimates of extraction.* The mines are opened by drifts, and as the coal is irregular, the hills have been entered at many points. A heavy sandstone occurs between the two seams worked, and pillars have not been drawn in the lower seam, because it has been feared that the upper seam would be damaged. In the upper seam, pillar drawing has not been practiced to any great extent, because when tried, it has resulted in breaks extending to the surface through which considerable water has entered. It is the intention of the operators to extract the pillars when the mines have been worked out, and the final percentage of extraction will probably be high.

28. *Tennessee.*—Little information is available on the percentage of recovery in Tennessee, and the statements obtained are not altogether in agreement. One operator,† formerly connected with the industry in Tennessee, states that a few years ago the mining practices were not good. On the first mining, about 50 per cent of the coal was taken, and the ultimate recovery was probably about 80 per cent. Because of the low value of coal lands, less effort is made to get a maximum recovery than in some other districts where coal lands are more valuable.

R. A. Shiflett,‡ Chief Mine Inspector, says that it would be difficult to give any general percentage for extraction since the coal measures vary in dip from horizontal to 40 degrees, and in some

* Butler, J. E., Personal Communication.
† Coxe, E. H., Personal Communication.
‡ Personal Communication.

cases from horitzontal to vertical. In a large number of mines it is impossible to map out any definite method of mining, and conditions have to be met as they are encountered.

Nearly all the drift mines are developed on the double-entry room-and-pillar system. Rooms are driven from 250 to 300 feet, and room pillars are drawn as soon as the rooms are finished. If conditions are favorable, the entry pillars and room stumps are recovered on the retreat; and where the coal is practically level, 40 to 50 inches thick and with good roof and bottom, about 90 per cent is extracted under careful management. This percentage is not reached in many mines because of lack of attention to high extraction. It is thought that the extraction in general does not exceed 65 per cent, but that this percentage could be greatly increased by proper methods and careful management.

One company,* whose method was to turn rooms on the advance and immediately to draw room pillars back to within 65 or 70 feet of the entry on completion of the room, obtained a recovery of nearly 90 per cent up to about the beginning of 1916. At that time four cross entries were lost from heaving of the soft bottom. The cover is 500 to 800 feet in thickness, the coal is 56 inches in thickness, and the bottom is of soft fire clay from 4 to 7 feet in thickness. This company is planning the introduction of the long-wall method. A face of about 300 feet will be formed by connecting the ends of two entries. It is thought that the single stick timbering, with perhaps an occasional crib, will be sufficient. It is expected that the bottom will heave and reach the roof as the latter bends down. The scarcity of labor and the irregularity of the car supply make the success of long-wall operations somewhat doubtful; and if it is necessary temporarily to abandon this method, another which is illustrated in Fig. 37 will be adopted. In this method apparently a little more than 50 per cent of the coal would be taken out from rooms, and the ultimate percentage of extraction should be almost complete. The method will permit concentrated working, and much of the trouble due to the conditions of the floor and roof will probably be avoided.

29. *Alabama.*—Although Alabama† is an important producer

* Hutcheson, W. C., Personal Communication.

† Strong, J. E., "Alabama Mining Methods," Mines and Minerals, Vol. 21, p. 195, and Personal Communication.

of coal, the working conditions are not so good as those of Pennsylvania, Kentucky, and some other states. One of the distinctive features of the Alabama coal field is that although there are five seams of coal, rarely more than one of them is workable at any one place; in one portion of Jefferson County, for instance, the Pratt seam, which is considered the topmost workable seam of the Alabama coal measures, may have a working thickness of 4 feet, while at a distance of two or three miles the same seam may not be more than

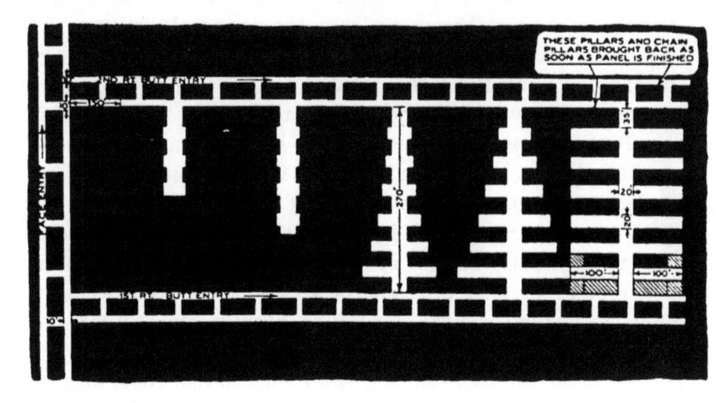

FIG. 37. PROPOSED PLAN OF WIND ROCK COAL COMPANY, TENNESSEE

2 feet thick. The coal beds, including the Pratt bed, vary rather abruptly within a few miles with regard to thickness, impurities, and character of roof.

The Pratt seam has been worked longer and more extensively than any other seam in this district. It is probable that the recovery from this seam, under the best methods of working with the room-and-pillar system, is about 87 per cent. This percentage applies only where the thickness is three feet or more; coal thinner than this cannot profitably be removed.

The Mary Lee seam, which is supposed to contain the thickest workable coal, lies about 300 feet below the Pratt seam, and is the one that the operators of the Birmingham district expect to work during the next twenty years. So far as it is known, the thickness of the bed ranges from 6 to 10 feet. It is difficult to get reliable estimates of the recovery of coal in this seam, but one operator reports, on the basis of an experience of twenty years, that the recovery has been about 90 per cent.

There are mines on lower seams in other localities, but little attention has been paid to the extraction of a high percentage of the coal. The operation of these mines depends largely upon market conditions, and they are probably not operated, on the average, more than half the time. Under these conditions the loss of pillar coal caused by falls of roof is necessarily large.

The Alabama mines, with the single exception of the Montavallo mine, where a change is being made to the long-wall system, are worked on the room-and-pillar system. The larger operations are in the neighborhood of Birmingham, and in this district the carboniferous measures are tilted and broken to a great extent. This condition affects the roof of the coal under cover for a considerable distance. Both top and bottom are of variable character.

In the larger operations at least, the triple entry system is used. Commencing at a distance of 800 feet from the surface, cross entries are usually driven about 350 feet apart. Until the entries have been driven a few hundred feet, it is not possible to determine whether they should be narrow or wide enough to provide storage for the impurities of the bed and the brushing of the roof. Rooms are generally opened narrow also, (30 feet wide) with 25-foot room pillars, until it is determined whether the character of the overlying strata and of the floor will permit the working of wider rooms.

Probably 75 per cent of the large operators in the district have adopted the plan of immediate pillar drawing in preference to that of driving the narrow work to the limit and pulling the pillars upon the retreat. In a number of instances, rooms are driven 40 feet wide with 30-foot pillars and are worked for a distance of 300 feet, or to the entry above; then a cut is taken across the end of the pillar, and the pillar is drawn back to the entry stump. When the room pillars are drawn on the advance, there is no difficulty in getting room stumps and air-course pillars after the entry work is complete.

Strong estimates the recovery in mines operated by the larger corporations to be from 87 to 90 per cent. Priestly Toulmin, another operator,* confirms these values by stating that the average extraction in Alabama is not less than 75 per cent and not more than 80 per cent, so that possibly 77½ per cent would be a fair value. In

* Personal Communication.

some mines the extraction is less than 60 per cent, and in others it is more than 95 per cent.

C. F. DeBardeleben* furnishes the following estimates:

At one operation where 768.6 acres had been worked over, assuming the average thickness of coal to be 4 feet and that 25 cubic feet of coal in place make a ton, the coal available was 5,356,834 tons

Tons mined 3,028,960
Extraction .... 56.5 per cent

Assuming that 12.5 per cent more will be obtained from pillars, the extraction will amount to about 63 per cent. At another operation where the average thickness is 5 feet and 316.6 acres have been worked over, the coal available was .... 2,758,219 tons

Tons produced 1,847,582
Extraction .... 67 per cent

Assuming a probable extraction of pillars, the final recovery will amount to about 70 per cent. At a third mine, where conditions were favorable because of level coal and the absence of gas, 25-foot rooms were driven on 75-foot centers; the pillar was drawn back half way to the entry as soon as a room was finished, the entry pillars and room stumps being extracted when the entry was abandoned. Under these conditions the recovery was about 85 per cent.

C. H. Nesbitt, Chief Mine Inspector of Alabama, estimates the average recovery to be 80 per cent, the highest percentages being reached in the Pratt and Montavallo beds.† There has been great improvement in the percentage of extraction in the past twenty years, and even in the past ten years. This improvement has been due largely to more nearly complete and accurate mapping, and to more improved and effective methods of controlling the water in slope and shaft mines.

30. *Indiana.*—Almost no information has been available concerning the percentage of coal extracted in Indiana mines. W. M. Zeller † reports that the extraction in the Brazil district is probably about 60 per cent. This estimate agrees fairly well with the estimates of operators in southern Illinois, and since such estimates have been found to be too high in almost all instances, it is probable that the

* Personal Communication.
† Personal Communication.

average extraction in Indiana, as in southern Illinois, will not exceed 50 per cent.

31. *Michigan.*—The coal beds in Michigan are irregular in extent and decrease in thickness with depth. Sometimes they are entirely cut out by erosion or replaced by sandstone and other materials. Usually the beds above the coal consist of black shale, and they are often weak. Owing to erosion, coal is sometimes found directly below clay, sand, or gravel, or below other unconsolidated rocks, where it is practically unworkable. At several mines the roof is of black bituminous limestone. In most instances the floor is of fire clay or shale, although sandstone is sometimes found. The thickness of coal varies from 2 feet, 6 inches to 3 feet, 10 inches, a fair average at Saginaw being 3 feet. In the Saginaw Valley the surface is level.*

R. M. Randall states † that the first company in the district operated within the city limits of Saginaw, and that because of the necessity of leaving pillars to protect the surface the recovery was only about 68 to 70 per cent. At present this company is operating in farming districts where it is not necessary to maintain the surface; and the recovery, within the last five years, has been about 90 per cent. The room-and-pillar system is used with rooms projected 40-feet wide on 50-foot centers and driven 150 feet, but the actual dimensions vary according to the conditions of the roof. Short-wall machines are used for undercutting. It is estimated that 75 per cent of the room pillars and 95 per cent of the entry pillars are recovered, and that the extraction on the advance is 70 per cent. The conditions at the old and at the new mines have been so different that it is impossible to give an average value for the extraction, but it is believed that the extraction in the new mines in the area actually worked will be from 85 to 90 per cent.

32. *Iowa.*—The physical conditions in the Iowa coal field are not uniform. The cover ranges in thickness from a few inches to 300 feet, and consists of the coal measure beds and glacial drift, the latter commonly constituting the larger part of the thickness. The workable coal beds generally have a top of draw shale varying

---

* Lane, A. C., Mich. Geol. Sur., Vol. 8; Mines and Minerals, Vol. 23, p. 148.
† Personal Communication.

from a few inches to two feet in thickness. Above this is a black shale, or sometimes a bituminous or argillaceous limestone. In the latter case this rock is often strong enough to permit a reduction in the amount of timber used and thereby to facilitate mining. This condition makes it possible in Appanoose County to work a bed in which the thickness of clean coal is only about 25 to 27 inches. The bottom everywhere consists of plastic fire clay, and in the Appanoose field the undercutting is done in this clay. Its occurrence is frequently the cause of creep in room-and-pillar mines. The most unfavorable conditions are found through the northern end of the Iowa coal fields.

Even in single districts the percentage of extraction varies between wide limits. The maximum extraction, estimated at 90 per cent or more, is reached in the Appanoose field, where the long-wall system is employed. An operator* familiar with conditions in these long-wall mines says that the extraction is complete, but that it is less than the previously calculated amount of coal in the ground because of the presence of faults.

In the room-and-pillar districts the extraction rarely if ever exceeds 75 per cent, and under especially bad conditions of bottom and top with an abundance of water, it may not exceed 50 per cent. Probably a fair average of recovery for the state is 70 per cent. The percentages given refer only to the bed mined and to the area of actual mining operations. When larger areas are considered, the percentage of recovery is less because of the loss of considerable quantities of coal through lack of coöperation between owners, a loss estimated to be at least 10 per cent.†

The engineer‡ of one of the operating companies says that in the room-and-pillar mines with which he is familiar the recovery will average about 75 per cent, and that a recovery of 80 per cent is expected in the newer mines.

33. *Missouri.*—The coal fields of Missouri may be roughly divided into three districts, the first district lying near the middle of the state in Macon and Randolph Counties where operations are conducted on the room-and-pillar method, the second district farther

* Taylor, H. N., Personal Communication.
† Beyer, Professor S. W., Personal Communication.
‡ Jorgensen, F. F., Personal Communication.

west along the Missouri River in the vicinity of Lexington where operations are conducted on the long-wall system, and the third district in the southwestern part of the state, where the conditions are similar to those of southeastern Kansas and northeastern Oklahoma and where the room-and-pillar system has generally been followed. Recently a considerable quantity of coal has been obtained in the southwest district by stripping.

In Randolph and Macon Counties, in the neighborhood of Bevier, the coal is considerably broken by faults and horsebacks, and recovery does not exceed 50 per cent.* In the long-wall district the extraction in the area worked out is practically complete, but most of the operations are conducted on a small scale and no estimates covering the probable extraction over the whole area are available.

In the southwestern part of the state the continuity of the coal is considerably broken by horsebacks, as in the neighboring parts of Kansas and Oklahoma. Mining methods have not been highly developed, and no great attention has been paid to completeness of extraction. It is not probable that the extraction in this district, within the areas worked, will be more than 50 per cent.

34. *Arkansas.*—Steel says† that the ordinary waste of coal in Arkansas is unusually great even for this country, a fact to be accounted for partly by unfavorable geological conditions. In addition to the wastes common to all coal producing states there are others due to local geological and physical conditions, which Steel considers unusually unfavorable in Arkansas.

There is considerable loss because of irregularities of entries, due to the varying dip of the bed. Entries which are turned from the slope at standard distances measured along the coal seam will have variable and perhaps severe grades if they are driven straight or will be very crooked if they are driven on grade. If the dip increases and the entries are driven on grade, the distance between entries decreases and sometimes the rooms between entries become so short that the entry from which they are turned is discontinued; then rooms from the entry below are driven long enough to take out all the coal, or part of it, which would have been taken out through the intermediate entry. Sometimes the length of rooms necessary to

* Taylor, H. N., Personal Communication.

† Steel, A. A., "Coal Mining in Arkansas," Ark Geol. Sur.

extract all the coal would be so great that part of it is left. Sometimes this is won through the upper entry, and sometimes it is lost.

There are also losses due to irregularities in the coal, and entries are frequently not extended through areas of low coal to get the good coal lying beyond. Areas of thin coal are commonly abandoned. The losses due to thin or poor coal are greater perhaps in Arkansas than in other states, because the dip of the beds makes the driving of entries around these poor areas considerably more expensive, and it is not profitable to take out good coal lying beyond poor coal unless the area of the good coal is large. There is often considerable loss from the abandonment of parts of beds. In many places the different benches of thick beds are separated by thick partings; if a single bench is thick enough to mine, it is worked separately, and sometimes the bench above it or below it is lost. The loss of coal in this form, though not so great as formerly, is probably greater in Arkansas than in any other state, with the possible exception of Colorado. Loss due to the need of protecting the surface is not serious, because the value of the surface is low, and the rough topography insures good drainage.

H. Denman,* an operator familiar with the district, expresses the opinion that the recovery in both Arkansas and Oklahoma does not exceed 50 per cent. In certain portions of a mine the recovery may be as high as 70 per cent, but he believes if the whole area of the mine is considered, the percentage of extraction will not, in any case, exceed 55 per cent. These statements are applicable to both Arkansas and the neighboring Oklahoma district, as the same system of mining is used in both.

The system of mining in the Arkansas-Oklahoma field is practically the same that was used when the field was first opened about forty years ago. There is no systematic attempt at laying out mines with the view of drawing pillars, but the general plan is to get as much coal as possible in the first working and to abandon the remainder. There is one mine in which an attempt is being made to plan the work so as to obtain the pillar coal, but this attempt is so recent that it is impossible to foretell the degree of its success. The widths of rooms and pillars are influenced by the charges for narrow work and for yardage, which are so high that neither narrow rooms nor long break-throughs can be driven. At present the average room

* Personal Communication.

neck is about 10 feet long, and if longer necks could be driven without increased cost, it might be possible to prevent squeezing of the entry and to obtain a considerable amount of coal from the entry pillars, but the yardage cost is so high that this procedure seems unprofitable.

35. *Kansas.*—The coal produced in Kansas comes from three districts, the one in the southeastern corner of the state being by far the most important. The others are the Leavenworth and the Osage districts. The Leavenworth district lies in the northeastern part of the state and may be considered as connected with the district of northwest Missouri, although the strata dip toward the west and the coal is found at greater depths in Kansas than in Missouri. All operations in the Leavenworth district are on the long-wall system, and the extraction, in the areas mined out, is practically complete. Coal in this district ranges from about 19 to about 24 inches in thickness. The depth is about 700 feet. A 3-foot bed lying at a depth of 1,000 feet was found at Atchison about ten years ago and was worked by the long-wall method, but the work was not commercially successful and was abandoned. The Osage district lies to the south of Topeka and is not important commercially. The coal is about 20 inches thick, and is mined entirely by the long-wall method. The extraction is practically complete within the area mined out. This is the thinnest bed of bituminous coal worked in the United States.

In the southeastern district of the state the coal beds lie on the west slope of the Ozark uplift and dip toward the west and northwest. The beds contain numerous horsebacks which interefere with systematic mining. The room-and-pillar method is followed, and little attempt is made to extract pillar coal. Practically all coal in Kansas, except that produced by the long-wall method and by stripping, is shot from the solid, a method which unquestionably leads to the production of small coal, especially where the holes are greatly overcharged as they usually are. The recovery is in the neighborhood of 50 per cent, although it may sometimes be greater in limited areas, because the horsebacks may be made to serve as pillars. H. N. Taylor, in a personal communication, confirms this estimate of extraction. He says that in places a considerable loss is experienced, because the rate for mining low coal is so high as to be considered prohibitive, and even if the rate is paid it is difficult to get men to work the low coal. A. C. Terrill reports* that the closest estimates of those most familiar with conditions place the recovery at about 50 per cent.

The approaching exhaustion of the shallower mines has necessitated the working of the northern part of the district where the

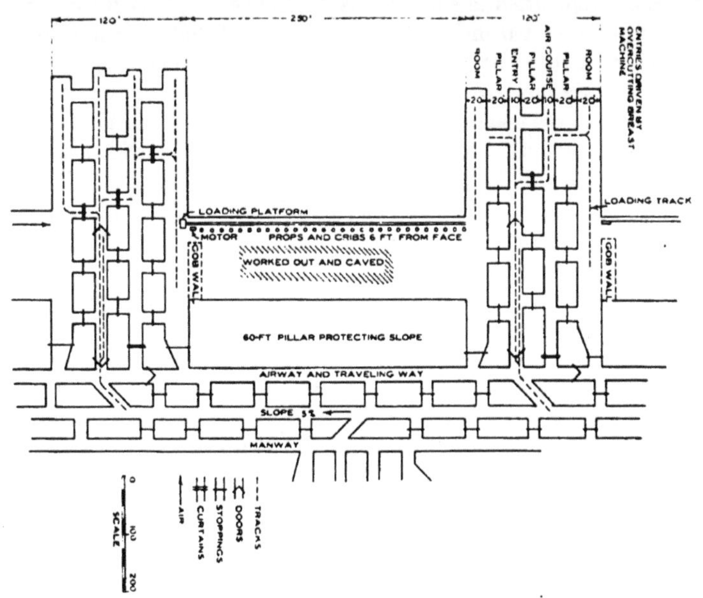

FIG. 38. PANEL LONG-WALL IN OKLAHOMA

cover is about 250 feet thick. At least one of the operators desired to work these deeper mines by the panel method, but it has not as yet been found possible to reach satisfactory arrangements with the mine workers. Since a makeshift adopted to prevent the spread of squeezes leaves two rooms out of seven unworked, the recovery has been reduced about 1,000 tons per acre. The operators still hope that they may be able to introduce the panel system and thus materially increase the recovery.†

In recent years a large amount of coal has been taken from this district and from the neighboring region in Missouri by extensive

* Personal Communication.

† Taylor, H. N., Personal Communication.

stripping operations. In the area worked over, the extraction by this method is practically complete.

36. *Oklahoma.*—In Oklahoma the coal is produced largely by individual operators, the land being owned by the Indian nations and leased to operators in small tracts. * Elaborate plans for mining are not to be expected under these conditions. Elliot estimated that the recovery of the entire bed worked was not more than 55 per cent. He said that the low percentage of recovery was due to the extravagant system of room-and-pillar mining adopted, and that this system could not be changed because of unfavorable labor conditions.

The Rock Island Coal Mining Company obtains an extraction of 48.18 per cent in the McAlester district. In the Hartshorne district this company has five mines, their percentages of extraction being 56.6, 52.6, 55.0, 51.5, and 47.8, respectively. The average percentage of extraction at these five mines is 52.7 and the average of all the mines of the company in Oklahoma is 51.8. †

This company is now trying a panel long-wall plan (Fig. 38) with the hope of increasing the extraction from about 57 to about 70 per cent. The coal is about 3 feet, 4 inches thick, and dips from 5 to 8 degrees. The working face is parallel with the dip. The roof along the face was at first supported by cribs built of 8-inch by 8-inch timbers about 4 feet long and these cribs were withdrawn and moved forward as the face advanced, the roof being allowed to fall. A row of props was also used to support the top above a conveyor used for carrying the coal along the face. At present the use of cribs has been discontinued, except along the ribs of the entries, and 10-inch by 10-inch props are used to support the roof. These are drawn and reset as the face advances. The necessity of using props on both sides of the conveyor constitutes one of the difficulties of the operation. The roof breaks as the face advances. There seems to be no great difficulty in the use of undercutting machines, but sometimes the coal falls too soon for convenience in loading, and large lumps clog the conveyor. While this operation must still be considered in the experimental stage, the working face has been advanced about 130 feet without serious difficulty. It is planned that the pillars flank-

* Elliot, James, "Conditions of the Coal Mining Industry of Oklahoma," Proc. Amer. Min. Cong., p. 221, 1911.

† Scholz, Carl, Personal Communication.

ing the long-wall panel, left during the advance, shall be taken out on the retreat.

37. *Texas.*—There are three bituminous coal fields and one lignite field in Texas. In one bitumionus field in the north central portion of the state, practically all the mines are operated on the long-wall plan, and the recovery is nearly complete. The two other bituminous fields are located in the southwestern part of the state along the Rio Grande. In this district the mines are operated on the room-and-pillar plan, and the recovery is said to be about 75 per cent. The lignite field extends entirely across the state from the northeast corner in a southwesterly direction to the Rio Grande. All lignite mines are worked on the room-and-pillar method, and the recovery varies greatly in different parts of the state, but 75 per cent is probably the average.*

38. *North Dakota.*—J. W. Bliss, State Engineer, estimates that the recovery in the coal mines of North Dakota is between 70 and 75 per cent. The manager of one mine claims a recovery of about 85 per cent. †

39. *Colorado.*—The principal producing districts of Colorado are the bituminous district in the southeastern part of the state, near Trinidad, and the lignite district just east of the mountains in the northern part of the state. The bituminous district is the more important. In the Trinidad district‡ the average thickness of the coal is about 6 feet. The top is strong, and the bottom is weak. Entries are driven to a fixed boundary, and the rooms which are needed to supply enough coal to keep the driver busy are turned. When the boundary is reached, rooms are turned at the inby end of the entry, and pillar drawing is commenced as soon as the rooms reach their limits. Nearly all the coal is taken out on the retreat. Rooms have a maximum width of 18 feet, and room pillars are 32 feet wide. All work is done with picks. The coal is soft and occasionally the pillars crush, but most of the difficulty encountered is due to heaving of the bottom. The cover averages more than 600

* Gentry, B. S., State Inspector of Mines, Personal Communication.
† Personal Communication.
‡ Weitzel, E. H., Personal Communication.

feet in thickness. The output from mines in this district is used largely in connection with steel making, and operations are very regular, most of the mines being worked every day in the year. The recovery at a typical mine in this district, calculated for operations over a period of several years, is 87.2 per cent.

In the domestic coal district in the neighborhood of Walsenburg, the average thickness of coal is 5 feet. It is stronger than the coal in the Trinidad district, and a squeeze is very unusual. The cover averages about 400 feet in thickness. As a rule the bottom in this district is stronger than the top, and little difficulty is experienced from heaving. The work is less regular than in the Trinidad district, although it is fairly regular except in March and April when the mines are usually worked about half time. Rooms are driven 25 feet wide on 50-foot centers. There is little difficulty in drawing pillars. The tendency in these districts has been to drive narrow rooms and to leave wide pillars, and this has assisted in increasing the percentage of recovery. The extraction in a typical mine in this district, calculated for operations over a period of several years, is 91.7 per cent. The chief engineer* of another company operating in this same district believes the extraction in certain portions of the mines of his company will reach 80 per cent. In the Canyon district the long-wall system is used, and the recovery is nearly complete.

40. *New Mexico.*—No information is available concerning the percentages of extraction in New Mexico.

41. *Utah.*—The principal coal fields of Utah are located in Carbon County.† The main coal horizon has from two to four workable beds, from 5 to 28 feet in thickness. The main workable bed, known as the Castle Gate, varies in thickness from 5 to 20 feet, and rests on a massive close-grained sandstone. The problem presented by these deposits is one of mining thick seams, comparatively level or slightly inclined. Formerly some seams 4½ to 8 feet in thickness were worked, but at present most of the mining is done in seams varying from 8 to 28 feet in thickness. The physical features to be taken into consideration in this district are: the number of workable seams, the thickness of seams and their relation to one another,

* Personal Communication.

† Watts, A. C., "Coal Mining Methods in Utah," Coal Age, Vol. 10, p. 214 and p. 258; and Personal Communication.

the character of the coal, the dip of seams, the character of roof and floor, cover, faults, dikes, wants, the flow of water, sometimes gas, and the burned-out coal beds in place with their residual heat.

Throughout most of the fields there are at least two workable seams, and these are generally found in one coal horizon. In several sections, however, there are three and sometimes four workable seams 5 feet or more in thickness. The distances between these seams vary considerably, so that in some sections there are no unusual problems involved while in others two or more workable seams are found with so little intervening strata that the problem of successful extraction has not yet been solved. There may be, for example, an 8- to 14-foot seam underlying a 6- to 10-foot seam with about 200 feet of intervening strata; in another instance a 5- to 8-foot bed lies 60 feet below one 5 to 11 feet thick, and this lies from 12 to 20 feet below a 22-foot seam, which in turn lies 30 feet below a 6-foot seam. In still another instance an 11-foot bed is found from 3 to 40 feet below a 6-foot bed. There is considerable variation in the physical characteristics of the beds, some being hard and brittle and others tough. In some instances the cleavage is good, while in others it is not pronounced. Almost without exception the coals are hard to cut, and some are hard to shoot. The average dip does not exceed 10 per cent, and in some places the beds are practically flat. As a rule the floor consists of hard smooth sandstone from which the coal parts rather readily. In many cases the roof is of shale varying in thickness from a few inches to several feet. Where a sandstone roof is found, it is generally too hard to break for easy mining. In some places the cover is more than 2,000 feet thick, and there are only a few localities in which it is less than 1,000 feet in thickness. This heavy cover makes the mining of these flat thick seams a serious problem in itself, but the additional complication of great irregularity in depth and the unyielding qualities of the thick beds of overlying sandstone make the problem still more serious. A condition which modifies, at least locally, the laying out and working of a mine is the fact that near the outcrop there are sometimes found large areas of burned coal. These sometimes extend 2,500 feet in from the outcrop. Mining in burned areas is often dangerous, if the burning has been at the top, because of the disintegration of the roof.

With one exception all the mines of the district are opened from the outcrop by means of slopes of drifts. Where conditions of topography and property permit, main slopes are driven directly on the pitch of the seams. All mining is by the room-and-pillar method. An attempt to use the long-wall method in one case failed because of the unyielding nature of the roof. The double-entry system is almost universal, although in one case a triple entry is used, and in some cases the double-entry system has been so modified by the connection of the first rooms on the cross entries that it has become practically a 4-entry system. In the earlier workings rooms were

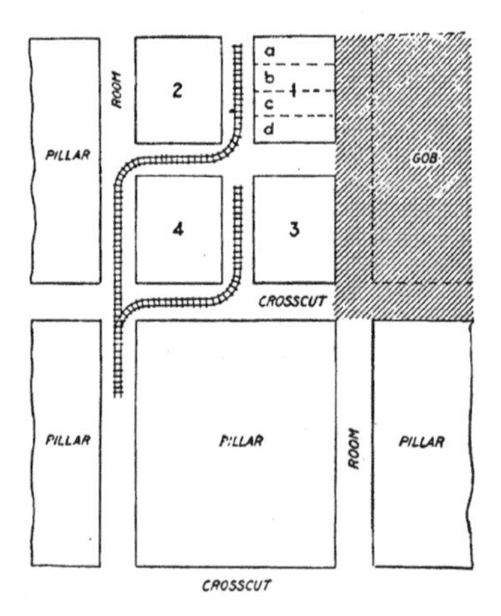

FIG. 39. PILLAR DRAWING IN UTAH

turned from the cross entries as these were driven, but the system resulted in the occurrence of bounces, which seem to take the place of the squeezes that occur with more yielding materials. In later operations the panel system has been used, and the pillars are drawn on the retreat.

Methods of drawing pillars are of particular interest, since they show how almost complete extraction can be attained under conditions which seem unfavorable. These are described by Watts substantially as follows:

In one method (Fig. 39), the block at the end of the pillar on the inby side of the cross-cut is divided by another cross-cut driven through its center, and from the center of this new cross-cut a narrow room which splits the stump into two parts is driven to the gob. The upper inside stump is taken out first by slices beginning at the inby end; then the remaining stump is removed. The lower half of the original block or pillar meanwhile is split by a narrow road, and the process is thus continued down the pillar, each block being divided into four parts.

Pillar drawing in a flat seam 12 to 14 feet thick under a cover 800 feet thick and under a roof which broke fairly well when posts were removed has been successfully accomplished by the following

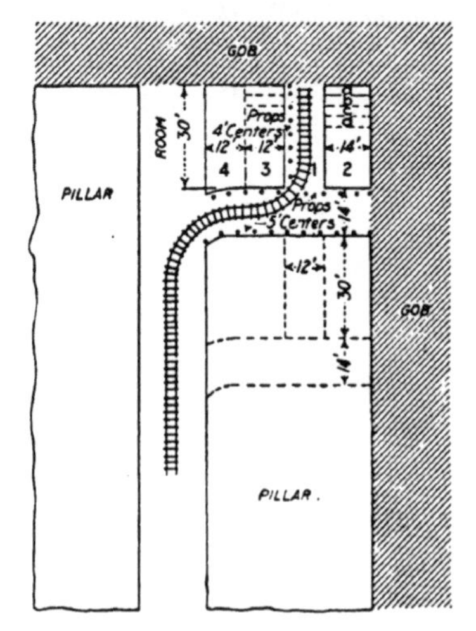

FIG. 40. PILLAR DRAWING IN UTAH

method: 20-foot rooms were driven with 50-foot pillars (Fig. 40), and a cross-cut was driven through the pillar; thus a 30-foot stump was left. This 30- by 50-foot stump was then split by a 12-foot room which left a 24- by 30-foot stump next to the room and a 14- by 30-foot stump on the other side. The latter stump was then taken out in slices which begin at the gob, and the roof was supported by props set every 4 feet. The coal was undercut by hand and shot with black powder. When this block had been removed, the track was taken up, and all props were drawn except a row adjacent to the rib of block No. 3. These blocks were numberd in the order of their extraction, 1, 2, 3, and 4. Block No. 3 was then taken out from the cross-cut to the gob, track was laid in the space, and block No. 4 was taken out in the reverse direction, that is, beginning at the gob. In this mine 3 to 6 feet of top coal are left up to protect the roof on the advance, but this coal is taken down on the retreat. When cross-cuts are made in pillars preparatory to drawing them the whole height of the seam is taken.

Pillar drawing in 16-foot coal with a cover of 400 to 1,000 feet, with no top seam and with the roof breaking well when props are drawn, is accomplished as follows: Rooms about 400 feet long are

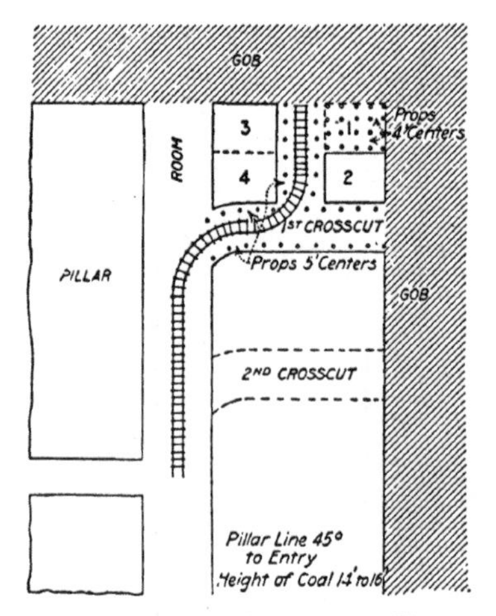

FIG. 41. PILLAR DRAWING IN UTAH

driven straight up the pitch which averages about 10 per cent. Pillars are drawn on the retreat, and the line of break is kept at an angle of 45 degrees with the entry; thus work is done on six or seven pillars at a time. Rooms are about 20 feet wide, and pillars are 50 feet wide. A cross-cut is driven through the pillar 30 to 35 feet from the end (Fig. 41); thus a block about 25 to 30 feet by 50 feet is cut off. This block is then split by a room about 12 feet wide. Blocks 1 and 2 are drawn by slicing which begins at the end next to the gob. The top is supported by means of props at 4-foot intervals, and after the two blocks have been removed and the track has been taken out, these props are pulled, and the area is allowed to cave. The track is then laid in the main room, and blocks 3 and 4 are taken out by end slicing from the room; then the track is taken out, and the props are pulled. Another cross-cut meanwhile has been made through the pillar nearer the entry, and a room has been driven through the stump so that by the time the first stumps have been extracted, work is being begun on the lower stumps.

In this mine the size and the systematic placing of props have an important bearing on the successful recovery of the pillars. Until heavy pine props were used, trouble was likely to occur at any time. Now props as large as 10 inches in diameter at the small end with

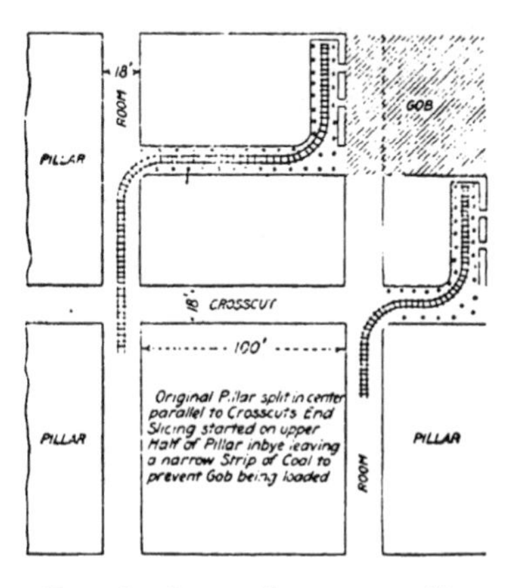

FIG. 42. PILLAR DRAWING IN UTAH

correspondingly heavy caps and, in some places, cross bars are used. Props are set at regular distances. Many of these props are recovered and re-used three or four times. By the adoption of this method, the safety factor is largely increased, the percentage of recovery is greater, and the product is of better quality. In some cases it is customary to mark pillars at regular distances so that the mine foreman or pillar boss may easily determine the progress of the pillar work daily and may keep the ends of the stumps in proper alignment.

Another method of pillar drawing sometimes used is similar to that last mentioned, although the stump left is a little shorter. This stump is then split into quarters, and the work of extraction proceeds from the cross-cut toward the gob, a thin section of coal being left to the last around the edges of the block to prevent the mixture of fallen roof with clean coal.

A plan which has proved satisfactory, partly because it does not split the pillar into many small stumps, is illustrated by Fig. 42. The original block of pillar coal 100 feet wide by 120 feet long is divided into equal parts by a cross-cut, and the upper half is taken out by slicing beginning at the inner side, a curtain of coal being left to prevent the loading of gob, and one or two rows of props being put along the side of the coal. Before the track and props are pulled, most of this section of coal is loaded out. In this case the rooms are 18 feet wide and the pillars 100 feet wide.

Little information is available regarding the percentage of extraction in Utah. There is probably only one mine in the state which has been worked out, and no reliable information can be obtained concerning this mine. It is believed, however, that the extraction was probably about 75 per cent. In coal 12 to 16 feet thick and under cover varying from 200 to 2,000 feet an extraction as high as 90 per cent has been made, if marketable coal alone is considered. If all the coal in the bed is considered, the recovery is about 80 per cent. In beds ranging from 15 to 30 feet in thickness, retreating work has hardly been started so that no information on total recovery is available. It is possible that it will be rather low. It could be made higher if the filling method could be used, but the price of coal does not warrant the use of this method.

A condition largely influencing the percentage of extraction is the presence of more than one workable seam with little intervening material. Under present conditions the percentage of extraction from an area containing seams with 3 to 12 feet of intervening rock is at best only 65 per cent of all the coal. In one mine an attempt was made to take out the coal from two beds, the lower being 11 feet thick and the upper 5 to 6 feet thick with intervening rock 2½ to 12 feet thick. The workings were in the lower bed, and frequently the roof caved as soon as the pillars were drawn and practically all the upper seam was lost.

A. B. Apperson* gives the percentage of extraction in two mines as nearly 95 per cent of the total seam, while the extraction at another mine is about 85 per cent. At the mines yielding the lower percentage of extraction, the cover is about 800 feet. At one of the mines

* Personal Communication.

yielding the higher percentage of extraction, pillar drawing was commenced at the middle of the mine under a cover of approximately 1,700 feet. A good break is obtained about 50 feet behind the pillar extending the full length of line. Only small areas have been worked out in these mines.

42. *Washington.*—No reliable information is available concerning the percentage of extraction in Washington. Conditions are somewhat unusual in that most of the coal has been badly folded and faulted and consequently crushed, and the deposits have been steeply tilted.* It is impossible to separate the refuse in the mines, and a large percentage of it has to be washed.

* Daniel, Professor Jos., Personal Communication.

APPENDIX

Development of Mining Methods in England and on the Continent

43. *Brief History of Coal Mining Practice in England.*—It is interesting to review briefly the history of the coal mining methods of England, because the mining methods employed in this country are largely applications of methods developed in England and brought over by miners.

The many methods of obtaining coal may be grouped on the basis of recovery under two main headings: one in which the whole of the coal seam is taken out in the first working, and another in which

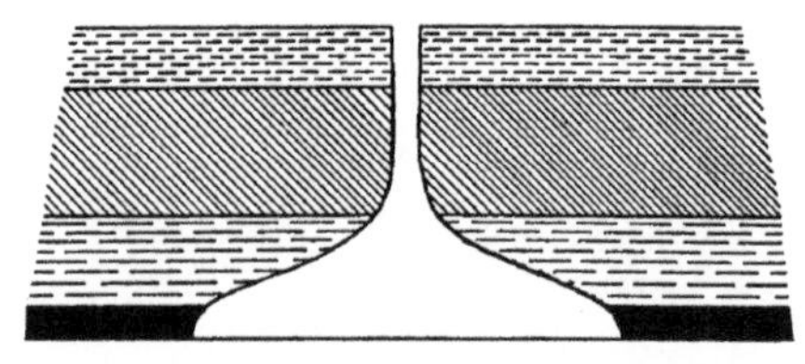

Fig. 43. Bell Pit

only a part of the seam is removed in the first working. These may be called the no-pillar, or long-wall, system and the pillar system.

The earliest mining was naturally done on the outcrop of the seams, and as this practice became difficult or impossible, the use of "bell-pits" (Fig. 43) was developed. These were holes or shafts, from 3 to 4 feet in diameter, which were sunk through the shallow overburden near the outcrop and widened out at the bottom in order to allow the excavation of as much coal as possible without permitting the roof to fall in. It was of course impossible to extract much coal from a pit of this kind, and in order to obtain the coal even from a small area it was necessary to dig a large number of pits. This method was gradually abandoned, and the coal was worked by means of galleries driven out from the bottom of the shaft, usually in an unsystematic manner; thus began the use of pillars to sustain the roof. The driving of galleries permitted the working of much greater areas than could be reached from the bell-pits; however, no areas of more than a few acres were worked from one shaft, nor were systematic ventilation and regularity in laying out the workings introduced until the exhaustion of the shallow coal made necessary a study of methods to be employed in deeper workings. Until the introduction of the Newcomen engine, when pumping by steam power became possible, shafts were rarely as deep as 200 feet; they

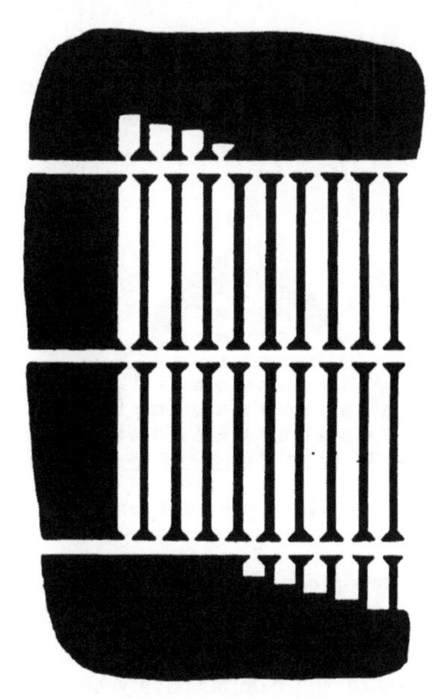

FIG. 44. BORD-AND-PILLAR

were 7 or 8 feet in diameter, and the area worked from one shaft was seldom more than 600 feet in radius.*

The structure of many coal seams is such that there are two directions, determined by the cleat of the coal, in which the seams can be most easily worked. The direction at right angles to the face cleats is known as "bordway," while the other direction approximately at right angles to the first is known as "headway." The excavations made in a direction at right angles to the principal or face cleats

* Bulman and Redmayne, "Colliery Working and Management," p. 3, 1906.

were called bords, and, as the coal was most easily taken out in this direction, these excavations were made wider than the connecting passages or headways. The coal left in place to sustain the roof was called pillars; thus originated the term, "bord-and-pillar" method, which in its various developments is commonly known in this country as the room-and-pillar method. This method was developed in different forms in England and was variously called "bord-and-pillar," "bord-and-wall," "post-and-stall," and "stoop-and-room."

In early times the "pillars" were probably made very small and square measuring from 3 to 6 feet each way. In the eighteenth century the bords were usually made 9 feet wide and the pillars 12 feet

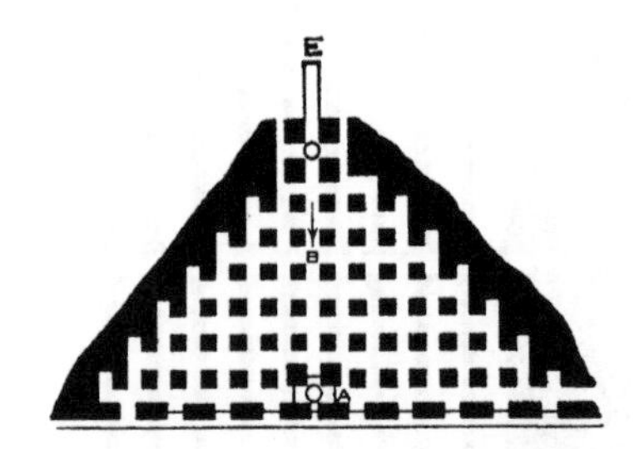

FIG. 45. STOOP-AND-ROOM

wide, though they were of course irregular. The bords were commonly widened out between the headways (Fig. 44), and the pillars were thus gouged to as great an extent as was considered safe, it being desirable, in view of the comparatively small area which could be reached from a single shaft and in view also of the inadequate ventilation, to extract as much coal as possible within the area worked. This method of working was essentially wasteful as not much more than 50 per cent of the coal was obtained, and since the pillars left were unable to bear the weight of the cover, they were soon crushed and further working was made impossible.* Possibly a larger percentage of coal was taken out in some places as Redmayne † says it was rare that more than 65 per cent of the available area could be

* Boulton, W. S., "Practical Coal Mining," Vol. 1, p. 296.
† Redmayne, R. A. S., "Modern Practice in Mining," Vol. 3, p. 82.

extracted; but he refers to John Buddle as saying, at a somewhat later period when the pillars were 18 by 66 feet, that not more than 45½ per cent of the contents of a fiery seam could be obtained under any method of working then known. It was not until much later that the exhaustion of the most easily worked deposits directed attention to the desirability of higher extraction.

A system with square pillars and working places of almost uniform width (Fig. 45) has continued in common use at Whitehaven and in Scotland down to the present time. In the north of England the pillars were usually oblong, probably because the highly developed face cleat of the coal made the extraction in one direction much easier than in others.* The lengthening of the pillars reached its greatest extent in South Wales where cross holing was so little employed as scarcely to form a part of the system of working.

The date at which the extraction of pillar coal was begun is not known, but it seems certain that pillars were removed in the north of England before 1740. The following statement† is made concerning the removal of pillars:— "The documentary evidence cited goes to show that, previous to 1708, the general practice was to leave small pillars of coal standing for the support of the roof; 30 years later pillars were being partially, sometimes entirely, removed; and during the remainder of that century, in mines free from gas, a second working of the pillars was frequently carried out. In the deeper and fiery collieries, which began to be developed about the middle of the eighteenth century, the risk of creep as well as of gas explosions prevented the removal of the pillars. The invention of the safety lamp, improvements in ventilation, and the formation of much larger pillars in the first working . . . . were introduced during the first 30 to 40 years of the present (nineteenth) century . . . . which enabled the pillars to be removed in a second working."

Concerning the size of pillars, Jars, a French engineer who published "Voyages Métallurgiques" in 1774, says in "A Journey Through the North of England," that underground pillars of coal were made from 39 to 54 feet square, and that working places were from 5 to 16 feet wide. At this time the pillars were left until all the coal was exhausted. Another traveler who made a tour of Scotland in

* Galloway, R. L., "Annals of Coal Mining and the Coal Trade," p. 181, 1898.
† Bulman and Redmayne, "Colliery Working and Management," p. 14, 1906.

1772 said that pillars 45 feet square were left and that not more than one-third of the coal was worked.*

The extraction of a portion of the pillars in gassy mines by a second working was just beginning to be a regular part of the bord-and-pillar system at this period. It could, however, be effected only in a very incomplete manner so long as the miners had to depend upon candles and steel mills for light. At this time also the extensive adoption of the long-wall system began.†

In the early part of the nineteenth century little change seems to have been made in the size of pillars used in the Newcastle district, according to a statement of an author who speaks of them as being 60 by 27 feet or, in some instances, 27 feet square. About this time the drawing of pillars seems to have become common in Northumberland, as Mackenzie, who wrote a "View of Northumberland" in 1825, speaks of the mode of working coal as being much improved in the last few years. He says (second edition, page 90), "from seven-eights to nine-tenths of the coal is at present raised, whilst formerly but one-half, and frequently less, was all that could be obtained." No doubt this statement refers to the general practice of removing pillars, which had been made practicable in gassy mines by the invention of the Davy lamp.

Conflicts of interests between coal producers and owners of the surface are of early record. It was, of course, the desire of the colliers to remove as much of the coal as possible, even where the surface was supposed to be maintained, and the result of making pillars too small was subsidence. There is probably no definite record of the first occurrence of subsidence, but one of the earliest mining leases written in the English language, dated 1447, indicates that it was the custom to leave pillars to sustain the surface and that subsidence had already taken place.‡

In the latter part of the eighteenth century the working of pillars in a fiery mine, such as Wallsend Colliery, was not considered practicable, and only about 39 per cent of the coal was obtained while 61 per cent was permanently lost. This coal was at a depth of 600 feet, and the workings represent the best practice of the bord-and-pillar system at that period.¶

* Galloway, R. L. "Annals of Coal Mining and the Coal Trade," p. 353, 1898.
† Ibid, p. 362.
‡ Ibid, p. 69.
¶ Ibid, p. 398.

Until about the end of the eighteenth century an extraction of 45.5 per cent was considered the maximum which could be obtained in the deep collieries of the Tyne.* The first person to offer a partial remedy for this very unsatisfactory condition was Thomas Barnes, viewer of Walker Colliery, who projected a scheme in 1795 for recovering a portion of the pillars without causing loss of the mine. This system provided for dividing the workings into small sections of 10 to 20 acres and isolating these with artificial barriers formed by filling the excavated spaces with stones and refuse for a breadth of 120 to 150 feet. By this method one-half of alternate pillars, or one-quarter of the remaining coal, was removed, and the percentage of extraction was increased from about 39 to about 54 per cent. Wherever pillars were thus removed, a squeeze was brought on, but the barriers kept it from spreading. This method proved to be successful, and it was adopted at other collieries.

Probably about this time the pillars at Wallsend Colliery were left larger as a preparatory step toward a second working. Buddle said that after about one-third of the colliery had been worked by means of 36-foot winnings (12 feet to the bord, 24 feet to the wall or pillar) in which no more coal was left in pillars than was considered sufficient to support the roof, the size of the winnings was increased to 45 feet (15 feet to the bord and 30 feet to the pillar). "This change of size," he said, "was not made for the purpose of obtaining a greater produce in the first working of the seam. But the notion of the future working of the pillars then began to be entertained, and the increased size of the winnings was considered a more favorable apportionment of the excavation and pillar for the attainment of this object." This is the first record found of a second working in the deep Tyne Collieries.† Pillars seem to have been worked in the northern part of England about the middle of the eighteenth century.

44. *Ventilation.*—The distance to which workings could be driven and the extent to which pillars could be drawn, especially in gassy mines, were found to depend largely upon ventilation. In the latter half of the eighteenth century improvements in ventilation, which had been used earlier in the Cumberland field, were introduced into

* "Trans. Nat. Hist. Soc. of Northumberland," Vol. 2, p. 323.
† Galloway, R. L., "Annals of Coal Mining and the Coal Trade," pp. 315–318, 1898.

the north of England, where the frequency of explosions made better ventilation necessary. Until this time it had been considered sufficient to conduct the air current along the working face, an arrangement known as " face-airing;" consequently, the worked-out places, which were behind the miners in advancing work, were left without ventilation. As long as the extent of workings was very limited, this method was not attended with great danger; but, as the mines became deeper and more gassy, workings were made larger and the danger from this inadequate ventilation increased, because the worked-out places became magazines for the accumulation of fire damp.

The improved method of ventilation, which was known as "coursing the air," consisted in so directing the air that the whole current passed through all the openings in the mine. While this method was effective in preventing the accumulation of standing gas, it introduced a great danger in that the air took up constantly increasing quantities of gas in its passage through the mine, and, since it was constantly exposed to the lights of miners, it became dangerous in the latter part of its course. It was, moreover, constantly contaminated by the breathing of men and animals and by the smoke from the candles. This method was introduced in the north of England about 1765 or 1766, and it was about this date that the steel mill also was introduced for the purpose of giving light.* Though this method was fairly satisfactory in small mines, it was very unsatisfactory in large ones. At Walker Colliery, although the pits were only half a mile apart, the air current traversed a line exceeding thirty miles in length. At Hebburn Colliery the air course was also said to be not less than the same length. Not only was it difficult to keep the air passages open and the doors and stoppings tight, but the friction of the air limited the velocity of the ventilating current, which would have been low at best since the force causing this current was supplied only by a furnace. At this colliery the circulation of five or six thousand cubic feet of air per minute was considered sufficient, and the velocity was about three feet per second.

45. *The Panel System.*—There was great difficulty in carrying on work in the deep collieries of the North, because squeezes occurred. A method of working described as common in the North at this period consisted in having bords 12 feet wide and 24 feet apart

*Galloway, R. L., "Annals of Coal Mining and the Coal Trade," p. 279, 1898.

connected by headways 60 feet apart, thus leaving pillars of coal 24 by 60 feet. Another size of pillar given is 30 by 72 feet.*

Early in the nineteenth century, John Buddle, Jr., who had succeeded his father as manager at Wallsend and who was responsible for important improvements in coal mining methods, devised and put into practice improved methods of working and ventilation whereby squeezes were effectually kept in check, and the existing system of ventilation was greatly improved. He effected these improvements by dividing the workings into independent districts or panels, as Barnes had done. Buddle's idea, however, was to provide for confining the movement by separating the districts or panels with barriers of solid coal left in the first working. This method was adopted in developing the Wallsend G pit in 1810. Buddle's improvement in ventilation involved dividing or splitting the current. This method of ventilation proved successful and was quickly adopted at other mines to which it could be applied, but the air currents employed were still very feeble.

From the preceding descriptions it will be seen that all the essentials of the room-and-pillar system as now practiced in this country had been developed in Great Britain prior to 1810.

46. *Square Work of South Staffordshire.*—In the Thick seam of South Staffordshire where the coal varies in thickness from 18 to 36 feet, a method which bears a close resemblance to the panel method was developed. The district had been greatly troubled with fires due to spontaneous combustion, and in order to extinguish these fires easily or to confine them within the immediate vicinity of their origin this method, known as "square work," was developed. It consists in dividing the area to be worked into a number of large chambers termed "sides-of-work," surrounded on all sides by panels of solid coal known as "fire ribs." The only openings in these panels are those necessary for the extraction of coal and for ventilation. The panels are nearly square, and from four to sixteen pillars, the number varying according to the size of the chamber, are left to support the roof. Fig. 46 shows an old form of square work. Under the system in its simple form and in the first working, only from 40 to 50 per cent of the available coal is recovered, but the larger portion of that left is recovered by second or even third

* Ibid, p. 395.

workings carried out after the lapse of some years. The final loss in working may, therefore, not exceed 10 per cent of the available coal; the coal recovered in these later workings, however, is frequently badly crushed.*

47. *The Long-wall System.*—The other general method of coal mining, the long-wall method, has been from early times prevalent in Shropshire, from which district it has spread into others. The date of the origin of this method is doubtful, but it is said to have been in general use in the Shropshire district about the middle of the nineteenth century.†

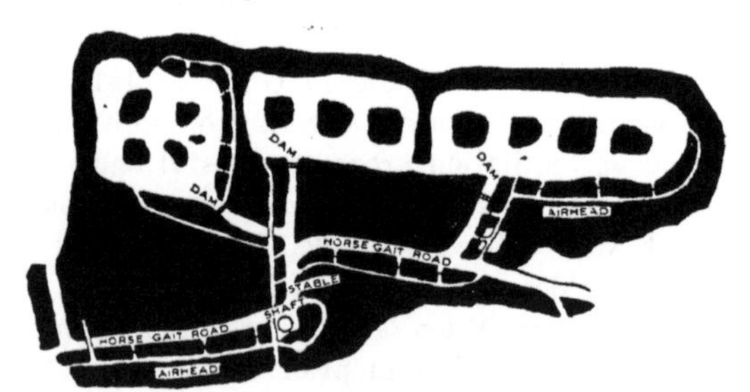

FIG. 46. OLD SQUARE WORK

The long-wall method of mining has been highly developed in England and Scotland and has been applied at greater depths and to thicker beds of coal than it has been in this country. Considered from the point of view of completeness of extraction, the system fulfills the highest requirements: It not only permits, but requires the excavation of the whole bed of coal. Whether all the coal shall be taken out of the mine depends of course on whether it is marketable.

This method of working has not been as generally applied in the United States as have the various forms of the room-and-pillar system. There are, however, certain districts in which it is used almost exclusively. Among the most prominent of these is the long-wall district of Illinois which has been described as District I in Chapter II. The other districts in which the long-wall method is used are those of northwest Missouri, northeast Kansas, the

* Redmayne, R. A. S., "Modern Practice in Mining," Vol. 3, p. 116.
† "Trans. North of Eng. Inst. Min. Engrs.," Vol. 2, p. 261.

Osage district of Kansas, the Appanoose district of Iowa, the north-central district of Texas, and the Canyon district of Colorado. Scattering applications of the method are found elsewhere, but the physical conditions in the regions mentioned have been best suited to its use. It seems probable that the long-wall system, with modifications perhaps, will be more widely used in this country in the future.

48. *Percentage of Recovery in England.*—The methods followed in England have not been developed with the purpose of obtaining a high percentage of recovery. It was not until 1854* that a special department for collecting and publishing mineral statistics was created, and not until 1861 that any systematic estimate of coal resources was made. About this time predictions forecasting the exhaustion of the coal supply within a century caused a great disturbance, and a royal commission was appointed in 1866. A report of this commission was made public in 1871.

The part of this report which deals with waste in working is of special interest. The commission estimated the "ordinary and unavoidable loss" to be about 10 per cent, though they said, "In a large number of instances, when the system of working practiced is not suited to the peculiarities of the seams, the ordinary waste and loss amount to sometimes as much as 40 per cent." The principal part of this unavoidable waste arises from the crushing of pillars.

In addition to this unavoidable loss, there is waste or loss, variable in amount, but sometimes very great, arising from the following causes:†

(1) The leaving below ground or consuming in large heaps of small coal on the surface (presumably the loss from this source is much less at present because of the greater consumption of small sized coal, as in this country).

(2) Undercutting, often wastefully made, in good coal.

(3) The leaving, either wholly or in part, of an adjoining or neighboring bed when it becomes crushed and unworkable, because it is not wanted at the time, or because if it should be worked, the cost per ton of the coal extracted is increased.

* Digest of the Evidence given before the Royal Commission on Coal Supplies, Vol. I., p. IX., 1905.

† Ibid, p. XXXIII.

(4) Existence of coal on properties which are too small to be worked alone or in small parts of colleries cut off by a large fault.

(5) Disputes over the cost of drainage.

(6) The breaking in of water from the sea or from a river estuary.

(7) The leaving of barriers around small properties or crooked boundaries.

(8) Lack of plans or records showing the extent of old workings, operations of seams not sufficiently proved to justify expenditure for sinking pits; sufficient information might have been obtained if records of previous explorations had been preserved in available form.

(9) The piercing of water-bearing strata by shafts and bore holes which are not protected by water-tight casings, or are not carefully filled and puddled when temporarily left or abandoned.

(10) The cutting through of main faults serving as natural barriers to keep back water and the consequent flooding of the coal.

(11) The leaving of large areas of coal in populous and manufacturing districts to support the surface and the buildings.

While some of the causes mentioned do not apply directly to conditions in this country, the list furnishes a complete synopsis of reasons for coal losses.

Since the issuance of the report of 1871, there have been great improvements in the methods of getting coal. At the present time the long-wall system is in general use, and the waste has been lowered; yet in some parts of the United Kingdom, notably Northumberland, the pillar-and-stall system is still in general use.

Among the factors contributing to a higher rate of recovery is the greatly increased value of small sizes of coal. It was computed in 1871 that the average value of the small coal mined in Great Britain was only 60 cents per ton, while in 1905 the small sizes of steam coal from the South Wales district brought about $1.90 per ton; in all the other coal fields the value has been doubled and even trebled. The principal cause of this change lies in the improved preparation of coal. The manufacture of producer gas on a large scale and the growth of the briquet industry have also increased the possible uses of the small sizes. One of the effects of the increase in the value of small coal has been some decrease of the comparative advantage of the long-wall system, since the production of a large amount of fine coal with the pillar-and-stall system is less objectionable than formerly.*

Interest in the subject continued, and another investigation, more exhaustive than the earlier one, was made by the Royal Commission on Coal Supplies which organized in 1902 and presented its report in 1905. The Royal Commission of 1905 adhered to the limit of depth, namely 4,000 feet, established by the earlier commission. It was thought that, although there might be no insuperable physical or mechanical difficulties in the working of beds at greater depths, the expense would be so great that imported coal could be obtained more cheaply.

With regard to thickness, the commission which reported in 1871 had included seams exceeding one foot in thickness as workable. The question is largely a commercial one, and thinner seams are being worked now than formerly. Mr. Gerrard, inspector of mines for the Manchester district, obtained from all the inspection districts returns which showed that in 1900 17.7 per cent of the entire output was taken from seams not exceeding three feet in thickness.† In the United States, limits of 3,000 feet in depth and of 14 inches in thickness have been decided upon by the Department of the Interior as factors determining what portions of the remaining public lands shall be considered coal lands.‡

The Royal Commission took evidence also on the cost of working, and gave figures which show how greatly the labor cost rises and the individual output declines as thinner beds are mined. Mr. Gerrard gave the underground wages as ranging from $1.68 to $2.28 per ton in seams up to 12 inches, and from 63 cents to $1.36 in all underground seams in his district from 1 foot, 1 inch to 3 feet, while the daily output ranged from one-half ton to 3¼ tons. It was estimated that the cost of digging, loading, and hauling in Scotland was $1.24 for a seam 14 to 15 inches thick, and 65 cents for one from 2 to 2½ feet thick, while the daily output varied from 22 hundredweight to 1½ tons. In Somersetshire the average cost of working thin seams has been about $1.92 per ton for a number of years, while

*Digest of the Evidence given before the Royal Commission on Coal Supplies, Vol. I., p. XXV., 1905.

†Digest of the Evidence given before the Royal Commission on Coal Supplies, Vol. I., p. XXXV., 1905.

‡Fisher, Cassius A., "Standards Adopted for Coal Lands of the Public Domain," U. S Geol. Sur., Bul. 424, Ashley and Fisher, p. 63, 1910.

in Yorkshire the cost in 1900 varied from about 96 cents to $1.68.* The Commission of 1905 finally decided to retain the figure of one foot as the limit of thickness.

In connection with the subjects of depth and thickness, it should be noted that it is not the practice in Europe to work single thin beds at great depths. The thin beds are worked in conjunction with thicker ones, and it is the lower cost of production in the latter which makes the working of the thin ones commercially possible. The high cost of working thin beds is partly responsible for the high cost of European coal. The American practice is distinctly different, for there are few, if any, districts in this country in which any bed of bituminous coal is worked unless it is believed that such working shows a profit without reference to other workings. Instances of the working of more than one bed of bituminous coal from the same shaft are rare in the United States.

49. *Percentage of Coal Lost.*—A detailed inquiry was made by the Royal Commission into the various sources of loss. The points of greatest interest in connection with the present study were covered as follows:†

"*Coal left for Support.*—It is evident, that, except in very special cases, it is not possible to remove all the coal. A certain amount must be left in order to maintain shafts, etc., and to support the surface —as, for instance, under houses, railway, canals and rivers—and there seems little hope under existing circumstances of avoiding this source of loss. The amount of coal left for support depends largely upon whether its value is greater than the damage, which would be caused by its removal. . . .

"*Barriers.*—We have evidence that much coal has been and is lost through the practice of leaving unnecessary barriers between royalties and properties; but the present tendency to take large areas under lease is reducing the loss from this cause, and in many cases barriers between properties are now worked out by mutual arrangements.

"*Thick Seams.*—Where the seams are of abnormal thickness much coal is, in some cases, wasted, and for various reasons. Sometimes it

* Digest of Evidence given before the Royal Commission on Coal Supplies, Vol. I., p. 178 et seq., 1905.

† Digest of the Evidence given before the Royal Commission on Coal Supplies, Vol. I., p. XXXVI., 1905.

is considered that the whole seam cannot be taken out with safety, and part is therefore left to form a roof. Further, such thick seams are more difficult to work, and when the whole of the seam is not of the same quality, there is a temptation to take out the best coal first and to leave the rest for possible future working. Suggestions have been made by some of the witnesses as to the best method of working such thick seams, and there is little doubt that improved methods combined with the increasing use of inferior coal will to a large extent obviate the difficulties mentioned.

"*Inferior and Small Coal Left in Mines.*—According to the evidence inferior coal is frequently left in the mine owing to its being unsalable, and in some districts considerable quantities of small coal are also left. In recent years there have been vast improvements in the methods of, and the appliances for, preparing and utilizing small and inferior coal, and the higher appreciation of such coal should go far to put an end to this waste."

Table 10 presents the conclusions of the commissioners of different districts regarding the deductions which should be made to cover losses in calculating the amount of coal remaining available.* It is to be understood that the values given do not refer merely to the losses within a definite mined-out area but to the total losses which are to be expected in extracting the total coal remaining available. Since both amounts of losses and reasons for them are governed largely by local conditions, it is unnecessary to go into details, especially since it was found impossible there, as it has been here, to arrive at definite statement for the losses in all cases. Values, however, are founded upon the opinions of men familiar with the practice in the districts, and they are at least approximately correct.

TABLE 10

PERCENTAGES OF COAL LOSSES AS ESTIMATED BY THE ROYAL COMMISSION OF 1905

District	Per Cent lost	District	Per Cent lost	District	Per Cent lost
South Wales and Monmouthshire	20.68	Warwickshire	2.20	Northumberland	22.61
Forest of Dean	15.50	Leicestershire	26.00	Durham	20.23
North Staffordshire	17.00	Lancashire	20.70	Cumberland	28.20
Shropshire	13.00	Cheshire	18.70	Scotland	26.20
South Staffordshire	27.50	Flintshire	27.60		
		Denbighshire	33.30	Average	21.28

* Ibid. p. XXV.

In the introduction to the report of the Commission, prepared by the editor of the "Colliery Guardian," it is stated (p. xxvi), "Much of the evidence goes to show that the more general adoption of the long-wall system in recent years has resulted in an increased yield of coal. But there are many localities where the conditions are not considered favorable to long-wall working, and where pillars are still left. In the worst cases, in exceptionally bad ground, as much as 50 per cent of the coal is often left behind for this purpose, only to be crushed and oxidized and rendered unfit for future recovery. In the undersea workings in Cumberland as much as 75 per cent is thus left behind. Perhaps the most interesting point brought out in the evidence is that which concerns thick seams. It certainly does seem unfortunate that where there are 9 feet of good coal in a single seam, nearly one-third of this should be left behind. Yet this happens in many of our thickest seams, and the loss threatens to be still more serious as the depth increases."

50. *Mining Conditions on the Continent.*—In the Franco-Belgian basin the beds are for the most part thin, and they are worked, to a considerable extent, at greater depths than those reached in the United States. In Westphalia the beds are mostly steeply dipping, and in Upper Silesia there are combinations of steep dip with great thickness of coal. The development of mining methods in the United States up to the present time has not been affected by practice in these districts.

51. *Percentage of Extraction on the Continent.*—In France it is the custom to extract as much coal as possible from the bed and to fill the resulting space with rock or other material. The filling material is usually transported to its destination in cars, and the method of packing depends largely on the inclination of the bed. In steeply dipping beds the material is allowed to run into place by gravity, but where the slope is not sufficient to permit this method of packing, it is packed by hand. This custom does not entirely prevent subsidence, but it permits the extraction of nearly all the coal without serious disturbance of the surface. While the method of packing followed in these districts permits the removal of nearly all the coal, the removal is accomplished at an expense which would be regarded as prohibitive in the United States in view of the narrow margin between

cost of production and selling price here. The method of filling by flushing is coming into use in France, but has not yet displaced dry filling in most of the mines. Whatever system of filling is used, and whether the coal is taken out by pillar or long-wall method, the extraction is nearly complete.

Some of the most difficult problems found in any coal mining district have been encountered in Belgium. There is no other country in which such thin seams are worked and in which coal is generally mined at such great depths. At Quaregnon a series of thirty-three seams is worked, the average useful thickness being 1 foot, 3½ inches, while the greatest thickness is 2 feet, 2 inches. These beds vary in dip from 8 to 90 degrees. The flatter portions of the bed are worked by long-wall, and the steeper parts by inverted steps forming an interrupted long-wall face. Other beds of nearly the same thickness are being worked, and it appears in all cases that those thin beds are attacked by some form of long-wall working in which, of course, the extraction is practically complete.* The discussion of these districts is much briefer than their importance as coal mining districts would warrant were it not for the fact that the methods used would not in general be adaptable to physical and commercial conditions in this country. They furnish interesting illustrations of high percentages of extraction under difficult conditions, but can hardly be regarded as indicative of what it would be possible to do in the United States.

In the Westphalian district in Germany large amounts of coal have been lost, not so much as the result of poor mining methods or lack of attention to completeness of extraction as because of the necessity of preventing subsidence of the surface. This region is one of great industrial activity, and surface values have so increased within the last half century that high extraction without filling has become impossible. At first, hand filling was employed, the material used being the waste produced in the large amount of rock excavation necessary in beds lying at various angles combined with slack from collieries where coke was not made. More recently the method of hydraulic filling has been introduced. Where the packing is well done and the mining conditions are favorable, the loss of coal is possibly not more than five per cent, which may be considered a fair estimate of the

* Digest of the Evidence given before the Royal Commission on Coal Supplies, Vol. I., pp. 41, 76, 393, 1905.

loss even where the long-wall method is followed. There is, however, a greater loss in some of the thicker steeply dipping beds, though it has not been possible to obtain estimates of the amount.

The Upper Silesian coal field,* situated in the southeast corner of Prussia and extending into Austria and into Russian Poland, has an area of 2,160 square miles. The character of the seam varies considerably both in composition and in thickness, and thick seams occur only in the northern portion of the field where they are very numerous and many of which are of great thickness.

In this coal field, the problem of removing coal beds of great aggregate and individual thicknesses without serious disturbance of the surface has been met by the development of sand flushing processes of filling. This method of filling was borrowed from the anthracite district of the United States where it had first been used.

The mines are worked with and without sand filling. In the method without sand filling much coal is left unworked in the form of pillars and as support under towns or villages. There is a considerable loss resulting from the difficulty of extracting coal left as barriers between the working places and in the old workings. There is also a considerable loss because of fire. The estimated total loss under this method is 25 per cent.

At present sand filling is being used more or less extensively in most of the mines in the thick beds. It is especially advantageous where spontaneous combustion is prevalent and where surface support is necessary. With sand filling when only a part of the coal is replaced by sand it is estimated that the loss of coal is 10 to 15 per cent; with complete replacement of coal by sand filling, the loss is only from 3 to 5 per cent. Smaller and cheaper timber is used in this case, and the greater portion of this timber is recovered for future use. In four mines in Upper Silesia in which sand filling is used extensively and in sufficient quantities to suit the conditions of the mines, the cost in the seams is between 12 and 18 cents per ton. The cost is variable, however, and is calculated in different ways. The average working cost per ton of coal at the surface in this district is $1.51, of which 37 cents is for underground labor.

A report by J. B. Hadesty† shows that the sand filling system has not yet been adopted on a large scale in the western part of Europe,

* Gullachsen, Berent Conrad, "The Working of the Thick Coal Seams in Upper Silesia," Trans. Inst. Min. Engrs., Vol. 42, p. 209, 1911.

† "Pennsylvania State Anthracite Mine Cave Commission Report," Journal Pa. Legislature, Appendix, 1913.

and the statements on cost of filling show that it would be impossible to adopt the process in the United States without materially increasing the cost of production.

It is unnecessary to go into the methods of mining and the results obtained in other coal producing districts. While there are great coal deposits in other parts of the world, and while large quantities of coal are produced, these regions have not been developed sufficiently to work out what may be called a settled practice. No other districts, moreover, are really large producers of coal in the same sense as those already considered. The problems to be considered in connection with districts only partially developed, or districts which though well developed supply only a limited market, are different from those in this country, and the results in such districts are no indication of what can be accomplished here.

BIBLIOGRAPHY

Books

Boulton, W. S. " Practical Coal Mining." Vol. 1, 1913.
Bulman and Redmayne. " Colliery Working and Management." 1906.
de la Goupillière, Haton. " Cours d' Exploitation des Mines." 1907.
Galloway, R. L. "Annals of Coal Mining." Series I., 1898.
Galloway, R. L. "Annals of Coal Mining." Series II., 1898.
Galloway, W. " Lectures on Mining." 1900.
Heise-Herbst. " Bergbaukunde."
Hughes, H. W. "A Text Book of Coal Mining." 1904.
Mayer, L. W. " Mining Methods in Europe." 1909.
Pamely, Caleb. " Colliery Managers' Handbook." 1891.
Redmayne, R. A. S. " Coal—Its Occurrence, Value, and Methods of Boring." " Modern Practice in Coal Mining." Vol. I., 1908.

General

Chance, H. M. "A New Method for Working Deep Coal-Beds." Trans. Amer. Inst. Min. Engrs., Vol. 30, p. 285, 1900.

Conner, Eli T. "Anthracite and Bituminous Mining." Coal Age, Vol. I., pp. 2, 42, 76, 170, 277, 1911-12.

Edwards, J. C. and Gibb, H. M. "An Ideal Method of Mining." Coll. Eng., Vol. 33, p. 665, 1913.

Edwards, J. C. " Machine Mining in Room Pillars." Coll. Eng., Vol. 34, p. 591, 1914.

Grady, Wm. H. " Cost Factors in Coal Production." Trans. Amer. Inst. Min. Engrs., Vol. 51, p. 138, 1915.

Grady, W. H. " Some Details of Mining Methods with Special Reference to the Maximum of Recovery." W. Va. Coal Min. Inst., Dec., 1913; Coal Age, Vol. 5, pp. 126, 156, 1914.

Gresley, W. S. " Rib Drawing with Machines." Mines and Minerals, Vol. 21, p. 82, 1900.

Hall, R. D. " Permanent Roof Sustention." Coal Age, Vol. I., p. 481, 1911.

Hall, R. D. " Squeezes in Mines and Their Causes." Mines and Minerals, Vol. 30, p. 286, 1909.

Hesse, A. W. " Maximum Coal Recovery." Coll. Eng., Vol. 35, p. 13, 1914.

Knox, George. " Hydraulic Filling as Roof Support." Coll. Eng., Vol. 34, p. 225, 1913.

Knox, George. " Relation of Subsidence to Packing." Coll. Eng., Vol. 34, p. 87, 1913.

Parker, E. W. " Conservation of Coal in the United States." Trans. Amer. Inst. Min. Engrs., Vol. 40, p. 596, 1909.

Payne, Henry M. "American Long-wall Mining Methods." Eng. and Min. Jour., Vol. 90, p. 1020, 1910.

Rice, George S. "Mining Costs and Selling Prices of Coal in the United States and Europe with Special Reference to Export Trade." Second Pan-American Scientific Congress.

Unsigned. "Width of Room and Pillar." Mines and Minerals, Vol. 26, p. 107, 1905. (Gives dimensions of workings, characteristics of beds, percentage of extraction, etc.)

Weeks, Joseph D. "Some Fuel Problems." Trans. Amer. Inst. Min. Engrs., Vol. 25, p. 943, 1895.

SUBSIDENCE

Knox, George. "The Relation between Subsidence and Packing, with Special Reference to Hydraulic Packing of Goaves." Trans. Inst. Min. Engrs., Vol. 44, p. 527, 1912–13.

UNITED STATES

ALABAMA

Hutchins, Neill. "Kellerman Mine, Kellerman, Alabama." Mines and Minerals, Vol. 31, p. 204, 1910.

Strong, J. E. "Alabama Mining Methods." Mines and Minerals, Vol. 21, p. 195, 1900.

Unsigned. "Mine of Birmingham Fuel Company at Townley, Alabama." Coal Age, Vol. 5, p. 524, 1914.

ALASKA

Payne, Henry M. "Coal Mining in Yukon Territory." Coll. Eng., Vol. 35, p. 133, 1914.

ARKANSAS

Steel, A. A. "Coal Mining in Arkansas." Ark. Geol. Sur.

CALIFORNIA

Horsewill, F. J. "Tesla Coal Mines." Mines and Minerals, Vol. 19, p. 146, 1898.

COLORADO

Herrick, R. L. "Coal Mining at Primero, Colorado." Mines and Minerals, Vol. 30, p. 598, 1910.

Whiteside, F. W. "The Delagua Mines." Mines and Minerals, Vol. 29, p. 317, 1909.

ILLINOIS

Andros, S. O. "Coal Mining in Illinois." Ill. Coal Min. Invest., Bulletin 13, 1915.

Andros, S. O. "Coal Mining Practice in District I. (Long-wall)." Ill. Coal Min. Invest., Bulletin 5, 1914.

Andros, S. O. "Coal Mining Practice in District II." Ill. Coal Min. Invest., Bulletin 7, 1914.

Andros, S. O. "Coal Mining Practice in District III." Ill. Coal Min. Invest., Bulletin 9, 1915.

Andros, S. O. "Coal Mining Practice in District IV." Ill. Coal Min. Invest., Bulletin 12, 1915.

Andros, S. O. "Coal Mining Practice in District V." Ill. Coal Min. Invest., Bulletin 6, 1914.
Andros, S. O. "Coal Mining Practice in District VI." Ill. Coal Min. Invest., Bulletin 8, 1914.
Andros, S. O. "Coal Mining Practice in District VII." Ill. Coal Min. Invest., Bulletin 4, 1914.
Andros, S. O. "Coal Mining Practice in District VIII. (Danville)." Ill. Coal Min. Invest., Bulletin 2, 1914.
Cady, G. H. "Coal Resources of District I. (Long-wall.)" Ill. State Geol. Sur., Bulletin 10, 1915.
Cady, G. H. "Coal Resources of District VI." Ill. State Geol. Sur., Bulletin 15, 1916.
Cartlidge, Oscar. "Mine No. 3, Saline County Coal Co." Mines and Minerals, Vol. 32, p. 387, 1912.
Kay, Fred H. "Coal Resources of District VII." Ill. State Geol. Sur., Bulletin 11, 1915.
Kay, Fred H., and White, K. D. "Coal Resources of District VIII. (Danville)." Ill. State Geol. Sur., Bulletin 14, 1915.
Rice, George S. "Mining-Wastes and Mining-Costs in Illinois." Trans. Amer. Inst. Min. Engrs., Vol. 40, p. 31, 1909.
Roberts, Warren, and Cartlidge, Oscar. "Buckner No. 2 Mine." Coll. Eng., Vol. 33, p. 121, 1912.

INDIANA

Ashley, Geo. H. "Coal Mining in Indiana." Mines and Minerals, Vol. 20, p. 202, 1899.
Hall, R. D. "Method of Working in Indiana." Coal Age, Vol. 7, p. 94, 1915.
Parsons, Floyd W. "Mining Coal in South Indiana." Eng. and Min. Jour., Vol. 90, p. 869, 1910.
Price, Wm. Z. "The J. K. Dering Coal Co." Coll. Eng., Vol. 35, p. 65, 1914.

IOWA

Price, Wm. Z. "The Electra Mine." Coll. Eng., Vol. 35, p. 11, 1914.

KENTUCKY

Alford, Newell G. "Problems Encountered in Kentucky Coal Mining." Coal Age, Vol. 5, p. 674, 1914.

MARYLAND

Hall, R. D. "Georges Creek Coalfield, Maryland." Coal Age, Vol. 1, p. 10, 1911-12.
Hesse, H. V. "Maximum Recovery of Coal." Mines and Minerals, Vol. 29, p. 373, 1909.
Jenkins, Jonathan. "Pillar Drawing." Mines and Minerals, Vol. 30, p. 151, 1909.
Rutledge, J. J. "Ocean No. 7, or 'Klondyke Mine,'" Mines and Minerals, Vol. 26, p. 5, 1905.

MICHIGAN

Lane, Alfred C. "Coal of Michigan." Mich. Geol. Sur., Vol. 8, Pt. 2, 1902; Mines and Minerals, Vol. 32, p. 148, 1911.

OHIO

Burroughs, W. G. "Black Diamond Concrete Tipple." Coll. Eng., Vol. 34, p. 475, 1914.

Burroughs, W. G. "The High Shaft Mine." Coll. Eng., Vol. 35, p. 69, 1914.

Burroughs, W. G. "Hisylvania Mine No. 23." Coll. Eng., Vol. 35, p. 421, 1914.

Burroughs, W. G. "The Panel System in Ohio." Coll. Eng., Vol. 34, p. 562,

OKLAHOMA

Elliot, James. "Conditions of the Coal Mining Industry of Oklahoma." Proc. Amer. Min. Cong., p. 221, 1911.

PENNSYLVANIA

Baton, G. S. "Rib Drawing in the Connellsville Coke Region." Mines and Minerals, Vol. 27, p. 561, 1907.

Beeson, A. C., and Parsons, F. W. "Model Coal Mine at Marianna." Eng. and Min. Jour., Vol. 91, p. 177 et seq., 1911.

Claghorn, Clarence R. "A Modified Long-wall System." Mines and Minerals, Vol. 22, p. 16, 1901.

Cunningham, F. W. "Best Methods of Removing Coal Pillars." Proc. Coal Min. Inst. Amer., p. 275, 1910; p. 35, 1911; Mines and Minerals, Vol. 31, p. 495, 1911.

Cunningham, J. S. "The Windber Mine." Mines and Minerals, Vol. 21, p. 340, 1901.

Dixon, Charlton. "A New Method of Coal Mining." Mines and Minerals, Vol. 27, p. 32, 1906.

Goodale, S. L. "Safety Through Systematic Timbering." Coll. Eng., Vol. 32, p. 195, 1911.

Howarth, H. W. "Mining by Concentration Method." Coal Age, Vol. 9, p. 125, 1916.

Jennings, S. J. "The Panel Retreating System." Mines and Minerals, Vol. 27, p. 532, 1907.

Keighley, Fred C. "Mining Coal with Friable Roof and Soft Floor." Coal Age, Vol. 7, p. 1008, 1915.

Mullen, Patrick. "New Mining Method in the Connellsville Region." Coal Age, Vol. 10, p. 700, 1916.

Phelps, H. M. "The Marianna Coal Mines." Mines and Minerals, Vol. 31, p. 523, 1911.

Phillips, Elias. "Late Methods of Rib Drawing." Mines and Minerals, Vol. 26, p. 380, 1906.

Report of Commission Appointed to Investigate the Waste of Coal Mining, 1893.

Schellenberg, F. Z. "Systematic Exploitation in the Pittsburgh Coal-Seam." Trans. Amer. Inst. Min. Engrs., Vol. 41, p. 225, 1910.

TENNESSEE

Hutchinson, W. S. "The Wind Rock Coal Mine." Mines and Minerals, Vol. 31, p. 65, 1910.

UTAH

Manley, B. P. "The Somerset Mines." Coll. Eng., Vol. 34, p. 667, 1914.

Turner, R. J. "Consolidated Fuel Co., Utah." Mines and Minerals, Vol. 31, p. 385, 1911.

Watts, A. C. "Coal-Mining Methods in Utah." Coal Age, Vol. 10, pp. 214, 258, 1916.

Washington

Ash, Simon H. "Working a Steep Coal Seam by the Long-wall Method." Coal Age, Vol. 9, p. 742, 1916.
Evans, Geo. W. "Working an Inclined Coal Bed." Coll. Eng., Vol. 35, p. 18, 1914.

West Virginia

Beebe, James C. "Protection of Mines and Miners." Mines and Minerals, Vol. 28, p. 554, 1908.
Cornet, F. C. "Proposed Method of Long-wall Mining." Coal Age, Vol. 4, p. 120, 1913.
Evans, Geo. D. "E. E. White Coal Co. Mines." Coll. Eng., Vol. 35, p. 401, 1915.
Gay, H. S. "A Single Room System." Mines and Minerals, Vol. 27, p. 325, 1907.
Hall, R. D. "The Fairmont, West Virginia Coal Region." Coal Age, Vol. 1, p. 138, 1911.
Hesse, A. W. "Pillar-Drawing Methods in Fairmont Region." Coal Age, Vol. 4, p. 762, 1913.
Hesse, H. V. "Mining Methods for Maximum Recovery of Coal." W. Va. Min. Inst., p. 75, 1908; Mines and Minerals, Vol. 29, p. 373, 1909.
Stoek, H. H. "Coal Fields of West Virginia." Mines and Minerals, Vol. 29, pp. 219, 283, 1908.
Stoek, H. H. "Coal Fields of Central W. Va." Mines and Minerals, Vol. 30, p. 188, 1909.
Stoek, H. H. "The Kanawha Region, W. Va." Mines and Minerals, Vol. 30, p. 70, 1909.
Stoek, H. H. "New River Coal Field, West Virginia." Mines and Minerals, Vol. 29, p. 509, 1909.
Stoek, H. H. "Pocahontas Region Mining Methods." Mines and Minerals, Vol. 29, p. 394, 1909.
Stoek, H. H. "Raleigh County Mining Methods." Mines and Minerals, Vol. 29, p. 471, 1909.
Stow, Audley H. "Mining in the Pocahontas Field." Coal Age, Vol. 3, p. 549, 1913.
Unsigned. "The Boissevain Plant." Mines and Minerals, Vol. 28, p. 497, 1908.

OTHER COUNTRIES

Canada

Jacobs, E. "The Blairmore-Frank Coal Field." Mines and Minerals, Vol. 25, p. 359, 1905.
Leblanc, E. E. "Long-wall Mining in Alberta." Coal Age, Vol. 7, p. 712, 1915.
Quigley, J. S. "Methods of Drawing Pillars in Pitching Seams." Trans. Amer. Inst. Min. Engrs., Vol. 17, p. 406, 1888–9.

France

Dean, Samuel. "A Tour Through European Coal Mines—III." Coal Age, Vol. 7, p. 110, 1915.

GERMANY

Annett, Hugh Clarkson. "Hydraulic Stowing of Gob at Shamrock I. and II., Colliery, Herne, Westphalia, Germany." Trans. Inst. Min. Engrs., Vol., 37, p. 257, 1908–09.

Young, Geo. J. "Brown Coal Mining in Germany." Trans. Amer. Inst. Min. Engrs., Vol. 54, p. 327, 1916.

GREAT BRITAIN

Charlton, Wm. "A Method of Working the Thick Coal Seam in Two Sections." Trans. Inst. Min. Engrs., Vol. 21, p. 264, 1900-01; p. 110, 1902.

Dean, Samuel. "Bentley Colliery." Coll. Eng., Vol. 35, p. 71, 1914.

Holland, Laurence. "Notes on Working the Thick Coal of South Staffordshire and Warwichshire." Trans. Inst. Min. Engrs., Vol. 37, p. 46, 1908–09.

Jackson, J. H. "Notes on Early Mining in Staffordshire and Worcestershire." Trans. Inst. Min. Engrs., Vol. 27, p. 98, 1903–4.

Meachem, Isaac. "Notes on the Loss of Mineral Areas in South Staffordshire." Trans. Inst. Min. Engrs., Vol. 43, p. 11, 1912.

O'Donahue, T. A. "Notes on the Valuation of Mineral Properties." Trans. Inst. Min. Engrs., Vol. 43, p. 19, 1912.

Smith, Alexander. "Brief History of Coal-Mining in Warwichsire." Trans. Inst. Min. Engrs., Vol. 34, p. 355, 1907–08.

INDIA

Adamson, T. "Working a Thick Coal Seam, Bengal." Trans. Inst. Min. Engrs., Vol. 25, p. 10, 1902–03.

McCale, C. H. "Pillar Working in the Raniganj and Jharia Coal Fields." Min. and Geol. Inst. of India, Vol. 7, p. 42, 1912.

Simpson, F. L. G. "Goaf Packing at the Mohpani Mines." Trans. Min. and Geol. Inst. of India, Coll. Guard., Vol. 112, p. 1213, 1916.

Stonier, Geo. A. "The Bengal Coal-Fields, and Some Methods of Pillar-Working in Bengal, India." Trans. Inst. Min. Engrs., Vol. 28, p. 537, 1904–05.

JAPAN

Yonekra, K. "Japanese Coal Mines." Mines and Minerals, Vol. 24, pp. 349, 508, 1904.

MEXICO

Brown, E. O. Forster. "Coal Mining in Mexico." Trans. Inst. Min. Engrs., Vol. 49, p. 381, 1915.

SILESIA

Gullachsen, Berent Conrad. "The Working of the Thick Coal Seams of Upper Silesia." Trans. Inst. Min. Engrs., Vol. 42, p. 209, 1911–12.

SPAIN

Louis, Henry. "Coal Mining in Asturias, Spain." Trans. Inst. Min. Engrs., Vol. 28, p. 420, 1904–05.

INDEX

PUBLICATIONS OF THE ILLINOIS COAL MINING INVESTIGATIONS

Bulletin 1. Preliminary Report on Organization and Method of Investigations, 1913. (Out of print.)

Bulletin 2. Coal Mining Practice in District VIII. (Danville), by S. O. Andros, 1914.

Bulletin 3. A Chemical Study of Illinois Coals, by Prof. S. W. Parr. (In press.)

Bulletin 4. Coal Mining Practice in District VII. (Mines in bed 6 in Bond, Clinton, Christian, Macoupin, Madison, Marion, Montgomery, Moultrie, Perry, Randolph, St. Clair, Sangamon, Shelby, and Washington Counties), by S. O. Andros, 1914.

Bulletin 5. Coal Mining Practice in District I. (Long-wall), by S. O. Andros, 1914. (Out of print.)

Bulletin 6. Coal Mining in District V. (Mines in bed 5 in Saline and Gallatin Counties), by S. O. Andros, 1914.

Bulletin 7. Coal Mining Practice in District II. (Mines in bed 2 in Jackson County), by S. O. Andros, 1914.

Bulletin 8. Coal Mining Practice in District VI. (Mines in bed 6 in Franklin, Jackson, Perry, and Williamson Counties), by S. O. Andros, 1914.

Bulletin 9. Coal Mining Practice in District III. (Mines in beds 1 and 2 in Brown, Calhoun, Cass, Fulton, Greene, Hancock, Henry, Jersey, Knox, McDonough, Mercer, Morgan, Rock Island, Schuyler, Scott, and Warren Counties), by S. O. Andros, 1915.

Bulletin 10. Coal Resources of District I. (Long-wall), by G. H. Cady, 1915.

Bulletin 11. Coal Resources of District VII. (Counties listed in Bulletin 4), by Fred H. Kay, 1915.

Bulletin 12. Coal Mining Practice in District IV. (Mines in bed 5 in Cass, DeWitt, Fulton, Knox, Logan, Macon, Mason, McLean, Menard, Peoria, Sangamon, Schuyler, Tazewell, and Woodford Counties), by S. O. Andros, 1915.

Bulletin 13. Coal Mining in Illinois, by S. O. Andros, 1915.

Bulletin 14. Coal Resources of District VIII. (Danville), by Fred H. Kay and K. D. White.

Bulletin 15. Coal Resources of District VI., by G. H. Cady, 1916.

Bulletin 16. Coal Resources of District II., by G. H. Cady, 1917.

Bulletin 17. Surface Subsidence in Illinois Resulting from Coal Mining, by L. E. Young, 1916.

Bulletin 18. Tests on Clay Materials Available in Illinois Coal Mines, by R. T. Stull and R. K. Hursh.

***Bulletin 72.** U. S. Bureau of Mines, Occurrence of Explosive Gases in Coal Mines, by N. H. Darton, 1915.

***Bulletin 83.** U. S. Bureau of Mines, The Humidity of Mine Air, by R. Y. Williams, 1914.

***Bulletin 99.** U. S. Bureau of Mines, Mine Ventilation Stoppings, by R. Y. Williams.

***Bulletin 102.** U. S. Bureau of Mines, Inflammability of Illinois Coal Dust, by J. K. Clement and L. A. Scholl, Jr.

***Bulletin 137.** U. S. Bureau of Mines, Use of Permissible Explosives in the Coal Mines of Illinois, by James R. Fleming and John W. Koster.

***Bulletin 138.** U. S. Bureau of Mines, Coking of Illinois Coals, by F. K. Ovitz.

Bulletin 91. Engineering Experiment Station, University of Illinois, Subsidence Resulting from Mining, by L. E. Young and H. H. Stoek.

Bulletin 100. Engineering Experiment Station, University of Illinois, The Percentage of Extraction of Bituminous Coal, with Special Reference to Illinois Conditions, by C. M. Young.

---

*Copies may be obtained by addressing the Director, U. S. Bureau of Mines, Washington, D. C.

THE UNIVERSITY OF ILLINOIS

THE STATE UNIVERSITY

Urbana

EDMUND J. JAMES, Ph. D., LL. D., President

THE UNIVERSITY INCLUDES THE FOLLOWING DEPARTMENTS:

The Graduate School

The College of Liberal Arts and Sciences (Ancient and Modern Languages and Literatures; History, Economics, Political Science, Sociology; Philosophy, Psychology, Education; Mathematics; Astronomy; Geology; Physics; Chemistry; Botany, Zoology, Entomology; Physiology; Art and Design)

The College of Commerce and Business Administration (General Business, Banking, Insurance, Accountancy, Railway Administration, Foreign Commerce; Courses for Commercial Teachers and Commercial and Civic Secretaries)

The College of Engineering (Architecture; Architectural, Ceramic, Civil, Electrical, Mechanical, Mining, Municipal and Sanitary, and Railway Engineering)

The College of Agriculture (Agronomy; Animal Husbandry; Dairy Husbandry; Horticulture and Landscape Gardening; Agricultural Extension; Teachers' Course; Household Science)

The College of Law (three years' course)

The School of Education

The Course in Journalism

The Courses in Chemistry and Chemical Engineering

The School of Railway Engineering and Administration

The School of Music (four years' course)

The School of Library Science (two years' course)

The College of Medicine (in Chicago)

The College of Dentistry (in Chicago)

The School of Pharmacy (in Chicago; Ph. G. and Ph. C. courses)

The Summer Session (eight weeks)

Experiment Stations and Scientific Bureaus: U. S. Agricultural Experiment Station; Engineering Experiment Station; State Laboratory of Natural History; State Entomologist's Office; Biological Experiment Station on Illinois River; State Water Survey; State Geological Survey; U. S. Bureau of Mines Experiment Station.

The library collections contain (July 1, 1917) 400,720 volumes and 102,029 pamphlets.

For catalogs and information address

THE REGISTRAR
URBANA, ILLINOIS

UNIVERSITY OF ILLINOIS BULLETIN
ISSUED WEEKLY

Vol. XV SEPTEMBER 10, 1917 No. 2

[Entered as second-class matter Dec. 11, 1912, at the Post Office at Urbana, Ill., under the Act of Aug. 24, 1912.]

COMPARATIVE TESTS OF SIX SIZES OF ILLINOIS COAL ON A MIKADO LOCOMOTIVE

BY

EDWARD C. SCHMIDT, JOHN M. SNODGRASS

AND

OTTO S. BEYER, JR.

BULLETIN No. 101

ENGINEERING EXPERIMENT STATION

PUBLISHED BY THE UNIVERSITY OF ILLINOIS, URBANA

PRICE: FIFTY CENTS

EUROPEAN AGENT

CHAPMAN & HALL, LTD., LONDON

THE Engineering Experiment Station was established by act of the Board of Trustees, December 8, 1903. It is the purpose of the Station to carry on investigations along various lines of engineering and to study problems of importance to professional engineers and to the manufacturing, railway, mining, constructional, and industrial interests of the State.

The control of the Engineering Experiment Station is vested in the heads of the several departments of the College of Engineering. These constitute the Station Staff and, with the Director, determine the character of the investigations to be undertaken. The work is carried on under the supervision of the Staff, sometimes by research fellows as graduate work, sometimes by members of the instructional staff of the College of Engineering, but more frequently by investigators belonging to the Station corps.

The results of these investigations are published in the form of bulletins, which record mostly the experiments of the Station's own staff of investigators. There will also be issued from time to time, in the form of circulars, compilations giving the results of the experiments of engineers, industrial works, technical institutions, and governmental testing departments.

The volume and number at the top of the title page of the cover are merely arbitrary numbers and refer to the general publications of the University of Illinois; *either above the title or below the seal is given the number of the Engineering Experiment Station bulletin or circular which should be used in referring to these publications.*

For copies of bulletins, circulars, or other information address the

ENGINEERING EXPERIMENT STATION,
URBANA, ILLINOIS

UNIVERSITY OF ILLINOIS
ENGINEERING EXPERIMENT STATION

BULLETIN No. 101 SEPTEMBER, 1917

COMPARATIVE TESTS OF SIX SIZES OF ILLINOIS COAL ON A MIKADO LOCOMOTIVE

BY
EDWARD C. SCHMIDT
PROFESSOR OF RAILWAY ENGINEERING
JOHN M. SNODGRASS
ASSISTANT PROFESSOR OF RAILWAY MECHANICAL ENGINEERING
AND
OTTO S. BEYER, JR.
FIRST ASSISTANT IN RAILWAY ENGINEERING, ENGINEERING EXPERIMENT STATION

ENGINEERING EXPERIMENT STATION
PUBLISHED BY THE UNIVERSITY OF ILLINOIS, URBANA

CONTENTS

CONTENTS (Continued)

LIST OF FIGURES

LIST OF FIGURES (Continued)

LIST OF TABLES

COMPARATIVE TESTS OF SIX SIZES OF ILLINOIS COAL ON A MIKADO LOCOMOTIVE

I. Introduction

1. *Preliminary Statement.*—Until a few years ago practically all of the coal used on locomotives was mine-run coal—the entire unscreened products of the mines. In recent years, however, increasing quantities of screened lump coal have been used in locomotive service. This increase in the consumption of lump coal has been due partly to economic factors, such as the increasing market for the screenings which result from the production of lump coal; and partly to the belief that lump coal, when burned on a locomotive, produces enough more steam than mine-run coal to compensate for its greater cost. Special considerations, such as the desire to lessen the amount of smoke formed, have also led in some instances to the use of lump coal, which is generally believed to require less skill in firing than mine-run coal. Because of the gradual adoption of mechanical stokers for locomotives, the railroads are also using constantly increasing amounts of various sizes of screenings for locomotive fuel. Thus far they have made comparatively little use of any except the sizes mentioned, although traffic and market conditions occasionally make it feasible and desirable to employ such sizes as egg, egg-run, and nut coal on locomotives, provided the prices are such as to warrant their use.

Under these circumstances railway purchasing departments are continually confronted with the problem of choosing between mine-run and lump coal, and occasionally with that of choosing between these and other sizes as well as between various sizes of screenings. For such a choice, information regarding the relative values of the various sizes of coal in locomotive service is obviously essential; but unfortunately very little such information is in existence. Nearly all locomotive laboratory tests have been made with mine-run coal, and what little information is available concerning the relative values of mine-run and lump coal has been derived from road tests, and is inadequate and conflicting. There are practically no data concerning the other sizes.

An appreciation of the situation thus briefly reviewed, and a recognition of the economic importance of reliable information on this subject led, in 1914, to the appointment by The International Railway Fuel Association of a special Committee on Fuel Tests. This committee was instructed to arrange tests in locomotive service for various sizes of coal in order to determine their steam-producing capacities and to define their relative values. This committee held its first meeting in November, 1914, at the University of Illinois, and arrangements then broached in conference between the committee and the representatives of the Engineering Experiment Station of the University later resulted in an agreement for co-operation between the Fuel Association, the University, and the United States Bureau of Mines in carrying on an investigation of the subject under consideration. The tests whose results are here presented constitute the beginning of this investigation. The agreement contemplates the continuation of the research on coals from various other fields. Under the terms of this agreement the University of Illinois has furnished the facilities of its locomotive laboratory, the services of the staff of its department of railway engineering, and a portion of the funds required for the tests; the Fuel Association has provided the remainder of the funds; and the Bureau of Mines has made all the chemical analyses and the heat determinations of the coal, ash, and cinders. In perfecting these arrangements, the Fuel Association was represented by the committee whose members are named in Section 2; and the Bureau of Mines, by Director Van. H. Manning, and Mr. O. P. Hood, Chief Mechanical Engineer.

The funds supplied by the Fuel Association were obtained by subscription and did not become available until the Spring of 1916; the locomotive, then under construction, was not delivered until the Fall of that year. The tests were begun in December, 1916, and were completed in February, 1917.

The body of this bulletin contains information concerning the test program, the coal, the locomotive, the laboratory, the test methods and conditions, and the results. Appendixes I, II, and III contain more detailed statements regarding the locomotive and the methods, and complete tabulated results. In Appendix IV, there are presented certain data relating to engine performance. In Appendix V, there are the results of a few of the tests which, in order to study the uniformity of conditions during their progress, were divided into three periods, and the data for each period were separately calculated.

The results of this investigation have already been presented in a report to the International Railway Fuel Association Convention held in Chicago in May, 1917; and, in somewhat different form, will appear in the Proceedings of the Association for this year.

2. *Acknowledgments.*—The Committee on Fuel Tests previously referred to, under whose direction the work was planned and the general program defined, was composed of the following:

J. G. Crawford, Fuel Engineer, Chicago, Burlington & Quincy Railroad, *Chairman*

H. B. Brown, General Fuel Inspector, Illinois Central Railroad

W. P. Hawkins, Fuel Agent, Missouri Pacific Railway System

O. P. Hood, Chief Mechanical Engineer, United States Bureau of Mines

L. R. Pyle, Fuel Supervisor, Minneapolis, St. Paul & Sault Ste. Marie Railroad

W. L. Robinson, Supervisor of Fuel Consumption, Baltimore & Ohio Railroad

E. C. Schmidt, Professor of Railway Engineering, University of Illinois

The locomotive used during the tests was loaned by the Baltimore and Ohio Railroad Company, through the interest and courtesy of Mr. J. M. Davis, Vice President; Mr. F. H. Clark, General Superintendent of Motive Power; and Mr. M. K. Barnum, Assistant to the Vice President. The tests came at a time when traffic demands were extraordinary, and the loan of the locomotive constituted as great a contribution to the work as that made by any other agency.

The funds provided by the International Railway Fuel Association were donated to the Association by the following railroads, coal companies, and railway supply manufacturers:

Atchison, Topeka, and Santa Fe Railway

Atlantic Coast Line Railway

Baltimore and Ohio Railroad

Chicago Great Western Railway

Chicago, Indianapolis, and Louisville Railway

Erie Railroad

Long Island Railroad

Minneapolis, St. Paul, and Sault Ste. Marie Railway

Norfolk and Western Railway

St. Louis Southwestern Railway
Seaboard Air Line Railway
Big Muddy Coal and Iron Company
T. C. Keller and Company
Old Ben Coal Corporation
W. P. Rend and Company
Southern Coal and Mining Company
Taylor Coal Company
United Coal Mining Company
American Arch Company
American Locomotive Company
Franklin Railway Supply Company
Locomotive Stoker Company
Locomotive Superheater Company
The Pilliod Company

The Locomotive Stoker Company and the Baltimore and Ohio, the Chicago and Northwestern, the Erie, and the Minneapolis, St. Paul & Sault Ste. Marie Railroad Companies each delegated to the laboratory a man to act as test observer and calculator for the entire period of the tests. Mr. L. R. Pyle, Fuel Supervisor of the road last named, was in charge of the cab operations and supervised the work of the fireman. The uniformity attained in the firing and in the conditions of combustion was due largely to the experience and skill of Mr. Pyle.

The department of mining engineering of the University of Illinois contributed the use of its laboratory facilities for crushing and sampling the coal and analysing the flue gas; and Professors H. H. Stoek and E. A. Holbrook of that department gave advice on many matters connected with the investigation. The laboratory coal screen used in the tests was designed by Professor Holbrook.

II. Purpose and Program

As has been stated, the ultimate purpose of the tests was to determine the relative values of different sizes of coal when burned on a locomotive. The immediate purpose was to find for each size, at two rates of evaporation, the number of pounds of water evaporated per pound of coal, in the expectation that these values of evaporation would provide a proper basis for comparing the performance of the sizes and for defining their relative values. The tests were made on a Mikado (2-8-2) type locomotive.

Six sizes of Franklin County, Illinois, coal were selected—mine run, 2-inch by 3-inch nut, 3-inch by 6-inch egg, 2-inch lump, 2-inch screenings, and 1¼-inch screenings. The general test program involved for each size of coal six tests, three of which were made at a medium rate of evaporation, and the remaining three at a high rate. The medium rate was chosen to represent an average rate of working the locomotive, in so far as it is possible to define such an average. During tests run at this medium rate about 23,000 pounds of water were evaporated per hour under the prevailing conditions of steam pressure, superheat temperature and feedwater temperature; from 3,100 to 4,300 pounds of coal were fired per hour; and the engine was worked at 33 per cent cut-off and at about 19 miles per hour, developing approximately 1,300 indicated horse power and about 22,500 pounds drawbar pull. During tests when the engine was worked at the high rate of evaporation, about 43,000 pounds of water were evaporated per hour, the hourly coal consumption varied from about 7,000 to 9,300 pounds, the cut-off and speed were respectively 55 per cent and 26 miles per hour, while the horse power was about 2,200, and the drawbar pull about 28,500 pounds.

The number of tests actually run with each size at each rate of evaporation was as follows:

Size of Coal	No. of Tests at the Medium Rate of Evaporation	No. of Tests at the High Rate of Evaporation
Mine Run	3	3
2" x 3" Nut	4	3
3" x 6" Egg	3	3
2" Lump	3	4
2" Screenings	3	2
1¼" Screenings	3	2

III. The Coal Used

3. *Source and Mining Methods.*—All coal used during the tests was secured from the United Coal Mining Company's Mine No. 1, located one mile east of Christopher, Franklin County, Illinois, on the Illinois Central and the Chicago, Burlington and Quincy Railroads. This mine was chosen by the Fuel Association Committee because western railroads draw a large fuel supply from this field, and because of its nearness to the locomotive laboratory.

The coal is derived from what is designated by the Illinois Geological Survey as bed No. 6 of the Carboniferous Age, Carbondale formation. The bed averages in thickness at this mine about 9 feet and carries almost throughout, at from 18 to 30 inches from the floor, a "blue band" variable in thickness and consisting of "bone," shaly coal, or gray shale. The mine is worked under the room-and-pillar system, and the coal is undercut with electric chain machines. It separates at a parting of mother coal about 14 to 30 inches from the top of the bed, and the coal above this parting is left for the roof. The coal face and the mine itself are quite uniformly dry.

All the coal was mined, screened, and loaded by the methods usually employed at the mine for supplying the ordinary commercial product. It was inspected during the process of loading by one of the regular fuel inspectors of the Chicago, Burlington and Quincy Railroad, who at the time of inspection took at the tipple samples for analysis, the results of which were later used to compare the moisture in the coal when loaded with its moisture content when used at the laboratory.

While it was originally planned to ship all test coal in box cars to protect it from the weather during transit and before it could be unloaded at the laboratory, only two cars of mine-run coal were so shipped. Under the prevailing conditions of business and car supply, the plan proved impracticable and had to be abandoned, and all coal except these two car loads was shipped in ordinary flat-bottomed gondola cars. As promptly as possible after its receipt at the laboratory—on the average 6 days, and in no instance more than 12 days after its arrival—the coal was unloaded into covered bins where it remained protected from the weather until used. The cars were unloaded by hand shoveling about as they would have been at some of the older types of railway coal pockets, and the coal was probably subjected to about the same amount of breakage in this process. The maximum time which elapsed between loading the coal at the mine and testing it was 37 days in one instance. Taking the tests as a whole the average time between loading and testing was about 25 days.

4. *Preparation.*—At the mine all coal was dumped from the mine cars into a hopper from which it was run out on a stationary deadplate where it spread out before reaching the shaking screens. The various sizes were prepared as follows:

The mine run coal was the entire unselected product of the mine.

The 2-inch lump was made by passing mine run coal over a screen having 144 square feet area with 2-inch round openings, and consisted of everything going over this screen.

The 2-inch by 3-inch nut consisted of what passed over the previously mentioned screen and through a screen having 72 square feet area with 3-inch perforations; and it was re-screened over a stationary screen having an area of 18 square feet with slots ⅞-inch wide and 8-inches long, and a stationary screen of 20 square feet area with 1¼-inch round perforations.

The 3-inch by 6-inch egg passed first over a screen having 144 square feet area with 2-inch round perforations and 72 square feet area with 3-inch round perforations, and then through a screen having 32 square feet area with 6-inch round perforations.

The 2-inch screenings were passed through the screen over which the 2-inch lump coal was made, namely, 144 square feet area with 2-inch round perforations.

The 1¼-inch screenings were made at the re-screening plant through a revolving screen having 411 square feet area of plate with ¾-inch round perforations and 188 square feet area of plate with 1¼-inch perforations.

5. *Chemical Analyses.*—During the progress of each test, while the coal was being loaded into the charging wagons to be taken to the firing platform, samples were taken for the purpose of analysis. These samples varied in amount from 500 to 1000 pounds, and they were taken according to methods prescribed by the American Society for Testing Materials as set forth in the year book of the society for 1915. The sampling process is described in Appendix II. Under arrangements made with the United States Bureau of Mines, all analyses of coal, ash, and cinders were made at the laboratories of the bureau in Pittsburgh, where the samples were shipped immediately upon the conclusion of each test.

The results of these analyses are given for each test in the tables in Appendix III. The averages of the coal analyses for all tests made with each grade of coal are presented in Table I. An inspection of this table reveals a rather unusual uniformity among the various sizes with regard to their composition and heating value. Considering all six sizes, the ash content varied from 8.06 per cent to 10.59 per cent and the heating value per pound of dry coal varied from 12,711 to 13,239 B. t. u. The analyses for the two sizes of screenings correspond very closely in all respects and their average heating value

TABLE 1

THE CHEMICAL ANALYSES AND HEATING VALUES OF THE COALS

(The table gives the averages for all tests for each size.)

Size of Coal	Proximate Analyses—Coal as Fired					Calorific Values			Ultimate Analyses—Coal as Fired				Moisture in Coal Determined from Sample taken at Mine per cent
	Moisture per cent	Volatile Matter per cent	Fixed Carbon per cent	Ash per cent	Sulphur Separately Determined per cent	Per lb. of Coal as Fired B. T. U.	Per lb. of Dry Coal B. T. U.	Per lb. of Combustible B. T. U.	Carbon per cent	Hydrogen per cent	Nitrogen per cent	Oxygen per cent	
Mine Run	8.14	34.18	47.92	9.76	0.95	11873	12926	14463	66.63	4.28	1.55	8.69	7.82
2" x 3" Nut	8.60	34.83	47.70	8.87	0.88	11957	13082	14487	67.50	4.36	1.38	8.42	8.48
3" x 6" Egg	8.82	34.57	48.56	8.06	0.94	12071	13239	14523	68.19	4.50	1.51	7.99	.....
2" Lump	9.27	34.46	47.49	9.07	0.88	11817	13023	14469	66.34	4.23	1.49	8.73	.....
2" Screenings	9.25	32.05	48.12	10.59	0.85	11550	12727	14408	65.74	4.43	1.48	7.66	.....
1¼" Screenings	9.09	32.34	48.01	10.57	0.97	11557	12711	14385	65.49	4.35	1.43	8.10	9.07

based on dry coal, was only about two per cent less than the average heating value of the four large sizes. Their average ash content was 10.58 per cent, and the average ash for the other sizes was 8.94 per cent—a difference of 1.64 per cent. As would be expected, the mine run occupies an intermediate position between the screenings and the egg, nut and lump, both with regard to ash content and to heating value. The uniformity of the analyses and of the heating values makes it clear that such differences in performance as developed between the various sizes are due chiefly to differences in their mechanical make-up, and only in small measure to differences in their chemical composition. This fact is further emphasized by discussion which appears later in the report.

6. *The Make-up of the Coals as Received.*—Because of differences in the nature of the coal, in mining methods, and in methods of preparation, there is frequently much uncertainty about the meaning of such terms as "mine run," "lump," etc. The mine run grade from a district where the coal is soft and friable, for example, is likely to contain a larger proportion of fine coal than mine run made from a harder coal. Similarly the methods of mining, the use of bar instead of plate screens, or square-hole instead of round-hole screens, all entail differences in the make-up of coals which are designated by identical names. For these reasons the laboratory has devised a method of screening samples of the coals used during tests for the

purpose of separating them into their size elements in order to be able to define and record the actual mechanical make-up of the various grades.* All the coals used in these tests were thus screened, and this screening process is referred to in the report as the mechanical analysis.

The samples for this purpose were taken while the cars were being unloaded, by methods which are described in Appendix II. Three carloads each of mine run and lump, and two carloads of each of the other four grades were received at the laboratory. For both the mine run and the lump coals, two of the three carloads of each size were sampled for screening. Samples were taken from each car of nut and each car of egg, whereas the two cars of 2-inch screenings and the two cars of 1¼-inch screenings were combined for each size, and one sample only was taken from each. There was thus taken for mechanical analysis a total of ten samples, each of which weighed about two tons.

These samples were screened by means of the specially designed shaker screen shown in Fig. 1. This consists of two inclined steel frames each of which is supported by four vertical wooden slab springs. These frames are shaken by connecting rods attached to the pulley-driven eccentrics which appear at the right of the figure, and which were run at a speed of 80 revolutions per minute. The frames carry removable plate screens provided with round perforations. Five such screens were used perforated respectively with 4-inch, 2-inch, 1-inch, ½-inch, and ¼-inch holes.

Starting with the 4-inch screen in the upper and the 2-inch screen in the lower frame, one of the samples—mine run, for example—was fed over the upper frame and the coal was separated in three parts; one containing what passed over the 4-inch screen, the other what passed through the 4-inch screen and over the 2-inch screen, and the screenings which passed through the 2-inch screen. The first two portions were then set aside for weighing, the screens were replaced by the plates with 1-inch and ½-inch holes, the screenings were again fed onto the upper plate and the process repeated, ending finally with the ¼-inch screen. In this way the sample was divided into six parts whose size limits were as designated by the headings of Columns 2 to 7 in Table 2. These parts were then weighed and the ratios of their weights to that of the original sample were calculated.

*The term "grade" is occasionally used throughout this bulletin instead of the word "size." It refers solely to one of the six sizes tested, and does not imply any difference in quality or kind.

Table 2 presents the average values of these ratios and it defines, therefore, for each grade the magnitude of the size. elements which went to make up the original coal and thus records definitely its composition. The significance of Table 2 is perhaps made clearer by

TABLE 2

THE SIZE ELEMENTS OF THE COALS AS RECEIVED AT THE LABORATORY

(This table gives the direct results of the separation made by the use of the laboratory screens)

SIZE OF COAL	Per Cent over 4" Screen	Per Cent through 4" over 2" Screen	Per Cent through 2" over 1" Screen	Per Cent through 1" over ½" Screen	Per Cent through ½" over ¼" Screen	Per Cent through ¼ "Screen	Total
1	2	3	4	5	6	7	8
Mine Run........	29.6	22.3[1]	16.8[1]	11.4	7.4	12.5	100.0
2" x 3" Nut	....	63.9	30.3	2.8	1.1	1.9	100.0
3" x 6" Egg	41.0	48.3	5.3	2.0	1.1	2.3	100.0
2" Lump..........	61.6	26.4	7.5	1.9	.9	1.7	100.0
2" Screenings......	....	....	33.2	25.7	14.2	26.9	100.0
1¼" Screenings ...	....	....	4.5	37.9	20.0	37.6	100.0

[1] Derived from plotted curves (Fig. 9).

Figs 3 to 8 inclusive. Each of these illustrations applies to one of the sizes and each figure is reproduced from a photograph of the various size elements which came from the screen and which, after weighing, were assembled side by side as shown in the cuts. These figures present graphically the same information as is given in Table 2. Fig. 2 is reproduced from photographs of the original coal samples and represents the six sizes as they were received at the laboratory.

The facts presented in Table 2 may be re-combined to permit tabular and graphical definitions of the grades in another form. Considering in Table 2 the 2-inch by 3-inch nut coal, if we add Columns 4 to 7 we find that 36.1 per cent of this coal passes through a 2-inch screen. Adding Columns 5, 6, and 7 we find that 5.8 per cent will pass through a 1-inch screen, etc. Obviously also 100 per cent of this grade passed a 3-inch screen in the original preparation at the mine. The total per cents passing the various sized screens determined in this manner from Table 2 are assembled in Table 3, where we find that for the 2-inch by 3-inch nut coal, 31.1 per cent, 5.8 per cent, 3.0 per cent, and 1.9 per cent passed respectively 2-inch, 1-inch, ½-inch, and ¼-inch screens. If now we plot as in Fig. 9 the percentages given in Table 3, together with the corresponding screen size, we get for the nut coal curve No. 3 there drawn, which serves to

Fig. 1. The Laboratory Coal Screen

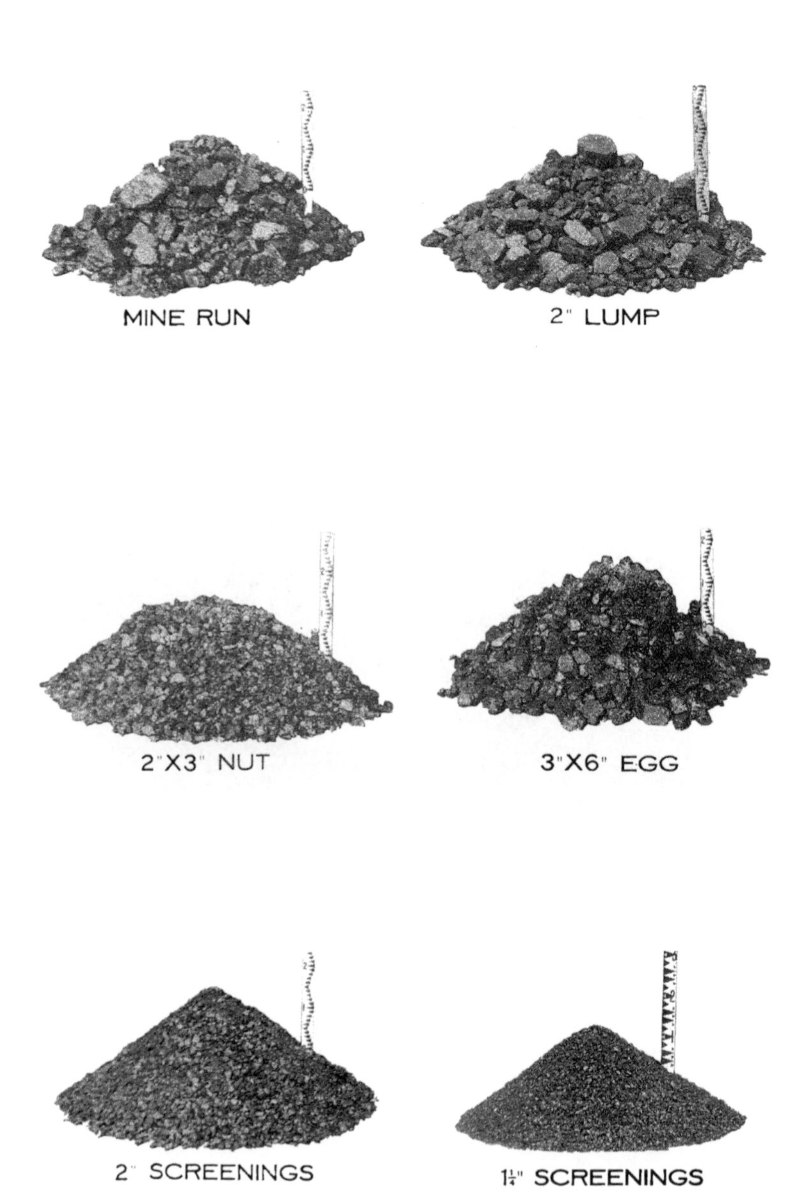

FIG. 2. THE SIX SIZES OF COAL USED DURING THE TESTS, IN THE CONDITION IN WHICH THEY WERE DELIVERED AT THE LABORATORY

Fig. 3. The Size Elements of the Mine Run Coal

Fig. 4. The Size Elements of the 2-inch by 3-inch Nut Coal

Fig. 5. The Size Elements of the 3-inch by 6-inch Egg Coal

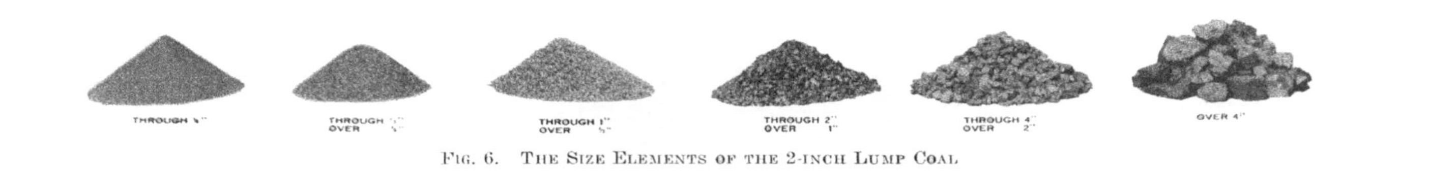

FIG. 6. THE SIZE ELEMENTS OF THE 2-INCH LUMP COAL

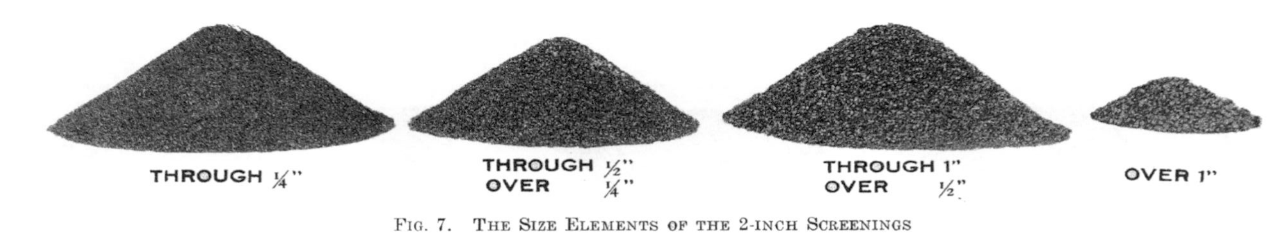

FIG. 7. THE SIZE ELEMENTS OF THE 2-INCH SCREENINGS

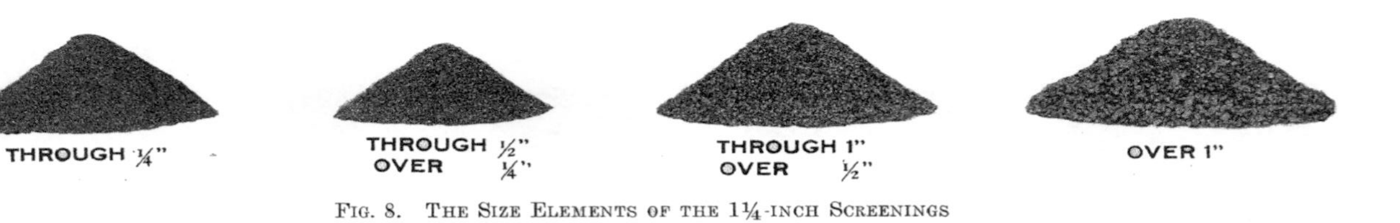

FIG. 8. THE SIZE ELEMENTS OF THE 1¼-INCH SCREENINGS

define its composition and which permits us to determine not only the percentages which successively pass through the screen openings marked in Fig. 9, but presumably to determine these percentages for screens of any intermediate size. The six curves drawn in Fig. 9 are plotted from the percentage values and the screen sizes given in Table 3 for each of the grades. Those portions of the curves drawn with broken lines are not supported by direct experimental data. The scale shown in the upper part of the diagram represents the screen sizes which are commonly used in the mines of southern Illinois.

TABLE 3

THE MAKE-UP OF THE COALS AS RECEIVED AT THE LABORATORY

(This table presents the results computed from Table 2)

SIZE OF COAL	Per Cent over 4" Screen	Per Cent through 4" Screen	Per Cent through 2" Screen	Per Cent through 1" Screen	Per Cent through ½" Screen	Per Cent through ¼" Screen
1	2	3	4	5	6	7
Mine Run......	29.6	70.4	48.1	31.3	19.9	12.5
2" x 3" Nut.....	....	....	36.1	5.8	3.0	1.9
3" x 6" Egg.....	41.0	59.0	10.7	5.4	3.4	2.3
2" Lump.......	61.6	38.4	12.0	4.5	2.6	1.7
2" Screenings ...	....	....	....	66.8	41.1	26.9
1¼" Screenings .	....	....	....	95.5	57.6	37.6

If in Fig. 9, we follow curve No. 5 pertaining to the mine run, we find that about 90 per cent of it will pass through a screen with 9-inch round holes; about 87 per cent of it will pass through a 7-inch screen; about 70 per cent through a 4-inch screen; 48 per cent through a 2-inch screen, and so on. It is interesting to note that the 2-inch by 3-inch nut, the 3-inch by 6-inch egg, and the 2-inch lump contain nearly the same proportions of coal which passes through holes 1-inch or less in diameter; whereas in these sizes the proportions of coarser coal differ materially. Other comparisons are rendered feasible by having all six sizes thus represented on one diagram. It should be borne in mind that the curves in Fig. 9 define the make-up of coals in the condition in which they were unloaded from the cars at the laboratory.

7. *The Make-up of the Coals as Fired.*—All grades except the mine run and lump were unloaded into the charging wagons from the bins without further preparation, and they were consequently fired in exactly the condition in which they arrived at the laboratory,

except for the breakage incident to unloading and the insignificant breakage due to shoveling into the charging wagons.

Since, however, the mine run and the lump coals contained as usual a considerable proportion of lumps too large for proper firing, the attempt was made to break these two sizes down to the extent to which, in the judgment of those in charge of the tests, these grades

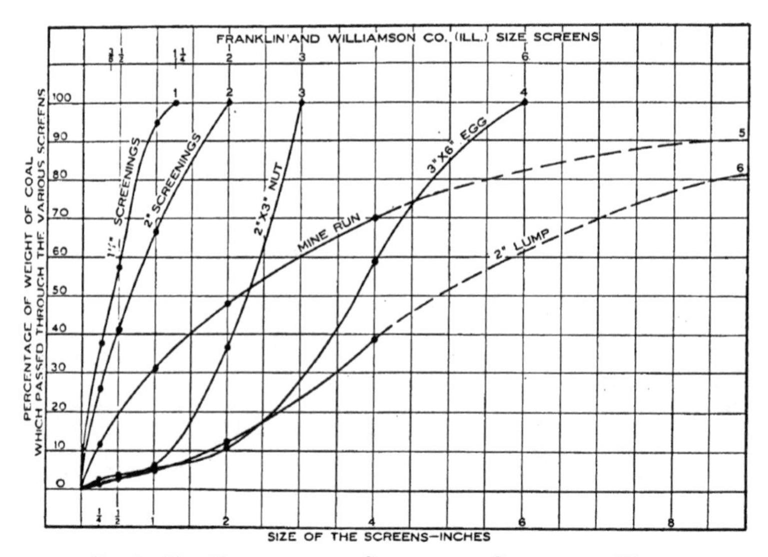

FIG. 9. THE MAKE-UP OF THE COALS IN THE CONDITION IN WHICH THEY WERE RECEIVED

are generally broken down at the coal chute. These two coals as fired contain, therefore, a smaller proportion of large lumps than when they were received and the extent to which this extra preparation modified the make-up of the coals is defined in the table, the figures, and the discussion which follow.

After the mechanical analysis samples taken to represent the mine run and lump coals as received at the laboratory had been screened and separated as described in the preceding section, the large lumps in each sample were broken down to the same extent as these sizes were broken before firing, and under the supervision of the same test operators who controlled this process during the progress of the tests.

These reduced samples for mine run and lump coal were accepted as representing these two grades in the condition in which they were fired, and they were subjected to the same screening process as has been previously described. The results of this mechanical analysis are presented for these two coals "as fired" in Table 4, and the size percentages of these grades in the condition in which they were received are also embodied for comparison's sake in the same table.

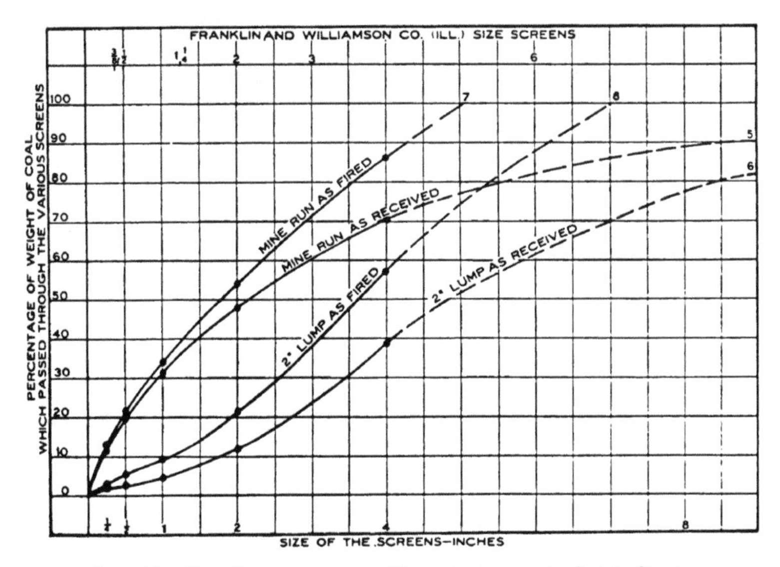

FIG. 10. THE MAKE-UP OF THE MINE RUN AND THE LUMP COALS, AS RECEIVED AND AS FIRED

The values in Table 4 are plotted in Fig. 10, in which curve No. 5 applies to mine run coal as received and No. 7 to the same grade as fired; while curves 6 and 8 apply to the 2-inch lump coal in the conditions as received and as fired, respectively. The extent to which these two sizes were broken down is revealed by an inspection of Table 4 and Fig. 10. Considering the curves in the figure, it is apparent that the largest lumps in the mine run were somewhat further broken down than those in the 2-inch Lump. After reduction all mine run passed through a 5-inch screen, whereas only about 74 per cent of the lump would pass a screen of this size.

TABLE 4

THE MAKE-UP OF THE MINE RUN AND THE LUMP COAL, BOTH AS RECEIVED AND AS FIRED

SIZE OF COAL	Per Cent over 4" Screen	Per Cent through 4" Screen	Per Cent through 2" Screen	Per Cent through 1" Screen	Per Cent through ½" Screen	Per Cent through ¼" Screen
1	2	3	4	5	6	7
Mine Run:						
As Received..	29.6	70.4	48.1[1]	31.3	19.9	12.5
As Fired	13.1	86.9	54.3[1]	34.3	22.0	13.8
2" Lump:						
As Received..	61.6	38.4	12.0	4.5	2.6	1.7
As Fired......	42.7	57.3	21.0	9.0	5.3	2.0

1 Derived from plotted curves (Fig. 10).

In order to permit comparisons of the mechanical make-up of all six sizes as fired, curves Nos. 1, 2, 3, and 4 from Fig. 9 (applying to the grades which were fired as received) and curves 7 and 8 from Fig. 10 are brought together in Fig. 11, which consequently shows the make-up of all the grades in the condition in which they were fired during the tests.

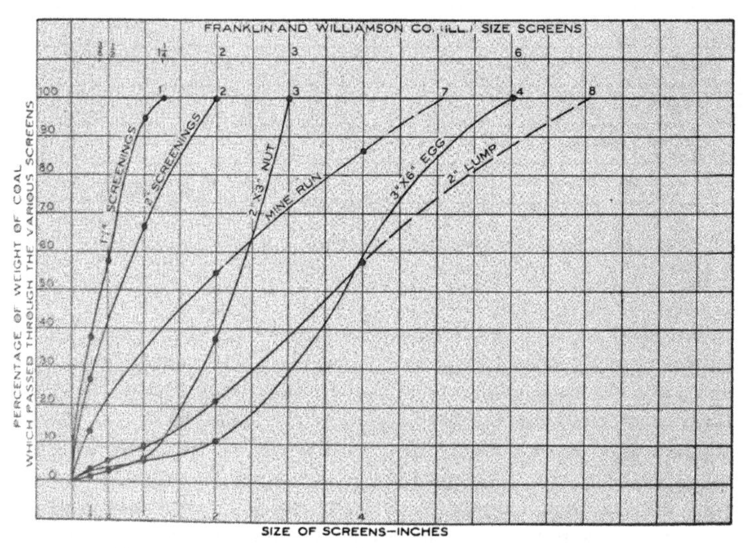

FIG. 11. THE MAKE-UP OF THE COALS, IN THE CONDITION IN WHICH THEY WERE FIRED

IV. The Locomotive

8. *Design and Main Dimensions.*—The locomotive used during the tests was loaned for the purpose by the Baltimore and Ohio Railroad Company. It was of the Mikado type (2-8-2); its road number was 4846, and its classification, Q-7-F. It was built by the Baldwin Locomotive Works during the summer of 1916, entered service in September, and upon its arrival at the laboratory, had run approximately 3400 miles. It arrived at the laboratory in excellent condition.

The principal dimensions of the locomotive are as follows:

Total weight, in working orders, lb.	284,500
Weight on drivers, lb.	222,000
Cylinders (simple), diameter and stroke, in.	26 x 32
Diameter of drivers, in.	64
Firebox, length and width, in.	120 x 84
Firebox volume, cu. ft.	348.6
Grate area, sq. ft.	69.8
Heating surface, 2¼-inch tubes (fire side), sq. ft.	2410
Heating surface, 5½-inch tubes (fire side), sq. ft.	973
Heating surface, firebox and tube sheets (fire side) sq. ft.	247
Heating surface, total (fire side) sq. ft.	3630
Heating surface, superheater (fire side) sq. ft.	1030
Boiler pressure, lb. per sq. in.	190
Tractive effort, lb.	54587

The boiler was of the wagon-top type with radial stays. It was equipped with a Schmidt top-header superheater consisting of 34 elements, a Street stoker, and a Security brick arch carried on four tubes. The front end was self-cleaning and was equipped with a plain 6-inch nozzle-tip without bridge or split, which was used throughout all tests.

The grates were of the box type, the design of which is shown in detail in Appendix I. The total air opening through the grates amounted to 17 square feet or 24.4 per cent of the grate area. The area of the air inlet to the ash pan amounted to 8.3 square feet or 49 per cent of the air opening through the grates.

The locomotive was regularly equipped with a hand-operated door which was replaced, however, during the period of the tests by a Franklin pneumatic door of the butterfly type. This was used during all tests except those with the two sizes of screenings, which were fired by means of the Street stoker.

The design of the locomotive is described in further detail in Appendix I.

9. *Inspection.*—In order to ensure uniformity of its condition as regards accumulations of scale, soot, etc., and to define these conditions, the locomotive, during its stay in the laboratory, was handled and inspected as follows:

The boiler water was changed every five or six days and the boiler was washed about every two weeks. More frequent washings were unnecessary because the laboratory water is relatively free from scale-forming salts and suspended matter. Monthly inspections in accordance with Interstate Commerce Commission regulations were made during the test period as during regular service.

During the progress of the earlier tests, in order to ensure uniformity in the condition of the heating surfaces, all tubes, flues, and superheater elements were blown free from soot and small "honey-comb" immediately before each alternate test. Although there was nothing in the test results to indicate that this was not being done often enough, to remove all uncertainty this blowing down process was gone through before each test during the latter part of the series. The front end netting was cleaned of all cinders, and the cinders were removed from the top of the arch at the same time. The arch was inspected after each test, and all defective bricks were immediately replaced.

Upon its arrival at the laboratory, the interior of the boiler was inspected and was found to be free from all scale except a very thin coating. The laboratory water itself is not only nearly free from scale-forming materials but it tends frequently to soften hard scale deposited by other water; and a final interior inspection revealed the fact that this softening had occurred in this instance, and that the original scale had been considerably disintegrated. There was nothing, however, in the change in the scale to indicate that the heat transfer through the surfaces had, in any significant degree, varied on account of scale during the progress of the tests. The facts that the boiler had run only 3400 miles before arriving at the laboratory and that its mileage during the tests was only 3600, are perhaps in themselves sufficient evidence that there could have been no accumulation of scale nor change in its thickness sufficient materially to affect the test results.

V. The Laboratory

The locomotive laboratory is fully described in Bulletin 82 of the Engineering Experiment Station of the University of Illinois; descriptions have been published also in the Proceedings of the American

Fig. 12. Baltimore and Ohio Railroad Locomotive, 4837, Identical in Design with the One Used in the Tests

Fig. 13. An Interior View of the Laboratory, with a Locomotive in Test Position.

Railway Master Mechanics' Association, Vol. 46, 1913, and in the Proceedings of the Western Railway Club for March, 1913. It is unneccessary, therefore, to include here any detailed statement concerning its design. Since, however, the amount of the cinder losses in this series of tests serves in large measure to account for the differences in performance of the various grades of coal, these losses have an especial significance; and it seems appropriate therefore to describe briefly the cinder separator by means of which they were determined.

All gases and exhaust steam are discharged across an open space above the locomotive stack into the mouth of a large steel elbow which carries them up and over to a horizontal duct running through the roof trusses to the rear of the building, terminating at an exhaust fan. This elbow and duct are illustrated in Fig. 13. The gases and steam are drawn through the duct by the fan and they are then passed through a breeching to the cinder separator itself which is located outside the building and forms the base of a stack through which the gases are finally discharged to the air. The separator and stack are shown in section in Fig. 14.

The cinder-laden gases enter the separator at *B* and, in order to leave, they must pass downward and around the sleeve *A*. This passage gives them a whirling motion, which causes the cinders by centrifugal force to move toward the outside wall along which they fall to the hopper below, while the gases pass out through the sleeve to the stack. The cinders collecting at the bottom of the hopper are drawn off, weighed, and analysed. This separator collects all solid matter which issues from the locomotive stack, except possibly the finest dust. The cinders taken from the separator during tests with mine-run coal have contained from 10 to 18 per cent of material which passed a screen with 200 meshes to the inch.

VI. Firing Methods

It is desired to present at this point only such information concerning the methods used in the laboratory as relates to the measurements of coal and to the firing. Information about methods relating to other test processes is given in Appendix II.

10. *Coal Measurements.*—Before starting the test, the engine was run at the desired load and speed long enough beforehand to permit the rate of feed through the injectors to be adjusted to the test conditions, and it was generally unnecessary to change this rate

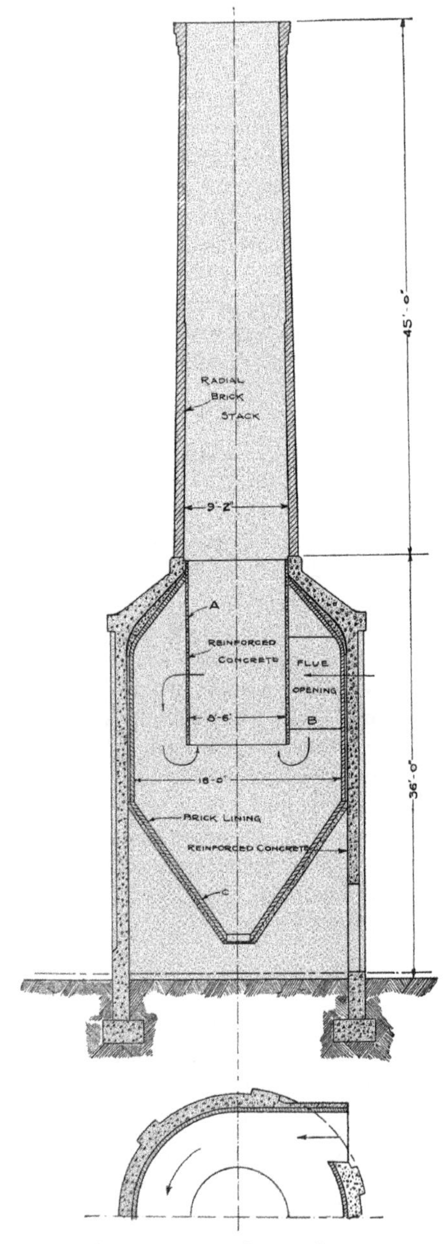

Fig. 14. Cross Section of the Cinder Collector and Stack.

during the progress of the test. The tests were not started until the desired conditions of load, speed, boiler pressure, and a proper condition of the fire had been established. The coal was weighed in one-thousand pound lots before being delivered to the firing platform. Throughout each medium rate test, the time of firing the last scoopful of each ton was recorded, together with the levels of the water in the main feed tank and in the boiler gauge glass. During the high rate tests, these facts were recorded at the time of firing the last scoopful of each two tons of coal. This procedure made it possible to control the regularity of the firing process and it also makes available facts which may be used to illustrate the regularity of feed of both the coal and the water. For this purpose tests 2405 and 2416, fairly characteristic of the series, have been chosen. During test 2416, run at a medium rate of evaporation, the time required to fire each of the ten successive lots of 2000 pounds, varied only from 34 to 36 minutes; and the amount of water fed per minute during these ten intervals varied only from 390 to 413 pounds. During test 2405, which was run at a high rate of evaporation, the times required to burn each of the five successive lots of 4000 pounds of coal were respectively 36, 33, 31, 32 and 31 minutes; and the water fed per minute during these intervals varied only from 693 pounds to 709 pounds.

11. *Firing Methods.*—The locomotive was fired throughout all tests by Mr. C. Welker, a skilled fireman, who was detailed by the Illinois Central Railroad from its regular force. Previous to the tests he had had about seven and one-half years' experience firing in service, and he had also had about a half year's experience as fireman in the laboratory. The supervision of the fireman, the control of the injectors, and other cab operations were, during all except the last three tests, in charge of Mr. L. R. Pyle, Fuel Supervisor of the Minneapolis, St. Paul and Sault Ste. Marie Railroad, and member of the Fuel Test Committee. During the last three tests Mr. Pyle's place was taken by Mr. B. F. Crolley, Supervisor of Locomotive Operation of the Baltimore and Ohio Railroad. During the tests of the two sizes of screenings, which were fired by the stoker, the firing was supervised by Mr. E. Prouty, Mechanical Expert of the Locomotive Stoker Company.

All hand-fired tests were fired by the "three-scoop system," that is, three scoopfuls of coal were fired at a time, both during the medium and the high rate tests, although during the latter it was necessary to cut down the interval between firings. The fireman was instructed to keep these intervals as regular as possible and in this he was aided from time to time by stop watch observations. The degree of regularity attained is evidenced by the figures concerning the rate of coal consumption which have already been cited, and by the graphical logs shown in Appendix III. During the stoker-fired tests, the same regularity of feed was sought and attained. No coal was fed by hand during any of these tests, and an inspection of the fire at the end of the test showed in each case a uniform layer with no holes and no banks.

The thickness of the fire was kept as nearly uniform as was practicable, and the grates were shaken as little as possible. The Lump coal proved the most difficult to fire, especially at the high rate of combustion; and in one high rate test with this grade the grates had to be shaken six times. This was unusual, however, and during the entire series the grates were shaken, on the average, only twice during each run. The approximate thicknesses of the fire carried are shown in Table 5.

TABLE 5

APPROXIMATE THICKNESSES OF FIRE CARRIED

Size of Coal	Rate	Approximate Fire Thickness—Inches		
		At the Beginning	Maximum	At the End.
Mine Run	Medium	6	9	6
	High	6	12	12
2" x 3" Nut	Medium	5	8	8
	High	7	12	12
3" x 6" Egg	Medium	7	11	10
	High	9	12	10
2" Lump	Medium	7	12	12
	High	9	13	12
2" Screenings	Medium	4	9	7
	High	6	12	10
1¼" Screenings	Medium	4	8	7
	High	5	10	9

VII. TEST CONDITIONS

Owing to the fact that only two sets of conditions as to speed and cut-off were employed throughout the tests, other conditions such as drafts, temperatures, and pressures were also in general quite uniform for all tests at a given rate. The degree of uniformity shown is in a large measure indicative of desirable test conditions and is therefore in some degree significant of the reliance which may be placed upon the determination of those variables, such as evaporative performance, which constituted the main purpose of the tests.

Fig. 15 and the discussion of the present section are intended to present the more important test conditions and the variation of these conditions as between different tests and different groups of tests, and as between the medium and high rate tests. Figs. 29 and 30 in Appendix III present graphical logs for tests 2416 and 2405, one a medium and the other a high rate test, during each of which approximately ten tons of coal were burned. These graphical logs are representative of the degree of uniformity in test conditions which existed throughout individual tests.

Fig. 15 presents in graphical form averages of test conditions for all grades of coal for both medium and high rate tests. The graphs have been so arranged as to show the variation in conditions for different grades of coal at a given rate, and also to show difference in conditions between medium and high rate tests.

12. *Drafts.*—The two upper graphs of Fig. 15 present averages for front-end and firebox draft. The extremes of front-end draft for the medium rate tests were 2.8 and 3.8 inches of water; the average for all medium rate tests was 3.4 inches of water. The extremes of front-end draft for the high rate tests were 8.4 and 10.1 inches of water and the average for all high rate tests was 9.3 inches of water. The averages for the different grades of coal vary but little from the common average for all of the tests at the corresponding rate. The mean firebox draft for all medium rate tests was 1.6 inches of water and for all high rate tests, 4.2 inches. The averages for the individual tests and for the various grades of coal do not vary greatly from these mean values. The high rate mine run tests show the greatest variation in front-end draft. The high rate Screening tests show a somewhat lower average firebox draft than is shown for the other grades, due probably to the fact that although the rate of combustion was relatively high, the fires were comparatively thin and open.

The data relative to drafts show uniformity of conditions between the tests of a group as well as between the different groups at a given rate of performance. In general the drafts were comparatively low in relation to rate of combustion, indicating satisfactory arrangement of the draft appliances.

13. *Temperatures.*—Front-end temperatures are shown to have been uniform for the medium and for the high rate tests, both by the graphs of Fig. 15 and by the tabulated values. The average firebox temperatures also are shown to have been fairly uniform for the high rate tests and slightly less so for the medium rate tests. The minimum, maximum, and average values of firebox temperature for all medium rate tests were respectively 1735, 2090, and 1893 degrees F.; and the corresponding values for the high rate tests were 2078, 2334, and 2228 degrees F.

14. *Superheat and Branch-pipe Pressure.*—The variations in, and the values for, averages of degrees of superheat and pressure in branch-pipe are shown graphically in Fig. 15. The minimum, maximum, and average values of degrees of superheat for all medium rate tests were respectively 187, 211, and 198 degrees; and the corresponding values for the high rate tests were 207, 265, and 243 degrees. Considerable lack of uniformity is shown by the results, particularly in view of the general uniformity which existed in other test conditions. The branch-pipe pressure for all medium rate tests averaged 179 pounds per square inch, which was 10 pounds lower than the average boiler pressure for the same tests. For the high rate tests the branch-pipe pressure averaged 166 pounds per square inch—22 pounds lower than the corresponding average boiler pressure.

15. *Rate of Combustion and Rate of Evaporation.*—The two lower graphs of Fig. 15 show the average rate of combustion and the average rate of evaporation for the different sizes of coal. Rate of combustion is shown in pounds of coal fired per square foot of grate per hour; and rate of evaporation, in pounds of equivalent evaporation per square foot of heating surface per hour. For the medium rate tests the minimum, maximum, and average values for rate of combustion were, respectively, 45.1, 61.9, and 51.7 pounds of coal per square foot of grate per hour; while for the high rate tests the corresponding values were 99.1, 133.7, and 109.6 pounds. Since the rate of evaporation per square foot of heating surface per hour was maintained approximately constant for all tests at a given rate and since the heating value of all sizes of the coal was about the same, it follows that the various rates of combustion should indicate closely the relative efficiencies with which the coal was burned; and they may be used as rough measures of the relative values of the different grades.

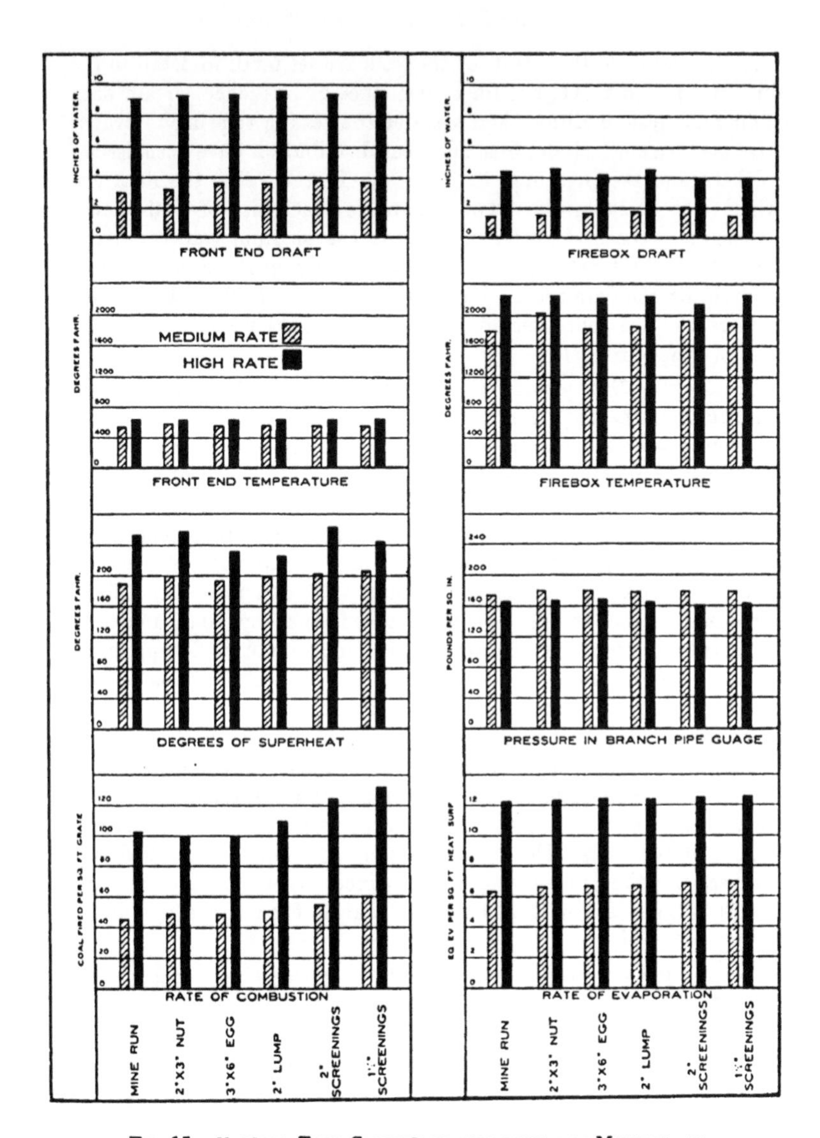

FIG. 15. VARIOUS TEST CONDITIONS, FOR BOTH THE MEDIUM AND HIGH RATE TESTS

VIII. The Results of the Tests

All the data and the results of the tests are set forth in detail in the tables of Appendix III; but for convenience of reference certain data defining the test conditions and some of the results relating to evaporative performance, cinder loss, and heat distribution have been assembled here and are presented in Table 6. In this table the data are divided into twelve groups; two groups for each size of coal tested. For each size the first group pertains to the medium rate tests; the second, to the high rate tests. Averages for each group appear immediately below the data for the individual tests. In Table 6 the column headings and the code item numbers are identical with those in the tables of Appendix III.

16. *Actual Evaporation per Pound of Coal.*—The number of pounds of superheated steam produced for each pound of coal in the condition in which it was fired appears in Column 33 of Table 6; and the number of pounds of steam produced by each pound of dry coal is given in Column 34. The averages of these values for each size of coal are brought together in Table 7, where they appear separately for the medium rate and the high rate tests. Inspection of Table 7 reveals the fact that the relative standing of the various sizes differs when based on coal as fired and when based upon dry coal; and that it differs also between the medium and the high rate tests. Comparisons between the sizes, however, should not be made on the basis of the values of evaporation shown in Table 7; because not only is the moisture content of the coal as fired variable; but during the tests there were slight variations in feed water temperature, in boiler pressure, and in the degree of superheat, which make it impracticable to compare values of actual evaporation in order to determine the relative standing of the coals. Further discussion of Table 7 is omitted for these reasons; the table is presented principally to exhibit the differences between evaporation based on coal as fired and evaporation based on dry coal.

17. *Equivalent Evaporation per Pound of Dry Coal.*—Because of the incidental variations just cited it became necessary to find another basis of comparison. Since the different sizes as they are loaded at the mine contain inherently different amounts of moisture, there would be some justice in trying to base comparisons on coal as loaded at the mine. The significance of this basis is, however, impaired by the fact that the various sizes are seldom or never fired in the same condition

Table 6
Test Conditions and Principal Results

1	2	3	4	5	6	7	8	9	10	11	12
Size	Rate	Test Number	Duration of Test in Hours	Speed in Miles per Hour	Cut-off, Per Cent of Stroke	Draw-bar Pull, lb.	Pressure lb. per sq. in. — Boiler Gauge	Pressure lb. per sq. in. — Branch Pipe Gauge	Temperatures Degrees F. — Front-end	Temperatures Degrees F. — Branch Pipe	Temperatures Degrees F. — Fire-Box
		Code Item ☛	**345**	**353**	**499**	**487**	**380**	**383**	**367**	**370**	**374**
Mine Run	Medium	2400	3.62	18.9	....	21970	190.4	179	535	573	1735
		2401	6.28	18.9	34.0	21727	190.0	172	535	566	1835
		2402	5.20	19.0	33.0	21822	190.2	174	539	564	1812
		Average	**5.03**	**18.9**	**33.5**	**21840**	**190.2**	**175**	**536**	**568**	**1794**
	High	2405	2.72	25.5	54.3	28771	187.8	168	627	628	2271
		2406	2.68	25.6	53.7	28718	187.8	167	631	631	2334
		2429	1.92	25.7	55.9	28672	189.4	164	624	618	2140
		Average	**2.44**	**25.6**	**54.6**	**28720**	**188.3**	**166**	**627**	**626**	**2248**
2"x 3" Nut	Medium	2408	3.77	19.0	33.0	22490	189.0	178	595	589	2090
		2409	2.33	18.9	32.4	22411	188.5	177	588	582	2034
		2410	4.33	19.0	31.5	22417	186.2	181	570	569	2008
		2426	4.50	18.9	31.7	22640	189.9	182	555	572	1967
		Average	**3.73**	**19.0**	**32.2**	**22490**	**188.4**	**180**	**577**	**578**	**1025**
	High	2412	2.67	25.8	....	28958	187.1	168	607	629	2293
		2413	3.00	25.7	55.7	29100	187.1	168	611	632	2267
		2414	2.00	25.7	59.3	29128	187.5	168	631	634	2174
		Average	**2.56**	**25.7**	**57.5**	**29062**	**187.2**	**168**	**616**	**632**	**2245**
3"x 6" Egg	Medium	2415	3.50	18.9	32.4	22840	189.9	180	543	576	1808
		2416	5.83	18.9	33.3	23115	189.5	180	540	574	1801
		2423	4.00	18.8	32.2	22533	190.0	182	539	571	....
		Average	**4.44**	**18.9**	**32.6**	**22829**	**189.8**	**181**	**541**	**574**	**1805**
	High	2420	2.00	25.7	57.2	29046	190.1	171	588	610	2210
		2422	2.17	25.9	58.2	29030	189.7	170	634	590	2278
		2424	1.98	25.8	56.6	29104	190.1	170	626	617	2183
		Average	**2.05**	**25.8**	**57.3**	**29060**	**190.0**	**170**	**616**	**606**	**2224**
2" Lump	Medium	2417	4.00	18.9	36.3	23026	189.9	180	546	578	1838
		2418	5.83	19.0	33.5	23085	190.1	180	545	578	1857
		2419	3.67	19.0	32.7	22983	190.0	180	553	578	1849
		Average	**4.50**	**19.0**	**34.2**	**23031**	**190.0**	**180**	**548**	**578**	**1848**
	High	2425	1.00	25.7	56.0	28530	190.0	168	618	595	2178
		2427	1.50	25.8	55.7	27909	183.4	162	625	578	2308
		2428	1.83	25.9	55.1	28441	186.8	166	635	603	2277
		2442	2.00	25.5	56.5	29266	188.9	166	637	616	2192
		Average	**1.58**	**25.7**	**55.8**	**28537**	**187.3**	**166**	**629**	**598**	**2239**
2" Screenings	Medium	2430	2.62	19.0	32.6	22906	188.6	181	549	583	2010
		2434	3.13	18.7	33.6	23091	190.0	181	541	584	1817
		2435	0.97	19.0	34.9	23268	191.1	182	549	578	1936
		Average	**2.24**	**18.9**	**33.7**	**23088**	**189.9**	**181**	**546**	**582**	**1921**
	High	2436	1.35	25.5	56.8	27976	185.3	161	631	634	2078
		2437	1.50	25.7	56.9	28938	189.1	162	634	637	2194
		Average	**1.43**	**25.6**	**56.9**	**28457**	**187.2**	**162**	**633**	**636**	**2136**
1¼" Screenings	Medium	2431	1.77	19.0	33.9	22332	184.5	178	551	589	2003
		2432	1.87	19.0	34.0	22912	189.1	181	544	591	1798
		2433	3.10	18.9	33.2	22588	187.7	180	543	572	1874
		Average	**2.25**	**19.0**	**33.7**	**22611**	**187.1**	**180**	**546**	**584**	**1892**
	High	2440	1.50	25.5	58.2	29061	186.6	163	634	604	2273
		2441	1.50	25.6	55.6	29392	191.0	166	639	629	2234
		Average	**1.50**	**25.6**	**56.9**	**29227**	**188.8**	**165**	**637**	**617**	**2254**

TABLE 6 (Continued)

TEST CONDITIONS AND PRINCIPAL RESULTS

1	2	3	13	14	15	16	17	18	19	20	21
			Draft, in. of Water			De-grees of Sup-er-heat	Coal as Fired, lb.			Dry Coal, lb.	
SIZE	Rate	Test Number	Front-end Front of Dia-phragm	Fire-box	Ash-pan		Total	Per Hour	Per Hour per Sq. Ft. of Grate Surface	Per Hour	Per Hour per Sq. Ft. of Grate Surface
		Code Item ☞	394	396	397	409	418	......	....	626	627
Mine Run	Medium	2400	2.8	1.2	0.2	194	11399	3151	45.1	2895	41.5
		2401	2.8	1.5	0.2	190	20000	3183	45.6	2934	42.0
		2402	3.0	1.6	0.2	187	17000	3269	46.8	3005	43.1
		Average	**2.9**	**1.4**	**0.2**	**190**	**16133**	**3201**	**45.8**	**2945**	**42.2**
	High	2405	8.4	4.2	0.4	254	20000	7361	105.5	6753	96.8
		2406	8.6	4.3	0.4	257	18630	6944	99.5	6377	91.4
		2429	10.1	4.5	0.4	246	14000	7303	104.6	6690	95.9
		Average	**9.0**	**4.3**	**0.4**	**252**	**17543**	**7203**	**103.2**	**6607**	**94.7**
2"x 3" Nut	Medium	2408	3.0	1.4	0.2	210	12955	3439	49.3	3146	45.1
		2409	2.9	1.3	0.2	204	7808	3347	48.0	3054	43.8
		2410	2.9	1.4	0.2	189	14731	3400	48.7	3102	44.4
		2426	3.6	1.8	0.1	192	16310	3624	51.9	3348	48.0
		Average	**3.1**	**1.5**	**0.2**	**199**	**12951**	**3453**	**49.5**	**3163**	**45.3**
	High	2412	9.2	4.6	0.5	255	18683	7005	100.4	6382	91.4
		2413	9.2	4.4	0.5	258	20811	6937	99.4	6313	90.4
		2414	9.3	4.5	0.5	260	13884	6942	99.5	6324	90.6
		Average	**9.2**	**4.5**	**0.5**	**258**	**17793**	**6961**	**99.8**	**6340**	**90.8**
3"x 6" Egg	Medium	2415	3.6	1.8	0.2	197	11888	3397	48.7	3107	44.5
		2416	3.6	1.7	0.2	195	19915	3414	48.9	3101	44.4
		2423	3.3	1.4	0.2	191	13520	3380	48.4	3079	44.1
		Average	**3.5**	**1.6**	**0.2**	**194**	**15108**	**3397**	**48.7**	**3096**	**44.3**
	High	2420	9.5	4.3	0.5	235	13882	6941	99.4	6333	90.7
		2422	9.2	4.0	0.5	215	14996	6920	99.1	6315	90.5
		2424	9.3	4.1	0.5	242	14000	7060	101.2	6438	92.2
		Average	**9.3**	**4.1**	**0.5**	**231**	**14293**	**6974**	**99.9**	**6362**	**91.1**
2" Lump	Medium	2417	3.6	1.7	0.2	199	13753	3438	49.3	3118	44.7
		2418	3.5	1.7	0.2	199	20537	3521	50.5	3173	45.5
		2419	3.5	1.7	0.2	199	13344	3639	52.1	3289	47.1
		Average	**3.5**	**1.7**	**0.2**	**199**	**15878**	**3533**	**50.6**	**3193**	**45.8**
	High	2425	9.3	4.3	0.5	221	7499	7499	107.4	6850	98.1
		2427	9.3	4.3	0.4	207	11775	7850	112.5	7133	102.2
		2428	9.4	4.4	0.3	230	14122	7704	110.4	6992	100.2
		2442	10.0	4.6	0.5	243	15279	7640	109.5	6951	99.6
		Average	**9.5**	**4.4**	**0.4**	**225**	**12169**	**7673**	**110.0**	**6982**	**100.0**
2" Screen-ings	Medium	2430	3.8	1.9	0.2	203	10000	3821	54.7	3460	49.6
		2434	3.6	......	0.2	204	11950	3814	54.6	3455	49.5
		2435	3.7	2.1	0.2	198	3822	3952	56.6	3597	51.5
		Average	**3.7**	**2.0**	**0.2**	**202**	**8591**	**3862**	**55.3**	**3504**	**50.2**
	High	2436	9.4	3.7	0.4	263	11556	8560	122.6	7771	111.3
		2437	9.2	3.8	0.3	265	13254	8836	126.6	8027	115.0
		Average	**9.3**	**3.8**	**0.4**	**264**	**12405**	**8698**	**124.6**	**7899**	**113.2**
1¼" Screen-ings	Medium	2431	3.5	1.9	0.2	210	7635	4321	61.9	3960	56.7
		2432	3.6	1.1	0.2	211	7813	4185	60.0	3818	54.7
		2433	3.6	1.3	0.2	193	13218	4264	61.1	3899	55.9
		Average	**3.6**	**1.4**	**0.2**	**205**	**9555**	**4257**	**61.0**	**3892**	**55.8**
	High	2440	9.5	4.0	0.5	232	14000	9333	133.7	8487	121.6
		2441	9.4	3.7	0.5	256	13750	9167	131.3	8183	117.2
		Average	**9.5**	**3.9**	**0.5**	**244**	**13875**	**9250**	**132.5**	**8335**	**119.4**

TABLE 6 (Continued)

TEST CONDITIONS AND PRINCIPAL RESULTS

1	2	3	22	23	24	25	26	27	28	29	30
			B.t.u. per lb. of		Cinder Loss						
Size	Rate	Test Number	Dry-Coal	Stack Cinders	Per Hour, lb.	Per Cent of Coal as Fired	Per Cent of Dry Coal	Per Cent of B.t.u. in Coal Fired	Indicated Horse-Power	Drawbar Horse-Power	Superheated Steam per I.H.P. Hour, lb.
		Code Item ☞	458	462	......	......	427	888	711	743	740
Mine Run	Medium	2400	12983	8399	99	3.2	3.4	2.2	......	1108.6	......
		2401	13012	8563	94	3.0	3.2	2.1	1224.5	1095.2	18.18
		2402	12929	8570	102	3.1	3.4	2.2	1223.6	1104.6	18.74
		Average	**12975**	**8511**	**98**	**3.1**	**3.3**	**2.2**	**1224.1**	**1102.8**	**18.46**
	High	2405	12811	11081	709	9.6	10.5	9.1	2157.7	1954.4	19.63
		2406	12933	11030	588	8.5	9.2	7.9	2151.5	1963.8	19.40
		2429	12888	10921	651	8.9	9.7	8.2	2191.0	1965.2	19.65
		Average	**12877**	**11011**	**649**	**9.0**	**9.8**	**8.4**	**2166.7**	**1961.1**	**19.56**
2"x 3" Nut	Medium	2408	13118	8023	63	1.8	2.0	1.2	1293.6	1138.4	17.93
		2409	13102	7585	72	2.2	2.4	1.4	1286.5	1129.7	17.98
		2410	13023	8231	72	2.1	2.3	1.5	1280.9	1133.5	17.93
		2426	12983	8458	107	3.0	3.2	2.1	1313.7	1139.9	18.72
		Average	**13057**	**8074**	**79**	**2.3**	**2.5**	**1.6**	**1293.7**	**1135.4**	**18.14**
	High	2412	13081	10728	413	5.9	6.5	5.3	......	1988.7	......
		2413	13176	10822	397	5.7	6.3	5.2	2200.3	1993.0	19.34
		2414	13090	10634	390	5.6	6.2	5.0	2221.8	1994.9	19.00
		Average	**13116**	**10728**	**400**	**5.7**	**6.3**	**5.2**	**2211.1**	**1992.2**	**19.17**
3"x 6" Egg	Medium	2415	13273	7987	78	2.3	2.5	1.5	1313.7	1150.1	18.01
		2416	13122	7999	76	2.2	2.5	1.5	1324.3	1167.6	17.86
		2423	13282	8329	70	2.1	2.3	1.4	1291.1	1131.0	18.41
		Average	**13226**	**8105**	**75**	**2.2**	**2.4**	**1.5**	**1309.7**	**1149.6**	**18.09**
	High	2420	13198	10771	500	7.2	7.9	6.5	2188.7	1991.6	19.83
		2422	13345	11234	468	6.8	7.4	6.2	2214.3	2001.4	19.40
		2424	13214	10584	538	7.6	8.4	6.7	2220.0	2003.4	19.63
		Average	**13252**	**10863**	**502**	**7.2**	**7.9**	**6.5**	**2207.7**	**1998.8**	**19.62**
2" Lump	Medium	2417	13043	7713	71	2.1	2.3	1.4	1323.2	1160.7	17.92
		2418	13086	7574	68	1.9	2.1	1.2	1329.6	1168.6	17.73
		2419	12836	7106	85	2.3	2.6	1.4	1321.0	1162.2	18.31
		Average	**12988**	**7464**	**75**	**2.1**	**2.3**	**1.3**	**1324.6**	**1163.8**	**17.99**
	High	2425	13205	10917	545	7.3	8.0	6.6	2201.4	1957.8	19.61
		2427	12958	10849	602	7.7	8.4	7.1	2169.7	1919.6	19.68
		2428	13061	10829	523	6.8	7.5	6.2	2200.5	1963.8	19.95
		2442	12974	10415	595	7.8	8.6	6.9	2198.5	1989.6	20.24
		Average	**13050**	**10753**	**566**	**7.4**	**8.4**	**6.7**	**2192.5**	**1957.7**	**19.87**
2" Screenings	Medium	2430	12748	9407	315	8.3	9.1	6.7	1304.9	1160.7	18.02
		2434	12782	9569	338	8.9	9.8	7.3	1314.5	1152.2	18.00
		2435	12710	9113	372	9.4	10.4	7.4	1365.2	1179.1	18.25
		Average	**12747**	**9363**	**342**	**8.9**	**9.8**	**7.1**	**1328.2**	**1164.0**	**18.09**
	High	2436	12769	10611	1127	13.2	14.5	12.1	2126.3	1905.6	19.86
		2437	12625	11018	1316	14.9	16.4	14.3	2243.5	1980.4	19.33
		Average	**12697**	**10815**	**1222**	**14.1**	**15.5**	**13.2**	**2184.9**	**1943.0**	**19.60**
1¼" Screenings	Medium	2431	12929	10505	579	13.4	14.6	11.9	1329.9	1133.4	18.32
		2432	12650	10157	551	13.2	14.4	11.6	1343.1	1161.6	18.22
		2433	12793	10784	461	10.8	11.8	10.0	1310.2	1141.0	18.71
		Average	**12791**	**10482**	**530**	**12.5**	**13.6**	**11.2**	**1327.7**	**1145.3**	**18.42**
	High	2440	12692	10870	1513	16.2	17.8	15.3	2215.6	1977.2	19.29
		2441	12492	11203	1457	15.9	17.8	16.0	2231.6	2006.7	19.48
		Average	**12592**	**11037**	**1485**	**16.1**	**17.8**	**15.7**	**2223.6**	**1992.0**	**19.39**

TABLE 6 (Concluded)

TEST CONDITIONS AND PRINCIPAL RESULTS

1	2	3	31	32	33	34	35	36	37	38	39
			Superheated Steam, lb.				Equivalent Evaporation, lb.				
SIZE	Rate	Test Number	Per Hour	Per Sq.Ft. of Heating Surface per Hour	Per Pound of Coal as Fired	Per Pound of Dry Coal	Per Hour	Per Sq. Ft. Heating Surface per Hour	Per Pound of Dry Coal	B.t.u. Absorbed by Boiler per Pound of Dry Coal	Boiler Efficiency Per Cent
		Code Item ☞	......	......	......	......	645	648	658	......	666
Mine Run	Medium	2400	22566	4.84	7.16	7.79	29764	6.39	10.28	9989	76.89
		2401	22328	4.79	7.01	7.61	29451	6.32	10.04	9756	74.95
		2402	22970	4.93	7.02	7.64	30183	6.48	10.04	9756	75.46
		Average	**22621**	**4.85**	**7.06**	**7.68**	**29799**	**6.40**	**10.12**	**9834**	**75.77**
	High	2405	42176	9.05	5.73	6.24	57022	12.24	8.44	8201	64.08
		2406	41946	9.00	6.04	6.58	56795	12.19	8.91	8658	66.93
		2429	42854	9.20	5.87	6.41	57767	12.40	8.64	8395	65.10
		Average	**42325**	**9.08**	**5.88**	**6.41**	**57195**	**12.28**	**8.66**	**8418**	**65.37**
2"x 3" Nut	Medium	2408	23327	5.00	6.78	7.42	31048	6.66	9.87	9591	73.11
		2409	23254	4.99	6.95	7.61	30881	6.63	10.11	9824	75.02
		2410	23059	4.95	6.78	7.43	30461	6.54	9.82	9542	73.05
		2426	24808	5.33	6.85	7.41	32796	7.04	9.80	9523	73.33
		Average	**23612**	**5.07**	**6.84**	**7.47**	**31297**	**6.72**	**9.90**	**9620**	**73.63**
	High	2412	42964	9.22	6.13	6.73	58130	12.47	9.11	8852	67.67
		2413	42720	9.17	6.16	6.77	57929	12.43	9.18	8920	67.67
		2414	42209	9.06	6.08	6.67	57236	12.28	9.05	8794	67.15
		Average	**42631**	**9.15**	**6.12**	**6.72**	**57765**	**12.39**	**9.11**	**8855**	**67.50**
3"x 6" Egg	Medium	2415	23944	5.14	7.05	7.71	31678	6.80	10.20	9911	74.66
		2416	23788	5.11	6.97	7.67	31448	6.75	10.14	9853	75.09
		2423	23970	5.14	7.09	7.78	31665	6.79	10.28	9989	75.25
		Average	**23901**	**5.13**	**7.04**	**7.72**	**31597**	**6.78**	**10.21**	**9918**	**75.00**
	High	2420	43219	9.28	6.22	6.82	58043	12.46	9.16	8901	67.46
		2422	42881	9.20	6.20	6.79	57160	12.27	9.05	8794	65.92
		2424	43302	9.29	6.13	6.73	58328	12.52	9.06	8804	66.61
		Average	**43134**	**9.26**	**6.18**	**6.78**	**57844**	**12.42**	**9.09**	**8833**	**66.66**
2" Lump	Medium	2417	23906	5.13	6.96	7.67	31652	6.79	10.15	9863	75.68
		2418	23778	5.10	6.75	7.49	31458	6.75	9.91	9630	73.57
		2419	24340	5.23	6.69	7.40	32227	6.92	9.80	9523	74.21
		Average	**24008**	**5.15**	**6.80**	**7.52**	**31779**	**6.82**	**9.95**	**9672**	**74.49**
	High	2425	42889	9.21	5.72	6.26	57214	12.28	8.35	8114	61.47
		2427	42254	9.07	5.38	5.92	56071	12.03	7.86	7638	58.92
		2428	44136	9.47	5.73	6.31	58403	12.53	8.35	8114	62.14
		2442	44910	9.64	5.88	6.46	60539	12.99	8.71	8464	65.19
		Average	**43547**	**9.35**	**5.68**	**6.24**	**58057**	**12.46**	**8.32**	**8083**	**61.93**
2" Screenings	Medium	2430	24122	5.18	6.31	6.97	32010	6.87	9.25	8988	70.55
		2434	24014	5.15	6.30	6.95	31914	6.85	9.24	8979	70.24
		2435	25147	5.40	6.36	6.99	33270	7.14	9.25	8988	70.75
		Average	**24427**	**5.24**	**6.32**	**6.97**	**32398**	**6.95**	**9.25**	**8985**	**70.51**
	High	2436	42287	9.07	4.94	5.45	57468	12.33	7.40	7191	56.25
		2437	43981	9.44	4.98	5.48	59814	12.84	7.45	7239	57.35
		Average	**43134**	**9.26**	**4.96**	**5.47**	**58641**	**12.59**	**7.43**	**7215**	**56.80**
1¼" Screenings	Medium	2431	24907	5.34	5.76	6.29	33102	7.10	8.36	8123	62.81
		2432	24714	5.30	5.90	6.47	32870	7.05	8.61	8366	66.08
		2433	24933	5.35	5.85	6.39	32961	7.07	8.45	8211	64.21
		Average	**24851**	**5.33**	**5.84**	**6.38**	**32978**	**7.07**	**8.47**	**8233**	**64.37**
	High	2440	43186	9.27	4.63	5.09	57956	12.44	6.83	6637	52.28
		2441	43941	9.43	4.80	5.37	59540	12.78	7.28	7074	56.64
		Average	**43564**	**9.35**	**4.72**	**5.23**	**58748**	**12.61**	**7.06**	**6856**	**54.46**

(as regards moisture) in which they were loaded, nor were they in the same condition in this instance; and furthermore the necessary mine samples were available for only three of the six sizes tested. Final comparisons were consequently drawn only on the usual basis of dry coal. The use of the customary "equivalent evaporation from and at 212 degrees" eliminates the effect of the remaining variations in test conditions; and the final comparison of the grades was therefore made by the use of the values of this equivalent evaporation per pound of dry coal.

TABLE 7

THE ACTUAL EVAPORATION PER POUND OF COAL AS FIRED AND ALSO PER POUND OF DRY COAL

SIZE OF COAL	Actual Evaporation per lb. of Coal as Fired—lb.		Actual Evaporation per lb. of Dry Coal—lb.	
	At the Medium Rate of Evaporation	At the High Rate of Evaporation	At the Medium Rate of Evaporation	At the High Rate of Evaporation
1	2	3	4	5
Mine Run	7.06	5.88	7.68	6.41
2" x 3" Nut	6.84	6.12	7.47	6.72
3" x 6" Egg	7.04	6.18	7.72	6.78
2" Lump	6.80	5.68	7.52	6.24
2" Screenings	6.32	4.96	6.97	5.47
1¼" Screenings	5.84	4.72	6.38	5.23

These values of equivalent evaporation per pound of dry coal are given for each test in Column 37 of Table 6. Inspection of these figures discloses great uniformity among the values applying to each size and each rate. Only in the case of the high rate tests with the 2-inch lump coal is there any considerable variation between the equivalent evaporation values for the individual tests; and even in this group the maximum variation from the average is only 5½ per cent. In view of this uniformity we are entirely warranted in using the average values for the various groups and in basing conclusions upon them. These averages of equivalent evaporation per pound of dry coal are therefore assembled in Table 8 together with the averages of the rate of evaporation per square foot of heating surface per hour taken from Column 36 of Table 6. Table 8 embodies consequently the final direct results of the tests.

In Table 8 the coals are arranged in the order of the evaporation at the medium rate as given there in Column 2. The egg coal heads the list with an equivalent evaporation of 10.21 pounds per pound of dry

TABLE 8

THE EQUIVALENT EVAPORATION PER POUND OF DRY COAL FOR BOTH THE MEDIUM AND THE HIGH RATE TESTS

Size of Coal	For the Medium Rate Tests		For the High Rate Tests	
	Equivalent Evaporation per lb. of Dry Coal lb.	Equivalent Evaporation per Hour per sq. ft. of Heating Surface lb.	Equivalent Evaporation per lb. of Dry Coal lb.	Equivalent Evaporation per Hour per sq. ft. of Heating Surface lb.
1	2	3	4	5
3" x 6" Egg	10.21	6.78	9.09	12.42
Mine Run	10.12	6.40	8.66	12.28
2" Lump	9.95	6.82	8.32	12.46
2" x 3" Nut	9.90	6.72	9.11	12.39
2" Screenings	9.25	6.95	7.43	12.59
1¼" Screenings	8.47	7.07	7.06	12.61

coal followed by the other grades in the order in which they appear in the table. For the high rate tests the nut coal gave the best performance, namely an equivalent evaporation of 9.11 pounds per pound of dry coal, while the other sizes stand in the order in which they are cited in the table. These relations stand out more clearly in Fig 16 which has been prepared by plotting values of equivalent evaporation and rate of evaporation given in Table 8. In Fig. 16 the vertical distances represent equivalent evaporation per pound of dry coal, whereas the horizontal distances represent the pounds of equivalent evaporation per hour on each square foot of heating surface. For the 3-inch by 6-inch egg coal these quantities are, for the medium rate tests 10.21 pounds and 6.78 pounds, respectively; and for the high rate tests 9.09 pounds and 12.42 pounds, respectively (see Table 8). These pairs of values are plotted in Fig. 16 where they appear as the two points which define the line marked 3-inch by 6-inch Egg. These points are connected by a straight line, which implies the assumption that the equivalent evaporation varies regularly and directly with the rate of evaporation. While there are, in this series, no tests at intermediate rates to support this assumption, it is amply warranted by the results of numerous other locomotive boiler tests. The other lines in Fig. 16 are similarly plotted from values given in Table 8; and they define the performance of the other five sizes.

Inspection of Fig. 16 reveals, as usual, for all grades a sharp decrease in evaporation as the rate of evaporation increases. The rate of this decrease is nearly alike for all sizes except the 2-inch by 3-inch nut for which it is roughly one-half of that for the other sizes. This change in evaporation with rate of evaporation makes it necessary to reduce the values of evaporation to a common rate before drawing final comparisons between the various grades. To effect this reduction the rates of evaporation for all the medium rate tests shown in Column 3 of Table 8 have been averaged and this average—6.79 pounds per square foot of heating surface per hour—has been represented

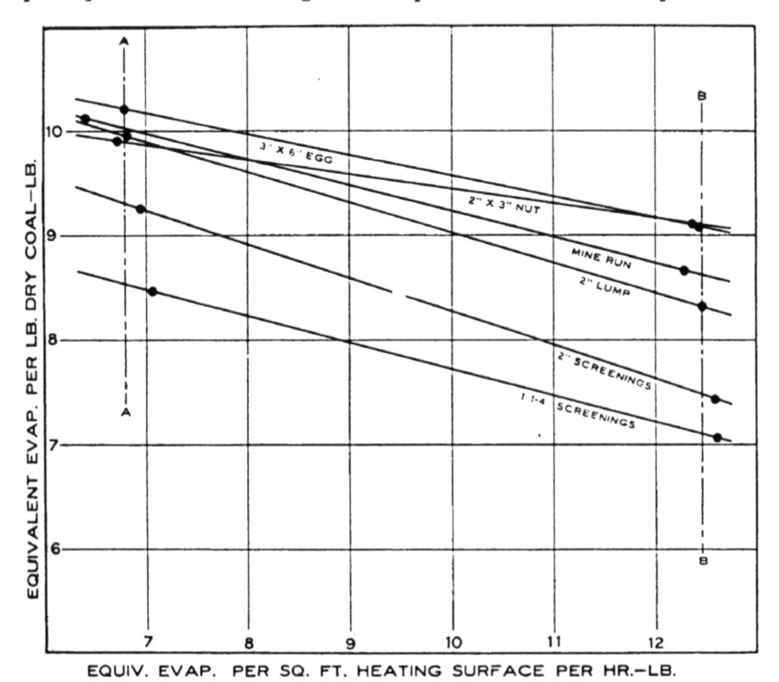

FIG. 16. THE RELATION BETWEEN EQUIVALENT EVAPORATION PER POUND OF DRY COAL AND THE RATE OF EVAPORATION, FOR EACH SIZE OF COAL TESTED

by the vertical line AA in Fig. 16. Similarly the average high rate—12.46 pounds per square foot of heating surface per hour—is found from Column 5 of Table 8 and is defined by the line BB in this figure. If now in Fig. 16 we measure off the vertical distances on AA at the points where this line is intersected by the performance lines for the various sizes, we obtain six values of equivalent evaporation per pound of dry coal, one for each size, and all pertaining to the common medium rate of evaporation defined by the line *AA*, namely, 6.79 pounds per square foot of heating surface per hour. These values are shown in Column 2 of Table 9 and since they pertain to the same rate of evaporation they are rigidly comparable. In like manner the evaporation values defined by the intersections with the line *BB* are given in Column 4 of Table 9, and pertain to the common high rate—12.46 pounds per square foot of heating surface per hour.

TABLE 9

THE RELATIVE STANDING OF THE VARIOUS SIZES, BASED ON CORRECTED VALUES OF THE EQUIVALENT EVAPORATION PER POUND OF DRY COAL

(This table gives the values of equivalent evaporation per pound of dry coal, corrected for a common medium rate of evaporation of 6.79 pounds per square foot of heating surface per hour, and for a common high rate of evaporation of 12.46 pounds per square foot of heating surface per hour)

GRADE OF COAL	For the Common **Medium** Rate of Evaporation—6.79 lb. per sq. ft. of Heating Surface per Hour		For the Common **High** Rate of Evaporation—12.46 lb. per sq. ft. of Heating Surface per Hour	
	Equivalent Evaporation per lb. of Dry Coal lb.	Relative Values, Based on Mine Run	Equivalent Evaporation per lb. of Dry Coal lb.	Relative Values, Based on Mine Run
1	2	3	4	5
3" x 6" Egg	10.21	1.02	9.08	1.05
Mine Run	10.02	1.00	8.62	1.00
2" Lump	9.96	0.99	8.32	0.97
2" x 3" Nut	9.89	0.98	9.10	1.06
2" Screenings	9.30	0.93	7.47	0.87
1¼" Screenings	8.54	0.85	7.10	0.82

Table 9 presents therefore average values of equivalent evaporation per pound of dry coal for each of the sizes of coal—first for a common medium rate of evaporation in Column 2, and next for a common high rate of evaporation in Column 4. These are the final results of the tests, and they may be compared to determine the relative value of the various sizes. Such a comparison of the values in Column 2 shows that when the boiler was worked at the medium rate the 3-inch by 6-inch egg coal gave the highest evaporation, with the other sizes following in the order in which they appear in the table; whereas at the high rate the 2-inch by 3-inch nut coal gave the highest evaporation followed by the egg, mine run, lump, 2-inch screenings, and 1¼-inch screenings in the order named. Further comparison is more conveniently made upon a percentage basis, for which purpose the performance of the mine run is taken as unity or 100 per cent, and the other sizes are represented in Columns 3 and 5 of Table 9 by numbers which define the relation of their performance to that of the mine run.

At the *medium rate* the four larger grades gave nearly the same performance, the maximum difference among them being but 4 per cent. The steam production per pound of egg coal was 2 per cent greater than with the mine run, while with the lump and the nut it was respectively 1 per cent and 2 per cent less than with mine run. The performance with 2-inch screenings was 7 per cent less and with 1¼-inch screenings 15 per cent less than with mine run. If we assume that mine run coal on the tender was worth $2.00 per ton, the relative worth on the tender of the other sizes during the medium rate tests was:

3-inch x 6-inch Egg . .	$2.04	2-inch Screenings . . .	$1.86
2-inch Lump	1.98	1¼-inch Screenings . .	1.70
2-inch x 3-inch Nut . .	1.96		

At the *high rate* the 2-inch by 3-inch nut coal gave the best performance, producing 6 per cent more steam than the mine run; the 3-inch by 6-inch egg came next with an evaporation 5 per cent more than that of the mine run; while the 2-inch lump evaporated 3 per cent less. At this rate of evaporation the 2-inch screenings and the 1¼-inch screenings produced per pound respectively 13 per cent and 18 per cent less steam than the mine run. If we again assume that mine run was worth on the tender $2.00 per ton, the relative worth of the other sizes during the high rate tests was as follows:

2-inch x 3-inch Nut . .	$2.12	2-inch Screenings . . .	$1.74
3-inch x 6-inch Egg . .	2.10	1¼-inch Screenings . .	1.64
2-inch Lump	1.94		

The facts presented in Table 9 and in the foregoing discussion are graphically presented in Figs. 17 and 18. Fig. 17 shows the relative steam producing capacity or the relative values of the six different sizes of fuel when used at the medium rate of evaporation; and Fig. 18 presents these relations for the high rate of evaporation. These two figures and Table 9 embody the final and principal results of the whole test series.

While the differences in performance of the sizes is due in some measure to inherent variations in heating value and in ash content, these variations are too small to account fully for the difference in

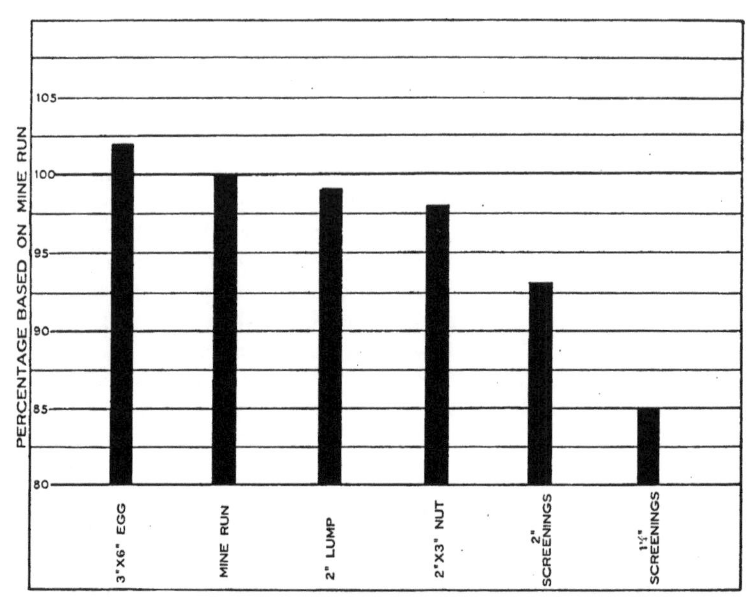

Fig. 17. The Relative Evaporative Efficiencies of the Coals, for the Medium Rate Tests

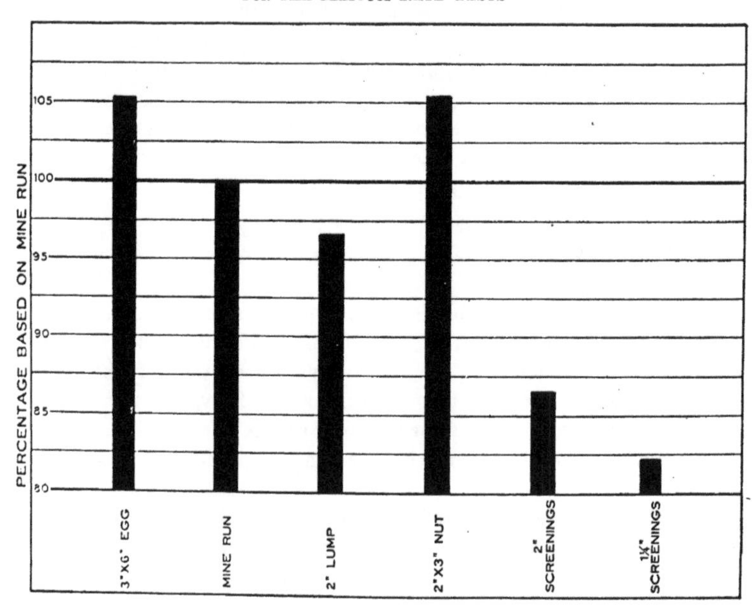

Fig. 18. The Relative Evaporative Efficiencies of the Coals, for the High Rate Tests

performance, nor is an explanation on these grounds applicable to all of the sizes. The difference in performance appears to be due chiefly to the variations in cinder loss and in the conditions of combustion which it was possible to maintain with the different sizes. This conclusion is supported by the discussion of cinder losses and of the heat distribution which follows in the next two sections.

18. *Cinder Losses.*—Information relative to the losses due to cinders passing out through the stack is given in Fig. 19. In considering the cinder losses as here presented it should be borne in mind that all of the coal tested was of one kind, that is, it came from one mine. Coals possessing other physical characteristics might show somewhat different results as to cinder losses under the conditions of the tests here considered. It should also be remembered that for a given rate, medium or high, the draft was, for all grades of coal, practically constant as shown in Fig. 15.

Fig. 19 shows the amount of the stack losses when the weight of the cinders collected from the stack is expressed as a percentage of the weight of the dry coal fired, and also the amount of such loss when the heat content of the cinders collected from the stack is expressed as a percentage of the British thermal units in the coal fired. The loss when expressed as per cent of B. t. u. is numerically less than when expressed as per cent of weight of dry coal, due to the fact that the cinders do not have so high a heat value per pound as the coal from which they originate. Also, due to the fact that cinders produced at high rates of combustion have higher heating values than cinders produced at low rates of combustion, the differences between percentages for medium rate and high rate tests are greater when expressed in terms of heat units than when expressed in terms of dry coal. The average heating value of the stack cinders for all medium rate tests was 8635 B. t. u. and the average value for all high rate tests was 10854 B. t. u. Column 23, Table 6, shows the heating value of stack cinders for each test and the average values for each of the twelve groups of tests. The heating values of the cinders from the medium rate tests with screenings were higher than corresponding values from other grades of coal.

In Fig. 19 it will be seen that, during the medium rate tests, from 2.3 to 13.6 pounds of cinders were collected from the stack for each 100 pounds of dry coal fired; while for the high rate tests from 6.3 to 17.8 pounds were collected for each 100 pounds of coal. The

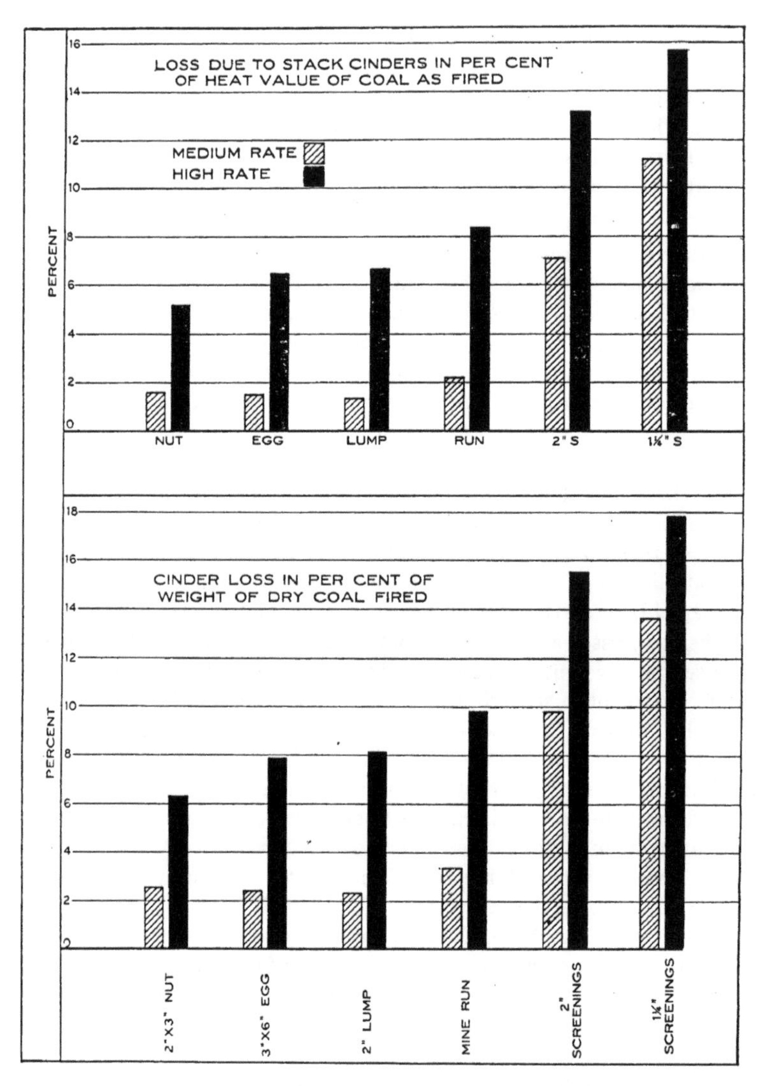

Fig. 19. The Cinder Losses, Expressed in Per Cent of the Heat in the Coal and as Per Cent of the Weight of the Dry Coal

screened coals in all cases produced fewer cinders than the mine run coal; and the screenings produced a materially greater quantity of cinders than any of the larger sizes.

When the losses are expressed as B. t. u. percentages, Fig. 19 shows that for the medium rate tests the loss on account of stack cinders was smallest in the case of the lump coal, amounting to 1.3 per cent of the heat content of the coal fired. The corresponding losses for the egg, nut and mine run coals were 1.5, 1.6, and 2.2 per cent, respec-

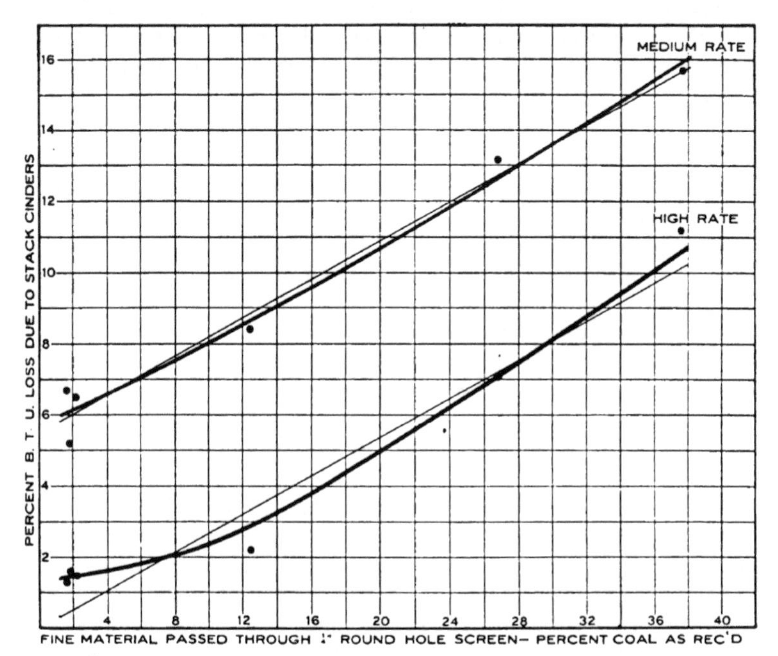

FIG. 20. THE RELATION BETWEEN CINDER LOSS AND THE PER CENT OF FINE MATERIAL IN THE COAL

tively. The loss, during medium rate tests, for the 2-inch screenings was 7.1 per cent and for the 1¼-inch screenings, 11.2 per cent. The average loss from the screenings was roughly five times as great as the average loss from the larger coals during the medium rate tests. For the high rate tests the smallest heat loss due to stack cinders occurred with the nut coal. The average losses for nut, egg, lump and mine run are 5.2, 6.5, 6.7, and 8.4 per cent, respectively. The corresponding loss for the 2-inch screenings was 13.2 per cent and for the 1¼-inch screenings 15.7 per cent. The average loss from the screenings during the high rate tests was more than twice as great as the average loss from the larger coals.

The figures and data indicate that with very fine coals such as screenings the cinder loss is large even at medium rates of combustion and with comparatively low front-end draft; but that under these conditions the cinder loss is not serious for the larger coals even when they contain a considerable amount of fine material as in mine run coal. For conditions involving high rates of combustion and strong drafts, however, the stack cinder loss is a serious one for all sizes of coal.

Fig. 20 shows the relation existing between the loss due to stack cinders and the amount of ¼-inch or smaller material in the coal as received. The data presented in Fig. 20 are also shown in Table 10. The curves in addition to showing the relative magnitude of the cinder losses for the two rates of operation, show that the cinder losses increased quite uniformly with the increase of fine material in the coal. At the medium rate about 1 per cent of the coal would apparently be lost as cinders if there were no ¼-inch fine material at all in the coal; while at the higher rate and without such material, the loss would be about 5.5 per cent. The curve for the high rate tests shows an increase in the cinder loss of very nearly one per cent for each increase of 3.7 per cent in the amount of ¼-inch material in the coal. The light straight lines in Fig. 20 show, for both rates, a uniform increase of one per cent in cinder loss for each 3.7 per cent increase in the ¼-inch material in the coal. The straight line represents the plotted points of the high rate tests closely but does not represent so well the points plotted for the medium rate tests. For the purposes of further discussion, however, the straight lines have been accepted as defining with sufficient accuracy the relations for both rates.

For conditions similar to those of the high rate tests, therefore, the percentage loss of fuel due to stack cinders may be expected to be approximately 5.5 plus the per cent of ¼-inch material in the coal divided by 3.7. Expressed as a formula this becomes

$$C = 5.5 + (F \div 3.7),$$

where C is the fuel loss on account of stack cinders expressed as per cent of B. t. u. in the coal, and F is the per cent of original coal passing

TABLE 10

PER CENT OF FINE MATERIAL IN COAL, AND LOSSES DUE TO STACK CINDERS

Size	Per Cent of Fine Material in Coal as Received, Passing Through			Rate	Loss Due to Stack Cinders		
	¼" Round Hole Screen	½" Round Hole Screen	1" Round Hole Screen		Per Cent of B.t.u. in Coal Fired	Wt. of Cinders in Per Cent of Coal as Fired	Wt. of Cinders in Per Cent of Dry Coal
1	2	3	4	5	6	7	8
2" Lump	1.72	2.62	4.53	Medium	1.3	2.1	2.3
				High	6.7	7.4	8.4
2" x 3" Nut	1.87	2.93	5.77	Medium	1.6	2.3	2.5
				High	5.2	5.7	6.3
3" x 6" Egg	2.28	3.40	5.40	Medium	1.5	2.2	2.4
				High	6.5	7.2	7.9
Mine Run	12.50	19.94	31.30	Medium	2.2	3.1	3.3
				High	8.4	9.0	9.8
2" Screenings	26.88	41.09	66.82	Medium	7.1	8.9	9.8
				High	13.2	14.1	15.5
1¼" Screenings	37.59	57.62	95.56	Medium	11.2	12.5	13.6
				High	15.7	16.1	17.8

through a ¼-inch round hole screen. The expression for conditions similar to those of the medium rate tests is:

$$C = F \div 3.7.$$

The per cent of coal passing through a ¼-inch round hole screen has been used in the foregoing analysis since that was the smallest screen used in testing the coal for size. Similar analyses making use of the per cent of coal passing through a ½-inch or 1-inch screen show similar relations and result in similar formulas, with only a change in the value of the divisor. When F is the per cent of coal passing through a ½-inch round-hole screen, these formulas become:

$C = 5.5 + (F \div 5.7)$, for the high rate conditions,
$C = F \div 5.7$, for the medium rate conditions; and when F is the per cent of coal passing through a 1-inch round-hole screen, the corresponding formulas are:
$C = 5.5 + (F \div 9.3)$, for the high rate conditions,
$C = F \div 9.3$, for the medium rate conditions.

It should be remembered that kind of coal, intensity of draft, firebox and front end arrangement and probably other factors may materially affect the relations existing between cinder losses and the amount of fine material in coal, and that in the tests under consideration these variables have a very limited range. The results, therefore, if applied to conditions other than those from which they were derived should be used with caution and with an understanding of their limitations.

During tests with the four larger grades, larger quantities of cinders were collected than there was ¼-inch, or smaller, material in the coal. For these coals a considerable portion of the cinders must therefore have come from comparatively large pieces of coal. In the Screenings tests the cinders collected were materially less in amount than the ¼-inch or smaller material that existed in the coal. At all comparatively high rates of combustion therefore, and probably also at lower rates, there must be factors determining the amount of cinders produced other than the original amount of fine material in the coal fired.

19. *Heat Distribution.*—Fig. 21 presents average heat balances for the tests with each grade of coal for both medium and high rate tests. The figures have been so constructed that the groups are arranged with relation to decreasing values of the per cent of heat absorbed by the boiler during the high rate tests. This places the grades in the following order: nut, egg, mine run, lump, 2-inch screenings, and 1¼-inch screenings, the nut coal having shown the highest boiler efficiency, followed by the other sizes in the order named. During the medium rate tests the mine run coal showed the highest average boiler efficiency followed by egg, lump, nut, 2-inch screenings and 1¼-inch screenings in the order named. The percentages of heat absorbed by the boiler, during the medium rate tests, for the four grades of coal other than screenings, were, however, very nearly the same, ranging only from 75.8 per cent for mine run to 73.6 per cent for nut coal.

Fig. 21, in addition to the per cent of heat absorbed by the boiler, shows the following items, all in percentages of the heat of the coal fired; loss due to stack cinders; loss due to hydrogen in the coal, moisture in the coal, and moisture in the air; loss due to combustible in the ash; loss due to heat of the escaping gases; loss due to incomplete combustion of gases; and the "radiation and unaccounted for" loss. The complete heat balances, tabulated in Appendix III, present the same information in more detail and for each test. Fig. 21 reveals various relations concerning the heat distribution. For the high rate tests the figures representing cinder losses increase in size to about the same extent as the figures representing heat absorbed by the

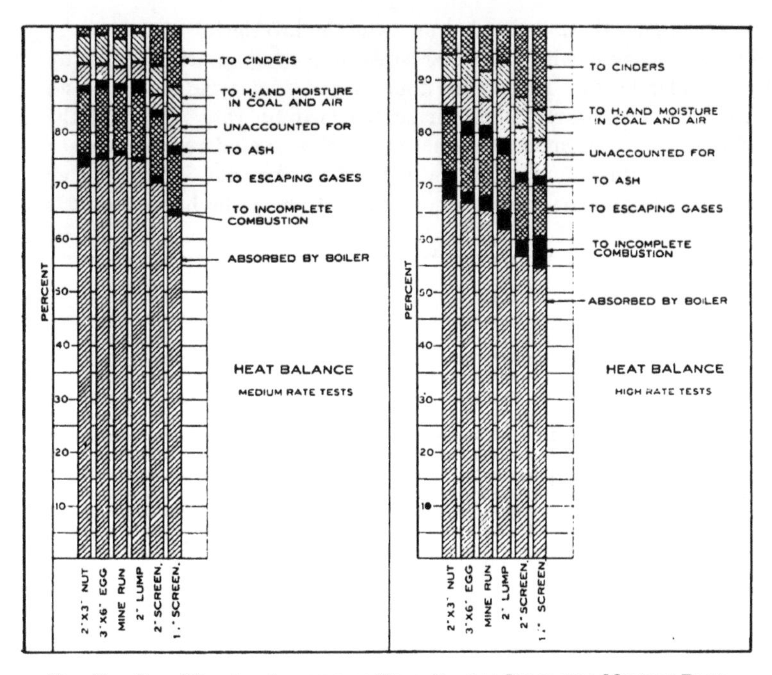

FIG. 21. THE DISTRIBUTION OF THE HEAT DURING BOTH THE MEDIUM RATE AND THE HIGH RATE TESTS

boiler decrease, in passing from nut to 1¼-inch screenings. In general also, all of the figures representing losses other than the cinder loss appear to be nearly equal. The principal exceptions to the general statements just made are found in the facts that losses due to incomplete combustion vary considerably, and that the heat distribution representing the lump coal tests shows a small cinder loss and a large "unaccounted for" loss. It may be said, however, that there was little variation in the losses due to escaping gases, to the ash-pan, to incomplete combustion, to moisture, and to radiation and unaccounted for, as between the different grades; and that the differences in the amounts of heat absorbed by the boiler are to be accounted for chiefly by the variation in the losses due to cinders.

This last statement is illustrated by Fig. 22, in which the height of each vertical hand is proportioned to the sum of the heat absorbed by the boiler and the heat carried away in the cinders. It is obvious from the figure that, at the medium rate, the inferiority of the screenings as compared with the other sizes is entirely accounted for by their cinder losses. Among the four larger grades the cinder losses were small and nearly alike during the medium rate tests, and the differences in performances are not chargeable to cinders, but to other factors. Of these four sizes the mine run shows the highest boiler efficiency, despite the largest cinder loss. During the high rate tests

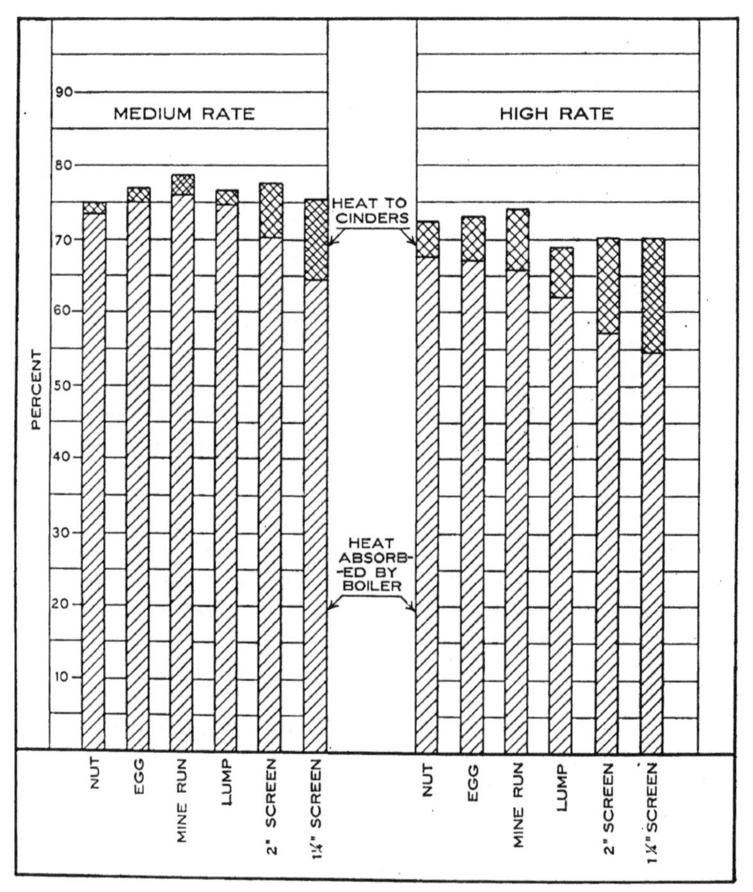

Fig. 22. The Sum of the Heat Absorbed by the Boiler and the Heat Lost in the Cinders, for Both the Medium and the High Rate Tests

the inferiority of the screenings is again almost entirely accounted for by the cinder loss. The difference in performance of the four other grades is also apparently due chiefly to the heat carried away in the cinders, except as regards the lump coal which, although its cinder loss was less than that of the mine run and about equal to that of the egg and nut, nevertheless gave a performance inferior to all of them. This fact reflects the difficulty experienced in firing the lump coal at the high rate, which has been previously alluded to.

The radiation and unaccounted for losses are, for all grades, quite uniform, the minimum and maximum values for the twelve groups being 3.0 per cent and 9.3 per cent. The minimum, maximum, and average values for the entire 36 tests are 2, 11.2, and 5.2 per cent respectively. If there were no unaccounted for loss the average value of 5.2 per cent should represent with some degree of exactness the radiation loss. In addition to the heat losses accounted for there are probably other losses not measured, such as those due to sensible heat carried away by the ash and cinders, unburned combustible gases not determined by the gas analysis, and unburned carbon in the smoke other than the cinders which are collected. Making an allowanco of 1 or 2 per cent for the losses just mentioned and deducting this from the total average unaccounted for—5.2 per cent—would leave the average value for the loss due to radiation at about 3 or 4 per cent. While we have no very reliable data as to the radiation loss under conditions similar to those of the tests, the figure 3 or 4 per cent is probably not greatly in error. There is, further, for all tests a comparatively small variation of the radiation and unaccounted for loss from the mean value. Because of these facts it is fair to conclude that, in general, the heat distributions as given in the tables account for practically all of the heat content of the coal; that the amounts actually unaccounted for are so small as not seriously to invalidate any portion of the balances; and finally that the approximately complete and correct accounting for of all the heat content of the coal makes it probable that values defining the heat distribution may safely be taken as a basis for conclusions concerning the test results.

IX. Conclusions

Such generalizations as follow seem warranted by the test results. They are presented as applicable only to the coal tested. How closely they apply to coals from other fields is not clear, although it is probable that they hold good for other coals of like mechanical make-up and similar physical properties. If it is desired to apply these conclusions to coals from other fields, the facts should be borne in mind that the six sizes tested were more nearly alike in chemical composition and heating value than is often the case, that the cinder losses account in large measure for the differences in performance, that the firing was unusually uniform and constantly supervised, that the large lumps in both the mine run and lump coals were broken before firing, and that the same exhaust nozzle was used throughout all tests.

The purpose of the tests and the general program are set forth in Chapter II.

The heating values and the chemical analyses of the six sizes of coal tested are given in Table 1 of Chapter III, and their mechanical make-up is defined in sections 6 and 7 of that chapter.

The final results of the tests, expressed in terms of equivalent evaporation per pound of dry coal, are presented in Columns 2 and 4 of Table 9 in Chapter VIII, and they are discussed at the end of section 17 in that chapter.

The relative values of the six sizes are defined by the percentage values given in Columns 3 and 5 of Table 9 in Chapter VIII, and are illustrated by Figs. 17 and 18. It should not be forgotten that these percentages define the relative values of the coals on the tender —not at the mine.

At the prices which prevailed when the tests were made, both sizes of screenings were slightly more economical than the mine run coal. Among the four larger grades the mine run was much more economical than either the egg, nut, or lump coals. Averaging the results at both rates of evaporation, the price differential between 2-inch screenings and 1¼-inch screenings was just offset by the superior performance of the former.

Except as regards the lump coal at the high rate of evaporation and the four larger grades at the medium rate, the heat lost in the cinders accounts almost entirely for the differences in performance among the various grades. These losses are shown in Figs. 21 and 22 and they are discussed in sections 18 and 19. For the Screenings they varied during the medium rate tests from 7.1 to 11.2 per cent, and in the high rate tests from 13.2 to 15.7 per cent. Among the four larger sizes the heat lost in the cinders varied during the medium rate tests from 1.3 to 2.2 per cent, and in the high rate tests from 5.2 to 8.4 per cent.

Inspection of Figs. 21 and 22 reveals the fact that, despite greater heat loss in the cinders, mine run coal at the medium rate of evaporation had a higher boiler efficiency than either the egg or the lump; and at the high rate its efficiency was greater than that of the lump, and only 1.3 per cent inferior to that of the egg. It is assumed that this is due to the better combustion of the smaller pieces of coal, which are more numerous in the mine run than in the two other sizes.

The inferiority of the performance of the nut coal at the medium rate was probably due to insufficient draft. Its superior performance at the high rate is considered to be due to its small cinder loss and to the evenness and uniformity of the fire which it was possible to maintain with this grade.

At the high rate of evaporation it was more difficult to handle the fire with lump coal than with mine run; and at both rates the evaporative efficiency of the lump was less than that of mine run. The test results offer, therefore, no support for the popular belief in the superiority of lump coal.

As stated in Chapter III the large lumps in both the mine run and lump coals were broken before firing—the former somewhat the more thoroughly, as is evidenced by the fact that after being thus cracked all of the mine run would pass a 5-inch round opening, whereas only 74 per cent of the lump would pass an opening of this size. As has been stated, the evaporative efficiency of the mine run was greater than that of the lump at both rates of evaporation. Since these two coals were not in other respects identical, the facts cited do not form a conclusive argument for the advantage of breaking the large lumps; but, taken in connection with the firing experience in the laboratory, they do offer support for the opinion expressed by the Fuel Test Committee that the cracking of coal to the point where it will all pass a 5-inch or 6-inch round-hole screen is worth more than it costs at well equipped coal chutes.

APPENDIX I

The Locomotive

The locomotive has been briefly described in the body of this report. For convenience of reference some of the facts there cited are repeated in this appendix which is intended to describe the locomotive in detail.

20. *General Design.*—Baltimore and Ohio locomotive 4846 is of the 2-8-2 type and is shown in general design in Figs. 12, 23, 24, and 25. It was built by the Baldwin Locomotive Works in the summer of 1916. It uses superheated steam at 190-pound boiler pressure, in simple cylinders, 26 inches in diameter by 32 inches stroke. Its principal general dimensions are as follows:—

Weight of locomotive, in working order	284500 lb.
Weight of tender, loaded	180000 lb.
Weight of locomotive and tender, in working order	464500 lb.
Weight on front drivers	55600 lb.
Weight on intermediate drivers	54900 lb.
Weight on main drivers	56200 lb.
Weight on back drivers	55300 lb.
Weight on drivers, total	222000 lb.
Weight on leading truck	19400 lb.
Weight on trailing truck	43100 lb.
Nominal maximum tractive effort	54587 lb.
Driving wheel base	16 ft.- 9 in.
Total wheel base of locomotive	35 ft.- 0 in.
Driving wheel diameter—nominal	64 in.
Driving wheel diameter—actual	63.92 in.
Leading truck wheel diameter	33 in.
Trailing truck wheel diameter	46 in.
Main driving journals	11½ x 21 in.
Other driving journals	9½ x 13 in.
Leading truck journals	6 x 6 in.
Trailing truck journals	8 x 14 in.

21. *The Boiler, Firebox, and Front End.*—The boiler, the general design of which is shown in Figs. 26 and 27, was of the wagon top radial stay type, composed of four ring courses and the back end. The main steam dome was mounted over an opening about 27 inches in diameter and an auxiliary dome was mounted on the back end, about one-third of the length of the firebox back of the flue sheet. Flexible staybolts were used throughout.

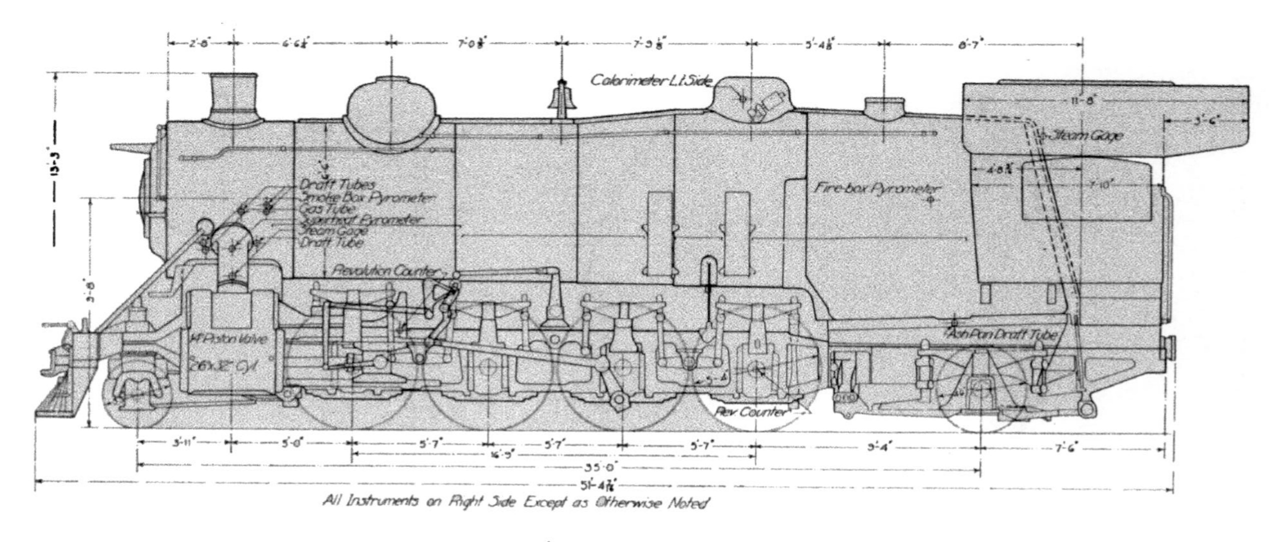

Fig. 23. Side Elevation of Baltimore and Ohio Railroad Locomotive, 4846.

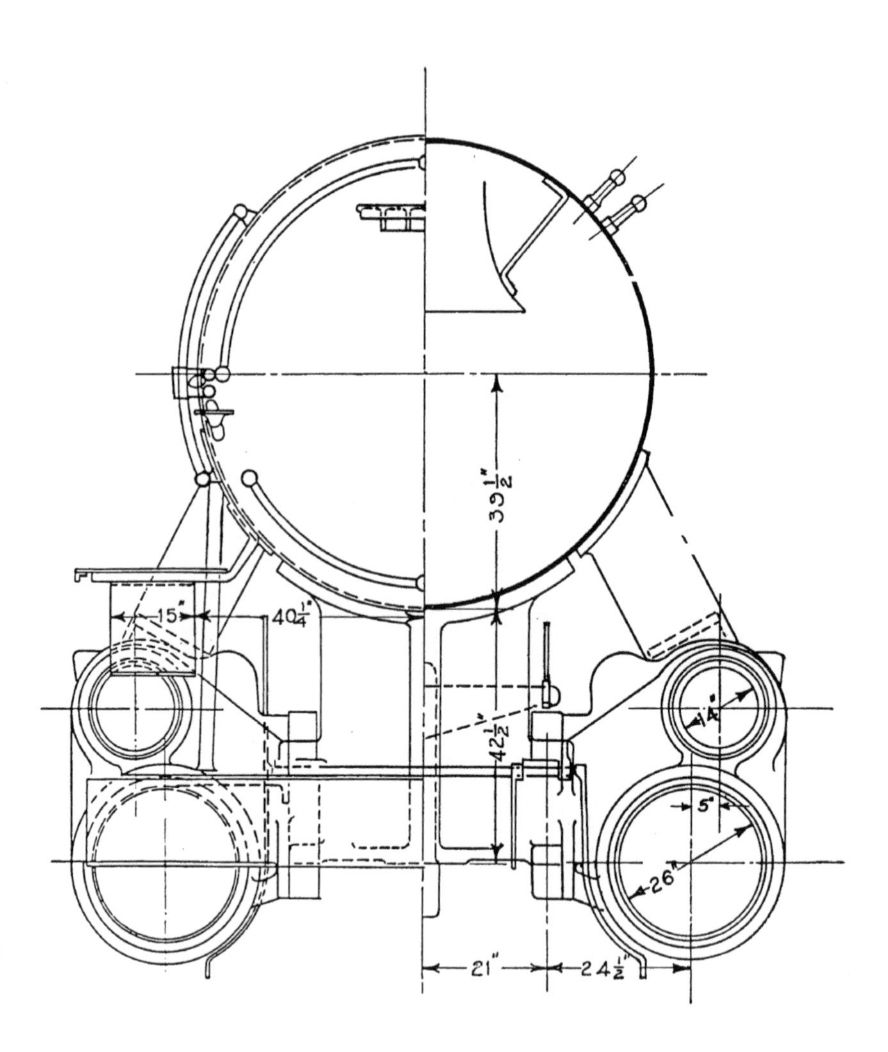

Fig. 24. Partial Front Elevation

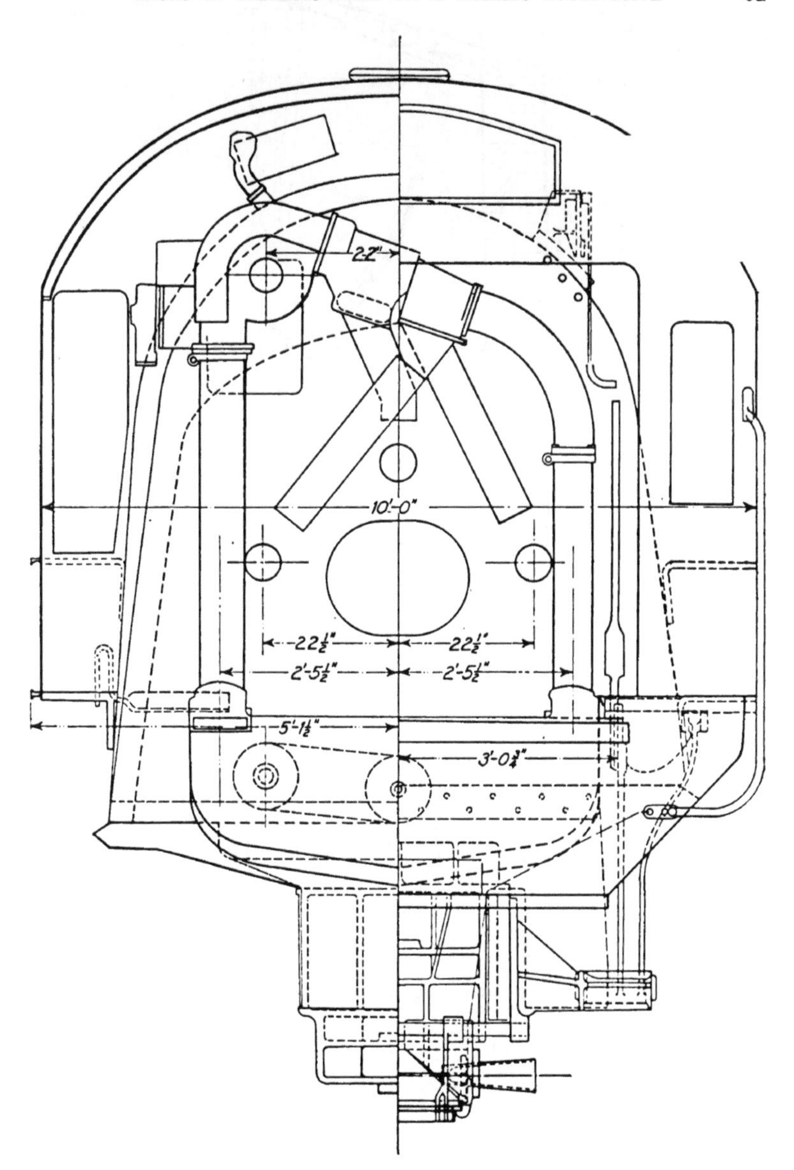

FIG. 25. REAR ELEVATION AND SECTION THROUGH THE CAB

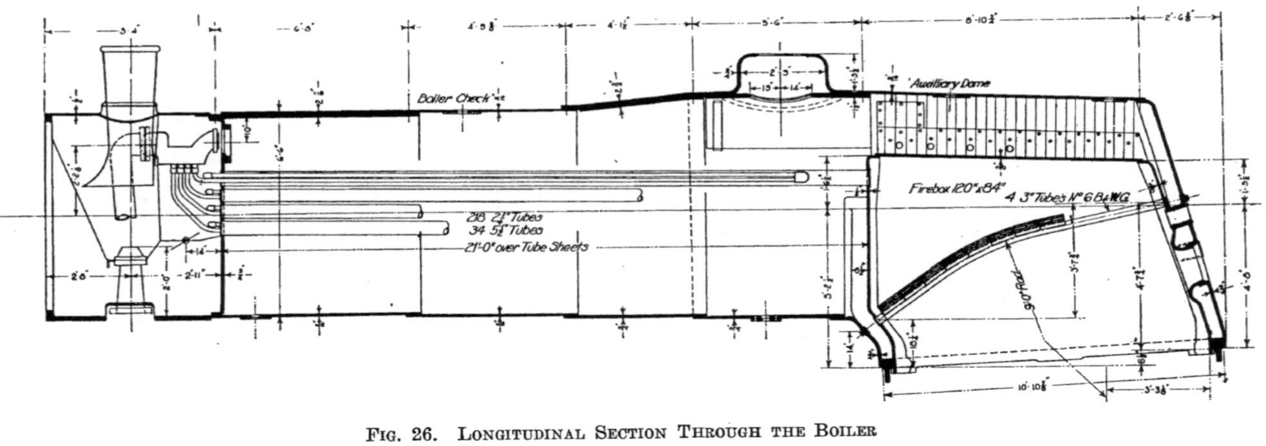

FIG. 26. LONGITUDINAL SECTION THROUGH THE BOILER

The firebox was provided with a "Security" brick arch carried on four 3-inch arch tubes. The grates—shown in Fig. 28—were of the box type with a total area of 69.8 square feet and a total area through the grate openings of 17.0 square feet—24.4 per cent of the grate area. In ordinary operation the firebox was fed by a Street mechanical stoker built by The Locomotive Stoker Company. Its general design is shown in Fig. 25. Three inlets of 5½-inch inside diameter were provided in the back head for the stoker nozzles.

The general design of the front end and the superheater appears in Figs. 26 and 27. While the locomotive was being broken in in service the front-end arrangements shown in the figures were tested by using coal similar to that to be used during the tests and were found to be satisfactory. They were not modified during the progress of the tests. The superheater was of the Schmidt top-header type and consisted of 34 elements. The principal boiler dimensions appear in the following list:

Outside diameter of first ring	78 in.
Cylindrical courses, thickness of sheet	$\frac{11}{16}$ and ¾ in.
Wrapper sheet, thickness	$\frac{9}{16}$ in.
Back flue sheet, thickness	½ in.
Front flue sheet, thickness	¾ in.
Firebox sheets, thickness	⅜ in.
Number of 2¼-inch tubes	218
Number of 5½-inch tubes	34
Number of 3-inch arch tubes	4
Length between tube sheets	21 ft.
Water space in the boiler	547.3 cu. ft.
Steam space in the boiler	144.7 cu. ft.
Heating surface of 2¼-inch tubes, fireside	2,410.0 sq. ft.
Heating surface of 5½-inch tubes, fireside	972.8 sq. ft.
Heating surface of 3-inch tubes, fireside	31.4 sq. ft.
Heating surface of front tube sheet, fireside	15.3 sq. ft.
Heating surface of firebox, fireside	200.4 sq. ft.
Total water heating surface, fireside	3,630.0 sq. ft.
Superheating surface, fireside	1,030.0 sq. ft.
Total water and superheating surface, fireside	4,660.0 sq. ft.
Number of superheater tubes	136
Outside diameter of superheater tubes	1$\frac{7}{16}$ in.
Total length of superheater tubes	2,733.5 ft.
Length of firebox, inside	120 in.
Width of firebox, inside	84 in.
Depth of firebox, at front	81 in.
Depth of firebox, at back	71½ in.
Volume of firebox	348.6 cu. ft.
Grate area	69.8 sq. ft.
Exhaust nozzle, tip diameter	6 in.

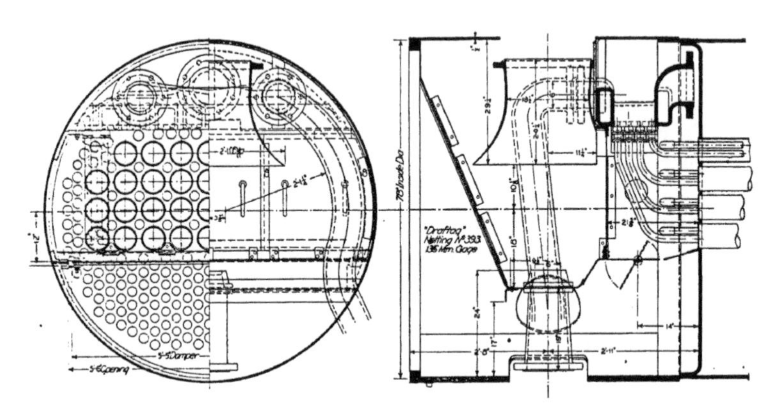

FIG. 27. THE FRONT-END ARRANGEMENT AND THE SUPERHEATER

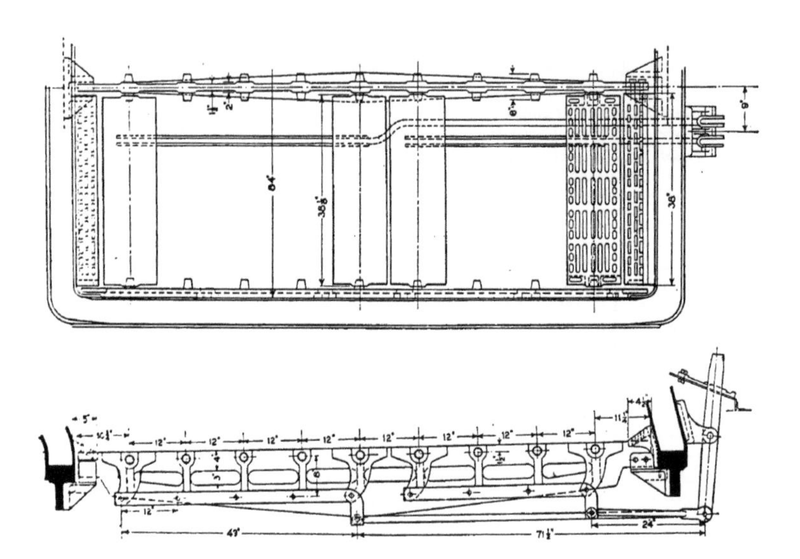

FIG. 28. THE GRATES

22. *The Cylinders and the Valves.*—The arrangement of the cylinders and the valves is shown in Fig. 24. The valves were driven by a Baker-Pilliod gear. The following list presents the principal cylinder and valve dimensions together with data useful in interpreting the indicator diagrams:

Cylinder diameter, right side	25.771 in.
Cylinder diameter, left side	25.767 in.
Valve chamber diameter, right side	14.0 in.
Valve chamber diameter, left side	14.0 in.
Stroke of piston, both sides	32.0 in.
Piston rod diameter, both sides	4.0 in.
Piston displacements:	
Right side, head end	9.660 cu. ft.
Right side, crank end	9.427 cu. ft.
Left side, head end	9.657 cu. ft.
Left side, crank end	9.424 cu. ft.

Clearance volumes—per cent of piston displacement:

Right side, head end	11.0 per cent
Right side, crank end	11.5 per cent
Left side, head end	11.1 per cent
Left side, crank end	11.4 per cent

APPENDIX II

Test Methods and Calculations

The test methods employed were, in general, those outlined in the "Method of Conducting Locomotive and Road Tests" as published in the Proceedings of the American Railway Master Mechanics' Association, Volume 47, page 538.

All tests were run under one of two sets of conditions as to speed and cut-off: The "Medium Rate" tests at a speed of 100 revolutions per minute and at 33 per cent cut-off, and the "High Rate" test at a speed of 135 revolutions per minute and at 55 per cent cut-off. The test methods employed were the same for all tests, and throughout each test all conditions subject to control were maintained as nearly constant as possible. The graphical logs in Appendix III show to what extent uniformity of test conditions was obtained during tests 2416 and 2405, and these logs may be taken as fairly representative of test conditions for all of the tests.

All instruments were known to be correct within reasonable limits or were calibrated at intervals and suitable corrections applied to the observed data. Observations were in general taken every ten minutes. Indicator diagrams were taken from each end of both cylinders at intervals varying from ten minutes on some tests to forty minutes on other tests. Owing to the uniformity of test conditions and to the fact that only two sets of conditions as to speed and cut-off were employed, the taking of indicator diagrams more frequently was unnecessary. The locations of the more important instruments and apparatus are indicated in the figures in Appendix I.

23. *Duration of Tests.*—The tests varied in length from 58 minutes for test 2435 to 6 hours and 17 minutes for test 2401. The general test program contemplated one medium and one high rate test for each size of coal during which approximately ten tons of coal should be burned per test; and two medium and two high rate tests for each grade of coal during which approximately 6⅔ tons of coal should be burned per test. An examination of the data shows that during five tests ten tons or more of coal were burned per test; that during 30 tests the amount of coal per test varied from 4 to 9 tons; and that in one test only 2 tons of coal were burned. For the entire 36 tests the average amount of coal burned was 7 tons per test. As an average therefore about 200 pounds of coal per square foot of grate were burned per test.

24. *Beginning and Closing a Test.*—Fires were built upon a clean grate for each test. With sufficient steam pressure, the locomotive was started and was gradually brought to the required conditions of speed and cut-off. It was then operated for a short time under the required conditions and until a satisfactory fire and a satisfactory boiler pressure were being maintained. On signal the ash pan and cinder separator were closed, observations of water levels and steam pressure were made, and the test was begun. In closing a test simultaneous observations were made upon water levels, steam pressure and condition of fire. The locomotive was then stopped as quickly as conditions warranted. As soon as possible after stopping, ashes were removed from ash pan, and cinders from the cinder separator.

In all cases it was endeavored to have the same amount of combustible matter upon the grate at the close as at the start of the test. The removal of ash from the fire in connection with the closing of the test was primarily for the purpose of judging the amount of combustible upon the grate and not for the purpose of collecting ash. The endeavor was made to have the boiler pressure and the water level in the boiler substantially the same at the close as at the beginning of the test. Corrections were made for such irregularities as occurred.

25. *Temperatures, Pressures, etc.*—Temperatures in the firebox were measured by a platinum and platinum-rhodium thermocouple; and front-end and superheated steam temperatures by copper and copper-constantan couples. Mercury thermometers were used at other points where temperature observations were made.

Boiler pressure observations were taken from a gauge located in the engine cab. Draft pressures were measured by means of "U" tubes with water or with differential draft gauges. Quality of steam was determined by means of a throttling calorimeter fitted with a suitable sampling tube. During portions of a few tests the moisture in the steam was so great that it could not be measured by means of the throttling calorimeter. Speed was measured by means of a stroke counter.

26. *Flue Gas Sampling and Analysis.*—Front-end gas samples were collected through a sampling pipe provided with numerous small holes along the pipe through which the gas was drawn. The time during which a single sample was collected varied from 20 to 60 minutes, depending mainly upon the total length of the test. The taking of samples covered in general the entire time of the test. All samples were collected over mercury and analyzed immediately after collection. The apparatus used for the analysis of the flue gases was

Burrell and Seibert's modification of Haldane's apparatus. The accuracy of this apparatus is sufficient to distinguish 0.01 per cent of carbon monoxide, of methane, or of hydrogen. In the present work, CO_2 percentages were checked to 0.02 per cent and unusual care was taken both in the collection of samples and in the analysis in order that reliable data might be secured regarding the percentages of carbon monoxide, of methane, and of hydrogen.

The tabulated data relating to the composition of the flue gases, as well as the heat losses due to methane and to hydrogen, indicate that under ordinary conditions very little of the original heat of the coal is lost because of the presence of these gases and that only a small error will be made if the volume of these gases which is present be treated as carbon monoxide instead of as methane and hydrogen.

27. *Samples of Coal, Ash, and Cinders for Chemical Analysis.*—Following the close of a test, the ashes collected in the ash pan and the cinders collected in the cinder separator were weighed and sampled. Samples weighing from 50 to 150 pounds were collected as the ash and cinders were being weighed, a small amount being taken from each barrow load after passing over the scales.

Ninety-five per cent or more of each cinder sample being smaller than $\frac{1}{16}$ inch, the large sample was thoroughly mixed and reduced by "quartering" to a five-pound sample. The ashes were mixed and crushed to ¼-inch size and reduced to a five-pound sample by "quartering."

The general practice of sampling the coal for chemical analysis was that outlined in the 1915 Year Book of the American Society for Testing Materials. During each test, as the coal was loaded from the bins into the wagons to be transferred to the firing platform, amounts weighing approximately 15 pounds (one scoopful) were placed in sampling cans. The number of these portions was so proportioned that a total sample of 1,000 pounds would be collected from the total amount of coal fired during one test. In the case of the 1¼-inch and the 2-inch screenings, because of their general uniformity and thorough mixture resulting from the process of screening and loading, the number of scoops of sample coal was so proportioned to the gross amount of coal burned that total samples weighing approximately 500 pounds instead of 1,000 pounds were collected. For test 2435 a sample of only 200 pounds was collected. For all other tests the samples weighed 500 pounds or more. The average weight of all samples collected for the grades larger than screenings was 885 pounds.

Special care was exercised to ensure that the coal selected for samples was in all respects representative of the coal being fired. In general, samples for the chemical laboratory were prepared from the large samples immediately after collection. The samples were prepared largely by mechanical means which produced results equivalent to the hand method described in the year book of the American Society for Testing Materials. The entire sample was crushed by rolls to less than 1-inch size, then mixed by "coning" and reduced by "long pile" mixing and "quartering" to from 125 to 250 pounds. This amount was then pulverized and, through quartering, was reduced to a five-pound sample.

The five-pound samples of coal, ash, and cinders were submitted to the chemical laboratory for analysis.

28. *Chemical Analysis of Coal, Ash, and Cinders.*—The chemical analysis and heat determinations were made by the United States Bureau of Mines at the Experiment Station Laboratory, Pittsburgh, Pa. The methods of analysis and details of the apparatus used by the Bureau of Mines in analyzing coal are fully described in Technical Paper 8 issued by the bureau in June, 1913, and all samples of coal, ash and stack cinders were analyzed in accordance with those methods.

Proximate analyses and direct B. t. u. determinations were made for the coal sample for each test. One ultimate analysis was made for each size of coal tested. The ultimate analyses were made from composite samples. Each composite sample was made by combining equal parts by weight from the air-dried samples representing the tests for each grade of coal. The ultimate analyses for the individual tests which appear in the report are based upon the percentages of moisture, ash, and sulphur as determined by the proximate analysis; and upon the assumption that the percentages of carbon, hydrogen, oxygen, and nitrogen as determined for the individual tests are proportional to the percentages of carbon, hydrogen, oxygen, and nitrogen as determined for the composite samples of that size by ultimate analysis.

Proximate analyses and direct B. t. u. determinations of the cinder samples were made for each test. Proximate analyses were made of the ash sample for each test and direct B. t. u. determinations were made for each ash sample for tests 2400 to 2427 inclusive. For the ash samples subsequent to test 2427, the B. t. u. values were calculated from an average B. t. u. value for one pound of moisture-free and ash-free content of the ash samples. The average moisture-free and ash-free B. t. u. value for all ash samples 2400 to 2424 inclusive, is 14,148 B. t. u. per pound, and all ash samples subsequent to test 2427 have B. t. u. values dependent upon this average value and proportional to the moisture-free and ash-free content of the individual samples.

29. *Samples of Coal for Mechanical Analysis.*—From each car of coal delivered, a sample was taken for mechanical analysis to determine the grade percentages in each size of coal. All samples were collected in uniform manner, the handling from car to separating screens being such that approximately the same amount of incidental breakage took place as occurred when the regular firing coal was transferred from the cars to the firing platform. As each car of the run of mine and the 2-inch lump coal was unloaded, every twentieth scoopful and every twentieth lump unloaded by hand were set aside. In the case of the other coals, which contained no large lumps, every fifteenth scoopful was set aside. The weight of each sample collected was about five per cent of the weight of the coal in each car.

30. *Smoke Records.*—The Ringelmann scale was used in making the smoke observations. Nos. 1, 2, 3, 4, and 5 of the Ringelmann chart represent respectively, 20, 40, 60, 80 and 100 per cent of black smoke. Owing to the large amount of steam escaping with the stack gases, changes in temperature and light greatly affect the appearance of the smoke as regards its apparent blackness. Due to these and other causes which affect the value of observations of this kind, the tabulated results regarding blackness of smoke should be accepted as only approximately correct.

31. *Methods of Calculation.*—The methods used in determining the calculated results are in general similar to the detailed methods of calculation published in Bulletin No. 82, University of Illinois, Engineering Experiment Station.

The calculations relating to heat losses due to the presence of hydrogen and methane in the escaping gases were based upon the determination of the amounts of these gases present and upon heat values of 62100 and 23842 B. t. u. per pound for hydrogen and methane, respectively.

The steam tables of G. A. Goodenough, presented in "Properties of Steam and Ammonia," have been used in all calculations pertaining to the properties of steam.

Certain methods of calculation relating to the determination of the amount of superheated steam produced and the amount used by the engine are as follows:

Item 409. Degrees of Superheat
(Branch-pipe Temperature)—
(Temperature of Saturated Steam at Branch-pipe pressure).

Item 644. Factor of Evaporation

$$\frac{Hs - h}{971.7}$$

H^s = Total heat of steam at branch-pipe pressure.
h = Heat of liquid due to feed water temperature.

Heat Transfer Across Water Heating Surface per Minute, B. t. u.

Item 633 $\times (q + xr - h) \div 60$
$q + xr - h$ = the heat added to each pound of water evaporated by the boiler exclusive of the superheater.

Heat Transfer Across Superheater Heating Surface per Minute, B. t. u.

(Pounds of steam to superheater per minute) $\times$ $(Hs - q - xr)$

$Hs - q - xr$ = the heat added to each pound of steam passing through the superheater.

Item 645. Equivalent Evaporation per Hour, Pounds.

[(Heat transfer per hour across water HS) + (Heat transfer per hour across Superheater *HS*)] ÷ 971.7

Superheated Steam per Hours, Pounds.
Item 645 ÷ Item 644

Superheated Steam to Engine per Hour, Pounds.
(Superheated steam per hour) —
(Superheated steam loss per hour due to Calorimeter leaks, Corrections, etc.)

APPENDIX III

Tabulated Data and Results

The purpose of this appendix is to present, for the sake of those who are interested in the details of the tests, all the data and the results. The appendix consists of sixteen tables and two figures.

Tables 11 to 26, inclusive, contain the results for each of the 36 tests arranged in six groups. Each of the six groups presents the data and the results for a particular size of fuel. Within each group the arrangement is such that the medium rate tests precede the high rate tests. The tests were numbered consecutively in the order in which they were run and their arrangement within the tables is, with few exceptions, also in this order. Under each size of fuel the results of all tests made at a common rate of evaporation have been averaged and these averages appear in the tables in bold face type. The columns headed "Test Number" and "Laboratory Designation" are repeated from table to table to facilitate cross reference. The first term of the column headed "Laboratory Designation" indicates the kind of fuel; the second, the nominal speed in revolutions per minute; and the third, the nominal cut-off in per cent of stroke. The abbreviations used in this column are: M. R. for mine run; 2 in. S. for 2-inch screenings; and 1¼ in. S. for 1¼-inch screenings. The data and the results are presented under 146 column headings. The numbers assigned to these columns are included between 344 and 900 and they appear in the tables in the order of these numbers, which are in general the same as those used in the code for testing locomotives published in the Proceedings of the American Railway Master Mechanics' Association, Vol. 47, p. 538.

In Fig. 29 and Fig. 30 are shown graphical logs for tests 2416 and 2405 respectively, which are fairly typical of all the tests. Test 2416 is a medium rate test, during which 19915 pounds of 3-inch by 6-inch egg coal were fired; whereas No. 2405 is a high rate test during which 20,000 pounds of mine run coal were fired. The lines plotted in these two figures afford a basis for judging of the uniformity of the test conditions which prevailed during these tests.

Table 11
General Conditions

Test Number	Laboratory Designation	Duration of Test, Hours	Speed: Revolutions: Total	Speed: Revolutions: Average per Minute	Speed: Equivalent: Speed in Miles per Hour	Speed: Equivalent: Piston Speed in Feet per Minute	Reverse Lever Notches from Center	Throttle
	Code Item ☞	345	351	352	353	354	360	363
2400	M. R.-100-33	3.62	21576	99.4	18.9	530.1	2	Full
2401	M. R.-100-33	6.28	37439	99.3	18.9	529.6	2	Full
2402	M. R.-100-33	5.20	31101	99.7	19.0	531.7	2	Full
	Average	....	....	**99.5**	**18.9**	**530.5**	....	....
2405	M. R.-135-55	2.72	21807	133.8	25.5	713.6	6	Full
2406	M. R.-135-55	2.68	21687	134.7	25.6	718.4	6	Full
2429	M. R.-135-55	1.92	15529	135.0	25.7	720.0	6	Full
	Average	....	....	**134.5**	**25.6**	**717.3**	....	....
2408	Nut-100-33	3.77	22528	99.7	19.0	531.7	2	Full
2409	Nut-100-33	2.33	13904	99.3	18.9	529.6	2	Full
2410	Nut-100-33	4.33	25886	99.6	19.0	531.2	2	Full
2426	Nut-100-33	4.50	26782	99.2	18.9	529.0	2	Full
	Average	....	....	**99.5**	**19.0**	**530.4**	....	....
2412	Nut-135-55	2.67	21643	135.3	25.8	721.6	6	Full
2413	Nut-135-55	3.00	24285	134.9	25.7	719.4	6	Full
2414	Nut-135-55	2.00	16189	134.9	25.7	719.4	6	Full
	Average	....	....	**135.0**	**25.7**	**720.1**	....	....
2415	Egg-100-33	3.50	20834	99.2	18.9	529.0	2	Full
2416	Egg-100-33	5.83	34822	99.5	18.9	530.6	2	Full
2423	Egg-100-33	4.00	23741	98.9	18.8	527.4	2	Full
	Average	....	....	**99.2**	**18.9**	**529.0**	....	....
2420	Egg-135-55	2.00	16214	135.1	25.7	720.5	6	Full
2422	Egg-135-55	2.17	17650	135.8	25.9	724.2	6	Full
2424	Egg-135-55	1.98	16133	135.6	25.8	723.2	6	Full
	Average	....	....	**135.5**	**25.8**	**722.6**	....	....
2417	Lump-100-33	4.00	23837	99.3	18.9	529.6	2	Full
2418	Lump-100-33	5.83	34894	99.7	19.0	531.7	2	Full
2419	Lump-100-33	3.67	21918	99.6	19.0	531.2	2	Full
	Average	....	....	**99.5**	**19.0**	**530.8**	....	....
2425	Lump-135-55	1.00	8111	135.2	25.7	721.0	6	Full
2427	Lump-135-55	1.50	12194	135.5	25.8	722.7	6	Full
2428	Lump-135-55	1.83	14963	136.0	25.9	725.3	6	Full
2442	Lump-135-55	2.00	16071	133.9	25.5	714.1	6	Full
	Average	....	....	**135.2**	**25.7**	**720.8**	....	....
2430	2 in. S.-100-33	2.62	15671	99.8	19.0	532.2	2	Full
2434	2 in. S.-100-33	3.13	18475	98.3	18.7	524.2	2	Full
2435	2 in. S.-100-33	0.97	5791	99.8	19.0	532.2	2	Full
	Average	....	....	**99.3**	**18.9**	**529.5**	....	....
2436	2 in. S.-135-55	1.35	10870	134.2	25.5	715.7	6	Full
2437	2 in. S.-135-55	1.50	12133	134.8	25.7	718.9	6	Full
	Average	....	....	**134.5**	**25.6**	**717.3**	....	....
2431	1¼ in. S.-100-33	1.77	10603	100.0	19.0	533.3	2	Full
2432	1¼ in. S.-100-33	1.87	11184	99.9	19.0	532.8	2	Full
2433	1¼ in. S.-100-33	3.10	18511	99.5	18.9	530.6	2	Full
	Average	....	....	**99.8**	**19.0**	**532.2**	....	....
2440	1¼ in. S.-135-55	1.50	12063	134.0	25.5	714.6	6	Full
2441	1¼ in. S.-135-55	1.50	12105	134.5	25.6	717.3	6	Full
	Average	....	....	**134.3**	**25.6**	**716.0**	....	....

TABLE 12

TEMPERATURES

Test Number	Laboratory Designation	Temperature, Degrees F.						
		Front-End	Laboratory		Branch Pipe	Feed Water	Fire-Box	Out-Door
			Dry Bulb	Wet Bulb				
	Code Item ☞	367	368	369	370	373	374	....
2400	M. R. -100-33	535	75	71	573	58.9	1735	51
2401	M. R. -100-33	535	73	70	566	56.0	1835	49
2402	M. R. -100-33	539	79	78	564	59.2	1812	62
	Average	**536**	**76**	....	**568**	**58.0**	**1794**	....
2405	M. R.-135-55	627	64	61	628	56.5	2271	35
2406	M. R.-135-55	631	60	58	631	55.6	2334	25
2429	M. R.-135-55	624	48	44	618	55.2	2140	37
	Average	**627**	**57**	....	**626**	**55.8**	**2248**	....
2408	Nut-100-33	595	51	54	589	55.2	2090	12
2409	Nut-100-33	588	54	55	582	55.2	2034	10
2410	Nut-100-33	570	58	56	569	54.1	2008	12
2426	Nut-100-33	555	49	48	572	54.9	1967	23
	Average	**577**	**53**	....	**578**	**54.9**	**2025**	....
2412	Nut-135-55	607	50	55	629	55.7	2293	28
2413	Nut-135-55	611	45	52	632	54.5	2267	18
2414	Nut-135-55	631	51	56	634	55.9	2174	28
	Average	**616**	**49**	....	**632**	**55.4**	**2245**	....
2415	Egg-100-33	543	65	64	576	55.9	1808	44
2416	Egg-100-33	540	62	63	574	56.2	1801	42
2423	Egg-100-33	539	46	47	571	54.8	....	8
	Average	**541**	**58**	....	**574**	**55.6**	**1805**	....
2420	Egg-135-55	588	58	56	610	55.9	2210	42
2422	Egg-135-55	634	56	57	590	54.7	2278	35
2424	Egg-135-55	626	51	53	617	55.0	2183	9
	Average	**616**	**55**	....	**606**	**55.2**	**2224**	....
2417	Lump-100-33	546	59	58	578	56.0	1838	46
2418	Lump-100-33	545	58	58	578	57.0	1857	36
2419	Lump-100-33	553	57	57	578	56.2	1849	36
	Average	**548**	**58**	....	**578**	**56.4**	**1848**	....
2425	Lump-135-55	618	46	47	595	56.3	2178	24
2427	Lump-135-55	625	50	46	578	55.3	2308	26
2428	Lump-135-55	635	43	42	603	54.4	2277	24
2442	Lump-135-55	637	49	42	616	54.4	2192	18
	Average	**629**	**47**	....	**598**	**55.1**	**2239**	....
2430	2 in. S.-100-33	549	56	48	583	56.4	2010	29
2434	2 in. S.-100-33	541	48	46	584	54.8	1817	—2
2435	2 in. S.-100-33	549	62	49	578	57.1	1936	25
	Average	**546**	**55**	....	**582**	**56.1**	**1921**	....
2436	2 in. S.-135-55	631	40	38	634	53.4	2078	5
2437	2 in. S.-135-55	634	44	40	637	54.1	2194	27
	Average	**633**	**42**	....	**636**	**53.8**	**2136**	....
2431	1¼ in. S.-100-33	551	63	59	589	57.0	2003	48
2432	1¼ in. S.-100-33	544	76	66	591	57.6	1798	60
2433	1¼ in. S.-100-33	543	59	54	572	55.3	1874	13
	Average	**546**	**66**	....	**584**	**56.6**	**1892**	....
2440	1¼ in. S.-135-55	634	42	38	604	54.1	2273	9
2441	1¼ in. S.-135-55	639	43	38	629	54.0	2234	10
	Average	**637**	**43**	....	**617**	**54.1**	**2254**	....

TABLE 13

PRESSURES

TEST NUMBER	Laboratory Designation	Pressure—lb. per sq. in. Boiler, Average Gauge	Branch Pipe Average Gauge	Laboratory Barometric	Draft, in. of Water: Front End, Front of Diaphragm	Front End, Back of Diaphragm, Below Damper	Front End, Back of Diaphragm, Above Damper	Fire Box	Ash Pan
	Code Item☞	**380**	**383**	**388**	**394**	**395**	....	**396**	**397**
2400	M. R.-100-33	190.4	179	14.3	2.8	2.3	2.1	1.2	0.2
2401	M. R.-100-33	190.0	172	14.3	2.8	2.4	2.1	1.5	0.2
2402	M. R.-100-33	190.2	174	14.2	3.0	2.5	2.3	1.6	0.2
	Average	**190.2**	**175**	**14.3**	**2.9**	**2.4**	**2.2**	**1.4**	**0.2**
2405	M. R.-135-55	187.8	168	14.3	8.4	6.6	5.9	4.2	0.4
2406	M. R.-135-55	187.8	167	14.2	8.6	6.7	6.0	4.3	0.4
2429	M. R.-135-55	189.4	164	14.3	10.1	7.8	7.0	4.5	0.4
	Average	**188.3**	**166**	**14.3**	**9.0**	**7.0**	**6.3**	**4.3**	**0.4**
2408	Nut-100-33	189.0	178	14.2	3.0	2.4	2.1	1.4	0.2
2409	Nut-100-33	188.5	177	14.3	2.9	2.3	2.0	1.3	0.2
2410	Nut-100-33	186.2	181	14.4	2.9	2.3	2.1	1.4	0.2
2426	Nut-100-33	189.9	182	14.5	3.6	2.8	2.5	1.8	0.1
	Average	**188.4**	**180**	**14.4**	**3.1**	**2.5**	**2.2**	**1.5**	**0.2**
2412	Nut-135-55	187.1	168	14.5	9.2	7.3	6.6	4.6	0.5
2413	Nut-135-55	187.1	168	14.6	9.2	7.2	6.4	4.4	0.5
2414	Nut-135-55	187.5	168	14.5	9.3	7.3	6.5	4.5	0.5
	Average	**187.2**	**168**	**14.5**	**9.2**	**7.3**	**6.5**	**4.5**	**0.5**
2415	Egg-100-33	189.9	180	14.2	3.6	3.1	2.6	1.8	0.2
2416	Egg-100-33	189.5	180	14.2	3.6	3.0	2.6	1.7	0.2
2423	Egg-100-33	190.0	182	14.4	3.3	2.7	2.2	1.4	0.2
	Average	**189.8**	**181**	**14.3**	**3.5**	**2.9**	**2.5**	**1.6**	**0.2**
2420	Egg-135-55	190.1	171	14.2	9.5	7.7	6.7	4.3	0.5
2422	Egg-135-55	189.7	170	14.2	9.2	7.4	6.4	4.0	0.5
2424	Egg-135-55	190.1	170	14.5	9.3	7.3	6.4	4.1	0.5
	Average	**190.0**	**170**	**14.3**	**9.3**	**7.5**	**6.5**	**4.1**	**0.5**
2417	Lump-100-33	189.9	180	14.2	3.6	3.0	2.5	1.7	0.2
2418	Lump-100-33	190.1	180	14.2	3.5	3.0	2.5	1.7	0.2
2419	Lump-100-33	190.0	180	14.4	3.5	3.0	2.5	1.7	0.2
	Average	**190.0**	**180**	**14.3**	**3.5**	**3.0**	**2.5**	**1.7**	**0.2**
2425	Lump-135-55	190.0	168	14.6	9.3	7.5	6.4	4.3	0.5
2427	Lump-135-55	183.4	162	14.4	9.3	7.1	6.3	4.3	0.4
2428	Lump-135-55	186.8	166	14.4	9.4	7.4	6.6	4.4	0.3
2442	Lump-135-55	188.9	166	14.4	10.0	8.0	7.2	4.6	0.5
	Average	**187.3**	**166**	**14.5**	**9.5**	**7.5**	**6.7**	**4.4**	**0.4**
2430	2 in. S.-100-33	188.6	181	14.3	3.8	3.1	3.1	1.9	0.2
2434	2 in. S.-100-33	190.0	181	14.5	3.6	3.0	2.5	....	0.2
2435	2 in. S.-100-33	191.1	182	14.3	3.7	3.1	2.7	2.1	0.2
	Average	**189.9**	**181**	**14.4**	**3.7**	**3.1**	**2.8**	**2.0**	**0.2**
2436	2 in. S.-135-55	185.3	161	14.4	9.4	7.6	6.5	3.7	0.4
2437	2 in. S.-135-55	189.1	162	14.4	9.2	7.4	6.3	3.8	0.3
	Average	**187.2**	**162**	**14.4**	**9.3**	**7.5**	**6.4**	**3.8**	**0.4**
2431	1¼ in. S.-100-33	184.5	178	14.4	3.5	2.9	2.5	1.9	0.2
2432	1¼ in. S.-100-33	189.1	181	14.1	3.6	3.0	2.6	1.1	0.2
2433	1¼ in. S.-100-33	187.7	180	14.4	3.6	3.1	2.6	1.3	0.2
	Average	**187.1**	**180**	**14.3**	**3.6**	**3.0**	**2.6**	**1.4**	**0.2**
2440	1¼ in. S.-135-55	186.6	163	14.4	9.5	7.5	6.6	4.0	0.5
2441	1¼ in. S.-135-55	191.0	166	14.4	9.4	7.5	6.4	3.7	0.5
	Average	**188.8**	**165**	**14.4**	**9.5**	**7.5**	**6.5**	**3.9**	**0.5**

TABLE 14

QUALITY OF STEAM, COAL, CINDERS AND ASH, AND AIR SUPPLY

TEST NUMBER	Laboratory Designation	Quality of Steam in Dome	Degrees of Super-Heat	Factor of Correction for Quality of Steam	Coal Fired Total lb.	Dry Coal Fired Total lb.	Combustible by Analysis Total lb.	Ash by Analysis Total lb.	Stack Cinders Total lb.	Air per lb. of Carbon Consumed lb.	Air per lb. of Coal as Fired lb.
	Code Item ☞	**407**	**409**	**412**	**418**	**419**	**420**	**421**	**423**	......	......
2400	M. R.-100-33	0.9815	194	0.987	11399	10472	9393	1079	360	19.5	12.5
2401	M. R.-100-33	0.9791	190	0.985	20000	18432	16512	1920	592	19.4	12.4
2402	M. R.-100-33	0.9801	187	0.986	17000	15628	13960	1668	529	18.3	11.7
	Average	**0.9802**	**190**	....	**16133**	....	....	....	**494**	**19.1**	**12.2**
2405	M. R.-135-55	0.9628	254	0.973	20000	18348	16346	2002	1926	15.6	8.9
2406	M. R.-135-55	0.9574	257	0.969	18630	17110	15353	1757	1579	15.7	9.2
2429	M. R.-135-55	0.9437	246	0.960	14000	12824	11390	1434	1247	16.6	9.5
	Average	**0.9546**	**252**	....	**17543**	....	....	....	**1584**	**16.0**	**9.2**
2408	Nut-100-33	0.9890	210	0.992	12955	11853	10702	1150	236	17.0	11.2
2409	Nut-100-33	0.9836	204	0.988	7808	7125	6423	702	169	16.7	10.8
2410	Nut-100-33	0.9840	189	0.989	14731	13442	12100	1342	311	16.6	10.8
2426	Nut-100-33	0.9853	192	0.989	16310	15064	13550	1514	482	16.5	10.8
	Average	**0.9855**	**199**	....	**12951**	....	....	....	**300**	**16.7**	**10.9**
2412	Nut-135-55	0.9516	255	0.965	18683	17022	15371	1652	1103	14.2	8.7
2413	Nut-135-55	0.9470	258	0.962	20811	18938	17196	1742	1191	15.4	9.6
2414	Nut-135-55	0.9460	260	0.961	13884	12647	11452	1195	780	15.8	9.8
	Average	**0.9482**	**258**	....	**17793**	....	....	....	**1025**	**15.1**	**9.4**
2415	Egg-100-33	0.9871	197	0.991	11888	10875	9922	953	273	19.7	12.9
2416	Egg-100-33	0.9885	195	0.992	19915	18087	16372	1715	445	20.5	13.3
2423	Egg-100-33	0.9873	191	0.991	13520	12317	11243	1073	281	18.0	11.8
	Average	**0.9876**	**194**	....	**15108**	....	....	....	**333**	**19.4**	**12.7**
2420	Egg-135-55	0.9466	235	0.962	13882	12666	11514	1152	1001	16.1	9.7
2422	Egg-135-55	0.9485	215	0.963	14996	13684	12564	1120	1014	15.3	9.5
2424	Egg-135-55	0.9466	242	0.962	14000	12767	11645	1121	1067	15.4	9.3
	Average	**0.9472**	**231**	....	**14293**	....	....	....	**1027**	**15.6**	**9.5**
2417	Lump-100-33	0.9882	199	0.992	13753	12470	11243	1227	285	20.0	12.6
2418	Lump-100-33	0.9861	199	0.990	20537	18508	16727	1781	396	19.5	12.3
2419	Lump-100-33	0.9852	199	0.989	13344	12060	10661	1400	312	18.8	11.5
	Average	**0.9865**	**199**	....	**15878**	....	....	....	**331**	**19.4**	**12.1**
2425	Lump-135-55	0.9440	221	0.960	7499	6850	6247	603	545	15.1	9.3
2427	Lump-135-55	0.9490	207	0.963	11775	10700	9617	1083	903	14.0	7.9
2428	Lump-135-55	0.9442	230	0.960	14122	12816	11566	1250	958	14.7	8.7
2442	Lump-135-55	0.9439	243	0.960	15279	13902	12481	1421	1189	15.4	9.0
	Average	**0.9453**	**225**	....	**12169**	....	....	....	**899**	**14.8**	**8.7**
2430	2 in. S.-100-33	0.9803	203	0.986	10000	9054	7991	1063	825	17.4	10.3
2434	2 in. S.-100-33	0.9668	204	0.976	11950	10826	9595	1231	1058	18.3	10.7
2435	2 in. S.-100-33	0.9714	198	0.979	3822	3478	3073	404	360	18.1	10.7
	Average	**0.9728**	**202**	....	**8591**	....	....	....	**748**	**17.9**	**10.6**
2436	2 in. S.-135-55	0.9442	263	0.960	11556	10491	9297	1194	1521	17.1	9.4
2437	2 in. S.-135-55	0.9439	265	0.960	13254	12041	10570	1471	1975	15.0	7.9
	Average	**0.9441**	**264**	....	**12405**	....	....	....	**1748**	**16.1**	**9.3**
2431	1¼ in. S.-100-33	0.9794	210	0.985	7635	6998	6281	718	1023	17.1	9.6
2432	1¼ in. S.-100-33	0.9801	211	0.986	7813	7129	6276	853	1029	18.7	10.3
2433	1¼ in. S.-100-33	0.9681	193	0.977	13218	12087	10750	1336	1430	18.5	10.5
	Average	**0.9759**	**205**	....	**9555**	....	....	....	**1161**	**18.1**	**10.1**
2440	1¼ in. S.-135-55	0.9442	232	0.960	14000	12730	11234	1497	2269	14.6	7.5
2411	1¼ in. S.-135-55	0.9437	256	0.960	13750	12275	10663	1612	2185	14.8	7.4
	Average	**0.9440**	**244**	....	**13875**	....	....	....	**2227**	**14.7**	**7.5**

TABLE 15

COAL, CINDERS, ASH, SMOKE AND HUMIDITY

TEST NUMBER	Laboratory Designation	Stack Cinder Loss Per Cent of Total Coal as Fired	Stack Cinder Loss Per Cent of Total Dry Coal Fired	Ash from Ash Pan				Smoke Per Cent of Blackness by Ringelmann Chart	Humidity Moisture per lb. of Dry Air lb.
				Total lb.	Per Cent of Total Dry Coal Fired	Per Cent of Total Coal as Fired	Per Cent of Ash by Analysis		
	Code Item ☞	....	427	428	429	....	430	431	435
2400	M. R.-100-33	3.2	3.4	335	3.2	2.9	31.1	29	0.016
2401	M. R.-100-33	3.0	3.2	1331	7.2	6.7	69.3	31	0.015
2402	M. R.-100-33	3.1	3.4	587	3.8	3.5	35.2	35	0.020
	Average	**3.1**	**3.3**	....	**4.7**	**4.4**	**45.2**	**32**	....
2405	M. R.-135-55	9.6	10.5	1248	6.8	6.2	62.3	59	0.011
2406	M. R.-135-55	8.5	9.2	1272	7.4	6.8	72.4	57	0.010
2429	M. R.-135-55	8.9	9.7	976	7.6	7.0	68.1	38	0.005
	Average	**9.0**	**9.8**	....	**7.3**	**6.7**	**67.6**	**51**	....
2408	Nut-100-33	1.8	2.0	555	4.7	4.3	48.3	41	0.008
2409	Nut-100-33	2.2	2.4	381	5.4	4.9	54.3	37	0.009
2410	Nut-100-33	2.1	2.3	1029	7.7	7.0	76.7	37	0.009
2426	Nut-100-33	3.0	3.2	789	5.2	4.8	52.1	29	0.007
	Average	**2.3**	**2.5**	....	**5.8**	**5.3**	**57.9**	**36**	....
2412	Nut-135-55	5.9	6.5	1037	6.1	5.6	62.8	58	0.008
2413	Nut-135-55	5.7	6.3	677	3.6	3.3	38.9	50	0.006
2414	Nut-135-55	5.6	6.2	865	6.8	6.2	72.4	49	0.008
	Average	**5.7**	**6.3**	....	**5.5**	**5.0**	**58.0**	**52**	....
2415	Egg-100-33	2.3	2.5	690	6.3	5.8	72.4	1	0.012
2416	Egg-100-33	2.2	2.5	1155	6.4	5.8	67.4	3	0.012
2423	Egg-100-33	2.1	2.3	637	5.2	4.7	59.4	18	0.007
	Average	**2.2**	**2.4**	....	**6.0**	**5.4**	**66.4**	**7**	....
2420	Egg-135-55	7.2	7.9	1041	8.2	7.5	90.4	22	0.009
2422	Egg-135-55	6.8	7.4	982	7.2	6.6	87.7	38	0.010
2424	Egg-135-55	7.6	8.4	1009	7.9	7.2	90.0	39	0.008
	Average	**7.2**	**7.9**	....	**7.8**	**7.1**	**89.4**	**33**	....
2417	Lump-100-33	2.1	2.3	1017	8.2	7.4	82.9	3	0.010
2418	Lump-100-33	1.9	2.1	1520	8.2	7.4	85.4	3	0.010
2419	Lump-100-33	2.3	2.6	1185	9.8	8.9	84.6	7	0.010
	Average	**2.1**	**2.3**	....	**8.7**	**7.9**	**84.3**	**4**	......
2425	Lump-135-55	7.3	8.0	199	2.9	2.7	33.0	45	0.006
2427	Lump-135-55	7.7	8.4	1349	12.6	11.5	124.6	53	0.006
2428	Lump-135-55	6.8	7.5	1008	7.9	7.1	80.6	....	0.005
2442	Lump-135-55	7.8	8.6	1319	9.5	8.6	92.8	48	0.004
	Average	**7.4**	**8.1**	....	**8.2**	**7.7**	**82.8**	**49**	....
2430	2 in. S.-100-33	8.3	9.1	388	4.3	3.9	36.5	22	0.005
2434	2 in. S.-100-33	8.9	9.8	578	5.3	4.8	47.0	23	0.006
2435	2 in. S.-100-33	9.4	10.4	123	3.5	3.2	30.5	28	0.005
	Average	**8.9**	**9.8**	....	**4.4**	**4.0**	**38.0**	**24**	....
2436	2 in. S.-135-55	13.2	14.5	587	5.6	5.1	49.2	52	0.004
2437	2 in. S.-135-55	14.9	16.4	637	5.3	4.8	43.3	58	0.004
	Average	**14.1**	**15.5**	....	**5.5**	**5.0**	**46.3**	**55**	....
2431	1¼ in. S.-100-33	13.4	14.6	378	5.4	5.0	52.7	27	0.010
2432	1¼ in. S.-100-33	13.2	14.4	330	4.6	4.2	38.7	26	0.011
2433	1¼ in. S.-100-33	10.8	11.8	634	5.3	4.8	47.5	30	0.008
	Average	**12.5**	**13.6**	....	**5.1**	**4.7**	**46.3**	**28**	....
2440	1¼ in. S.-135-55	16.2	17.8	740	5.8	5.3	49.4	52	0.004
2441	1¼ in. S.-135-55	15.9	17.8	453	3.7	3.3	28.1	62	0.004
	Average	**16.1**	**17.8**	....	**4.8**	**4.3**	**38.8**	**57**	....

TABLE 16

COAL ANALYSIS

Test Number	Laboratory Designation	Proximate Analysis—Coal as Fired					Calorific Value per lb. of Coal as Fired B.t.u.	Ultimate Analysis Coal as Fired			
		Fixed Carbon, Per Cent	Volatile Matter, Per Cent	Moisture, Per Cent	Ash, Per Cent	Sulphur Separately Determined, Per Cent		Carbon, Per Cent	Hydrogen, Per Cent	Nitrogen, Per Cent	Oxygen, Per Cent
	Code Item ☞	437	438	440	441	442	443	449	450	451	452
2400	M. R.-100-33	48.08	34.32	8.13	9.47	0.93	11929	66.90	4.29	1.56	8.73
2401	M. R.-100-33	48.51	34.05	7.84	9.60	0.95	11992	67.01	4.30	1.56	8.74
2402	M. R.-100-33	46.96	35.16	8.07	9.81	0.98	11885	66.62	4.28	1.55	8.69
	Average	**47.85**	**34.51**	**8.01**	**9.63**	**0.95**	**11935**	**66.84**	**4.29**	**1.56**	**8.72**
2405	M. R.-135-55	47.83	33.90	8.26	10.01	0.81	11752	66.44	4.26	1.55	8.67
2406	M. R.-135-55	48.56	33.85	8.16	9.43	0.92	11876	66.91	4.29	1.56	8.73
2429	M. R.-135-55	47.55	33.81	8.40	10.24	1.09	11806	65.91	4.23	1.53	8.60
	Average	**47.98**	**33.85**	**8.27**	**9.89**	**0.94**	**11811**	**66.42**	**4.26**	**1.55**	**8.67**
2408	Nut-100-33	47.10	35.51	8.51	8.88	0.83	12002	67.60	4.37	1.38	8.43
2409	Nut-100-33	47.14	35.12	8.75	8.99	1.01	11956	67.16	4.34	1.37	8.38
2410	Nut-100-33	46.50	35.91	8.48	9.11	0.85	11918	67.42	4.36	1.38	8.41
2426	Nut-100-33	48.69	34.39	7.64	9.28	0.82	11992	68.00	4.39	1.39	8.48
	Average	**47.36**	**34.23**	**8.35**	**9.07**	**0.88**	**11967**	**67.55**	**4.37**	**1.38**	**8.43**
2412	Nut-135-55	48.28	33.99	8.89	8.84	0.86	11918	67.29	4.35	1.38	8.39
2413	Nut-135-55	48.01	34.62	9.00	8.37	0.92	11990	67.54	4.36	1.38	8.42
2414	Nut-135-55	48.20	34.28	8.91	8.61	0.87	11923	67.46	4.36	1.38	8.41
	Average	**48.16**	**34.30**	**8.93**	**8.61**	**0.88**	**11944**	**67.43**	**4.36**	**1.38**	**8.41**
2415	Egg-100-33	47.89	35.57	8.52	8.02	1.29	12143	68.18	4.49	1.50	7.99
2416	Egg-100-33	48.34	33.87	9.18	8.61	0.93	11918	67.45	4.45	1.49	7.90
2423	Egg-100-33	48.39	34.77	8.90	7.94	0.73	12100	68.40	4.51	1.51	8.01
	Average	**48.21**	**34.74**	**8.87**	**8.19**	**0.98**	**12054**	**68.01**	**4.48**	**1.50**	**7.97**
2420	Egg-135-55	48.65	34.29	8.76	8.30	0.86	12042	68.11	4.49	1.50	7.98
2422	Egg-135-55	49.24	34.54	8.75	7.47	0.92	12175	68.76	4.53	1.52	8.05
2424	Egg-135-55	48.83	34.35	8.81	8.01	0.91	12049	68.27	4.50	1.51	8.00
	Average	**48.91**	**34.39**	**8.77**	**7.93**	**0.90**	**12089**	**68.38**	**4.51**	**1.51**	**8.01**
2417	Lump-100-33	46.84	34.91	9.33	8.92	1.06	11826	66.26	4.22	1.48	8.72
2418	Lump-100-33	46.36	35.09	9.88	8.67	0.83	11794	66.21	4.22	1.48	8.72
2419	Lump-100-33	46.54	33.35	9.62	10.49	1.00	11601	64.78	4.13	1.45	8.53
	Average	**46.58**	**34.45**	**9.61**	**9.36**	**0.96**	**11740**	**65.75**	**4.19**	**1.47**	**8.66**
2425	Lump-135-55	48.06	35.24	8.66	8.04	0.93	12062	67.64	4.31	1.52	8.90
2427	Lump-135-55	47.66	34.01	9.13	9.20	0.89	11776	66.34	4.22	1.49	8.73
2428	Lump-135-55	46.38	35.52	9.25	8.85	0.82	11853	66.58	4.24	1.49	8.76
2442	Lump-135-55	48.58	33.11	9.01	9.30	0.64	11806	66.56	4.24	1.49	8.76
	Average	**47.67**	**34.47**	**9.01**	**8.85**	**0.82**	**11874**	**66.78**	**4.25**	**1.50**	**8.79**
2430	2 in. S.-100-33	47.98	31.93	9.46	10.63	0.84	11542	65.54	4.42	1.47	7.64
2434	2 in. S.-100-33	48.12	32.17	9.41	10.30	0.89	11579	65.81	4.44	1.48	7.67
2435	2 in. S.-100-33	48.33	32.08	9.01	10.58	0.74	11565	66.04	4.45	1.48	7.70
	Average	**48.14**	**32.06**	**9.29**	**10.50**	**0.82**	**11562**	**65.80**	**4.44**	**1.48**	**7.67**
2436	2 in. S.-135-55	48.70	31.75	9.22	10.33	0.79	11592	66.03	4.45	1.48	7.70
2437	2 in. S.-135-55	47.45	32.30	9.15	11.10	0.97	11470	65.30	4.40	1.47	7.61
	Average	**48.05**	**32.03**	**9.19**	**10.72**	**0.88**	**11531**	**65.67**	**4.43**	**1.48**	**7.66**
2431	1¼in.S.-100-33	48.65	33.61	8.34	9.40	0.99	11851	67.06	4.45	1.46	8.30
2432	1¼in.S.-100-33	48.13	32.20	8.75	10.92	1.07	11543	65.40	4.34	1.43	8.09
2433	1¼in.S.-100-33	48.49	32.84	8.56	10.11	0.90	11698	66.36	4.41	1.45	8.21
	Average	**48.42**	**32.88**	**8.55**	**10.14**	**0.99**	**11697**	**66.27**	**4.40**	**1.45**	**8.20**
2440	1¼in.S.-135-55	47.85	32.39	9.07	10.69	0.97	11542	65.41	4.34	1.43	8.09
2441	1¼in.S.-135-55	46.91	30.64	10.73	11.72	0.91	11151	63.24	4.20	1.38	7.82
	Average	**47.38**	**31.52**	**9.90**	**11.21**	**0.94**	**11347**	**64.33**	**4.27**	**1.41**	**7.96**

TABLE 17

CALORIFIC VALUE OF COAL AND CINDERS, ANALYSIS OF FRONT-END GASES

Test Number	Laboratory Designation	Calorific Value B.t.u. per lb.				Analysis of Front End Gases—per cent					
		Dry Coal	Combustible	Stack Cinders	Ash	Oxygen O_2	Carbon Monoxide CO	Carbon Dioxide CO_2	Nitrogen N_2	Hydrogen H_2	Methane CH_4
	Code Item ☞	**458**	**459**	**462**	**463**	**466**	**467**	**468**	**469**	**470**	**471**
2400	M. R.-100-33	12983	14476	8399	3488	6.47	0.095	12.52	80.95	0.020	0.005
2401	M. R.-100-33	13012	14526	8563	3141	6.32	0.148	12.53	81.02	0.009	0.003
2402	M. R.-100-33	12929	14474	8570	2695	5.55	0.220	13.21	81.02	0.000	0.015
	Average	**12975**	....	**8511**	**3108**	**6.11**	**0.154**	**12.75**	**81.00**	**0.010**	**0.008**
2405	M. R.-135-55	12811	14380	11081	3935	2.79	0.601	15.19	81.26	0.093	0.068
2406	M. R.-135-55	12933	14414	11030	3852	2.15	1.100	14.71	82.00	0.020	0.020
2429	M. R.-135-55	12888	14510	10921	4410	3.70	0.307	14.54	81.38	0.033	0.033
	Average	**12877**	....	**11011**	**4066**	**2.88**	**0.669**	**14.81**	**81.55**	**0.049**	**0.040**
2408	Nut-100-33	13118	14528	8023	2633	4.59	0.440	13.98	80.89	0.070	0.035
2409	Nut-100-33	13102	14535	7585	3204	4.28	0.476	14.22	80.92	0.047	0.053
2410	Nut-100-33	13023	14461	8231	2187	4.11	0.376	14.45	80.97	0.058	0.034
2426	Nut-100-33	12983	14434	8458	1985	4.03	0.420	14.44	80.99	0.087	0.033
	Average	**13057**	....	**8074**	**2502**	**4.25**	**0.428**	**14.27**	**80.94**	**0.066**	**0.039**
2412	Nut-135-55	13081	14485	10728	3416	1.85	1.575	15.65	80.54	0.180	0.205
2413	Nut-135-55	13176	14512	10822	3184	2.41	0.610	15.34	81.00	0.265	0.190
2414	Nut-135-55	13090	14456	10634	3409	2.09	0.645	15.09	81.94	0.155	0.040
	Average	**13116**	....	**10728**	**3336**	**2.12**	**0.943**	**15.36**	**81.16**	**0.200**	**0.145**
2415	Egg-100-33	13273	14549	7987	3173	6.85	0.200	12.21	80.69	0.028	0.028
2416	Egg-100-33	13122	14495	7999	3357	7.12	0.118	11.85	80.92	0.002	0.000
2423	Egg-100-33	13282	14551	8329	3827	5.09	0.184	13.53	81.18	0.008	0.010
	Average	**13226**	....	**8105**	**3452**	**6.35**	**0.167**	**12.53**	**80.93**	**0.013**	**0.013**
2420	Egg-135-55	13198	14519	10771	4651	2.91	0.390	15.01	81.64	0.035	0.025
2422	Egg-135-55	13345	14535	11234	3343	2.43	0.400	15.71	81.30	0.112	0.040
2424	Egg-135-55	13214	14486	10584	4426	2.65	0.450	15.53	81.27	0.060	0.040
	Average	**13252**	....	**10863**	**4140**	**2.66**	**0.413**	**15.42**	**81.40**	**0.069**	**0.035**
2417	Lump-100-33	13043	14467	7713	3469	6.94	0.112	12.16	80.78	0.002	0.008
2418	Lump-100-33	13086	14479	7574	3598	6.49	0.102	12.50	80.92	0.003	0.002
2419	Lump-100-33	12836	14521	7106	3436	5.88	0.170	12.90	80.98	0.015	0.010
	Average	**12988**	....	**7464**	**3501**	**6.44**	**0.128**	**12.52**	**80.89**	**0.007**	**0.007**
2425	Lump-135-55	13205	14479	10917	3487	2.39	0.695	15.61	81.07	0.140	0.110
2427	Lump-135-55	12958	14418	10849	4869	0.90	1.630	15.96	80.94	0.350	0.203
2428	Lump-135-55	13061	14472	10829	4297	1.57	0.865	15.89	81.48	0.140	0.060
2442	Lump-135-55	12974	14450	10415	4527	2.67	0.343	15.66	81.26	0.038	0.033
	Average	**13050**	....	**10753**	**4295**	**1.88**	**0.883**	**15.78**	**81.19**	**0.167**	**0.102**
2430	2 in. S.-100-33	12748	14443	9407	3674	5.01	0.330	13.74	80.81	0.057	0.043
2434	2 in. S.-100-33	12782	14422	9569	3710	5.40	0.157	13.28	81.14	0.003	0.013
2435	2 in. S.-100-33	12710	14382	9113	3677	5.25	0.280	13.34	81.12	0.010	0.000
	Average	**12747**	....	**9363**	**3687**	**5.22**	**0.256**	**13.45**	**81.02**	**0.023**	**0.019**
2436	2 in. S.-135-55	12769	14409	10611	4343	4.25	0.503	13.88	81.20	0.073	0.073
2437	2 in. S.-135-55	12625	14382	11018	4021	2.08	0.650	15.76	81.33	0.137	0.057
	Average	**12697**	....	**10815**	**4182**	**3.17**	**0.577**	**14.82**	**81.27**	**0.105**	**0.065**
2431	1¼ in. S.-100-33	12929	14407	10505	4109	4.90	0.340	13.97	80.80	0.000	0.000
2432	1¼ in. S.-100-33	12650	14371	10157	4504	5.67	0.227	12.91	81.17	0.017	0.007
2433	1¼ in. S.-100-33	12793	14384	10784	4252	5.50	0.228	13.05	81.19	0.050	0.013
	Average	**12791**	....	**10482**	**4288**	**5.36**	**0.265**	**13.31**	**81.05**	**0.022**	**0.007**
2440	1¼ in. S.-135-55	12692	14384	10870	3838	2.11	1.380	15.37	80.42	0.313	0.443
2441	1¼ in. S.-135-55	12492	14380	11203	4469	1.92	0.760	15.86	81.15	0.177	0.137
	Average	**12592**	....	**11037**	**4154**	**2.02**	**1.070**	**15.62**	**80.79**	**0.245**	**0.290**

TABLE 18

WATER AND DRAWBAR PULL

Test Number	Laboratory Designation	Water						
		Delivered to Boiler by Injectors lb.	Weight of Water in Boiler at Start of Test Minus Weight in Boiler at Close of Test, lb.	Correction for Change of Water Level and Steam Pressure in Boiler, Start to Close, lb.	Loss from Boiler lb.	Loss from Boiler Corrected lb.	Presumably Evaporated lb.	Drawbar Pull lb.
	Code Item☞	**476**	**477**	**478**	**479**	**480**	**481**	**487**
2400	M. R.-100-33	80868	+1110	+ 794	0	0	81662	21970
2401	M. R.-100-33	141151	—1110	— 791	0	0	140360	21727
2402	M. R.-100-33	119334	+ 220	+ 157	0	0	119491	21822
	Average	....	....	....	....	....	**113838**	**21840**
2405	M. R.-135-55	113133	+2060	+1377	0	0	114510	28771
2406	M. R.-135-55	112494	+ 160	+ 114	0	0	112608	28718
2429	M. R.-135-55	80329	+2640	+1860	172	121	82068	28672
	Average	....	....	....	....	....	**103062**	**28720**
2408	Nut-100-33	88009	0	— 110	0	0	87899	22490
2409	Nut-100-33	54574	0	— 22	325	232	54320	22411
2410	Nut-100-33	99995	0	— 43	75	54	99898	22417
2426	Nut-100-33	111980	+ 180	+ 128	495	353	111755	22640
	Average	....	....	....	....	....	**88468**	**22490**
2412	Nut-135-55	112833	+2950	+1856	160	113	114576	28958
2413	Nut-135-55	127417	+1380	+ 973	180	127	128263	29100
2414	Nut-135-55	83950	+ 880	+ 573	120	85	84438	29128
	Average	....	....	....	....	....	**109092**	**29062**
2415	Egg-100-33	84089	0	0	315	225	83864	2284
2416	Egg-100-33	139492	0	— 325	525	375	138792	23115
2423	Egg-100-33	96044	+ 180	+ 128	360	257	95915	22533
	Average	....	....	....	....	....	**106190**	**22829**
2420	Egg-135-55	84550	+2740	+1920	120	85	86385	29046
2422	Egg-135-55	91315	+2420	+1707	130	92	92930	29030
2424	Egg-135-55	83485	+3320	+2330	120	85	85730	29104
	Average	....	....	....	....	....	**88348**	**29060**
2417	Lump-100-33	96335	— 590	— 422	360	257	95656	23026
2418	Lump-100-33	138919	+ 340	+ 220	525	375	138764	23085
2419	Lump-100-33	89522	0	0	330	236	89286	22983
	Average	....	....	....	....	....	**107902**	**23031**
2425	Lump-135-55	41743	+1610	+1136	60	42	42837	28530
2427	Lump-135-55	60757	+3860	+2639	135	96	63300	27909
2428	Lump-135-55	77730	+3130	+2125	165	116	79739	28441
2442	Lump-135-55	89495	+ 740	+ 543	90	63	89975	29266
	Average	....	....	....	....	....	**68963**	**28537**
2430	2 in. S.-100-33	63153	+ 350	+ 293	150	107	63339	22906
2434	2 in. S.-100-33	74750	+1070	+ 759	141	100	75409	23091
2435	2 in. S.-100-33	24008	+ 500	+ 376	45	32	24342	23268
	Average	....	....	....	....	....	**54363**	**23088**
2436	2 in. S.-135-55	55754	+1970	+1392	60	42	57104	27976
2437	2 in. S.-135-55	66936	—1040	— 732	68	48	66156	28938
	Average	....	....	....	....	....	**61630**	**28457**
2431	1¼ in. S.-100-33	43907	+ 360	+ 257	105	75	44089	22332
2432	1¼ in. S.-100-33	44696	+2250	+1585	110	79	46202	22912
2433	1¼ in. S-100-33	76873	+ 940	+ 689	180	128	77434	22588
	Average	....	....	....	....	....	**55908**	**22611**
2440	1¼ in. S.-135-55	64217	+ 980	+ 725	68	48	64894	29061
2441	1¼ in. S.-135-55	65872	+ 300	+ 211	68	48	66035	29392
	Average	....	....	....	....	....	**65465**	**29227**

TABLE 19

BOILER PERFORMANCE —COAL AND EVAPORATION

Test Number	Laboratory Designation	Coal as Fired lb.		Dry Coal Fired—lb.		Evaporation					
							Superheated Steam—lb.				Super-Heated Steam to Engine per Hour lb.
		Per Hour	Per Hour per Sq. Ft. of Grate Surface	Per Hour	Per Hour per Sq. Ft. of Grate Surface	Moist Steam per Hour lb.	Per Hour	Per Hour per Sq. Ft. of Heating Surface	Per lb. of Dry Coal	Per lb. of Coal as Fired	
	Code Item ☞	......	......	626	627	633	......	......	......	......	......
2400	M. R.-100-33	3151	45.1	2895	41.5	22577	22565	4.84	7.79	7.16	22466
2401	M. R.-100-33	3183	45.6	2934	42.0	22340	22328	4.79	7.61	7.01	22257
2402	M. R.-100-33	3269	46.8	3005	43.1	22979	22970	4.93	7.64	7.02	22936
	Average	**3201**	**45.8**	**2945**	**42.2**	**22632**	**22621**	**4.85**	**7.68**	**7.06**	**22533**
2405	M. R.-135-55	7361	105.5	6753	96.8	42145	42176	9.05	6.24	5.73	42365
2406	M. R.-135-55	6944	99.5	6377	91.4	41971	41946	9.00	6.58	6.04	41733
2429	M. R.-135-55	7303	104.6	6690	95.9	42811	42854	9.20	6.41	5.87	43059
	Average	**7203**	**103.2**	**6607**	**94.7**	**42309**	**42325**	**9.08**	**6.41**	**5.88**	**42386**
2408	Nut-100-33	3439	49.3	3146	45.1	23334	23327	5.00	7.42	6.78	23195
2409	Nut-100-33	3347	48.0	3054	43.8	23283	23254	4.99	7.61	6.95	23129
2410	Nut-100-33	3400	48.7	3102	44.4	23055	23059	4.95	7.43	6.78	22962
2426	Nut-100-33	3624	51.9	3348	48.0	24834	24808	5.33	7.41	6.85	24588
	Average	**3453**	**49.5**	**3163**	**45.3**	**23627**	**23612**	**5.07**	**7.47**	**6.84**	**23469**
2412	Nut-135-55	7005	100.4	6382	91.4	42960	42964	9.22	6.73	6.13	42951
2413	Nut-135-55	6937	99.4	6313	90.4	42754	42720	9.17	6.77	6.16	42561
2414	Nut-135-55	6942	99.5	6324	90.6	42219	42209	9.06	6.67	6.08	42224
	Average	**6961**	**99.8**	**6340**	**90.8**	**42644**	**42631**	**9.15**	**6.72**	**6.12**	**42579**
2415	Egg-100-33	3397	48.7	3107	44.5	23961	23944	5.14	7.71	7.05	23665
2416	Egg-100-33	3414	48.9	3101	44.4	23795	23788	5.11	7.67	6.97	23651
2423	Egg-100-33	3380	48.4	3079	44.1	23979	23970	5.14	7.78	7.09	23774
	Average	**3397**	**48.7**	**3096**	**44.3**	**23912**	**23901**	**5.13**	**7.72**	**7.04**	**23697**
2420	Egg-135-55	6941	99.4	6333	90.7	43193	43219	9.28	6.82	6.22	43393
2422	Egg-135-55	6920	99.1	6315	90.5	42884	42881	9.20	6.79	6.20	42953
2424	Egg-135-55	7060	101.2	6438	92.2	43233	43302	9.29	6.73	6.13	43581
	Average	**6974**	**99.9**	**6362**	**91.1**	**43103**	**43134**	**9.26**	**6.78**	**6.18**	**43309**
2417	Lump-100-33	3438	49.3	3118	44.7	23914	23906	5.13	7.67	6.96	23714
2418	Lump-100-33	3521	50.5	3173	45.5	23790	23778	5.10	7.49	6.75	23570
2419	Lump-100-33	3639	52.1	3289	47.1	24348	24340	5.23	7.40	6.69	24190
	Average	**3533**	**50.6**	**3193**	**45.8**	**24017**	**24008**	**5.15**	**7.52**	**6.80**	**23825**
2425	Lump-135-55	7499	107.4	6850	98.1	42837	42889	9.21	6.26	5.72	43160
2427	Lump-135-55	7850	112.5	7133	102.2	42200	42254	9.07	5.92	5.38	42698
2428	Lump-135-55	7704	110.4	6992	100.2	43502	44136	9.47	6.31	5.73	43892
2442	Lump-135-55	7640	109.5	6951	99.6	44988	44910	9.64	6.46	5.88	44441
	Average	**7673**	**110.0**	**6982**	**100.0**	**43382**	**43547**	**9.35**	**6.24**	**5.68**	**43548**
2430	2 in. S.-100-33	3821	54.7	3460	49.6	24203	24122	5.18	6.97	6.31	23516
2434	2 in. S.-100-33	3814	54.6	3455	49.5	24069	24014	5.15	6.95	6.30	23656
2435	2 in. S.-100-33	3952	56.6	3507	51.5	25173	25147	5.40	6.99	6.36	24918
	Average	**3862**	**55.3**	**3504**	**50.2**	**24482**	**24428**	**5.24**	**6.97**	**6.32**	**24030**
2436	2 in. S.-135-55	8560	122.6	7771	111.3	42299	42287	9.07	5.45	4.94	42232
2437	2 in. S.-135-55	8836	126.6	8027	115.0	44104	43981	9.44	5.48	4.98	43376
	Average	**8698**	**124.6**	**7899**	**113.2**	**43202**	**43134**	**9.26**	**5.47**	**4.96**	**42804**
2431	1¼in.S.-100-33	4321	61.9	3960	56.7	24951	24907	5.34	6.29	5.76	24370
2432	1¼in.S.-100-33	4185	60.0	3818	54.7	24747	24714	5.30	6.47	5.90	24469
2433	1¼in.S.-100-33	4264	61.1	3899	55.9	24979	24933	5.35	6.39	5.85	24514
	Average	**4257**	**61.0**	**3892**	**55.8**	**24892**	**24851**	**5.33**	**6.38**	**5.84**	**24451**
2440	1¼in.S.-135-55	9333	133.7	8487	121.6	43263	43186	9.27	5.09	4.63	42746
2441	1¼in.S.-135-55	9167	131.3	8183	117.2	44024	43941	9.43	5.37	4.80	43478
	Average	**9250**	**132.5**	**8335**	**119.5**	**43644**	**43564**	**9.35**	**5.23**	**4.72**	**43112**

TABLE 20

BOILER PERFORMANCE — EVAPORATION AND EQUIVALENT EVAPORATION

Test Number	Laboratory Designation	Steam Used at Calorimeter, Safety Valve, Leaks, etc.	Superheated Steam Loss per Hour due to Calorimeter Leaks, Corrections etc. lb.	Dry Coal Loss per Hour Equivalent to Steam Loss lb.	Factor of Evaporation	Equivalent Evaporation from and at 212 Degrees F.—lb.					
						Per Hour	Per Hour, Boiler Ex. cluding Super-Heater	Per Hour, Super-Heater Alone	Per Hour, per Sq. Ft. of Total Heating Surface	Per Hour, per Sq. Ft. of Total Heating Surface Excluding Super-Heater	Per Hour, per Sq. Ft. of Heat-Surface Super-Heater Alone
	Code Item ☞	638	....	643	644	645	646	647	648	649	655
2400	M. R.-100-33	717	+101	+13	1.319	29764	26867	2897	6.39	7.40	2.81
2401	M. R.-100-33	202	+ 78	+10	1.319	29451	26607	2844	6.32	7.33	2.76
2402	M. R.-100-33	288	+ 32	+ 4	1.314	30183	27299	2884	6.48	7.52	2.80
	Average	......	......	......	**1.317**	**29799**	......	......	**6.40**	......	......
2405	M. R.-135-55	87	—191	—31	1.352	57022	49563	7459	12.24	13.65	7.24
2406	M. R.-135-55	685	+199	+30	1.354	56795	49190	7605	12.19	13.55	7.38
2429	M. R.-135-55	426	—212	—33	1.348	57767	49703	8064	12.40	13.69	7.83
	Average	......	......	......	**1.351**	**57195**	......	......	**12.28**	......	......
2408	Nut-100-33	632	+122	+16	1.331	31048	28000	3048	6.66	7.71	2.96
2409	Nut-100-33	613	+128	+17	1.328	30881	27823	3058	6.63	7.66	2.97
2410	Nut-100-33	504	+ 87	+12	1.321	30461	27597	2864	6.54	7.60	2.78
2426	Nut-100-33	1511	+218	+29	1.322	32796	29727	3069	7.04	8.19	2.98
	Average	......	......	......	**1.326**	**31297**	......	......	**6.72**	......	......
2412	Nut-135-55	871	+ 4	+ 1	1.353	58130	50135	7995	12.47	13.81	7.76
2413	Nut-135-55	1115	+154	+23	1.356	57929	49766	8163	12.43	13.71	7.93
2414	Nut-135-55	383	— 8	— 1	1.356	57236	49059	8177	12.28	13.51	7.94
	Average	......	......	......	**1.355**	**57765**	......	......	**12.39**	......	......
2415	Egg-100-33	1259	+270	+35	1.323	31678	28705	2973	6.80	7.91	2.89
2416	Egg-100-33	1537	+136	+18	1.322	31448	28530	2918	6.75	7.86	2.83
2423	Egg-100-33	1130	+184	+24	1.321	31665	28751	2914	6.79	7.92	2.83
	Average	......	......	......	**1.322**	**31597**	......	......	**6.78**	......	......
2420	Egg-135-55	505	—165	—24	1.343	58043	50233	7810	12.46	13.84	7.58
2422	Egg-135-55	657	— 75	—11	1.333	57160	49960	7200	12.27	13.76	6.99
2424	Egg-135-55	384	—292	—43	1.347	58328	50323	8005	12.52	13.86	7.77
	Average	......	......	......	**1.341**	**57844**	......	......	**12.42**	......	......
2417	Lump-100-33	890	+183	+24	1.324	31652	28673	2979	6.79	7.90	2.89
2418	Lump-100-33	1778	+200	+27	1.323	31458	28453	3005	6.75	7.84	2.92
2419	Lump-100-33	816	+145	+20	1.324	32227	29121	3106	6.92	8.02	3.02
	Average	......	......	......	**1.324**	**31779**	......	......	**6.82**	......	......
2425	Lump-135-55	193	—284	—45	1.334	57214	49691	7523	12.28	13.69	7.30
2427	Lump-135-55	570	—436	—74	1.327	56071	49164	6907	12.03	13.54	6.71
2428	Lump-135-55	406	—334	—53	1.341	58403	50549	7854	12.53	13.93	7.63
2442	Lump-135-55	1354	+475	+74	1.348	60539	52276	8263	12.99	14.40	8.02
	Average	......	......	......	**1.338**	**58057**	......	......	**12.46**	......	......
2430	2 in. S.-100-33	1961	+611	+88	1.327	32010	28826	3184	6.87	7.94	3.09
2434	2 in. S.-100-33	1704	+360	+52	1.329	31914	28426	3488	6.85	7.83	3.39
2435	2 in. S.-100-33	412	+226	+32	1.323	33270	29780	3490	7.14	8.20	3.39
	Average	......	......	......	**1.326**	**32398**	......	......	**6.95**	......	......
2436	2 in. S.-135-55	711	+ 62	+11	1.359	57468	49194	8274	12.33	13.55	8.03
2437	2 in. S.-135-55	832	+614	+112	1.360	59814	51249	8565	12.84	14.12	8.32
	Average	......	......	......	**1.360**	**58641**	......	......	**12.59**	......	......
2431	1¼ in. S.-100-33	1206	+526	+84	1.329	33102	29692	3410	7.10	8.18	3.31
2432	1¼ in. S.-100-33	1262	+249	+38	1.330	32870	29448	3422	7.05	8.11	3.32
2433	1¼ in. S.-100-33	1818	+418	+65	1.322	32961	29525	3436	7.07	8.13	3.34
	Average	......	......	......	**1.327**	**32978**	......	......	**7.07**	......	......
2440	1¼ in. S.-135-55	1079	+439	+86	1.342	57956	50272	7684	12.44	13.85	7.46
2441	1¼ in. S.-135-55	955	+457	+85	1.355	59540	51156	8384	12.78	14.09	8.14
	Average	......	......	......	**1.349**	**58748**	......	......	**12.61**	......	......

TABLE 21

BOILER PERFORMANCE—HEAT TRANSFER, EQUIVALENT EVAPORATION, HORSE POWER AND EFFICIENCY

Test Number	Laboratory Designation	Heat Transfer across Water H. S. per Min. B.t.u.	Heat Transfer across Super-Heater H. S. per Min. B.t.u.	Per Cent of Evaporation by		Equivalent Evaporation from and at 212 degrees F.—lb.				Boiler Horse-Power	Efficiency of Boiler Per Cent
				Water Heating Surface	Super-Heating Surface	Per Hour per Sq. Ft. of Grate Area	Per lb. of Coal as Fired	Per lb. of Dry Coal	Per lb. of Combustible		
	Code Item ☞	......	......	......	......	656	657	658	659	660	666
2400	M. R.-100-33	435096	46917	90.3	9.7	426.4	9.44	10.28	11.46	863	76.89
2401	M. R.-100-33	430864	46072	90.3	9.7	421.9	9.25	10.04	11.21	854	74.95
2402	M. R.-100-33	442231	46713	90.4	9.6	432.4	9.23	10.04	11.24	875	75.46
	Average	**436064**	**46567**	......	......	......	**9.31**	**10.12**	......	**864**	**75.77**
2405	M. R.-135-55	802652	120811	86.9	13.1	817.0	7.75	8.44	9.48	1653	64.08
2406	M. R.-135-55	796819	123182	86.6	13.4	813.7	8.18	8.91	9.93	1646	66.93
2429	M. R.-135-55	804918	130612	86.0	14.0	827.6	7.91	8.64	9.72	1674	65.10
	Average	**801463**	**124868**	......	......	......	**7.95**	**8.66**	......	**1658**	**65.37**
2408	Nut-100-33	453535	49367	90.2	9.8	444.8	9.03	9.87	10.93	900	73.11
2409	Nut-100-33	450798	49535	90.1	9.9	442.4	9.23	10.11	11.22	895	75.02
2410	Nut-100-33	446844	46383	90.6	9.4	436.4	8.96	9.82	10.91	883	73.05
2426	Nut-100-33	481531	49709	90.6	9.4	469.9	9.05	9.80	10.89	951	73.33
	Average	**458177**	**48749**	......	......	......	**9.07**	**9.90**	......	**907**	**73.63**
2412	Nut-135-55	812016	129497	86.2	13.8	832.8	8.30	9.11	10.09	1685	67.67
2413	Nut-135-55	806198	132223	85.9	14.1	829.9	8.35	9.18	10.11	1679	67.67
2414	Nut-135-55	794562	132443	85.7	14.3	820.0	8.24	9.05	10.00	1659	67.15
	Average	**804259**	**131388**	......	......	......	**8.30**	**9.11**	......	**1674**	**67.50**
2415	Egg-100-33	464843	48158	90.6	9.4	453.9	9.33	10.20	11.17	918	74.66
2416	Egg-100-33	461940	47263	90.7	9.3	450.6	9.21	10.14	11.20	912	75.09
2423	Egg-100-33	465632	47191	90.8	9.2	453.7	9.37	10.28	11.27	918	75.25
	Average	**464138**	**47537**	......	......	......	**9.30**	**10.21**	......	**916**	**75.00**
2420	Egg-135-55	813324	126491	86.5	13.5	831.6	8.36	9.16	10.08	1682	67.46
2422	Egg-135-55	809436	116617	87.4	12.6	818.9	8.26	9.05	9.86	1657	65.92
2424	Egg-135-55	814798	129653	86.3	13.7	835.7	8.26	9.06	9.93	1691	66.61
	Average	**812519**	**124254**	......	......	......	**8.29**	**9.09**	......	**1677**	**66.66**
2417	Lump-100-33	464290	48258	90.6	9.4	453.5	9.21	10.15	11.26	917	75.68
2418	Lump-100-33	460773	48672	90.4	9.6	450.7	8.93	9.91	10.97	912	73.57
2419	Lump-100-33	471540	50315	90.4	9.6	461.7	8.86	9.80	11.08	934	74.21
	Average	**465534**	**49082**	......	......	......	**9.00**	**9.95**	......	**921**	**74.49**
2425	Lump-135-55	804835	121855	86.9	13.1	819.7	7.63	8.35	9.16	1658	61.47
2427	Lump-135-55	796173	111869	87.7	12.3	803.3	7.14	7.86	8.75	1625	58.92
2428	Lump-135-55	818563	127214	86.6	13.4	836.7	7.58	8.35	9.26	1693	62.14
2442	Lump-135-55	846524	133841	86.4	13.6	867.3	7.92	8.71	9.70	1755	65.19
	Average	**816524**	**123695**	......	......	......	**7.57**	**8.32**	......	**1684**	**61.93**
2430	2 in. S.-100-33	466957	51578	90.1	9.9	458.6	8.38	9.25	10.48	928	70.55
2434	2 in. S.-100-33	460520	56498	89.1	10.9	457.2	8.37	9.24	10.42	925	70.24
2435	2 in. S.-100-33	482315	56522	89.5	10.5	476.7	8.42	9.25	10.47	964	70.75
	Average	**469931**	**54866**	......	......	......	**8.39**	**9.25**	......	**939**	**70.51**
2436	2 in. S.-135-55	796631	134016	85.6	14.4	823.3	6.71	7.40	8.34	1666	56.25
2437	2 in. S.-135-55	830111	138731	85.7	14.3	857.0	6.77	7.45	8.49	1734	57.35
	Average	**813371**	**136374**	......	......	......	**6.74**	**7.43**	......	**1700**	**56.80**
2431	1¼in.S.-100-33	480723	55239	89.7	10.3	474.3	7.66	8.36	9.31	959	62.81
2432	1¼in.S.-100-33	476916	55422	89.6	10.4	470.9	7.85	8.61	9.78	953	66.08
2433	1¼in.S.-100-33	478008	55647	89.6	10.4	472.2	7.73	8.45	9.51	955	64.21
	Average	**478579**	**55436**	......	......	......	**7.75**	**8.47**	......	**956**	**64.37**
2440	1¼in.S.-135-55	814282	124462	86.7	13.3	830.3	6.21	6.83	7.74	1680	52.28
2441	1¼in.S.-135-55	828679	135796	85.9	14.1	853.0	6.50	7.28	8.38	1726	56.64
	Average	**821481**	**130129**	......	......	......	**6.36**	**7.06**	......	**1703**	**54.46**

TABLE 22

ENGINE PERFORMANCE

Test Number	Laboratory Designation	Cut-Off Per Cent of Stroke Average	Least Back-Pressure lb. per Sq. in. Average	Mean Effective Pressure lb. per Sq. in. Average	Indicated Horse Power				
					Right Side		Left Side		Total
					Head End	Crank End	Head End	Crank End	
	Code Item ☞	....	....	678	707	708	709	710	711
2400	M. R.-100-33	....	....	....	....	....	....	....	....
2401	M. R.-100-33	34.0	1.9	74.1	308.0	321.3	296.9	298.3	1224.5
2402	M. R.-100-33	33.0	2.1	73.7	309.3	320.7	296.1	297.5	1223.6
	Average	**33.5**	**2.0**	**73.9**	....	....	....	....	**1224.1**
2405	M. R.-135-55	54.3	11.7	96.9	531.8	567.2	529.2	529.5	2157.7
2406	M. R.-135-55	53.7	12.0	95.9	528.7	562.8	527.0	533.0	2151.5
2429	M. R.-135-55	55.9	12.9	97.5	529.9	572.3	546.2	542.6	2191.0
	Average	**54.6**	**12.2**	**96.8**	....	....	....	....	**2166.7**
2408	Nut-100-33	33.0	2.1	77.9	327.9	342.0	309.0	314.7	1293.6
2409	Nut-100-33	32.4	2.6	77.8	327.6	338.2	310.6	310.1	1286.5
2410	Nut-100-33	31.5	2.3	77.2	326.6	338.3	304.9	311.1	1280.9
2426	Nut-100-33	31.7	2.7	79.6	332.7	347.1	312.7	321.2	1313.7
	Average	**32.2**	**2.4**	**78.1**	....	....	....	....	**1293.7**
2412	Nut-135-55	....	....	....	....	....	....	....	....
2413	Nut-135-55	55.7	11.7	98.0	547.6	576.3	536.8	539.6	2200.3
2414	Nut-135-55	59.3	11.9	98.9	551.5	581.5	542.0	546.8	2221.8
	Average	**57.5**	**11.8**	**98.5**	....	....	....	....	**2211.1**
2415	Egg-100-33	32.4	2.7	79.6	331.7	350.0	311.7	320.3	1313.7
2416	Egg-100-33	33.3	2.4	80.0	334.2	351.5	317.4	321.2	1324.3
2423	Egg-100-33	32.2	2.4	78.4	330.1	336.5	312.6	311.9	1291.1
	Average	**32.6**	**2.5**	**79.3**	....	....	....	....	**1309.7**
2420	Egg-135-55	57.2	13.0	97.3	540.4	571.5	536.3	540.5	2188.7
2422	Egg-135-55	58.2	13.0	97.9	552.5	580.2	533.2	548.4	2214.3
2424	Egg-135-55	56.6	12.9	98.3	547.6	585.8	544.0	542.6	2220.0
	Average	**57.3**	**13.0**	**97.8**	....	....	....	....	**2207.7**
2417	Lump-100-33	36.3	2.2	80.0	335.5	350.3	320.1	317.3	1323.2
2418	Lump-100-33	33.5	2.6	80.1	338.8	350.8	320.0	320.0	1329.6
2419	Lump-100-33	32.7	2.8	79.7	335.0	351.9	314.4	319.7	1321.0
	Average	**34.2**	**2.5**	**79.9**	....	....	....	....	**1324.6**
2425	Lump-135-55	56.0	12.2	97.8	546.5	572.3	535.5	547.1	2201.4
2427	Lump-135-55	55.7	12.5	96.2	533.3	566.8	532.1	537.5	2169.7
2428	Lump-135-55	55.1	12.3	97.8	541.3	571.4	543.1	544.7	2200.5
2442	Lump-135-55	56.5	12.0	98.5	546.1	576.5	535.4	537.5	2198.5
	Average	**55.8**	**12.3**	**97.6**	....	....	....	....	**2192.5**
2430	2 in. S.-100-33	32.6	2.5	78.5	328.8	345.5	313.6	317.0	1304.9
2434	2 in. S.-100-33	33.6	2.8	80.3	337.3	345.0	321.0	311.2	1314.5
2435	2 in. S.-100-33	34.9	3.0	82.1	352.9	360.0	331.9	320.4	1365.2
	Average	**33.7**	**2.8**	**80.3**	....	....	....	....	**1328.2**
2436	2 in. S.-135-55	56.8	12.4	95.2	527.2	559.3	524.4	515.4	2126.3
2437	2 in. S.-135-55	56.9	12.7	100.0	563.6	585.5	550.9	543.5	2243.5
	Average	**56.9**	**12.6**	**97.6**	....	....	....	....	**2184.9**
2431	1¼ in. S.-100-33	33.9	2.9	79.9	337.2	353.8	321.9	317.0	1329.9
2432	1¼ in. S.-100-33	34.0	3.3	80.8	339.4	356.3	323.3	324.1	1343.1
2433	1¼ in. S.-100-33	33.2	3.0	79.1	334.3	342.6	318.7	314.6	1310.2
	Average	**33.7**	**3.1**	**79.9**	....	....	....	....	**1327.7**
2440	1¼ in. S.-135-55	58.2	11.4	99.3	544.5	580.1	546.7	544.3	2215.6
2441	1¼ in. S.-135-55	55.6	12.2	99.7	550.5	587.3	544.8	549.0	2231.6
	Average	**56.9**	**11.8**	**99.5**	....	....	....	....	**2223.6**

TABLE 23

GENERAL LOCOMOTIVE PERFORMANCE

Test Number	Laboratory Designation	Consumed per I. H. P. per Hour		Drawbar Horse-Power	Consumed per D. H. P. per Hour		Tractive Force Based on M.E.P. lb.	Machine Friction in Terms of Horse-Power	Machine Efficiency of Locomotive Per Cent	Efficiency of Locomotive Per Cent
		Dry Coal lb.	Super-Heated Steam lb.		Dry Coal lb.	Super-Heated Steam lb.				
	Code Item ☞	**734**	**740**	**743**	**744**	**747**	**764**	**770**	**778**	**779**
2400	M. R.-100-33	......	......	1108.6	2.60	20.27	......	......	......	7.55
2401	M. R.-100-33	2.39	18.18	1095.2	2.67	20.32	24288	129.3	89.4	7.33
2402	M. R.-100-33	2.46	18.74	1104.6	2.72	20.76	24166	119.0	90.3	7.25
	Average	**2.43**	**18.46**	**1102.8**	**2.66**	**20.45**	**24277**	**124.2**	**89.9**	**7.38**
2405	M. R.-135-55	3.14	19.63	1954.4	3.48	21.68	31766	203.3	90.6	5.72
2406	M. R.-135-55	2.95	19.40	1963.8	3.23	21.25	31457	187.7	91.3	6.09
2429	M. R.-135-55	3.07	19.65	1965.2	3.42	21.91	31962	225.8	89.7	5.78
	Average	**3.05**	**19.56**	**1961.1**	**3.38**	**21.61**	**31728**	**205.6**	**90.5**	**5.86**
2408	Nut-100-33	2.42	17.93	1138.4	2.75	20.38	25563	155.2	88.0	7.07
2409	Nut-100-33	2.36	17.98	1129.7	2.69	20.47	25524	156.8	87.8	7.24
2410	Nut-100-33	2.41	17.93	1133.5	2.73	20.26	25333	147.4	88.5	7.16
2426	Nut-100-33	2.53	18.72	1139.9	2.91	21.57	26088	173.8	86.8	6.73
	Average	**2.43**	**18.14**	**1135.4**	**2.77**	**20.67**	**25627**	**158.3**	**87.8**	**7.05**
2412	Nut-135-55	......	......	1988.7	3.21	21.60	......	......	......	6.07
2413	Nut-135-55	2.86	19.34	1993.0	3.16	21.36	32134	207.3	90.6	6.12
2414	Nut-135-55	2.85	19.00	1994.9	3.17	21.17	32438	226.9	89.8	6.14
	Average	**2.86**	**19.17**	**1992.2**	**3.18**	**21.38**	**32286**	**217.1**	**90.2**	**6.11**
2415	Egg-100-33	2.34	18.01	1150.1	2.67	20.58	26091	163.6	87.6	7.19
2416	Egg-100-33	2.33	17.86	1167.6	2.64	20.26	26208	156.7	88.2	7.34
2423	Egg-100-33	2.36	18.41	1131.0	2.70	21.02	25725	160.1	87.6	7.10
	Average	**2.34**	**18.09**	**1149.6**	**2.67**	**20.62**	**26008**	**160.1**	**87.8**	**7.21**
2420	Egg-135-55	2.90	19.83	1991.6	3.19	21.79	31922	197.1	91.0	6.04
2422	Egg-135-55	2.86	19.40	2001.4	3.17	21.46	32123	212.9	90.4	6.03
2424	Egg-135-55	2.92	19.63	2003.4	3.23	21.75	32256	216.6	90.2	5.96
	Average	**2.89**	**19.62**	**1998.8**	**3.20**	**21.67**	**32100**	**208.9**	**90.5**	**6.01**
2417	Lump-100-33	2.34	17.92	1160.7	2.67	20.43	26257	162.5	87.7	7.31
2418	Lump-100-33	2.37	17.73	1168.6	2.70	20.17	26257	161.0	87.9	7.22
2419	Lump-100-33	2.48	18.31	1162.2	2.81	20.81	26116	155.8	88.0	7.05
	Average	**2.40**	**17.99**	**1163.8**	**2.73**	**20.47**	**26210**	**159.8**	**87.9**	**7.19**
2425	Lump-135-55	3.13	19.61	1957.8	3.52	22.05	32076	243.6	88.9	5.47
2427	Lump-135-55	3.32	19.68	1919.6	3.76	22.24	31554	250.1	88.5	5.23
2428	Lump-135-55	3.20	19.95	1963.8	3.59	22.35	31869	236.7	89.2	5.44
2442	Lump-135-55	3.14	20.24	1989.6	3.45	22.34	32339	208.9	90.5	5.68
	Average	**3.20**	**19.87**	**1957.7**	**3.58**	**22.25**	**31960**	**234.8**	**89.3**	**5.46**
2430	2 in. S.-100-33	2.58	18.02	1160.7	2.90	20.26	25743	144.2	89.0	6.88
2434	2 in. S.-100-33	2.59	18.00	1152.2	2.96	20.53	26342	162.3	87.7	6.74
2435	2 in. S.-100-33	2.61	18.25	1179.1	3.02	21.13	26932	186.1	86.4	6.63
	Average	**2.59**	**18.09**	**1164.0**	**2.36**	**20.64**	**26339**	**164.2**	**87.7**	**6.75**
2436	2 in. S.-135-55	3.64	19.86	1905.6	4.07	22.16	31207	220.7	89.6	4.90
2437	2 in. S.-135-55	3.53	19.33	1980.4	3.99	21.90	32780	263.1	88.3	5.05
	Average	**3.59**	**19.60**	**1943.0**	**4.03**	**22.03**	**31994**	**241.9**	**89.0**	**4.98**
2431	1¼ in. S.-100-33	2.92	18.32	1133.4	3.42	21.50	26213	196.5	85.2	5.77
2432	1¼ in. S.-100-33	2.81	18.22	1161.6	3.26	21.06	26498	181.5	86.5	6.18
2433	1¼ in. S.-100-33	2.93	18.71	1141.0	3.36	21.48	25937	169.2	87.1	5.92
	Average	**2.89**	**18.42**	**1145.3**	**3.35**	**21.35**	**26216**	**182.4**	**86.3**	**5.96**
2440	1¼ in. S.-135-55	3.79	19.29	1977.2	4.25	21.62	32568	238.4	89.2	4.73
2441	1¼ in. S.-135-55	3.63	19.48	2006.7	4.04	21.67	32682	224.9	89.9	5.05
	Average	**3.71**	**19.39**	**1992.0**	**4.15**	**21.65**	**32625**	**231.7**	**89.6**	**4.89**

TABLE 24

ANALYSIS OF ASH AND STACK CINDERS

Test Number	Laboratory Designation	Analysis of Ash				Analysis of Stack Cinders			
		Fixed Carbon, Per Cent	Volatile Matter, Per Cent	Ash, Per Cent	Moisture, Per Cent	Fixed Carbon, Per Cent	Volatile Matter, Per Cent	Ash, Per Cent	Moisture, Per Cent
	Code Item ☞	....	....	832	833	....	....	847	848
2400	M. R.-100-33	19.92	4.63	74.11	1.34	53.88	4.87	40.27	0.98
2401	M. R.-100-33	18.30	2.75	78.85	0.10	55.49	4.28	39.48	0.75
2402	M. R.-100-33	15.42	3.16	81.07	0.35	55.47	4.05	39.78	0.70
	Average	**17.88**	**3.51**	**78.01**	**0.60**	**54.95**	**4.40**	**39.84**	**0.81**
2405	M. R.-135-55	24.48	3.34	71.92	0.26	73.75	4.08	21.98	0.19
2406	M. R.-135-55	25.34	2.11	71.77	0.78	73.86	2.91	22.80	0.43
2429	M. R.-135-55	27.83	3.35	67.97	0.85	72.91	3.37	23.47	0.25
	Average	**25.88**	**2.93**	**70.55**	**0.63**	**73.51**	**3.45**	**22.75**	**0.29**
2408	Nut-100-33	16.64	2.26	76.82	4.28	51.75	4.73	42.27	1.25
2409	Nut-100-33	21.25	2.50	73.68	2.57	48.87	4.47	42.69	3.97
2410	Nut-100-33	14.03	2.08	82.56	1.33	52.04	5.57	40.94	1.45
2426	Nut-100-33	13.65	1.70	77.29	7.36	56.35	3.55	39.08	1.02
	Average	**16.39**	**2.14**	**77.59**	**3.89**	**52.25**	**4.58**	**41.25**	**1.92**
2412	Nut-135-55	21.86	1.50	76.39	0.25	74.99	1.87	23.02	0.12
2413	Nut-135-55	21.35	1.83	75.32	1.50	73.36	2.15	24.37	0.12
2414	Nut-135-55	22.43	1.45	75.92	0.20	71.77	2.80	25.39	0.04
	Average	**21.88**	**1.59**	**75.88**	**0.65**	**73.37**	**2.27**	**24.26**	**0.09**
2415	Egg-100-33	19 88	2.51	77.26	0.35	51.03	4.10	44.84	0.03
2416	Egg-100-33	21.28	2.62	73.93	2.17	52.30	4.33	43.00	0.37
2423	Egg-100-33	24.21	2.70	72.38	0.71	55.04	3.94	39.13	1.89
	Average	**21.79**	**2.61**	**74.52**	**1.08**	**52.79**	**4.12**	**42.32**	**0.76**
2420	Egg-135-55	31 55	2.00	66.23	0.22	74.44	2.67	22.78	0.11
2422	Egg-135-55	22.16	1.32	76.26	0.26	77.32	2.04	20.60	0.04
2424	Egg-135-55	28.48	2.27	69.20	0.05	71.82	3.31	24.70	0.17
	Average	**27.40**	**1.86**	**70.56**	**0.18**	**74.53**	**2.67**	**22.69**	**0.11**
2417	Lump-100-33	22 39	1.95	75.55	0.11	48.94	5.02	45.49	0.55
2418	Lump-100-33	21.64	4.14	72.54	1.68	48.27	4.88	46.26	0.59
2419	Lump-100-33	21.31	2.88	74.06	1.75	44.59	5.11	48.59	1.71
	Average	**21.78**	**2.99**	**74.05**	**1.18**	**47.27**	**5.00**	**46.78**	**0.95**
2425	Lump-135-55	22.45	2.39	67.76	7.40	73.00	2.84	24.05	0.11
2427	Lump-135-55	32.52	1.42	65.98	0.08	73.56	2.88	23.28	0.28
2428	Lump-135-55	28.99	1.38	69.58	0.05	73.61	2.26	23.93	0.20
2442	Lump-135-55	29 61	2.38	66.77	1.24	69.58	3.22	26.92	0.28
	Average	**28.39**	**1.89**	**67.52**	**2.19**	**72.44**	**2.80**	**24.55**	**0.22**
2430	2 in. S.-100-33	20.33	5.64	65.86	8.17	61.08	5.43	32.37	1.12
2434	2 in. S.-100-33	19.71	6.52	71.22	2.55	61.57	5.85	31.75	0.83
2435	2 in. S.-100-33	19.30	6.69	57.81	16.20	58.19	6.24	34.14	1.43
	Average	**19.78**	**6.28**	**64.96**	**8.97**	**60.28**	**5.84**	**32.75**	**1.13**
2436	2 in. S.-135-55	26 39	4.31	68.14	1.16	66.06	8.14	25.64	0.16
2437	2 in. S.-135-55	23.16	5.26	70.09	1.49	63.32	13.67	19.88	3.13
	Average	**24.78**	**4.79**	**69.12**	**1.33**	**64.69**	**10.91**	**22.76**	**1.65**
2431	1¼ in. S.-100-33	21.38	7.66	69.41	1.55	65.98	7.13	26.44	0.45
2432	1¼ in. S.-100-33	21.66	10.17	61.67	6.50	63.95	7.06	28.48	0.51
2433	1¼ in. S.-100-33	23 01	7.04	67.85	2.10	69.86	5.90	23.75	0.49
	Average	**22 02**	**8.29**	**66.31**	**3.38**	**66.60**	**6.70**	**26.22**	**0.48**
2440	1¼ in. S.-135-55	21.23	5.89	69.56	3.32	67.35	8.54	23.87	0.24
2441	1¼ in. S.-135-55	24.50	7.09	66.24	2.17	69.57	8.47	21.50	0.46
	Average	**22.87**	**6.49**	**67.90**	**2.75**	**68.46**	**8.51**	**22.69**	**0.35**

TABLE 25

HEAT BALANCE—BRITISH THERMAL UNITS

Test Number	Laboratory Designation	B.t.u. Absorbed by Boiler per lb. of Coal as Fired	B.t.u. Loss per Pound of Coal as Fired									
			Due to Moisture in Coal	Due to Moisture in Air	Due to Hydrogen in Coal	Due to Escaping Gases	Due to Incomplete Combustion			Due to Combustible in Stack Cinders	Due to Combustible in Ash	Due to Radiation, and Unaccounted for
							CO	H_2	CH			
	Code Item ☞	**851**	**852**	**853**	**854**	**855**	......	......	......	**858**	**860**	**869**
2400	M. R.-100-33	9173	102	42	483	1419	49	11	8	265	103	275
2401	M. R.-100-33	8988	99	41	487	1408	76	5	5	254	209	421
2402	M. R.-100-33	8969	102	52	482	1331	107	0	23	267	93	460
	Average	**9043**	**101**	**45**	**484**	**1386**	**77**	**5**	**12**	**262**	**135**	**385**
2405	M. R.-135-55	7531	109	25	487	1239	222	35	78	1067	246	714
2406	M. R.-135-55	7949	108	24	506	1286	416	8	24	934	263	358
2429	M R.-135-55	7686	112	13	498	1342	120	14	40	973	307	701
	Average	**7722**	**110**	**21**	**497**	**1289**	**253**	**19**	**47**	**991**	**272**	**591**
2408	Nut-100-33	8775	112	23	503	1505	204	33	51	146	113	539
2409	Nut-100-33	8969	114	24	498	1430	214	20	74	164	156	292
2410	Nut-100-33	8706	110	24	497	1367	168	27	47	174	153	646
2426	Nut-100-33	8794	99	18	499	1355	189	39	46	250	96	608
	Average	**8811**	**109**	**42**	**499**	**1414**	**194**	**30**	**55**	**184**	**130**	**521**
2412	Nut-135-55	8065	117	17	475	1199	568	66	230	633	190	358
2413	Nut-135-55	8114	119	16	471	1336	241	106	234	619	104	630
2414	Nut-135-55	8007	118	21	504	1384	258	63	50	598	212	707
	Average	**8062**	**118**	**18**	**483**	**1306**	**356**	**78**	**171**	**617**	**169**	**565**
2415	Egg-100-33	9066	109	36	506	1526	108	16	47	184	184	362
2416	Egg-100-33	8949	117	35	512	1564	66	0	0	178	195	301
2423	Egg-100-33	9105	115	18	521	1437	90	5	15	173	180	440
	Average	**9040**	**114**	**30**	**513**	**1509**	**88**	**7**	**21**	**178**	**186**	**368**
2420	Egg-135-55	8123	114	22	520	1259	178	14	31	777	349	656
2422	Egg-135-55	8026	116	25	526	1355	157	44	49	759	219	899
2424	Egg-135-55	8026	117	20	527	1324	173	23	48	807	319	665
	Average	**8058**	**116**	**22**	**524**	**1313**	**169**	**27**	**43**	**781**	**296**	**740**
2417	Lump-100-33	8949	120	29	486	1519	59	0	13	160	256	236
2418	Lump-100-33	8677	127	29	482	1480	52	3	3	146	266	529
2419	Lump-100-33	8609	128	27	493	1411	82	7	15	166	305	358
	Average	**8745**	**125**	**28**	**487**	**1470**	**64**	**3**	**10**	**157**	**276**	**374**
2425	Lump-135-55	7414	115	16	489	1309	266	54	131	791	92	1382
2427	Lump-135-55	6938	121	12	499	1115	530	12	21	831	558	1139
2428	Lump-135-55	7366	124	13	493	1274	311	51	67	734	307	1113
2442	Lump-135-55	7696	120	10	502	1305	127	15	38	810	391	792
	Average	**7354**	**120**	**13**	**496**	**1251**	**309**	**33**	**64**	**792**	**337**	**1107**
2430	2 in. S.-100-33	8143	122	13	500	1252	141	24	57	776	143	372
2434	2 in. S.-100-33	8133	121	15	513	1303	70	2	18	847	180	378
2435	2 in. S.-100-33	8182	115	11	512	1283	124	4	0	858	118	357
	Average	**8153**	**119**	**13**	**508**	**1279**	**112**	**10**	**25**	**827**	**147**	**369**
2436	2 in. S.-135-55	6520	124	11	520	1357	195	29	88	1396	221	1131
2437	2 in. S.-135-55	6578	122	9	514	1150	212	45	58	1642	193	946
	Average	**6549**	**123**	**10**	**517**	**1264**	**204**	**37**	**73**	**1519**	**207**	**1039**
2431	1¼ in. S.-100-33	7443	107	21	513	1160	136	0	0	1408	203	860
2432	1¼ in. S.-100-33	7628	111	25	492	1180	97	6	9	1337	190	468
2433	1¼ in. S.-100-33	7511	110	19	502	1252	100	22	18	1167	204	793
	Average	**7527**	**109**	**22**	**502**	**1197**	**111**	**9**	**9**	**1304**	**199**	**707**
2440	1¼ in. S.-135-55	6034	122	8	451	1082	424	98	423	1762	203	934
2441	1¼ in. S.-135-55	6316	144	8	480	1082	231	54	129	1780	147	780
	Average	**6175**	**133**	**8**	**466**	**1082**	**328**	**76**	**276**	**1771**	**175**	**857**

TABLE 26

HEAT BALANCE —PERCENTAGE

Test Number	Laboratory Designation	Per Cent of Heat of Coal as Fired: Absorbed by Boiler	To Moisture in Coal	To Moisture in Air	To Hydrogen in Coal	To Escaping Gases	To Incomplete Combustion: CO	H_2	CH_4	To Combustible in Stack Cinders	To Combustible in Ash	To Radiation and Unaccounted for
	Code Item ☞	**881**	**882**	**883**	**884**	**885**	......	......	......	**888**	**890**	**899**
2400	M. R.-100-33	76.9	0.9	0.4	4.0	11.9	0.4	0.1	0.1	2.2	0.9	2.2
2401	M. R.-100-33	75.0	0.8	0.3	4.1	11.7	0.6	0.0	0.0	2.1	1.7	3.7
2402	M. R.-100-33	75.5	0.9	0.4	4.1	11.2	0.9	0.0	0.2	2.2	0.8	3.8
	Average	**75.8**	**0.9**	**0.4**	**4.1**	**11.6**	**0.6**	**0.0**	**0.1**	**2.2**	**1.1**	**3.2**
2405	M. R.-135-55	64.1	0.9	0.2	4.1	10.5	1.9	0.3	0.7	9.1	2.1	6.1
2406	M. R.-135-55	66.9	0.9	0.2	4.3	10.8	3.5	0.1	0.2	7.9	2.2	3.0
2429	M. R.-135-55	65.1	1.0	0.1	4.2	11.4	1.0	0.1	0.3	8.2	2.6	6.0
	Average	**65.4**	**0.9**	**0.2**	**4.2**	**10.9**	**2.1**	**0.2**	**0.4**	**8.4**	**2.3**	**5.0**
2408	Nut-100-33	73.1	0.9	0.2	4.2	12.5	1.7	0.3	0.4	1.2	0.9	4.6
2409	Nut-100-33	75.0	1.0	0.2	4.2	12.0	1.8	0.2	0.6	1.4	1.3	2.3
2410	Nut-100-33	73.1	0.9	0.2	4.2	11.5	1.4	0.2	0.4	1.5	1.3	5.3
2426	Nut-100-33	73.3	0.8	0.1	4.2	11.3	1.6	0.3	0.4	2.1	0.8	5.1
	Average	**73.6**	**0.9**	**0.2**	**4.2**	**11.8**	**1.6**	**0.3**	**0.5**	**1.6**	**1.1**	**4.3**
2412	Nut-135-55	67.7	1.0	0.1	4.0	10.1	4.8	0.6	1.9	5.3	1.6	2.9
2413	Nut-135-55	67.7	1.0	0.1	3.9	11.1	2.0	0.9	2.0	5.2	0.9	5.2
2414	Nut-135-55	67.2	1.0	0.2	4.2	11.6	2.2	0.5	0.4	5.0	1.8	5.9
	Average	**67.5**	**1.0**	**0.1**	**4.0**	**10.9**	**3.0**	**0.7**	**1.4**	**5.2**	**1.4**	**4.7**
2415	Egg-100-33	74.7	0.9	0.3	4.2	12.6	0.9	0.1	0.4	1.5	1.5	2.9
2416	Egg-100-33	75.1	1.0	0.3	4.3	13.1	0.6	0.0	0.0	1.5	1.6	2.5
2423	Egg-100-33	75.3	1.0	0.2	4.3	11.9	0.7	0.0	0.1	1.4	1.5	3.6
	Average	**75.0**	**1.0**	**0.3**	**4.3**	**12.5**	**0.7**	**0.0**	**0.2**	**1.5**	**1.5**	**3.0**
2420	Egg-135-55	67.5	1.0	0.2	4.3	10.4	1.5	0.1	0.3	6.5	2.9	5.3
2422	Egg-135-55	65.9	1.0	0.2	4.3	11.1	1.3	0.4	0.4	6.2	1.8	7.4
2424	Egg-135-55	66.6	1.0	0.2	4.4	11.0	1.4	0.2	0.4	6.7	2.7	5.4
	Average	**66.7**	**1.0**	**0.2**	**4.3**	**10.8**	**1.4**	0.2	0.4	**6.5**	**2.5**	**6.0**
2417	Lump-100-33	75.7	1.0	0.2	4.1	12.8	0.5	0.0	0.1	1.4	2.2	2.0
2418	Lump-100-33	73.6	1.1	0.2	4.1	12.5	0.4	0.0	0.0	1.2	2.3	4.6
2419	Lump-100-33	74.2	1.1	0.2	4.2	12.2	0.7	0.1	0.1	1.4	2.6	3.2
	Average	**74.5**	**1.1**	**0.2**	**4.1**	**12.5**	**0.5**	**0.0**	**0.1**	**1.3**	**2.4**	**3.3**
2425	Lump-135-55	61.5	1.0	0.1	4.1	10.9	2.2	0.5	1.1	6.6	0.8	11.2
2427	Lump-135-55	58.9	1.0	0.1	4.2	9.5	4.5	0.1	0.2	7.1	4.7	9.7
2428	Lump-135-55	62.1	1.0	0.1	4.2	10.8	2.6	0.4	0.6	6.2	2.6	9.4
2442	Lump-135-55	63.2	1.0	0.1	4.3	11.0	1.1	0.1	0.3	6.9	3.3	6.7
	Average	**61.9**	**1.0**	**0.1**	**4.2**	**10.6**	**2.6**	**0.3**	**0.6**	**6.7**	**2.9**	**9.3**
2430	2in. S.-100-33	70.6	1.1	0.1	4.3	10.8	1.2	0.2	0.5	6.7	1.2	3.3
2434	2in. S.-100-33	70.2	1.1	0.1	4.4	11.3	0.6	0.0	0.2	7.3	1.6	3.2
2435	2in. S.-100-33	70.8	1.0	0.1	4.4	11.1	1.1	0.0	0.0	7.4	1.0	3.1
	Average	**70.5**	**1.1**	**0.1**	**4.4**	**11.1**	**1.0**	**0.1**	**0.2**	**7.1**	**1.3**	**3.2**
2436	2in. S.-135-55	56.3	1.1	0.1	4.5	11.7	1.7	0.3	0.8	12.1	1.9	9.5
2437	2in. S.-135-55	57.4	1.1	0.1	4.5	10.0	1.9	0.4	0.5	14.3	1.7	8.1
	Average	**56.9**	**1.1**	**0.1**	**4.5**	**10.9**	**1.8**	**0.4**	**0.7**	**13.2**	**1.8**	**8.8**
2431	1¼in.S.-100-33	62.8	0.9	0.2	4.3	9.8	1.2	0.0	0.0	11.9	1.7	7.2
2432	1¼in.S.-100-33	66.1	1.0	0.2	4.3	10.2	0.8	0.1	0.1	11.6	1.7	3.9
2433	1¼in.S.-100-33	64.2	0.9	0.2	4.3	10.7	0.9	0.2	0.2	10.0	1.8	6.6
	Average	**64.4**	**0.9**	**0.2**	**4.3**	**10.2**	**1.0**	**0.1**	**0.1**	**11.2**	**1.7**	**5.9**
2440	1¼in.S.-135-55	52.3	1.1	0.1	3.9	9.4	3.7	0.9	3.7	15.3	1.8	7.8
2441	1¼in.S.-135-55	56.6	1.3	0.1	4.3	9.7	2.1	0.5	1.2	16.0	1.3	6.9
	Average	**54.5**	**1.2**	**0.1**	**4.1**	**9.6**	**2.9**	**0.7**	**2.5**	**15.7**	**1.6**	**7.4**

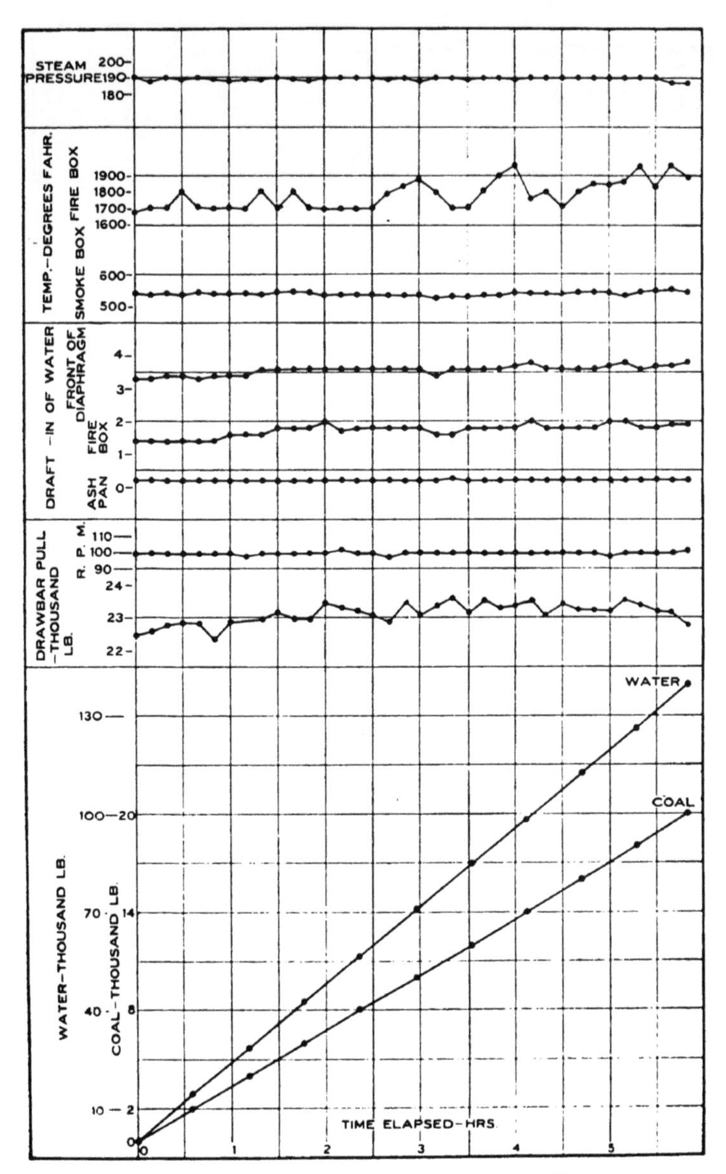

FIG. 29. GRAPHICAL LOG FOR MEDIUM RATE TEST NO. 2416

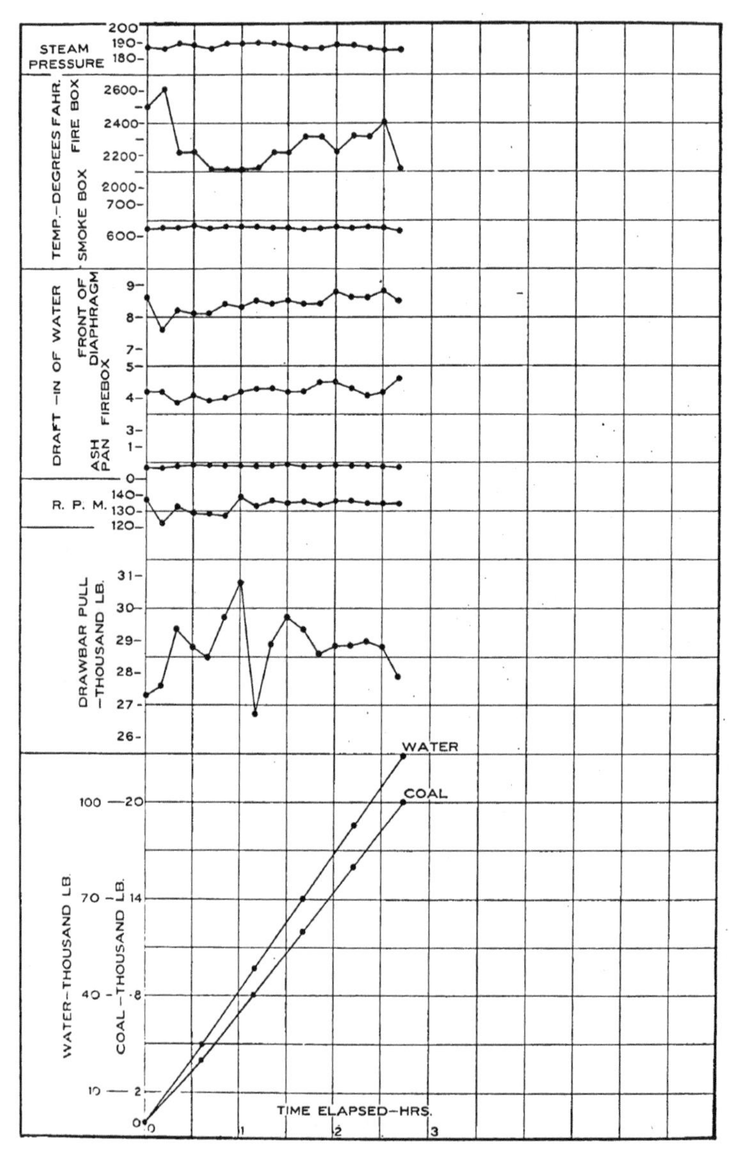

FIG. 30. GRAPHICAL LOG FOR HIGH RATE TEST NO. 2405

APPENDIX IV

Cylinder Performance

The main purpose of the tests entailed the definition of the boiler performance at only two rates of evaporation. This necessitated tests under only two conditions of speed and cut-off, which presented but little opportunity to gather data concerning the performance of the cylinders. Indicator cards were taken during all tests, under these conditions; but it proved impracticable to hold the locomotive for the determination of cylinder performance at other speeds and cut-offs, as had been originally intended. The available facts concerning cylinder performance are presented in detail for each of the tests in Tables 17, 21 and 22 of Appendix III, and their average values at both combinations of speed and cut-off are stated in the following sections. Representative indicator cards are reproduced in Fig. 31, and the data relating to them appear in Table 27.

32. *Medium Rate Tests.*—All medium rate tests were run with the reverse-lever in the second notch from the center of the quadrant, giving a cut-off of about 33 per cent. The speed was maintained as nearly as possible at 100 revolutions per minute, which is equivalent to 19.0 miles per hour on the road.

The average indicated horse power was 1305. It varied from 1224 to 1365.

The average drawbar pull was 22640 pounds. The average indicated horse power consumed in machine friction was 160.

The steam consumed per indicated horse power per hour varied from a minimum of 17.73 pounds to a maximum of 18.74 pounds; and the average for all medium rate tests was 18.18 pounds.

33. *High Rate Tests.*—All high rate tests were run with the reverse-lever in the sixth notch from the center, giving a cut-off of about 55 per cent. The speed was maintained as nearly as possible at 135 revolutions per minute, which is equivalent to 25.7 miles per hour.

The average indicated horse power was 2196. It varied from 2126 to 2243.

The average drawbar pull was 28826 pounds. The average indicated horse power consumed in machine friction was 223.

The steam consumed per indicated horse power per hour varied from a minimum of 19.00 pounds to a maximum of 20.24 pounds; the average for all high rate tests was 19.58 pounds.

34. *Variations in Power.*—Despite the fact that both the reverse-lever position and the speed were maintained constant during all medium rate and during all high rate tests, there was during the progress of the work considerable variation in indicated horse power in both groups. In general the power increased as time went on. An almost identical variation occurs in the areas of the indicator cards. Neither the variations in water rate nor in superheat offer an adequate explanation for these facts. It is assumed that they are due chiefly to changes in steam distribution brought about by wear in the valve gear.

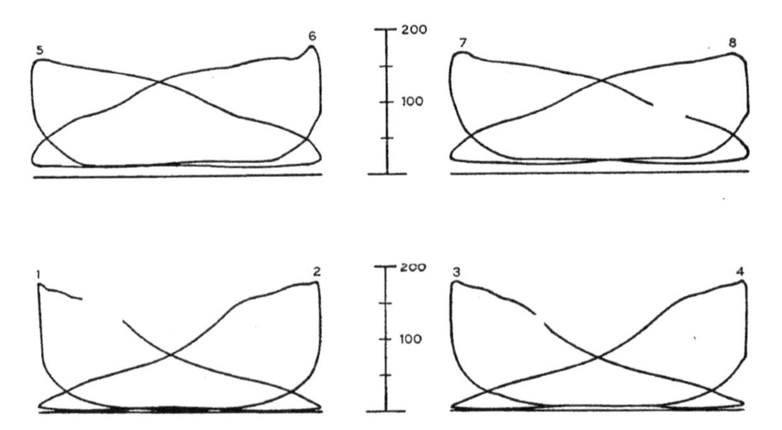

FIG. 31. REPRESENTATIVE INDICATOR DIAGRAMS FOR BOTH THE MEDIUM AND THE HIGH RATE TESTS

TABLE 27

INFORMATION CONCERNING THE INDICATOR DIAGRAMS SHOWN IN FIG. 37

Test Number	Laboratory Designation	Diagram No.	Right or Left Side	Head or Crank End	Average Cut-off for Test Per Cent	Average Speed for Test	
						Miles per Hour	Revolutions per Minute
2416	Egg-100-33	1	R	H	31.0	18.94	99.5
2416	Egg-100-33	2	R	C	33.8	18.94	99.5
2416	Egg-100-33	3	L	H	36.0	18.94	99.5
2416	Egg-100-33	4	L	C	32.3	18.94	99.5
2405	M. R.-135-55	5	R	H	49.9	25.47	133.8
2405	M. R.-135-55	6	R	C	58.8	25.47	133.8
2405	M. R.-135-55	7	L	H	55.0	25.47	133.8
2405	M. R.-135-55	8	L	C	53.4	25.47	133.8

APPENDIX V

Comparison of Long and Short Tests

In connection with this series of tests, the question arose as to the desirability or necessity of running tests of such a length that the amount of coal burned would be about the same as a freight locomotive would burn in making a trip over an ordinary railroad division. This included questions such as that concerning the relative reliability of short tests burning from 4 to 6 tons of coal, and of longer tests burning from 10 to 12 tons; and that of the relative pérformance during the first, middle and last parts of a test burning 10 or 12 tons of coal.

In order to determine to some extent the difference in boiler performance which might occur during short and long tests and during different parts of a test, the observations were so taken that the exact amount of water evaporated by the boiler could be determined for intervals corresponding to the firing of each 2000 pounds of coal during the medium rate tests, and corresponding to the firing of each 4000 pounds of coal during the high rate tests.

For the purpose of making comparisons, six tests—three at the medium rate and three at the high rate—have been selected and divided into shorter tests. From 8½ to 10½ tons of coal were burned during each of these six tests. The test data for each entire test were divided into three parts, resulting in data for eighteen comparatively short tests together with the data for the six original tests. With the data and results of the 24 tests thus obtained it is possible to make the following comparisons: First, between long and short tests; second, between the first, middle and last portions of a given test; and third, between the different portions and the entire test.

Table 28 presents the significant data and results for the 24 tests. Six groups of four tests each appear in the table. In each group sections *a*, *b* and *c* present, respectively the first, middle, and last portion of the entire test, developed as separate and distinct units. Immediately following the data for these three portions appear the corresponding results for the entire test.

The length of the different tests is shown in Columns 3, 4, 5 and 6 of Table 28. The entire medium rate tests were approximately 6 hours long and the entire high rate tests 3 hours long. The divided tests varied in length from about one-half hour to two and one-half hours. From 224 to 271 pounds of coal were burned per square foot of grate during each entire test while for the divided tests the corresponding amounts vary from 52 to 115 pounds of coal per square foot of grate. The equivalent evaporation per square foot of heating surface varied for the entire tests from 33 to 40 pounds and for the divided tests from 6 to 16 pounds. At the University of Illinois locomotive laboratory about 150 pounds of coal burned per square foot of grate, or from 15 to 20 pounds of equivalent evaporation per square foot of heating surface, have been considered as sufficient to avoid any serious errors in test results arising from inaccuracies in coal and water measurements.

TABLE 28

TEST CONDITIONS AND PRINCIPAL RESULTS FOR SIX TESTS, WHICH HAVE BEEN DIVIDED INTO THREE TESTS EACH

1	2	3	4	5	6	7	8	9	10	11	12
		Length of Test					Pressure lb. per sq. in.		Temperature Degrees F.		
Test Number Rate and Size	Section of Test or Entire Test	Minutes	Hours	Dry Coal Burned per Sq. Ft. of Grate During the Test lb.	Equivalent Evaporation per Sq. Ft. of H.S. During the Test lb.	Speed in Miles per Hour	Boiler Gauge	Branch-pipe Gauge	Front-end	Branch-pipe	Fire-box
Code Item ☞		......	345	......	......	353	380	383	367	370	374
2401	a	114	1.90	79.2	12.0	18.9	190.0	172	539	566	1774
Medium	b	151	2.52	105.6	16.4	19.0	190.0	172	533	564	1835
Mine	c	112	1.87	79.2	11.3	18.8	190.3	172	532	566	1907
Run	Entire	**377**	**6.28**	**264.1**	**39.7**	**18.9**	**190.0**	**172**	**535**	**566**	**1835**
2402	a	107	1.78	79.0	11.7	19.1	190.4	176	537	565	1772
Medium	b	115	1.92	79.0	12.4	18.9	190.1	172	542	563	1847
Mine	c	90	1.50	65.9	9.6	18.9	190.1	173	537	565	1825
Run	Entire	**312**	**5.20**	**223.9**	**33.7**	**19.0**	**190.2**	**174**	**539**	**564**	**1812**
2416	a	106	1.77	78.1	12.0	19.0	189.3	180	541	574	1852
Medium	b	141	2.35	104.1	15.7	18.8	189.5	180	536	572	1759
3" x 6"	c	103	1.72	76.9	11.7	19.0	189.5	180	544	577	1794
Egg	Entire	**350**	**5.83**	**259.1**	**39.4**	**18.9**	**189.5**	**180**	**540**	**574**	**1801**
2405	a	69	1.15	105.2	14.0	25.1	189.1	168	628	628	2284
High	b	31	0.52	52.6	6.4	25.2	188.7	169	627	628	2238
Mine	c	63	1.05	105.2	12.9	26.0	185.9	168	625	628	2279
Run	Entire	**163**	**2.72**	**262.9**	**33.3**	**25.5**	**187.8**	**168**	**627**	**628**	**2271**
2406	a	72	1.20	105.3	14.7	25.6	187.8	167	631	629	2257
High	b	34	0.57	52.6	7.0	25.7	188.5	169	632	630	2320
Mine	c	55	0.92	87.2	11.0	25.7	187.1	168	630	634	2467
Run	Entire	**161**	**2.68**	**245.1**	**32.7**	**25.6**	**187.8**	**167**	**631**	**631**	**2334**
2413	a	72	1.20	104.3	14.8	25.7	187.3	168	616	628	2227
High	b	34	0.57	52.2	7.3	25.7	186.5	168	615	632	2271
2" x 3"	c	74	1.23	114.9	15.2	25.7	187.3	168	602	636	2305
Nut	Entire	**180**	**3.00**	**271.3**	**37.3**	**25.7**	**187.1**	**168**	**611**	**632**	**2267**

TABLE 28 (Continued)

TEST CONDITIONS AND PRINCIPAL RESULTS FOR SIX TESTS, WHICH HAVE BEEN DIVIDED INTO THREE TESTS EACH

1	2	13	14	15	16	17	18	19	20	21	22
		Draft in. of Water					Coal as Fired lb.			Dry Coal lb.	
Test Number Rate and Size	Section of Test or Entire Test	Front-end Front of Diaphragm	Fire-box	Ash-pan	Draw-bar Pull lb.	Degrees of Super-heat	Total	Per Hour	Per Hour per Sq. ft. of Grate Surface	Per Hour	Per Hour per Sq. ft. of Grate Surface
Code Item ☞		394	396	397	487	409	418	....	....	626	627
2401 Medium Mine Run	a	2.7	1.3	0.2	21463	190	6000	3158	45.2	2911	41.7
	b	2.9	1.6	0.2	21833	188	8000	3178	45.5	2929	42.0
	c	2.8	1.6	0.2	21864	190	6000	3214	46.1	2962	42.4
	Entire	**2.8**	**1.5**	**0.2**	**21727**	**190**	**20000**	**3183**	**45.6**	**2934**	**42.0**
2402 Medium Mine Run	a	2.9	1.4	0.2	21873	187	6000	3365	48.2	3094	44.3
	b	3.0	1.7	0.2	21885	187	6000	3130	44.8	2877	41.2
	c	3.0	1.7	0.2	21684	189	5000	3333	47.8	3065	43.9
	Entire	**3.0**	**1.6**	**0.2**	**21822**	**187**	**17000**	**3269**	**46.8**	**3005**	**43.1**
2416 Medium 3" x 6" Egg	a	3.4	1.5	0.2	22782	195	6000	3396	48.7	3084	44.2
	b	3.6	1.8	0.2	23262	193	8000	3404	48.8	3092	44.3
	c	3.7	1.9	0.2	23231	198	5915	3445	49.4	3129	44.8
	Entire	**3.6**	**1.7**	**0.2**	**23115**	**195**	**19915**	**3414**	**48.9**	**3101**	**44.4**
2405 High Mine Run	a	8.2	4.1	0.4	28879	254	8000	6957	99.7	6382	91.4
	b	8.5	4.3	0.4	28708	253	4000	7737	110.9	7099	101.7
	c	8.6	4.4	0.4	28687	254	8000	7619	109.2	6990	100.2
	Entire	**8.4**	**4.2**	**0.4**	**28771**	**254**	**20000**	**7361**	**105.5**	**6753**	**96.8**
2406 High Mine Run	a	8.6	4.2	0.4	28600	255	8000	6667	95.5	6122	87.7
	b	8.8	4.5	0.4	29183	255	4000	7055	101.1	6480	92.8
	c	8.7	4.1	0.4	28642	260	6630	7230	103.6	6640	95.1
	Entire	**8.6**	**4.3**	**0.4**	**28718**	**257**	**18630**	**6944**	**99.5**	**6377**	**91.4**
2413 High 2" x 3" Nut	a	8.9	4.0	0.5	28791	253	8000	6667	95.5	6067	86.9
	b	9.2	4.4	0.5	29407	257	4000	7055	101.1	6420	92.0
	c	9.6	4.7	0.5	29294	261	8811	7146	102.4	6503	93.2
	Entire	**9.2**	**4.4**	**0.5**	**29100**	**258**	**20811**	**6937**	**99.4**	**6313**	**90.4**

Examination of the data shows that test conditions with regard to speed, pressures, temperatures, drafts, drawbar pull, and quality of steam were very uniform as between the first, middle, and final sections of each test, with the single exception of fire-box temperature. In general the temperature in the fire-box increased somewhat as the test proceeded.

Columns 22 and 29 show, respectively, the rate of combustion expressed in dry coal per square foot of grate per hour, and the rate of evaporation expressed in equivalent evaporation per square foot of heating surface per hour. During the medium rate tests these rates were quite uniform through the 3 sections of each entire test. During the high rate tests the rate of combustion increased as each test proceeded, the greater part of this increase occurring during the first

TABLE 28 (Concluded)

TEST CONDITIONS AND PRINCIPAL RESULTS FOR SIX TESTS, WHICH HAVE BEEN DIVIDED INTO THREE TESTS EACH

1	2	23	24	25	26	27	28	29	30	31	32
			Superheated Steam lb.				Equivalent Evaporation lb.				
Test Number Rate and Size	Section of Test or Entire Test	Quality of Steam in Dome	Per Hour	Per Sq. ft. of Heating Surface per Hour	Per Pound of Coal as Fired	Per Pound of Dry Coal	Per Hour	Per Sq. ft. of Heating Surface per Hour	Per Pound of Dry Coal	B.t.u. Absorbed by Boiler per lb. of Coal as Fired	Boiler Efficiency Per Cent
Code Item ☞		407	....	....	....	....	645	648	658	....	666
2401	a	0.9773	22405	4.81	7.09	7.70	29507	6.33	10.14	9076	75.68
Medium	b	0.9798	22973	4.93	7.23	7.85	30278	6.50	10.34	9260	77.22
Mine	c	0.9802	21390	4.59	6.66	7.23	28213	6.05	9.53	8532	71.14
Run	**Entire**	**0.9791**	**22328**	**4.79**	**7.01**	**7.61**	**29451**	**6.32**	**10.04**	**8988**	**74.95**
2402	a	0.9789	23217	4.98	6.90	7.51	30440	6.53	9.84	8794	73.99
Medium	b	0.9803	22961	4.92	7.33	7.98	30217	6.48	10.50	9377	78.90
Mine	c	0.9807	22710	4.87	6.82	7.41	29886	6.41	9.75	8716	73.34
Run	**Entire**	**0.9801**	**22970**	**4.93**	**7.02**	**7.64**	**30183**	**6.48**	**10.04**	**8969**	**75.46**
2416	a	0.9888	24037	5.16	7.08	7.80	31729	6.81	10.29	9076	76.15
Medium	b	0.9883	23557	5.05	6.92	7.62	31143	6.68	10.07	8891	74.60
3" x 6"	c	0.9886	23952	5.14	6.95	7.65	31736	6.81	10.14	8949	75.09
Egg	**Entire**	**0.9885**	**23788**	**5.11**	**6.97**	**7.67**	**31448**	**6.75**	**10.14**	**8949**	**75.09**
2405	a	0.9612	41962	9.01	6.03	6.58	56649	12.16	8.88	7910	67.30
High	b	0.9622	42155	9.09	5.48	5.96	57306	12.30	8.07	7200	61.27
Mine	c	0.9660	42176	9.05	5.54	6.03	57107	12.25	8.17	7288	62.01
Run	**Entire**	**0.9628**	**42176**	**9.05**	**5.73**	**6.24**	**57022**	**12.24**	**8.44**	**7531**	**64.08**
2406	a	0.9579	42309	9.08	6.35	6.91	57201	12.27	9.34	8337	70.20
High	b	0.9607	42321	9.08	6.00	6.53	57345	12.31	8.85	7900	66.52
Mine	c	0.9549	41200	8.84	5.70	6.20	55909	12.00	8.42	7498	63.14
Run	**Entire**	**0.9574**	**41946**	**9.00**	**6.04**	**6.58**	**56795**	**12.19**	**8.91**	**7949**	**66.93**
2413	a	0.9449	42374	9.09	6.36	6.98	57332	12.30	9.45	8357	69.70
High	b	0.9478	44272	9.50	6.28	6.90	60077	12.89	9.36	8279	69.05
2" x 3"	c	0.9489	42381	9.09	5.93	6.52	57553	12.35	8.85	7822	65.24
Nut	**Entire**	**0.9470**	**42720**	**9.17**	**6.16**	**6.77**	**57929**	**12.43**	**9.18**	**8114**	**67.67**

section. During the high rate tests the rate of evaporation increased with the increasing rate of combustion during the first part of the test, but decreased during the latter part. This indication of somewhat poorer performance during the latter part of a long test is shown more exactly by the values relating to efficiency.

Columns 30, 31 and 32 show the efficiency of the locomotive boiler as a heat transferring device. For the medium rate tests the results indicate that the efficiency during the first and middle sections of the tests was higher than during the last section. For two out of three of the medium rate tests the middle section showed a materially higher efficiency than either the first or last sections. For the high rate tests efficiency decreased in general from the first to the last section of each test, showing the best performance during the first part of the test and poorer performance as the test proceeded.

The differences in performance which have been pointed out as existing between different parts of a long test are in general small. Where test conditions are not unusual and are under control, as is the case in a testing laboratory, and where it is possible to maintain uniformly good fire-box conditions, short tests should give almost as reliable and almost the same results regarding evaporative performance and efficiency as much longer tests.

Boiler efficiency in general decreases as a test proceeds, so that, in so far as differences in efficiency exist, the average result for a long test would be lower than for a short test corresponding to the first part of the long test, and higher than for a short test corresponding to the last part of the long test. Boiler efficiency is more apt to be uniform throughout long tests at medium rates of combustion than at high rates of combustion.

The coal used during these tests gave little trouble in the firebox, a very small amount of clinkers being formed and the ash being readily removed. With coal which clinkers badly or which produces excessive honeycombing, the variations in performance between different parts of a long test might be much greater than those here shown.

LIST OF
PUBLICATIONS OF THE ENGINEERING EXPERIMENT STATION

Bulletin No. 1. Tests of Reinforced Concrete Beams, by Arthur N. Talbot, 1904. *None available.*

Circular No. 1. High-Speed Tool Steels, by L. P. Breckenridge. 1905. *None available.*

Bulletin No. 2. Tests of High-Speed Tool Steels on Cast Iron, by L. P. Breckenridge and Henry B. Dirks. 1905. *None available.*

Circular No. 2. Drainage of Earth Roads, by Ira O. Baker. 1906. *None available.*

Circular No. 3. Fuel Tests with Illinois Coal (Compiled from tests made by the Technological Branch of the U. S. G. S., at the St. Louis, Mo., Fuel Testing Plant, 1904–1907), by L. P. Breckenridge and Paul Diserens. 1909. *Thirty cents.*

Bulletin No. 3. The Engineering Experiment Station of the University of Illinois, by L. P. Breckenridge. 1906. *None available.*

Bulletin No. 4. Tests of Reinforced Concrete Beams, Series of 1905, by Arthur N. Talbot. 1906. *Forty-five cents.*

Bulletin No. 5. Resistance of Tubes to Collapse, by Albert P. Carman and M. L. Carr. 1906. *None available.*

Bulletin No. 6. Holding Power of Railroad Spikes, by Roy I. Webber, 1906. *None available.*

Bulletin No. 7. Fuel Tests with Illinois Coals, by L. P. Breckenridge, S. W. Parr, and Henry B. Dirks. 1906. *None available.*

Bulletin No. 8. Tests of Concrete: I, Shear; II, Bond, by Arthur N. Talbot. 1906. *None available.*

Bulletin No. 9. An Extension of the Dewey Decimal System of Classification Applied to the Engineering Industries, by L. P. Breckenridge and G. A. Goodenough. 1906. Revised Edition 1912. *Fifty cents.*

Bulletin No. 10. Tests of Concrete and Reinforced Concrete Columns, Series of 1906, by Arthur N. Talbot. 1907. *None available.*

Bulletin No. 11. The Effect of Scale on the Transmission of Heat through Locomotive Boiler Tubes, by Edward C. Schmidt and John M. Snodgrass. 1907. *None available.*

Bulletin No. 12. Tests of Reinforced Concrete T-Beams, Series of 1906, by Arthur N. Talbot. 1907. *None available.*

Bulletin No. 13. An Extension of the Dewey Decimal System of Classification Applied to Architecture and Building, by N. Clifford Ricker. 1907. *None available.*

Bulletin No. 14. Tests of Reinforced Concrete Beams, Series of 1906, by Arthur N. Talbot. 1907. *None available.*

Bulletin No. 15. How to Burn Illinois Coal Without Smoke, by L. P. Breckenridge. 1908. *None available.*

Bulletin No. 16. A Study of Roof Trusses, by N. Clifford Ricker. 1908. None available.

Bulletin No. 17. The Weathering of Coal, by S. W. Parr, N. D. Hamilton, and W. F. Wheeler. 1908. *None available.*

Bulletin No. 18. The Strength of Chain Links, by G. A. Goodenough and L. E. Moore. 1908. *Forty cents.*

Bulletin No. 19. Comparative Tests of Carbon, Metallized Carbon and Tantalum Filament Lamps, by T. H. Amrine. 1908. *None available.*

Bulletin No. 20. Tests of Concrete and Reinforced Concrete Columns, Series of 1907, by Arthur N. Talbot. 1908. *None available.*

Bulletin No. 21. Tests of a Liquid Air Plant, by C. S. Hudson and C. M. Garland. 1908. *Fifteen cents.*

Bulletin No. 22. Tests of Cast-Iron and Reinforced Concrete Culvert Pipe, by Arthur N. Talbot. 1908. *None available.*

Bulletin No. 23. Voids, Settlement, and Weight of Crushed Stone, by Ira O. Baker. 1908. *Fifteen cents.*

**Bulletin No. 24.* The Modification of Illinois Coal by Low Temperature Distillation, by S. W. Parr and C. K. Francis. 1908. *Thirty cents.*

Bulletin No. 25. Lighting Country Homes by Private Electric Plants, by T. H. Amrine. 1908 *Twenty cents.*

---

*A limited number of copies of bulletins starred is available for free distribution.

Bulletin No. 26. High Steam-Pressures in Locomotive Service. A Review of a Report to the Carnegie Institution of Washington, by W. F. M. Goss. 1908. *Twenty-five cents.*

Bulletin No. 27. Tests of Brick Columns and Terra Cotta Block Columns, by Arthur N. Talbot and Duff A. Abrams. 1909. *Twenty-five cents.*

Bulletin No. 28. A Test of Three Large Reinforced Concrete Beams, by Arthur N. Talbot. 1909. *Fifteen cents.*

Bulletin No. 29. Tests of Reinforced Concrete Beams: Resistance to Web Stresses, Series of 1907 and 1908, by Arthur N. Talbot. 1909. *Forty-five cents.*

**Bulletin No. 30.* On the Rate of Formation of Carbon Monoxide in Gas Producers, by J. K. Clement, L. H. Adams, and C. N. Haskins. 1909. *Twenty-five cents.*

**Bulletin No. 31.* Fuel Tests with House-heating Boilers, by J. M. Snodgrass. 1909. *Fifty-five cents.*

Bulletin No. 32. The Occluded Gases in Coal, by S. W. Parr and Perry Barker. 1909. *Fifteen cents.*

Bulletin No. 33. Tests of Tungsten Lamps, by T. H. Amrine and A. Guell. 1909. *Twenty cents.*

**Bulletin No. 34.* Tests of Two Types of Tile-Roof Furnaces under a Water-Tube Boiler, by J. M. Snodgrass. 1909. *Fifteen cents.*

Bulletin No. 35. A Study of Base and Bearing Plates for Columns and Beams, by N. Clifford Ricker. 1909. *Twenty cents.*

Bulletin No. 36. The Thermal Conductivity of Fire-Clay at High Temperatures, by J. K. Clement and W. L. Egy. 1909. *Twenty cents.*

Bulletin No. 37. Unit Coal and the Composition of Coal Ash, by S. W. Parr and W. F. Wheeler. 1909. *Thirty-five cents.*

**Bulletin No. 38.* The Weathering of Coal, by S. W. Parr and W. F. Wheeler. 1909. *Twenty-five cents.*

**Bulletin No. 39.* Tests of Washed Grades of Illinois Coal, by C. S. McGovney. 1909. *Seventy-five cents.*

Bulletin No. 40. A Study in Heat Transmission, by J. K. Clement and C. M. Garland. 1910. *Ten cents.*

Bulletin No. 41. Tests of Timber Beams, by Arthur N. Talbot. 1910. *Thirty-five cents.*

**Bulletin No. 42.* The Effect of Keyways on the Strength of Shafts, by Herbert F. Moore. 1910. *Ten cents.*

Bulletin No. 43. Freight Train Resistance, by Edward C. Schmidt. 1910. *Seventy-five cents.*

Bulletin No. 44. An Investigation of Built-up Columns Under Load, by Arthur N. Talbot and Herbert F. Moore. 1911. *Thirty-five cents.*

**Bulletin No. 45.* The Strength of Oxyacetylene Welds in Steel, by Herbert L. Whittemore. 1911. *Thirty-five cents.*

**Bulletin No. 46.* The Spontaneous Combustion of Coal, by S. W. Parr and F. W. Kressman. 1911. *Forty-five cents.*

**Bulletin No. 47.* Magnetic Properties of Heusler Alloys, by Edward B. Stephenson, 1911. *Twenty-five cents.*

**Bulletin No. 48.* Resistance to Flow Through Locomotive Water Columns, by Arthur N. Talbot and Melvin L. Enger. 1911. *Forty cents.*

**Bulletin No. 49.* Tests of Nickel-Steel Riveted Joints, by Arthur N. Talbot and Herbert F. Moore. 1911. *Thirty cents.*

**Bulletin No. 50.* Tests of a Suction Gas Producer, by C. M. Garland and A. P. Kratz. 1912. *Fifty cents.*

Bulletin No. 51. Street Lighting, by J. M. Bryant and H. G. Hake. 1912. *Thirty-five cents.*

**Bulletin No. 52.* An Investigation of the Strength of Rolled Zinc, by Herbert F. Moore. 1912. *Fifteen cents.*

**Bulletin No. 53.* Inductance of Coils, by Morgan Brooks and H. M. Turner. 1912. *Forty cents.*

**Bulletin No. 54.* Mechanical Stresses in Transmission Lines, by A. Guell. 1912. *Twenty cents.*

**Bulletin No. 55.* Starting Currents of Transformers, with Special Reference to Transformers with Silicon Steel Cores, by Trygve D. Yensen. 1912. *Twenty cents.*

**Bulletin No. 56.* Tests of Columns: An Investigation of the Value of Concrete as Reinforcement for Structural Steel Columns, by Arthur N. Talbot and Arthur R. Lord. 1912. *Twenty-five cents.*

**Bulletin No. 57.* Superheated Steam in Locomotive Service. A Review of Publication No. 127 of the Carnegie Institution of Washington, by W. F. M. Goss. 1912. *Forty cents.*

* A limited number of copies of bulletins starred is available for free distribution.

**Bulletin No. 58.* A New Analysis of the Cylinder Performance of Reciprocating Engines, by J. Paul Clayton. 1912. *Sixty cents.*

**Bulletin No. 59.* The Effect of Cold Weather Upon Train Resistance and Tonnage Rating, by Edward C. Schmidt and F. W. Marquis. 1912. *Twenty cents.*

**Bulletin No. 60.* The Coking of Coal at Low Temperatures, with a Preliminary Study of the By-Products, by S. W. Parr and H. L. Olin. 1912. *Twenty-five cents.*

**Bulletin No. 61.* Characteristics and Limitation of the Series Transformer, by A. R. Anderson and H. R. Woodrow. 1913. *Twenty-five cents.*

Bulletin No. 62. The Electron Theory of Magnetism, by Elmer H. Williams. 1913. *Thirty-five cents.*

Bulletin No. 63. Entropy-Temperature and Transmission Diagrams for Air, by C. R. Richards. 1913. *Twenty-five cents.*

**Bulletin No. 64.* Tests of Reinforced Concrete Buildings Under Load, by Arthur N. Talbot and Willis A. Slater. 1913. *Fifty cents.*

**Bulletin No. 65.* The Steam Consumption of Locomotive Engines from the Indicator Diagrams, by J. Paul Clayton. 1913. *Forty cents.*

Bulletin No. 66. The Properties of Saturated and Superheated Ammonia Vapor, by G. A. Goodenough and William Earl Mosher. 1913. *Fifty cents.*

Bulletin No. 67. Reinforced Concrete Wall Footings and Column Footings, by Arthur N. Talbot. 1913. *Fifty cents.*

**Bulletin No. 68.* Strength of I-Beams in Flexure, by Herbert F. Moore. 1913. *Twenty cents.*

Bulletin No. 69. Coal Washing in Illinois, by F. C. Lincoln. 1913. *Fifty cents.*

Bulletin No. 70. The Mortar-Making Qualities of Illinois Sands, by C. C. Wiley. 1913. *Twenty cents.*

Bulletin No. 71. Tests of Bond between Concrete and Steel, by Duff A. Abrams. 1914. *One dollar.*

**Bulletin No. 72.* Magnetic and Other Properties of Electrolytic Iron Melted in Vacuo, by Trygve D. Yensen. 1914. *Forty cents.*

Bulletin No. 73. Acoustics of Auditoriums, by F. R. Watson. 1914. *Twenty cents.*

**Bulletin No. 74.* The Tractive Resistance of a 28-Ton Electric Car, by Harold H. Dunn. 1914. *Twenty-five cents.*

Bulletin No. 75. Thermal Properties of Steam, by G. A. Goodenough. 1914. *Thirty-five cents.*

Bulletin No. 76. The Analysis of Coal with Phenol as a Solvent, by S. W. Parr and H. F. Hadley. 1914. *Twenty-five cents.*

**Bulletin No. 77.* The Effect of Boron upon the Magnetic and Other Properties of Electrolytic Iron Melted in Vacuo, by Trygve D. Yensen. 1915. *Ten cents.*

**Bulletin No. 78.* A Study of Boiler Losses, by A. P. Kratz. 1915. *Thirty-five cents.*

**Bulletin No. 79.* The Coking of Coal at Low Temperatures, with Special Reference to the Properties and Composition of the Products, by S. W. Parr and H. L. Olin. 1915. *Twenty-five cents.*

**Bulletin No. 80.* Wind Stresses in the Steel Frames of Office Buildings, by W. M. Wilson and G. A. Maney. 1915. *Fifty cents.*

**Bulletin No. 81.* Influence of Temperature on the Strength of Concrete, by A B. McDaniel. 1915. *Fifteen cents.*

Bulletin No. 82. Laboratory Tests of a Consolidation Locomotive, by E. C. Schmidt, J. M. Snodgrass and R. B. Keller. 1915. *Sixty-five cents.*

**Bulletin No. 83.* Magnetic and Other Properties of Iron-Silicon Alloys. Melted in Vacuo, by Trygve D. Yensen. 1915. *Thirty-five cents.*

Bulletin No. 84. Tests of Reinforced Concrete Flat Slab Structure, by A. N. Talbot and W. A. Slater. 1916. *Sixty-five cents.*

**Bulletin No. 85.* Strength and Stiffness of Steel Under Biaxial Loading, by A. J. Becker. 1916. *Thirty-five cents.*

**Bulletin No. 86.* The Strength of I-Beams and Girders, by Herbert F. Moore and W. M. Wilson. 1916. *Thirty cents.*

**Bulletin No. 87.* Correction of Echoes in the Auditorium, University of Illinois, by F. R. Watson and J. M. White. 1916. *Fifteen cents.*

Bulletin No. 88. Dry Preparation of Bituminous Coal at Illinois Mines, by E. A. Holbrook. 1916. *Seventy cents.*

---

* A limited number of copies of bulletins starred is available for free distribution.

**Bulletin No. 89.* Specific Gravity Studies of Illinois Coal, by Merle L. Nebel. 1916. *Thirty cents.*

**Bulletin No. 90.* Some Graphical Solutions of Electric Railway Problems by A. M. Buck. 1916. *Twenty cents.*

**Bulletin No. 91.* Subsidence Resulting from Mining, by L. E. Young and H. H. Stoek. 1916.

**Bulletin No. 92.* The Tractive Resistance on Curves of a 28-Ton Electric Car, by E. C. Schmidt and H. H. Dunn. 1916. *Twenty-five cents.*

**Bulletin No. 93.* A Preliminary Study of the Alloys of Chromium, Copper, and Nickel, by D. F. McFarland and O. E. Harder. 1916. *Thirty-five cents.*

**Bulletin No. 94.* The Embrittling Action of Sodium Hydroxide on Soft Steel, by S. W. Parr. 1917. *Thirty cents.*

**Bulletin No. 95.* Magnetic and Other Properties of Iron-Aluminum Alloys Melted in Vacuo, by Trygve D. Yensen and Walter A. Gatward. 1917. *Twenty-five cents.*

**Bulletin No. 96.* The Effect of Mouthpieces on the Flow of Water Through a Submerged Short Pipe, by Fred B. Seely. 1917. *Twenty-five cents.*

**Bulletin No. 97.* Effects of Storage Upon the Properties of Coal, by S. W. Parr. 1917. *Twenty cents.*

**Bulletin No. 98.* Tests of Oxyacetylene Welded Joints in Steel Plates, by Herbert F. Moore. 1917. *Ten cents.*

Circular No. 4. The Economical Purchase and Use of Coal for Heating Homes with Special Reference to Conditions in Illinois. 1917. *Ten cents.*

**Bulletin No. 99.* The Collapse of Short Thin Tubes, by A. P. Carman. 1917. *Twenty cents.*

**Circular No. 5.* The Utilization of Pyrite Occurring in Illinois Bituminous Coal, by E. A. Holbrook. 1917. *Twenty cents.*

**Bulletin No. 100.* Percentage of Extraction of Bituminous Coal with Special Reference to Illinois Conditions, by C. M. Young. 1917.

**Bulletin No. 101.* Comparative Tests of Six Sizes of Illinois Coal on a Mikado Locomotive, by E. C. Schmidt, J. M. Snodgrass, and O. S. Beyer, Jr. 1917. *Fifty cents.*

---

* A limited number of copies of bulletins starred is available for free distribution.

THE UNIVERSITY OF ILLINOIS

THE STATE UNIVERSITY

Urbana

EDMUND J. JAMES, Ph. D., LL. D., President

THE UNIVERSITY INCLUDES THE FOLLOWING DEPARTMENTS:

The Graduate School

The College of Liberal Arts and Sciences (Ancient and Modern Languages and Literatures; History, Economics, Political Science, Sociology; Philosophy, Psychology, Education; Mathematics; Astronomy; Geology; Physics; Chemistry; Botany, Zoology, Entomology; Physiology; Art and Design)

The College of Commerce and Business Administration (General Business, Banking, Insurance; Accountancy; Railway Administration, Foreign Commerce; Courses for Commercial Teachers and Commercial and Civic Secretaries)

The College of Engineering (Architecture; Architectural, Ceramic, Civil, Electrical Mechanical, Mining, Municipal and Sanitary, and Railway Engineering)

The College of Agriculture (Agronomy; Animal Husbandry; Dairy Husbandry; Horticulture and Landscape Gardening; Agricultural Extension; Teachers' Course; Household Science)

The College of Law (three years' course)

The School of Education

The Course in Journalism

The Courses in Chemistry and Chemical Engineering

The School of Railway Engineering and Administration

The School of Music (four years' course)

The School of Library Science (two years' course)

The College of Medicine (in Chicago)

The College of Dentistry (in Chicago)

The School of Pharmacy (in Chicago; Ph. G. and Ph. C. courses)

The Summer Session (eight weeks)

Experiment Stations and Scientific Bureaus: U. S. Agricultural Experiment Station; Engineering Experiment Station; State Laboratory of Natural History; State Entomologist's Office; Biological Experiment Station on Illinois River; State Water Survey; State Geological Survey; U. S. Bureau of Mines Experiment Station.

The library collections contain (July 1, 1917) 400,720 volumes and 102,029 pamphlets.

For catalogs and information address

THE REGISTRAR
URBANA, ILLINOIS

UNIVERSITY OF ILLINOIS BULLETIN

ISSUED WEEKLY

Vol. XV NOVEMBER 19, 1917 No. 12

[Entered as second-class matter Dec. 11, 1912, at the Post Office at Urbana, Ill., under the Act of Aug. 24, 1912.]

A STUDY OF THE HEAT TRANSMISSION OF BUILDING MATERIALS

BY

A. C. WILLARD

AND

L. C. LICHTY

BULLETIN No. 102

ENGINEERING EXPERIMENT STATION

PUBLISHED BY THE UNIVERSITY OF ILLINOIS, URBANA

PRICE: TWENTY-FIVE CENTS

EUROPEAN AGENT

CHAPMAN & HALL, LTD., LONDON

THE Engineering Experiment Station was established by act of the Board of Trustees, December 8, 1903. It is the purpose of the Station to carry on investigations along various lines of engineering and to study problems of importance to professional engineers and to the manufacturing, railway, mining, constructional, and industrial interests of the State.

The control of the Engineering Experiment Station is vested in the heads of the several departments of the College of Engineering. These constitute the Station Staff and, with the Director, determine the character of the investigations to be undertaken. The work is carried on under the supervision of the Staff, sometimes by research fellows as graduate work, sometimes by members of the instructional staff of the College of Engineering, but more frequently by investigators belonging to the Station corps.

The results of these investigations are published in the form of bulletins which record mostly the experiments of the Station's own staff of investigators. There will also be issued from time to time, in the form of circulars, compilations giving the results of the experiments of engineers, industrial works, technical institutions, and governmental testing departments.

The volume and number at the top of the title page of the cover are merely arbitrary numbers and refer to the general publications of the University of Illinois: *either above the title or below the seal is given the number of the Engineering Experiment Station bulletin or circular which should be used in referring to these publications.*

For copies of bulletins, circulars, or other information address the

ENGINEERING EXPERIMENT STATION,
URBANA, ILLINOIS.

UNIVERSITY OF ILLINOIS
ENGINEERING EXPERIMENT STATION

BULLETIN No. 102 NOVEMBER, 1917

A STUDY OF THE HEAT TRANSMISSION OF BUILDING MATERIALS

BY

A. C. WILLARD

PROFESSOR OF HEATING AND VENTILATION

AND

L. C. LICHTY

RESEARCH FELLOW, ENGINEERING EXPERIMENT STATION

ENGINEERING EXPERIMENT STATION

PUBLISHED BY THE UNIVERSITY OF ILLINOIS, URBANA

CONTENTS

CONTENTS (CONTINUED)

LIST OF FIGURES

LIST OF TABLES

A STUDY OF THE HEAT TRANSMISSION OF BUILDING MATERIALS*

I. Introduction

1. *Purpose of Investigation.*—The object of this investigation is to determine the coefficients of heat transmission of standard building materials for exterior building walls under conditions similar to those commonly found in practice. In applying these coefficients to actual problems it is necessary to take account of the temperature of the air inside and outside the building; therefore the data must be determined under similar or comparable conditions. Since it is impossible to test all types and combinations of exterior wall construction, some method must be devised for applying the data obtained from the tests of a relatively small number of simple walls. This method should permit, by a simple calculation, the determination of the coefficient of transmission in the case of the various compounds and special walls. It is necessary, therefore, only to apply the simple principles underlying the transfer of heat to, through, and from a building wall, and to determine the basic coefficients required in making the calculations of the actual transmission coefficient. In order to explain this procedure for the general case it is necessary to analyze the process by which heat transmission through a wall takes place.

2. *Acknowledgments.*—The writers acknowledge their indebtedness to Professor L. A. Harding, formerly in charge of the work in Experimental Mechanical Engineering at the University of Illinois, and to Dean C. R. Richards, Dean of the College of Engineering and formerly head of the Department of Mechanical Engineering at the University of Illinois. This investigation was begun under Professor Harding's supervision, and a large part of the test data was secured before his resignation. Dean Richards contributed valuable suggestions during the progress of the tests.

*The results here reported do not constitute results of a complete series of experiments. It is the present purpose to continue the study of heat transmission of building materials and to release results for publication from time to time as the work proceeds.

II. Principles of Heat Transmission

3. *Conduction.*—Heat passes through a wall by conduction provided the temperature on one side of the wall is higher than that on the other. The amount of heat (H) passing through the wall depends upon the following factors:

(1) The coefficient of conductivity (C) which varies with the material of the wall. This coefficient is practically constant* for any given wall for the ranges of surface temperatures found in heating buildings; it is not constant, however, at high temperatures if there is much variation in the mean temperature of the wall.†

(2) The thickness of the wall in inches (x).

(3) The difference in temperature in degrees F. between the two surfaces ($t_1 - t_2$).

(4) The time in hours (D).

(5) The mean area of the cross section in square feet through which the heat passes (A).

The nature of the two surfaces is of no consequence provided the faces are parallel. The amount of heat transferred is also independent of the actual temperatures of the two surfaces but it depends upon the difference in temperatures between the two surfaces; thus the heat transmitted by conduction is, in B. t. u.,

$$H_C = \frac{C}{x}(t_1 - t_2)\ D\ A. \quad . \quad . \quad . \quad . \quad (1)$$

It is at once apparent that all these factors except the surface temperatures are easily ascertainable, and it should be noted that the inside and outside air temperatures are of no value in determining the amount of heat transmitted by conduction.

Heat reaches the surface of a wall both by radiation and by convection provided the objects and the air are at a higher temperature than the surface of the wall. By a reverse process ‡ of radiation and convection, the heat which has been received by the inner surface of a wall and transmitted through it by conduction may be given off

*Tests by Grober and Nüsselt in "Mechanical Refrigeration," Macintire, J. H., p. 155, 1914.

†Clement, J. K., and Egy, W. L., "The Thermal Conductivity of Fire-Clay at High Temperatures." Univ. of Ill. Eng. Exp. Sta., Bul. 36, 1909.

‡Preston, Thomas, "The Theory of Heat," 1904.

by the outer surface. In refrigeration, the direction of flow is reversed, but the coefficient is the same for ordinary ranges of temperature. The air and objects on the cooler side of the wall must, of course, remain at a lower temperature than the surface of the wall.

4. *Radiation.*—Newton seems to have been the first to consider the "law of cooling" of a body by radiation, and, as stated by Preston, "he supposed that the rate of cooling was proportional to the excess of the temperature of the body above that of the medium in which it was immersed." The same authority expresses the heat lost by radiation in a unit of time as the difference between the total heat given off by radiation *from* the body and the total heat gained by radiation *to* the body or $(ft - ft_1)$, in which

ft = the total heat loss per second by radiation when the the body is at the temperature (t), and

ft_1 = the total heat received per second by radiation when the surroundings are at the temperature (t_1).

This is equivalent to assuming that for the absolute scale of temperature the function $ft = Rt - B$ so that $ft - ft_1 = R\ (t - t_1)$ and the heat lost by radiation is

$$H_R = R\ (t - t_1)\ D\ A \text{ in B. t. u.,} \quad . \quad . \quad . \quad (2)$$

in which R is the coefficient of radiation for the surface of the material under consideration, t and t_1 are Fahrenheit temperature readings, D is the time in hours, and A is the area in square feet. Preston says, "This formula has been found to represent the facts fairly well for small differences of temperature. For differences exceeding 40 or 50 degrees C. this law was found to deviate seriously from the truth."*

Where the differences in temperature become large and when "black bodies," which are perfect radiators and perfect absorbers of heat, are used, the total energy transferred by radiation has been shown by Stefan† and Boltzmann‡ to depend: first, upon a factor called coefficient of radiation (R), and secondly, upon the difference

*Preston, Op. Cit.

†Joseph Stefan, physicist, was born in Austria-Hungary in 1835 and died in 1893. He was Professor of Higher Mathematics and Physics at the University of Vienna, and carried on extensive investigations on the transmisson of light, heat and sound, and the diffusion of gases.

‡Ludwig Boltzmann, physicist, was born in Vienna, Austria, in 1844 and died in 1906. He was Professor of Experimental and Theoretical Physics at the Universities of Gratz and Vienna, and made many valuable researches in the field of thermodynamics. He published a great many books and papers on various physical subjects and problems, including "Der Zweite Hauptsatz der Mechanischen Wärmtheorie."

between the fourth powers of the absolute temperatures of the surfaces between which the transfer takes place. For English units, therefore, the heat transmitted by radiation is, in B. t. u.,

$$H_R = R\ (T_1^4 - T_2^4)\ D\ A \quad . \quad . \quad . \quad . \quad . \qquad (2a)$$

in which, as previously, D is the time in hours, and A is the area of the surface in square feet, T_1 and T_2 are degrees F. on the absolute scale and in the case of sooted surfaces with the cooler entirely surrounding the warmer surface, which must not "see" anything but the cooler surface, $R = 1.6 \times 10^{-9}$. The experimental determination of values of R for either small or great temperature differences for commercial building materials would require rather elaborate equipment.

It is evident, from equation (2a) that for large temperature differences, a most rapid increase in radiation will occur: first, with an increase in the temperature of the warmer body; and secondly, if, for the same temperature differences, both temperatures T_1 and T_2 are increased.* It should be noted that the energy transferred by radiation must first be transformed at the surface of the warmer body from the form of sensible heat or kinetic energy into radiant energy, so that it may pass, like light,† through space to the surface of the cooler body, and there be retransformed into heat energy.‡

5. *Convection.*—The amount of heat transferred by convection, or air movement, over the surface of a body has been shown by Peclet¶ to depend upon:

(1) A factor, coefficient of convection (N), which varies with the form and arrangement of the surface.

(2) Some power of the velocity of flow across the surface in feet per second $V^{\frac{1}{n}}$.

(3) The difference in temperature between the surface and the surrounding air $(t_1 - t_2)$ in degrees F.

The heat transferred by convection is independent of the nature of the surface, and the absolute temperature of the surface and the surrounding air; consequently for English units, the heat transferred from or to a wall by convection is, in B. t. u.,

$$H_N = N\ V^{\frac{1}{n}}\ (t_1 - t_2)\ D\ A \quad . \quad . \quad . \quad . \quad . \qquad (3)$$

---

*Burgess, G. K., and Le Chatelier, H., "The Measurement of High Temperatures." p. 247, 1912.

†"Heat Transmission," Engineering, Vol. 100, p. 499, November 12, 1915.

‡Bolton, R. P., "The Establishment of a Standard for Transmission Losses from Buildings of All Constructions," Journal A. S. H. and V. E., Vol. 21, p. 30, July, 1915.

¶"Traité de la Chaleur." Vol. I, Livre VI, 1860.

in which D is the time in hours and A is the area of the surface in square feet. The values of N and n must be determined experimentally. Elaborate testing equipment is required for the determination of these values.

6. *Heat Transmission to, through, and from a Simple Wall.*—In view of the foregoing analysis, it will be apparent from the diagram, Fig. 1, that the temperature gradient through a wall, which

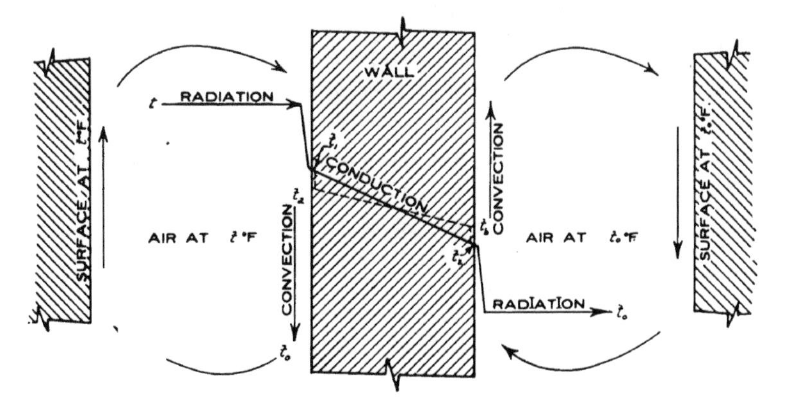

FIG. 1. TEMPERATURE GRADIENT THROUGH A WALL

separates air at a higher temperature (t) from air at a lower temperature (t_0), can be represented by the line t, t_1, t_2, t_0, provided it is recognized that the surface temperature of the wall and the air in contact with it are not the same. This difference in temperature occurs within a thin layer or film of air very close to the surface. That the wall surface must be at a lower or at a higher temperature than the surrounding air and the objects is evident; otherwise there would be no "flow" or transfer of heat to or from the wall.

It has also been suggested by Dalby* that there is probably a further drop of temperature head, represented by the distance between t_1 and t_a, which is required to force the flow across the surface where the gas film is in contact with the wall, corresponding to a po-

*"Heat Transmission," Inst. M. E. (British), p. 921, October, 1909.

tential difference at a joint in an electric circuit. This is usually referred to as contact resistance.* Contact resistance cannot be measured by thickness, but requires an appreciable amount of potential difference, whether the potential difference be in volts or degrees, and whether the current flow be in amperes or thermal units.

If the air and surface temperatures are known, it is possible, for small temperature differences, to express the amount of heat entering the wall per square foot per hour in terms of equations (2) and (3), in which D and A become unity, as:

$$H_R+H_N=R\left[(t+460)-(t_1+460)\right]+N\times V^{\frac{1}{n}}(t-t_1) \quad . \quad . \quad (4)$$

For walls of fixed height, possibly from nine to ten feet, standing in still air so that only natural or gravity convection currents exist, the factor $\left(V^{\frac{1}{n}}\right)$, moreover, becomes a constant and can be included in N making it N_1, so that the combined or total heat transfer at the surface of the wall can be expressed as:

$$H_{R+N}=R\ (t-t_1)+N_1\ (t-t_1),$$
$$=(R+N_1)\ (t-t_1). \quad . \quad . \quad . \quad . \quad . \quad (5)$$

or, letting $K_1=(R+N_1)$,

$$H_{R+N}=K_1\ (t-t_1)\dagger \quad . \quad . \quad . \quad . \quad (6)$$

The heat passing through the wall per square foot per hour can be expressed by equation (1) as:

$$H_C=\frac{C}{X}\ (t_1-t_2), \quad . \quad . \quad . \quad . \quad . \quad (7)$$

if the two wall surface temperatures are known.

The heat leaving the wall for the limited range of temperatures found in practice can be expressed, furthermore, by equations (2) and (3) simplified to read as follows:

$$H_{R+N}=K_2\ (t_2-t_0), \quad . \quad . \quad . \quad (8)$$

in which $(R+N_2)=K_2$ are new coefficients determined under outside conditions of temperature and average wind movement.

Finally, if the overall transmission coefficient from air inside to air outside is U, in B. t. u. per square foot per hour, the heat transmitted is

$$H_U=U\ (t-t_0). \quad . \quad . \quad . \quad . \quad . \quad (9)$$

Since the heat H transmitted in B. t. u. per square foot per hour is the same in each of the above cases,

---

*Refrigerating World, p. 29, September, 1914.

†Harding and Willard, "Mechanical Equipment of Buildings." Vol. I, p. 55, 1916.

$$H = K_1 (t - t_1)$$
$$= \frac{C}{X} (t_1 - t_2)$$
$$= K_2 (t_2 - t_0)$$
$$= U (t - t_0)$$

$$\text{and} \frac{1}{K_1} + \frac{X}{C} + \frac{1}{K_2} = \frac{t - t_1}{H} + \frac{(t_1 - t_2)}{H} + \frac{(t_2 - t_0)}{H} = \frac{t - t_0}{H}$$

$$\text{But} \frac{1}{U} = \frac{t - t_0}{H};$$

$$\text{hence} \frac{1}{U} = \frac{1}{K_1} + \frac{1}{K_2} + \frac{X}{C}$$

$$\text{and } U = \frac{1}{\frac{1}{K_1} + \frac{1}{K_2} + \frac{X}{C}}$$

In the case of compound walls (See Fig. 2 A, B, and C) with no air spaces and with a variety of materials of different conductivities, C_1, C_2, C_3, etc., and of various thicknesses, X_1, X_2, X_3, etc., the equation assumes the form,

$$U = \frac{1}{\frac{1}{K_1} + \frac{1}{K_2} + \frac{X_1}{C_1} + \frac{X_2}{C_2} + \frac{X_3}{C_3} + \text{, etc.}} \quad . \quad . \quad . \quad (11)$$

If the compound wall contains air spaces, (See Fig. 2 D) the value of

$$U = \frac{1}{\frac{1}{K_1} + \frac{1}{K_2} + \frac{1}{K_3} + \frac{1}{K_4} + \frac{X_1}{C_1} + \frac{X_2}{C_2} + \frac{X_3}{C_3} + \text{, etc.}} \quad . \quad . \quad (12)$$

in which the proper surface coefficients for each material and for each surface must be used. All coefficients will be taken for "still air" conditions, except that for the outside wall the surface coefficient must be increased, as indicated under "Applications" (Fig. 2 D), to correspond with the average wind movement.

It must, therefore, be apparent that, if values of the combined surface coefficients (K_1) and (K_2) and of the coefficient of conductivity (C) can be determined experimentally by measuring the heat transmitted through typical wall materials, the very practical and useful coefficient of transmission (U) can be readily calculated by equation (10) for a wall of any given thickness (x). As will be shown later under methods of testing, it is a fairly simple matter to determine the surface coefficients of walls standing in still air and the coefficients of conductivity at the same time that the actual determination of the coefficient of transmission (U) is being made.

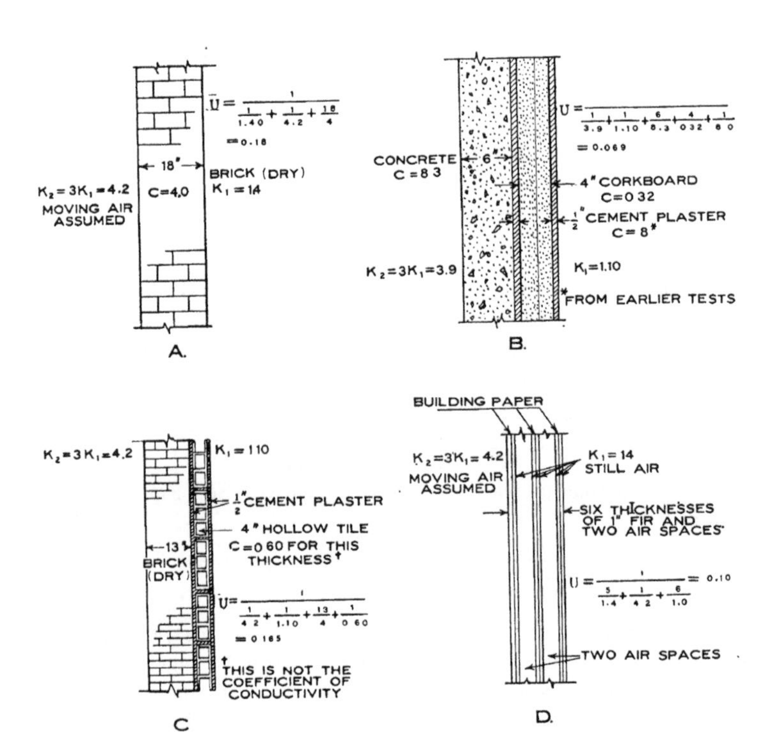

FIG. 2. APPLICATION OF TEST DATA TO SIMPLE AND COMPOUND WALLS, EXAMPLES IN THE CALCULATION OF U

III. METHODS OF TESTING FOR HEAT TRANSMISSION OF BUILDING MATERIALS

7. *General Conditions.*—There are two general cases to be considered in determining the value of the coefficient (U) for walls: first, the case of walls standing in perfectly still air, except for the vertical convection currents set up by the wall itself, and secondly, that of walls standing in moving air where the velocities cover the range found in atmospheric air during the average heating season. Both cases have been investigated in these tests so that the data reported are divided into two parts, one for "still air tests" and the other for "moving air tests."

8. *Investigations.*—Heat transmission tests previously made have in many cases been confined to small specimens so that the data secured have proved unsatisfactory when applied to walls of practical proportions. All investigators in this field have profited by the pioneer experimental work of the French physicist, Peclet.* His work was followed by investigations (which included the work of Rietschel† and Grashof‡), conducted under the auspices of the German and Austrian governments. The results of this work have been translated in part by J. H. Kinealy¶ and still form the basis for many transmission coefficients in use to-day.

The most prominent American investigator has been Prof. C. L. Norton§ of the Massachusetts Institute of Technology. The best

* Jean Claude Eugène Peclet was born at Besançon, France, in 1793 and died in Paris in 1857. He was Professor of Industrial Physics at the Central School of Arts and Manufacturers in Paris. Among his most important works is his "Traité de la Chaleur et de son Application aux Arts et aux Manufactures," published in two volumes in 1829, a second edition being issued in 1843 and translated into German.

† Herman Iman Rietschel was born at Dresden, Germany, in 1847. He was a professor in the Königliche Technische Hochschule, Berlin. He wrote "Leitfaden zum Berechnen und Entwerfen von Luftüngs—und Heitzungs-Anlagen. Ein Hand und Lehrbuch für Ingenieure und Architecken," which was published in 1894 (second edition). He also wrote "Theorie und Praxis der Bestimmung der Rohrweiten von Warmwasserheizungen," published in 1897.

‡ Franz Grashof was born at Düsseldorf, Germany, in 1826 and died in 1893. He was one of the founders of the Verein Deutscher Ingenieure and was president of this society for thirty-five years. He was Professor of Applied Mechanics and Director of the Department of Machine Construction in the Polytechnical High School, Karlsruhe. His writings are very numerous and include the "Resultate der Mechanischen Wärmetheorie," published in 1870.

¶ John Henry Kinealy was born at Hannibal, Missouri, in 1864. Among his other works is the translation of the Prussian tables for heat transmission. This translation was published in 1899 and is known as "Formulas and Tables for Heating."

§ Charles Ladd Norton was born at Springfield, Massachusetts, in 1870. He is Professor of Heat Measurements in the Massachusetts Institute of Technology. His investigations have included the transmission of heat through various building materials as well as the thermal conductivity of earthy materials and cold storage insulation.

equipped thermal-transmission testing plant in this country has been erected by the Armstrong Cork Company, at Beaver Falls, Pennsylvania. A similar plant is located at the Pennsylvania State College at State College, Pennsylvania. In the tests run at the former plant little attention has been given to surface temperatures, since only actual or overall transmission air to air coefficients were desired. In the plant at Pennsylvania State College both air and surface temperatures are measured by means of platinum resistance pyrometers, and the Engineering Experiment Station at State College has been studying the effect produced on the heat transmission by varying the relative humidity and velocity of the air passing over the outside surface of a building wall.

The Worcester Polytechnic Institute* has recently conducted a series of tests on the heat transmission of various types of ice house construction. Prof. J. R. Allen,† University of Michigan, has recently reported the results of tests on transmission coefficients for glass made under a variety of conditions.

The latest heat transmission tests of importance are those of L. B. McMillan made at the University of Wisconsin.‡ Steam pipe coverings were investigated to determine their heat insulating properties. For determining the temperature of the air in the test room, high grade mercury thermometers were used, and after considerable experimenting it was decided to use constantan-copper thermocouples for the pipe temperatures. In this connection the potentiometer method of measuring the electro motive force of the couples was used. To imbed the thermocouple junction in the pipe, a chip was raised on the surface of the pipe, the thermocouple junction was held underneath, and the chip was forced down; thus the couple was held in contact with the metal pipe.

Similar work is also in progress at the Mellon Institute.

---

*Refrigerating World, June, 1915.

†"Heat Transmission Through Building Materials," Journal A. S. H. and V. E., Vol. 22, p. 1, July, 1916.

‡Journal, Am. Soc. of M. E., January, 1916.

IV. Testing Methods and Equipment

9. *Methods.*—The equipment or apparatus for making heat transmission tests in building materials varies according to the method of testing and the data desired. Some excellent laboratory plants have been designed for making heat transmission tests, and some of the most elaborate of these have been used abroad.* The methods most commonly employed in this country may be classified, according to principle at least, as follows:

(1) Ice Box method.
(2) Oil Box method.
(3) Cold Air Box method.
(4) Hot Air Box method.
(5) Flat or Hot Plate method.

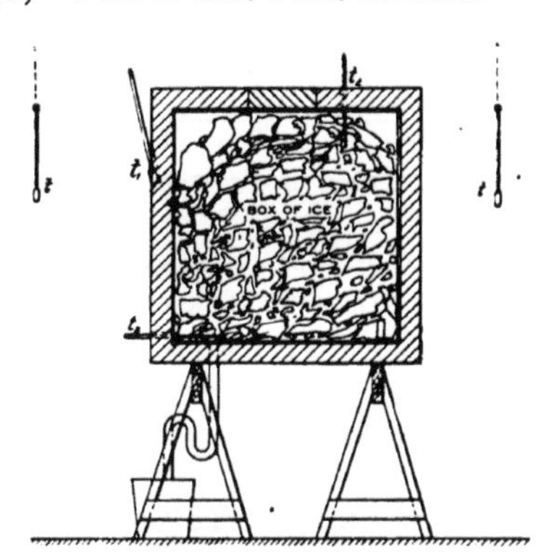

Fig. 3. The Ice Box Method of Testing Heat Transmission

10. *The Ice Box Method.*—This method, illustrated by Fig. 3, is the simplest one employed in making heat transmission tests. Ice is placed inside a metal box, or cube, and the material to be tested is placed outside. If the rate at which the ice melts, the temperature of the melting ice, and the outside air temperature are known, the heat transmission is readily obtainable.

This method may prove unsatisfactory in the following respects:

* Ohmes, Arthur K., "A Notable Institution for the Advancement of the Heating and Ventilating Art," Journal A. S. H. and V. E., Vol. 22, January, 1916.

(a) The melting of ice in pockets and its retention in the box after melting cause low results.

(b) Frequent additions of ice must be made to keep the box as full as possible.

(c) The inside temperature reading is not the true inside tem-

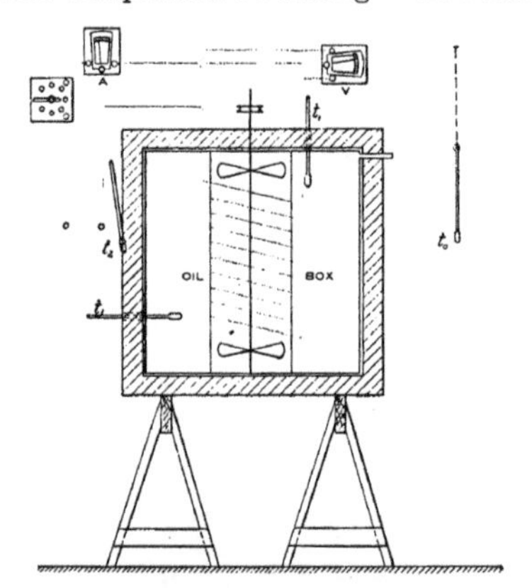

FIG. 4. THE OIL BOX METHOD OF TESTING HEAT TRANSMISSION

perature because of the temperature gradient through the walls of the metal box, which results in low coefficients.

(d) The range of temperature drop, through which the material is to be tested, is fixed unless some means are employed for regulating the outside air temperature.

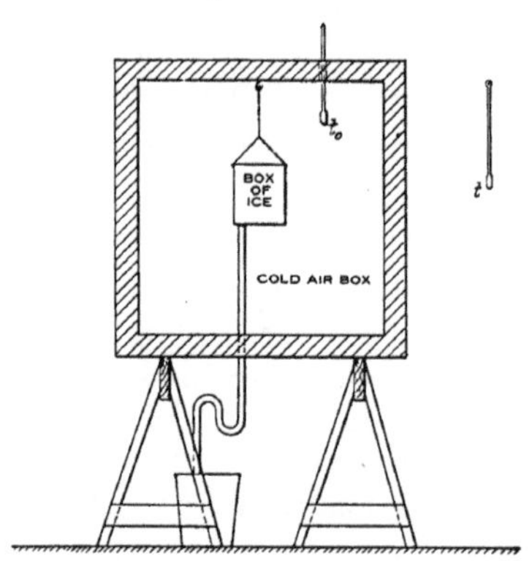

FIG. 5. THE COLD AIR BOX METHOD OF TESTING HEAT TRANSMISSION

11. *The Oil Box Method.*—In this method, illustrated by Fig. 4, a metal box is covered with the material to be tested. Oil is placed inside the box and is kept at the desired temperature by means of an electrical heater immersed in the oil. A stirring device keeps the oil at uniform temperature. The amount of heat transmitted through the material under test is determined from the electrical input.

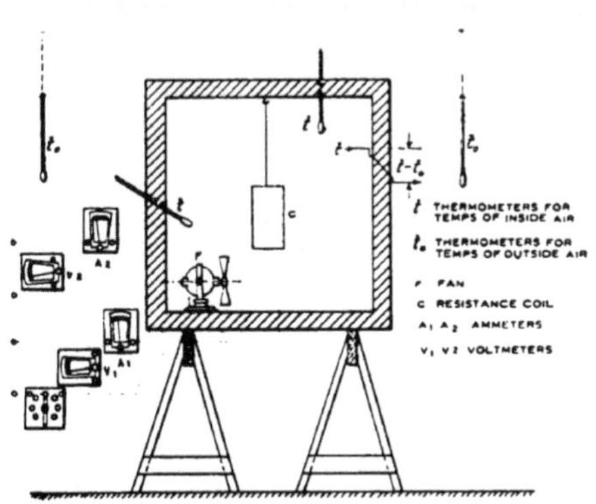

FIG. 6. THE HOT AIR BOX METHOD OF TESTING HEAT TRANSMISSION

In this method the range of temperature through which the tests may be run is much larger than in the ice box method, and it is not necessary to add to the material in the metal box. In other respects, however, this method presents the same disadvantages as the ice box method.

12. *The Cold Air Box Method.*—In this method, illustrated by Fig. 5, a box of cracked ice, hung near the top on the center line of the testing box, is substituted for the electrical heating element. The melting of the ice maintains the temperature of the air in the box at a lower degree than the air outside. The suspension of the ice box near the top of the test box supposedly causes natural circulation to maintain the inside air temperature nearly uniform. The heat transmitted is determined by weighing the amount of ice melted. Since it is not possible to control satisfactorily the temperature inside the box, this method is generally inferior to the hot air box method.

13. *The Hot Air Box Method.*—The test box for the hot air box method (Fig. 6) is made entirely of the material to be tested, unless the material is such that a skeleton frame work is necessary to provide

strength for the structure. The heat is supplied by electrical means, a resistance coil or a bank of lamps being used. A fan is usually employed inside the test box to circulate the air and maintain a uniform air temperature throughout the interior of the box. The amount of heat transmitted through the material is determined from the combined electrical input to the heater and to the fan motor.

This method, or a modification of it, is considered, according to Prof. C. L. Norton, the best to employ in testing materials for heat transmission. The use of the fan, however, is objectionable if a determination of the inside coefficient (K_1) is to be made, since the

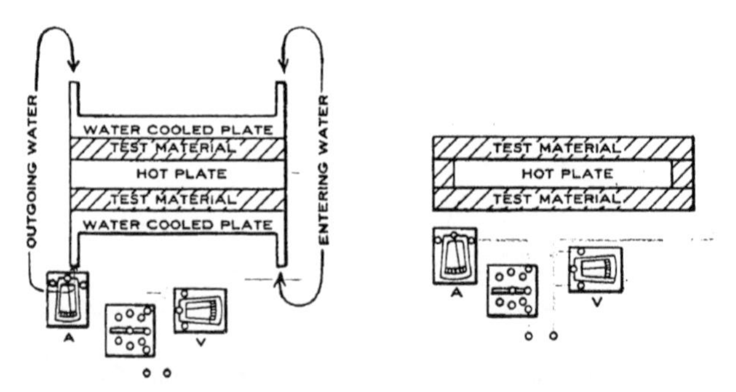

FIG. 7. THE HOT PLATE METHOD OF TESTING HEAT TRANSMISSION

velocity of the air over the inside surface tends to increase the value of K_1. In test specimens of large dimensions the fan is necessary if the interior air is to be maintained at a uniform temperature throughout.

14. *The Flat or Hot Plate Method.*—In this method (Fig. 7), the heating element consists of an electric grid, which is made of resistance wire placed between asbestos sheets. The material to be tested is placed on both sides of the grid and in contact with it. Outside the test material are placed two hollow flat plates, which are kept at constant temperature by means of water circulation through the plates. All the heat, except that lost from the edges in some plates, goes through the test material into the water-cooled outside plates. In some hot plates no water-cooled outside plates are used, the edges being covered with the test material; thus the heat passes through only the test material. The heat lost is measured by the

electrical input, and the temperature difference between the inside and outside surfaces is usually determined by electrical means. Given the dimensions of the plate, the conductivity of the specimen may be readily determined.

The amount of the heat loss from the edges, in the one case, is unknown. According to one authority, this loss is not only considerable but varies in amount, the variation depending upon the nature and the thickness of the material being tested. A correction, the accuracy of which is rather uncertain, for this edge loss must be made. When the water-cooled plates are used, only the conductivity can be determined; without them the outside surface coefficient may also be obtained. In neither case can the transmission coefficient (air to air) be obtained.

15. *The Determination of the Heat Transmission Coefficient under the Foregoing Methods.*—In any of the box methods of testing, the determination of the heat transmission coefficient (U) is a comparatively simple matter, but in order to obtain any of the other coefficients an accurate determination of surface temperatures must be made. Various means of temperature measurement have been employed for this purpose. Oil wells for mercury thermometers have been sunk in the surface of the material, the center line of the oil well lying in the plane of the material being tested. Mercury thermometers have been fully imbedded in the material, half-way imbedded, or fastened on the surface in attempts to determine the true surface temperature. Experiments show that the accuracy with which mercury thermometer determinations of surface temperatures can be made depends almost entirely upon the dimensions and the nature of the test specimen; the temperature gradient through the material being the governing factor. If the material being tested is thick with a correspondingly small temperature drop per inch through the material, the displacement of the center line of the thermometer with regard to the surface of the material is much less important than with a comparatively thin test specimen and its correspondingly steeper temperature gradient.

Dalby further complicates the matter by his suggestion, previously mentioned, that there is probably a further drop of temperature head occurring just at the surface of the material which is required to force the flow of heat across the surface. The temperature head required to cause the flow of heat through the material, however, is the difference between the temperatures at each end of the temperature gradient through the material itself.

The temperature gradient through the material is usually assumed to be a straight line. This assumption suggests the possibility of determining the surface temperatures at two different points along this gradient and then solving graphically or analytically for the surface temperatures, but most building materials are not homogeneons enough to warrant the straight line assumption. The density of the material must be considered in determining the homogeneousness of the material.

Attempts to determine surface temperatures by means of platinum discs held against the surface of the material have been made at the Engineering Experiment Station of Pennsylvania State College.

V. Description of Specimens, Testing Apparatus, and Method of Conducting Tests

16. *The Testing Plant.*—Since practically full size specimens were to be used in determining the heat transmission of all the walls investigated and in ascertaining the values of the surface and conductivity coefficients, the hot air box method of testing was chosen. The box was built of the material to be tested except where means of support were needed. In order to get convection conditions similar to those in actual practice, the boxes were built about the height of a room and were supported on small piers; thus air was allowed to circulate around practically all parts of the box.

The heating element was composed of "Yankee Silver" resistance wire,* helically wound with increasing pitch from bottom to top upon a wooden frame support placed inside the box. With this method of heating it was found after considerable experimenting that the same variation in air temperature was maintained as is found in the average room, that is, the air was warmer near the ceiling than at the floor line.

A voltmeter across the terminals of the resistance heating coil, an ammeter in the line, and a water resistance box, or rheostat, to control the current constituted the apparatus necessary to control and determine the heat input into the box. All electrical instruments were accurately calibrated against standard meters of the Electrical Engineering Department.

Air and surface temperatures were measured with mercury thermometers and thermocouples or with thermocouples alone. All mercury thermometers were carefully calibrated against a standard Centigrade thermometer of known accuracy.

A diagrammatic sketch of the apparatus used for determining the various coefficients is presented as Fig. 8. Only one thermocouple circuit is shown in the diagram.

For the moving air tests, in addition to the apparatus illustrated by Fig. 8, a hood shown in Fig. 9 was placed over the column. This hood, placed so that the column was centrally located inside, was connected to a No. 4 Sirocco multivane fan by means of a 24-inch duct about thirty feet long. A variable speed direct-current motor,

* The wire used for the resistance coil is known as a "Yankee Silver" No. 16 B. and S. gage, a patented alloy manufactured by Driver-Harris Company, Harrison, New Jersey. The resistance per mil foot is 200 ohms at 75 degrees F. The temperature coefficient is 0.000086 per degree. The specific gravity is 8.6. The weight per cubic inch is 0.31 pounds.

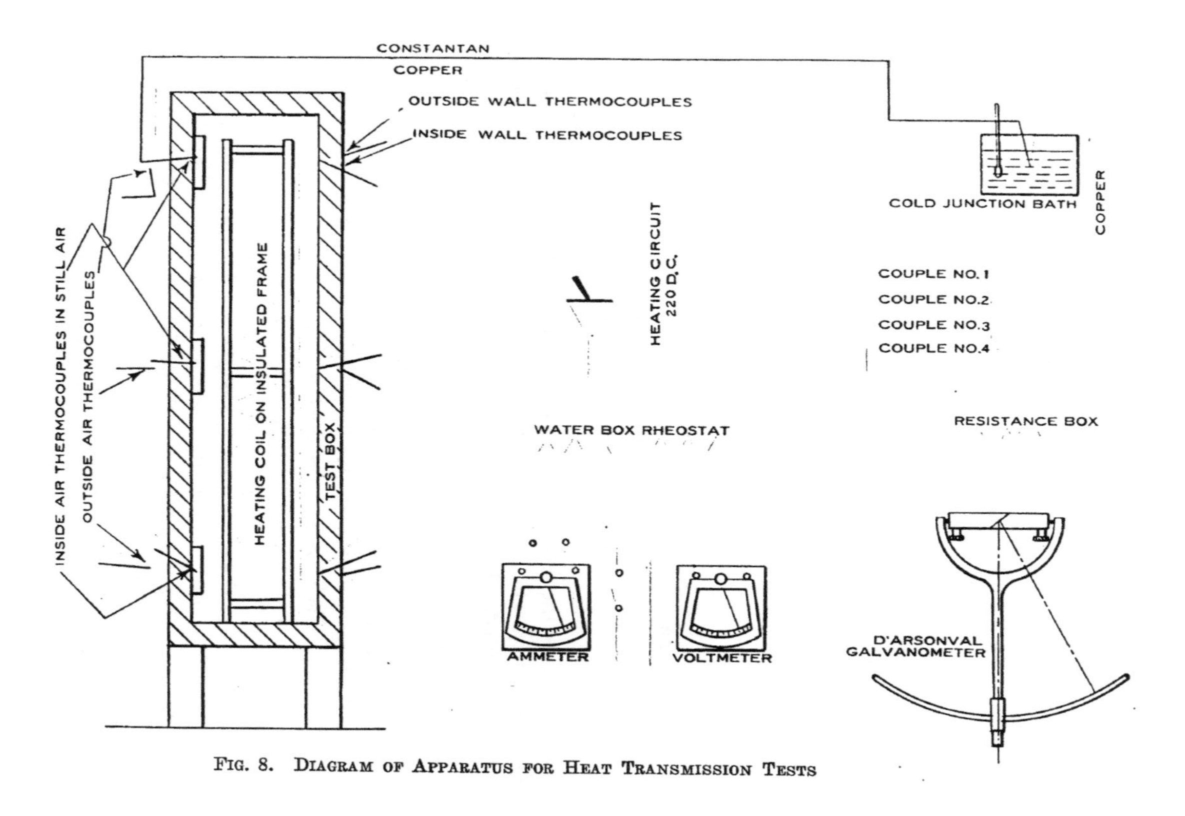

FIG. 8. DIAGRAM OF APPARATUS FOR HEAT TRANSMISSION TESTS

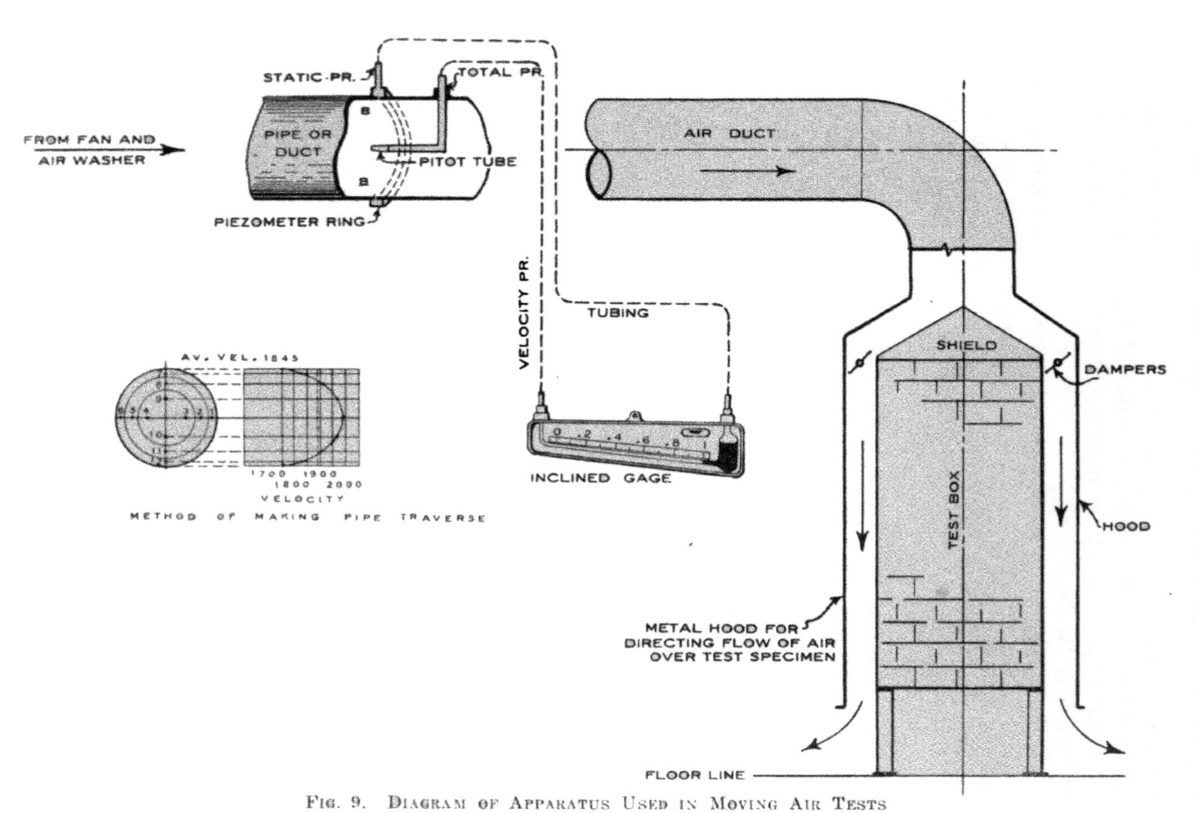

Fig. 9. Diagram of Apparatus Used in Moving Air Tests

belt-connected to the Sirocco fan, furnished the means for varying the velocities of air flow over the surface of the column being tested. By means of dampers in the hood the velocity of the air was kept uniform over the four sides.

A Pitot tube and a piezometer ring were used to determine the quantity of air discharged by the fan; consequently the velocity of the air over the surface of the test column was readily calculated.

Photographic views of the test apparatus are presented in Figs. 10, 11, and 12.

17. *Calibration of Thermocouples.*—The thermocouple has been found to be better suited to the determination of surface temperatures than the thermometer. The junction, when embedded just in the surface of the material, indicates the surface temperature as nearly as it is possible, at the present time, to make such measurements. The accuracy of the temperatures determined in this manner is also dependent upon the temperature gradient through the material, but owing to the difference in dimensions between a thermocouple junction and the thickness of the material to be tested, this factor is not significant.

The thermocouples used in these tests were made of copper and a nickel alloy of copper called constantan, No. 25 B. and S. gage, double silk covered wire being used. The junctions were made by fusing the ends of the two different wires together in the flame of a blast lamp. Each junction was placed in a glass tube closed at one end and filled with oil, and the two wires were insulated from each other by a small glass tube slipped around one of the wires.

Calibrations were made with a cold junction temperature of 70 degrees F., the hot junction temperature being controlled by means of a hot-water bath. In series with each couple, by means of switches, were placed a resistance and a Leeds and Northrup Type H D'Arsonval galvanometer for indicating by angular deflections the current generated and therefore the temperature in each thermocouple circuit. The deflection method of measuring the temperature difference between the cold and hot junctions was used. In order to make this method as accurate as possible, all calibrations were made with the galvanometer balanced and set in position on a concrete pier; in this position it remained without any change through the set of tests on any specimen with which a given set of thermocouples was connected. The deflection method is considered entirely reliable, if proper care

Fig. 10. View of Test Columns Showing Brick Column in Place and Other Columns as Made Ready for Testing

FIG. 11. VIEW OF AIR WASHER AND FAN WITH COLUMNS IN BACKGROUND

FIG. 12. VIEW SHOWING AIR DUCT AND BOX SHIELD AROUND COLUMN FOR AIR VELOCITY TESTS

This view also shows the rheostat and the electrical instruments for controlling and measuring the heating current supplied to the column in the right foreground. The concrete pier and leads to thermocouples are shown at the left. *Note*—Small galvanometer shown on pier was not used during the tests reported.

is taken, and has been used by other investigators in this field. The Babcock and Wilcox Company have recently applied this method in a series of tests run at Bayonne, N. J., on the rate of heat transfer through boiler tubes.

In calibrating thermocouples it is customary to hold the temperature, of the cold junction constant at a predetermined point, to vary the hot junction temperature, and to note the deflection of the galvanometer. Owing to the slight inconvenience of keeping the cold junction at the temperature at which it was calibrated, some experimenters prefer to run a series of calibrations with the temperature of the cold junction varying over the range expected during the test.

Another method of procedure is to calibrate, as in the first case, by holding the cold junction at a certain temperature and then, with a fixed hot junction temperature, the cold junction temperature is changed to a point above and to a point below the original temperature during the calibration. The three readings or deflections are plotted, the equation of the curve determined, and the correction for the cold junction temperature is readily made.

The first method was followed during these experiments. A typical set of thermocouple calibration curves is presented in Fig. 13. It will be noted that the curve A, of the lower temperature differences, is obtained with 300 ohms external resistance in the circuit while 1,300 ohms were used to obtain the curve of higher temperature differences. This method gives a larger deflection for the lower temperatures than would otherwise be obtained with a single fixed resistance in the circuit, yet the deflection, corresponding to the highest temperature to be measured, remains within the range of the galvanometer.

After calibration, the so-called hot junctions were removed from the glass containers and fastened in position on the surface of the material to be tested or in the air about one inch from the surface. All thermocouples and mercury thermometers used for determining air temperatures were shielded against direct radiation by means of paper shields.

Thermocouples for determining surface temperatures were attached to the surfaces in the manner described in the following: For wood, a thin shaving was glued over the junction, the junction being somewhat imbedded in the surface of the material before the application of the thin shaving; for the materials composed of asbestos, the junction was covered by a thin sheet of asbestos or held against the

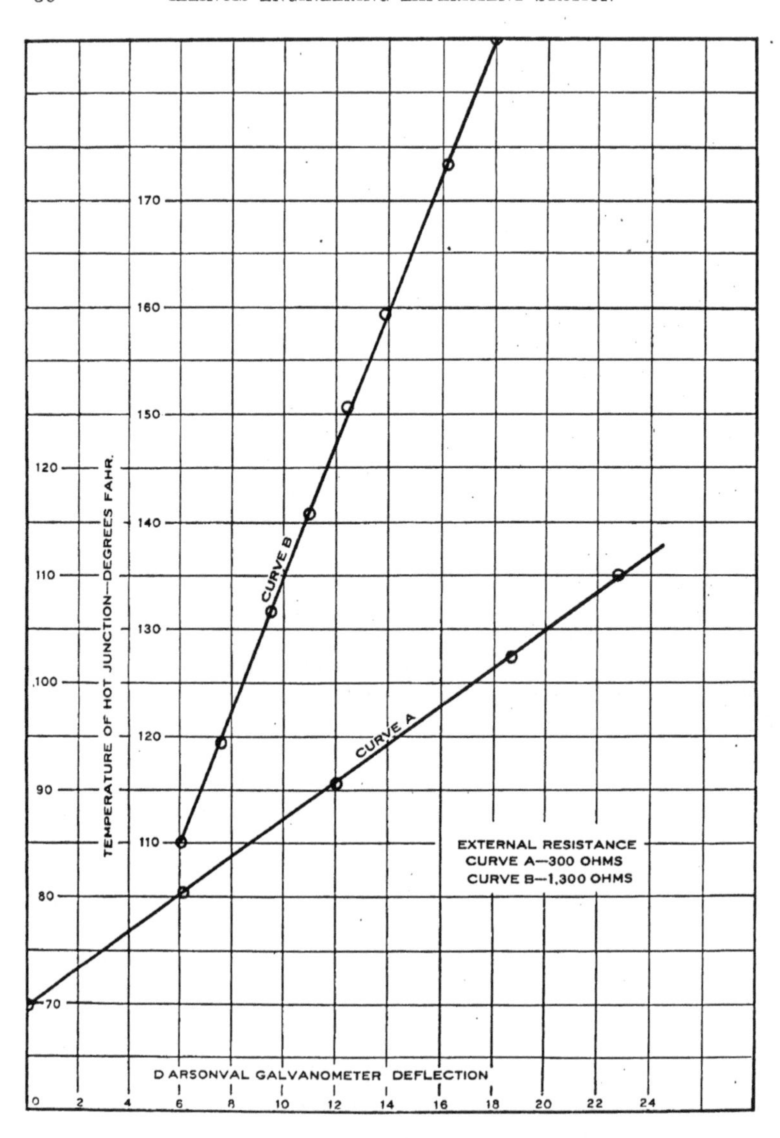

FIG. 13. TYPICAL THERMOCOUPLE CALIBRATION CURVES

surface with a mixture of powdered asbestos and water; for the vitreous materials, the junction was fastened to the surface with a thin layer of plaster of Paris.

In determining surface temperatures with mercury thermometers, it was found that if the thermometer was imbedded in a trench just deep enough to permit the thermometer to lie entirely below the surface of the material, the temperature recorded was higher than that given by the thermocouple, the difference being dependent upon the thickness of the material. If the thermometer were placed against the surface, all the thermometer being out of the surface, yet covered, and stuck to the surface, as previously described, it was found that the temperature recorded was somewhat lower than that recorded by the thermocouple. With comparatively thick materials the temperatures indicated by the thermometers and the thermocouples were practically the same when the thermometers were so imbedded that their axes lay in the plane of the surface of the material.

18. *The Method of Conducting Tests.*—The temperature of the air inside the test column was brought to the desired point by allowing a relatively large current to flow through the heating coil. As the rising air temperature approached the desired point, the current was decreased until the right amount was flowing to maintain the required temperature. In all cases sufficient time, ranging from twenty-four to seventy-two hours, according to the material and its thickness, was then allowed to elapse in order to insure constant heat flow. Readings of thermometers, thermocouples, and electrical instruments were then taken at intervals of thirty minutes. From five to seven readings were taken during each test, local conditions determining the duration of the tests. Usually three or four tests were run on each material, various air temperature differences being maintained for each test. In most cases both mercury thermometers and thermocouples were installed for each test. This installation made it possible to run both tests at the same time, and provided the data for a direct comparison.

The moving air tests were conducted in a manner similar to that described in the preceding paragraph, and in addition a traverse of the air duct was made during each test, the fan speed was recorded, and the exit velocity of the air over the four sides of the column was checked with an anemometer at the regular 30-minute intervals. The relative humidity of the air was also determined for each test.

In the combined moving air and humidity tests, the humidity of the air as it left the air washer was also recorded. The air washer made it possible to supply air at practically one hundred per cent relative humidity for the combined moving air and humidity tests.

19. *Calculations for Finding Coefficients.*—The readings, taken during a test, for each thermometer, thermocouple, or electrical meter were averaged and then corrected according to the calibration curves. The corrected averages for any one section, such as the outside surface temperatures, were then averaged, the result being the outside surface temperature in the case mentioned. From these final temperatures the various drops, air to air, wall to wall, air to wall, and wall to air, were determined.

The heat transmitted was determined from the electrical input by means of the following relation:

Volts × Amperes × 3.412 = B. t. u. loss per hour.

Since the heat input was known, the various coefficients were determined from the following equations:

$$U = \frac{H}{S_m \ (t_1 - t_4)}$$

$$K_1 = \frac{H}{S_1 \ (t_1 - t_2)}$$

$$K_2 = \frac{H}{S_2 \ (t_3 - t_4)}$$

$$C = \frac{Hx}{S_m \ (t_2 - t_3)}$$

in which,

S_m is the mean area of the inside and outside surfaces in square feet and is taken as the arithmetical mean of S_1 and S_2.

S_1 is the inside area of the test box in square feet.

S_2 is the outside area of the test box in square feet.

H is the total heat transmission of the box in B. t. u. per hour.

t_1 is the inside air temperature.

t_2 is the inside wall temperature.

t_3 is the outside wall temperature.

t_4 is the outside air temperature.

In the moving air tests, flow of air occurred over the four sides of the test box only; thus it is necessary to make a correction to allow for the difference in loss of heat from the ends and sides. It is assumed that the loss of heat from the ends is the same during the moving air tests as during the still air tests, and the correction is accordingly made in the following manner:

$$H_s = H_t - u\,S_e\ (t_1 - t_4)$$

in which,

H_s is the heat loss through the sides in B. t. u.

H_t is the total heat loss from the test box in B. t. u.

u is the unit transmission of the test box, obtained from the still air tests.

S_e is the mean area of the ends of the box in square feet.

t_1 is the inside air temperature in degrees F.

t_4 is the outside air temperature in degrees F.

In determining the coefficients for the moving air tests, the relations given on page 32 are used; H_8 is substituted for the heat loss, and the respective areas of the sides are used instead of those in the entire box.

In making a traverse to determine the quantity of air flowing, the duct was divided into five concentric zones of equal area, and readings were taken on the circle which equally divided the area of each zone. The traverse was made across on one diameter only, thus giving ten readings in inches of water, which will be called h. To calculate the mean velocity, these values of h were substituted in the following equation:

$$V_m = \frac{18.27}{d^{1/2}} \left[\frac{(\sqrt{h_1} + \sqrt{h_2} + \sqrt{h_3} + \text{etc.})}{n} \right]$$

in which,

V_m is the mean velocity in feet per second.

d is the density of the air at the mean temperature pounds per cubic foot.

h is the velocity pressure in inches of water.

n is the number of readings taken.

Knowing the mean velocity of the air in the duct, the relative areas of the duct, and the space between the test column and the surrounding hood, the air velocity over the surface of the test column can readily be determined.

The space between the hood and the test column was changed in going from the low to the high velocity tests in order to run the velocities as high as desired.

TABLE 1

HEAT TRANSMISSION TESTS—STILL AIR

Material	Density	Thickness	Areas: Mean / Inside / Outside	Test	Heat Loss Per Hour B. T. U.	Temperature Differences: Air to Air	Wall to Wall	Inside Air to Wall	Outside Wall to Air	Coefficients: U	C	K_1	K_2
2-in. Tile ½-in. Cement Plaster on Both Sides	119.9 Lb. / Cu. Ft.	2.02"	46.16 Sq. Ft. 36.06 Sq. Ft. 56.26 Sq. Ft.	Thermometer	1744.5	77.57	19.18	28.08	30.31	0.49	1.97[1]	1.72	1.02
					1039.0	55.41	12.41	23.02	19.98	0.41	1.81[1]	1.25	0.93
					690.3	32.93	7.26	13.87	11.71	0.45	2.06[1]	1.38	1.05
				Thermocouple	1744.5	84.69	38.05	14.95	31.69	0.45	0.99[1]	3.24	0.98
					1039.0	57.38	26.75	10.25	21.88	0.39	0.84[1]	2.81	0.85
					690.3	35.99	14.65	8.80	11.29	0.42	1.02[1]	2.18	1.09
4-in. Tile ½-in. Cement Plaster on Both Sides	127.0 Lb. / Cu. Ft.	3.84"	44.33 Sq. Ft. 24.28 Sq. Ft. 64.39 Sq. Ft.	Thermometer	1133.5	57.74	20.07	24.31	13.36	0.44	1.27[1]	1.92	1.32
					687.7	41.08	14.94	14.72	11.42	0.38	1.04[1]	1.92	0.94
					400.5	25.11	8.05	9.94	7.12	0.36	1.12[1]	1.66	0.87
				Thermocouple	1133.5	59.78	43.30	3.65	12.83	0.43	0.59[1]	12.80	1.37
					687.7	45.12	25.65	7.85	11.62	0.34	0.61[1]	3.61	0.92
					400.5	29.19	13.60	8.15	7.44	0.31	0.67[1]	2.02	0.84
6-in. Tile ½-in. Cement Plaster on Both Sides	124 3 Lb. / Cu. Ft.	6.77"	57.61 Sq. Ft. 31.11 Sq. Ft. 84.12 Sq. Ft.	Thermometer	1243.5	64.29	24.71	24.67	14.91	0.34	0.87[1]	1.62	0.99
					798.5	39.49	14.54	16.79	8.16	0.35	0.95[1]	1.53	1.16
					519.5	23.37	8.33	11.03	4.09	0.39	1.08[1]	1.51	1.51
				Thermocouple	1243.5	68.30	46.24	6.06	16.00	0.32	0.47[1]	6.60	0.92
					798.5	39.54	28.00	2.93	8.61	0.35	0.50[1]	8.76	1.10
					519.5	24.07	17.56	1.90	4.61	0.38	0.51[1]	8.79	1.34

[1] Conductivity for total thickness and construction as stated.

TABLE 1 (Continued)
HEAT TRANSMISSION TESTS—STILL AIR

Material	Density	Thickness	Areas: Mean, Inside, Outside	Test	Heat Loss Per Hour B. T. U.	Temperature Differences: Air to Air	Wall to Wall	Inside Air to Wall	Outside Wall to Air	Coefficients: U	C		K_2
1-in. Wood (Fir)	33.37 Lb. / Cu. Ft.	1.06"	51.04 Sq. Ft. 46.94 Sq. Ft. 55.14 Sq. Ft.	Thermometer	1602.5	72.48	24.62	27.24	20.62	0.46	1.35	1.25	1.41
					1041.0	49.20	16.49	18.44	14.27	0.44	1.31	1.20	1.32
					466.5	22.74	6.84	9.16	6.74	0.43	1.42	1.08	1.26
				Thermocouple	1602.5	70.15	33.03	16.80	20.32	0.47	1.01	2.03	1.43
					1041.0	48.20	24.37	9.93	13.90	0.45	0.89	2.23	1.36
					466.5	20.92	10.80	3.73	6.39	0.46	0.90	2.66	1.33
2-in. Cork Board (Nonpareil)	9.74 Lb. / Cu. Ft.	2.03"	39.34 Sq. Ft. 31.72 Sq. Ft. 46.96 Sq. Ft.	Thermometer	368.2	66.27	48.59	10.52	7.16	0.29	0.39	1.10	1.09
					213.3	41.83	30.10	7.06	4.67	0.26	0.37	0.95	0.98
					144.9	28.51	20.57	5.16	2.78	0.26	0.36	0.89	1.11
				Thermocouple			58.90	5.67	6.37	0.27	0.32	2.05	1.23
							36.80	3.87	4.18	0.25	0.30	1.74	1.09
							25.67	2.86	2.19	0.24	0.29	1.60	1.41
1½-in. Magnesia Board							44.36	26.71	13.63	0.37	0.71	0.86	1.35
							30.89	22.21	10.38	0.36	0.75	0.76	1.29
							20.77	14.07	6.71	0.37	0.73	0.78	1.31
										0.38	0.51		1.46
										0.37	0.50		1.37
										0.38	0.52		1.30

TABLE 1 (Continued)
HEAT TRANSMISSION TESTS—STILL AIR

Material	Density	Thickness	Areas: Mean, Inside, Outside	Test	Heat Loss Per Hour B. T. U.	Temperature Differences: Air to Air	Wall to Wall	Inside Air to Wall	Outside Wall to Air	Coefficients: U	C	K_1	K_2
2-Course Brick	131.9 Lb. / Cu. Ft.	8.77″	72.08 Sq. Ft. 39.35 Sq. Ft. 104.80 Sq. Ft.	Thermometer	1548.0	78.89	33.21	30.54	15.14	.273	5.67	1.29	0.98
					1231.5	61.10	25.09	26.41	9.60	.281	5.97	1.18	1.22
					858.7	35.65	12.08	19.08	4.49	.335	8.64	1.14	1.82
				Thermocouple	1548.0	77.92	47.93	.....	16.89	.276	3.93	3.00	1.13
					1231.5	60.39	36.59	.....	8.40	.284	4.09	2.03	1.40
					858.7	35.10	20.58	9.80	4.72	.340	5.08	2.23	1.73
3-in. Concrete 1-2-4 Mixture	139.7 Lb. / Cu. Ft.	3.19″	33.76 Sq. Ft. 23.58 Sq. Ft. 43.94 Sq. Ft.	Thermometer	2088.0	78.27	18.47	26.39	33.41	.790	10.68	3.36	1.42
					1527.5	57.81	12.59	22.66	22.58	.782	11.46	2.86	1.54
					1066.0	40.14	8.81	16.44	14.89	.787	11.44	2.75	1.63
				Thermocouple	2088.0	80.95	23.77	20.83	36.35	.765	8.31	4.25	1.31
					1527.5	58.97	16.52	18.36	23.99	.767	8.74	3.53	1.45
					1066.0	39.61	12.77	12.23	14.61	.797	7.89	3.70	1.66
3-in. Concrete Roofing Covered	Concrete 139.7 Lb./Cu. Ft. Roofing 1.34 Lb./Sq. Ft. Gravel 0.83 Lb. / Sq. Ft.	3.34″	35.03 Sq. Ft. 23.58 Sq. Ft. 46.47 Sq. Ft.	Thermometer	1261.5	65.33	22.74	19.19	13.40	0.55	1.58[1]	2.79	2.03
					690.0	31.25	8.42	13.17	9.66	0.63	2.34[1]	2.22	1.54
					356.3	17.60	4.11	7.15	6.34	0.58	2.48[1]	2.11	1.21
				Thermocouple	1261.5	64.64	27.15	14.30	13.19	0.56	1.33[1]	3.74	2.06
					690.0	28.55	10.68	8.17	9.70	0.69	1.84[1]	3.58	1.53
					356.3	17.07	3.88	5.84	7.35	0.60	2.62[1]	2.59	1.04

[1] Conductivity for total thickness and construction as stated.

TABLE 1 (Continued)

HEAT TRANSMISSION TESTS—STILL AIR

Material	Density	Thickness	Areas: Mean	Areas: Inside	Areas: Outside	Test	Heat Loss Per Hour B. T. U.	Temperature Differences: Air to Air	Temperature Differences: Wall to Wall	Temperature Differences: Inside Air to Wall	Temperature Differences: Outside Wall to Air	Coefficients: U	Coefficients: C	Coefficients: K_1	Coefficients: K_2
1-in. Asbestos Board (Corrugated Interior)	20.42 Lb. / Cu. Ft.	1.00"	50.10 Sq. Ft.	46.58 Sq. Ft.	53.61 Sq. Ft.	Thermometer	1248.0	73.72	32.13	25.80	15.79	0.34	0.78	1.04	1.47
							796.0	51.28	23.17	17.82	10.29	0.31	0.69	0.96	1.44
							496.2	32.10	12.91	12.15	7.04	0.31	0.77	0.88	1.32
						Thermocouple	1248.0	71.16	51.57	5.10	14.49	0.35	0.48	5.26	1.61
							796.0	50.53	36.96	3.34	10.23	0.32	0.43	5.12	1.45
							496.2	31.07	22.10	2.17	6.80	0.32	0.45	4.91	1.36
Sheet Asbestos 60 Sheets 1-64-in. Thick	48.25 Lb. / Cu. Ft.	1.1"	50.96 Sq. Ft.	48.75 Sq. Ft.	53.16 Sq. Ft.	Thermometer	856.2	77.33	35.12	31.23	10.98	0.24	0.53	0.56	1.47
							619.7	56.41	24.40	23.15	8.86	0.24	0.55	0.55	1.32
							371.9	32.93	13.46	14.57	4.90	0.24	0.60	0.52	1.43
						Thermocouple	856.2	78.43	63.00	4.00	11.43	0.24	0.29	4.39	1.41
							619.7	57.10	44.60	3.53	8.97	0.23	0.27	3.57	1.30
							371.9	33.05	25.97	1.86	5.22	0.24	0.31	4.10	1.34
Single Strength Glass 76.3% Glass	141.1 Lb. / Cu. Ft.	0.085"	43.23 Sq. Ft.	42.93 Sq. Ft.	43.53 Sq. Ft.	Thermocouple	2240.5	67.64	6.00	27.67	33.97	1.00	0.96	2.44	1.99
							1500.0	48.02	2.57	20.10	25.35	0.95	1.51	2.25	1.78
							832.5	32.20	2.22	14.56	15.42	0.78	0.97	1.72	1.63
2 Panes Glass ½" Air Space 69.3% Glass	141.1 Lb. / Cu. Ft.	0.085" and 0.127"	45.46 Sq. Ft.	42.93 Sq. Ft.	47.98 Sq. Ft.	Thermocouple	1636.5	75.11	34.57	14.97	25.57	0.69	1.50	3.68	1.93
							1104.0	54.43	24.53	13.57	16.33	0.64	1.43	2.73	2.03
							731.5	38.09	17.10	9.70	11.29	0.61	1.36	2.54	1.95

TABLE 1 (Concluded)

HEAT TRANSMISSION TESTS—STILL AIR

Material	Density	Thickness	Areas Mean Inside Outside	Test	Heat Loss Per Hour B. T. U.	Temperature Differences				Coefficients			
						Air to Air	Wall to Wall	Inside Air to Wall	Outside Wall to Air	U	C	K_1	K_2
Brick, One-Course	131.9 Lb. / Cu. Ft.	3.79"	25.78 Sq. Ft. 17.48 Sq. Ft. 34.08 Sq. Ft.	Thermo-couple	1033.0	77.71	38.15	16.35	23.21	.516	3.98	3.61	1.31
					688.7	60.15	27.15	13.80	19.20	.443	3.73	2.85	1.05
					303.7	26.14	9.80	9.65	6.69	.452	4.55	1.80	1.33
Glass Single Pane 91.4% Glass	141.1 Lb. / Cu. Ft.	0.085"	10.14 Sq. Ft. 10.02 Sq. Ft. 10.25 Sq. Ft.	Thermo-couple	751.2	72.85	3.05	29.00	40.80	1.11	2.26	2.83	1.97
					444.0	48.27	2.65	19.55	26.07	0.99	1.53	2.48	1.82
					147.1	19.35	1.65	7.95	9.75	0.82	0.82	2.02	1.62
2-in. Tile ½-in. Plaster (Both Sides) Outside Roofing Covered	Tile 119.86 Lb./Cu. Ft. Roofing 1.34 Lb./Sq. Ft. Gravel 0.83 Lb. / Sq. Ft.	Tile 2.02" Roofing 0.15"	47.61 Sq. Ft. 36.06 Sq. Ft. 59.15 Sq. Ft.	Thermometer	1942.0	87.71	28.29	35.43	23.99	0.47	1.44	1.52	1.37
					1486.5	66.04	20.45	28.53	17.06	0.47	1.53	1.45	1.47
					1039.0	45.53	13.75	19.92	11.86	0.48	1.59	1.45	1.48
				Thermocouple	1942.0	83.39	48.55	11.70	28.14	0.49	0.84	4.60	1.17
					1486.5	69.07	38.32	11.95	18.80	0.45	0.82	3.45	1.34
					1039.0	51.13	26.67	10.65	13.81	0.43	0.82	2.71	1.27

VI. Results and Test Data

20. *Still Air Tests.*—The material, density, thickness, and heat transmitting areas of the various walls investigated are given in Table 1. This table also presents the average test data for heat transmitted and temperature differences which are required in calculating the coefficients U, C, K_1, and K_2. The calculated values for each coefficient are given in the last four columns of the table. In most cases there are two separate sets of tests for each material, the first being the mercury thermometer tests, and the second the thermocouple tests. For every thermometer test there is a corresponding thermocouple test, the two having been run at the same time.

21. *Coefficients of Heat Transmission.*—In each case, U is the heat, in B. t. u., transmitted by one square foot of wall surface per hour, per degree of difference in air temperature, inside to outside, for the thickness of wall actually tested. For the solid walls, such as brick, concrete, or cork, C is the conductivity of one square foot of wall surface per hour, per degree of difference in the surface temperatures, per inch in thickness of the material tested. For walls made of other than solid materials, such as the tile walls, the value of C is given for the thickness of wall actually tested. The surface coefficient K_1 is the heat received by one square foot of wall surface per hour, per degree difference in temperature between the inside air and inside wall surface. K_2 is the heat emitted by one square foot of wall surface per hour, per degree of difference in temperature between the outside wall surface and the outside air. The thickness or nature of the wall does not affect the determination of surface coefficients.

22. *Discussion of Results.*—The values of the transmission coefficients, U, have undoubtedly been affected by the abnormally high values of inside surface coefficients, K_1, when based on thermocouple readings. The radiation from the heating coil to the inside wall of the test column was apparently great enough to make the inside surface coefficient larger than the outside one. It was thought that with this method of heating, that is, eliminating the fan inside the box (See Hot Air Box method of testing), that surface coefficients corresponding approximately to still air conditions would be obtained for both the inside and outside surfaces. In this case, the air-to-air heat

transmission of the walls would have corresponded to still air conditions.

It will be noted from the result sheets, Table 1, that the K_2 values of each set of tests roughly check each other; this agreement indicates that mercury thermometers imbedded half-way in the material give fairly close readings of the outside surface temperatures.

TABLE 2

COEFFICIENTS BASED ON HEAT TRANSMISSION TESTS

(See "Applications of Test Data to Typical Walls.")

NOTE: These values are selected from Table 1 and are based on the tests run under most satisfactory conditions

No.	Material	C per 1" Thickness per Sq. Ft. per 1° F.	K Still Air per 1° F.
1	Brick Wall (Mortar Bond & Dry Conditions) . . .	4.00	1.40
2	Concrete, 1-2-4 Mixture	8.30	1.30
3	Wood (Fir, one surface finished)	1.00	1.40
4	Corkboard	0.32	1.25
5	Magnesia Board	0.50	1.45
6	Glass (actual glass 91.4% of total area)	2.06[3]	2.00
7	2-in. Tile, ½-in. plaster on both surfaces	1.00[4]	1.10
8	4-in. Tile, ½-in. plaster on both surfaces	0.60[4]	1.10
9	6-in. Tile, ½-in. plaster on both surfaces	0.47[4]	1.10
10	2-in. Tile, plastered as above and roofing covered . .	0.84	1.25
11	Asbestos Board	0.50	1.60
12	Sheet Asbestos	0.30	1.40
13	Double Glass, ½-in. air space (glass 69.3% of total area)	1.50[3] [4]	2.00
14	Roofing[1]	5.30[4]	1.25
15	Air Space[2]	1.00-1.70[4]	....

[1] Calculated from values of C for 2-in tile with and without roofing. $\left(\frac{1}{C}=\frac{1}{C_1}+\frac{1}{C_2}\right)$

[2] See "Air Spaces."

[3] Per sq ft. of actual glass set in wood frame but based on total heat transmitted.

[4] For thickness and construction stated, not per 1" of thickness.

The values of the transmission coefficients, U, show that for the air-to-air coefficient mercury thermometers give fairly accurate results. For the inside surface temperatures, however, where oil wells imbedded in the surface were used, the results from the mercury thermometer tests are far from correct; thus there is no checking of the K_1 and C values of the two sets of tests. Only the thermocouple temperatures are to be relied upon for inside wall temperatures.

For calculating heat transmission coefficients of simple or compound walls, it is, fortunately, only necessary to have the coefficients of conductivity and the surface coefficients of the materials composing the wall; consequently, in selecting the coefficients for Table 2 only the outside surface coefficients were considered in getting the values found under the column headed K. These values are the surface coefficients for still air conditions, and are in B. t. u. per square foot of surface per hour, per degree difference between the surface temperature and the temperature of the air in contact with it. Values of C are in B. t. u. per one square foot of surface per hour per degree difference between the inside and outside surface temperatures, per inch in thickness for solid walls, and for the actual thickness of hollow walls of simple or compound construction. In selecting these coefficients, the greatest significance has been attached to the values obtained from the tests of greatest air temperature differences, and only the tests in which thermocouples were used were considered.

Attention is called to the value of C for roofing, deduced from the values of C obtained in the 2-inch tile tests with and without roofing. This value was calculated from the relation:

$$\frac{1}{C}=\frac{1}{C_1}+\frac{1}{C_2}$$

where C_1 and C_2 are the conductivities of the tile and the roofing respectively. In the table of coefficients values will also be found for the so-called conductivity of air spaces, an explanation of which will be found under the section on air spaces.

23. *Moving Air Tests.*—Results of the moving air tests on brick are presented in Table 3.

Curves showing the effect of the velocity of air on the various coefficients for a 4-inch brick wall are given in Figs. 14 and 15. Just why K_1, the inside surface coefficient, should increase with an increasing air velocity over the outside of the box is difficult to explain. Air moving over the outside surface causes the outside surface temperatures to approach that of the outside air with an increase in velocity. The loss of heat being practically the same in all but a few of the tests, the temperature of the inside surface would have to drop until the temperature gradient through the material was the same as in the still air tests. On account of the lowering of the inside wall temperature, a drop of the inside air temperature, of such magnitude that the coefficient for the inside surface would be the same as under still air conditions, would be expected. According to the tests, however, the temperature difference is less than this amount, resulting in

TABLE 3

RESULTS OF THE MOVING AIR TESTS ON BRICK

PARTLY SATURATED AIR

No.	Total Heat Loss	Corrected Heat Loss	Air Temperature Difference	Wall Temperature Difference	Air to Wall Difference	Wall to Air Difference	U	C	K_1	K_2	Air Velocity	Relative Humidity
1	1278.0	1117.2	73.96	47.40	19.05	7.51	0.70	4.15	3.89	5.30	1563	....
2	1227.5	1061.0	76.55	49.15	19.45	7.95	0.64	3.98	3.63	4.76	985	80.0
3	1126.5	967.2	73.25	45.55	17.80	9.90	0.61	3.91	3.61	3.48	461	43.5
4	1175.5	1019.3	71.81	46.70	17.95	7.16	0.66	4.02	3.77	5.07	1955	50.5
5	1413.0	1244.2	77.62	53.95	18.95	4.72	0.74	4.25	4.36	9.40	3100	50.0
6	1434.5	1241.5	88.73	56.00	19.25	5.48	0.65	4.09	4.29	8.08	4378	....
7	1485.0	1317.0	77.23	55.60	16.40	5.23	0.79	4.37	5.34	9.16	5200	71.0
8	1191.0	1039.5	69.54	48.55	16.50	4.49	0.69	3.95	4.19	8.26	3135	51.0
9	1171.5	1024.4	67.66	45.95	17.40	4.31	0.70	4.11	3.91	8.47	2570	61.0
10	1147.0	991.4	73.30	42.70	19.40	11.20	0.63	4.08	3.40	3.15	373	49.5
11	1144.5	977.0	78.90	43.95	25.90	9.05	0.57	3.91	2.51	3.84	701	35.0
12	1150.0	992.5	74.15	45.00	21.00	8.15	0.62	3.88	3.14	4.34	98[illegible]	44.0
13	1148.0	991.5	73.70	46.55	19.35	7.50	0.62	3.75	3.40	4.71	1175	32.0

an increase of K_1, with the velocity of the air. The conductivity curve is practically horizontal, as would be expected, since the mean temperature of the material was about the same for all the tests. The transmission curve increases with the velocity of the air in a manner similar to the increase of the outside surface coefficient.

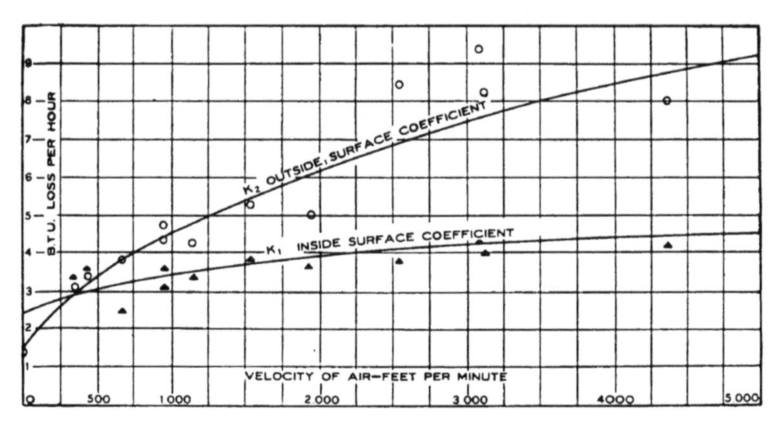

FIG. 14. K_1 AND K_2 CURVES, BRICK BOX MOVING AIR TEST (AVERAGE RELATIVE HUMIDITY, SEE TABLE 3)

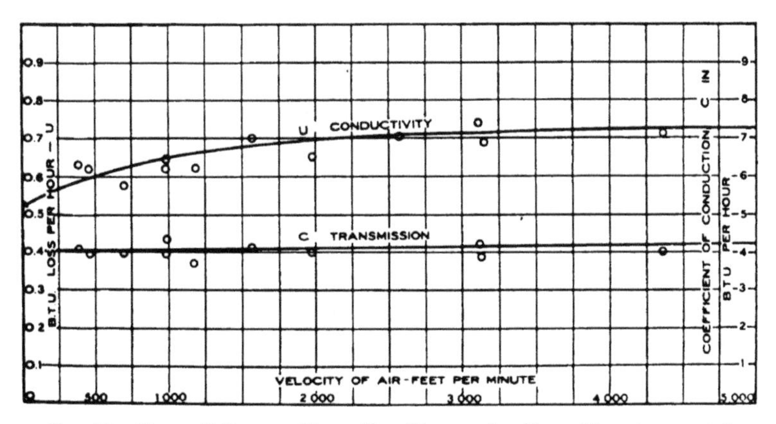

FIG. 15. U AND C CURVES, BRICK BOX MOVING AIR TEST (AVERAGE RELATIVE HUMIDITY, SEE TABLE 3)

TABLE 4

RESULTS OF THE MOVING AIR TESTS ON WOOD

PARTLY SATURATED AIR

No.	Total Heat Loss	Corrected Heat Loss	Air Temperature Difference	Wall Temperature Difference	Air to Wall Difference	Wall to Air Difference	U	C	K_1	K_2	Air Velocity	Humidity
1	1369.5	1176.5	79.58	60.40	9.45	9.73	0.59	0.82	5.37	4.47	2322	53.5
2	1401.5	1212.2	78.05	60.65	9.10	8.30	0.62	0.84	5.75	5.40	3138	61.0
3	1619.0	1422.3	81.08	64.00	8.20	8.88	0.70	0.94	7.44	5.92	4015	45.0
4	1452.0	1250.7	83.02	63.05	9.10	10.87	0.60	0.84	5.93	4.25	5665	79.5
5	1405.5	1212.8	79.44	60.95	8.25	10.24	0.61	0.85	6.35	4.38	....	77.0
6	1379.0	1186.7	79.25	61.00	9.25	9.00	0.60	0.82	5.54	4.87	3875	81.0
7	1283.0	1098.8	75.82	57.95	9.10	8.77	0.58	0.80	5.21	4.63	2889	75.5
8	1341.5	1144.6	81.18	60.20	9.55	11.43	0.56	0.80	5.16	3.70	1905	87.5
9	1321.0	1143.1	73.33	54.25	9.75	9.33	0.62	0.89	5.16	4.53	4925	77.0
10	1237.0	1058.1	73.76	56.35	8.70	8.71	0.57	0.79	5.25	4.49	1668	70.0
11	1314.0	1139.0	77.14	58.15	9.85	9.14	0.59	0.83	4.99	4.60	1083	79.0
12	1238.0	1058.7	73.95	55.40	9.65	8.90	0.57	0.81	4.74	4.40	1236	72.5
13	1258.5	1072.5	76.65	57.20	9.80	9.65	0.56	0.79	4.72	4.12	882	71.5
14	1320.0	1120.0	82.43	59.40	11.00	13.03	0.54	0.79	4.39	3.17	548	68.5
15	1348.5	1145.4	83.77	60.05	10.55	13.17	0.54	0.81	4.69	3.21	41	....

24. *Discussion of Results.*—Results of the moving air tests on wood are presented in Table 4. These data are plotted on the curve sheets, Figs. 16 and 17, and show the rate of change of the various coefficients with air velocity. The remarks regarding K_1 for the brick box tests apply equally well to the K_1 values for the wood box tests. In the still air tests the value of K_1 for wood is evidently incorrect and has been disregarded in drawing the K_1 curve for this material.

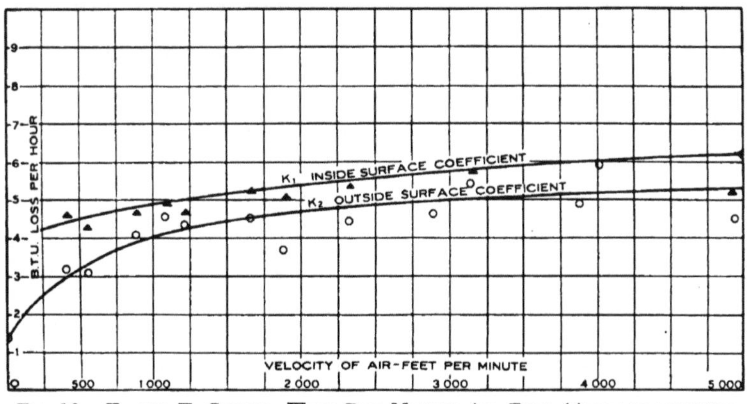

FIG. 16. K_1 AND K_2 CURVES, WOOD BOX MOVING AIR TEST (AVERAGE RELATIVE HUMIDITY, SEE TABLE 4)

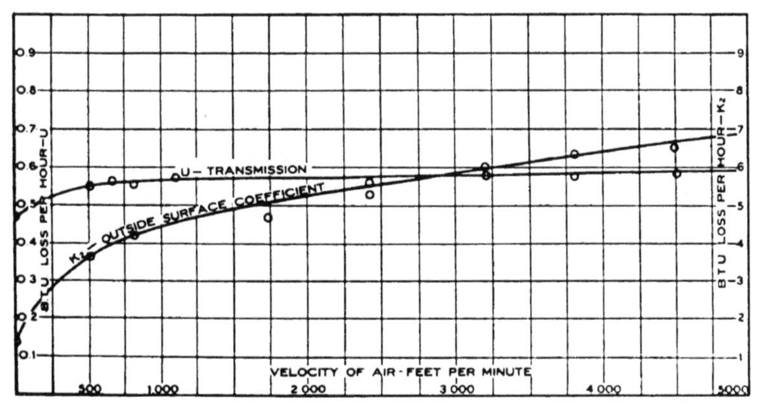

FIG. 17. U AND K_2 CURVES, FOR WOOD BOX MOVING AIR TEST (AVERAGE RELATIVE HUMIDITY, SEE TABLE 4)

TABLE 5

RESULTS OF THE SATURATED AIR TESTS ON THE BRICK BOX

No.	Total Heat Loss	Corrected Heat Loss	Air Temperature Difference	Wall Temperature Difference	Air to Wall Difference	Wall to Air Difference	U	C	K_1	K_2	Air Velocity	Humidity
1	1141.5	980.0	74.35	47.35	18.60	8.40	0.61	3.64	3.50	4.15	882	95.0
2	1147.0	989.0	72.75	46.00	17.45	9.30	0.63	3.78	3.77	3.79	698	98.0
3	1149.5	989.1	73.80	47.50	19.10	7.20	0.62	3.66	3.44	4.89	1338	92.0
4	1148.0	979.9	77.40	50.50	14.75	12.15	0.59	3.41	4.42	2.88	472	99.5

TABLE 6

RESULTS OF THE SATURATED AIR TESTS ON THE WOOD BOX

No.	Total Heat Loss	Corrected Heat Loss	Air Temperature Difference	Wall Temperature Difference	Air to Wall Difference	Wall to Air Difference	U	C	K_1	K_2	Air Velocity	Humidity
1	1420.5	1208.8	87.23	63.80	11.40	12.03	0.55	0.80	4.57	3.71	528	98.0
2	1405.0	1200.2	84.40	62.80	11.70	9.90	0.57	0.81	4.43	4.48	678	98.5
3	1457.0	1243.0	88.26	65.80	11.60	10.86	0.56	0.80	4.63	4.23	837	98.5
4	1465.0	1255.3	86.42	64.45	11.10	9.87	0.58	0.82	4.88	4.70	1108	98.0
5	1239.0	1052.0	77.15	47.55	21.30	8.30	0.54	0.93	2.13	4.69	1763	99.8
6	1275.0	1093.0	75.20	49.45	18.95	6.80	0.58	0.93	2.49	5.94	3210	98.0
7	1275.5	1087.6	77.25	51.30	18.35	7.60	0.56	0.89	2.56	5.29	2438	98.0
8	1275.5	1093.2	75.15	50.00	18.75	6.40	0.58	0.92	2.52	6.31	3795	97.5
9	1277.0	1096.2	74.55	49.80	18.55	6.20	0.59	0.93	2.55	6.53	4510	86.0

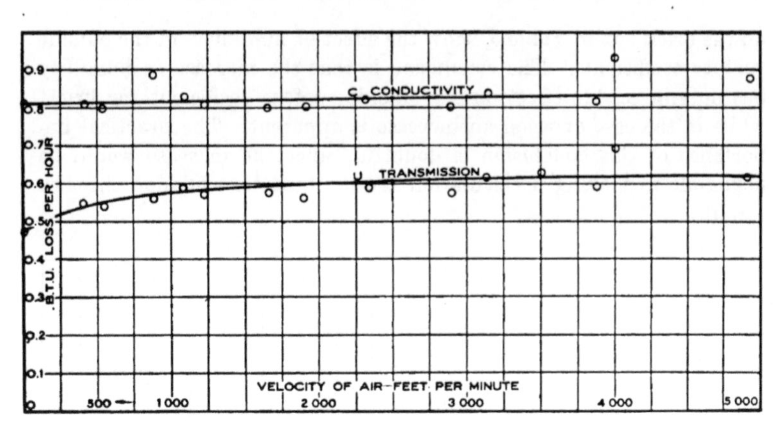

FIG. 18. U AND C CURVES, WOOD BOX MOVING AIR TEST

25. *The Effect of Variations in Relative Humidity on Surface Coefficients.*—Results of the saturated air tests on the brick box are presented in Table 5, and those on the wood box in Table 6. Curves of the various coefficients determined from the saturated air tests on the wood box are shown in Fig. 18.

The three curves shown in Fig. 19, which are taken from the K_2 curves in Figs. 14 and 16, at the lower velocities, and the four points plotted, which are obtained from the data of the saturated air tests

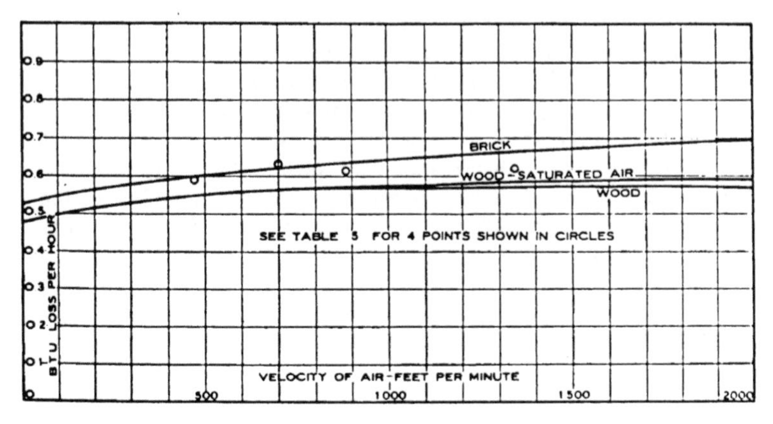

FIG. 19. K_2 CURVES, AIR VELOCITY BELOW 2000 FEET PER MINUTE

on the brick box in Table 5, show the effect of humidity on the outside surface coefficients. The conclusion is that the increase in humidity has no appreciable effect on the outside surface coefficient for brick, while in the case of wood an increase is apparent. The practical importance of this conclusion is doubtful, since the increase would be negligible in making a calculation for the heat transmission through a wall.

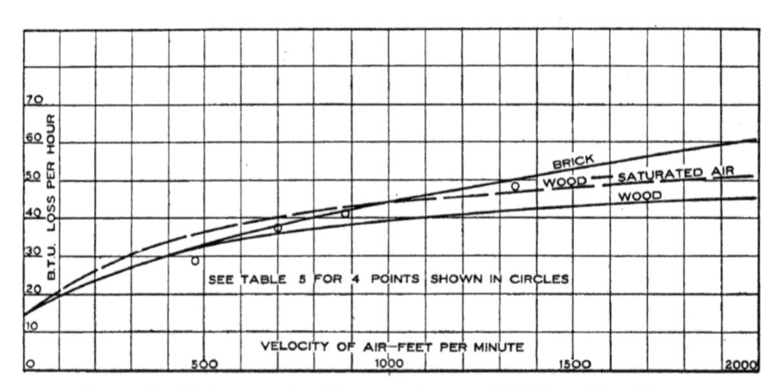

FIG. 20. *U* CURVES, AIR VELOCITY BELOW 2000 FEET PER MINUTE

The three curves shown in Fig. 20, taken from the *u* curves in Figs. 15, 17, and 18, at the lower velocities, and the four points plotted, which are obtained from the results of the saturated air tests on the brick box Table 5, show the effect of humidity on the heat transmission coefficient. The plotted points lie close enough to the curve for brick with partially saturated air to justify the conclusion that increasing the relative humidity from an average of 51.6 per cent to an average of 95.9 per cent does not increase the heat transmission through a brick wall four inches thick. In the case of the wood box, the effect of increasing the relative humidity of the air from an average of 71.3 per cent to an average of 96.9 per cent is also practically negligible.

26. *The Value of Air Spaces.*—Heat is transferred across an air space by means of all three methods of heat transfer, radiation, convection, and conduction.

With a large drop in temperature across the air space the circulation of the air will obviously be more rapid, and the convection loss will therefore be greater than with a small drop. Removing the air from the space has no appreciable effect until a very high vacuum is reached, as Nüsselt* found that a 29.96-inch vacuum (referred to a 30-inch barometer) had little effect on the loss by convection. In order, therefore, to reduce the loss by convection to an appreciable amount a very high vacuum is necessary. Convection loss, then, depends on the velocity of circulation of the air, which varies with the temperature difference of the two containing walls.

Air is a poor conductor of heat, and this fact accounts for the general belief that no matter what the structure air spaces built into walls will reduce the loss of heat to a great extent. The double glass box with a half-inch air space between the panes illustrates the value of an air space in constructions of this nature. A comparison of the brick and tile tests shows favorable results for the air space.

When higher temperatures than those met in ordinary building wall construction are encountered, however, the transfer of heat across an air space assumes a different aspect. For the lower temperature differences the radiation factor is not of very great importance. But since the quantity of heat which passes across an air space by means of radiation is proportional to the difference of the fourth powers of the absolute temperatures of the surfaces† enclosing the air space, it is evident that the radiation loss will increase rapidly with the temperatures of the two surfaces, althought the difference between these surfaces remains constant. On the other hand, if a solid material such as used in building walls should be used instead of the air space, the heat would be transferred through it by means of conduction alone. The amount of heat lost by conduction, moreover, would increase only slightly with an increase in temperature if the temperature difference between the two surfaces remained constant.

An air space may thus be as effective a heat insulator as a solid insulating material, at the lower temperatures and with the same temperature difference, but with a higher mean temperature and the same temperature difference, the air space would prove to be very inefficient.

Ray and Kreisinger‡ state that the amount of heat passing through furnace walls would be much reduced if the air spaces were

* Wilhelm Nüsselt, a private lecturer at the University of Dresden in Mechanical and Electrical Engineering. He has written on the subject of heat.

† This is exactly true only for perfect radiators and perfect absorbers.

‡ "The Flow of Heat through Furnace Walls," U. S. Bureau of Mines, Bul. 8, 1911.

filled with brick or preferably with materials of poor conductivity, such as ash, sand, or mineral wool. In other words, because of the radiation factor, when heat at low temperatures is to be insulated, use air spaces; when heat is at high temperatures as in the case of furnace walls, use solids of poor conductivity.

From the results of the single glass and the double glass box tests, the so-called conductivity of the ½-inch air space, meaning by "so-called conductivity" the B. t. u. loss per square foot per hour per degree of difference in the surface temperatures of the two containing walls, is found to be 1.77.

Prof. L. A. Harding, in a Pennsylvania State College Experiment Station Bulletin, gives a value of 1.66 for the so-called conductivity of air spaces ranging from one to six inches in thickness. From tests reported in Ice and Refrigeration* a value of 1.25 is deduced for a 1-inch air space. From tests on a steel mail-car side, reported by Prof. A. C. Willard,† a value of 1.59 is deduced for a 4-inch air space. Nüsselt states that air spaces greater than ¾ inch in thickness give no additional value for heat insulating purposes, a statement substantiated by the foregoing data. An average of the previously mentioned values gives a value of 1.57 for the "so-called conductivity" of an air space.

From the data already presented, the temperature drop across the air space was calculated, and a curve was plotted with air-temperature differences as abscissae and the so-called conductivity of air spaces as ordinates. The curve, indicating the variation of the so-called conductivity with the temperature difference across the space, is shown in Fig. 21.

In making calculations for heat transmission coefficients of compound walls, an air space may be treated in either of the following ways: the air space may be regarded as a solid insulating material through which the heat passes according to the so-called conductivity theory or considering the transfer by the three methods, radiation, convection, and conduction, the radiation and convection action may be combined into a single surface coefficient and the true conductivity of the air neglected. For every air space two surface coefficients, accordingly, would be considered. If different surfaces enclosed the air space, different surface coefficients would be used for the two walls.

---

*Refrigerating World, October, 1914.

†Railway Age Gazette, Vol. 56 n. s., p. 1572, June 26, 1914.

It has been customary to assume still air conditions inside the air space, and consequently to use the surface coefficients corresponding to this condition. This assumption gives values for U, when calculated which are generally in accord with the actual transmission values obtained in tests of hollow building walls. It would probably not prove true for air spaces at high temperatures, as in furnace work.

VII. Applications of Data to Typical Simple and Compound Walls

27. *Types of Wall Construction.*—Reference has already been made to the fact that it is manifestly impossible to test all types and combinations of building materials used in actual wall construcion. It will, therefore, generally be necessary to calculate from test data, similar to those given in this bulletin and amplified by further tests, the values of the overall or heat transmission coefficients (U) for many walls. Two cases exist, one involving solid walls of one or more materials and the other including hollow walls of simple or compound construction with one or more air spaces.

28. *Solid Wall Construction.*—In a simple wall or a compound wall without air spaces there are two surfaces which enter into the calculations for the heat transmission coefficient. For the inside surface a coefficient corresponding to still air conditions is used. This is obtained from the list of coefficients, Table 2. For the outside surface a coefficient of three times that corresponding to still air conditions is used; this allows for a wind velocity of practically fifteen miles per hour. The conductivity values for the materials involved are obtained from the same table of coefficients. With these values substituted in the heat transmission formula, the heat loss in B. t. u., per square foot, per hour, per degree of difference in air temperatures is obtained and the total loss through the walls of a building for any given air temperature difference may be readily calculated, as shown by Fig. 2.

The walls shown involve materials for which tests have already been run to determine K_1, K_2 and C· Additional tests will furnish data for a variety of materials which are not so generally used as those listed in Table 2.

29. *Air Space Construction.*—For walls containing an air space or spaces, the accepted method of determining the heat transmission of the wall is the same as that for a simple or compound wall with the addition of two surface coefficients for each space included in the construction. The surface coefficients used for air spaces are those used for still air conditions. As mentioned before, however, this assump-

tion of still air conditions in an air space is probably not true in all cases.

To determine the heat transmission of a wall containing an air space according to this alternate method, the following tentative solution may be adopted: The temperature drop across the air space is assumed, the so-called conductivity value is determined from the curve, Fig. 21, calculation for the transmission coefficient is made, and the loss for the overall air temperature difference determined. Dividing this value by the conductivity value used for the air space gives the

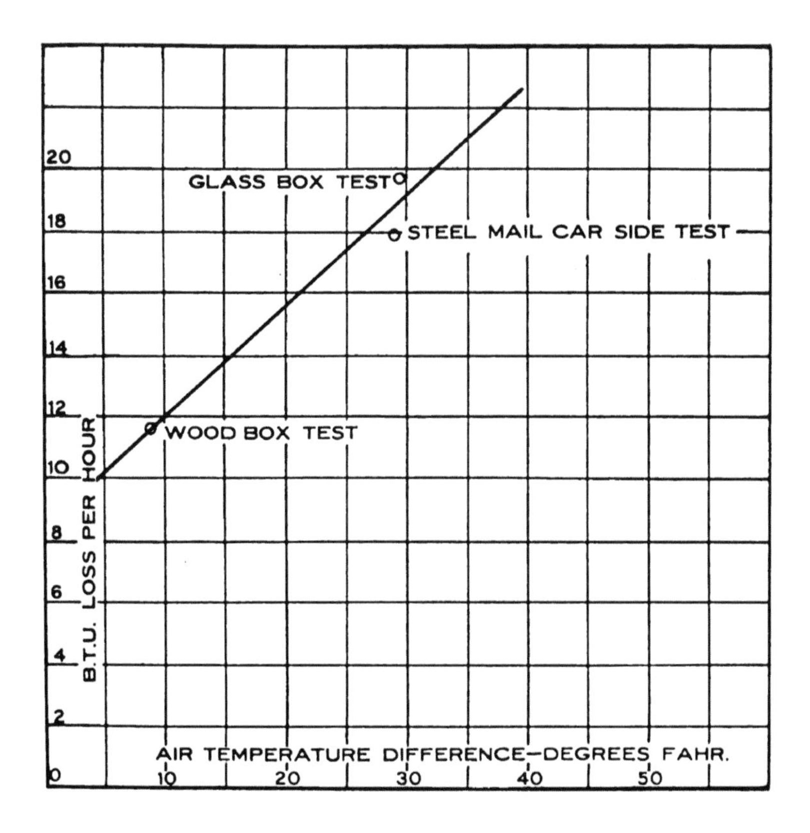

FIG. 21. THE SO-CALLED CONDUCTIVITY CURVE FOR AIR SPACE CONSTRUCTION, BASED ON TEMPERATURE DIFFERENCE BETWEEN ENCLOSING WALLS

temperature drop across the air space: This result should check the assumed temperature drop if the assumption is correct. If the resulting temperature is greater than the assumed one, an assumption of temperature drop larger than the previous one is made, and calculations are made again, and so on until the calculated drop checks the assumed one.

VIII. Conclusions

The following summary includes the more important conclusions which have been drawn from the results of the investigation:

(1) For determining the transmission coefficient, air to air, and the outside surface coefficient, mercury thermometers may be used and if properly installed will give results of practically the same accuracy as those determined with the thermocouples. Mercury thermometer wells were shown to be of no value in determining surface temperatures. The method finally adopted was to imbed the thermometer in the surface so that its center line would lie in the plane of the surface of the material being tested. Thermocouples are to be preferred for determining temperatures in this work.

(2) A variation in the air velocity over the surface affects both the air-to-air coefficient and the outside surface coefficient. The outside surface coefficients of brick and wood surfaces are affected in practically the same manner at the lower air velocities, while the coefficient for brick gradually rises above that of wood for the higher velocities.

(3) The coefficient of transmission, U, for a simple or compound wall, floor, or roof can be readily computed provided the surface coefficients for the building materials used in the surfaces are known for both still and moving air conditions. It is also necessary to know the coefficients of conductivity for all the materials used in the wall. These coefficients should be determined on full size walls for the temperature ranges usually encountered. (See Applications Fig. 2). The effect of moving air is to increase the value of the outside surface coefficient, and this increase involves the combined action of heat transfer by radiation and convection. While this effect is variable, according to the velocity of wind, it seems evident that it can always be expressed as some function of the still air coefficient for the same wall; thus where the average wind movement during the heating season is approximately fifteen miles per hour, a value of $K_2 = 3\,K_1$ is recommended. In any case, the test data presented herewith indicate the manner and degree in which the factor varies for different velocities so that the value of K_2 can be modified to conform with the conditions of wind movement ex-

isting in the locality where the heat transmission data are to be applied.

(4) The effect of the relative humidity of the air on the heat transmission of an ordinary building wall is so small that it can safely be neglected in determining the total heat loss from a building. An increase in the relative humidity of the air causes a slight increase in the value of the surface coefficient; hence the overall transmission coefficient becomes larger. If the walls are actually wet so that they absorb moisture, the coefficient of conduction, C, may be seriously affected, as in the case of brick work where an increase of at least twenty-five per cent may easily occur.

(5) Air space construction is of value at low temperatures, but not at high temperatures. In all cases the heat transmission across an air space will rapidly increase not only with an increase in the temperature difference in the enclosing walls but also with the same difference of temperature, if the absolute values of the enclosing wall temperatures are increased.

LIST OF

PUBLICATIONS OF THE ENGINEERING EXPERIMENT STATION

Bulletin No. 1. Tests of Reinforced Concrete Beams, by Arthur N. Talbot, 1904. *None available.*

Circular No. 1. High-Speed Tool Steels, by L. P. Breckenridge. 1905. *None available.*

Bulletin No. 2. Tests of High-Speed Tool Steels on Cast Iron, by L. P. Breckenridge and Henry B. Dirks. 1905. *None available.*

Circular No. 2. Drainage of Earth Roads, by Ira O. Baker. 1906. *None available.*

Circular No. 3. Fuel Tests with Illinois Coal (Compiled from tests made by the Technological Branch of the U. S. G. S., at the St. Louis, Mo., Fuel Testing Plant, 1904–1907), by L. P. Breckenridge and Paul Diserens. 1908. *Thirty cents.*

Bulletin No. 3. The Engineering Experiment Station of the University of Illinois, by L. P. Breckenridge. 1906. *None available.*

Bulletin No. 4. Tests of Reinforced Concrete Beams, Series of 1905, by Arthur N. Talbot. 1906. *Forty-five cents.*

Bulletin No. 5. Resistance of Tubes to Collapse, by Albert P. Carman and M. L. Carr. 1906. *None available.*

Bulletin No. 6. Holding Power of Railroad Spikes, by Roy I. Webber. 1906. *None available.*

Bulletin No. 7. Fuel Tests with Illinois Coals, by L. P. Breckenridge, S. W. Parr, and Henry B. Dirks. 1906. *None available.*

Bulletin No. 8. Tests of Concrete: I, Shear; II, Bond, by Arthur N. Talbot. 1906. *None available.*

Bulletin No. 9. An Extension of the Dewey Decimal System of Classification Applied to the Engineering Industries, by L. P. Breckenridge and G. A. Goodenough. 1906. Revised Edition 1912. *Fifty cents.*

Bulletin No. 10. Tests of Concrete and Reinforced Concrete Columns, Series of 1906, by Arthur N. Talbot. 1907. *None available.*

Bulletin No. 11. The Effect of Scale on the Transmission of Heat through Locomotive Boiler Tubes, by Edward C. Schmidt and John M. Snodgrass. 1907. *None available.*

Bulletin No. 12. Tests of Reinforced Concrete T-Beams, Series of 1906, by Arthur N. Talbot. 1907. *None available.*

Bulletin No. 13. An Extension of the Dewey Decimal System of Classification Applied to Architecture and Building, by N. Clifford Ricker. 1907. *None available.*

Bulletin No. 14. Tests of Reinforced Concrete Beams, Series of 1906, by Arthur N. Talbot. 1907. *None available.*

Bulletin No. 15. How to Burn Illinois Coal Without Smoke, by L. P. Breckenridge. 1908. *None available.*

Bulletin No. 16. A Study of Roof Trusses, by N. Clifford Ricker. 1908. *None available.*

Bulletin No. 17. The Weathering of Coal, by S. W. Parr, N. D. Hamilton, and W. F. Wheeler. 1908. *None available.*

Bulletin No. 18. The Strength of Chain Links, by G. A. Goodenough and L. E. Moore. 1908. *Forty cents.*

Bulletin No. 19. Comparative Tests of Carbon, Metallized Carbon and Tantalum Filament Lamps, by T. H. Amrine. 1908. *None available.*

Bulletin No. 20. Tests of Concrete and Reinforced Concrete Columns, Series of 1907, by Arthur N. Talbot. 1908. *None available.*

Bulletin No. 21. Tests of a Liquid Air Plant, by C. S Hudson and C. M. Garland. 1908. *Fifteen cents.*

Bulletin No. 22. Tests of Cast-Iron and Reinforced Concrete Culvert Pipe, by Arthur N. Talbot. 1908. *None available.*

Bulletin No. 23. Voids, Settlement and Weight of Crushed Stone, by Ira O. Baker. 1908. *Fifteen cents.*

**Bulletin No. 24.* The Modification of Illinois Coal by Low Temperature Distillation, by S. W. Parr and C. K. Francis. 1908. *Thirty cents.*

Bulletin No. 25. Lighting Country Homes by Private Electric Plants, by T. H. Amrine. 1908. *Twenty cents.*

*A limited number of copies of bulletins starred is available for free distribution.

Bulletin No. 26. High Steam-Pressures in Locomotive Service. A Review of a Report to the Carnegie Institution of Washington, by W. F. M. Goss. 1908. *Twenty-five cents.*

Bulletin No. 27. Tests of Brick Columns and Terra Cotta Block Columns, by Arthur N. Talbot and Duff A. Abrams. 1909. *Twenty-five cents.*

Bulletin No. 28. A Test of Three Large Reinforced Concrete Beams, by Arthur N. Talbot. 1909. *Fifteen cents.*

Bulletin No 29. Tests of Reinforced Concrete Beams: Resistance to Web Stresses, Series of 1907 and 1908, by Arthur N. Talbot. 1909. *Forty-five cents.*

**Bulletin No. 30.* On the Rate of Formation of Carbon Monoxide in Gas Producers, by J. K. Clement, L. H. Adams, and C. N. Haskins. 1909. *Twenty-five cents.*

**Bulletin No. 31.* Tests with House-Heating Boilers, by J. M. Snodgrass. 1909. *Fifty-five cents.*

Bulletin No. 32. The Occluded Gases in Coal, by S. W. Parr and Perry Barker. 1909. *Fifteen cents.*

Bulletin No. 33. Tests of Tungsten Lamps, by T. H. Amrine and A. Guell. 1909. *Twenty cents.*

**Bulletin No. 34.* Tests of Two Types of Tile-Roof Furnaces under a Water-Tube Boiler, by J. M. Snodgrass. 1909. *Fifteen cents.*

Bulletin No. 35. A Study of Base and Bearing Plates for Columns and Beams, by N. Clifford Ricker. 1909. *Twenty cents.*

Bulletin No. 36. The Thermal Conductivity of Fire-Clay at High Temperatures, by J. K. Clement and W. L. Egy. 1909. *Twenty cents.*

Bulletin No. 37. Unit Coal and the Composition of Coal Ash, by S. W. Parr and W. F. Wheeler. 1909. *Thirty-five cents.*

**Bulletin No. 38.* The Weathering of Coal, by S. W. Parr and W. F. Wheeler. 1909. *Twenty-five cents*

**Bulletin No. 39.* Tests of Washed Grades of Illinois Coal, by C. S. McGovney. 1909. *Seventy-five cents.*

Bulletin No. 40. A Study in Heat Transmission, by J. K. Clement and C. M. Garland. 1910. *Ten cents.*

Bulletin No. 41. Tests of Timber Beams, by Arthur N. Talbot. 1910. *Thirty-five cents.*

**Bulletin No. 42.* The Effect of Keyways on the Strength of Shafts, by Herbert F. Moore. 1910. *Ten cents.*

Bulletin No. 43. Freight Train Resistance, by Edward C. Schmidt. 1910. *Seventy-five cents.*

Bulletin No. 44. An Investigation of Built-up Columns Under Load, by Arthur N. Talbot and Herbert F. Moore. 1911. *Thirty-five cents.*

**Bulletin No. 45.* The Strength of Oxyacetylene Welds in Steel, by Herbert L. Whittemore. 1911. *Thirty-five cents.*

**Bulletin No. 46.* The Spontaneous Combustion of Coal, by S. W. Parr and F. W. Kressman. 1911. *Forty-five cents*

**Bulletin No 47.* Magnetic Properties of Heusler Alloys, by Edward B. Stephenson, 1911. *Twenty-five cents.*

**Bulletin No 48.* Resistance to Flow Through Locomotive Water Columns, by Arthur N. Talbot and Melvin L. Enger. 1911 *Forty cents.*

**Bulletin No. 49.* Tests of Nickel-Steel Riveted Joints, by Arthur N. Talbot and Herbert F. Moore. 1911. *Thirty cents.*

**Bulletin No. 50.* Tests of a Suction Gas Producer, by C. M. Garland and A. P. Kratz. 1912 *Fifty cents.*

Bulletin No. 51. Street Lighting, by J. M. Bryant and H. G. Hake. 1912. *Thirty-five cents.*

**Bulletin No. 52.* An Investigation of the Strength of Rolled Zinc, by Herbert F. Moore. 1912. *Fifteen cents.*

**Bulletin No. 53.* Inductance of Coils, by Morgan Brooks and H. M. Turner. 1912. *Forty cents.*

**Bulletin No. 54.* Mechanical Stresses in Transmission Lines, by A. Guell. 1912. *Twenty cents.*

**Bulletin No. 55.* Starting Currents of Transformers, with Special Reference to Transformers with Silicon Steel Cores, by Trygve D. Yensen. 1912. *Twenty cents.*

**Bulletin No. 56.* Tests of Columns: An Investigation of the Value of Concrete as Reinforcement for Structural Steel Columns, by Arthur N. Talbot and Arthur R. Lord. 1912. *Twenty-five cents.*

**Bulletin No 57.* Superheated Steam in Locomotive Service. A Review of Publication No. 127 of the Carnegie Institution of Washington, by W. F. M. Goss. 1912. *Forty cents.*

*A limited number of copies of bulletins starred is available for free distribution.

**Bulletin No. 58.* A New Analysis of the Cylinder Performance of Reciprocating Engines, by J. Paul Clayton. 1912. *Sixty cents.*

**Bulletin No. 59.* The Effect of Cold Weather Upon Train Resistance and Tonnage Rating, by Edward C. Schmidt and F. W. Marquis. 1912. *Twenty cents.*

**Bulletin No. 60.* The Coking of Coal at Low Temperatures, with a Preliminary Study of the By-Products, by S. W. Parr and H. L. Olin. 1912. *Twenty-five cents.*

**Bulletin No. 61.* Characteristics and Limitation of the Series Transformer, by A. R. Anderson and H. R. Woodrow. 1913. *Twenty-five cents.*

Bulletin No. 62. The Electron Theory of Magnetism, by Elmer H. Williams. 1913. *Thirty-five cents.*

Bulletin No. 63. Entropy-Temperature and Transmission Diagrams for Air, by C. R. Richards. 1913. *Twenty-five cents.*

**Bulletin No. 64.* Tests of Reinforced Concrete Buildings Under Load, by Arthur N. Talbot and Willis A. Slater. 1913. *Fifty cents.*

**Bulletin No. 65.* The Steam Consumption of Locomotive Engines from the Indicator Diagrams, by J. Paul Clayton. 1913. *Forty cents.*

Bulletin No. 66. The Properties of Saturated and Superheated Ammonia Vapor, by G. A. Goodenough and William Earl Mosher. 1913. *Fifty cents.*

Bulletin No. 67. Reinforced Concrete Wall Footings and Column Footings, by Arthur N. Talbot. 1913. *Fifty cents.*

**Bulletin No. 68* The Strength of I-Beams in Flexure, by Herbert F. Moore. 1913. *Twenty cents.*

Bulletin No. 69. Coal Washing in Illinois, by F. C. Lincoln. 1913. *Fifty cents.*

Bulletin No. 70. The Mortar-Making Qualities of Illinois Sands, by C. C. Wiley. 1913. *Twenty cents.*

Bulletin No. 71. Tests of Bond between Concrete and Steel, by Duff A. Abrams. 1913. *One dollar.*

**Bulletin No. 72.* Magnetic and Other Properties of Electrolytic Iron Melted in Vacuo, by Trygve D. Yensen. 1914. *Forty cents.*

Bulletin No. 73. Acoustics of Auditoriums, by F. R. Watson. 1914. *Twenty cents.*

**Bulletin No. 74.* The Tractive Resistance of a 28-Ton Electric Car, by Harold H. Dunn. 1914. *Twenty-five cents.*

Bulletin No. 75. Thermal Properties of Steam, by G. A. Goodenough. 1914. *Thirty-five cents.*

Bulletin No. 76. The Analysis of Coal with Phenol as a Solvent, by S. W. Parr and H. F. Hadley. 1914. *Twenty-five cents.*

**Bulletin No. 77.* The Effect of Boron upon the Magnetic and Other Properties of Electrolytic Iron Melted in Vacuo, by Trygve D. Yensen. 1915. *Ten cents.*

**Bulletin No. 78.* A Study of Boiler Losses, by A. P. Kratz. 1915. *Thirty-five cents.*

**Bulletin No. 79.* The Coking of Coal at Low Temperatures, with Special Reference to the Properties and Composition of the Products, by S. W. Parr and H. L. Olin. 1915. *Twenty-five cents.*

**Bulletin No. 80.* Wind Stresses in the Steel Frames of Office Buildings, by W. M. Wilson and G. A. Maney. 1915. *Fifty cents.*

**Bulletin No. 81.* Influence of Temperature on the Strength of Concrete, by A. B. McDaniel. 1915. *Fifteen cents.*

Bulletin No. 82. Laboratory Tests of a Consolidation Locomotive, by E. C. Schmidt, J. M. Snodgrass, and R. B. Keller. 1915. *Sixty-five cents.*

**Bulletin No. 83.* Magnetic and Other Properties of Iron-Silicon Alloys. Melted in Vacuo, by Trygve D. Yensen. 1915. *Thirty-five cents.*

Bulletin No. 84. Tests of Reinforced Concrete Flat Slab Structure, by A. N. Talbot and W. A. Slater. 1916. *Sixty-five cents.*

**Bulletin No. 85.* The Strength and Stiffness of Steel Under Biaxial Loading, by A. T. Becker. 1916. *Thirty-five cents.*

**Bulletin No. 86.* The Strength of I-Beams and Girders, by Herbert F. Moore and W. M. Wilson. 1916. *Thirty cents.*

**Bulletin No. 87.* Correction of Echoes in the Auditorium, University of Illinois, by F. R. Watson and J. M. White. 1916. *Fifteen cents.*

Bulletin No. 88. Dry Preparation of Bituminous Coal at Illinois Mines, by E. A. Holbrook. 1916. ***Seventy cents.***

---

*A limited number of copies of bulletins starred is available for free distribution.

**Bulletin No. 89.* Specific Gravity Studies of Illinois Coal, by Merle L. Nebel. 1916. *Thirty cents.*

**Bulletin No. 90* Some Graphical Solutions of Electric Railway Problems, by A. M. Buck. 1916. *Twenty cents.*

Bulletin No. 91. Subsidence Resulting from Mining, by L. E. Young and H. H. Stoek. 1916. *None available.*

**Bulletin No. 92.* The Tractive Resistance on Curves of a 28-Ton Electric Car, by E. C. Schmidt and H. H. Dunn. 1916. *Twenty-five cents.*

**Bulletin No 93.* A Preliminary Study of the Alloys of Chromium, Copper, and Nickel, by D. F. McFarland and O. E. Harder. 1916. *Thirty cents.*

**Bulletin No. 94.* The Embrittling Action of Sodium Hydroxide on Soft Steel, by S. W. Parr. 1917. *Thirty cents.*

**Bulletin No. 95.* Magnetic and Other Properties of Iron-Aluminum Alloys Melted in Vacuo, by T. D. Yensen and W. A. Gatward. 1917. *Twenty-five cents.*

**Bulletin No. 96.* The Effect of Mouthpieces on the Flow of Water Through a Submerged Short Pipe, by Fred B. Seely. 1917. *Twenty-five cents.*

**Bulletin No. 97.* Effects of Storage Upon the Properties of Coal, by S. W. Parr. 1917. *Twenty cents.*

**Bulletin No. 98.* Tests of Oxyacetylene Welded Joints in Steel Plates, by Herbert F. Moore. 1917. *Ten cents.*

Circular No 4. The Economical Purchase and Use of Coal for Heating Homes, with Special Reference to Conditions in Illinois. 1917. *Ten cents.*

**Bulletin No. 99.* The Collapse of Short Thin Tubes, by A P. Carman. 1917. *Twenty cents.*

**Circular No. 5.* The Utilization of Pyrite Occurring in Illinois Bituminous Coal, by E. A. Holbrook. 1917. *Twenty cents.*

**Bulletin No. 100.* Percentage of Extraction of Bituminous Coal with Special Reference to Illinois Conditions, by C. M. Young. 1917.

**Bulletin No. 101.* Comparative Tests of Six Sizes of Illinois Coal on a Mikado Locomotive, by E. C. Schmidt, J. M. Snodgrass, and O. S. Beyer, Jr. 1917. *Fifty cents.*

**Bulletin No. 102.* A Study of the Heat Transmission of Building Materials, by A. C. Willard and L. C. Lichty. 1917. *Twenty-five cents.*

---

*A limited number of copies of bulletins starred is available for free dis ribution.

THE UNIVERSITY OF ILLINOIS

THE STATE UNIVERSITY

Urbana

EDMUND J. JAMES, Ph. D., LL. D., President

THE UNIVERSITY INCLUDES THE FOLLOWING DEPARTMENTS:

The Graduate School

The College of Liberal Arts and Sciences (Ancient and Modern Languages and Literatures; History, Economics, Political Science, Sociology; Philosophy, Psychology, Education; Mathematics; Astronomy; Geology; Physics; Chemistry; Botany, Zoology, Entomology; Physiology; Art and Design)

The College of Commerce and Business Administration (General Business, Banking, Insurance, Accountancy, Railway Administration, Foreign Commerce; Courses for Commercial Teachers and Commercial and Civic Secretaries)

The College of Engineering (Architecture; Architectural, Ceramic, Civil, Electrical, Mechanical, Mining, Municipal and Sanitary, and Railway Engineering)

The College of Agriculture (Agronomy; Animal Husbandry; Dairy Husbandry; Horticulture and Landscape Gardening; Agricultural Extension; Teachers' Course; Household Science)

The College of Law (three years' course)

The School of Education

The Course in Journalism

The Courses in Chemistry and Chemical Engineering

The School of Railway Engineering and Administration

The School of Music (four years' course)

The School of Library Science (two years' course)

The College of Medicine (in Chicago)

The College of Dentistry (in Chicago)

The School of Pharmacy (in Chicago; Ph. G. and Ph. C. courses)

The Summer Session (eight weeks)

Experiment Stations and Scientific Bureaus: U. S. Agricultural Experiment Station; Engineering Experiment Station; State Laboratory of Natural History; State Entomologist's Office; Biological Experiment Station on Illinois River; State Water Survey; State Geological Survey; U. S. Bureau of Mines Experiment Station.

The library collections contain (December 1, 1917) **411,737 volumes and 104,524 pamphlets.**

For catalogs and information address

THE REGISTRAR
URBANA, ILLINOIS

THE Engineering Experiment Station was established by act of the Board of Trustees, December 8, 1903. It is the purpose of the Station to carry on investigations along various lines of engineering and to study problems of importance to professional engineers and to the manufacturing, railway, mining, constructional, and industrial interests of the State.

The control of the Engineering Experiment Station is vested in the heads of the several departments of the College of Engineering. These constitute the Station Staff and, with the Director, determine the character of the investigations to be undertaken. The work is carried on under the supervision of the Staff, sometimes by research fellows as graduate work, sometimes by members of the instructional staff of the College of Engineering, but more frequently by investigators belonging to the Station corps.

The results of these investigations are published in the form of bulletins which record mostly the experiments of the Station's own staff of investigators. There will also be issued from time to time, in the form of circulars, compilations giving the results of the experiments of engineers, industrial works, technical institutions, and governmental testing departments.

The volume and number at the top of the title page of the cover are merely arbitrary numbers and refer to the general publications of the University of Illinois: *either above the title or below the seal is given the number of the Engineering Experiment Station bulletin or circular which should be used in referring to these publications.*

For copies of bulletins, circulars, or other information address the

ENGINEERING EXPERIMENT STATION,
URBANA, ILLINOIS.

UNIVERSITY OF ILLINOIS
ENGINEERING EXPERIMENT STATION

BULLETIN NO. 103 NOVEMBER, 1917

AN INVESTIGATION OF TWIST DRILLS

BY

BRUCE W. BENEDICT
DIRECTOR OF THE SHOP LABORATORIES

AND

W. PENN LUKENS
RESEARCH FELLOW IN THE ENGINEERING EXPERIMENT STATION

ENGINEERING EXPERIMENT STATION
PUBLISHED BY THE UNIVERSITY OF ILLINOIS, URBANA

CONTENTS

LIST OF FIGURES

LIST OF FIGURES (Continued)

AN INVESTIGATION OF TWIST DRILLS

I. Introduction

1. *Purpose.*—During recent years the economic importance of increasing production in modern manufacturing has stimulated development and improvement in shop practice, in machinery, and in metal working process. Practice governing the cutting of metals has had an important bearing upon the rate of production, and competition among manufacturers to cut down costs has resulted in great improvements in the practical application of metal cutting tools. In this development the engineer-scientist and the organizations devoted to experimental research have not had an important part, and the design of metal cutting tools has not been based upon an exact knowledge of performance. The investigation of twist drills, the results of which are here reported, is the first of a series of investigations of the design and performance of metal cutting tools planned by the Engineering Experiment Station.

Unlike many of the tools employed with machinery companion of the drill-press, the twist drill is normally the product of tool making establishments especially equipped for its manufacture. One result of this plan of manufacturing has been a standardization of design and of methods of production to a degree greater than is found in the manufacture of other metal-working tools. So closely have manufactures followed a common practice that twist drills of different makes are very similar or practically identical in appearance. Only after minute inspection can variations in design be detected; and these are slight. An extra thousandth of an inch more in the web, a few degrees more or less in the helix angle, and different standards of machining and finishing are the characteristics distinguishing one drill from another. This similarity in form and the general excellence of performance have helped to remove the twist drill from the field of the investigator.

There is little information regarding the comparative performance of drills of different makes, and the producer who is constantly facing a multitude of unsolved problems has not concerned himself with the twist drill. Although a number of investigators in this country and abroad have made studies of the drilling of metals during the past decade and have contributed much to the knowledge of the subject, the existing data emphasize the gaps yet to be filled rather than the extent of the ground already covered. This statement does not reflect in any manner whatever on the work which has already been done. The fullest recognition is recorded the work of those investigators who have contributed all the existing knowledge of drills and drilling, but the subject is large and investigators have usually worked independently on certain individual phases of the problem without reference to a coördinate program.

The experiments reported in this bulletin were undertaken primarily to solve certain problems relating to drilling which arose in the University laboratories and in the shops of manufacturers, and the purpose has been to establish facts relating to every day problems of shop practice. Data have been obtained on the basis of which the influence of the angle of helix and of the methods of point grinding upon the power required by the drill might be studied. The absorption of power by the drill-press has been noted and the subject of drill endurance for different angles of grinding has been briefly investigated. All experiments have been made upon cast iron of fairly soft composition.

The fact is emphasized that the investigations here reported are essentially preliminary. It is the present purpose to extend the work in the experimental study of drilling through a series of several investigations, and it is probable that many of the data and conclusions here presented will be shown in their true value only after the projected program of investigations is further advanced.

2. *Acknowledgments.*—The investigations were conducted by the Department of Mechanical Engineering as the work of the Engineering Experiment Station of the University of Illinois. Dean C. R. Richards, Dean of the College of Engineering and formerly head of the Department of Mechanical Engineering at the University of Illinois, exercised general direction over the work. Director B. W. Benedict supervised the tests, the laboratory work of which was conducted by W. P. Lukens. Upon the latter fell the burden of the work involved in the making of the tests, and the compilation of data for publication. G. H. Radebaugh and the other members of the staff of the Shop Laboratories gave valuable assistance in constructing and reconstructing some portions of the apparatus.

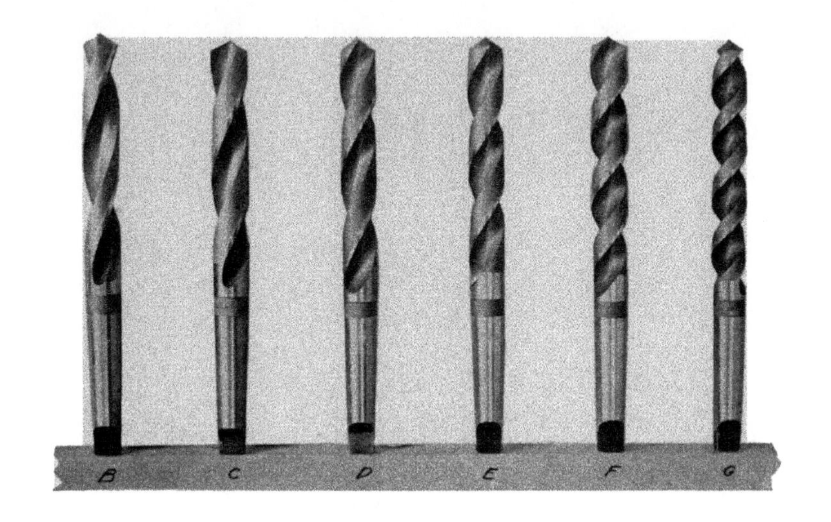

FIG. 1. STANDARD TWIST DRILLS, "B" TO "G," USED IN THE TESTS

0.125 inches. It will be

II. Methods Employed and Apparatus Used in Conducting the Tests

3. *Drills Used.*—One-inch high-speed drills were used throughout the tests. This size was chosen on account of the strength of the drills and the small risk of breakage under heavy loads. It was also felt that this size was representative and that the data obtained would form the basis for correct deductions respecting drill performance in general. Drills of many prominent makers were purchased and were tested under similar operating conditions. Little difference was observed in the character of the drills made by various manufacturers, but their design was shown to have a marked influence on performance.

Drills with eight different helix angles (measured with the axis of the drill) were used in the greater number of these tests. The drills with the 10-, 35-, 40-, and 45-degree angles were made especially for the tests by a well-known twist drill manufacturer, while the 22-, 26-, and 32-degree and the 32-degree flat-twist drills were obtained on the market from standard stock. For convenience in recording test results, these drills were designated by the letters "A" to "H," inclusive, as indicated in the following list:

"A"—Helix Angle	10 degrees
"B"—Helix Angle	22 degrees
"C"—Helix Angle	26 degrees
"D"—Helix Angle	27 to 32 degrees
"E"—Helix Angle	35 degrees
"F"—Helix Angle	40 degrees
"G"—Helix Angle	45 degrees
"H"—Helix Angle	32 flat twist

A photograph of drills "B" to "G" inclusive is reproduced as Fig. 1. Cross-sections of all the drills are shown by Fig. 2. The variation in the form of drills, even when the product of the same manufacturer, is well shown, while the greater chip space afforded by the larger helix angle may be noted. The method of point thinning used on drill "H" is shown in Fig. 56. In Fig. 3 are shown graphically the variations in web thickness from point to hilt of all test drills except "H," which was given the uniform thickness of 0.125 inches. It will be noted that drill "A" in being thinner toward the tang is made contrary to the usual practice, probably because of an error in the set-up of the milling machine. This drill ultimately broke in testing and was replaced by another 10-degree drill with a thicker web. Drill "B," as shown in Figs. 1 and 2 has the heel of the flute cut away for the entire length; yet it is apparently amply strong and has the advantages of giving greater chip space. It is also easier to grind and possibly uses slightly less metal.

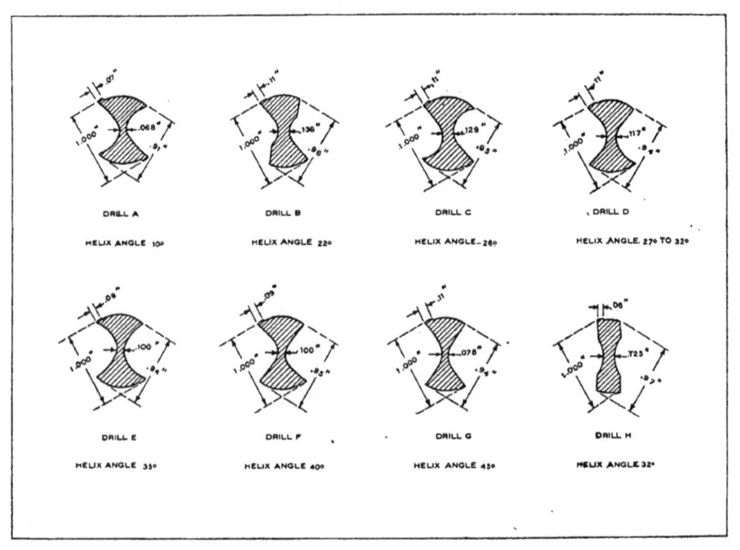

FIG. 2. SECTIONS OF THE VARIOUS TYPES OF DRILLS USED IN THE TESTS

Other drills used in some of the point angle and chisel edge tests or in the endurance tests were standard drills of various manufacturers.

4. *Test Blocks.*—The test material consisted of soft cast iron blocks made in the foundry of the University. Every effort was made to produce blocks of uniform structure and chemical composition. To insure uniformity the blocks were cast in bars as shown by Fig. 4 and were then planed on the bottom. Since most of the

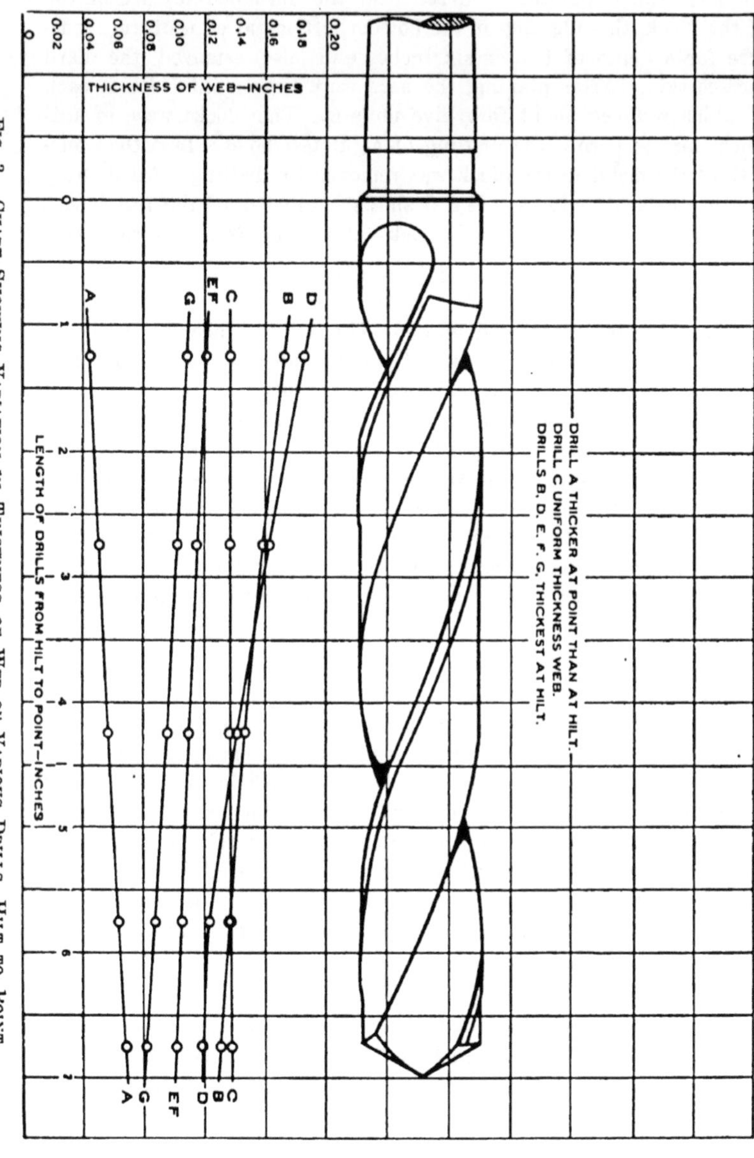

FIG. 3. CHART SHOWING VARIATION IN THICKNESS OF WEB OF VARIOUS DRILLS, HILT TO POINT

air holes and impurities occurred near the unplaned surface or top of the block, this planing of the bottom left metal of uniform structure for a depth of five or six inches and also removed the hard surface skin. After planing, the bars were broken into blocks, each of which weighed about forty-five pounds. The blocks were of sufficient size to permit the drilling of eight test holes. Less than one-fifth of the metal in the block was removed by drilling. All drilling was in finished surfaces. Fig. 5 shows a number of the test blocks ready for use, and also used blocks and chips from several days' testing.

5. *The Drill-Press and Motor.*—The drilling machine used was a 22-inch all geared drill-press. It was considered essential for purposes of these tests that no slippage occur between the motor and the drill, and all speeds and rates of feed were controlled by gears, so that the operation was positive and exact. All gears and bearings, except the spindle sleeve, were oiled automatically, and most of the gears were run in a bath of oil. The bearings throughout were of babbitt metal, except the ball thrust bearing on the spindle. Most of the gears were of chrome-nickel steel, and the final feed drive to the spindle was by rack and pinion. All speed and feed change levers were accessible.

The motor used for driving the drill-press, shown in Fig. 5, was a 25-horse-power alternating current, two-phase, 440-volt Westinghouse motor running at 1200 r. p. m. Although it was frequently necessary to operate the motor at loads approximating one hundred per cent overload, the regulation was very satisfactory; the speed variation being usually less than one per cent; in many of the tests, therefore, it was unnecessary to calculate the speed independently. The highest recorded load during a test was 54 h. p., 15 h. p. being delivered to the drill point. The power was transmitted through double helical cast iron gears to the drill-press, the gears running in fibre gear grease in a tight case.

6. *The Dynamometer.*—A view of the dynamometer under working conditions is presented as Fig. 7. At A (see also Fig. 6) is shown a cord which runs through a pulley system from the drill spindle to rotate the torque and thrust cards at a rate which permits

Fig. 4. Test Blocks Cast in Bars

Fig. 5. Drill-press, Motor, and Test Blocks

Fig. 6. The Drill-press and Dynamometers

the circumferential traverse of the pen on the record to be read in inches of spindle advance. At B is shown the torque gage and at C the thrust gage, recording the pressures in the reservoirs D and E respectively. At F is shown the casting for holding the test block, and at G a circular pan to catch the chips. The cord A passing around pulleys on the dial shafts is held taut by the weight K, so that the motion of the cards is sufficiently uniform for test purposes. By means of the pendulum shown in Fig. 8 a time record is obtained on the torque card, the pendulum being timed to complete an electrical contact each second. The wires, L, lead to an electro-magnet, which actuates the recording pen, and yields a record which may be correlated with torque coördinate of the torque card to give the drill penetration in inches per minute. These seconds' marks may be clearly seen on the torque card shown in Fig. 7 and it may be noted that the drill took practically 16 seconds to penetrate 2.5 inches, or the rate of penetration was 9.4 inches per minute. Since the feed per revolution in this instance was known to be exactly 0.0256 inches, the r. p. m. may be calculated by dividing 0.0256 into 9.4. The arrangement of the time recorder was changed slightly between the taking of the photographs for Figs. 6 and 7.

The dynamometer is clamped to the fixed table of the drill base and in effect consists of two distinct dynamometers of similar type, one recording torque and the other thrust. Both dynamometers are of the hydraulic type and employ a mixture of alcohol and glycerine as the transmitting medium. Heavy rubber diaphragms were used for taking the load and retaining the fluid. The pressure on the diaphragms is recorded upon the chart of a Bristol recording gage, the whole system being calibrated as a unit to insure accurate results.

The thrust dynamometer consists of a cast iron cylinder of $11\frac{21}{32}$ inches internal diameter, containing practically one gallon of fluid. Upon the rubber diaphragm covering this chamber rests a disc having a diameter slightly smaller than the inner diameter of the reservoir. Resting on this disc and supporting another disc on which the test piece is clamped is a large ball-thrust bearing which permits the upper disc to rotate with a minimum amount of friction. Projecting from the upper disc is a radial arm which transmits the torque of the drill to a vertical reservoir at the left. From the fluid reservoir a pipe leads to the standard Bristol recording gage, which records the pressure in the usual manner. The torque arm (J, Fig. 7) is of such length that the distance from the drill center to the point of application of the torque at the reservoir is exactly one foot; consequently readings of the torque are given in foot-pounds on the cards. The torque dynamometer is exactly similar in principle to the thrust dynamometer but is smaller, having a capacity for about a pint of fluid. The dynamometer is made of cast iron, except for the ball-bearing. There are few machined surfaces, and the apparatus is simple and strong. By the use of gages having different pressure ranges the degree of sensitiveness may be varied to suit the requirements of the test. Figs. 9 and 10 show the general principles of the construction.

7. *Calibration.*—As has been stated, it was necessary to calibrate the apparatus for the torque and thrust pressures and for the correlation of card rotation with spindle motion. This latter process was accomplished with the apparatus in place by advancing the spindle a certain measured amount and marking on the cards the advance of the recording pens. Enough test records were made to insure accurate average results and permanent cards were prepared showing this relation. The calibration of the thrust gage was accomplished by the direct application of pressure in a Riehle testing machine; permanent average records were thus established. The thrust calibration was made with gages giving a 2500-and a 6000-pound maximum total pressure, and in each case the radial motion of the recording pen was in direct ratio with the load increase. The torque calibration was made by means of a harness which permitted the application of the horizontal load to the vertical disc. In one calibration known weights were applied, the direction of the force being changed by means of a pulley. In order to obviate errors due to the friction of the pulley and the bending of the cord around the pulley a second calibration test was made in which the load was applied by a previously calibrated spring balance. The two calibrations agreed fairly well, but the latter was accepted as the more nearly accurate. For the range of pressures used in these tests the radial motion of the recording pen was in almost direct ratio with the load increase. There is an initial discrepancy which seems to be due to the fact that the fluid is in a vertical plane and thus requires a load of about two pounds to equalize the pressure throughout the fluid. Since no test recorded less than ten foot-pounds of torque,

FIG. 7. THE DYNAMOMETERS WITH DRILL AND TEST BLOCK IN PLACE

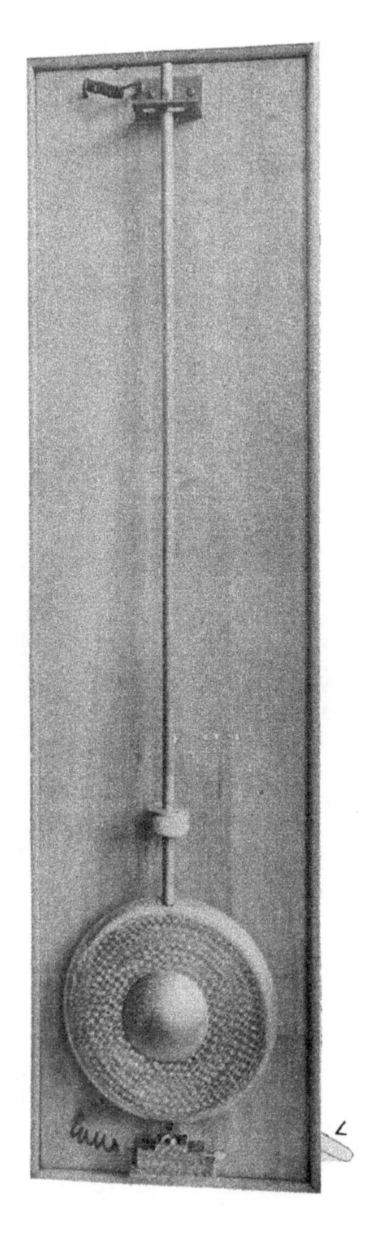

Fig. 8. Pendulum Used for Recording Seconds on the Torque Card

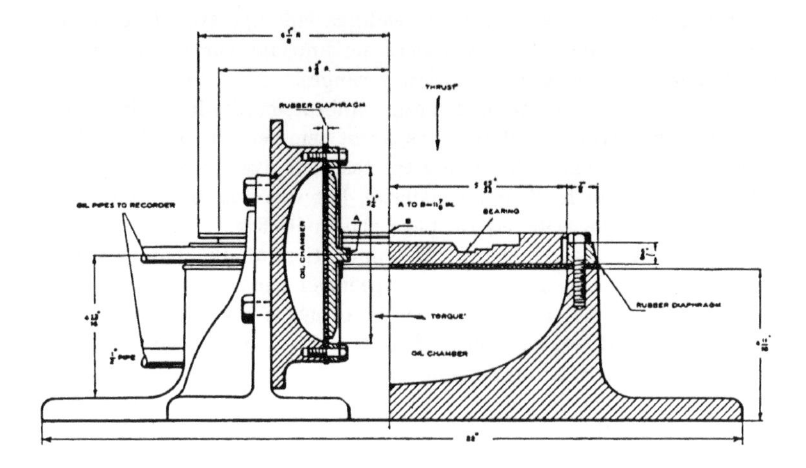

FIG. 9. SECTIONAL ELEVATION OF DRILL DYNAMOMETERS

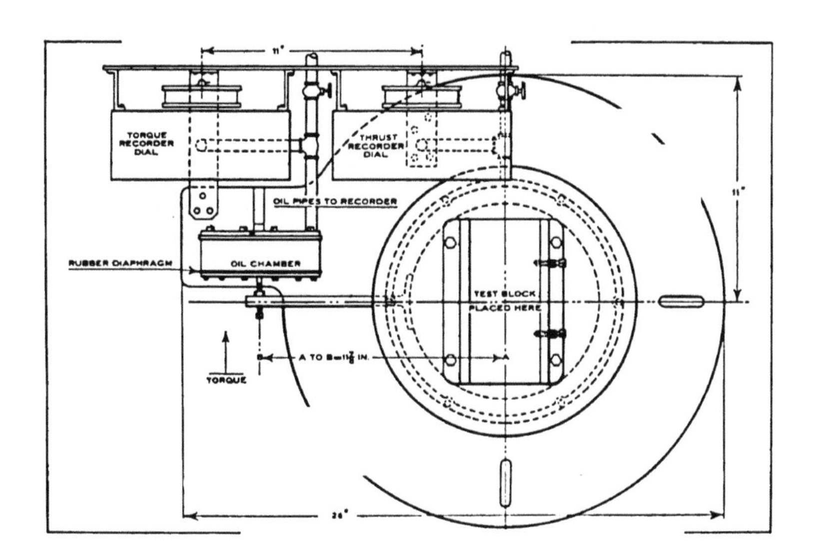

FIG. 10. PLAN VIEW OF DRILL DYNAMOMETERS

this initial discrepancy affects all readings equally, even if due to errors in calibration. The tests show no unusual variations which would suggest inaccuracy in the lower ranges.

When the apparatus had been calibrated satisfactorily, the calibration was transferred to transparent templets from which the cards could be evaluated by laying the templets over them. Tracing cloth was first used for templet material, but was not found entirely satisfactory; so the celluloid templets shown in Fig. 11 were made. This illustration clearly shows the application of the templets to the cards, while Fig. 12 shows two specimen cards as obtained in tests. Values of the cards are taken as an eye average between the depths of one and two and one-half inches, this average being found to give more nearly uniform results than an exact average over a certain range. The torque may be estimated within one-half foot-pound, and the thrust within fifty pounds. Absolute exactness of values is not only unnecessary, but practically impossible.

The thrust card illustrated shows that the block was harder for a depth of about one inch than further down, while at a depth of several inches the difficulty of chip disposal may have caused an increase in both torque and thrust. The time record and the method of calculating r. p. m. and h. p. are shown.

8. *Drill Grinder.*—The drill grinder used for grinding the drills for all tests was suited to drills varying in size from No. 60 to 3½ inches, and was fitted with an adjustable holder for variable point angles. The machine is illustrated by Fig. 13, the shadow showing the swing of the drill-holder. No drills for these tests were hand ground, but to compare the average accuracy of hand grinding with this machine grinding a few one-inch drills were ground by good mechanics and measured. In no case was the hand ground drill perfectly ground, the contour of the two sides being different even when the angles and lip length were nearly the same. It was interesting to note that almost invariably the included point angle was made about 110 or 112 degrees instead of the standard 118 degrees. This fact may account for the seemingly greater endurance of hand ground drills frequently noted by mechanics.

9. *Test Procedure.*—In all tests, except a few to which special reference is made, holes were drilled in cast iron with the skin re-

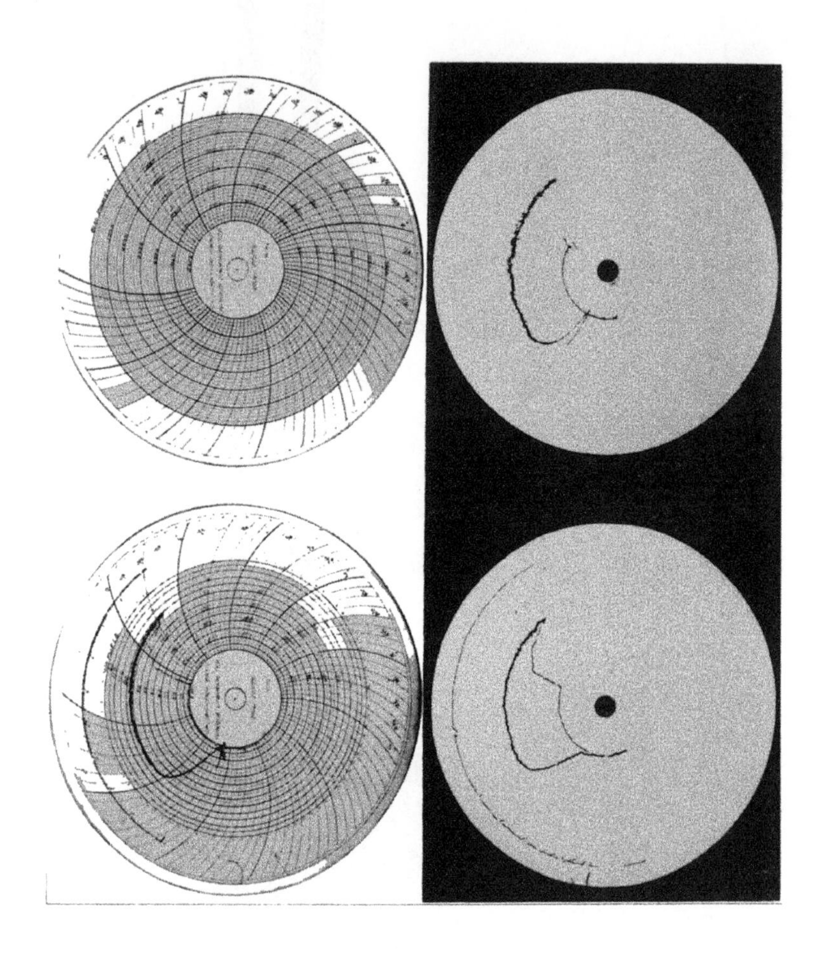

FIG. 11. SPECIMEN DIAGRAMS AND CELLULOID TEMPLETS USED FOR READING TORQUE AND THRUST DIAGRAMS

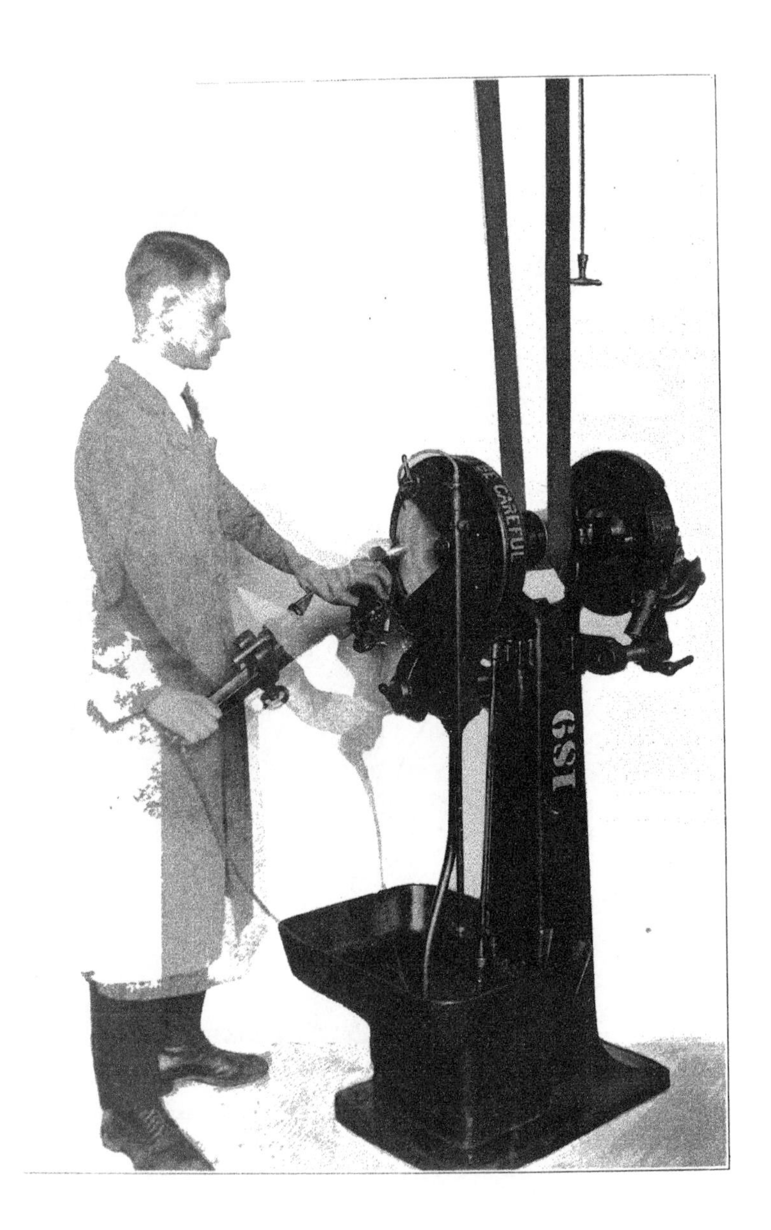

Fig. 13. The Drill Grinder

moved. For power tests the holes were drilled to a depth of three or four inches, but for endurance tests the depth was about five inches. For power tests the drill was sharpened after drilling four or five holes at moderate rates and after one or two holes at the higher rates. All drills were ground dry and care was taken not to overheat the edges on the wheel. Where comparative tests were necessary, as in the chisel point tests and point angle tests, the holes were all

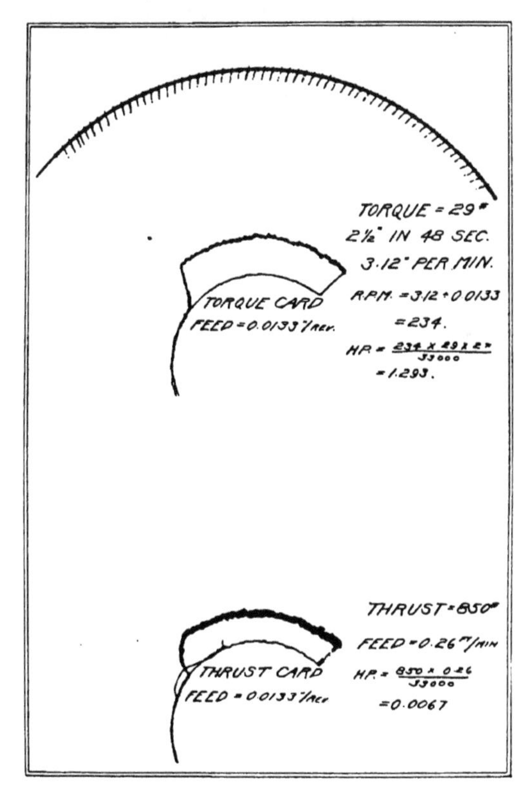

FIG. 12. SPECIMEN TORQUE AND THRUST CARDS

drilled in the same block, or in blocks which were tested with preliminary holes to insure the same quality of test material. In the tests to determine the effect of speed, holes were usually drilled at a given speed in the same block, and similarly in the tests to determine the effect of the rate of feed, all holes were drilled at a given feed in the same block. In tests in which the number of holes was too great to permit the use of a single test block, the variation in test material was partially compensated for by averaging five or more holes; thus in some of the helix angle curves each point is the average of thirty test holes. The variation in hardness of blocks has more troublesome in the endurance tests, since a single hard block might destroy the value of the whole test. Where only two drills were to be compared, half of each test block was devoted to each drill, although even this plan was not wholly satisfactory, since one drill would often penetrate five or six times as much metal as the other.

III. Power Tests of Twist Drills

10. *The Influence of Helix Angle Variations.*—In these tests eight drills were used, and all drills were resharpened after drilling five holes. Tests were run with point angles of 98, 118, and 138 degrees.* For the tests with the 118-degree point angle, six different combinations of peripheral clearance angle were also used, but the influence of this angle was found to be so slight that for the 98- and 138-degree tests only one clearance angle was used.† For convenience in recording the data the drills were designated "A," "B," "C," etc., the 98-degree point angle was designated "X," the 118 "Y," and the 138 "Z," and the clearance angles of 2, 4½, 7, 10, 12½, and 15 degrees, respectively, were designated as "a," "b," "c," "d," "e," "f," respectively. Twenty holes were drilled with each combination of angles, and four rates of feed and five speeds were used. For instance, holes 1 to 5, inclusive, were drilled with a feed of 0.0056 inches per revolution and respective speeds of 92, 233, 366, 457, and 570 revolutions per minute; thus the combination "C"—"Y"—"b"—"4" would mean a hole drilled with the speed of 457 r. p. m., feed of 0.0056 inches per revolution, by a 26-degree helix angle drill with a point angle of 118 degrees and a clearance angle of 4½ degrees. Holes 6 to 10, inclusive, were likewise drilled with a feed of 0.0133 inches per revolution and with the same speeds as holes 1 to 5. Holes 11 to 15 were drilled with a feed of 0.0256 inches per revolution, and holes 16 to 20 were drilled with a feed of 0.041 inches per revolution. Hereinafter the holes of this series of tests will be referred to by these symbols, as, for examples, "E-Y-c-12" and "D-X-e-7." As applied to the tests to determine the effect of helix angle, the values of torque and thrust for holes 1 to 5, 6 to 10, 11 to 15, and 16 to 20 were added and a flat average taken, so that each value used is the average of five different holes drilled in five different test blocks.

Figs. 14 and 15 present in diagrammatic form the results of the tests with the standard 118-degree point angle and different helix angles, each plotted point being the average for thirty holes (six

---

*In the discussions in this bulletin, the value of the total or "included" point angle will be used instead of the half angle more generally used in practice.

† For a definition of terms see Appendix II, p. 109.

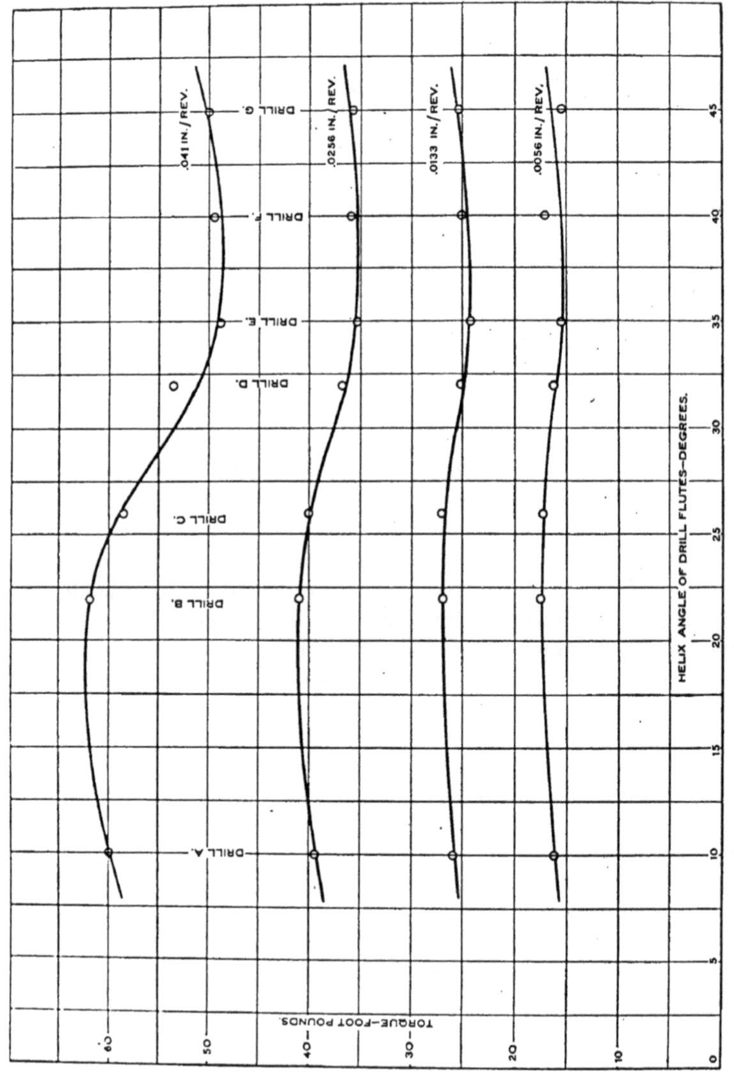

FIG. 14. CURVES SHOWING EFFECT OF HELIX ANGLE VARIATION ON TORQUE FOR DIFFERENT FEEDS

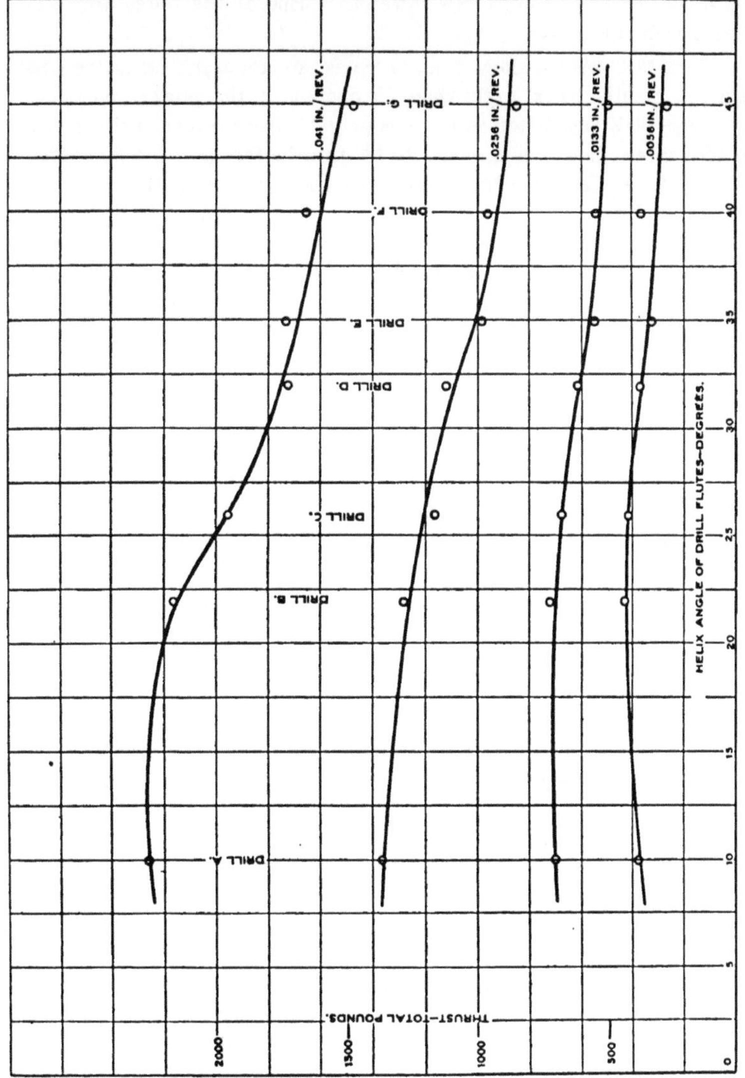

FIG. 15. CURVES SHOWING EFFECT OF HELIX ANGLE VARIATION ON THRUST FOR DIFFERENT FEEDS

clearance angles and five speeds). The irregularity shown by the curves is probably due to the different forms of the flutes and different web thicknesses.

The test results show that there is no apparent advantage in using a greater helix angle than 35 degrees if the power consumed by the drill is the factor to be considered. The uniform helix angle of 35 degrees, however, appears to be even better than the 32-degree angle. A satisfactory explanation why the torque is greater for the 40-and 45-degree drills is not apparent, though possibly it is due to the flute construction, which gives a convex cutting edge, to a rapid dulling of the thin edge which may affect results before the completion of a single hole, or possibly to less effective removal of chips. The performance of drill "A" is inconsistent, but this is probably explained by the very thin chisel point and the shape of the flute (see Fig. 18), which is very similar to that of drill "E." No general law can be stated, but the results seem to indicate the advisability of using a greater helix angle than is now standard.

The thrust variation with different helix angles seems to follow a more nearly uniform law than the torque variation. If drill "A" is excepted, a nearly straight line may be drawn for the other drills, the thrust decreasing uniformly as the helix angle increases.

Figs. 16 and 17 show the results of tests of drills having different point angles and various helix angles. Although the results for the 98- and 118-degree angles are approximately parallel, the results for the 138-degree angle drills "F" and "G" are appreciably better than the others in power consumption. In fact the drill performance for the 138-degree point angle is more regular than for the other points, although drills "A" and "G" show inconsistently low values for both torque and thrust. In general the results seem to indicate that the larger helix angles give the best performance. It may be noted from Fig. 18 that the larger point angles give a more nearly straight cutting edge for drills "F" and "G" and may partly account for the better performance of these drills under that grinding. It is, however, difficult to note separately the effect of the helix angle and that of flute form; the two are interrelated in these drills, because they were shaped in the same milling cutter and the flute form is partly a product of helix angle variation. Each point in Figs. 16 and 17 is the average of five test holes drilled at different speeds but with a constant clearance angle of 7 degrees.

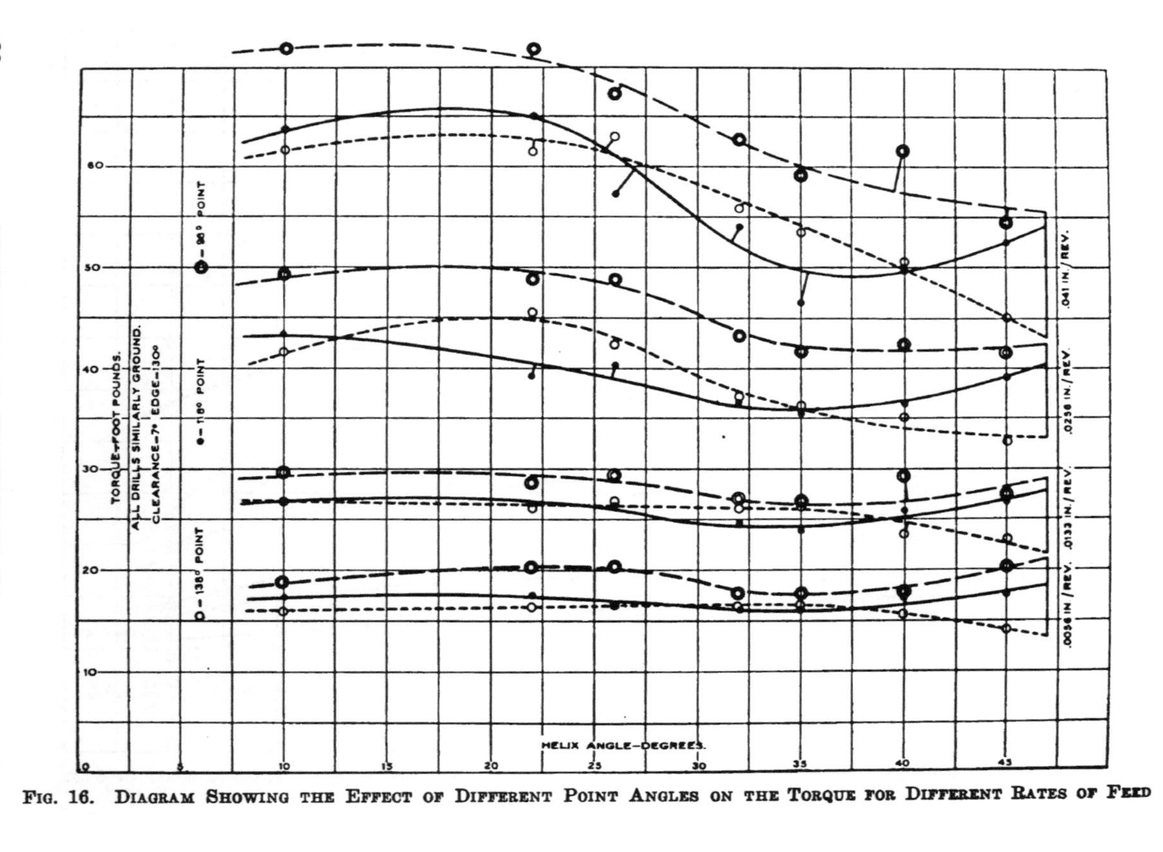

FIG. 16. DIAGRAM SHOWING THE EFFECT OF DIFFERENT POINT ANGLES ON THE TORQUE FOR DIFFERENT RATES OF FEED

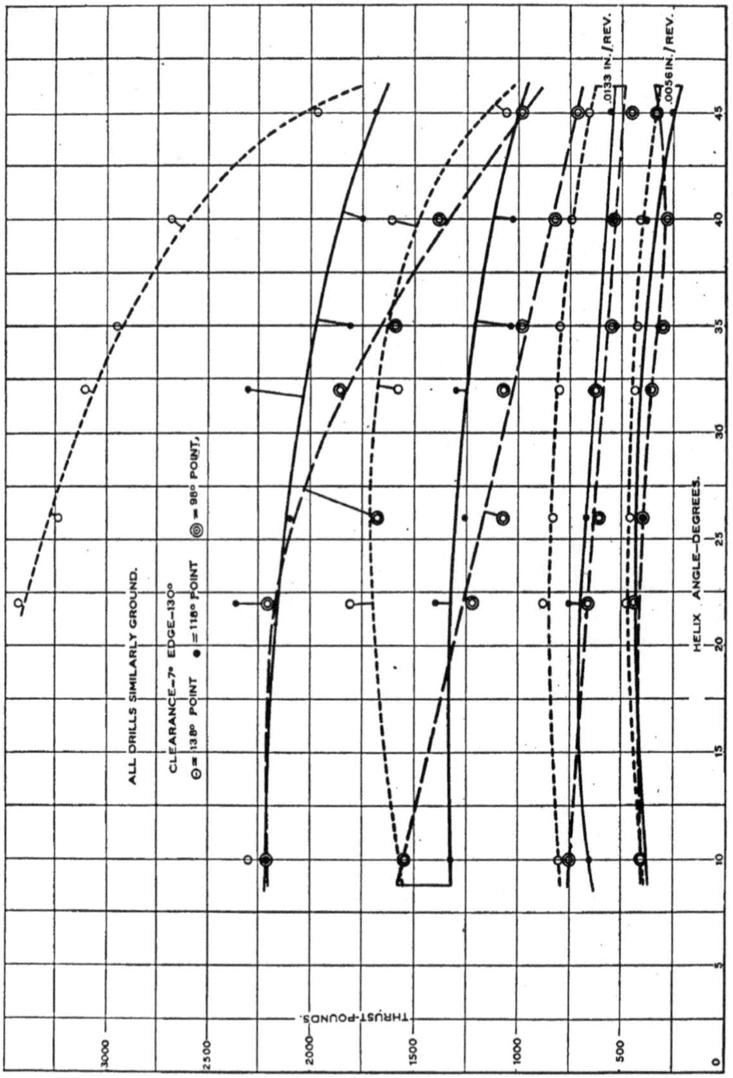

Fig. 17. Diagram Showing Effects of Different Point Angles on the Thrust for Different Rates of Feed

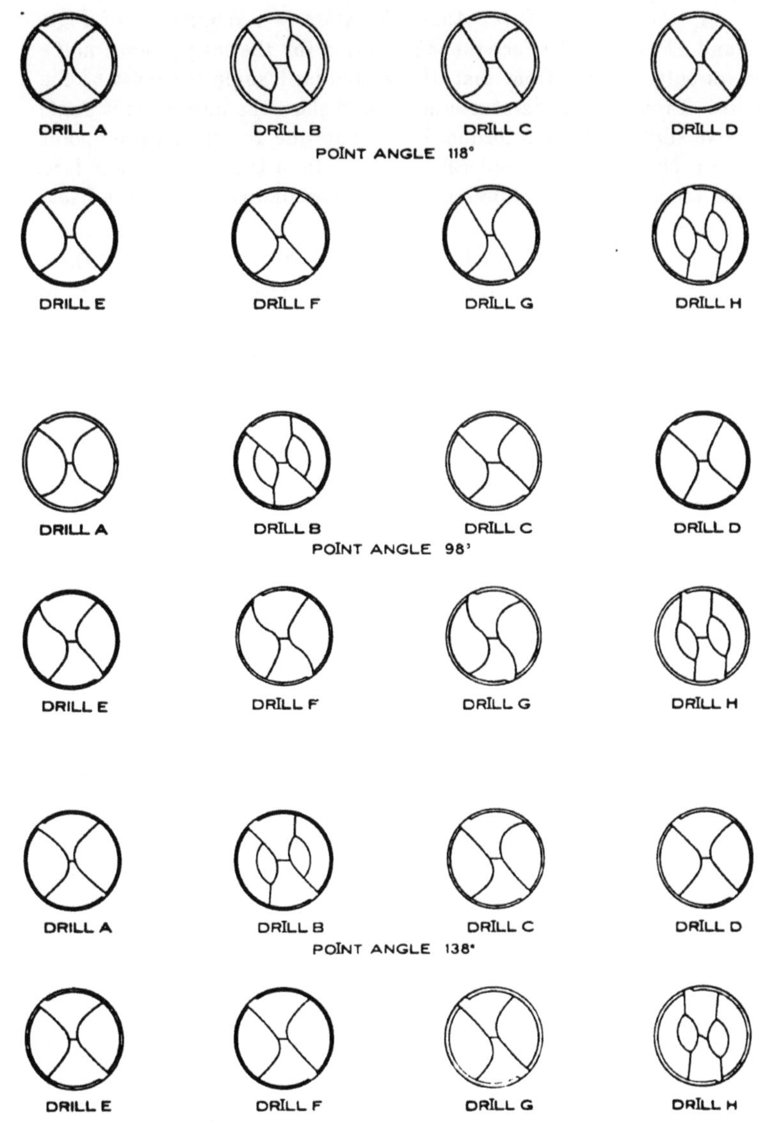

FIG. 18. END VIEWS OF THE 118-, 98-, AND 138-DEGREE POINT TEST DRILLS

11. *Influence of Point Angle Variations.*—Examination of Figs. 16 and 17 reveals the fact that the torque for the larger point angle varies only slightly from that of the standard angle for most of the drills. Probably a general average would show the larger point angle to give slightly greater torque, but the torque for the smaller point angle is shown to be considerably greater than that for the standard angle for all drills, the difference being most noticeable for the heavier feeds.

Since the tests recorded in Figs. 16 and 17 covered only three point angles, a more comprehensive set of tests was conducted in which drills having point angles varying by increments of 5 from 88 degrees to 148 degrees were used. The thrust results from these tests are more nearly uniform and corroborate most previous experiments, since the thrust is greatest with the large angles and least with the small angles. Fig. 19 shows the results of these point angle tests. The curves are self-explanatory, but inspection shows that it would be possible to grind a drill to a 110- or 112-degree included point angle without increasing the torque, but with a slight diminution in thrust and the possible advantage of increased endurance. The reason is not apparent why the thrust should "hump" suddenly as it does for about a 125-degree included point angle, but a similar series of tests with a different drill showed an even more pronounced hump and completely corroborated the results shown by the thrust curves. The slight change in torque between 118 and 138 degrees and the slight change in thrust between 118 and 98 degrees tend to explain the peculiarities referred to in connection with the discussion of Figs. 16 and 17. Different drills show different families of curves for this series of tests, a possible explanation for the disagreement of previous experimenters concerning the effect of point angle on torque change. One drill, for example, showed an absolutely uniform torque for all point angles, although the general tendency is for torque to increase as the angle becomes smaller.

Fig. 18 shows the variations in end contour of the different drills for changes in point angle. It will be noticed that the smaller angles give a convex cutting edge while the larger angles give a slightly concave cutting edge. Whether the shape of the edge has any effect on the power consumption is not known, but it is seen that drill "A" which was almost equally efficient for all point angles had a concave edge in all cases. Drill "G" had a convex edge for both the 98- and

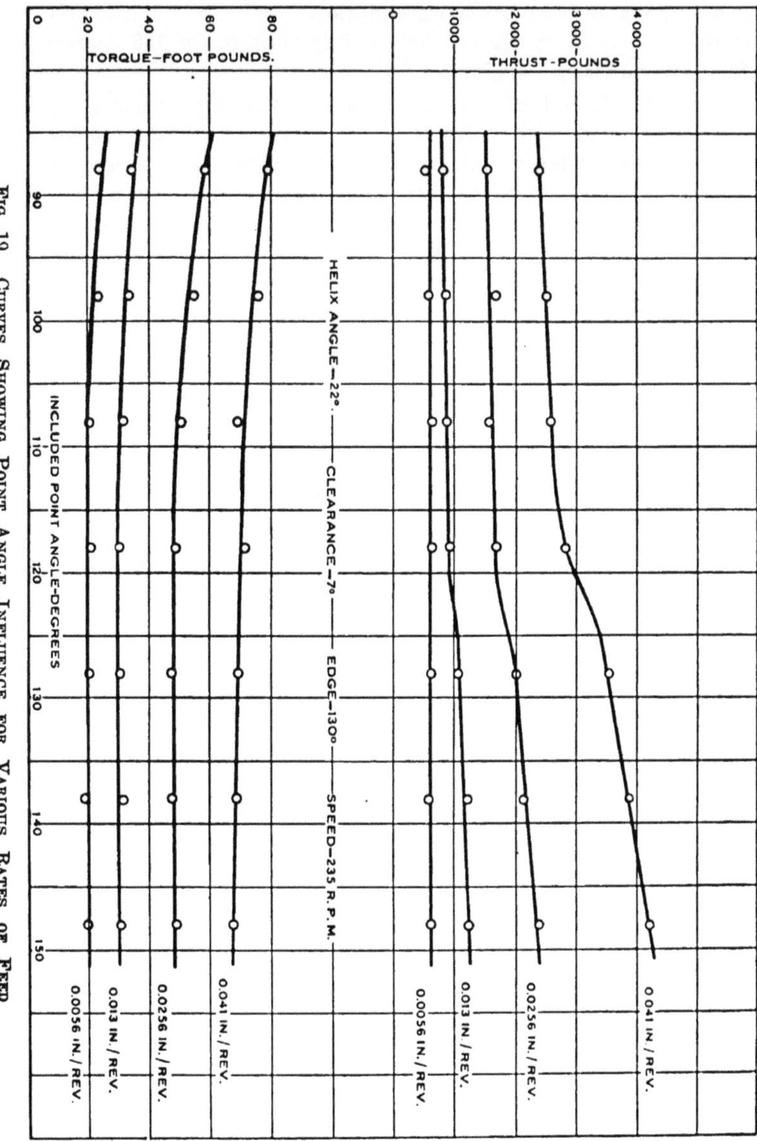

Fig. 19. Curves Showing Point Angle Influence for Various Rates of Feed

118-degree points, but a straight edge for the 138-degree point with which it shows its least power consumption. Drill "F" is similarly convex for 118 and 98 degrees, but slightly concave for 138 degrees with power variations similar to "G."

Figs. 20, 21, and 22 are illustrations of the chips removed by the different drills with different point angles. Fig. 22 probably shows the true tendency of helix angle change, which produces increasingly longer chip helixes with the smaller cutting angles made by the larger helix angles. This tendency was not apparent in the case of the 118-degree test by the convex edge of drill "G," which produced distorted and easily broken chips. The form of the cutting edge apparently influences the chip form, for the concave edges rolled a long, freely coiled chip having a slightly convex outline, while the convex edge rolled a tightly coiled chip with a concave outline which was so stressed as to break easily. The straight edge turned nicely coiled chips which had a straight outline and nested more closely than those from the concave edge. The drills which produced the longest chip coils also used the least power, an indication which may be employed in comparing drills in commercial work without the use of a dynamometer.

Fig. 23 shows the chips and the point form of the drill used in the tests, results of which are shown by Fig. 19. The variations in chip form with the variations in edge contour produced by different point angles are clearly shown. It would seem logical to assume that more power would be used in producing the tightly coiled chips of the 88-degree angle than in producing the loosely coiled chips of the 148-degree angle.

12. *Influence of Clearance Angle Variations.*—In the series of tests for helix angle influence and for point angle influence, the group of tests of drills having the 118-degree point angle also included six different clearances at the drill periphery. These clearances were 2, 4½, 7, 10, 12½, and 15 degrees, respectively, and the corresponding edge angles were held at 120, 124, 128, 133, 137, and 140 degrees, respectively, these being about the normal combinations given by the drill grinder and presenting the best edge contour. It is thus possible to plot curves showing the effect of the clearance variation for the different drills. Such results are shown by Figs. 24 and 25. Considering only the torque, which is practically in

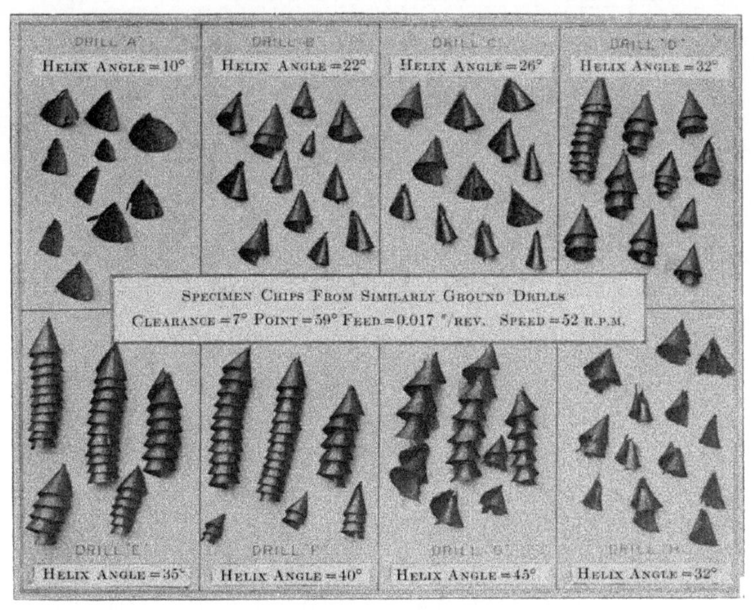

FIG. 20. CHIPS REMOVED BY SIMILARLY GROUND 118-DEGREE POINT ANGLE DRILLS HAVING DIFFERENT HELIX ANGLES

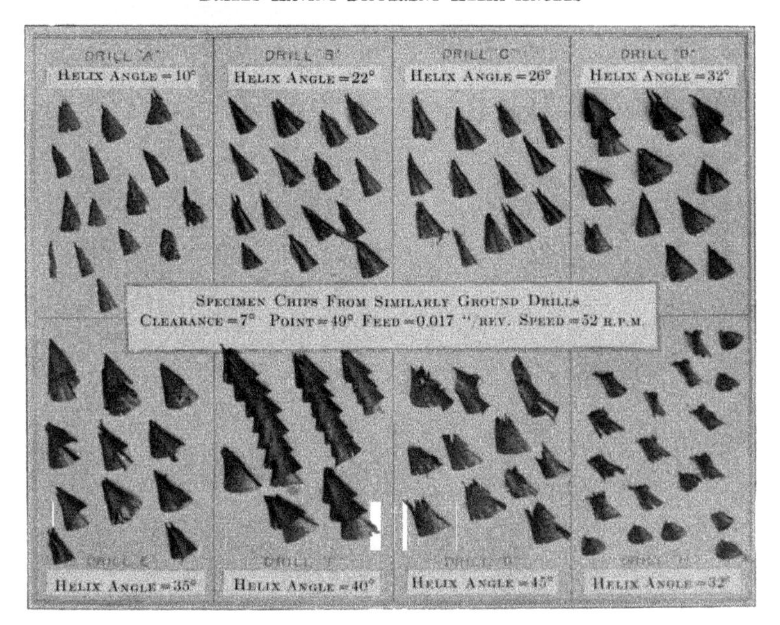

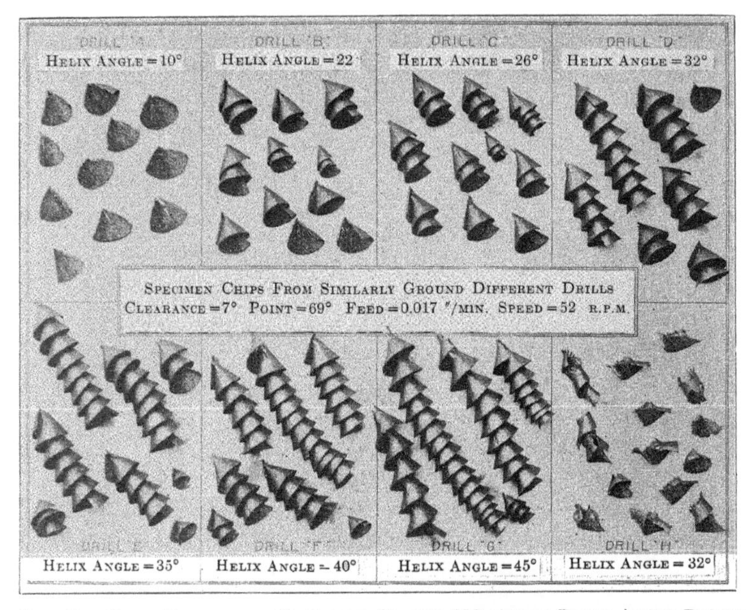

FIG. 22. CHIPS REMOVED BY SIMILARLY GROUND 138-DEGREE POINT ANGLE DRILLS HAVING DIFFERENT HELIX ANGLES

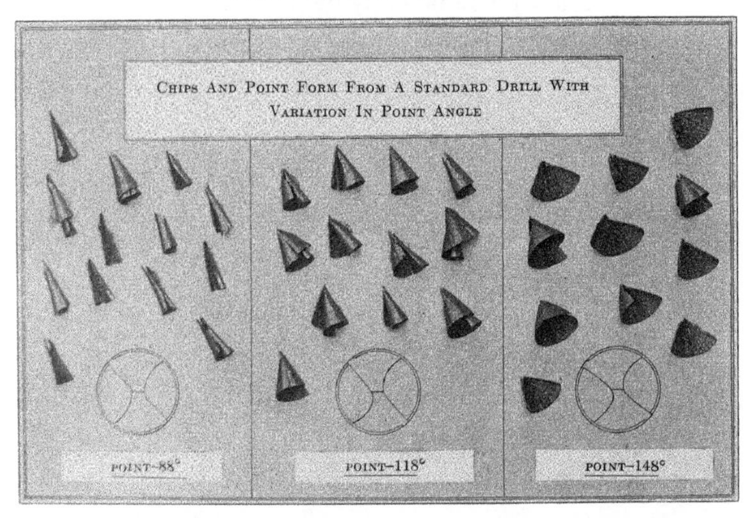

FIG. 23. POINT FORM AND CHIPS REMOVED BY A STANDARD DRILL WITH DIFFERENT POINT ANGLES

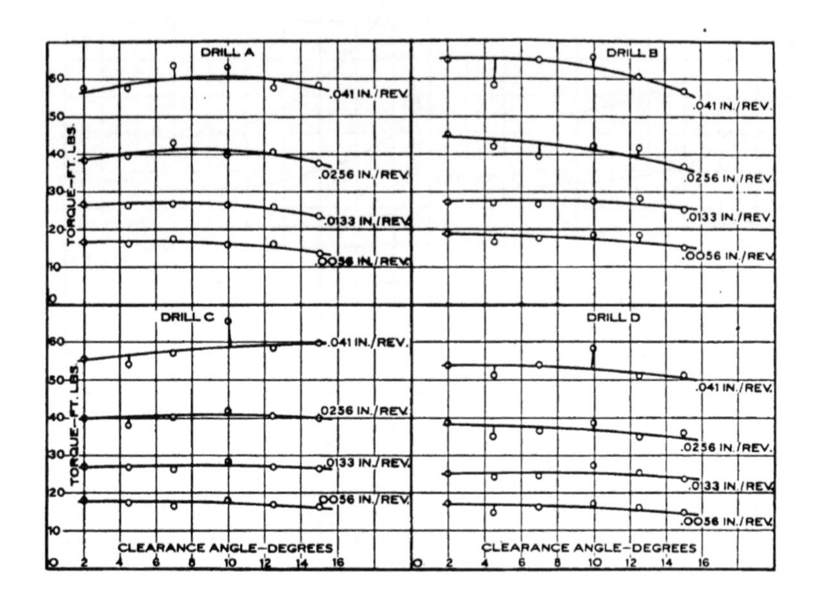

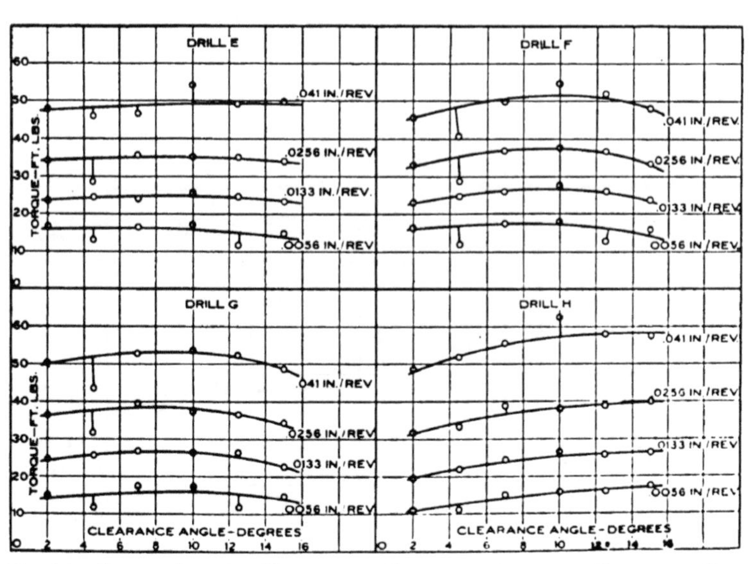

FIG. 24. CURVES SHOWING EFFECTS OF CLEARANCE ANGLE ON TORQUE AT DIFFERENT RATES OF FEED

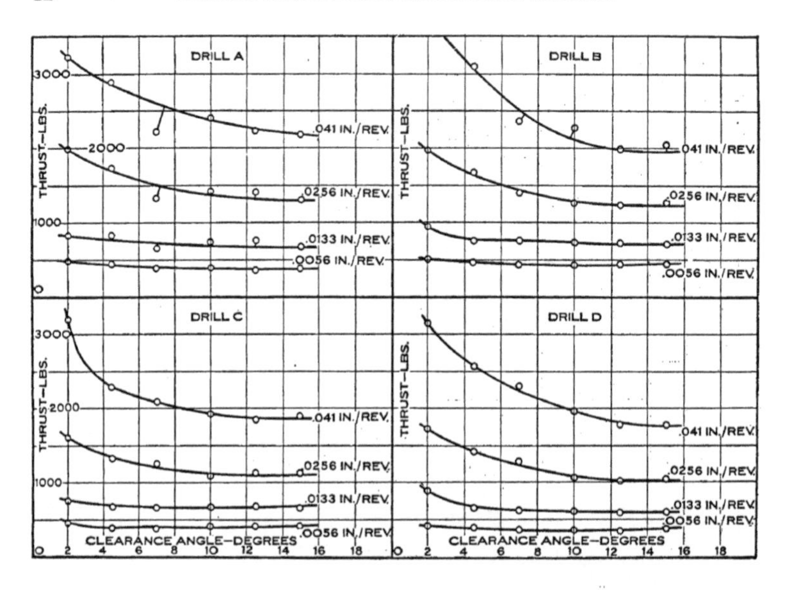

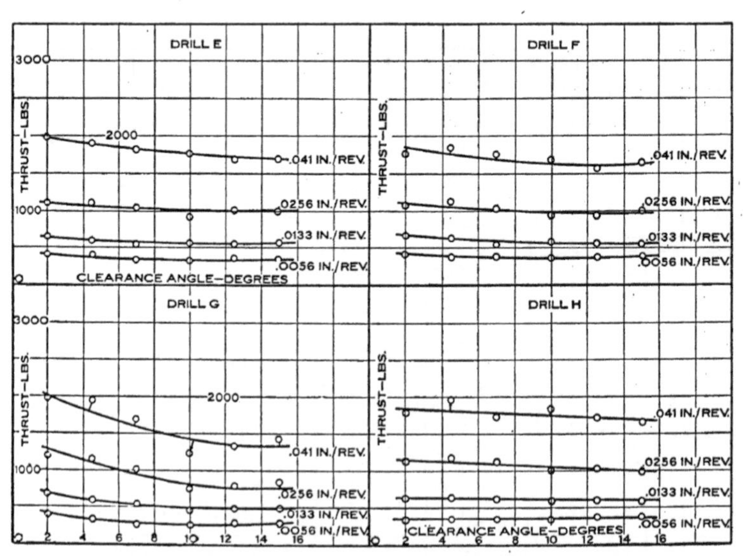

FIG. 25. CURVES SHOWING THE EFFECTS OF CLEARANCE ANGLE ON THRUST AT DIFFERENT RATES OF FEED

identical ratio with total power, it will be seen that there is no pronounced general tendency. Whatever variation appears might well be charged to differences in the test blocks; thus all the points having 4½ degrees clearance, except those for the 0.013-inch feed and for drill "H," show perceptibly less power than the others. All these holes were drilled in a set of twenty blocks, which were from a single day's pouring in the foundry and apparently softer than usual. The 4½-degree tests of drill "H" were made in a different set of blocks, as were the tests for the 0.013-inch feed which were re-run for this purpose, and these latter points appear to be normal. Drill "B" shows a distinct lessening of the power for an increase in clearance angle, while drill "H" shows exactly the opposite tendency. In most of the other drills a horizontal line would represent the variation fairly well.

Considering the thrust, uniform and pronounced tendencies are observed, the thrust being high for the lesser clearances and gradually decreasing as the clearance increases. Most of the sets of curves show a certain resemblance to the family of curves in Fig. 58; thus they present a basis for the supposition that the variation was due to insufficient clearance at the drill center rather than at the periphery. The flatness of the curves of drill "H" is probably due to the fact that the edge angle was larger than for the other drills, it being difficult to measure accurately the peculiar form of the angle. The flatness of the curves for "E" and "F," however, would seem to indicate some peculiar action due to the helix angle, since even poor clearance grinding on these drills seems not to affect the good drilling results. The thrust values for drill "G" are lower than those for any other drill, but they show the same general tendencies as drills "A," "B," "C," and "D." Comparison of these curves with the curves of Fig. 58 led to the making of a separate test in which the edge angle influence and the variation due to differences in the test blocks could be eliminated. The results of this test are shown by Fig. 26. The six holes at the different clearances were drilled in the same test block, although it was necessary to use a separate block for each different feed. The edge angle was kept constant at 130 degrees since this angle was thought sufficiently large to eliminate the influence of small center clearance. The included point angle was kept constant at 188 degrees and the speed at 235 r. p. m.

Torque decreases slightly as the clearance angle increases at the

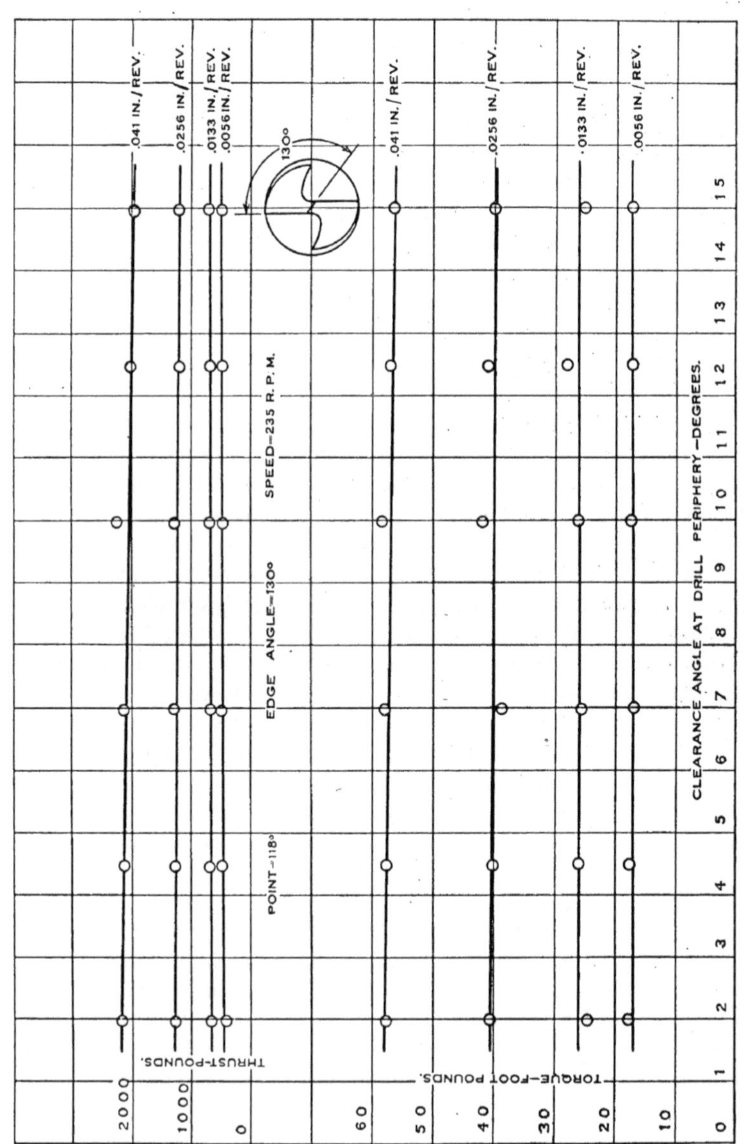

Fig. 26. Curves Showing Effects of Varying the Clearance Angle at the Drill Periphery on Thrust and Torque

two heavier feeds, although this variation is not perceptible for the lesser feeds, and the greatest variation between any two holes at a given feed is only three pounds, as compared with the eight and ten pound differences shown in Fig. 24.

Variations in the thrust almost identical with those in the torque may be observed. The conclusion is that a clearance angle of two degrees at the periphery is ample for all ordinary work provided the edge angle is 130 degrees or more. It seems probable also that the lesser clearance angles would give a greater drill endurance and are, therefore, to be preferred.*

13. *Influence of Edge Angle Variations.*—To determine more definitely the relation of the edge angle to the center clearance angle of the drill, a series of tests was run in which the point angle, the speed, and the peripheral clearance angle were kept constant, while the chisel edge angle was varied by 5-degree increments between 110 and 140 degrees. The seven holes for the different edge angles were drilled in the same test block, a different block being used for each of the four different feeds. The results show that the torque variation (see Fig. 27) is slight for any variation in edge angle, but that the thrust variation may be very large. From the ratio of increase it was deemed useless to attempt drilling the 110-degree hole at 0.041-inch feed, the dynamometer capacity being only 6000 pounds. It will be seen that an edge angle of about 130 degrees is as large as is necessary for average drilling, but that this may be increased to 135 or 140 degrees for very heavy feeds. The 135-degree angle is recommended by most twist drill makers. *The edge angle should vary according to the feed used.*

14. *Influence of Cutting Angle Variations.*—The tests run to determine the influence of varying the helix angle were put to another use in connection with a study of the cutting angle. The cutting angles, measured at the drill periphery, are plotted in Fig. 28, the different drills being designated by different point symbols and the average curves of torque and thrust drawn. All holes were drilled at the same feed per revolution and with the same point angle, each point on the diagram, Fig. 28, being the average of the results of five holes drilled at different speeds in different blocks.

* See pages 82 and 114.

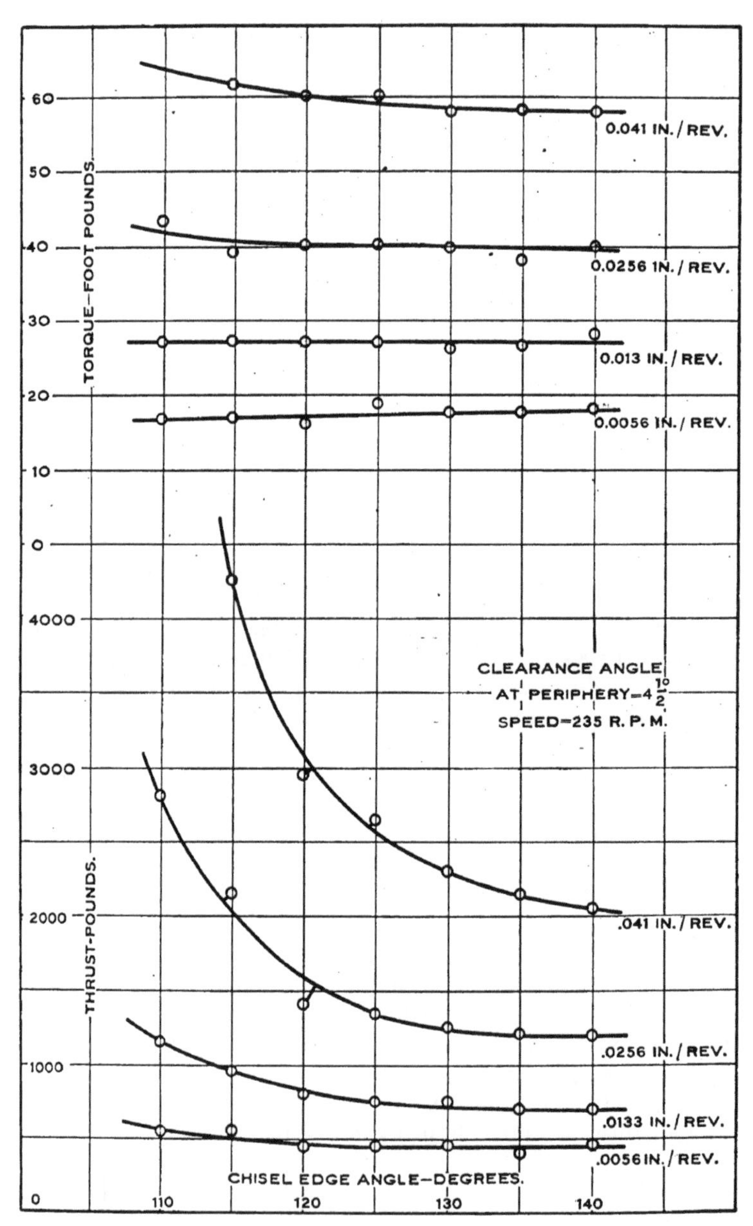

FIG. 27. CURVES SHOWING THE INFLUENCE OF THE EDGE ANGLE ON TORQUE AND

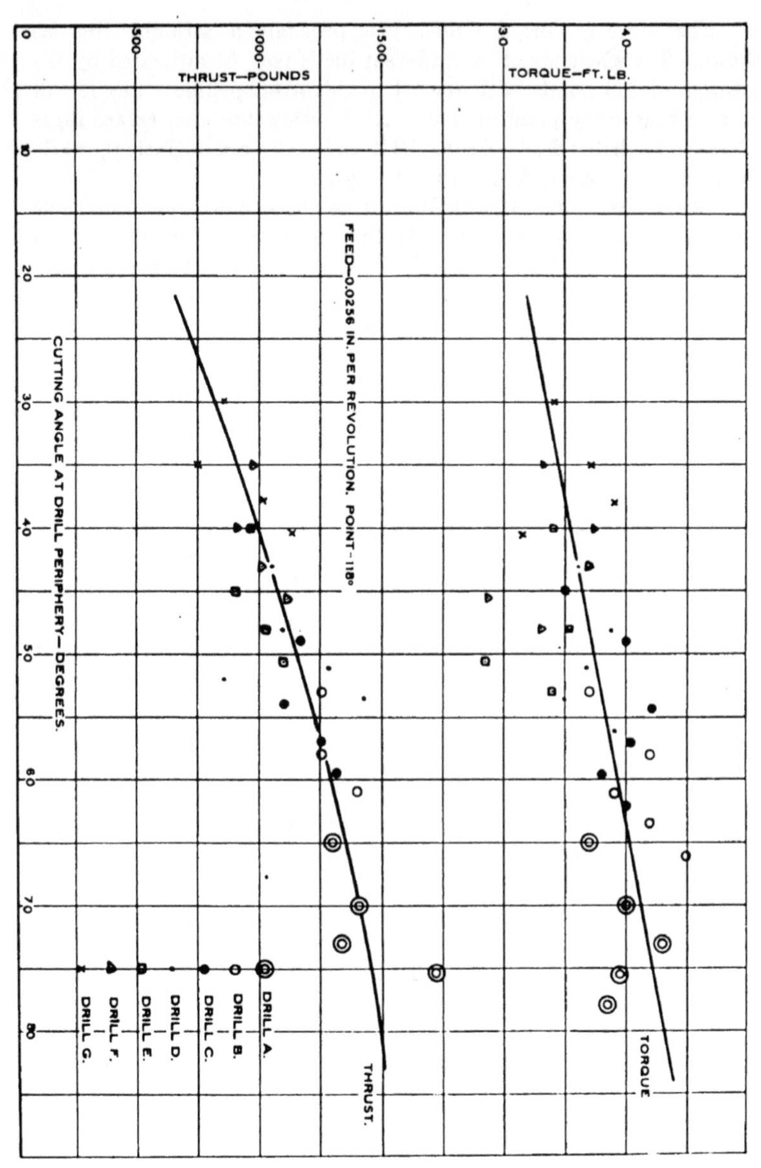

Fig. 28. Curves Showing Effect of Varying the Cutting Angle on Torque and Thrust

The curve of torque tendency is probably a straight line as shown. The efficiency of the different drills may be estimated by the position of the points with regard to this average line; thus it will be seen that every point of drill "E" is below the line, as are three of those of drill "F." Drills "B" and "C" are inefficient, while drill "D" is average and drill "A" good.

The thrust curve is less difficult to place and shows the same general tendency as the torque. In this curve also, all points of drill "E" lie below the line, while "C," "F" and "G" are average and "A" good.

A study of the two curves presents interesting possibilities. That the thrust value apparently descends to zero when the cutting angle is zero is logical enough, but the torque curve apparently descends to a value of about twenty-eight foot-pounds when the cutting angle is zero. Whether it really descends only to this value or suddenly curves down to zero is of course uncertain, but it is reasonable to believe that it has a real value.

The photograph, Fig. 29, illustrates the sort of drilling which may be accomplished with a carefully ground drill. The illustration shows drill "E" in a block of cast iron bringing up coiled chips from a depth of about three-fourths inches. The spindle was stopped when the chips had emerged this far and the photograph taken after blowing away a few of the finer chips which had been made as the drill entered the block. The emerging chips were still attached to the block at the bottom of the hole. The included point angle was 118 degrees, the clearance angle 7 degrees, feed 0.017 inches, and speed 58 r. p. m.

15. *Influence on Torque and Thrust of Varying the Feed and Speed.*—The torque of the drill varies with the rate of feed per revolution in a manner which is fairly uniform. Previous experiments show that the torque does not increase so rapidly as the feed for any given drill, and while these experiments confirm this fact, they show slighter differences from a direct variation or relationship. Fig. 30 shows the results of a series of tests run especially for the purpose of determining this relationship. It is to be noted that the curves are nearly straight. Fig. 31 shows comparative curves in cast iron from the Smith experiments and from the Bird and Fairfield experiments,* each of which shows a greater degree of curvature, the latter much more than the former.

*See Appendix V for a discussion of the results of experiments by other investigators.

FIG. 29. DRILL "E" AT WORK

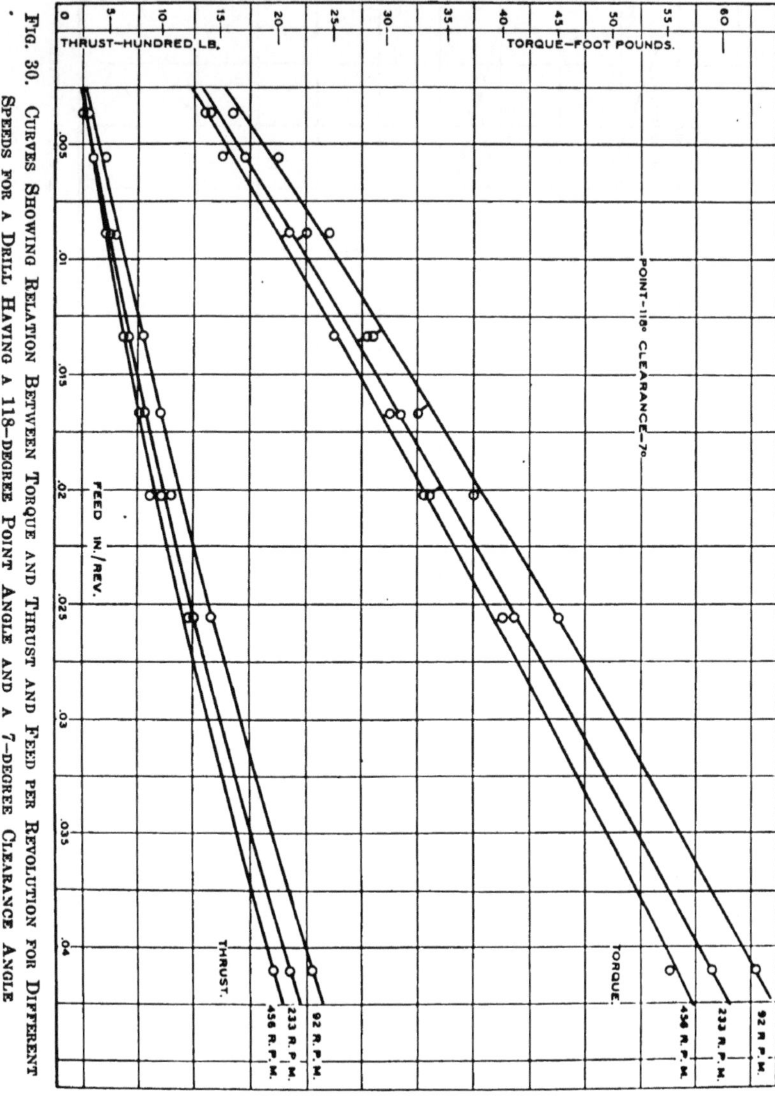

FIG. 30. CURVES SHOWING RELATION BETWEEN TORQUE AND THRUST AND FEED PER REVOLUTION FOR DIFFERENT SPEEDS FOR A DRILL HAVING A 118-DEGREE POINT ANGLE AND A 7-DEGREE CLEARANCE ANGLE

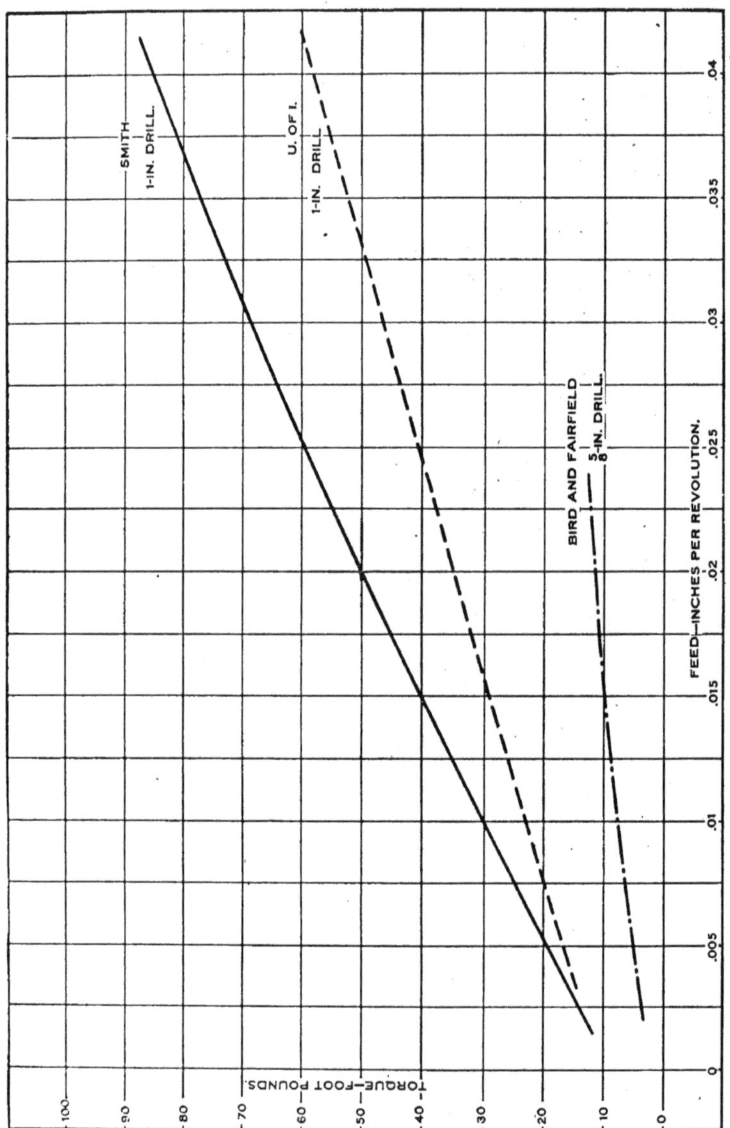

FIG. 31. COMPARATIVE CURVES OF THE EFFECT OF FEED ON TORQUE FROM THE SMITH, BIRD AND FAIRFIELD, AND UNIVERSITY OF ILLINOIS EXPERIMENTS IN CAST IRON

The law governing the effect of feed variation on the thrust was not so easily established, since drills having insufficient center clearance seem to cause a rise in thrust for the heavier feeds (see Fig. 27). Possibly owing to this cause or to an increasing drill dullness, the tests of Bird and Fairfield show the thrust rising more rapidly than the feed. Smith states that the thrust is directly proportional to the feed for tests in soft cast iron. The tests plotted in Fig. 30 show the thrust rising at a slightly faster rate than the feed, although this increase is probably due to the insufficient center clearance mentioned.

From the curves of Fig. 32 it will be seen that the torque decreases as the speed increases for a given feed per revolution, although in most cases no decrease in torque appears for speeds greater than 400 r. p. m. For speeds less than 150 r. p. m. the results are not uniform. In nearly all the tests this decrease in torque at the higher speeds was noted, particularly when the drill was a trifle dull. Occasionally it amounted to as much as twenty per cent. Each of the eight holes drilled at different speeds with a given feed was drilled in the same block, so that variation in test blocks would have no influence on the values.

The thrust observations from these same tests are presented in Fig. 32, the same general characteristics being shown as for the torque.

Dempster Smith* states that there is no pronounced variation of torque or thrust with speed variation, but his tests were all run at less than 150 r. p. m., and the irregularities he observed are similar to those shown in Fig. 32. His curves for the equivalent of a one-inch drill in cast iron together with the torque curves of the present tests are shown in Fig. 33. Bird and Fairfield show no pronounced variation in torque or thrust between speeds of 140 and 600 r. p. m., but possibly the small size of drill used and the lower feeds render the results not comparable. Hallenbeck, from his tests for Baker Brothers of Toledo, presents a family of curves which show a wide variation in thrust with increasing speeds, although all his tests were run in hard steel with a large drill. The general tendency of the Hallenbeck curves is contrary to that of these tests, as shown in Fig. 33.

---

*See Appendix V for discussion of the results of experiments by other investigators

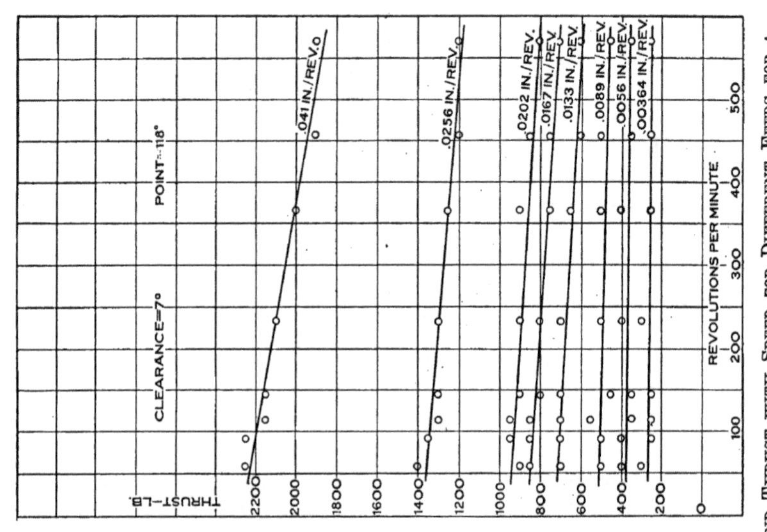

FIG. 32. CURVES SHOWING THE VARIATION OF TORQUE AND THRUST WITH SPEED FOR DIFFERENT FEEDS FOR A DRILL HAVING A 118-DEGREE POINT ANGLE AND A 7-DEGREE CLEARANCE ANGLE

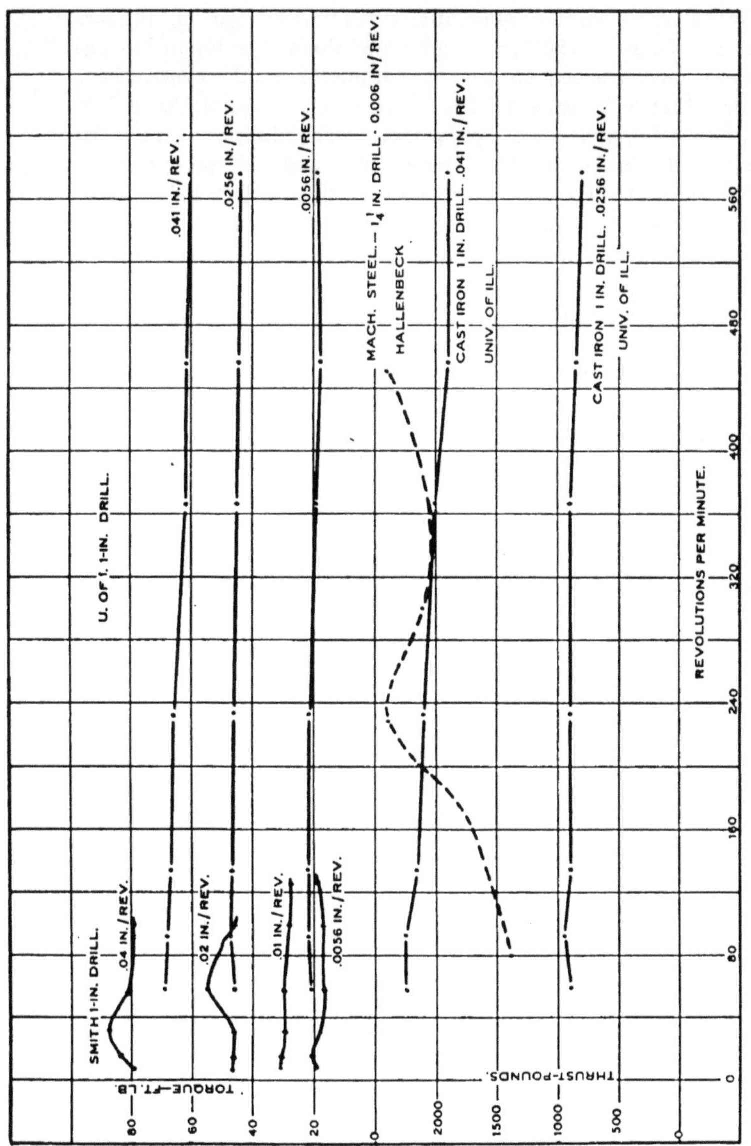

FIG. 33. CURVES SHOWING TORQUE AND THRUST VARIATION WITH SPEED FOR DIFFERENT TESTS

16. *Influence of Chisel Edge Variations on Torque and Thrust.*—In Fig. 34 are shown the results of tests obtained by varying the form of the chisel edge. Whether the chisel was thinned or blunted seemed to have no very marked effect on the torque, the total variation being less than that to be expected from differences in test blocks, but the effect on the thrust was more pronounced. The blunting of the chisel increased the thrust, and the thinning of it reduced the thrust, but the bluntness of the chisel in ordinary drilling will probably not be so pronounced as the artificial bluntness of the drills tested.

When the chisel was thinned to about three-quarters of its normal thickness, the effect on thrust was scarcely distinguishable. The thinning was purposely poorly done to observe the effect of thinning by an inexperienced person. Thinning the point to about half its normal thickness produced a decrease in thrust of about thirteen per cent, which is probably not so great a saving as can be obtained if more time and attention are given to the thinning. In the Smith trials, for example, it seems that the half-chisel point reduced the thrust from five to twenty-five per cent for various drills. It is probable, however, that the importance of thinning the chisel is exaggerated and that the possible saving is more than offset by the danger of weakening the drill.

While the effect of thinning the chisel edge is variable and not important, the total effect of chisel thickness on thrust is pronounced. Smith gives the total chisel effect as about twenty-five per cent, but his standard drill was one in which the chisel edge was half the web thickness. Some of his tests, however, show that the removal (by preliminary drilling) of a cylinder of metal equal to that which would be removed by the chisel may account for from thirty-seven to fifty-seven per cent of the total thrust. In the discussion of those tests Brooks gave it as his opinion that the chisel effect was about fifty per cent of the total thrust.

In the results plotted in Fig. 35, an attempt has been made to disclose the actual influence of the chisel edge on both torque and thrust. The web thickness of the drill used was 0.120 inches, and the chisel edge length was 0.136 inches. In the solid metal the drill showed a thrust of eleven hundred pounds for the speed and feed used. Pilot holes of successively larger diameter were drilled in the test block and in another block which showed the same normal thrust when drilled in the solid. These pilot holes were opened out with

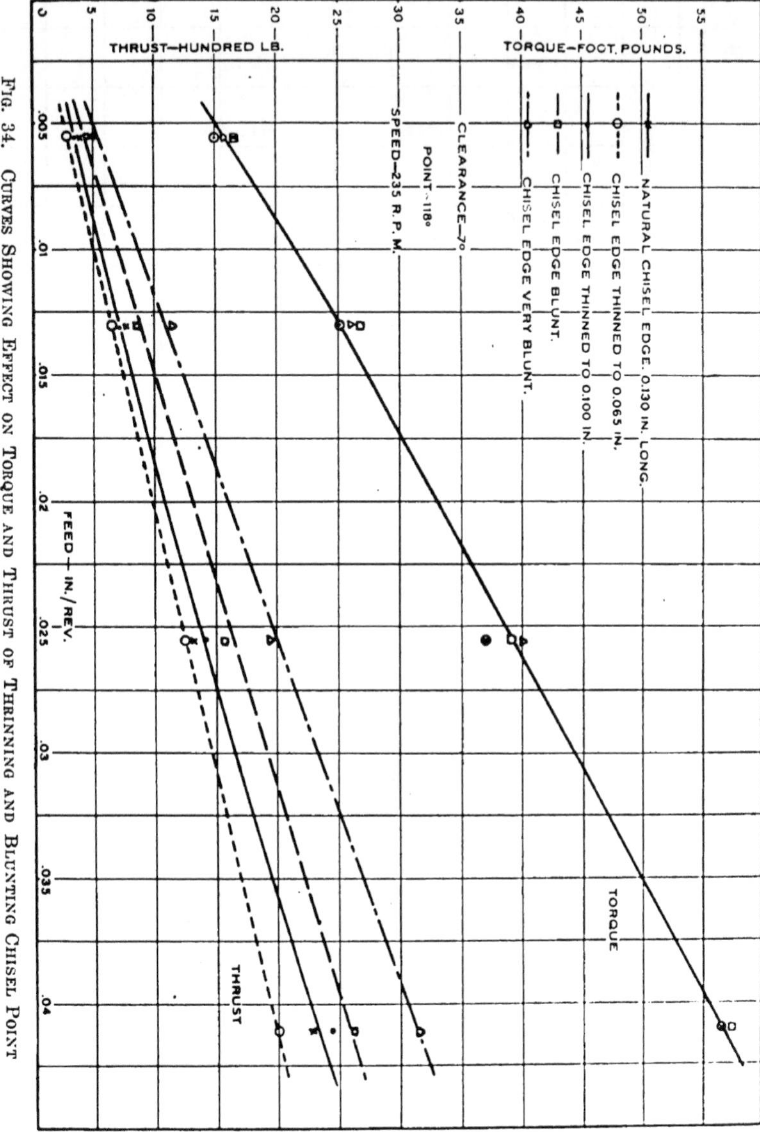

Fig. 34. Curves Showing Effect on Torque and Thrust of Thinning and Blunting Chisel Point

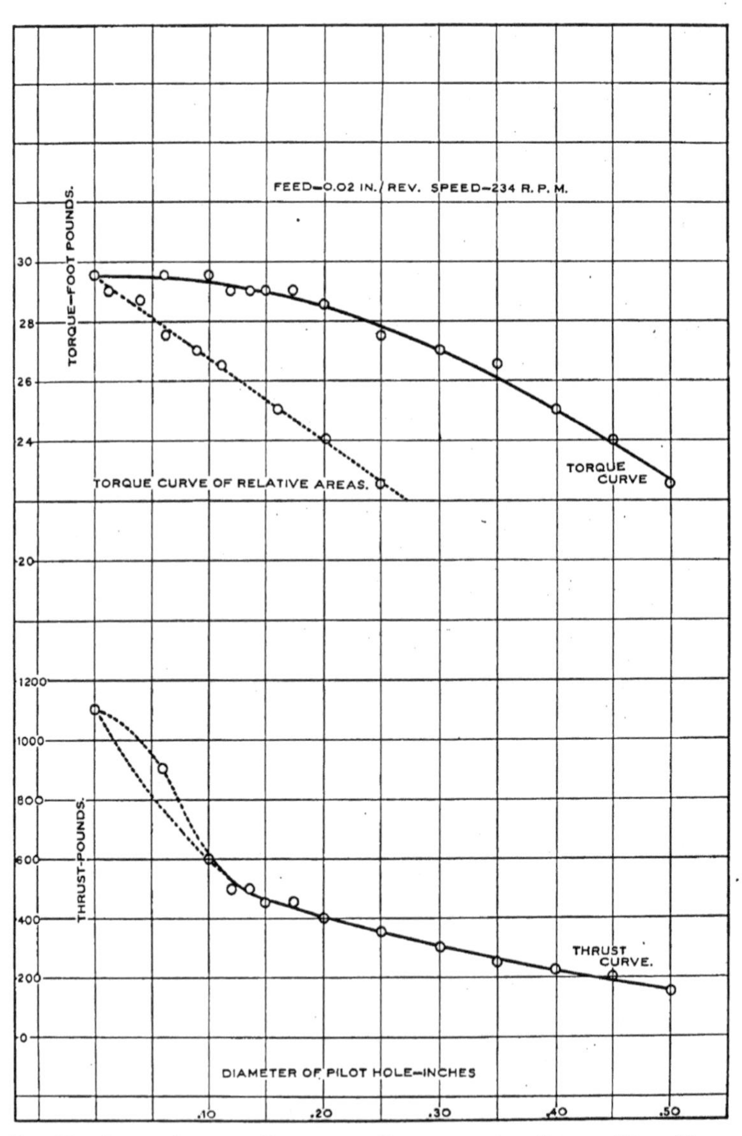

FIG. 35. CURVES SHOWING RESULTS OF INCREASING THE SIZE OF PILOT HOLES FOR CONSTANT SPEED AND FEED

the standard one-inch drill and the thrust and torque recorded as shown in Fig. 35. With the diameter of the pilot hole equal to the web thickness, the thrust was decreased fifty-nine per cent. With pilot holes of less than this diameter the indications were not satisfactory, the small pilot holes becoming filled with fine particles of iron to such an extent that their value vanished when the large drill had penetrated one-half to three-quarters of an inch, although the small holes were at least two inches deep. The observation shown for the 0.10-inch diameter drill extended to a depth of one inch, but that of the 0.06-inch drill, being taken before the drill was cutting to its full diameter, was only relative. For the larger holes the observed data were quite uniform. It is seen that when the pilot hole is larger than the web thickness, the decrease in thrust is almost in a direct ratio with the increase in drill diameter. If this lower portion of the curve be extended to the ordinate of zero diameter pilot hole, i. e., drilling in the solid with the one-inch drill, it indicates a thrust value of about sixty per cent of the actual value; whereas, if the line ran straight to the point of one-inch pilot hole, therefore zero thrust, its extrapolation to the point of zero pilot hole would indicate a thrust value of fifty per cent of the actual value.

It will be noted from the results that the torque is scarcely affected by the pilot hole until the diameter of the hole becomes larger than fifteen per cent of the diameter of the large drill. For pilot hole diameters of more than fifteen per cent the torque decreases in a curve which was identified as corresponding to the diameter of the pilot hole squared or as the relative areas of the larger and smaller holes. This curve of relative areas was plotted as shown in the straight dashed line. If the pilot holes have diameters of 0.5 inches, the areas are to each other as 1 to 0.25. In other words, the pilot hole affects the torque only as it removes a certain percentage of the total metal.

Smith reports a series of tests in which a ¾-inch hole was opened out to 1½ inches at different rates of feed in medium cast iron, and examination of the results confirms the given indication. Since the diameters have a ratio of two to one, the areas have a ratio of four to one; therefore the power consumed by the opening-out process should be three-fourths that required to drill the 1½-inch hole in the solid. As given for a feed of 0.01 inches, the values are 74.6 and 76.8 per cent, for a feed of 0.013 inches the values are 73.9 and 83.0

per cent, for a feed of 0.02 inches, 65.8 and 75.5 per cent and for 0.04 inches 80.3 per cent. The average of these values is 75.7 per cent, which compares with the 75 per cent theoretical value. For a pilot hole having a diameter equal to the web thickness Smith reports a torque reduction of only one per cent in steel and no noticeable reduction in cast iron. Since the average ratio of chisel edge to drill diameter in the drills used by Smith was 0.096, the area ratios are 0.0092 to 1, or a relation of 0.92 per cent, which compares well with the 1 per cent reduction in power which he noted. For thrust values on the opening-out tests, Smith reports that the percentage of reduction apparently increases with the feed, but a comparison of his values for the same feed with those plotted in Fig. 35 shows corresponding values of 82 and 86.4 per cent, respectively. For feeds of 0.01 inches, 0.013 inches, and 0.04 inches Smith shows reductions in thrust of 73, 76.5, and 85 per cent, respectively. This variation, together with values determined by opening out a hole with a diameter of 0.15 of an inch, is shown in Fig. 36. The decrease in proportional thrust with increased feed is evident in both curves.

17. *Total and Partial Power Consumption.*—In order to determine the gross power required for drilling at different rates, and the drill-press efficiencies and general characteristics, alternating current wattmeters were connected to each power phase to measure the total input to the motor. These instruments were checked for accuracy, and all readings were taken practically simultaneously for the two phases. These readings, with the torque and thrust cards taken from the drill point, made it possible to note the gross and net values of power for any hole drilled as well as the effect of torque and thrust variation on the power absorbed by the motor and press.

Fig. 37 shows the power consumed by the motor and drill-press when no work is being performed. When the drill spindle is not revolving, all the power input is consumed in the motor, in the main drive shaft of the press, and in the main shaft of the press transmission. When the transmission gears are engaged, they rotate the drill spindle at different speeds, involving losses in the transmission gears and bearings, in the gears and bearings of the drive to the spindle, and in a portion of the feed gearing. If the feed gears be thrown in so that the spindle is advanced, but no drilling done, no increase in the total power input is to be noted. The spindle speeds are

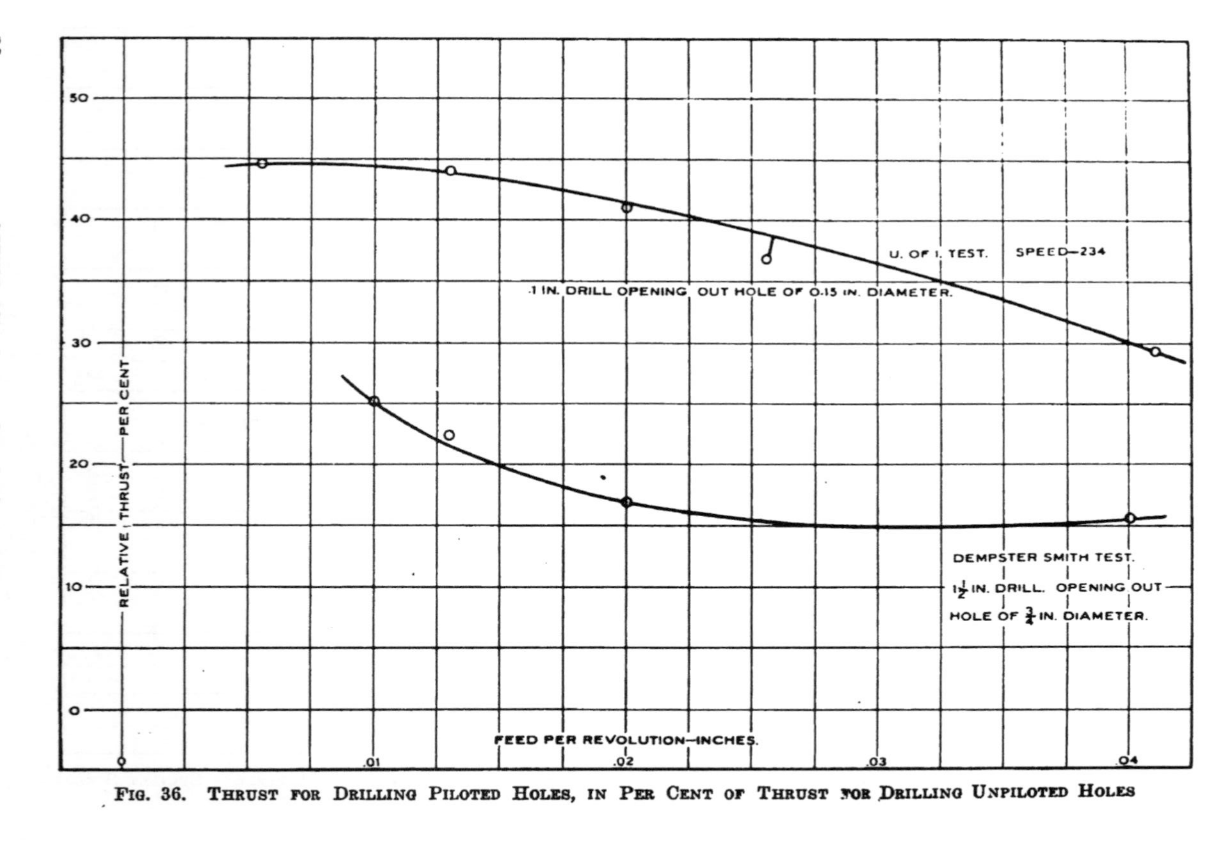

FIG. 36. THRUST FOR DRILLING PILOTED HOLES, IN PER CENT OF THRUST FOR DRILLING UNPILOTED HOLES

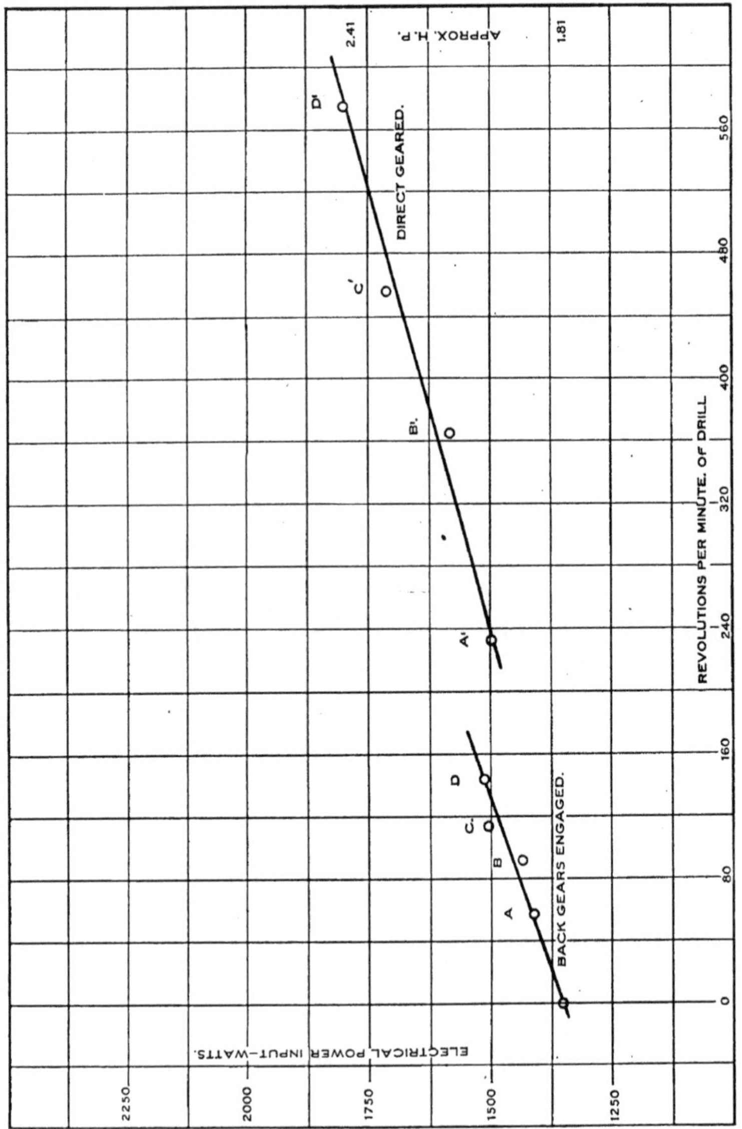

FIG. 37. THE POWER CONSUMPTION OF THE MOTOR AND THE DRILL-PRESS, RUNNING IDLE, FOR DIFFERENT SPINDLE SPEEDS

obtained through four gear combinations and the back gears; thus eight spindle speeds are possible. The effect of the back gearing is clearly shown by the curves of Fig. 36, as the points A, B, C, and D have the same gear combinations as points A,′ B,′ C,′ and D,′ the only difference being that for the first set the drive was through the back gears, while for the second the drive was direct. The frictional loss with increased spindle speed is well shown, the increase amounting to 0.6 h. p. for a speed of 575 r. p. m.

In the tests recorded in Fig. 38, holes were drilled at four different feeds and six different speeds. For the two lower the back gears were engaged. It is interesting to note that the two lines of direct and geared drive are practically parallel for the different feeds and show constant power absorption for the back gears for all different spindle speeds, since the speed of the back gears is constant, but the power absorption of these gears varies almost directly with the rate of feed, being about 400 watts for 0.041-inch feed, 270 watts for 0.0256-inch feed, and 160 watts for 0.013-inch feed. It will be recalled that both the torque and thrust increased in practically a direct ratio with the feed; so this increase in power consumption of the back gears is due to the increased pressure imposed upon them by the increased torque or thrust, or both. The purpose these results serve is to indicate the total power required for drilling or for removing a certain amount of metal under varying conditions of feed and speed. These values, of course, apply only to this specific set-up, each apparatus for drilling having its own power characteristics.

In Fig. 39 are shown the efficiencies for the different points illustrated by Fig. 37. These values show the influence of the back-gearing on efficiency, and also give an idea as to the absolute values which may be obtained for the higher rates of drilling.

The values of Fig. 38 have been used in still another manner as shown by Fig. 40 in which power is plotted against rate per minute for both total and net powers of drilling. The lower curves of net power are interesting as a basis for studying the question of the best feed and speed combination for the removal of metal. If a vertical line of constant rate of drilling is taken, it will be found to intersect the curves of power and feed relation, and to show varying power requirements for removing the same amount of metal per minute. Instead of rate per minute, the abscissae could as well be cubic inches of metal removed per minute, since the drill diameters are the same.

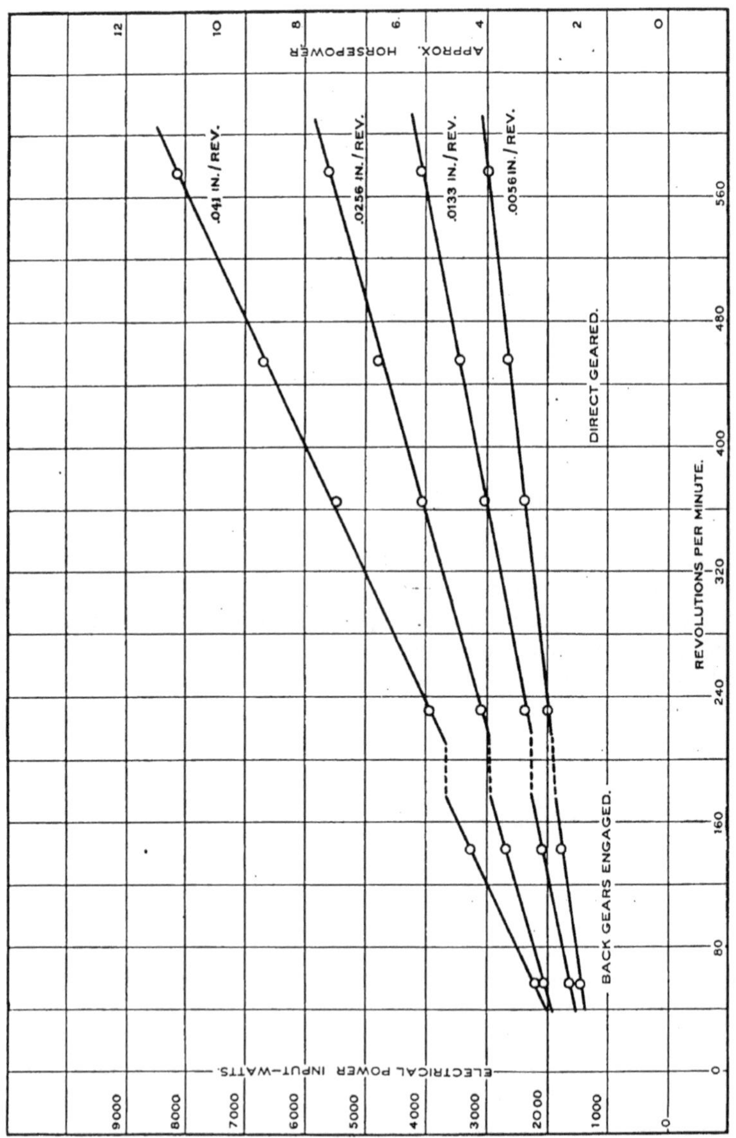

Fig. 38. Diagram Showing Relation Between the Total Power Input and the Speed for Different Rates of Feed

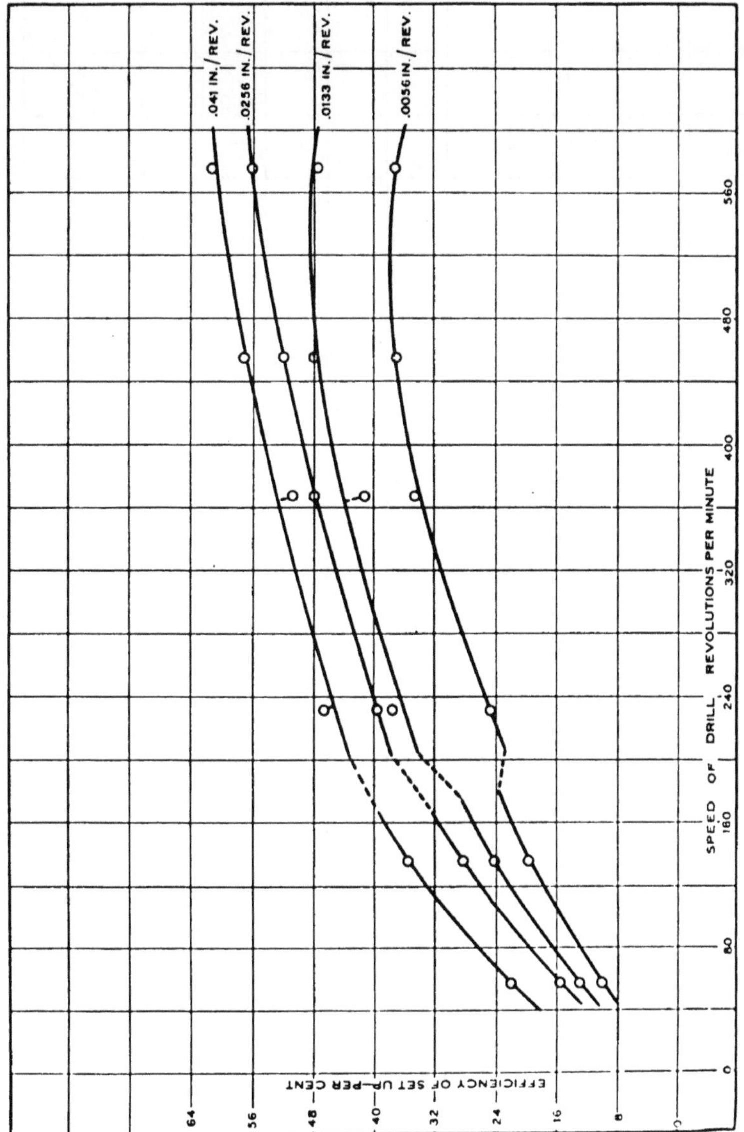

Fig. 39. Diagram Showing Efficiency of the Apparatus for Various Speeds and Feeds

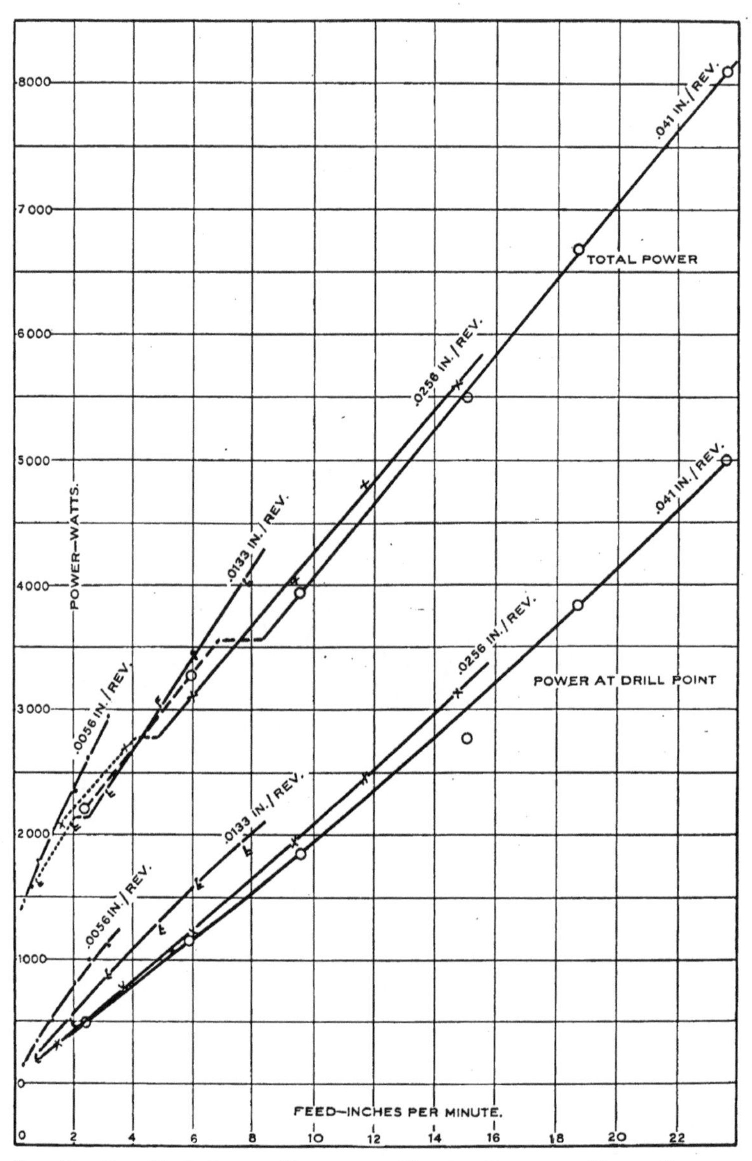

FIG. 40. THE VARIATION OF POWER FOR VARYING RATES OF METAL REMOVAL

and the same relative curves would be given. These curves show clearly that it requires less net power to remove metal with a heavy feed and slow speed than with light feed and high speed, the power difference running as high as one hundred per cent for the lower rates of feed. Other tests, which were run for a different purpose, were used in checking this observation and will be found plotted in Fig. 41. In Fig. 40 the curvature of the lines of feed and power seems to be due to irregularity in test material, since the curves of Fig. 41 are practically straight lines and show this power difference very pronouncedly. Still another set of test values is shown plotted in a different manner in Fig. 42. In this case the horse-power consumed in removing one cubic inch of metal with a one-inch drill at different rates of drilling has been plotted against the rate per minute. The families of curves for different feeds per revolution, therefore, represent holes drilled at different speeds. The three points marked with crosses, almost on the 0.041-inch curve, represent holes drilled at a feed of 0.061 inches per revolution, and show that the gain in efficiency by using a heavier feed than 0.041 inches is slight. In fact there is but little gain in using a feed heavier than 0.0256 inches per revolution. It will be noted that there is a slight decrease in power due to the increased speed; thus the findings shown in Fig. 32 are corroborated although the principal saving is due to the increased feed. This variation of power with feed is shown by the dashed line and heavy points, which is a curve of constant speed drawn through the average value for the power at each feed. It will be noted that the total power saving by the use of the heavier feeds may amount to 0.8 h. p. in drilling at the rate of three inches per minute.

Although the drill itself uses less power when drilling with the heavier feeds and slower speeds, the rule for the total power is not so simple. If it were possible to use the direct gearing to obtain the whole range of speed, the relations for gross power would probably be the same as for net power. The introduction of the back gears, however, varies the relation considerably for the lower rates of drilling, as will be seen from Fig. 40. The gross values given on this chart have been transferred to different coördinates in Fig. 43, which shows a family of curves of power for different rates of drilling and which signifies at what feed per revolution it is most economical to remove metal. A number of these values are taken from extrapola-

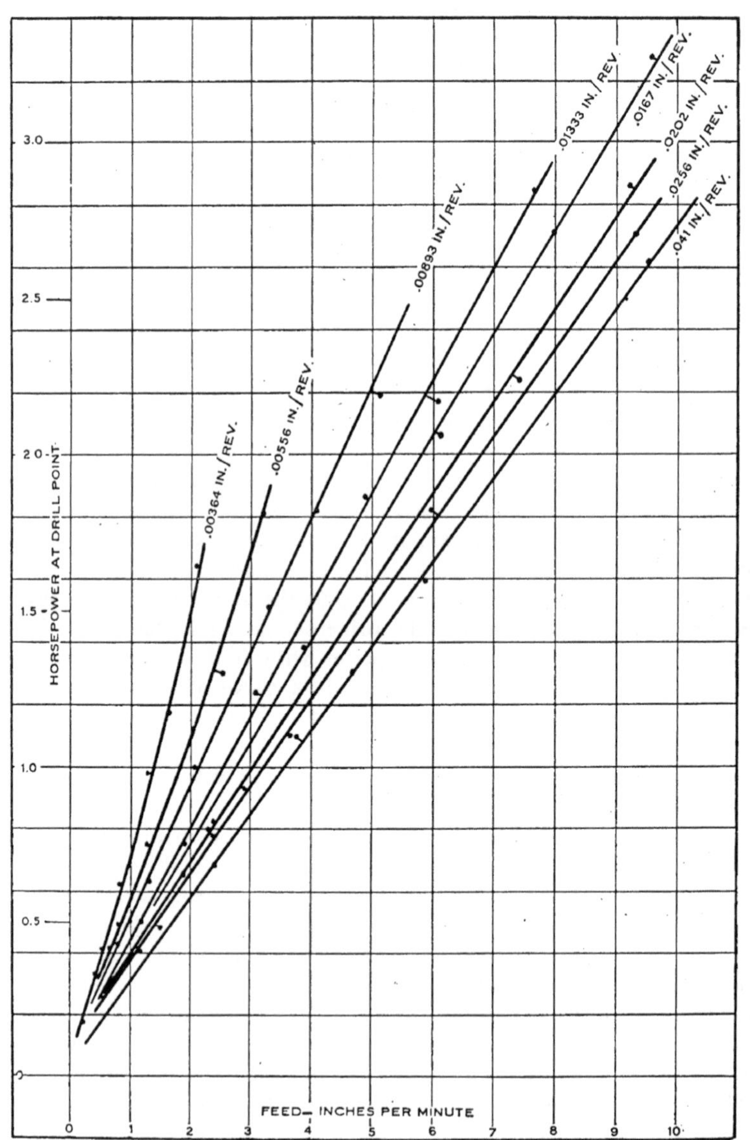

FIG. 41. CURVES SHOWING THE VARIATION IN NET HORSE-POWER WITH VARYING RATES OF DRILLING FOR DIFFERENT SPEEDS

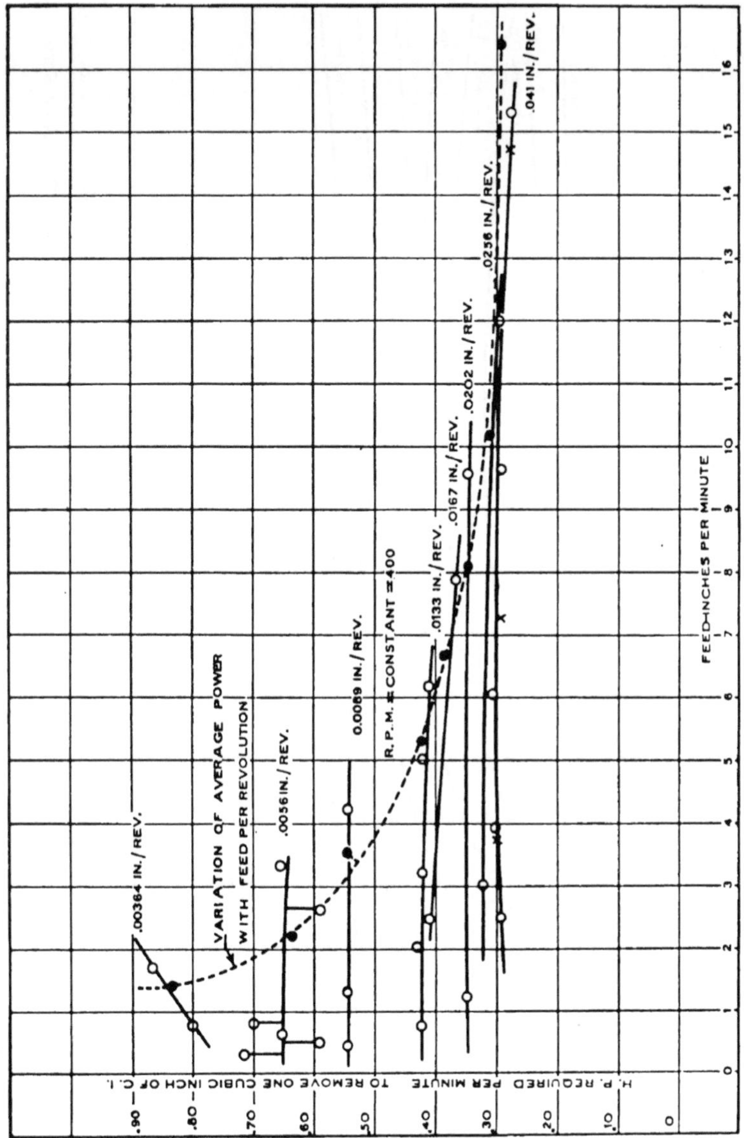

FIG. 42. VARIATION OF NET HORSE-POWER REQUIRED TO REMOVE ONE CUBIC INCH OF SOFT CAST IRON AT DIFFERENT RATES

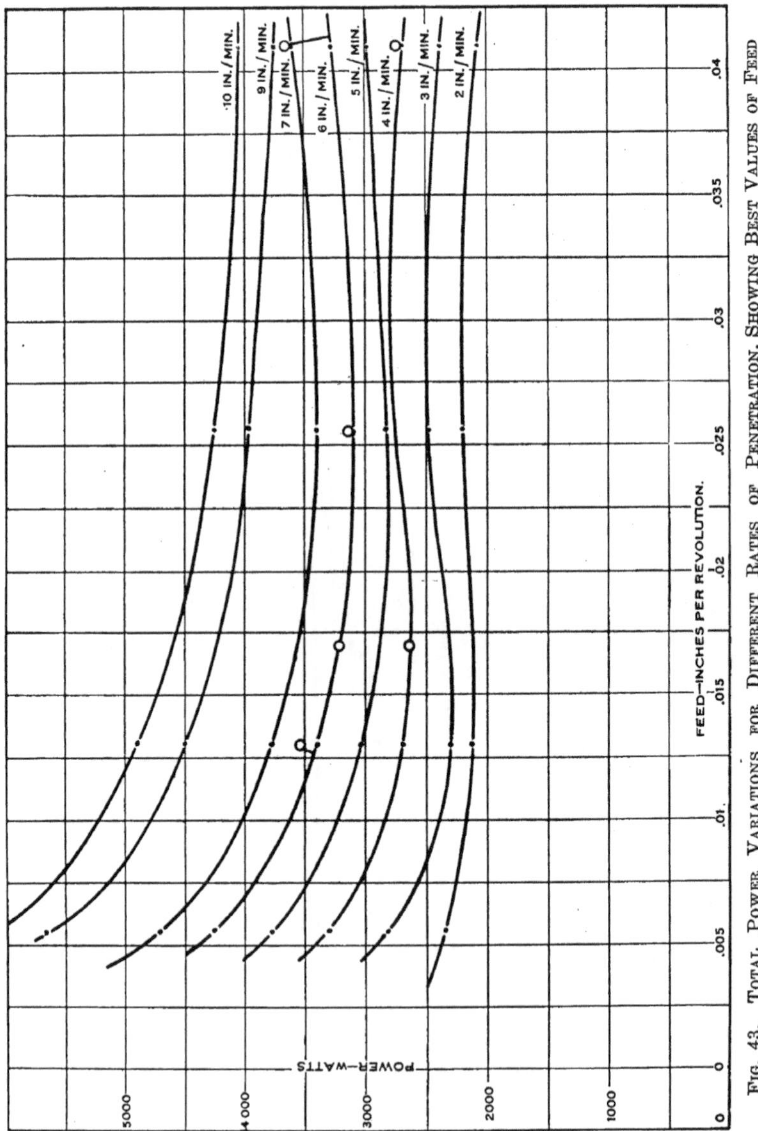

FIG. 43. TOTAL POWER VARIATIONS FOR DIFFERENT RATES OF PENETRATION, SHOWING BEST VALUES OF FEED PER REVOLUTION

tions of the curves of Fig. 40, and represent combinations beyond the range of the press. The circled points shown are actual tests which were drilled at a rate closely approximating the rates of the curves, and seem to check fairly well the tendencies shown. A general rule for drilling economically with this set-up seems to be that the feed per revolution shall increase with the rate per minute, being never less than 0.0133 inches per revolution when drilling in cast iron.

The individual characteristic power and feed relations of different drilling apparatus are shown by Fig. 44, in which a number of test curves from different experimenters is presented. It will be noted that the power curves on the same set-up do not necessarily show the same trend; thus, for example, the two Westinghouse curves take opposite curvatures, and a few indications by Hallenbeck show that for the heavier feeds the curvature would also be reversed. The Bickford and University of Illinois tests are both at comparatively low powers; so the straight lines shown might change their tendencies in the higher regions.

The variations of power and efficiency with varying feeds per revolution may be observed by referring to Fig. 45. These tests were drilled at a constant speed and with eight variations of feed. The effect of the back gearing in the feed system is perceptible, although of no practical importance, since the power absorption of these gears is relatively slight. The power used at the drill point, which was determined from the test cards, varies almost in direct ratio with the feed per revolution. This power was subtracted from the total input to determine the power absorbed by the motor and drill-press. The change in total efficiency, ratio of power absorbed in useful work at the drill point to the total electrical input, is a function of the varying power absorption of the press and of the relation of the total power input to the power absorbed in running the press idle. This latter effect is most pronounced at the lower rates of drilling, when the useful power is small as compared with the absorbed or "overhead" power. If the drill is dulled so that it absorbs more power, the efficiency of the apparatus will increase, owing to the increase in net power consumption. This efficiency curve is a strong argument for the higher rates of drilling, or, at least, for having the apparatus so proportioned that it will be working at nearly full capacity most of the time.

As has been mentioned, the increase in torque and thrust affects

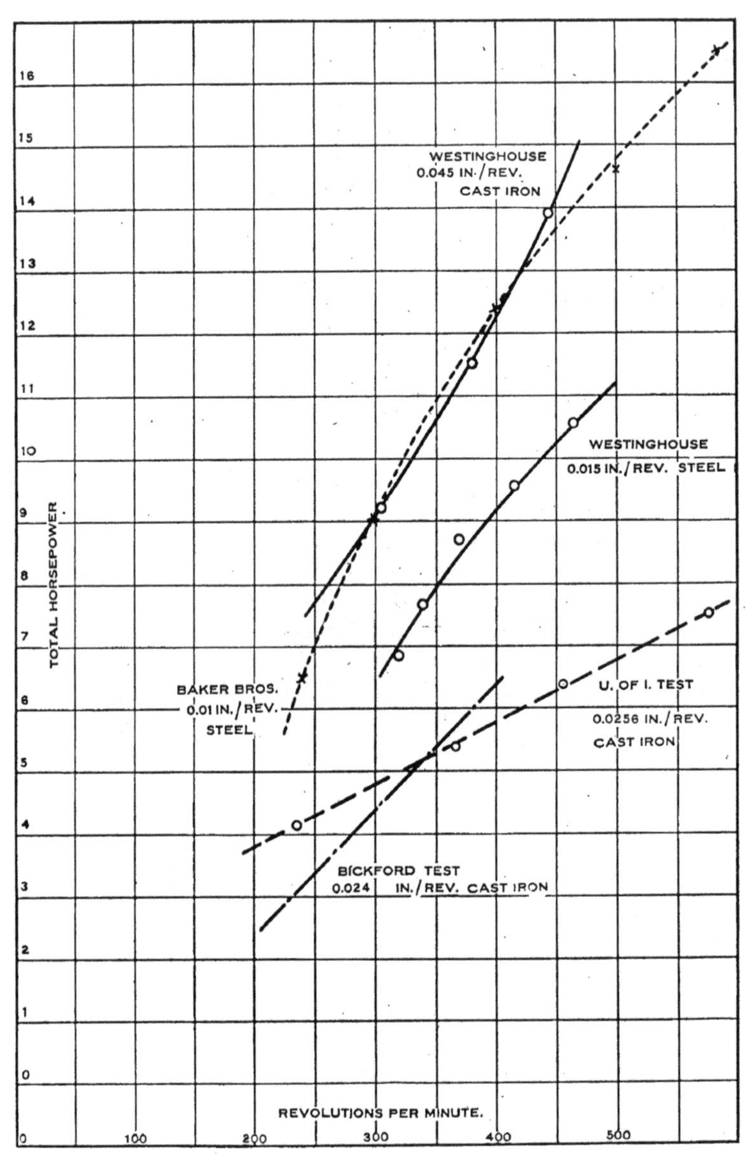

FIG. 44. THE POWER AND FEED RELATIONS FOR VARIOUS TESTS

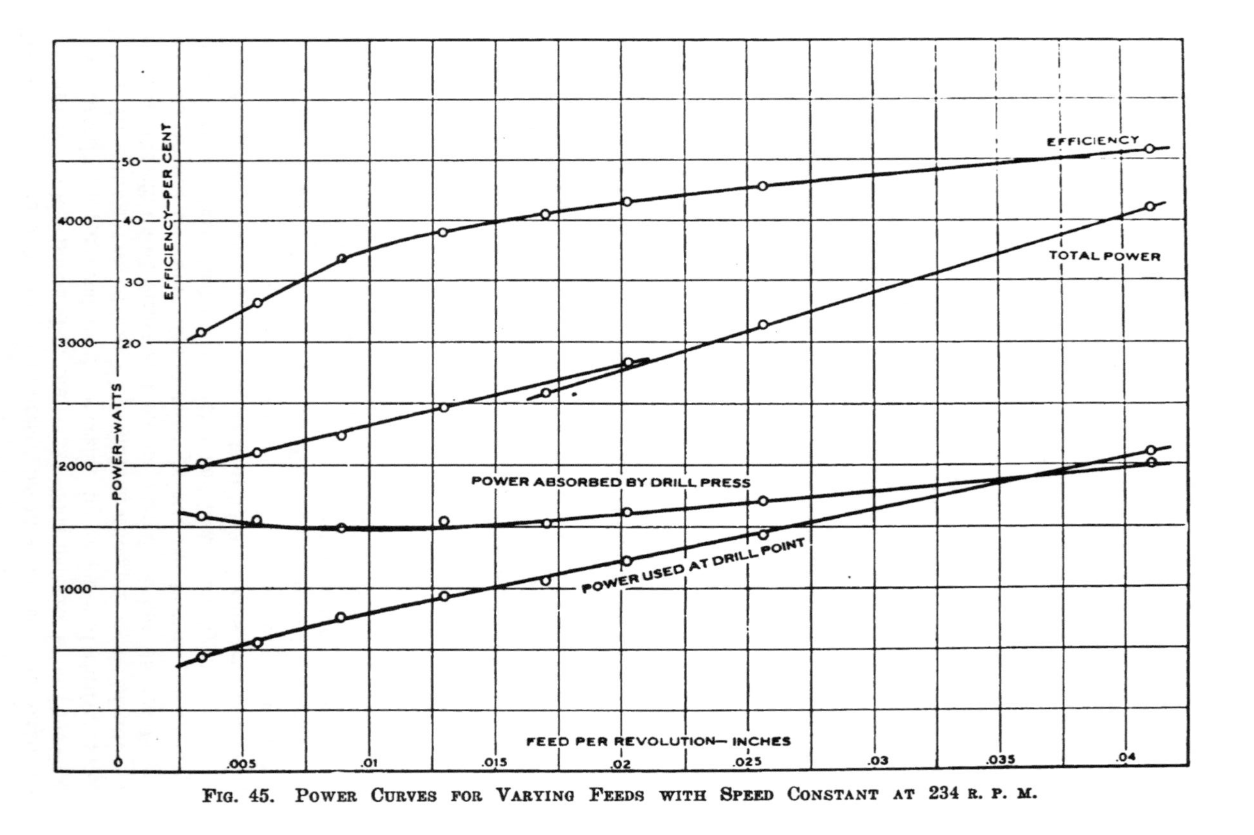

Fig. 45. Power Curves for Varying Feeds with Speed Constant at 234 R. P. M.

the power absorbed by the drill-press, because of the increased pressure on the bearings and in the gears. This increase is not, however, directly proportional to the increase in torque and thrust, as is seen from the curves of Fig. 45. Various attempts were made to segregate the effects of thrust and torque variations, but the results were too diverse to admit of analysis. It was not known just how much power was absorbed by the motor and how much by the press, and the very slight variations below feeds of 0.0202 inches per revolution gave few values for the development of curves. Some of the general results are shown in Fig. 46, which presents the results of tests in which an effort was made to vary the thrust independently of the torque by dulling the chisel point, by drilling into a pilot hole, and by using a drill with an insufficient center clearance. By these means the thrust was varied as much as 450 per cent while the torque was varied only about five or six per cent. When the drill edges were ground straight, it was possible to vary the torque about twenty per cent while the thrust varied only ten or twelve per cent. Satisfactory analysis was, however, not possible, and the only general indication was that the increase in power was about as the cube root of the increase in thrust times the torque. Probably the curves of Fig. 46 will be useful as indicating the possible power saving through the use of pilot holes, which at the higher feeds reduce the total power by ten or twelve per cent. The extra power required when the drill is dull, is improperly ground, or has the cutting angle increased through poor point thinning is shown in curves B, D, and E respectively. For the different curves at the 0.041-inch feed, the thrust and torque values are:

Torque	*Thrust*
A — 58 foot-pounds	2900 pounds
B — 56 foot-pounds	3900 pounds
C — 54 foot-pounds	850 pounds
D — 68 foot-pounds	2700 pounds
E — 72 foot-pounds	2950 pounds

Test "D" was scarcely comparable with the other tests, but was drilled for comparison with a drill having a 135-edge angle, which gave values for this feed and speed of 65 foot-pounds torque and 2850 pounds thrust. The difference in torque was partly due to the use of a different drill from that used in tests A, B, C, and D and partly to the different test block used. The lower curves showing the

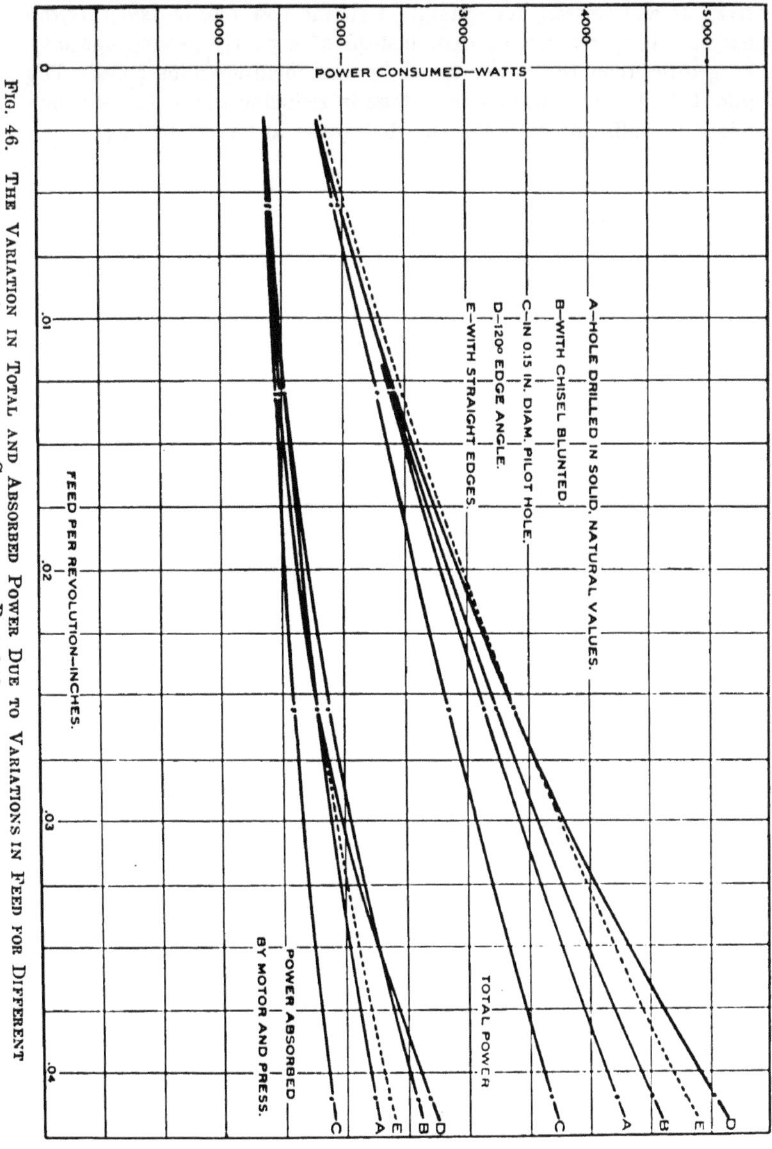

Fig. 46. The Variation in Total and Absorbed Power Due to Variations in Feed for Different Conditions of Drilling

power absorbed by the drill-press are, however, partially comparative. It will be seen, for example, that the power to be saved by the use of a properly ground drill instead of a poorly ground one may be greater than the power saved by first drilling pilot holes. The pilot hole, however, has its advantage in reducing thrust forces which might be sufficient to break the drill-press in heavy drilling.

IV. Tests to Determine Drill Endurance

18. *Influence of the Helix Angle on Drill Endurance.*—A large number of holes was drilled to determine the effect of helix angle on drill power. The last five of each set of twenty holes were drilled at a feed of 0.041 inches per revolution, at speeds of 92, 233, 367, 456, and 575 r. p. m. The two highest of these speeds are excessive for a one-inch drill in cast iron, and the behavior of the drills in drilling at these speeds is considered in the study of endurance. At the lower rates of feed at these speeds the attempt was made to run all five holes without sharpening the drill, and this was possible in almost every case with no increase in the torque which might have been due to dullness, although one or two indications of dulling are noted, and once or twice it was thought advisable to sharpen the drill after the 456 r. p. m. test before drilling the 575 r. p. m. test. Several times in the 0.041-inch tests it was necessary to sharpen the drills, and once or twice it was impossible to drill the last hole, because the drill burned as it entered the block.

With the 118-degree point grinding drill "A" did not need to be re-sharpened for any hole. Drill "B" was dulled by two of the 575 r. p. m. holes (with the 10-and 12½-degree clearance). Drill "C" was dulled by one of the holes, drill "D" by none, drill "E" by one, and drill "F" by two. Drill "G" was so badly burned on one hole that it was impossible to drill more than an inch deep with it. Two other high-speed holes had already dulled this drill badly.

With the 138-degree point grinding the indications of dullness were more numerous. Drill "A," however, was not dulled by any hole drilled. Drill "B" was dulled by three of the holes (only twenty holes were drilled with this point angle as against 120 for each drill with the 118-degree point angle). Drill "C" was dulled by two holes, and so badly burned at the last two holes that they could not be drilled more than an inch deep. Drill "D" was dulled by two holes, burned on one hole, and failed to complete its set by two holes. Drill "E" was dulled by one hole and slightly burned by another. Drill "F" was dulled by three holes, being slightly burned in one of these. Drill "G" was dulled by four holes, and failed to complete its set by one hole.

In general, drill "A" showed up remarkably well for endurance and consumed a relatively small amount of power. The other drills showed somewhat uniform tendencies. For the 118-degree point the drills showed progressively better with increasing helix angles until the 32-degree angle was reached and then they became progressively worse with further increases in the helix angle. For the 138-degree point the drills showed progressively better until the 35-degree angle was reached and then progressively worse with further increases. Other tests indicated that the drill "C" was not so well tempered as the other drills, since it showed poorer endurance than other drills exactly similar in form. It will be noted that the 35-degree helix angle performed well under all conditions, averaging slightly better than drill "F" and considerably better than the 32-degree drill. Drills "E" and "F" gave the best average performance of all the drills in both power and endurance indications, although endurance results are by no means conclusive. The differences in test materials, in the quality and tempering of the drill metal, and possibly in the slight burning of the drills during grinding render the degree of reliability of the endurance tests open to question. If the drills had been run at normal speeds and worn to dullness instead of being burned, the results might have been different.

19. *Influence of Clearance Angle on Endurance.*—The results of the tests referred to in the preceding discussion of the effect of the helix angle on endurance may again be employed in obtaining indications of the effects of the clearance angle on endurance. Six different clearance angles were used on all drills for the tests with the 118-degree point angle.

With the 2-degree clearance angle no drill was dulled in drilling any hole. Likewise with the 4½- and 7-degree clearances all the drills stood up. With a 10-degree clearance, drills were dulled by three of the holes. With a 12½-degree clearance, drills were slightly dulled by four holes. With a 15-degree clearance, drills were slightly dulled in several holes, badly dulled in two holes, and burned in two more. The endurance, consequently, lessens with increase in the clearance angle, although the performance of the drills in the 12½-degree clearance tests was, if anything, a trifle better than in the 10-degree clearance tests.

A partial series of tests was run with a standard drill at a speed of 456 r. p. m. and a feed of 0.041 inches per revolution to determine the effect of clearance on endurance. In each case the object was to run the drill to a degree of dullness approximating an increase of ten pounds in torque. This rapid testing is faulty, however, because the greatest damage is apparently done just as the drill enters the block at high speed. A drill may finish one hole in good shape, but burn upon entering the block on the next test. The total number of inches drilled is so small that this entrance effect is important. At a slower rate, the damage at entrance would be less and the total amount of drilling greater; so more nearly correct values would be obtained. Indications from a number of tests of the rapid drilling, however, gave the following average results: for a 4-degree clearance the total distance drilled was 14¼ inches, for a 6-degree clearance 25½ inches, for a 9-degree clearance 18¾ inches, and for a 12-degree clearance 17½ inches. The 6-degree and 12-degree clearances and the 4-degree and 9-degree clearances were drilled in the same test blocks. Apparently a clearance of 6 degrees is preferable to 12, and 9 degrees to 4.

A series of endurance tests at moderate speeds was started, involving different point and clearance angles with the same drill. These were, however, interrupted and not resumed, so that the results are fragmentary and may be applied to point or clearance angle analysis without certainty of the proportionate effect of either angle. With 15-degree clearance and 124-degree point, the drill went through about 240 inches of soft cast iron with no perceptible increase in dullness. It was then started on an unusually hard block, in which it broke down after drilling ten inches. With a 4½-degree clearance and 112-degree point this drill went through 240 inches of the soft cast iron (half of each block was used for each drill), drilled 40 inches in the hard block mentioned, and then continued until it had drilled through about three thousand lineal inches of iron, at which point the power required had increased about 35 per cent. At 2400 inches the power had increased about 25 per cent, which was the extent of dullness adopted as standard. With a 12-degree clearance and 118-degree point, the drill went through about 1400 inches of iron before attaining the 25 per cent increase in power. Practically all these tests were unsatisfactory and the results probably would have been more nearly uniform if a uniform grade of steel had been used instead of cast iron. On the basis of the test data, however, the indications favor a combination of a small clearance angle with a small point angle.

Incidentally a portion of these tests, with the 124-degree—7-degree combination and the 118-degree—12-degree combination, was run parallel with tests of a drill having a 26-degree helix angle as against the 32-degree angle of this drill. In the first case 85 inches were drilled by the 26-degree drill against 720 inches by the 32-degree drill. In the second test the 26-degree drill went through 360 inches against the 1400 inches of the 32-degree drill.

20. *Influence of Point Angle on Endurance.*—In addition to the combination point angle and clearance angle tests referred to, the group of tests employed for studying the effects of helix angle variations was used for a study of the effects of different point angles. For the group of tests having a 7-degree clearance and a 118-degree point, only a few holes dulled the drill slightly. For the same clearance and 138-degree point angle four holes burned the drill so that they could not be advanced more than a fraction of an inch in depth. Eight other holes dulled the drill, and in three others the drill lips were perceptibly burned, although the hole was advanced to a depth of two or three inches. For the group having a 7-degree clearance and a 98-degree point angle no drill was perceptibly dulled by any hole, although there was a tendency to chip the lip edges and the chisel point, because of the insufficient backing of metal. In the 98-degree tests drill "A" broke under a feed of 0.041 in. per rev. and a speed of 92 r. p. m., although it had drilled this same hole and four others at the same feed with points of 118 and 138 degrees. This failure probably was due to a weakening of the drill by the removal of chisel edge support.

21. *Influence of Rounded Corner on Endurance.*—A number of rapid endurance tests was run at a speed of 575 r. p. m. and a feed of 0.041 in. per rev. in soft cast iron, with pairs of drills, one of which had rounded corners. The results of two of these tests are shown graphically in Fig. 47. On the basis of the number of inches drilled for a given increase in torque an endurance ratio of about four to one in favor of the rounded corners is shown in one case, and about eight to one in the other. Other tests were run in blocks of varying hardnesses which showed a ratio in favor of the rounded corners of about ten to one in the harder materials, and three and

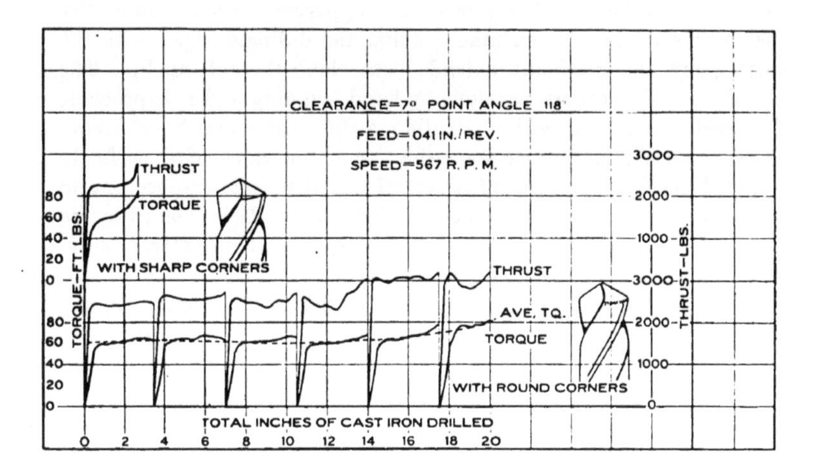

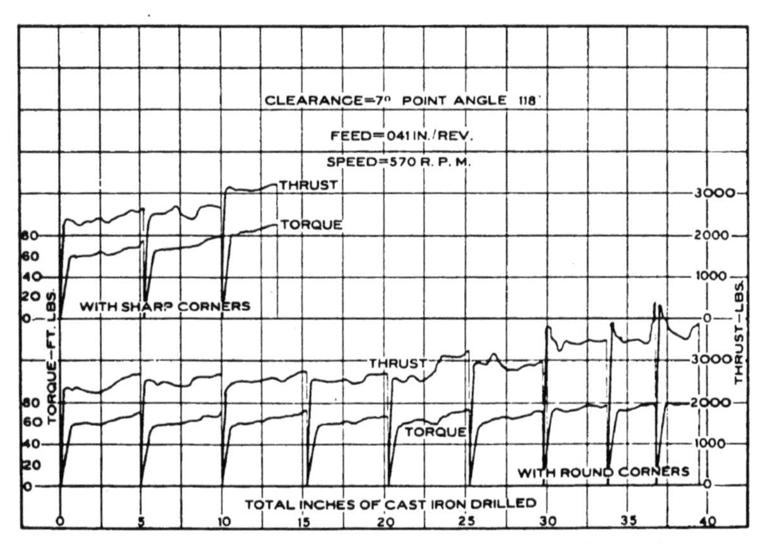

FIG. 47. TEST CARDS SHOWING REMARKABLE INCREASE IN DRILL ENDURANCE OBTAINED BY ROUNDING DRILL CORNERS

four to one in the softer materials. The apparent effect was to render the corners immune to burning, so that the dullness began with the chisel point and gradually extended out along the cutting lip. The rounding of the corners was done by hand grinding and it is probable that each corner removed in drilling a different quantity of metal, so reducing the endurance somewhat from the value of a perfectly rounded corner. This modification of the drill corner does not in the least affect the size of the hole drilled. The hole is cleaned out to size by the cutting edge above the corners, so that no extreme accuracy of corner grinding is necessary. There should, however, be sufficient clearance behind the rounded corner to prevent rubbing action.

22. *Comparative Endurance Tests of Different Kinds of Drills.*—A short series of comparative endurance tests was run in soft cast iron at high speed and heavy feed on several milled drills, a forged twist drill, and a flat-twist drill. The results are plotted in Fig. 48. The test points for the forged drill are shown, and the average curves are given for the other drills. It will be seen from the results of the forged drill test that the power indications as the drill begins to burn are irregular, although it is possible to plot fairly regular curves. The reversed curves for the three drills which showed the greatest endurance are pronounced, while the curves for the drills with the least endurance show fairly regular increases of power at the drill point as drilling advances. A general similarity of behavior of the drills, however, is to be noted in that, following initial drilling, the tool "settles to a cutting edge" and does its most effective work after the initial sharpness is worn off. This effect was evident also in the longer endurance tests previously mentioned, the drill very often running through five to eight hundred inches of cast iron with no variation in the average power curve, after the first gradual dulling had taken place. The rapid rise in power after the drill has once started to burn is logical, as each increment of dullness adds to the friction of cutting thereby heating the drill and destroying its resistance.

Comparative endurance tests of this sort are more satisfactory than endurance tests in which different drill angles are tested, for in the latter there is likely to be a variation in the drill endurance, even when the same drill is used for each test, because drills tend to

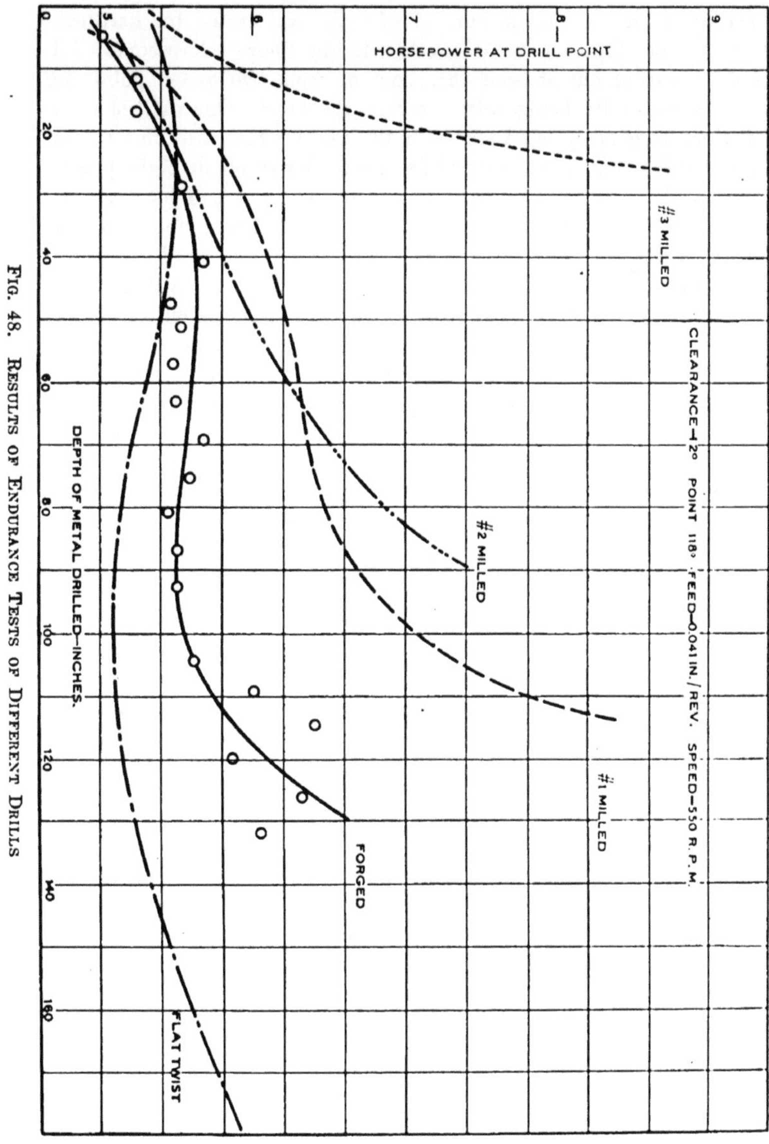

FIG. 48. RESULTS OF ENDURANCE TESTS OF DIFFERENT DRILLS

vary in character from point to tang through uneven tempering or variation in the metal. A number of drills was tested for hardness at various points along their length with the Shore scleroscope. All the drills except one showed variations of from two to ten points in hardness, generally being softer toward the tang. One drill showed a distinct ring about an inch from the point which was five to six points harder than the point, while one drill was about eight points softer in the middle than at either end. The average hardness at the drill points was about 86, only two drills being above 90, of which one was 92 and the other 91. One drill which produced average results in endurance tests had the surprisingly low value of 78. Some of the drills were so skillfully tempered that the entire length of the flutes tested about 84, but the hardness dropped to 40 or 45 at the point at which the drill attained its full diameter at the end of the flutes. Other drills were hard almost to the tang. In 1908 the Lincoln-Williams Twist Drill Company* made a number of endurance tests to show the relation between hardness and endurance (drilling in tool steel), but found no regularity in results. They found that the average hardness of high-speed drills was about 86, while that of carbon drills was about 89. The best high-speed drill tested was the hardest, and the poorest was the softest, but the variations between these extremes showed almost no regularity.

The variations in drill hardness practiced by the different makers are not more pronounced than the variations in helix angles and angles of point grinding of the drills. A number of standard drills was measured and their characteristics recorded as shown in the following:

MANUFACTURER	Type	Clearance Degrees	Point Degrees	Helix Angle Degrees
Detroit Twist Drill Co.	Milled	13½	60	32 to 27
Detroit Twist Drill Co.	Milled	7	61	25 to 20
Union Twist Drill Co.	Milled	4½	58	22
Morse Twist Drill Co.	Milled	2	58	26
Gelfor Twist Drill Co.	Flat-twist	6	63	32
Whitman & Barnes	Flat-twist	5	59½	29 to 26
Pratt & Whitney	Flat-twist	5	59½	30 to 25
Lincoln-Williams	Flat-twist	6	59	30
Lincoln-Williams	Milled	5	58	26 to 21
Union Twist Drill Co.	Milled	5	60	24
New Process	Milled	4	59½	22
Whitman & Barnes	Milled	6	59	26

*See Appendix V for discussion of experiments by other investigators.

On the special drills used in tests to determine the influence of the helix angle the clearances were 4½, 5, 5½, and 6 degrees, and the point angles were all 120 degrees. The edge angles of various drills showed variation between 117 and 135 degrees.

V. Minor Investigations

There has been some discussion concerning the effect of different helix angles on chip removal in deep holes; so from the number of tests run for the power determinations with different helixes a set of four cards was chosen. The values from these are shown transferred to rectangular coördinates in Fig. 49, the torque and thrust values being plotted side by side. All these holes were drilled with a speed of 570 r. p. m. and a feed of 0.041 in. per rev. All drills were similarly ground, and all but the first were advanced to a depth of about 4½ inches in cast iron. The drill with the 10-degree helix seemed to give the best performance, although this may have been due more to its large chip space than to the helix influence, since the drill with the 22-degree helix shows a steady rise in torque as the depth increased. The 32-degree drill had an increase twist to give it better chip clearance at the greater depths, but its performance was not appreciably better than that of the 40-degree drill. The increase in torque as the depth increases is in every case more pronounced than the increase in thrust. The sudden rise in the thrust values at the beginning of each hole may be due to the greater hardness of the metal near the surface or to a surge of fluid in the dynamometer resulting from the sudden application of the load.

The fact that all these holes were run at high speed seems to have been a factor affecting chip disposal, since results of tests run at a speed of 233 r. p. m. and feed of 0.041 in. per rev. show a much more rapid increase in torque as depth increases (see Fig. 50). In these 233 r. p. m. tests the 10- and 22-degree drills showed rapid increases in torque with increases in depth. The 32-degree increase-twist drill showed well to a depth of about two inches, while beyond this the torque increased rapidly. With the uniform helix of 40 degrees there was a steady increase in torque with depth, but this drill averaged better than any other. These results seem to indicate in general that all drilling should be done at as high speeds as possible and that the larger values of helix angle are better for deep hole drilling than the more nearly straight drills.

That the effect of depth on torque is governed by other conditions than the helix angle and the drill speed is shown by the results presented in Fig. 51, cards "A" and "B," the first of which shows

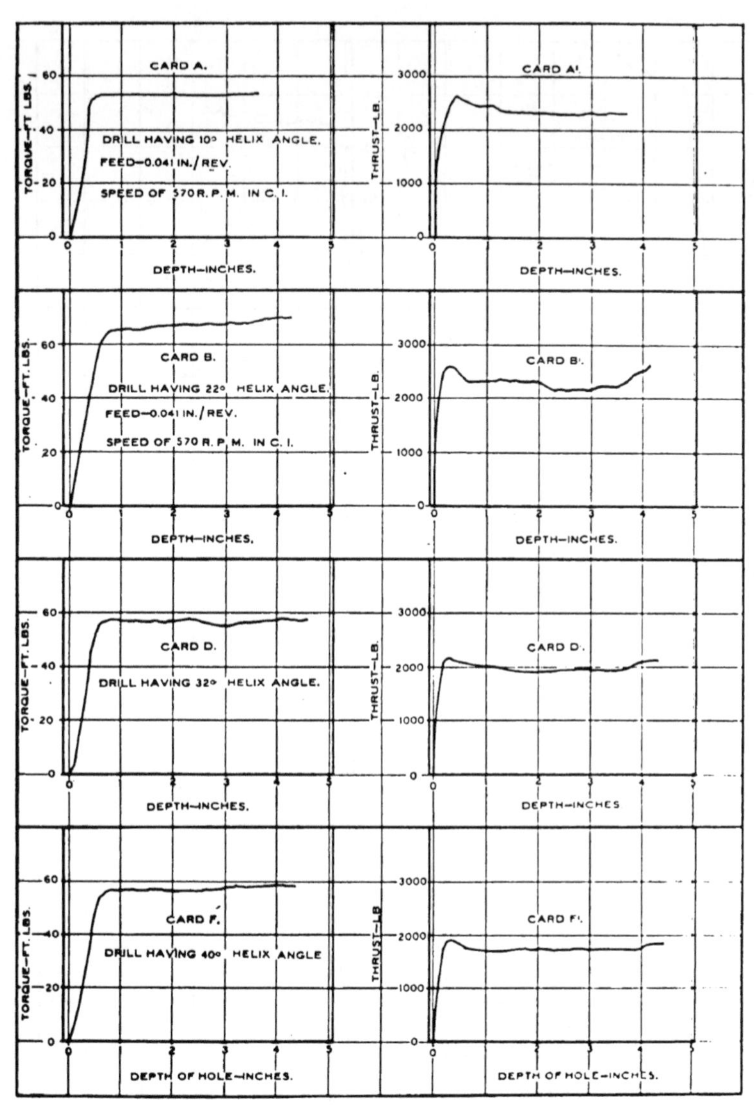

FIG. 49. TEST CARDS SHOWING CORRESPONDING TORQUE AND THRUST FOR DIFFERENT HELIX ANGLES

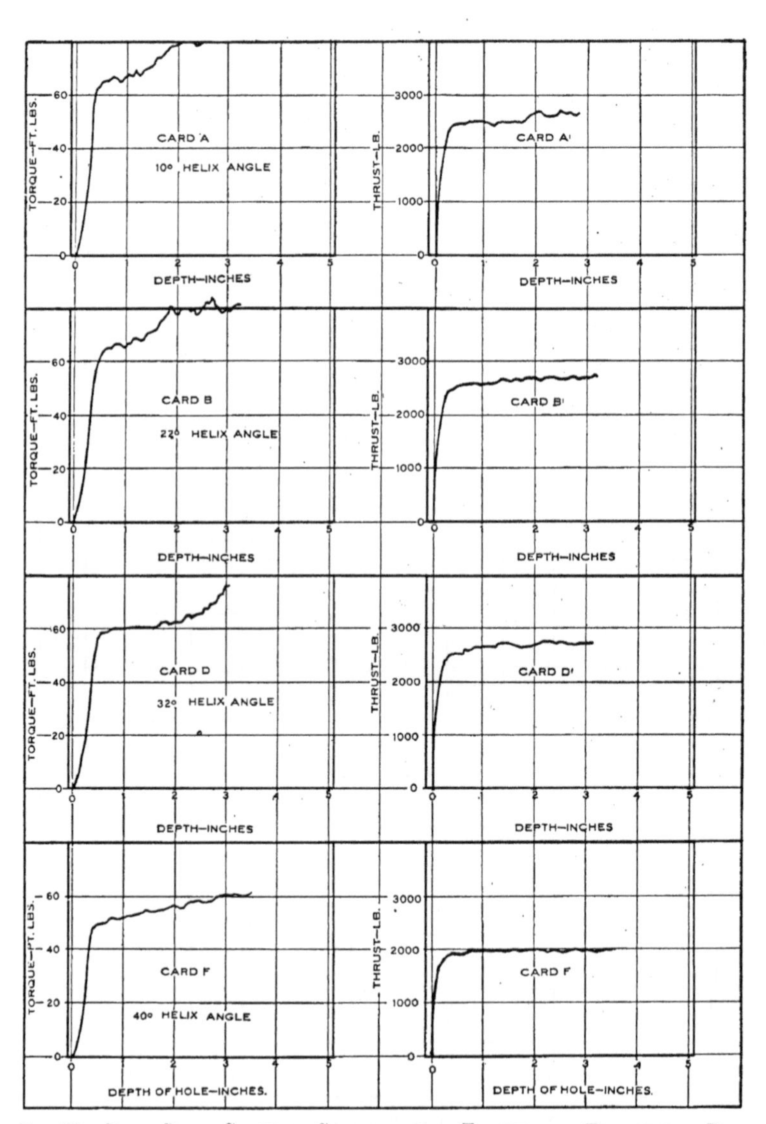

FIG. 50. TEST CARDS SHOWING CORRESPONDING TORQUE AND THRUST FOR DIFFERENT HELIX ANGLES

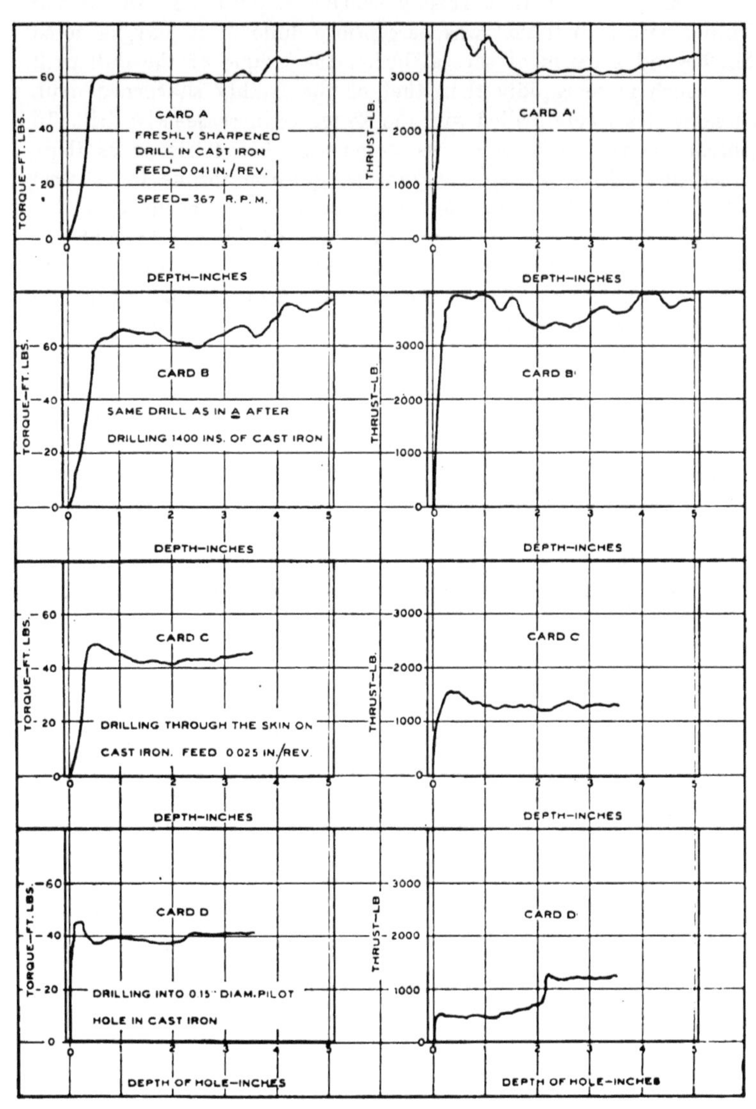

Fig. 51. Test Cards Showing Corresponding Torque and Thrust for Different Conditions for Drilling

results of drilling with a freshly sharpened drill, and the second drilling with a drill that was perceptibly dulled. It may be noted that beyond a depth of several inches the torque of the dull drill rises much more rapidly than that of the freshly sharpened drill. These two tests were drilled with the 32-degree increase-twist drill. In general the results show the same general characteristics as those presented in Figs. 49 and 50, that is, uniform torque for several inches, after which the torque beings to rise fairly abruptly.

Card "C" of Fig. 51 shows the effect of drilling through the skin on cast iron. The thrust at first rises almost instantaneously to a little above 1000 pounds and then drops to about 700 pounds before the drill shows perceptible penetration. With the lower feeds this effect was still more pronounced; thus for a feed of 0.013 inches per revolution the thrust rose to a value of 800 pounds and then dropped to 600 pounds. For a feed of 0.00364 in. per rev. the thrust rose to 550 pounds, dropped to 250 pounds, and then rose to about 450 pounds. Apparently the hard elastic skin resists penetration at first, but the instant it is punctured the thrust drops in value. The advantage of removing the skin before running a drill into the work is shown by the fact that after one of these test holes had been drilled, examination of the drill showed that the lips were scarred and chipped, although all holes were drilled at the slowest speed; so the destructive effect of shooting a drill through this hard surface at a high rate of speed may be understood.

Card "D" shows a typical record obtained when drilling into a small pilot hole for about two inches, striking the bottom of this hole, and then drilling in the solid for about an inch. It will be seen that the torque takes practically its full value instantaneously in a hole of this sort, giving a peak that was probably due to fluid surge in the dynamometer. The effect of the pilot on thrust begins to lessen after an inch or so, owing to filling up this small hole with the finer particles of iron removed by the large drill. For a hole about 0.25 inches in diameter this filling is not so marked, and the thrust rises almost instantaneously at the bottom of the smaller hole.

VI. Conclusions

A study of the results presented in the preceding pages relating to drilling in cast iron seems to justify the following conclusions:

(1) With the usual angles of grinding the power used at the drill point decreases as the helix angle increases until the value of the helix angle reaches 35 degrees, above which there is no further decrease in power. Above 40 degrees the power begins to increase. A helix angle of 30 to 35 degrees seems also to give the best average endurance in cast iron.

(2) For corresponding speeds and rates of feed the variation in torque for point angles between 108 and 138 degrees is slight, but torque increases for point angle values below 108 degrees. The variation in thrust for point angles between 98 and 118 degrees is slight, but the thrust increases rapidly as the point angle increases above 118 degrees. Drills with point angles greater than 118 degrees show less endurance than drills with point angles smaller than 118 degrees, while drills with point angles less than 118 degrees show less susceptibly to burning of the corners, although a greater liability to chipping of the cutting edge and a general weakness of the drill due to the removal of metal backing the chisel point. In general, a point angle of about 110 degrees seems more satisfactory than an angle of 118 degrees.

(3) Variations in the clearance angle at the drill periphery and variations in the clearance at the drill center (as indicated by the value of the edge angle) seem to affect the net power very slightly. Unless the peripheral clearance is less than two degrees, variations in its value have very slight effect on the values of torque or thrust, but if the edge angle is less than 120 degrees, the thrust tends to increase, showing a more rapid rate of increase for the heavier rates of feed. A peripheral clearance of about six or seven degrees seems to be the most satisfactory value as affecting drill endurance. An edge angle of 130 degrees is recommended for all but the heaviest feeds (above 0.041 in. per rev.). This angle should not vary with variations in peripheral clearance, but may be varied with the feed per revolution, decreasing for light feeds and increasing for heavy feeds.

(4) The power required at the drill point and the drill endurance seem to be much affected by the form of the chip space and the form of the cutting lip. Drills having the greatest chip spaces and concave cutting edges seem to give good results under all conditions of work and point grinding. Attempts to increase chip space near the hilt of the drill by decreasing the helix angle are apparently of small value, and the decreased helix angle as the drill is ground away causes increases in the power required and lessened endurance.

(5) In soft cast iron the torque does not increase so rapidly as the feed per revolution, although the variation of torque with feed is so nearly direct that it may for all practical purposes be so considered. For the correct angles of drill grinding the thrust should increase directly with the feed per revolution.

(6) In soft cast iron the torque decreases as the speed increases, although there is little advantage in running a one-inch drill faster than 350 r. p. m. The saving in power resulting from increasing speed may amount to ten or fifteen per cent. Thrust decreases as speed increases for corresponding rates of feed, although the advantages to be gained by running a one-inch drill faster than 400 r. p. m. are slight. The decrease in thrust may amount to fifteen per cent.

(7) Drilling a pilot hole the diameter of which is equal to the width of the chisel edge reduces the thrust of the large hole by sixty or seventy per cent, the reduction being greatest for the heavier feeds. The torque of the larger drill varies according to the area of metal removed by the pilot drill. For the average drill this means a saving in net power of one per cent.

(8) The power loss in the back gears of a drill-press increases directly as the feed per revolution and may be from five to eight per cent of the total power consumed. The efficiency of the drilling apparatus increases with the rate of drilling, and for these tests varied between eight and sixty per cent. The net power for a given rate of drilling varies with the feed used, being greater for the light feeds and less for the heavy feeds, although there is very slight advantage in using a feed greater than 0.041 in. per rev. The gross power is a function of several variables, including the variation in net power, the power con-

sumption of the drill when running idle, the increase in torque and thrust for the heavier feeds, and the means of gearing for different speeds. For general good efficiency of drilling with this apparatus in cast iron, the feed should not be less than 0.013 in. per rev. and should increase with the rate of drilling to 0.041 in. per rev. for a rate of nine inches per minute. So far as power consumption is concerned, whether net or gross, the best way to remove metal with a given drill is with as high a speed and as heavy a feed as the drill will stand.

(9) By first drilling a pilot hole with a diameter equal to the thickness of the drill web, the gross power when drilling the larger hole may be reduced ten or twelve per cent. A blunt drill increases the gross power by reason of the greater thrust which causes greater losses in the drill-press. A drill with a point angle greater than 118 degrees or a drill with an edge angle less than 130 degrees will increase the gross power for the same reason.

(10) The endurance of a drill may be increased 300 to 1000 per cent by the simple process of rounding the sharp corners at the drill periphery.

(11) For drilling holes at the depths of more than three inches a drill with a large helix angle gives best results. Such holes should be drilled at as high a speed as possible, for then the screwing action of the drill flutes removes the chips more nearly completely.

APPENDIX I

Development of the Twist Drill

To produce holes in hard materials seems to have been one of the early needs of the human race, since evidences are found of many early and crude devices for drilling. Almost invariably the primitive drill took the form of a sharp iron or stone point attached to a wooden shaft. The drilling action of this device was one of scraping or crushing rather than of true cutting. The motion was imparted directly by rolling the wooden shaft between the palms of the hands or indirectly through the medium of a bow or a stick-and-string arrangement. In any case the motion was a reciprocating one, and the whole process required patience and care. The usual requirement was to produce a hole in wood, but not infrequently the pump and bow drills were used in operations on glass, stone, and crockery. Among a few tribes was found a heavy two-man drill capable of piercing iron.

23. *Flat Drills.*—The direct progenitor of the present twist drill was the old style flat drill, made usually of carbon steel in a blacksmith shop and shaped according to the ideas of the maker. Such a drill could usually be forged in a few minutes. A length of square stock was flattened at one end and given a rough diamond shape, possibly with the edges turned over a bit to form a sort of hooked lip and then by means of a file or grindstone, the cutting lips and point were sharpened and the drill was brought to its final form. Drills of this type may be found in use at the present time, and although very inefficient they possess certain advantages due to the fact that they may be made with an extra long shank for deep holes or made in odd sizes.

The usual flat drill was made of square stock, but sometimes flat stock was used, probably because it was easier to forge and to grind and provided more space for the disposal of chips. Then some artisan tried twisting the drill while it was hot from the forge and discovered that he had a tool which gave noticeably better service. The idea of the twisted drill led in time to the development of a tool more or less similar to the modern flat-twist drill, but it was still in a crude state. Only when drills began to be made as a special product (The

Morse Twist Drill Company in 1862 was the first manufacturer), did they assume any uniformity of design or attain any real degree of efficiency. Then experimentation began with the different forms of milled flutes, with the helix angle of the flutes, with forms of point grinding, and with the composition of the steel used. Practically only in the last decade has any other than carbon steel been used, but previous to that the drill had been given a certain uniformity of design. The disadvantages of the drills used previous to the last decade lay in the fact that they had to be kept cool to prevent drawing the temper; this condition in turn meant either that the drill had to be run slowly or that some special provision had to be made for supplying a cooling and lubricating medium. Such a tool did not meet the requirements of modern large and rapid production; so after high-speed steels had proved their value in lathe and planer tools, attempts were made to use them in the form of twist drills.

24. *The High-Speed Steel Drill.*—Twist drills of high-speed steel are still in the process of development, both with reference to the metals used and the form of the drill. The principles of design determined for carbon steel drills have to a large extent governed in the making of the high-speed drill. Experiments are still necessary to establish the proper helix angle, flute space, web thickness, and point grinding. The steel used varies according to the formulas of the different drill makers. It has been determined by analysis that the composition varies widely and that even the carbon content shows no regularity. The carbon content varies from 0.25 to 0.60 per cent, manganese from 0.25 to 0.60 per cent, chromium from 0 to 5.0 per cent, tungsten from 10.0 to 20.0 per cent, and cobalt, vanadium, and silicon, etc. within narrow limits according to the ideas of the different steel and drill makers. This matter of the most desirable alloy has not been thoroughly investigated, nor have the equally important questions of forging and heat treatment. The large number of variables in the different alloys and the possibility of many methods of treatment greatly complicate the matter of testing. The testing itself is tedious, since it is necessary to make a complete drill for each material tested and to make provision for some special test apparatus, whereas the testing of lathe tools requires only a small piece of the material and only a standard lathe as a test machine. Unfortunately the qualities required in drills cannot

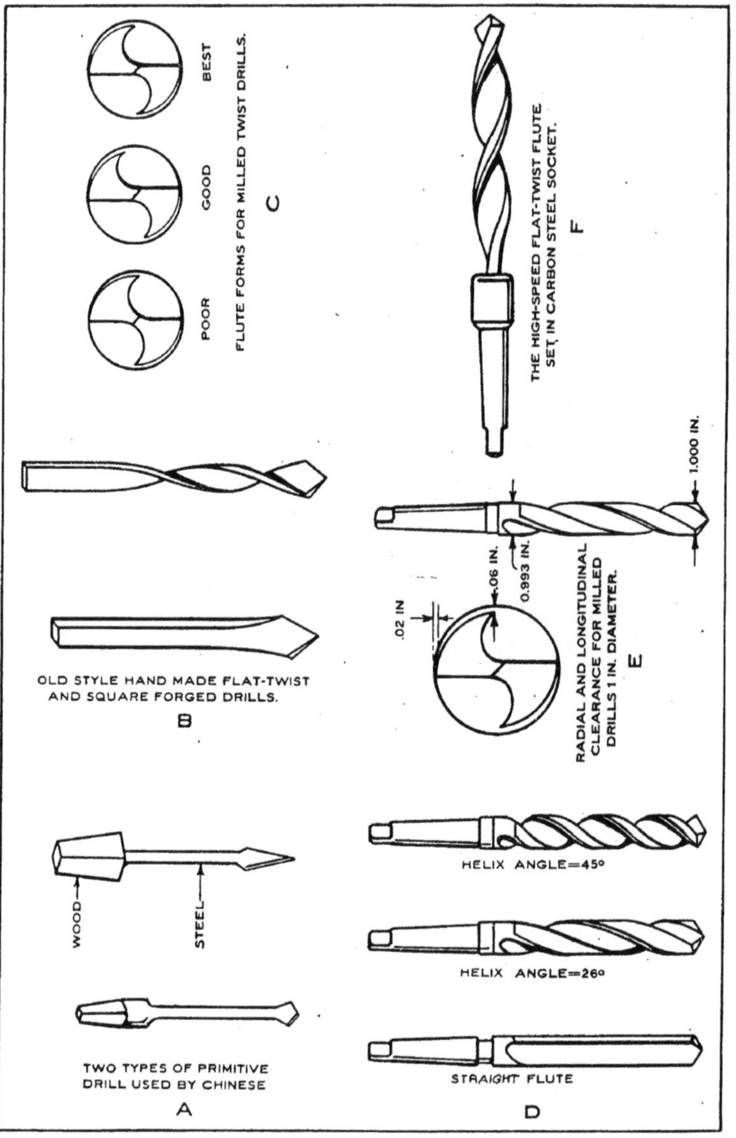

FIG. 52. TYPES, FORMS, AND CHARACTERISTICS OF DRILLS

be readily determined from lathe tool tests, because of the form of the cutting edges, the different conditions affecting the point, and the difficulties of proper cooling. Even at the present stage of the development, however, the advantages of the high-speed drill are apparent especially where rapid production necessitates high speeds and heavy feeds.

25. *Analysis of Drill Form.*—The modern drill consists of a straight metal shank carrying at one end two equal cutting lips and at the other some means of attaching the drill to a device which will impart the necessary rotary motion to it. The earlier types of drills were simple affairs, as may be seen by reference to Fig. 52-A, with the lips usually so shaped that a reciprocating rotary motion could be used. Small drills of this type may be made of wire nails or other pieces of round scrap. The only requirement is a sharp point which will make a hole bigger than the drill shank, without particular uniformity in hole diameter or outline. The drills shown in Fig. 52-B are a logical development from these early types, being merely larger and stronger, made of better material, probably a little more carefully forged and sharpened, and capable of producing a hole approximately true. The flat-twist drill shown is twisted by hand while hot from the forge, and the twist is neither uniform nor directed to any particular angle. This design was supposed to facilitate removal of the chips by screwing them out of the hole, but it was only approximately successful, because the diameter at the cutting lips is appreciably greater than the shank diameter. Efforts to make the helix of a uniform angle, to provide adequate chip space, and to combine strength with economy of material resulted in the production of the milled-twist drill. The first of these drills, made about 1850, proved disappointing in its weakness and lack of endurance, and it was not until Morse in 1864 produced a drill having more obtuse point and cutting edge angles that the milled-twist drill demonstrated its worth.

The form of flutes necessary to give best results is still a subject for discussion, various manufacturers using different forms. Three forms of flute are shown in Fig. 52-C. The requirements are that this space shall shape the chips to a compact helix when drilling steel and that it shall afford adequate space for their disposal. Usually the milling cutters are so shaped that in connection with a 118-degree included point angle the cutting lip will be a straight edge.

Apparently there is no reason to believe that the straight cutting edge is more efficient than a convex or concave edge, but the sentiment in favor of the straight edge is firmly established. Practice with regard to the helix angle also varies within certain limits, there being standard drills with a uniform helix angle of 20 degrees and others with a helix angle of 32 degrees. General practice seems to favor a drill with a helix of about 25 degrees, sometimes uniform from tang to point and sometimes varying from 20 degrees at the tang to 25 at the point. The theory of this variable or "increase" twist is that it provides a greater chip space at the tang, where this space would otherwise be decreased by the increasing thickness of the web. Drills for work in brass or similar soft material give best results when the drill flute is straight* (helix angle equals zero degrees), as shown in Fig. 52-D. In fact it is logical to suppose that this helix angle could well have different values, according to the material worked, just as the grinding angles used for lathe tools operating on steel are different from those used on cast iron. Fig. 52-D shows the straight flute drill, the standard 25- or 26-degree drill, and a special drill having a helix angle of 45 degrees.

The workmanship on modern twist drills is of a high order, and it is not unusual to find the cutting diameters guaranteed to within .0005 inches, but this diameter is true for only a small portion of the drill. Back of each advancing edge the drill may be concentric for a tenth of an inch, at which point the radial clearance begins, ranging from .01 to .05 inches. Usually this clearance is eccentric as shown in Fig. 52-E, and is produced by milling or grinding with the drill slightly off center. The drill has also a longitudinal clearance to prevent its binding in deep holes. This clearance amounts to about .001 inches per inch of length so that an old drill is almost invariably undersize. In all probability there is no need for such exactness in the cutting diameters, since a drill will usually cut a hole from .002 to .006 inches oversize, because of slight inaccuracies of grinding, eccentricity in the drill shank, or expansion from temperature rise.

While the standards for flute form and helix angle are neither so well established nor so necessary as are those for cutting diameters, the forms of drill shank have come to definite standards. This uniformity is necessary, in order that a drill purchased from any maker

*Cleveland Twist Drill Co., Catalogue.

will fit the spindle of any lathe, boring mill, or drill-press. There are, however, several types of drill-shank taper. Of these the Morse taper is the oldest and most widely used. Originally it was probably conceived as a uniform taper of five-eighths inch per foot, but inaccuracy in the first set of gages seems to have led to slight inaccuracies in the resulting product. The Brown and Sharpe taper of one-half inch per foot is optional with most drill makers. There are various forms of special sockets on the market, employing keys, pins, and other positive locks for driving the drill, but the taper shank with driving tongue is well established and convenient.

With the greater use of high-speed steels an old type of drill has been revived, namely the flat-twist drill. This form of drill has the advantage of requiring an appreciably smaller quantity of high-speed metal, since the twisted flutes may be firmly welded or brazed into a soft steel shank, or the flat flute may be extended to form a special shank. This drill has a further advantage in that the metal is hammered or rolled to form and then twisted; thus a closer and more homogeneous structure is given to the steel and a stronger and tougher product results. It will be found that such drills require more power than the milled drills because of the crude form of the chip space and the poorly shaped cutting edge. Their greater endurance and strength may, however, more than offset the power loss where heavy production in hard metals is desired. Fig. 52-F gives an idea of the appearance of the drill.

A further improvement in this drill consists in shaping the bar of high-speed metal so that the twisting process will give a resultant form approximating the milled drill. With modern facilities for this forging and twisting process, an excellently shaped drill is made, the final finish being given by a careful grinding. Such a drill is more expensive to make than the milled drill, but it seems to combine all the advantages of the milled and flat-twist drills.

26. *The Helix Angle.*—The standard lathe tool for operating on cast iron has a back slope or rake of eight degrees, a front slope or clearance of six degrees, and a side slope of fourteen degrees. This combination results in a lip angle of sixty-eight degrees. For operation on steel the rake and clearance angles are the same, but the side slope is twenty-two degrees which gives a lip angle of sixty-one degrees. Fig. 53-C shows the lathe tool as applied to a flat surface

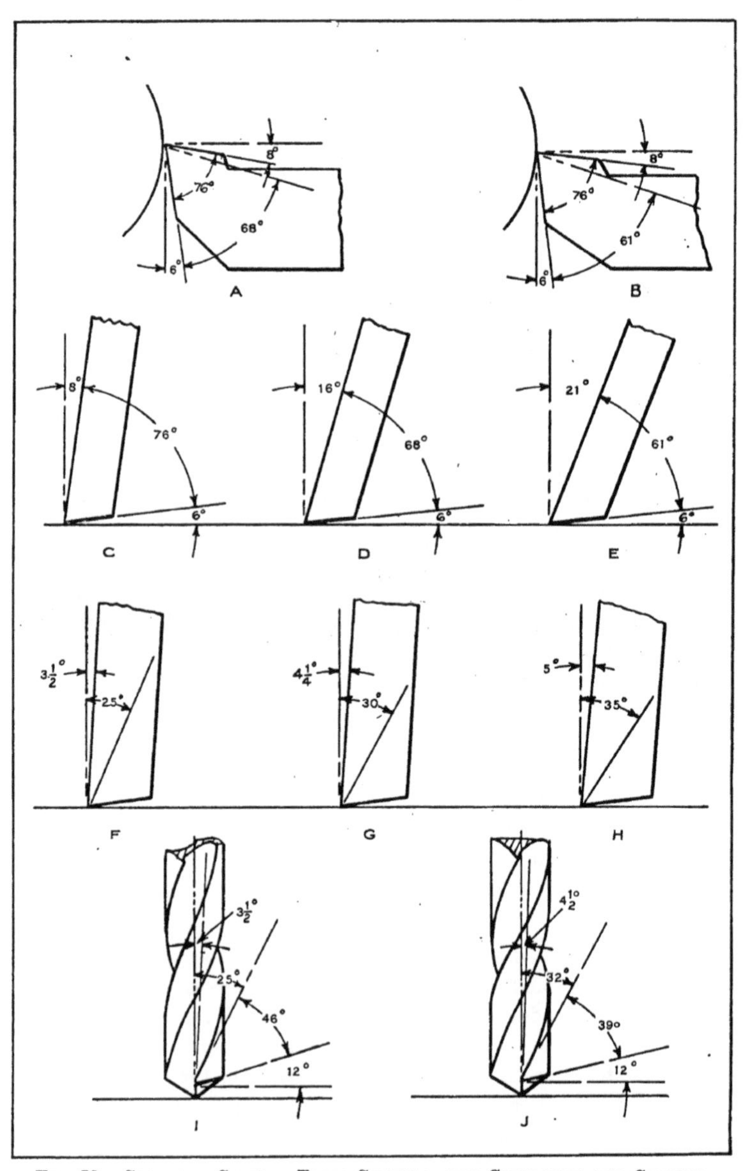

FIG. 53. STANDARD CUTTING TOOLS SHOWING THE CLEARANCE AND CUTTING ANGLES

on the principle of a drill or planer tool without consideration of the side slope. Figs. 53-D and 53-E show the theoretical application of the cutting angles of the tools for operating on cast iron and steel, respectively. It will be noted that the angles of back slope or rake now correspond to the helix angle of the ordinary milled or forged twist drill. Whereas the accepted helix angle for twist drills is about twenty-six degrees at the drill periphery, this application of lathe tool design would indicate angles of sixteen and twenty-one degrees for cast iron and steel, respectively. That this interpretation is not necessarily a true one will be seen from an examination of Figs. 53-F, G, and H.

Here the 25-, 30-, and 35-degree angles represent the helix angle of the drill at the periphery, while the 3½-, 4½-, and 5-degree angles represent the corresponding helix angles at the edge of the chisel point of the drill. The averages of the angles along the cutting edge are approximately fourteen, seventeen, and twenty degrees.

Assuming that this average angle should correspond to the cutting angle of a lathe tool, it will be found that the correct drill for cast iron will have a helix angle of twenty-eight degrees at the periphery, while the correctly formed drill for steel will have a corresponding helix angle of forty degrees. The drill, therefore, which would, according to this theory, produce the best results on both cast iron and steel would have a helix angle of thirty-four or thirty-five degrees at the periphery. This theoretical deduction cannot, however, be relied upon for practical application, since the entire operation of the twist drill is different from that of the lathe tool. No part of the cutting edge has the same speed as any other part. The drill point is kept cool only as the heat is conducted away by the metal being cut. The support given the drill by the metal upon which it is operating, the varying clearance of the cutting edge, and the difficulty of cooling and of chip disposal cause drill design to present its own peculiar problems, although it is possible that much may be learned from the lathe tool design. It is reasonable to suppose that if the carbon steel drill would give satisfactory service with an angle of twenty-five degrees, the new high-speed drills should show equal endurance with a sharper cutting edge, that is, with a helix angle of from thirty to thirty-five degrees at the drill periphery.

Fig. 53-I shows the regular carbon steel drill, and Fig. 53-J a drill of high-speed steel which has lately been put on the market.

In this high-speed steel drill the helix angle at the periphery is thirty-two degrees and this gives approximately a $4\frac{1}{2}$-degree rake or back slope at the edge of the chisel point. The clearance of twelve degrees shown in each drill is excessive, but is put at this figure by the drill manufacturers in order to assure sufficient clearance for even poor hand or machine grinding, just as lathe tool practice recommends a clearance of from ten to twelve degrees where grinding machines are not available, or where the tool gages are likely to be inaccurate or carelessly used. The usual clearance of six degrees would be sufficient if properly designed drill grinders were always used. The tendency is to allow too little clearance at the drill center; so this large clearance angle is designed to cover possible inaccuracies.

In tests conducted at the University of Illinois, drills were used having helix angles of from ten to forty-five degrees, which in conjunction with clearance angles of from two to fifteen degrees gave cutting lip angles of from thirty to seventy-eight degrees at the periphery. Such a range of cutting angles permitted a partial examination of the attendant phenomena, although it is always difficult to separate angle influence from variations induced by varying flute characteristics, variations in drill material, structure and temper, or variations in testing material.

27. *The Angle of the Cutting Edge.*—In the preceding pages it has been shown how the helix angle of the drill flutes influences the angle of the cutting edge, but this angle is not the only factor which influences the form and action of the cutting edge. Other factors include the point angle, the flute form, and the web thickness of the drill, as well as the peculiarities of point grinding. Of course it is to be expected that an increase in clearance angle, by forming a sharper edge, may result in a slight decrease in driving power. The larger helix will also act in the same way to produce a sharper edge, and may possess advantage also in its effect upon the formation of better chip helixes.

Considering the actual form of the edge as shaped by the average drill grinder, it will be observed that, as viewed from the side, it has a slight convex curve, shown to an exaggerated extent in Fig. 54-A. The end view of the standard ground drill shows, however, that the advancing edge is straight. This is true, because the drill makers have designed the flute millers in order to produce this straight edge

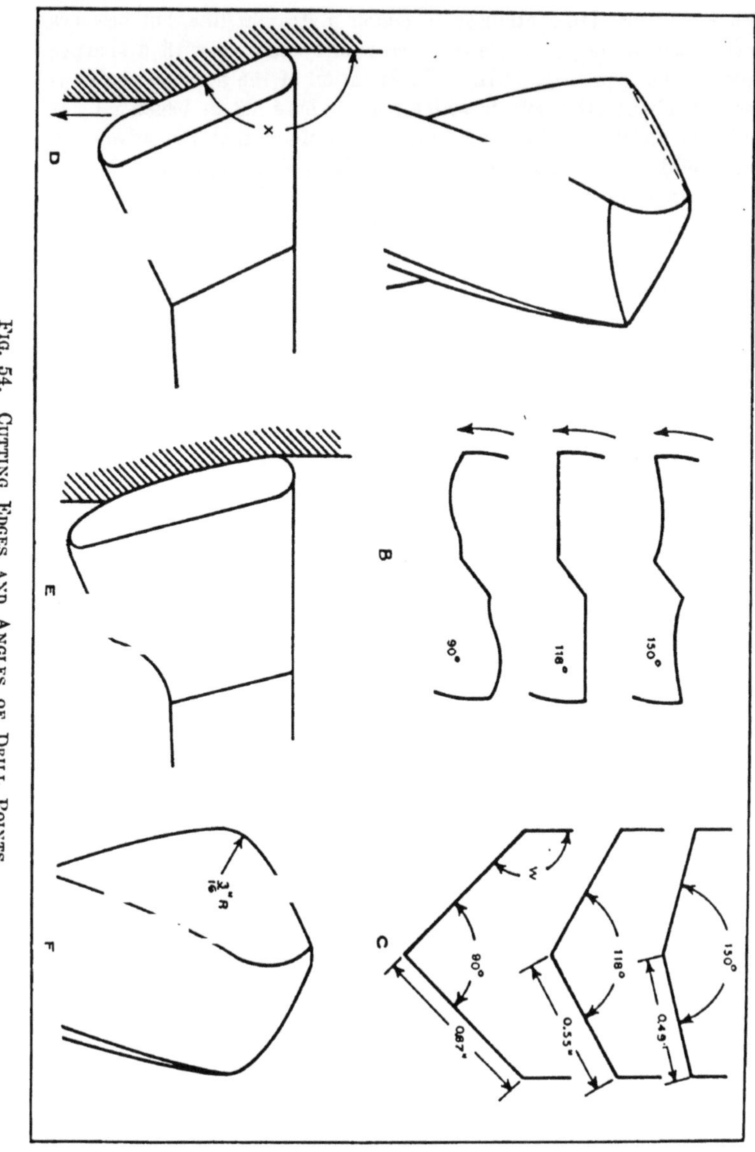

Fig. 54. Cutting Edges and Angles of Drill Points

in combination with the 118-degree point angle. If this point angle is changed, the edge contour is no longer a straight line, but becomes concave with a larger angle and irregularly convex with a sharper angle, as shown in Fig. 54-B. The variation of the point angle, however, introduces other effects which may or may not be beneficial. It would be possible so to design the flute millers that the advancing edge would present a straight line for any point angle; so this may be neglected for the present. Let the included point angle be increased from the standard 118 degrees to 150 degrees and then the length of cutting edge will be decreased by about twelve per cent. This decrease means that less length of cutting edge is removing the same amount of metal; so the wear should be proportionately increased. The more nearly flat point also removes a thicker chip proportionate to the decreased length of edge, which also should cause increased wear. The angles at the drill periphery, indicated by "W" in Fig. 54-C, also become more acute and should be more likely to wear since they have the highest peripheral speed of any part of the cutting edge. The angle of the chisel point, furthermore, is more obtuse and the action more nearly approaches rubbing, rather than the scraping cut of the standard point. This chisel edge at best merely scrapes off the metal but the further debasing of the action is best seen if it is supposed that the point angle is increased to 180 degrees. Then the chisel no longer even scrapes. It must be forced through the metal as a flat surface, crushing, if possible, but under a tremendous stress; so the logical conclusion is that the larger point angles are unsatisfactory from several standpoints.

If, on the other hand, the point angle is decreased, the length of cutting edge is increased, and this increase is proportionately greater than the decrease in cutting edge when the point angle is increased; thus the length of edge will be about twenty-two per cent greater for a decrease of thirty degrees of point angle; whereas an increase of thirty degrees over the standard angle gives a decrease in cutting edge length of only twelve per cent. The thickness of the chip also is decreased, the angles at the periphery are increased, and the action of the chisel edge is improved. The disadvantage of the sharper point angle lies in the fact that the chisel point has less support and may chip in hard material. This blunting and chipping of the chisel point may also lead to the whole drill splitting, because of the increased pressure necessary for cutting and danger of these small nicks providing a starting point for fracture.

The value of the longer cutting edge and of the thinner chip is shown by F. W. Taylor in his treatise, "On the Art of Cutting Metals," in which he states that if the chip thickness is reduced to one-half, the cutting speed may be increased to 1½ times its former value. Taylor's analysis of the proper form of cutting edge may in some respects be correlated to twist drill practice. He states that the straight edged tool as shown in Fig. 54-D is the most efficient form and would be widely used but for the tendency to chatter. The tool shown in Fig. 54-E is a modification of the flat-edge tool, which is given a slight curvature which produces a chip of variable thickness and thus reduces the chatter. This idea of the curved edge he carried still further by producing the standard round nosed tool so widely used in all roughing work. The edge curvature reduces the efficiency, however, and the flat tool shows an endurance 1.3 times greater than the round tool. It will be noted that the edges of all tools terminate in rounded corners. Even the parting or cutting-off tools have the corners slightly rounded. The great disadvantage of the old style diamond point tool lay in its sharp corner. Examination of Fig. 54-D will disclose the analogy to drill form: here the tool is advancing in the direction shown by the arrow, and although the tool is stationary, the work turns in such manner as to give the effect of a left-handed drill. The cutting edge corresponds to the lip of the drill, and the angle "x" corresponds to the angle "W" of Fig. 54-C. The rounded corner according to Taylor produces a chip which thins to zero, makes the tool less liable to destruction, and prevents injury to the work even if the tool is broken down in service. According to these lathe tool principles, it will be seen that the slight curvature of the drill lip noted in Fig. 54-A is beneficial in reducing the possibility of chatter when light feed and high speed are used. Theoretically, then, the ideal form of drill lip will be variable for different materials to be drilled; it will be ground to a more acute point angle for the soft materials and to a more obtuse one for hard materials. The matter of edge contour is not of great importance since all points of this edge lie in the same advancing plane. The corners at the drill periphery should be rounded, as shown exaggerated in Fig. 54-E, but the rounding of the whole drill end would not be desirable.

28. *Composition of High-Speed Steels.*—With few exceptions the primary metal of all high-speed steels is iron, and since the ulti-

mate cost of the steel is high by reason of the alloys used, the best grades of iron, low in sulphur and phosphorus, are used. Carbon was for a long time the principal alloy used, and the product was known as steel, soft steel, hard steel, carbon steel, or tool steel according to the variation in carbon content. Although steel has been in use for many centuries, it was not until 1860 that alloys other than carbon were used intelligently in its production. Previous to that time the famous Damascus steels are known to have contained tungsten, nickel, and possibly manganese, although these were present in a natural alloyed ore and were not added by the makers. In 1860 Robert Mushet discovered that the addition of tungsten and manganese in fairly large percentages would produce a tool steel hard enough for use without the usual quenching processes but needed only to be cooled in the air. Steels of this composition are known as "self hardening" or "air hardening" steels and were the immediate progenitors of the modern high-speed steels. Since 1880, much experimenting has been done in ascertaining the effect on steel of adding such substances as tungsten, chromium, manganese, silicon, vanadium, cobalt, and even copper, and aluminium. The possibility of so many alloys in so many percentages and the peculiarities produced by variations in the forging and heat treatment open a wide field for investigation.*

It is true that there is still much to be done in investigating the effect of alloys on steel, but enough has been done to warrant certain definite statements. Taylor, in 1907, laid down certain rules for composition and treatment for lathe tools which still are generally followed. He recommended a low manganese content, a carbon content of not more than 0.86 per cent, about six per cent chromium, nineteen per cent tungsten, low silicon and less than one per cent vanadium. His rules for heat treatment were:

(a) Heat slowly to 1500 degrees F.

(b) Heat rapidly from that temperature to just below the melting point.

(c) Cool rapidly to below 1550 degrees.

(d) Cool rapidly or slowly to room temperature.

(e) Reheat to 1150 degrees for about five minutes.

(f) Cool to room temperature rapidly or slowly.

* See Bibliography, page 138, for references to such investigations.

He further stated that this treatment would apply to any high-speed steel and that the method of obtaining these temperatures was of small importance, provided only that they could be determined with exactness and that the tool was not allowed to oxidize. A fresh coal fire was recommended for the initial heating and a lead bath for the reheating. More recent writers suggest gas muffle furnaces for the initial heating and salt baths for tempering.

That Taylor's formula may not hold for twist drills is suggested by some of the later investigators. In drills the conditions are such as to call for slightly different qualities in the metal, notably a greater toughness and strength. This logically leads to a lower percentage of carbon. Other considerations lead to a somewhat lower chromium content, and recent experiments suggest the use of a cobalt alloy and an increase in the percentage of vanadium. C. A. H. Lantsberry* suggests as the result of a study of a series of drill tests a steel containing not more than fourteen per cent tungsten, about four per cent chromium, one per cent vanadium, and carbon less than 0.60 per cent. Analyses of a number of stock American drills as determined by Norris† of the Westinghouse Electric and Manufacturing Company, in 1911, are given in the accompanying table.

ANALYSES OF STOCK DRILLS OF AMERICAN MAKES—NORRIS

DRILL	*C*	*Si*	*S*	*P*	*Mn*	*W*	*Cr*
A	0.48	0.20	0.014	0.010	0.40	15.86	3.46
B	.41	trace	.017	trace	.57	17.29	3.24
C	.41	trace	.012	.015	.50	17.84	3.12
D	.25	.19	.014	trace	.50	16.18	3.05
E	.38	.05	.010	.012	.28	19.98	4.95
F	.43	.32	.012	.018	.27	15.80	2.28
G	.49	.19	.018	.012	.28	19.45	2.89
H	.44	.14	.016	trace	.32	15.54	3.14
I	.24	trace	.023	.012	.27	19.43	4.07
J	.48	.14	.017	.037	.21	13.88	2.96
K	.46	trace	.010	.025	.50	14.98	2.65
L	.32	trace	.018	.045	.35	18.00	3.05
M	.45	.09	.015	.017	.55	17.80	2.89
N	.54	trace	.014	.015	.28	19.43	3.10
O	.44	.09	.015	.012	.52	17.76	2.90
P	.57	.48	.017	.020	.60	13.50	4.08
Q	.43	.19	.018	.038	.28	19.51	3.00
R	.49	.10	.019	.015	.48	18.79	4.70
S	.55	trace	.012	.017	.48	18.95	3.01
T	.41	.19	.011	.037	.34	10.31	3.20
U	.45	.19	.015	.017	.55	17.84	2.88
W	.49	.09	.010	.032	.24	18.39	2.35

*American Machinist, Vol. 34, p. 719, 1911.
† Journal West of Scotland Iron and Steel Institute, Jan. Feb., 1915.

These analyses show both the wide variation of values for the different variables and the tendency toward a low carbon and chromium content. The generally larger manganese content is noticeable, giving, according to Lantsberry, a greater hardness and a resistance to fire-cracking or cracks in hardening.

Lantsberry gives analyses of three of the most satisfactory of the English tool steels as follows:

C	*Si*	*Mn*	*W*	*Cr*	*V*	*Mo*
0.60	0.30	0.04	14.62	3.58	1.04	0.54
.064	0.00	0.00	14.76	4.27	1.00	1.22
.063	0.00	0.16	17.77	2.51	0.95	0.00

The chemical composition of the steel is, however, not the only factor which affects its performance, even when the heat treatments are identical. It has been established that the forging processes may be regulated to give a harder and tougher product. For this reason forged flute drills and many flat twist drills give better results than milled drills; thus of the drills analyzed by Norris the low tungsten drill *T* gave greater endurance than drill *E* which is high in tungsten, both being milled drills, but drill *I*, also a high tungsten drill but of forged flute design, was practically as good as drill *T*.

APPENDIX II

Drill Grinding

29. *Drill Grinding Machines.*—In 1885 a paper was presented before the American Society of Mechanical Engineers by W. H. Thorne in which were discussed the tendencies and practices of that time with regard to twist drills. Special attention was paid to the proper form of point grinding, incidental mention being made of the high prices of drill grinding machines. In the course of the discussion one engineer remarked that the man who would market a drill grinder for fifty dollars would be a public benefactor. Twenty-four years later, in the discussion of the experiments by Dempster Smith, the following statement is found:

"With regard to the question of grinding twist drills, drill grinding machines were, with the exception of common hack-saws, perhaps among the cheapest machines to be found in engineering shops. Was there any good reason why this should be the case? Why should twist drill grinding machines which had not even enough weight of metal in them to be rigid, and some of which were sold at ridiculously low prices, be used at all? The twist drill itself might probably be used in a machine costing from ten to twenty times as much as the grinding machine, and yet the efficiency of the drill and the drill-press was limited by the inefficient grinding machine employed."

It will be seen, therefore, that some great changes in drill grinding machines have been made in the last quarter-century. It may be stated, however, that it has normally been possible to procure a remarkably efficient grinder, capable of grinding drills up to 1¼ inches, for less than fifty dollars. Between fifty and one hundred dollars there is a wide choice of the highest grade machines capable of grinding the largest and smallest drills. Even the small machine shop can, therefore, profitably have as part of its equipment an efficient drill grinder.

The average grinding machine is extremely simple in its operation. Several types are automatic in their adjustment for the different diameters of drills, and those employing a calipering method require but a moment for adjustment. Usually an adjustment for the clearance angle is unnecessary, the point angle is fixed at the 59-degree standard, and the whole design of the machine is such as to give the most efficient form of point. With hand grinding of drills larger than three-eighths of an inch it is almost impossible to get accurate results. The length of each cutting edge must be the same, the lip angles must be identical, and the clearance should increase from the periphery to the drill center. A drill without these requirements will drill a larger hole, will use more power, will introduce strain on the drill-press, and is likely to split or break. A perfectly ground drill will perform forty per cent more work between grindings than a hand ground drill, and the amount of metal removed at each grinding will be less. Machine grinding eliminates the necessity for the removing of metal to correct initial inaccuracies. Perhaps no tool is so difficult to grind by hand or so easy to grind accurately by machine.

The drill grinder used in these tests is illustrated by the sketches shown in Fig. 55. The principles of operation briefly are as follows:

"The drill to be ground is laid in the V-rest, *E* (Fig. 55-A and B), the drill backing against the stop in the tail block. Near the front end of the 'V' is a slot by which the drill may be held while grinding or lifted out of the trough as required. At the extreme front end of the V-piece is the lip rest *D* (Fig. 55-C) against which the lip of the drill is rolled, thus bringing it and holding it in correct position.

"In Fig. 55-B is seen the fixed stud *A*, projecting from the front of the frame and parallel with the wheel shaft. On this stud is the split sleeve *C* with a screw to clamp it, and an arm extending backward with a bearing pointing obliquely forward and upward in which swings the bracket which carries the tool-holder. The grinding of the drill is done by the swinging of the drill-holder in this inclined bearing. The apex of the V-shaped groove in the drill-holder is in line with the inclined axis of the journal, so that when the drill-holder swings from side to side, it turns on this point as a pivot. The face of the grinding wheel being at an oblique angle to the axis of the journal supporting the holder, any parts supported in the drill-holder will be ground to correspond to the portion of the surface of a cone; consequently a small drill laid in the 'V' will come nearer the apex of the cone, and will be ground to conform to that portion of the cone, while a larger drill will not set so deep in the 'V' and will be ground to correspond to a cone with a larger base, thus the size of the drill laid in the holder automatically determines and secures the correct curvature."

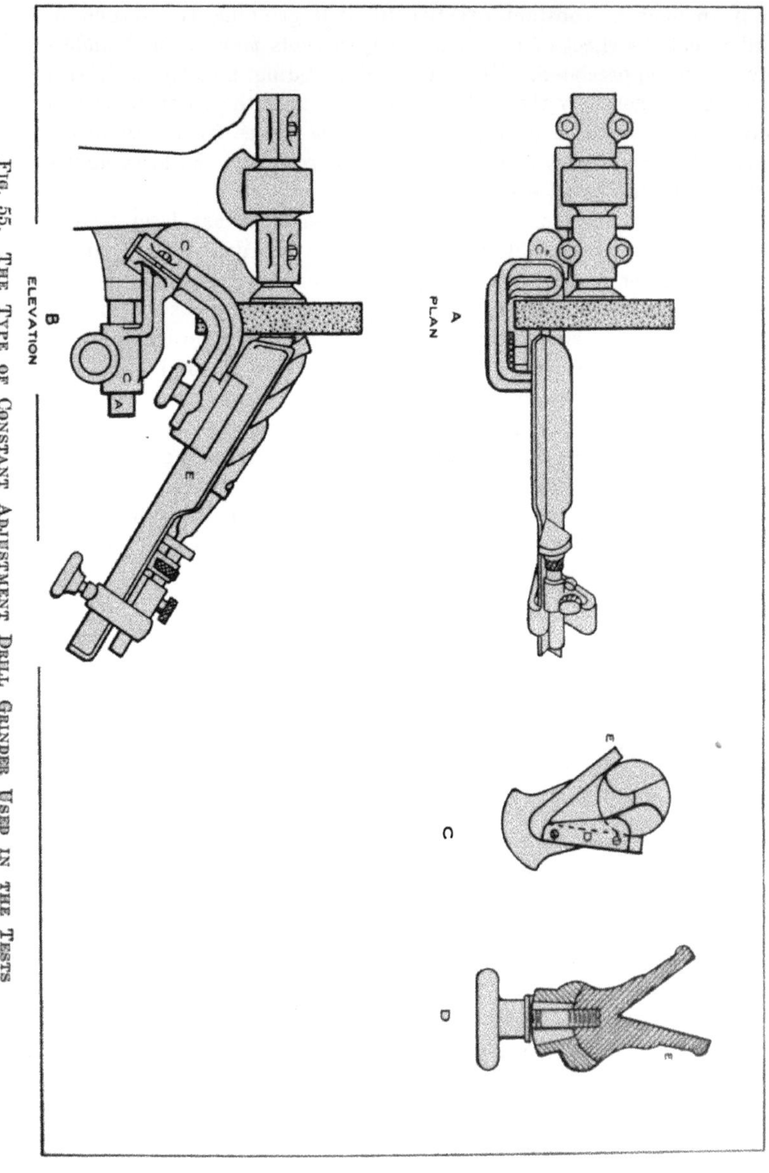

Fig. 55. The Type of Constant Adjustment Drill Grinder Used in the Tests

It may be remarked that although the grinder is accurate and easy to operate, constant practice in drill grinding is required in order that the effect of the various adjustments may be fully understood and appreciated. The range of grinding possible with this machine is shown by the fact that in these tests the point angle was varied from 44 to 74 degrees, the clearance angles from 2 to 15 degrees, the edge angles from 110 to 145 degrees, and the helix angles from 10 degrees to 45 degrees.

Carbon drills should be wet-ground, since it has been proved that it is almost impossible to dry-grind a carbon drill without drawing the temper on the cutting edge. While this change may be imperceptible to the naked eye, a strong glass will bring out a thin blue edge. Tests have shown this condition to diminish the efficiency of a drill at least five per cent, with consequent loss of production and increase in power consumption.

High-speed drills are usually dry-ground, although there is some discussion concerning the best practice. If the drills are ground dry, there is a possibility of heating the cutting edge to such a point as to injure the steel, especially if the grinding wheel has become glazed by use. Taylor in his work, "On the Art of Cutting Metals," says of the grinding of high-speed lathe tools:

"The writer trusts that he has made the fact clear that the property of the red hardness of tools is seriously impaired by even temporarily raising their temperature beyond 1240 degrees Fahr. He ventures to say that fully half of the high-speed tools now in use in the average machine shops have been more or less injured, and are therefore lacking in uniformity, owing to their having been overheated in the operation of grinding. Even when a heavy stream of water is thrown upon the nose of the tool throughout the operation of grinding, tools can be readily overheated by forcing the grinding or by allowing the tool to fit closely against the flat surface of the grindstone. The injury is the more serious because there is no way of detecting it except through finding by actual use that the tool has become of inferior quality. The writer has frequently seen tools which were ground under a heavy stream of water heated so that the metal close to their cutting edges showed a visible red heat."

This discussion shows that even the red-hard steels may be injured in the same manner as carbon steels and that care is required in the process of grinding. Taylor recommends further that a heavy stream of water at low velocity be played on the tool while grinding to afford ample cooling without splash. The principles of lathe tool grinding may be also applied to drill grinding, although the drill has an advantage in that it is constantly moved over the face of the wheel and has only a line contact therewith. Of course high-speed drills should never be ground under a small stream of water or cooled by being plunged in water after being heated in grinding or drilling. Such a procedure will inevitably start small surface cracks which will reduce the efficiency of the drill and may lead to deeper cracks which will destroy the drill.

One thing which is often overlooked in the grinding of high-speed steels is the degree of sharpness of the edge. An expert machinist will recommend that after the drill has been run a few minutes the edge should be touched up with a stone to give it an even polish. The drill will then cut well, and the process of edge honing seems to increase the endurance. Whether this advantage is real or imaginary is uncertain, although experiments show that the first sharpness of the tool is transitory and the real serviceable cutting edge shows slightly rounded under the glass. The tool seems to settle to a cutting edge after drilling for some minutes, and the edge honing probably helps by removing irregularities rather than by disturbing the rounded edge. This rounded edge is well known in lathe tools and constitutes one difficulty in the determination of relative dullness.

30. *Point Angles.*—In a communication Prof. W. W. Bird of the Worcester Polytechnic Institute states: "As far as we know, the original drill grinder was made to duplicate results secured by good hand work. The point angle of 59 degrees was simply arbitrary; 60 degrees would do just as well. Sentiment seems to call for a straight edge or lip, and sentiment is a factor in selling drills. Standard cutters having been designed for the 59-degree angle and a straight lip, a change of angle would mean either new cutters or a curved lip. . . . As it is, our experiments have shown that a few degrees difference in the point angle makes but very little difference in results." This statement of Professor Bird's concisely expresses the present status of the point angle, which is practically the same for the modern high-speed steels as it was for the old carbon steel drills thirty or forty years ago. As indicated under the discussion of cutting edge form, there are certain theoretical advantages in using a sharper point angle and it is probable that a change of a few degrees in grinding would result in a more durable edge without

materially affecting the contour of the edge or necessitating new flute millers.

Some of the possible variations from standard form are shown in Fig. 56. Sketch A shows how the angle of the cutting edge may be made more obtuse for drilling in a hard material. The form shown is rather extreme, and good results may often be obtained by merely going over the edges with a hand stone. This form is sometimes found efficient in drilling soft material, such as brass, where the regular point has a tendency to "hog in" or "grab." Sketch B shows the standard method of point thinning, which is an operation requiring care and experience, since the tendency is to spoil the cutting edges by giving them the obtuse angle shown in Sketch A. This thinning if improperly done will cause the center to be eccentric and make the drill cut too large, or may so weaken the web as to cause it to split in service. Sketch C shows how a drill was purposely blunted for the purpose of investigation, the cutting action of the chisel being destroyed and reduced to merely a crushing action. Sketch D shows a crude attempt at point thinning on a flat-twist drill, one of those used in tests at the University of Illinois. No attempt was made to preserve the cutting edge in good form, but the semicircular groove was ground down each side and the drill used in this shape. Rather unexpectedly the drill gave very good results and in a number of test performed better than a high grade standard milled drill.

31. *Clearance Angle.*—The question of clearance angle at the drill periphery has been touched upon in the discussion of the helix angle,* in which it was shown that this angle influences the actual cutting angle, and attention was called to the fact that the manufacturers specify an angle of clearance far in excess of the actual need in order to compensate for poor drill grinding.

That the clearance angle of the drill should increase from the periphery toward the center is a fact generally well-understood. An attempt has been made in Fig. 57 to show the reasons for this. At the scale used the sketch represents a feed of 0.25 inches per revolution,—about ten or twenty times the feed used in drilling. The increase toward the center is, however, proportional, the difference between angles Z and Z' shows the necessity for this form of grind-

* See page 29.

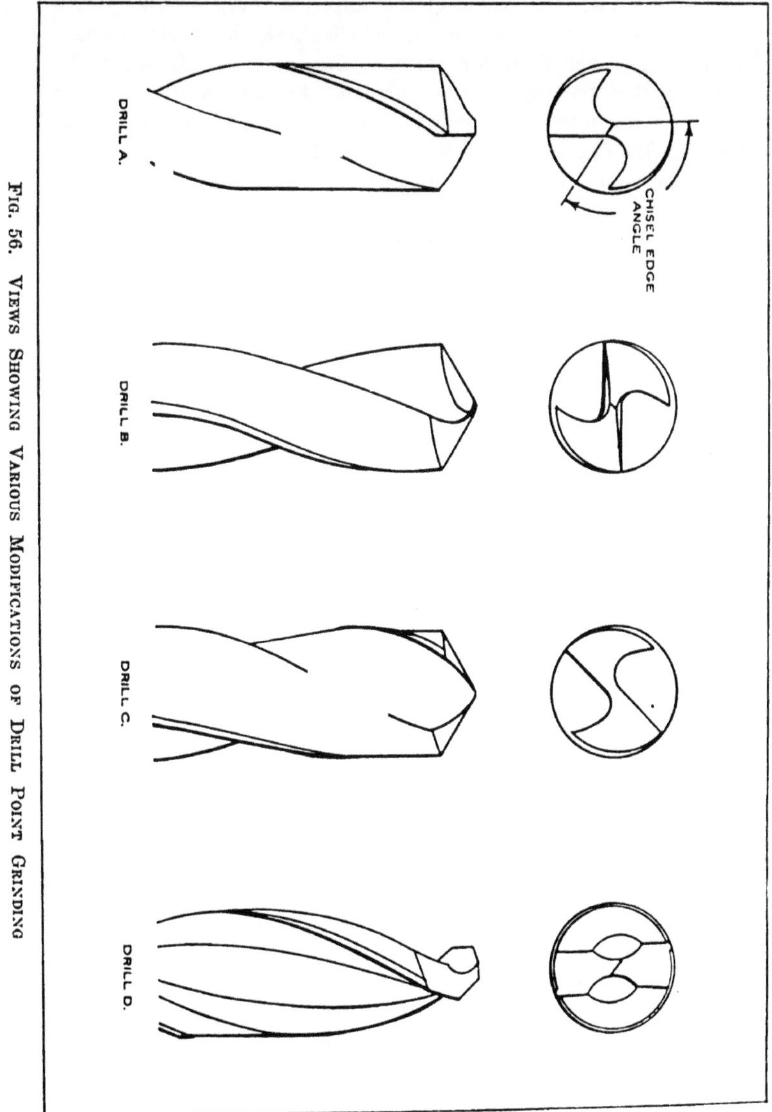

Fig. 56. Views Showing Various Modifications of Drill Point Grinding

ing. Most drill grinders are designed to give this increased clearance at the center. The large value of this angle for even ordinary work may be seen by reference to Fig. 58, which shows the clearances necessary on a one-inch drill for various rates of feed. It should be remembered that the angles shown are only the angles necessary for the drill to screw itself down into the work and are usually increased by four or five degrees to allow for a working clearance.

It is the value of the clearance angle at the center which largely determines the value of the "chisel edge angle" or "edge angle." As the clearance at the center increases, this edge angle also increases. The manufacturers recommend an angle of 135 degrees. Just what the proper value for the peripheral clearance may be is difficult

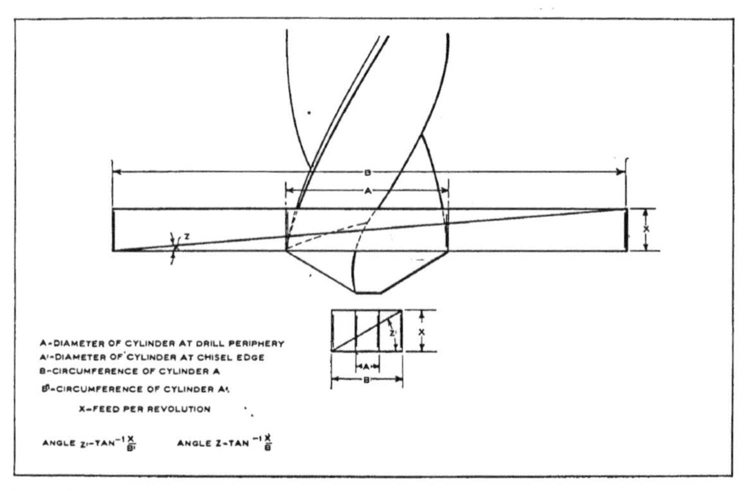

FIG. 57. DEMONSTRATING THE NEED FOR INCREASED CLEARANCE AT THE DRILL CENTER

to decide, for while the manufacturers recommend a twelve-degree angle they clearly state that it is excessive for a well ground drill, and drills come from their hands with clearance angles of about four and five degrees. In a recent publication* the superintendent of Baker Brothers, Toledo, Ohio, gives it as his experience that drills perform better if they are given a clearance of two degrees more than that given by the drill makers, or approximately six to seven

* Hallenbeck, G. E., Iron Trade Review.

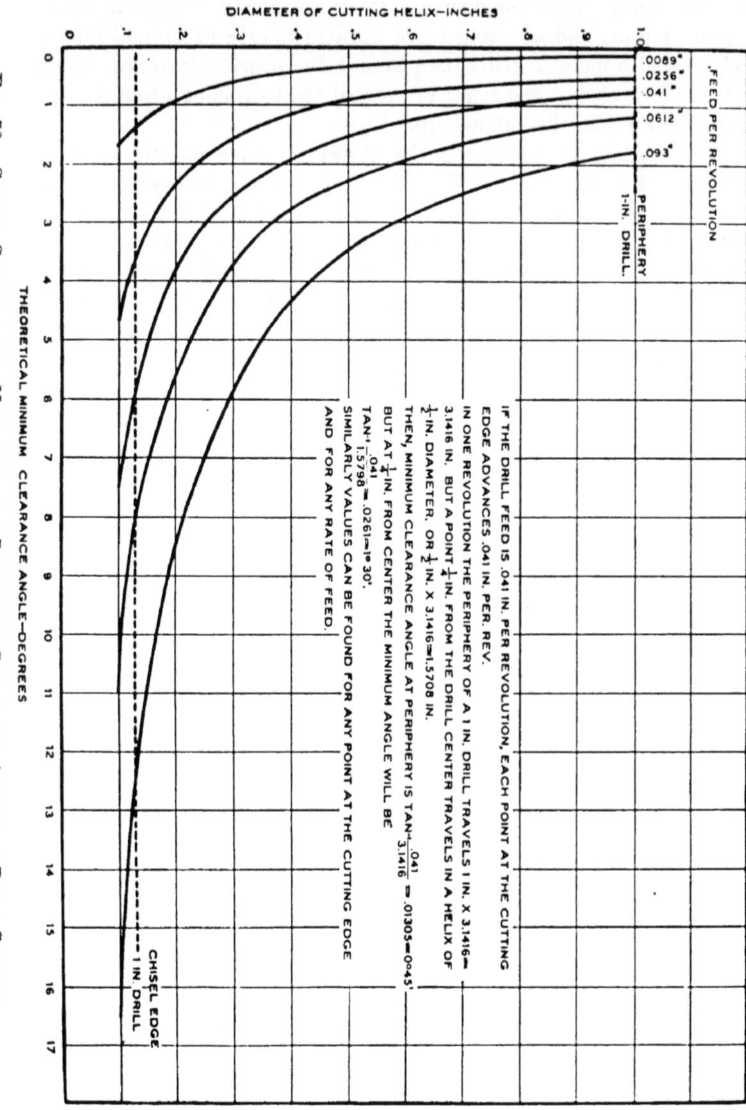

Fig. 58. Curves Showing the Necessity for Increased Clearance Angles at Drill Centers

degrees. This is about in accord with the findings of F. W. Taylor for lathe tools, who recommends a clearance of six degrees as using no more power, and as being more durable than a greater clearance angle and also more durable than even a five degree clearance, which he found just enough too small to prevent the tool being ground away on the flank below the cutting lip.

The method of measuring the clearance angle exactly is illustrated by the photograph, Fig. 59. This method is described and illustrated by Corneil Ridderhoff.* He states that "the outer ends of the paper are slid by each other until its top edge is parallel with the line of clearance immediately back of the cutting edge. A mark is then made at *A*, the paper removed and, being laid flat, a line is drawn from mark *A* to corner *B* at the other end. The angle intercepted between this line and the edge of the sheet is the angle of clearance on the drill." It may be further stated that if the sheet of paper is 8½ inches long, the mark *A* should be one inch from the edge to indicate an angle of seven degrees, while a mark 1¾ inches from the edge will indicate an angle of twelve degrees.

---

* American Machinist, Jan. 1, 1903, p. 5.

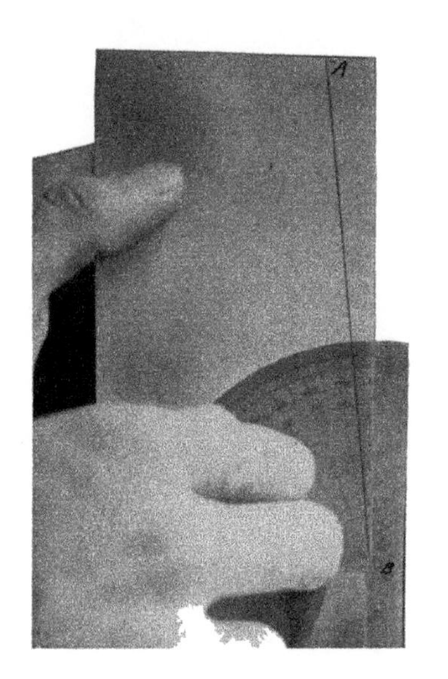

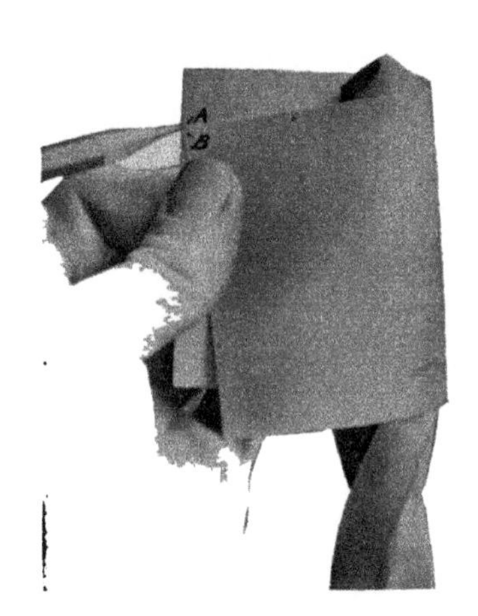

Fig. 59. Method of Measuring the Clearance Angle

APPENDIX III

Lubrication of Drills

32. *Application of Lubricant.*—As early as 1892 there is mention of a twist drill having oil-holes in the metal, these being formed by first drilling small holes longitudinally through the drill blank and then twisting and milling it. Previous to this date there had been attempts to use a lubricant at the drill point by carrying it through small brass tubes soldered or wired to the drill flutes or through channels milled in the outer surface of the flutes. Most of these lubricating drills were used in a turret lathe or similar machine in which the work revolved and the drill remained stationary, since this permitted the easy introduction of the lubricating fluid. The modern drill, however, with its oil holes can be used in an ordinary drill-press very satisfactorily with a special socket now made by drill manufacturers. The feature of this socket is a floating sleeve, having ground joints, which remains stationary while the drill revolves and conveys the fluid to the oil-hole openings in the drill shank. The use of such a drill necessitates a constant circulation pump system for the lubricant with means for catching the drip, and straining the oil.

Where oil-hole drills and their adapting sockets are not used, the problem of lubrication is a more difficult one. If a stream of fluid is played on the drill where it enters the work there is an excessive use of fluid, which tends to cause splashing and makes the work more difficult and messy. It is also probably true that the fluid is of very little value after the drill has penetrated to a depth of several inches. For slow speed drilling and for shallow holes this method may prove satisfactory. A recent scheme for lubrication has been suggested in which jigs are used. The jig is fitted with oil leads to carry the fluid to the spot where the drill enters the work, and this exact directing of the stream tends to give better results than the haphazard "flooding" so often seen.

Where the drill consists of the duplicating of the same process for a great number of pieces, as in some of the large automobile factories, lubrication may be reduced to a system. Special lubricating jigs, baths of fluid, or special positioning of the work may be employed. Some manufacturers have found that very good results may be obtained by inverting the work—the drill running up into the work instead of down, as is more usual. This permits the lubricating fluid to drain out of the holes quickly and completely, carrying with it the chips and giving a clean hole and a long-running drill. This method has the further advantage that the work does not rest in a pool of the fluid, but is drained clean as soon as the drill stops and allows quick removal with no slop or waste. Some manufacturers place the work horizontally, as in the standard boring mill, and find that this has the advantage of easy positioning of the work, quick draining, and complete chip removal.

It is the usual practice to use a liquid as the cooling and lubricating medium, but sometimes, and particularly when the material to be worked is cast iron, compressed air is used. This can be applied only through the use of drills with longitudinal holes and special sockets, for drill-press work, and is not a very general custom. Air serves to cool the drill, though probably very slightly, and also blows out the small particles of iron. Where an insufficient stream of water or other liquid only causes these small particles of iron to form a swarf which grinds away the drill, the air removes them in a dry state, clearing the hole and giving a free cut. If a heavy stream of water is used, it is probably superior to the air and has the further advantage of requiring no elaborate device for catching the dust and chips. There seems to be a tradition that a liquid lubricant should not be used on cast iron, but the probability is that a sufficient amount of liquid, supplied under proper pressure, will be found superior to air. More recently there has been some use of compressed air, as cooling fluid, in the drilling of deep holes in steel. In this event the drill lips must be gashed so that they will break up the chips, a fairly heavy feed and slow speed should be used, and the air should be supplied through an oil-hole drill at a considerable pressure (seventy or eighty pounds per square inch). It is reported that in machine steel this process gives a smooth hole and the drill keeps quite cool, the chips being removed by the air blast as soon as formed.

33. *Action of Lubricant.*—Among the users of drills, or in fact of any of the machine tools, there seems to be a considerable difference of opinion whether the use of a lubricant decreases the power necessary to drive the tool. That it increases the endurance of the tool

is undeniable, but that it also decreases the power is not so generally believed. In his lathe tool experiments F. W. Taylor* found that the use of a heavy stream of soda-water on the chip at the tool point permitted an increase in cutting speed in steel of about forty per cent. When cutting hard cast iron, the increase in speed through the use of water was sixteen per cent. Apparently the only action here was one of cooling the tool and chip, since the soda-water fluid has small lubricating effect. Unfortunately these experiments of Taylor's are practically the only ones on any type of tool which show the effect on endurance of the use of a cooling fluid. The use of compressed air in some drilling work shows that the combined effect of cooling and efficient chip removal is beneficial under certain conditions.

In drill tests run by Smith and Poliakoff,† no attempt was made to ascertain the effect of lubrication on drill endurance, but various tests were run to find the effect of lubrication on power consumption when drilling in steel. The results show that when using a copious supply of a solution of water and turning oil the power consumed was from seventy-two to ninety-two per cent of that when drilling dry, the greater saving being shown for the lower feeds. The end thrust on the drill was from twenty-five to thirty-five per cent less for soft and medium steels and twelve per cent less for hard steel. The lessened end thrust has no effect on the power consumption of the drill itself, but lessens the power input to the drill-press. The saving in power may be attributed to two or possibly three causes, namely, the lubrication of the cutting edge, the lubrication of the chips to reduce their friction on the drill and to facilitate their removal, and the washing out of the chips by the stream of fluid.

34. *Lubricants.*—There is almost as little uniformity in the lubricants used in commercial work as there is in the methods of application. A common lubricant is, of course, lard oil, but the expense of this has led to numerous substitutes. In the catalogues of drill makers the following lubricants are recommended:

For Tool Steel: Lard oil, machine oil, turpentine, soda-water, kerosene.

For Soft Steel: Lard oil, machine oil, soda-water.

---

*Taylor, F. W., "On the Art of Cutting Metals," Folder 15, N. Y., 1906.

† American Machinist, Vol. 32, pp. 739, 830.

For Cast Iron: Dry, or compressed air.

For Brass: Dry or paraffin-oil.

For Aluminum: Kerosene, soda-water, and aqualine and soda-water.

The same lubricants are generally recommended for wrought iron as for soft steel. That the lubricants recommended are not the only desirable ones is shown by an investigation of general practice. Here are suggested various compounds, such as:

For Steel: Boil together lard oil, 1/3, and soda-water, 2/3.

For Steel: Heat together borax, seven pounds, and water, forty gallons. When cool, add seven gallons of lard oil.

For Steel: A thin drilling compound — almost water — enough compound to make smooth and to prevent rust.

For Aluminum: Half kerosene and half lard oil.

In many places the most used fluid is a solution of about one gallon of soluble cutting oil to fifteen gallons of water. It is probable that this solution would prove satisfactory for most work, since a drilled hole does not usually require the perfect surface that can be given only by a solution rich in lard oil. Almost any machinist knows the value of turpentine as a lubricant in drilling very hard steel, although just why it has this value is not well known. It is suggested that it exercises a local softening effect or that the fluid may possibly penetrate the pores of the metal and afford a more intimate contact of lubricant with drill edges and chip.

APPENDIX IV

Drill Testing

35. *Methods of Testing.*—Drills and drilling processes may be tested to determine the following characteristics:

(1) Endurance of drills
(2) Power
  Input at the motor
  Used at the drill point
  Absorbed by the driving mechanism
(3) Composition of steel used in drills
(4) Forms of drills

The tests to determine the endurance of drills are simple and easy to conduct. They consist merely of driving the drill through steel or iron at desired speeds and rates of feed until failure, and noting the effects of the process upon the drill. By using a uniform material for the test drilling and by arranging a definite series of increasing rates of feed, it is possible to compare a number of different drills. A difficulty to be met in this sort of test lies in the fact that the influence of the different characteristics cannot be determined; thus the whole drill may be quite strong, yet an increasing bluntness, possibly made more pronounced through improper tempering, may cause the drill to break sooner than a weaker drill which holds its edge better, or one of the drills may run into a bit of segregated manganese or carbon in the test block and fail, but without giving any indication whether the cause lay in the drill or in the block. This test requires a powerful drill-press and considerable power. Modifying the test by running the drills merely to bluntness has its objections in the difficulty of estimating exactly the degree of bluntness from the appearance of the drill or of bringing two drills to the same appearance at different times.

The second method of testing involves the use of meters for recording or indicating the electrical input to the motor driving the drill-press. This method also presents difficulties in connection with the determination of the definite effects of the different influencing factors. It has been customary in such tests to run the motor and drill-press idle at different spindle speeds to determine the power absorbed by the driving mechanism and then tests are made for the total power consumption. Subtracting the first set of readings from the second set is supposed to give the power consumed by the drill alone. The fact is, however, that the result of this subtraction also includes extra power absorbed by the bearings of the motor and press due to the increase in torque and thrust, so that the drill is represented as consuming more power than it really does. This method of testing, therefore, is most useful when confined to the total power consumption, the results then being accurate, but the results obtained from such tests are specific rather than general, for different drill-presses will consume different amounts of power in doing the same work at the drill point, because of different arrangements of the drive and feed gears, of the number and type of bearings, and of the variations in feed and speed of the drill.

Tests to determine the power delivered to the drill point eliminate discrepencies due to mechanical differences in drill-presses and yield results which are of definite practical value and which are to a large extent comparable with results of other tests conducted in a similar manner. Power variations in the drill due to differences in drill form, sharpness, angles of grinding, speed, feed and other factors may, in these tests, be noted by means of dynamometers mounted to record the power delivered to the point. Exact indications of drill endurance may be obtained, and if failure is due to faults in the test blocks this fact may be detected from the test records obtained. The fact is to be recognized, however, that tests which involve the determination only of the power at the drill point do not yield results on the basis of which the efficiency or performance of the drill-press may be studied. There is a tendency also to lay down exact laws and formulas for drill performance based on such tests, since indications obtained may be exact, and to overlook the fact that there may not be uniformity in the metals drilled. Particularly is this true if cast iron is used, for what may be termed soft cast iron in England may be medium cast iron in America. The differences in methods of different experimenters will also influence results, even in so slight a matter as the thickness of the test blocks, or the removal of the casting skin, or in the depth to which the hole is drilled.

Tests in which only point dynamometers were used have yielded results which give the best insight into the matter of drill design, and it is to be expected that the intelligent development of this method will give further useful results. When this method of testing is combined with the method in which the power input into the drill-press is noted and records are kept of the total power consumption, the consumption of the motor and press running idle at different speeds, and the power consumed in useful work at the drill point, the analysis of the entire process of drilling becomes more nearly complete.

While such a set-up would give a close analysis of the processes of drilling, it would not completely investigate the effect of increased torque and thrust on the power consumed by the drill-press, as these two quantities tend to maintain a fixed relation to each other and cannot be separated without difficulty. For an investigation solely of the drill-press there would be needed an apparatus whereby the torque and thrust could be varied independently of each other from zero to a maximum. For the torque variation it might be possible to fit up something resembling a Prony brake on the drill spindle, thus varying the torque, measuring its value exactly, and showing the power absorbed at the spindle, while electrical readings would give the total power input. To vary the thrust it would be desirable that the spindle advance as in actual drilling; so some system would be necessary which would allow for this feed while applying a steady thrust of known value. Possibly this could be done by allowing the spindle to compress a series of springs or by lifting a certain weight through a system of pulleys. Such an apparatus, correctly and carefully designed, would allow of complete drill-press investigation, for it would duplicate the forces of actual drilling and keep them uniform, separate, and completely under control.

36. *Tests to Determine the Composition of Drill Steel.*—Tests to determine the structure and composition of the steel of which drills are made occupy an important place in the commercial world. Every drill manufacturer maintains a laboratory in which analyses of the chemical composition of the steels are made and the structure of the material determined by microscopic examination of etched surfaces and breaks. The task of testing certain percentages of the finished drills, driving them at certain speeds and feeds through a uniform test material, and judging of their quality by the number of inches of material they penetrate before they become too dull for further

work devolves upon the laboratory of the manufacturer. Some manufacturers also use a form of point dynamometer to get more nearly accurate data on the drill performance. It has been found, however, that the chemical composition of the steel is not a positive indication as to its fitness for drill making. Two steels exactly alike in composition may vary widely in performance when made into drills. The microscopic examination is more reliable if made by an expert. Examination of the break in steel is very useful, and the sclerescope test gives useful indications, but the ultimate criterion seems to be the performance of the drill under actual working conditions.

37. *Tests to Determine the Influence of Drill Form on the Power Required and on Endurance.*—Tests to determine the influence of drill form on the power required and on endurance include a study of the effects of the helix angle, the point and clearance angles, the shape of the drill flutes, and the method of making the drill, whether milled, forged, or flat-twisted. The general investigation of the forged, milled, and twisted drills is mainly one of comparative strength; so the method of endurance testing is mostly used. Where the effect of drill form upon power input is desired the use of meters is employed. Slight variations in power may be observed, since the flat-twist drill usually puts a greater load on the drill-press than other types. For the investigation of the effects of the angles of grinding upon the power required and upon endurance some form of point dynamometer must be used. It will be seen, therefore, that each method of drill testing has some particular field, depending upon the feature of drilling that is to be dealt with.

APPENDIX V

Experiments on Twist Drills by Other Investigators

38. *Tests to Determine the Power Input Required.*—The most notable series of tests on drills in which the electrical power input was determined was that conducted by H. M. Norris of the Cincinnati Bickford Tool Company, Cincinnati, Ohio. In 1902, when this concern was known as the Bickford Drill and Tool Company, Norris first began his tests, and his most recent contribution was in August, 1914. Most of these tests have been reported in the American Machinist and are noted specifically in the Bibliography,* although the most recent test is the only one which need be discussed here, it being the most comprehensive of the series.

Drilling was done with drills of from one to three inches in diameter in steel and cast iron. Graphic records of the total power input were obtained, from which was subtracted the power consumption of the motor at different speeds. The resultant power was taken as representing that consumed by the drill and drill-press. The purpose seems to have been to establish a basis of comparing these gross values with the net values obtained by a previous experimenter, Dempster Smith, although there were some difficulties to be met because of the different trains of gears used in obtaining different speeds. The essential differences of results are slight, however, and the work of Norris may be taken as affording a very practical interpretation of Smith's results. The tests of Norris also possess the added value of showing the important losses due to the use of back gears and of giving a general idea of the gross power needed to drive a drill-press in an average machine shop. In one test Norris notes a power saving of about twenty-three per cent which he suggests may be due to the use of ball-thrust bearings behind the gears of one machine. His conclusion is that the gross power absorbed by a machine of this (radial) type driving various sized drills may be calculated fairly exactly from the formula:

$$H.\,P. = c\left[r - 348\left(\frac{1}{d \mp 0.6}\right) 3.08\right]$$

in which "c" is a constant varying with drill diameter and feed, and obtainable from Norris' formulas for different sized drills and

* See page 138.

for different materials, "r" is the number of revolutions per minute, and "d" is the drill diameter. This formula gives powers varying from 1.84 for a one-inch drill in cast iron at 0.013 in. per rev. feed and 229 r. p. m. to 19.90 for a three-inch drill in machinery steel at 0.018 in. per rev. feed and 127 r. p. m.

In 1909 some tests were run by Baker Brothers, of Toledo, Ohio,* the results of which showed the power consumption of motor, press, and drill for varying feeds and speeds in steel. Some observations of thrust are recorded, although no mention is made of the apparatus used in recording them, and some work was done in observing the most advantageous speeds of the drill for different feeds. The conclusions are that the thrust tends to increase in direct ratio with the feed per revolution, that the thrust first increases and then decreases with increasing speed; thus the conclusions conform with the observation that the permissible feed is higher if the speed is increased beyond 300 r. p. m. It is also shown that the total power increases in a decreasing ratio for feeds lower than 0.010 in. per rev.; whereas with heavier feeds the tendency seems to be for the power to increase in an increasing ratio for increasing speed. The conclusion is that best results are to be obtained by drilling at comparatively high speeds and moderate feeds.

In 1910 the American Locomotive Works† tested three radial drill-presses, using flat-twist drills from one inch to one and one-half inches in diameter and recording only the total power input. All drilling was done in hammered steel billets of about 0.70 carbon content, and the drills driven to destruction by burning or breaking without injuring the drill-press. One of these radial drills required an average of about 14 h. p. to remove one pound of metal per minute. Another required an average of about 17 h. p. and the third required about 19 h. p. to do the same work. The first press reached a maximum h. p. of 38.3 while driving a 1½-inch drill at 300 r. p. m. and 0.0207 in. per rev. feed. The second press reached a maximum of 57.6 h. p. while driving a 1⁵⁄₁₆-inch drill at 312 r. p. m. and 0.0323 in. per rev. feed. The third press reached a maximum of 67.3 h. p. while driving a 1½-inch drill at 284 r. p. m. and 0.40 in. per rev. feed. The motor was rated at 20 h. p.

The general conclusion is that best results in endurance and

* Iron Trade Review, Apr. 29, 1909, p. 797.

† Bocorselski, F. E., American Machinist, Vol. 33, Part I, p. 481.

power in steel, with high-speed drills of the flat twist type may be obtained by running a 1-inch drill at 300 r. p. m. and 0.015 in. per rev. feed and a 1½-inch drill at 225 r. p. m. and 0.02 in. per rev. feed. For lubricating the drills, a thin cutting compound was found superior to a good cutting oil containing about forty per cent lard oil.

In 1911 some tests were run by E. R. Norris of the Westinghouse Electric and Manufacturing Company. A large number of drills was given comparative tests, their chemical composition having been previously determined. They were given power tests in cast iron and steel and finally tested to destruction in steel, their power efficiency and endurance efficiency both being noted. The total electrical power was noted, and from this was subtracted the power required to run the motor and press idle at the different speeds. This resultant power was assumed to be the power consumed by the drill. In addition to determining the most efficient drill (for both power and endurance) the tests yielded results which may be summarized as follows: More power is needed to drive flat twist drills than milled drills. Milled drills fail by breaking and flat twist drills by burning. The drill endurance is affected by the method of manufacture, since forged drills stand up best. In steel the power increases more rapidly as the amount of metal removed per minute increases. In general, with a properly constructed machine it is better to run at high speed and small feed, thereby prolonging the life of the drill, as the total power consumption is nearly constant, regardless of whether heavy feed and low speed or light feed and high speed are used. In the first case the drill consumes more power and the machine less, while in the second case the drill consumes less power and the machine more, but in both cases the power consumed and the work done are about the same.

Other experiments have been made to determine the power required to drive certain drill-presses when doing work at various rates, and some of these data may be useful in the design of drill-presses or in determining the power installation required in shops. The chief value of this method of testing lies in the disclosures regarding drill endurance. They show that production is generally run "to a power limit rather than the tool limit," although very often the limitation is found in the design of the drill-press, as an individual motor drive permits carrying a great overload for the short time required to drill a hole.

39. *Tests Employing Point Dynamometers.*—The first attempt to determine the forces at the drill point by means of dynamometers was made by Prof. L. P. Breckenridge of Lehigh University,* in 1888. In this test only the thrust of the drill was determined. The dynamometer consisted of a vertical cylinder closed at the bottom and fitted with a ram having an area of ten square inches. On this ram the work to be drilled was mounted and the thrust, transmitted to a liquid in the cylinder, was recorded by a pressure gage on an ordinary steam engine indicator card. The results are interesting historically, and this type of apparatus has been elaborated by later experimenters who have studied the whole question of thrust variation.

The next experiments of any note in which a point dynamometer was used were those of M. Codron, published in "Bulletin de la Société D'Encouragement pour l'Industrie Nationale, 1903," and also in book form as "Experiences sur le Travail des Machines Outils pour les Metaux." These experiments constituted a complete and careful series, but since no full translation has been published in English, the work has, in this country, not been accorded the importance it deserves. The apparatus was in the form of a horizontal boring mill, with the drill held in a rotating spindle and the work held on a circular plate. The thrust was recorded on one scale, by the compression of a spring, while on another scale the depth of the hole was noted. The torque was transmitted by a rope couple around the circular base-plate and recorded on two dynamometers. A ball-bearing transmitted the thrust without affecting the torque. The entire process of drilling a hole was analysed and formulas were developed for the coefficients of resistance to thrust and cutting pressures. These coefficients, R for thrust and R' for torque, were ultimately expressed in actual values and general formulas stated for drilling different metals as follows:

For cast iron, $R = 121 + \frac{1}{4}t$ tons per square inch.
$R' = 103.5 + \frac{3}{8}t$ tons per square inch.
For tool steel, $R = 254 + \frac{4}{11}t$ tons per square inch.
$R' = 387 + \frac{5}{42}t$ tons per square inch.
For hard steel, $R = 171 + \frac{14}{25}t$ tons per square inch.
$R' = 286 + \frac{1}{4}t$ tons per square inch.

*Journal, Lehigh University Eng. Society, 1888.

In which t is the feed per revolution in inches. The summary of conclusions states that a drill worked well with an included point angle varying from 90 to 180 degrees, although the thrust increases with the angle and the torque decreases. The power variations with increases in speed and feed were found to follow a "straight line" law, and the torque and thrust were unaffected by variation in the speed of the drill.

In America, in 1904, Bird and Fairfield, of the Worcester Polytechnic Institute* conducted a series of tests in which the horizontal type of apparatus was used, the dynamometer being separate from the milling machine which was used to drive the drill. All experiments were made on cast iron with a 5/8-inch drill. Conclusions were reached to the effect that the torque does not increase as rapidly as the feed, while the thrust increases faster than the feed; and that increasing the point angle results in a rapid increase in the thrust, but does not appreciably affect the torque value. The torque values for drilling brass, tool steel, and machine steel are given, respectively, as 0.715, 1.67, and 2.44 times that required to drill cast iron. The corresponding thrust values are 0.575, 1.7, and 2.6.

In 1907 a series of experiments was made with this same apparatus, slightly modified as to the means of recording torque, by Frary and Adams of the Worcester Polytechnic Institute.† Tests were made with drills from 1/2 to 7/8 inches in diameter in cast iron, at feeds varying from 0.0045 in. per rev. to 0.0225 in. per rev. In the analysis of results it is stated that the thrust increases in practically a direct ratio with both feed per revolution and drill diameter. Torque increases almost proportionally with the feed. Thrust decreases with decreasing point angle from 150 degrees to 90 degrees and then increases with further angle decrease. The minimum torque was given by a point angle of 130 degrees.

Probably the most careful and comprehensive tests which have been conducted on twist drills were those by Smith and Poliakoff at the Manchester Municipal School of Technology, Manchester, England.‡ Drills from 3/4 to 3 inches in diameter were run in soft, medium, and hard cast iron and in soft, medium, and hard steel under varying conditions of feed, speed, point angle, lubrication, etc.

---

*Trans. A. S. M. E. Vol. 26, p. 355, 1905.
† American Machinist, Vol. 30, pp. 210, 598.
‡ Proceedings, Institution of Mechanical Engineers, Paris 1 and 2, 1909.

Most of the tests were run at ten revolutions per minute, and the greatest speed used was about 130 r. p. m. The apparatus used consisted of a reconstructed milling machine fitted with a 120 h. p. motor and a point dynamometer. A speed variation of 5 to 150 r. p. m. and a feed of 0.0025 in. per rev. to 0.05 in. per rev. were possible. The thrust was transmitted through a thin metallic diaphragm and distilled water reservoir to a carefully calibrated gage. Torque was measured by an arm bolted to the ball-bearing face plate to which the work was secured and by an accurate platform scale. Every part of the apparatus was carefully made and calibrated, all observations were as accurate as possible, and the whole series of tests was conducted with precision and care. No attempt was made to measure the gross power, and all values represented the net effect at the drill point. Equations to cover the different conditions of drilling were developed. The experiments were run in two series and the results published.* The conclusions as stated in the original paper are as follows:

PART I

(a) The net horse-power for a given diameter of drill and feed is proportional to the revolutions or the cutting speed.

(b) The net horse-power is proportional to the torque, and for a given drill and speed does not increase as fast as the feed.

(c) Since the torque is practically proportional to the diameter of the drill squared, the horse-power for a given feed and cutting speed is directly proportional to the diameter of the drill squared.

(d) The net horse-power per cubic inch of metal removed per minute is inversely proportional to the feed and independent of the size of drill and cutting speed.

(e) The work required to drill a given hole, when one drill only is used, is greater than that required to drill the same hole in two operations with drills of different diameters. The greater the difference in the drill diameters, the greater is the saving in work, speed and feed remaining the same throughout. This is due to the fact that the mean cutting angle of the single drill is greater than the average angle in use for the two drills, and that the cutting pressure is proportional to the angle.

(f) With twist drills having the usual proportions, the

* Proceedings, Institution of Mechanical Engineers, 1909, pp. 378-381.

cutting angle is not sufficiently keen to drag the drill into the work when enlarging a hole in cast iron or steel.

PART II

Conclusions:

(g) Conclusion (b) confirmed. The net horse-power when operating on soft cast iron or medium steel varies as $t^{0.7}$ for a given drill and speed.

(h) The net horse-power for a given feed and speed does not increase as fast as the diameter but varies as $d^{0.8}$.

(i) The torque and horse-power when drilling medium steel is about 2.1 times that required to drill soft cast iron with the same drill speed and feed.

(j) The net horse-power per cubic inch of metal removed is inversely proportional to $d^{0.2}\, t^{0.3}$ and independent of the revolutions; thus the net horse-power per cubic inch of metal $= \dfrac{4 \times 2\pi N c d^{1.8}\, t^{0.7}}{33000 \times \pi d^2\, t\, N} = \dfrac{k}{d^{0.2}\, t^{0.3}}$ where c and k are constants.

If the diameter of the drill remains constant and feeds of 0.0025, 0.01, and 0.04 inches be taken, the corresponding horse-power will be in the order of 1, 0.66, and 0.435, which is the ratio of the cutting pressures. If t remains constant and values of $d = \frac{1}{2}$, 2 and 4 inches be taken, then the horse-power for each successive drill will be in the order of 1, 0.76, and 0.66.

(k) The net power required to enlarge a hole may be estimated from the cutting pressures given by the equation $f = \dfrac{0.44}{t^{.33}} \times$ cutting angle, where f = tons per sq. in. for cast iron, and 2.1 times that for medium steel.

(l) In a two-lipped drill the actual depth of cut taken by each lip is $t/2$; in a three-lipped drill is $t/3$ and so on.

If the number of lips is increased and t kept the same, the cutting pressure is equivalent to that for a proportionately decreased feed.

If the lips in a two-lipped drill are unequally ground, so that one lip does all the work, the cutting pressure is the same as that obtained by doubling the feed.

By gashing the cutting lips of a two-lipped drill so that the cut taken by one lip is the metal left by the other, the cutting pressure is the same as that given for twice the feed.

It is shown that the finer the feed the greater the proportionate cutting pressure, and consequently the greater the net horse-power per cubic inch of metal removed.

(m) The end thrust when operating on cast iron or steel does not increase in proportion to the feed for a given diameter of drill or in proportion to the diameter for a given feed.

(n) Whilst the chisel point scarcely affects the torque, it is responsible for about twenty per cent of the end thrust. (It should be noted here that the width of chisel point used by Smith was only half the web thickness, the drills having been point-thinned for all these tests.)

(o) The lubricated trials on steel when compared with the dry tests show a diminution in the torque and net horse-power, varying from 28 per cent with the 0.0025-inch feed to 8 per cent with the 0.0286-inch feed. This may be due to the lubricant washing away the small metal chips, which tend to jam between the walls of the hole and the drill, and to the preserved cutting edge. The diminished frictional resistance of the shaving across the lip together with the lubricant reduces the end thrust by about 25 per cent for all feeds.*

(p) The drill most commonly adopted in practice has an included angle at the point of 120 degrees. If this angle is increased, the torque diminishes but the end thrust increases, whilst if this angle is decreased the reverse is the result. So far as economy in power is concerned, the torque is the factor to consider, as the feeding horse-power is only about one per cent of the whole in small drills and very much less for the large sizes. From this point of view the drill with the larger point angle is to be preferred. The accompanying increased end thrust, however, strains the machine parts in proportion. When the point of the drill breaks through the metal at the bottom of the hole, a considerable portion of the end load is removed. The strain due to the load is partly released, thereby causing the drill to advance more than its rated feed and possibly break the drill. The drill with the greater included angle will be most likely to give trouble

*The lubricant was a mixture of turning oil and water in the proportions of one to fifty.

in this direction, both on account of the increased strain and cut, and conversely.

(q) By decreasing the helix of the drill a keener cutting angle with a decreased end thrust and torque can be obtained without altering the point angle above the accepted standard of 120 degrees. This would, however, affect the durability of the drill.

(r) With a small included point angle there is little metal to support the cutting edge of the drill at the chisel point, and trouble due to the blunting of this part is to be expected.

(s) In estimating the time required to drill a hole of given depth, the length of the point must be taken into account. The length of the point for different point angles is:

$$90 \text{ degrees} = 0.5\,d$$
$$120 \text{ degrees} = 0.29\,d$$
$$150 \text{ degrees} = 0.134\,d$$

The discussion following the presentation of the paper before the Institution of Mechanical Engineers brought out much interesting information and many suggestions. J. T. Nicholson remarked that what was now needed was an investigation for gross horse-power and investigation concerning the most economical speeds and feeds when the durability of the drill, time of grinding, and gross power were all considered.

In addition to the point-dynamometer experiments noted there have been minor investigations carried on by different manufacturing concerns. A point dynamometer was used by Baker Brothers of Toledo in their investigations, and in 1901 a dynamometer was used by the Cleveland Twist Drill Company in some tests of their drills for the determination of correct flute form. The apparatus of the Worcester Polytechnic Institute has been often used in testing for commercial purposes.

BIBLIOGRAPHY

Tests Using Point Dynamometers

BRECKENRIDGE, L. P. — American Machinist, Feb. 14, 1889. Abstract
American Machinist, Nov. 23, 1899. Discussion
Journal Lehigh University Engineering Society, August, 1888. Original

BIRD AND FAIRFIELD. — American Machinist, Vol. 27, pp. 1629, 1677. Abstract
Transactions A. S. M. E., Vol. 26, p. 355, 1905

CODRON, MONS. C. — Bulletin de la Societé d'Encouragement pour l'Industrie Nationale, 1903

CLEVELAND TWIST DRILL CO. — American Machinist, Vol. 24, p. 594, 1901

FRARY AND ADAMS. — American Machinist, Vol. 30, pp. 210, 598

HALLENBECK, G. E. — Iron Trade Review, p. 797, Apr. 29, 1909

SMITH AND POLIAKOFF. — American Machinist, Vol. 32, pp. 739, 830. Abstract.
Proceedings Institution Mechanical Engineers, Parts 1 and 2, 1909

Tests Using Electrical Input

BOCORSELSKI, F. E. — American Machinist, Vol. 33, Part I, p. 481. For American Locomotive Works, 1910

HALLENBECK, G. E. — Iron Trade Review, Apr. 29, 1909. For Baker Bros., Toledo, Ohio

NORRIS, E. R. — American Machinist, Vol. 34, p. 719, 1911. For Westinghouse Electric & Manufacturing Co.

NORRIS, H. M. — American Machinist, Vol. 27, pp. 52, 74, 1902
American Machinist, Vol. 27, pp. 52, 74, 1904
American Machinist, Vol. 33, p. 910, 1910
American Machinist, Vol. 35, p. 116, 1912

NORRIS, H. M. (*Cont'd*). American Machinist, Vol. 41, p. 256, 1914

General Tests and Discussions

RAND DRILL CO. American Machinist, Vol. 28, Feb. 16, 1905

LINCOLN-WILLIAMS TWIST DRILL CO. American Machinist, Vol. 31, p. 709, 1908

SEARS, W. T. American Machinist, Vol. 35, p. 209, 1911

American Machinist, Vol. 37, p. 382, 1912

NORRIS AND SEWARD. American Machinist, Vol. 35, p. 116, 1911

VAUCLAIN AND WILLE. American Machinist, Vol. 37, p. 719, 1912

LANTSBERRY, C. A. H. Journal West of Scotland Iron and Steel Institute, Jan.-Feb., 1915

Alloys of High-Speed Steels

BÖHLER. Wolfram und Rapid Stahl, 1903

CARPENTER. Journal Iron and Steel Institute, 1905 and 1906

DENIS. Revue de Metallurgie, 1914

EDWARDS. Carnegie Memoirs of the Iron and Steel Institute, 1908

GLEDHILL. Journal Iron and Steel Institute, 1904

HERBERT. Journal Iron and Steel Institute, 1910 and 1912

LANTSBERRY. Journal West of Scotland Iron and Steel Institute, 1915

NICHOLSON. Engineering, 1903

TAYLOR. American Society Mechanical Engineers, 1907

. . . . Zeitschrift des Verein Deutschen, Ing. Sept., 1901

LIST OF
PUBLICATIONS OF THE ENGINEERING EXPERIMENT STATION

Bulletin No. 1. Tests of Reinforced Concrete Beams, by Arthur N. Talbot, 1904. *None available*

Circular No. 1. High-Speed Tool Steels, by L. P. Breckenridge. 1905. *None available.*

Bulletin No. 2. Tests of High-Speed Tool Steels on Cast Iron, by L. P. Breckenridge and Henry B. Dirks. 1905. *None available.*

Circular No. 2. Drainage of Earth Roads, by Ira O. Baker. 1906. *None available.*

Circular No. 3. Fuel Tests with Illinois Coal (Compiled from tests made by the Technological Branch of the U. S. G. S., at the St. Louis, Mo., Fuel Testing Plant, 1904–1907), by L. P. Breckenridge and Paul Diserens. 1908. *Thirty cents.*

Bulletin No. 3. The Engineering Experiment Station of the University of Illinois, by L. P. Breckenridge. 1906. *None available.*

Bulletin No. 4. Tests of Reinforced Concrete Beams, Series of 1905, by Arthur N. Talbot. 1906. *Forty-five cents.*

Bulletin No. 5. Resistance of Tubes to Collapse, by Albert P. Carman and M. L. Carr. 1906. *None available.*

Bulletin No. 6. Holding Power of Railroad Spikes, by Roy I. Webber, 1906. *None available.*

Bulletin No. 7. Fuel Tests with Illinois Coals, by L. P. Breckenridge, S. W. Parr, and Henry B. Dirks. 1906. *None available.*

Bulletin No. 8. Tests of Concrete: I, Shear; II, Bond, by Arthur N. Talbot. 1906. *None available.*

Bulletin No. 9. An Extension of the Dewey Decimal System of Classification Applied to the Engineering Industries, by L. P. Breckenridge and G. A. Goodenough. 1906. Revised Edition 1912. *Fifty cents.*

Bulletin No. 10. Tests of Concrete and Reinforced Concrete Columns, Series of 1906, by Arthur N. Talbot. 1907. *None available.*

Bulletin No. 11. The Effect of Scale on the Transmission of Heat through Locomotive Boiler Tubes, by Edward C. Schmidt and John M. Snodgrass. 1907. *None available.*

Bulletin No. 12. Tests of Reinforced Concrete T-Beams, Series of 1906, by Arthur N. Talbot. 1907. *None available.*

Bulletin No. 13. An Extension of the Dewey Decimal System of Classification Applied to Architecture and Building, by N. Clifford Ricker. 1907. *None available.*

Bulletin No. 14. Tests of Reinforced Concrete Beams, Series of 1906, by Arthur N. Talbot. 1907. *None available.*

Bulletin No. 15. How to Burn Illinois Coal Without Smoke, by L. P. Breckenridge. 1908. *None available.*

Bulletin No. 16. A Study of Roof Trusses, by N. Clifford Ricker. 1908. None available.

Bulletin No. 17. The Weathering of Coal, by S. W. Parr, N. D. Hamilton, and W. F. Wheeler. 1908. *None available.*

Bulletin No. 18. The Strength of Chain Links, by G. A. Goodenough and L. E. Moore. 1908. *Forty cents.*

Bulletin No. 19. Comparative Tests of Carbon, Metallized Carbon and Tantalum Filament Lamps, by T. H. Amrine. 1908. *None available.*

Bulletin No. 20. Tests of Concrete and Reinforced Concrete Columns, Series of 1907, by Arthur N. Talbot. 1908. *None available.*

Bulletin No. 21. Tests of a Liquid Air Plant, by C. S. Hudson and C. M. Garland. 1908. *Fifteen cents.*

Bulletin No. 22. Tests of Cast-Iron and Reinforced Concrete Culvert Pipe, by Arthur N. Talbot. 1908. *None available.*

Bulletin No. 23. Voids, Settlement and Weight of Crushed Stone, by Ira O. Baker. 1908. *Fifteen cents.*

**Bulletin No. 24.* The Modification of Illinois Coal by Low Temperature Distillation, by S. W. Parr and C. K. Francis. 1908. *Thirty cents.*

Bulletin No. 25. Lighting Country Homes by Private Electric Plants, by T. H. Amrine. 1908 *Twenty cents.*

---

*A limited number of copies of bulletins starred is available for free distribution.

Bulletin No. 26. High Steam-Pressures in Locomotive Service. A Review of a Report to the Carnegie Institution of Washington, by W. F. M. Goss. 1908. *Twenty-five cents.*

Bulletin No. 27. Tests of Brick Columns and Terra Cotta Block Columns, by Arthur N. Talbot and Duff A. Abrams. 1909. *Twenty-five cents.*

Bulletin No. 28. A Test of Three Large Reinforced Concrete Beams, by Arthur N. Talbot. 1909. *Fifteen cents.*

Bulletin No. 29. Tests of Reinforced Concrete Beams: Resistance to Web Stresses, Series of 1907 and 1908, by Arthur N. Talbot. 1909. *Forty-five cents.*

**Bulletin No. 30.* On the Rate of Formation of Carbon Monoxide in Gas Producers, by J. K. Clement, L. H. Adams, and C. N. Haskins. 1909. *Twenty-five cents.*

**Bulletin No. 31.* Fuel Tests with House-Heating Boilers, by J. M. Snodgrass. 1909. *Fifty-five cents.*

Bulletin No. 32. The Occluded Gases in Coal, by S. W. Parr and Perry Barker. 1909. *Fifteen cents.*

Bulletin No. 33. Tests of Tungsten Lamps, by T. H. Amrine and A. Guell. 1909. *Twenty cents.*

**Bulletin No. 34.* Tests of Two Types of Tile-Roof Furnaces under a Water-Tube Boiler, by J. M. Snodgrass. 1909. *Fifteen cents.*

Bulletin No. 35. A Study of Base and Bearing Plates for Columns and Beams, by N. Clifford Ricker. 1909. *Twenty cents.*

Bulletin No. 36. The Thermal Conductivity of Fire-Clay at High Temperatures, by J. K. Clement and W. L. Egy. 1909. *Twenty cents.*

Bulletin No. 37. Unit Coal and the Composition of Coal Ash, by S. W. Parr and W. F. Wheeler. 1909. *Thirty-five cents.*

**Bulletin No. 38.* The Weathering of Coal, by S. W. Parr and W. F. Wheeler. 1909. *Twenty-five cents.*

**Bulletin No. 39.* Tests of Washed Grades of Illinois Coal, by C. S. McGovney. 1909. *Seventy-five cents.*

Bulletin No. 40. A Study in Heat Transmission, by J. K. Clement and C. M. Garland. 1910. *Ten cents.*

Bulletin No. 41. Tests of Timber Beams, by Arthur N. Talbot. 1910. *Thirty-five cents.*

**Bulletin No. 42.* The Effect of Keyways on the Strength of Shafts, by Herbert F. Moore. 1910. *Ten cents.*

Bulletin No. 43. Freight Train Resistance, by Edward C. Schmidt. 1910. *Seventy-five cents.*

Bulletin No. 44. An Investigation of Built-up Columns Under Load, by Arthur N. Talbot and Herbert F. Moore. 1911. *Thirty-five cents.*

**Bulletin No. 45.* The Strength of Oxyacetylene Welds in Steel, by Herbert L. Whittemore. 1911. *Thirty-five cents.*

**Bulletin No. 46.* The Spontaneous Combustion of Coal, by S. W. Parr and F. W. Kressman. 1911. *Forty-five cents.*

**Bulletin No. 47.* Magnetic Properties of Heusler Alloys, by Edward B. Stephenson, 1911. *Twenty-five cents.*

**Bulletin No. 48.* Resistance to Flow Through Locomotive Water Columns, by Arthur N. Talbot and Melvin L. Enger. 1911. *Forty cents.*

**Bulletin No. 49.* Tests of Nickel-Steel Riveted Joints, by Arthur N. Talbot and Herbert F. Moore. 1911. *Thirty cents.*

**Bulletin No. 50.* Tests of a Suction Gas Producer, by C. M. Garland and A. P. Kratz. 1912. *Fifty cents.*

Bulletin No. 51. Street Lighting, by J. M. Bryant and H. G. Hake. 1912. *Thirty-five cents.*

**Bulletin No. 52.* An Investigation of the Strength of Rolled Zinc, by Herbert F. Moore. 1912. *Fifteen cents.*

**Bulletin No. 53.* Inductance of Coils, by Morgan Brooks and H. M. Turner. 1912. *Forty cents.*

**Bulletin No. 54.* Mechanical Stresses in Transmission Lines, by A. Guell. 1912. *Twenty cents.*

**Bulletin No. 55.* Starting Currents of Transformers, with Special Reference to Transformers with Silicon Steel Cores, by Trygve D. Yensen. 1912. *Twenty cents.*

**Bulletin No. 56.* Tests of Columns: An Investigation of the Value of Concrete as Reinforcement for Structural Steel Columns, by Arthur N. Talbot and Arthur R. Lord. 1912. *Twenty-five cents.*

**Bulletin No. 57.* Superheated Steam in Locomotive Service. A Review of Publication No. 127 the Carnegie Institution of Washington, by W. F. M. Goss. 1912. *Forty cents.*

---

*A limited number of copies of bulletins starred is available for free distribution.

**Bulletin No. 58.* A New Analysis of the Cylinder Performance of Reciprocating Engines, by J. Paul Clayton 1912. *Sixty cents.*

**Bulletin No. 59.* The Effect of Cold Weather Upon Train Resistance and Tonnage Rating, by Edward C. Schmidt and F. W. Marquis. 1912. *Twenty cents.*

**Bulletin No. 60.* The Coking of Coal at Low Temperatures, with a Preliminary Study of the By-Products, by S. W. Parr and H. L. Olin. 1912. *Twenty-five cents.*

**Bulletin No. 61.* Characteristics and Limitation of the Series Transformer, by A. R. Anderson and H. R. Woodrow. 1913. *Twenty-five cents.*

Bulletin No. 62. The Electron Theory of Magnetism, by Elmer H. Williams. 1913. *Thirty-five cents.*

Bulletin No. 63. Entropy-Temperature and Transmission Diagrams for Air, by C. R. Richards. 1913. *Twenty-five cents.*

**Bulletin No. 64.* Tests of Reinforced Concrete Buildings Under Load, by Arthur N. Talbot and Willis A. Slater. 1913. *Fifty cents.*

**Bulletin No. 65.* The Steam Consumption of Locomotive Engines from the Indicator Diagrams, by J. Paul Clayton. 1913. *Forty cents.*

Bulletin No. 66. The Properties of Saturated and Superheated Ammonia Vapor, by G. A. Goodenough and William Earl Mosher. 1913. *Fifty cents.*

Bulletin No. 67. Reinforced Concrete Wall Footings and Column Footings, by Arthur N. Talbot 1913. *Fifty cents.*

**Bulletin No. 68.* The Strength of I-Beams in Flexure, by Herbert F. Moore. 1913. *Twenty cents.*

Bulletin No. 69. Coal Washing in Illinois, by F. C. Lincoln. 1913. *Fifty cents.*

Bulletin No. 70. The Mortar-Making Qualities of Illinois Sands, by C. C. Wiley. 1913. *Twenty cents.*

Bulletin No. 71. Tests of Bond between Concrete and Steel, by Duff A. Abrams. 1913. *One dollar.*

**Bulletin No. 72.* Magnetic and Other Properties of Electrolytic Iron Melted in Vacuo, by Trygve D. Yensen. 1914. *Forty cents.*

Bulletin No. 73. Acoustics of Auditoriums, by F. R. Watson. 1914. *Twenty cents.*

**Bulletin No. 74.* The Tractive Resistance of a 28-Ton Electric Car, by Harold H. Dunn. 1914. *Twenty-five cents.*

Bulletin No. 75. Thermal Properties of Steam, by G. A. Goodenough. 1914. *Thirty-five cents*

Bulletin No. 76. The Analysis of Coal with Phenol as a Solvent, by S. W. Parr and H. F. Hadley. 1914. *Twenty-five cents.*

**Bulletin No. 77.* The Effect of Boron upon the Magnetic and Other Properties of Electrolytic Iron Melted in Vacuo, by Trygve D. Yensen. 1915. *Ten cents.*

**Bulletin No. 78.* A Study of Boiler Losses, by A. P. Kratz. 1915. *Thirty-five cents.*

**Bulletin No. 79.* The Coking of Coal at Low Temperatures, with Special Reference to the Properties and Composition of the Products, by S. W. Parr and H. L. Olin. 1915. *Twenty-five cents.*

**Bulletin No. 80.* Wind Stresses in the Steel Frames of Office Buildings, by W. M. Wilson and G. A. Maney. 1915. *Fifty cents.*

**Bulletin No. 81.* Influence of Temperature on the Strength of Concrete, by A B. McDaniel. 1915. *Fifteen cents.*

Bulletin No. 82. Laboratory Tests of a Consolidation Locomotive, by E. C. Schmidt, J. M. Snodgrass, and R. B. Keller. 1915. *Sixty-five cents.*

**Bulletin No. 83.* Magnetic and Other Properties of Iron-Silicon Alloys, Melted in Vacuo, by Trygve D. Yensen. 1915. *Thirty-five cents.*

Bulletin No. 84. Tests of Reinforced Concrete Flat Slab Structure, by A. N. Talbot and W. A. Slater. 1916. *Sixty-five cents.*

**Bulletin No. 85.* The Strength and Stiffness of Steel Under Biaxial Loading, by A. T. Becker. 1916. *Thirty-five cents.*

**Bulletin No. 86.* The Strength of I-Beams and Girders, by Herbert F. Moore and W. M. Wilson. 1916. *Thirty cents.*

**Bulletin No. 87.* Correction of Echoes in the Auditorium, University of Illinois, by F. R. Watson and J. M. White. 1916. *Fifteen cents.*

Bulletin No. 88. Dry Preparation of Bituminous Coal at Illinois Mines, by E. A. Holbrook. 1916. ***Seventy cents.***

---

* A limited number of copies of bulletins starred is available for free distribution.

**Bulletin No. 89.* Specific Gravity Studies of Illinois Coal, by Merle L. Nebel. 1916. *Thirty cents.*

**Bulletin No. 90.* Some Graphical Solutions of Electric Railway Problems by A. M. Buck. 1916. *Twenty cents.*

Bulletin No. 91. Subsidence Resulting from Mining, by L. E. Young and H. H. Stoek. 1916. *None available.*

**Bulletin No. 92.* The Tractive Resistance on Curves of a 28-Ton Electric Car, by E. C. Schmidt and H. H. Dunn. 1916. *Twenty-five cents.*

**Bulletin No. 93.* A Preliminary Study of the Alloys of Chromium, Copper, and Nickel, by D. F. McFarland and O. E. Harder. 1916. *Thirty-five cents.*

**Bulletin No. 94.* The Embrittling Action of Sodium Hydroxide on Soft Steel, by S. W. Parr. 1917. *Thirty cents.*

**Bulletin No. 95.* Magnetic and Other Properties of Iron-Aluminum Alloys Melted in Vacuo, by Trygve D. Yensen and Walter A. Gatward. 1917. *Twenty-five cents.*

**Bulletin No. 96.* The Effect of Mouthpieces on the Flow of Water Through a Submerged Short Pipe, by Fred B. Seely. 1917. *Twenty-five cents.*

**Bulletin No. 97.* Effects of Storage Upon the Properties of Coal, by S. W. Parr. 1917. *Twenty cents.*

**Bulletin No. 98.* Tests of Oxyacetylene Welded Joints in Steel Plates, by Herbert F. Moore. 1917. *Ten cents.*

Circular No. 4. The Economical Purchase and Use of Coal for Heating Homes with Special Reference to Conditions in Illinois. 1917. *Ten cents.*

**Bulletin No. 99.* The Collapse of Short Thin Tubes, by A. P. Carman. 1917. *Twenty cents.*

**Circular No. 5.* The Utilization of Pyrite Occurring in Illinois Bituminous Coal, by E. A. Holbrook. 1917. *Twenty cents.*

**Bulletin No. 100.* Percentage of Extraction of Bituminous Coal with Special Reference to Illinois Conditions, by C. M. Young. 1917.

**Bulletin No. 101.* Comparative Tests of Six Sizes of Illinois Coal on a Mikado Locomotive, by E. C. Schmidt, J. M. Snodgrass, and O. S. Beyer, Jr. 1917. *Fifty cents.*

**Bulletin No. 102.* A Study of the Heat Transmission of Building Materials, by A. C. Willard and L. C. Lichty. 1917. *Twenty-five cents.*

**Bulletin No. 103.* An Investigation of Twist Drills, by Bruce W. Benedict and W. Penn Lukens 1917. *Sixty cents.*

---

*A limited number of copies of bulletins starred is available for free distribution.

THE UNIVERSITY OF ILLINOIS

THE STATE UNIVERSITY

Urbana

EDMUND J. JAMES, Ph. D., LL. D., President

THE UNIVERSITY INCLUDES THE FOLLOWING DEPARTMENTS:

The Graduate School

The College of Liberal Arts and Sciences (Ancient and Modern Languages and Literatures; History, Economics, Political Science, Sociology; Philosophy, Psychology, Education; Mathematics; Astronomy; Geology; Physics; Chemistry; Botany, Zoology, Entomology; Physiology; Art and Design)

The College of Commerce and Business Administration (General Business, Banking, Insurance, Accountancy, Railway Administration, Foreign Commerce; Courses for Commercial Teachers and Commercial and Civic Secretaries)

The College of Engineering (Architecture; Architectural, Ceramic, Civil, Electrical, Mechanical, Mining, Municipal and Sanitary, and Railway Engineering)

The College of Agriculture (Agronomy; Animal Husbandry; Dairy Husbandry; Horticulture and Landscape Gardening; Agricultural Extension; Teachers' Course; Household Science)

The College of Law (three years' course)

The School of Education

The Course in Journalism

The Courses in Chemistry and Chemical Engineering

The School of Railway Engineering and Administration

The School of Music (four years' course)

The School of Library Science (two years' course)

The College of Medicine (in Chicago)

The College of Dentistry (in Chicago)

The School of Pharmacy (in Chicago; Ph. G. and Ph. C. courses)

The Summer Session (eight weeks)

Experiment Stations and Scientific Bureaus: U. S. Agricultural Experiment Station; Engineering Experiment Station; State Laboratory of Natural History; State Entomologist's Office; Biological Experiment Station on Illinois River; State Water Survey; State Geological Survey; U. S. Bureau of Mines Experiment Station.

The library collections contain (December 1, 1017) 411,737 volumes and 104,524 pamphlets.

For catalogs and information address

THE REGISTRAR
URBANA, ILLINOIS

NIVERSITY OF ILLINOIS BULLETIN
ISSUED WEEKLY
l. XV DECEMBER 3, 1917 No. 14
Entered as second-class matter Dec. 11, 1912, at the Post Office at Urbana, Ill., under the Act of Aug. 24, 1912.]

TESTS TO DETERMINE THE RIGIDITY OF RIVETED JOINTS OF STEEL STRUCTURES

BY
WILBUR M. WILSON
AND
HERBERT F. MOORE

BULLETIN No. 104

ENGINEERING EXPERIMENT STATION
PUBLISHED BY THE UNIVERSITY OF ILLINOIS, URBANA.

PRICE: TWENTY-FIVE CENTS

THE Engineering Experiment Station was established by act of the Board of Trustees, December 8, 1903. It is the purpose of the Station to carry on investigations along various lines of engineering and to study problems of importance to professional engineers and to the manufacturing, railway, mining, constructional and industrial interests of the State.

The control of the Engineering Experiment Station is vested in the heads of the several departments of the College of Engineering. These constitute the Station Staff and, with the Director, determine the character of the investigations to be undertaken. The work is carried on under the supervision of the Staff, sometimes by research fellows as graduate work, sometimes by members of the instructional staff of the College of Engineering, but more frequently by investigators belonging to the Station corps.

The results of these investigations are published in the form of bulletins, which record mostly the experiments of the Station's own staff of investigators. There will also be issued from time to time, in the form of circulars, compilations giving the results of the experiments of engineers, industrial works, technical institutions, and governmental testing departments.

The volume and number at the top of the front cover page are merely arbitrary numbers and refer to the general publications of the University of Illinois: *either above the title or below the seal is given the number of the Engineering Experiment Station bulletin or circular which should be used in referring to these publications.*

For copies of bulletins, circulars, or other information address the

ENGINEERING EXPERIMENT STATION,
URBANA, ILLINOIS.

UNIVERSITY OF ILLINOIS
ENGINEERING EXPERIMENT STATION

BULLETIN No. 104 DECEMBER, 1917

TESTS TO DETERMINE THE RIGIDITY OF RIVETED JOINTS OF STEEL STRUCTURES

BY

WILBUR M. WILSON
ASSISTANT PROFESSOR OF CIVIL ENGINEERING

AND

HERBERT F. MOORE
RESEARCH PROFESSOR OF ENGINEERING MATERIALS

ENGINEERING EXPERIMENT STATION
PUBLISHED BY THE UNIVERSITY OF ILLINOIS, URBANA

CONTENTS

LIST OF FIGURES

LIST OF TABLES

TESTS TO DETERMINE THE RIGIDITY OF RIVETED JOINTS OF STEEL STRUCTURES

I. Introduction

1. *Object of Tests.*—The object of these tests was to determine the rigidity of riveted joints which connect the members of steel-framed structures.

If the geometrical element of a frame is a triangle, the members may be joined with frictionless hinges and the frame will resist external forces. If, however, the geometrical element of a frame is a rectangle, shear in the frame due to external loads produces turning moments on the connections, and in order to prevent the frame from collapsing the connection must be designed to resist moment. The distribution of the stresses in a rectangular frame depends upon the rigidity of the connections. In analyzing the stresses in such a frame it is customary to assume that the connections are perfectly rigid. If the connections are not perfectly rigid, it is apparent that the actual stress may not be equal to the computed stress. In addition to determining the rigidity of riveted connections, analyses have been made to determine the effect of lack of rigidity upon the distribution of stresses in a frame.

Although there are other structures in which the rigidity of the connections affects the distribution of the stresses, the discussion in this bulletin is limited to a discussion of the effect of the rigidity

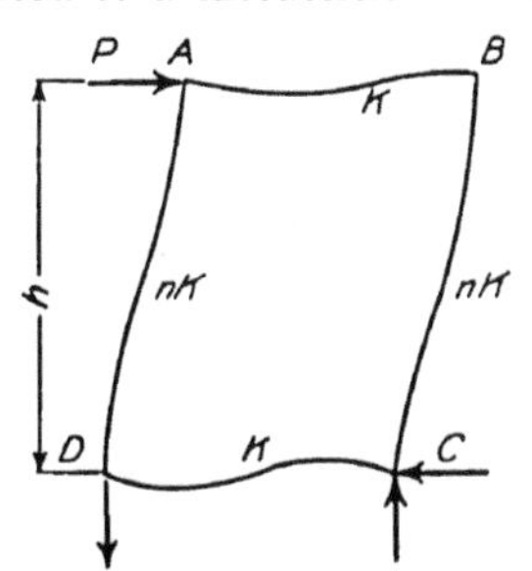

Fig. 1. Diagram of Deformation of Rectangular Frame

of connections upon the distribution of the stresses in a rectangular frame subjected to shear (Fig. 1). The most important example of this type of frame is the wind brace in steel skeleton buildings.

2. *Acknowledgments.*—The tests described in this bulletin were made under the general supervision of Dr. F. H. Newell, head of the Department of Civil Engineering, and of Dr. A. N. Talbot, head of the Department of Theoretical and Applied Mechanics. The tests formed a part of the regular research work of these Departments.

The specimens designated as *A*1, *A*2, and *A*3 were donated by the American Bridge Company.

Camillo Weiss, Research Fellow in Civil Engineering, Bernard Pepinsky, Research Fellow in Theoretical and Applied Mechanics, and W. L. Parish, Graduate Student in Architectural Engineering, rendered valuable service by assisting with the tests and with the calculations.

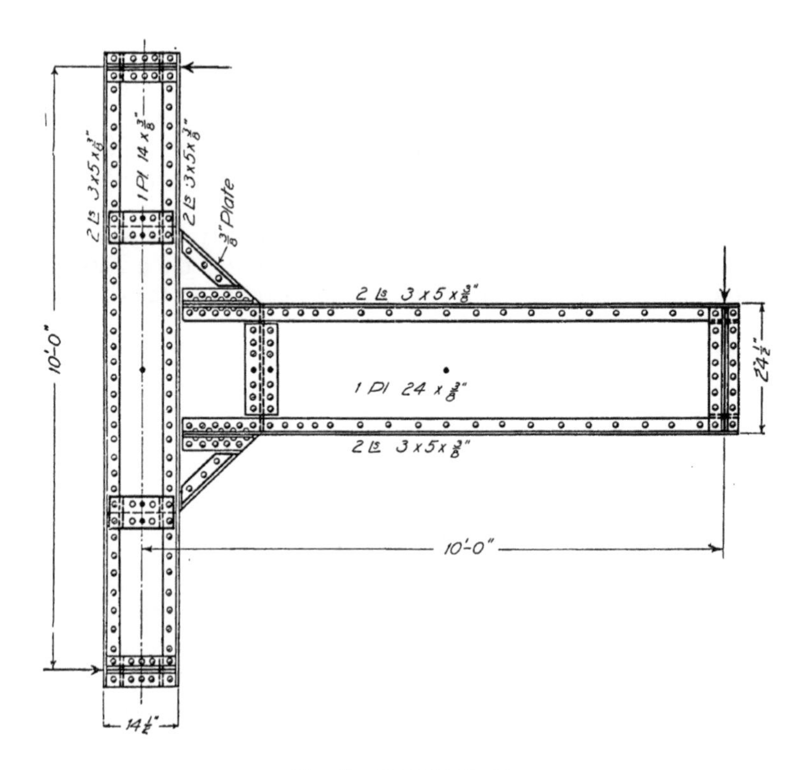

FIG. 2. TEST PIECE *A*1

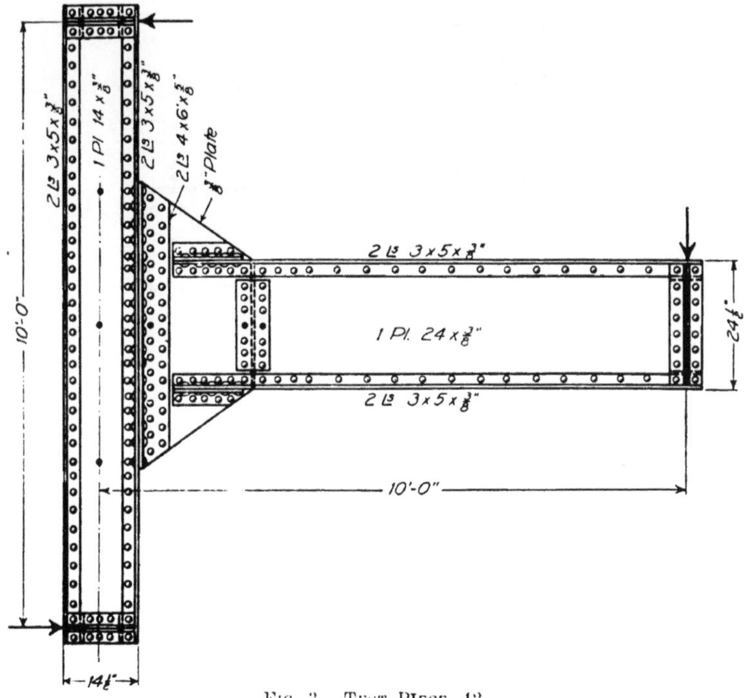

FIG. 3. TEST PIECE A2

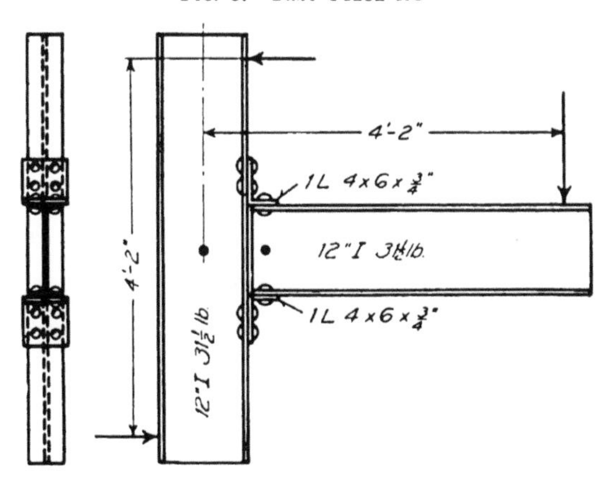

FIG. 4. TEST PIECE A3

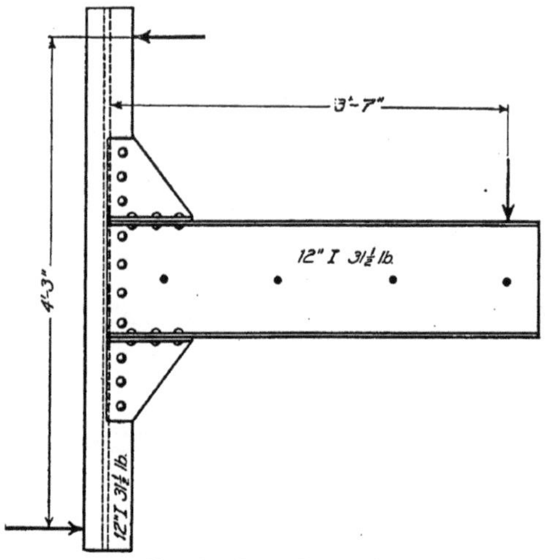

Fig. 5. Test Piece *A*4

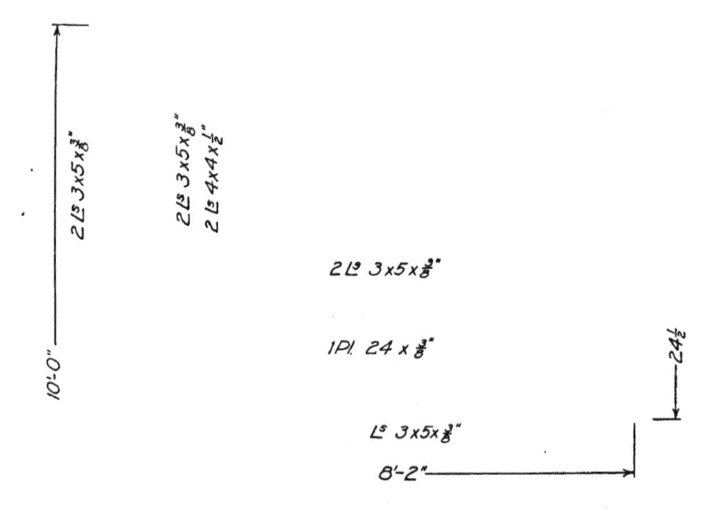

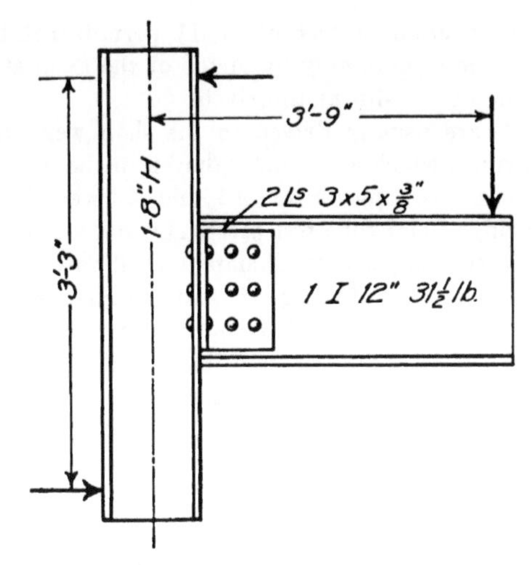

FIG. 7. TEST PIECE *A*6

II. DESCRIPTION OF TESTS

3. *Description of Test Pieces.*—The test pieces are shown in Figs. 2 to 7, inclusive. Two pieces of each type were used, tests on similar pieces being made simultaneously.

In selecting the test pieces it was the aim to use connections of types which are common in engineering structures and which resist loads and moments by methods which are fundamentally different. Test pieces *A*1 and *A*2 are types of connections used when the moment to be resisted is large; *A*4 is a type used when the moment is comparatively small and when the center lines of the columns and girdles do not intersect; *A*3, *A*5, and *A*6 are designed primarily to resist shear and are not intended to resist moment. For all specimens except *A*1, relative rotation between the columns and girders can occur by virtue of the deflection of a comparatively thin member in cross-bending. In the case of *A*3, for example, the portion of the vertical leg of the top lug angle between the lower rivet and the horizontal leg of the lug angle acts as a cantilever, and by bending the vertical leg the angle pulls away from the column. In a similar manner for *A*2, *A*4, *A*5, and *A*6, bending of the outstanding legs of the connection angles permits the girders to rotate relatively to the columns. The connection of test piece *A*1 permits rotation of the girder relative to the column only by virtue of the axial strain in the metal or by virtue of the slip of the rivets.

Rivets which are usually driven in the shop were driven with a press riveter; rivets which are usually driven in the field were driven with an air gun. Test pieces *A*1, *A*2, and *A*3 were fabricated by the American Bridge Company; test pieces *A*4, *A*5, and *A*6 were fabricated by the Burr Company of Champaign, Illinois. The I-beams of *A*3 were used in making *A*4 and *A*6, and the girders and columns of *A*2 were used in making *A*5. The rotation of the girders relative to the columns in *A*4, *A*5, and *A*6 was due to the strain in the connection angles, and these connection angles had not been previously stressed. It is therefore improbable that the rotation of the girders relative to the columns in *A*4, *A*5, and *A*6 was affected in any way by the fact that the main members of which the test pieces were composed had been stressed as members of pieces *A*2 and *A*3.

The structural details of the girders and columns of the test pieces are given in Table 1.

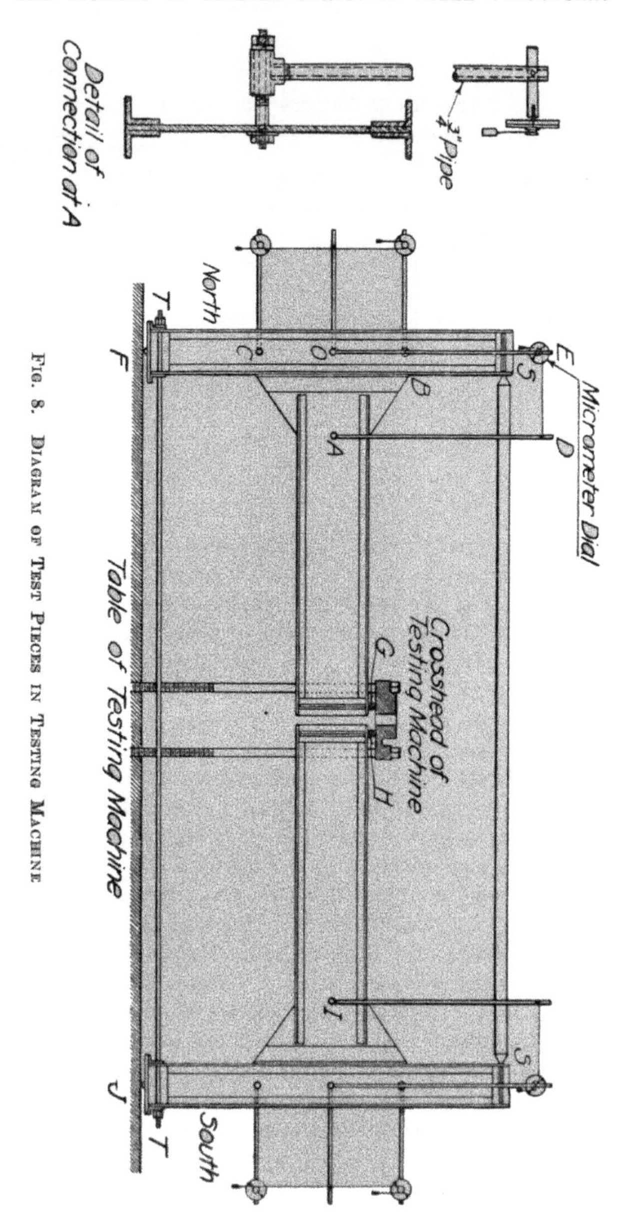

FIG. 8. DIAGRAM OF TEST PIECES IN TESTING MACHINE

TABLE 1

PROPERTIES OF COLUMNS AND GIRDERS

Test Piece	Column		Girder		For Frame 15 Feet High and 20 Feet Wide. $K=\frac{I}{l}$ for Girder, $nK=\frac{I}{l}$ for Column		
	Section	I for Gross Section	Section	I for Gross Section	nK (for col.)	K (for girder)	n
*A*1	1—*Pl*—14x3/8 4—*Ls* 5x3x3/8	584	1—*Pl*—24x3/8 4—*Ls* 5x3x3/8	1966	3.25	8.20	.4
*A*2	Same as *A*1	584	Same as *A*1	1966	3.25	8.20	.4
*A*3	12″—*I*—31.5	215.8	12—*I*—31.5	215.8	1.20	.90	1.33
*A*4	12″ Flat—*I*—31.5	9.5	12—*I*—31.5	215.8	.05	.90	.055
*A*5	Same as *A*1	584	Same as *A*1	1966	3.25	8.20	.4
*A*6	8″—*H*—39	139.5	12—*I*—31.5	215.8	.78	.90	.86

4. *Apparatus and Methods of Testing.*—The test pieces described were tested in pairs in a 300,000-pound Olsen four-screw testing machine. Fig. 8 shows the arrangement of a pair of test specimens, and Fig. 9 is a reproduction of a photograph of a pair of test specimens in the machine. The column of each test specimen was supported by a half-round piece of steel, *F* and *J*, (Fig. 8) which in turn was supported by the table of the testing machine. Tie rods at the bottom *T, T,* and a strut at the top, *S, S,* held the test pieces in the position shown. Load was applied by the cross-head of the testing machine through rollers, *G* and *H*, to the ends of the beams. The half-round bearings, *F* and *J*, the knife-edge ends of the strut, *S, S,* the flexibility of the long, slender tie rods, *T, T,* and the rollers, *G* and *H*, permitted free rotation of the specimens and insured an equal distribution of load between the two specimens.

The strain measurements include the rotation of the beam relative to the column, the slip of rivets, and the deformation of angles used in making the joint. The rotation of a beam with respect to a column was measured by determining for one or more points the change of slope of the beam with respect to the column and then comparing this change of slope with the change in slope due to elastic strain. The apparatus for measuring this change of slope is shown in Fig. 8. The arm, *OE*, is rigidly attached at *O*, the junction of the axis of the beam with the axis of the column, and the arm, *AD*, is attached

Fig. 9. General View of Test Pieces A1 in Testing Machine

rigidly at A, some point on the neutral axis of the beam. AD and OE are equal in length. If no rotation of A with respect to O takes place, the ends D and E will remain a constant distance apart, but if rotation does take place, the change of distance between D and E will be a measure of this rotation, and if the magnitude of this change of distance is denoted by ΔS, the angle of rotation is denoted by θ, and the length of arm by q, then $\tan\theta = \left\{\frac{\Delta S.}{q}\right.$

The motion of D with respect to E is measured by means of a micrometer fastened at E. This micrometer consists of a drum one inch in circumference carrying a pointer moving over a dial. Around the drum is wound a No. 36 insulated copper wire kept taut by a weight and fastened at D. The motion of the pointer over the dial indicates the magnitude of movement of D with respect to E.* A similar apparatus is used to measure the rotation of points in the column. The arms carrying the dials are clamped to the webs of the beams, or the columns, with bolts and pipe fittings as shown in Fig. 8. Slip of rivets, change of form of connecting angles, and other small deformations were measured by means of strain gages and attached micrometer dials.† Figs. 10 to 15, inclusive, show the location of the instruments for measuring changes of slope on the specimens.

*For a more detailed description of the wire-wound dial micrometer see Proc. A. S. T. M., p. 607, 1907.

†For a description of the strain gage and directions for its use see "Tests of Reinforced Concrete Buildings under Load," Univ. of Ill., Eng. Exp. Sta., Bul. 64, 1913, and Proc. A. S. T. M., p. 1019, 1913.

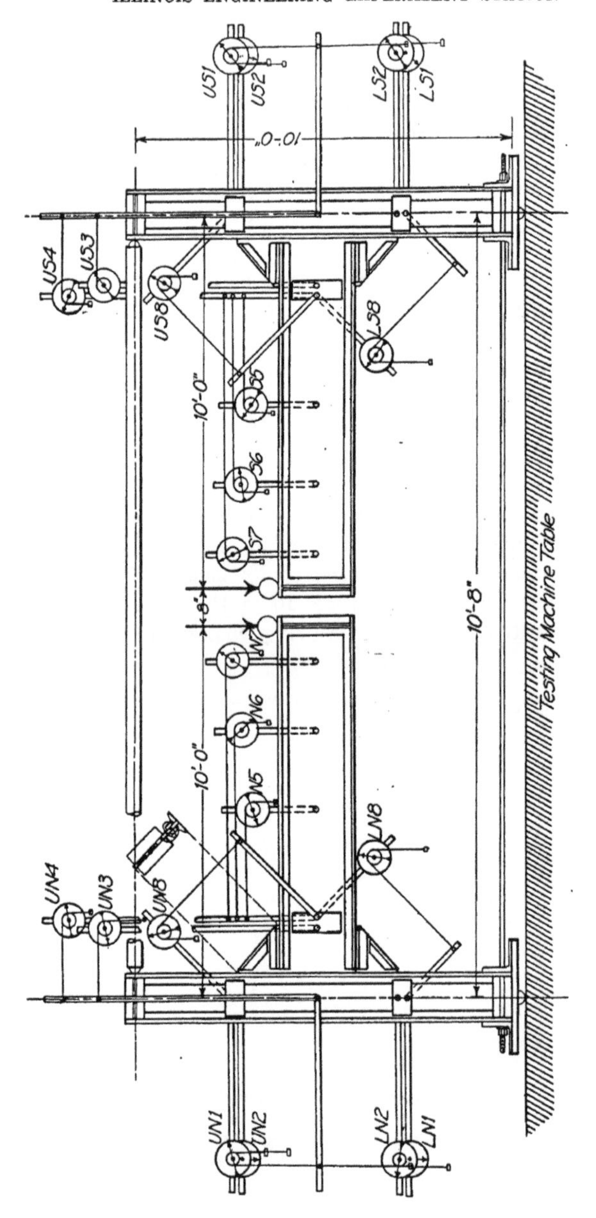

FIG. 10. MEASUREMENTS OF ROTATION FOR TEST PIECES *A1*

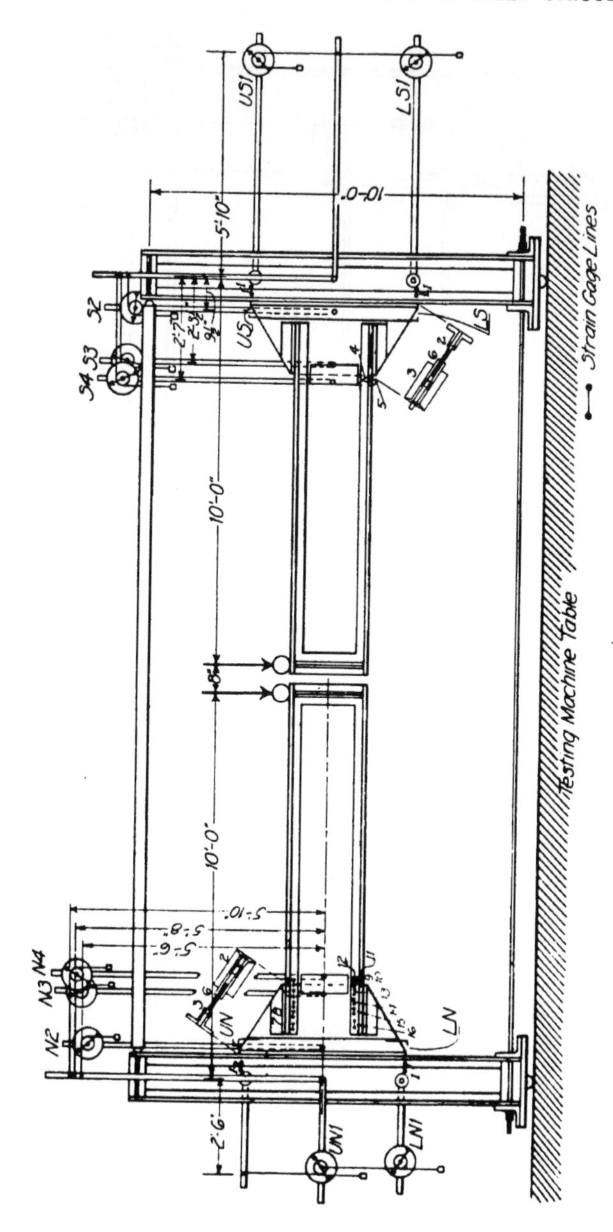

FIG. 11. MEASUREMENTS OF ROTATION FOR TEST PIECES A2

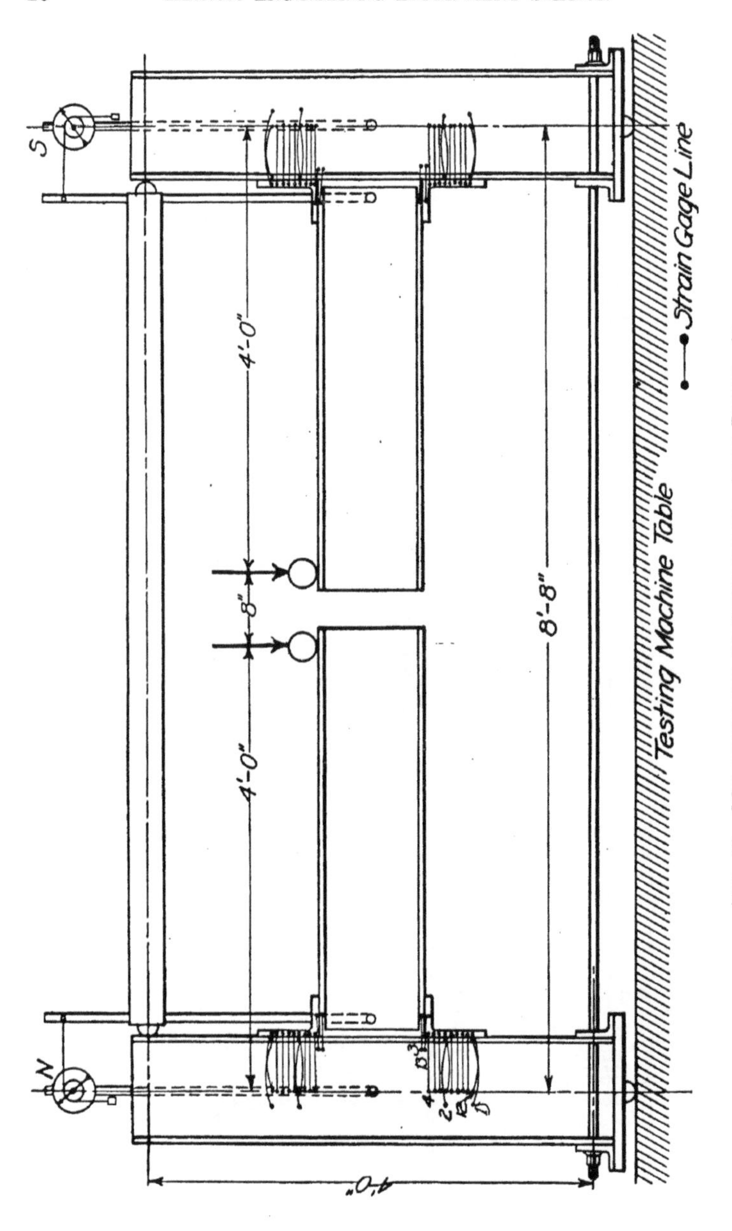

Fig. 12. Measurements of Rotation for Test Pieces A3

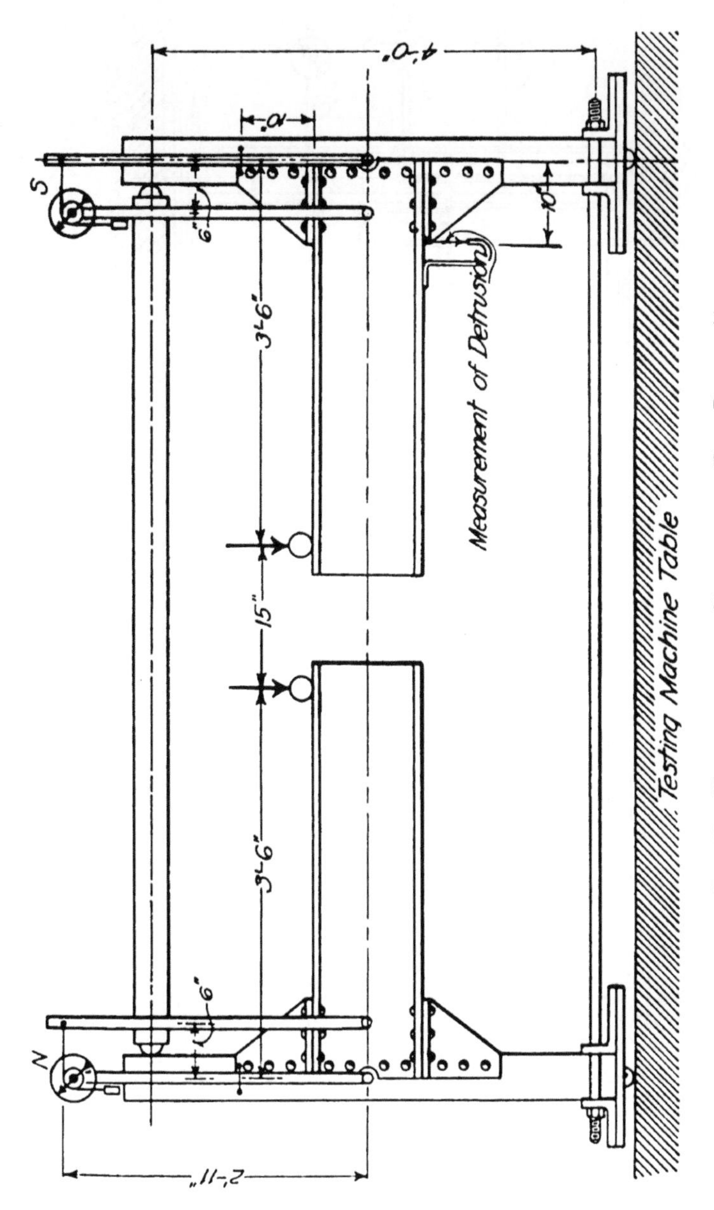

FIG. 13. MEASUREMENTS OF ROTATION FOR TEST PIECES A4

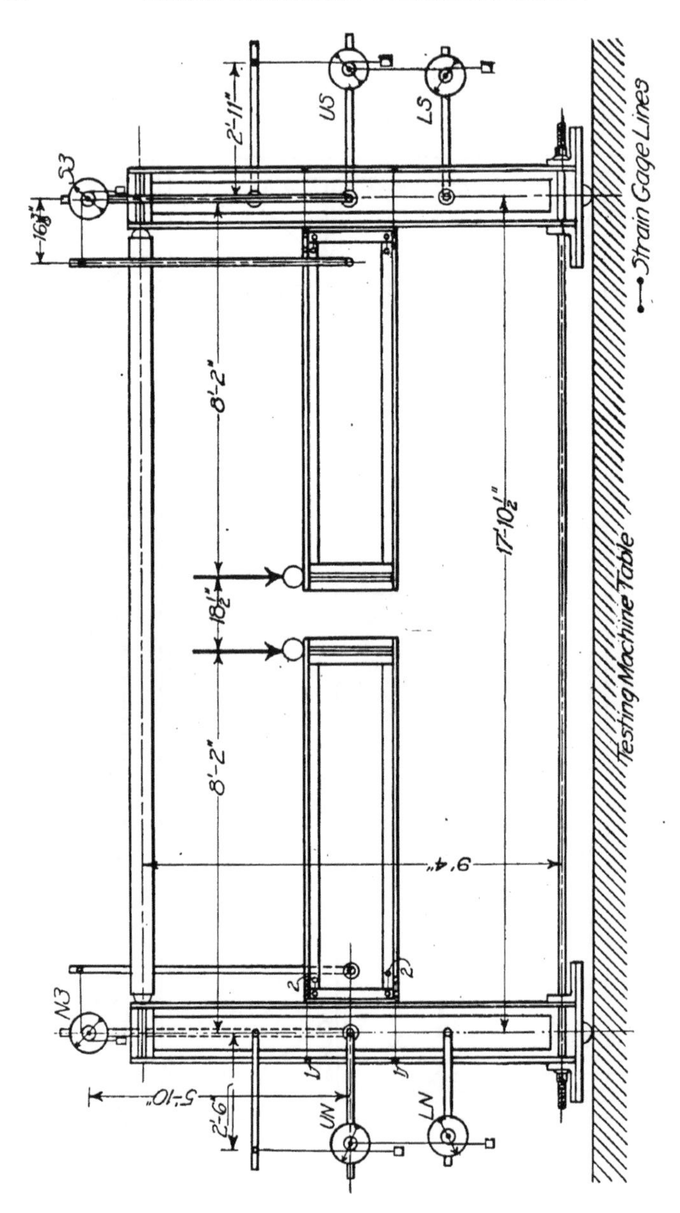

FIG. 14. MEASUREMENTS OF ROTATION FOR TEST PIECES A5

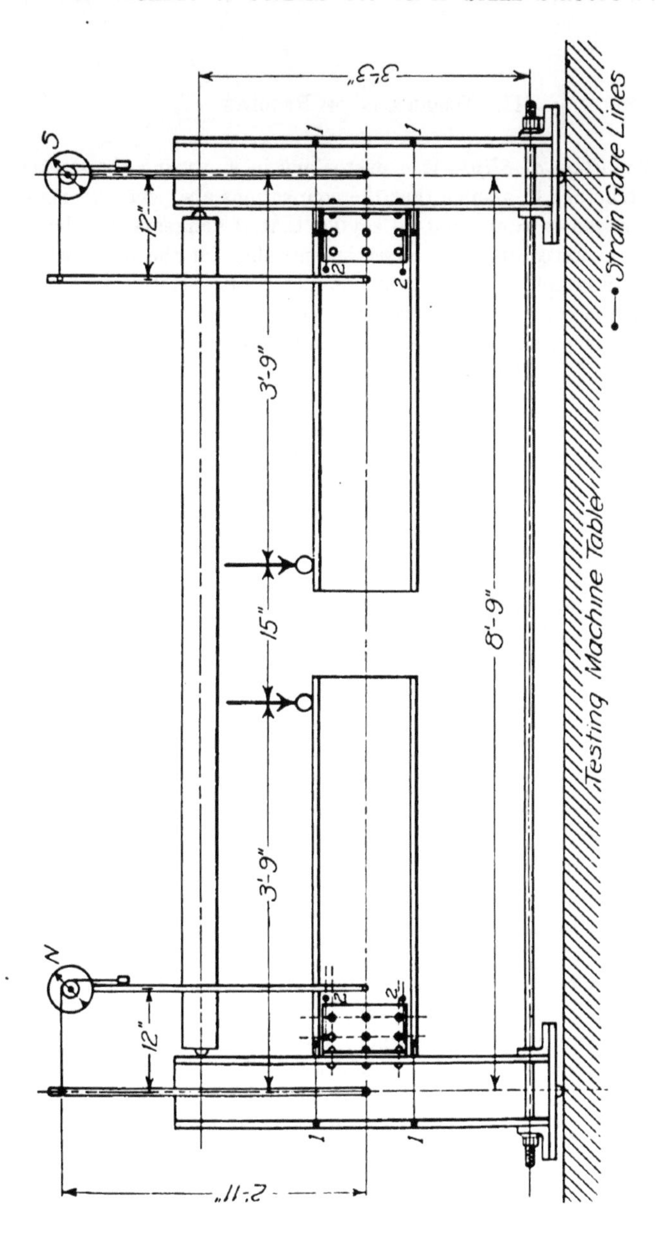

FIG. 15. MEASUREMENTS OF ROTATION FOR TEST PIECES A6

III. Discussion of Results

5. *Definition of Slip.*—It is customary in analyzing the stresses in a stiff structure to assume that the connections are perfectly rigid. This assumption is equivalent to saying that if tangents are drawn to the elastic curves of two intersecting members at the point where the curves intersect, these tangents do not rotate relatively to each other when the connection is stressed. If the tangents to the elastic curves at their point of intersection rotate relatively to each other, the connection is said to slip.

It is also customary in analyzing the stresses in a stiff structure to consider that where a column and girder intersect, both members maintain a constant cross-section up to the point of intersection of the elastic curves of the members, a condition impossible to obtain exactly, except for members having a width of zero.

If a column or a girder is subjected to flexure, a point in the elastic curve rotates relatively to other points in the elastic curve by virtue of the strain in the material even if the material is continuous. For this reason, when the connection between a column and a girder is subjected to a moment, a point in the elastic curve of the column rotates relatively to the point of intersection of the elastic curves of the two members. Similarly a point in the elastic curve of a girder rotates relatively to the point of intersection of the elastic curves of the two members; therefore the rotation of a point in the elastic curve of the girder relative to a point in the elastic curve of the column may be due partly to the slip in the joint and partly to the strain in the material. In determining the slip in the connections from the tests, the computed rotation of a point in the girder relative to a point in the column due to the strain in the material was subtracted from the measured rotation of one point relative to the other. In computing the relative rotation of the two points due to the strain of the material both members were considered as having constant cross-sections up to the point where the elastic curves of the two members intersect; thus what is really obtained in the analysis is not the error due to slip alone but rather the combined error resulting from two assumptions, one that the connection is perfectly rigid, and the other that both members maintain a constant cross-section up to the point of intersection of the two elastic curves.

The rotation of a point in the elastic curve of a member relative to another point in the elastic curve of the same member due to the strain of the material can be determined from the following proposition: For a member in flexure the change in slope between two points in the elastic curve of the member is equal to the area of the $\frac{M}{EI}$ diagram for the portion of the member between the two points.

6. *The Effect of the Slip in Connections upon a Rectangular Frame.*—Fig. 16 represents a rectangular frame subjected to a shear

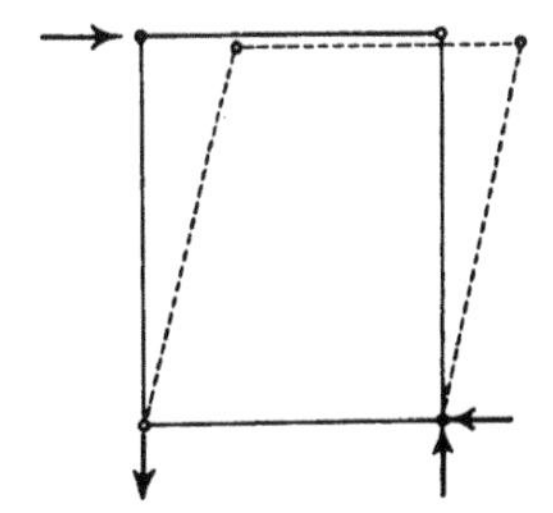

FIG. 16. DIAGRAM OF RECTANGULAR FRAME WITH PIN JOINTS

parallel with one side of the frame. If the connections at the corners of the frame are frictionless hinges, the frame will collapse as shown by the dotted lines. If, however, the connections at the corners are capable of resisting a moment, the frame will take the form shown in Fig. 1. The rigid connections between the vertical and horizontal members hold the ends of the vertical members in a nearly vertical position and thus prevent the frame from collapsing.

If the connections at the corners of the frame represented by Fig. 1 are not rigid but permit the vertical members to rotate slightly relatively to the horizontal members, because of the slip in the connections, a vertical member will be deflected from a vertical through its lower end more than if the connections are perfectly rigid. When all the connections slip, moreover, by the same angular amount, the stresses in the frame are the same as when all the connections are perfectly rigid;* but if the angular slip is greater in one or more connections than in the other connections, the stresses in the frame

* See page 30.

are not the same as when all the connections are perfectly rigid, that is, slip in the connections of a rectangular frame increases the deflection of the frame and, unless all connections slip by the same amount, makes the stresses in the frame different from the stresses in a similar frame having rigid connections.

7. *Method of Determining the Magnitude of the Change in the Deflection and in the Distribution of the Stresses in a Rectangular Frame Due to Slip in the Connections.*—In order to determine the magnitude of the change in the deflection and in the distribution of the stresses in a rectangular frame due to slip in the connections, the following formulas are derived:

Let

M_{AB} = moment at A in member AB.

θ_A = change in slope of the elastic curve of the member at the point A.

θ_B = change in slope of the elastic curve of the member at the point B.

$R = \frac{d}{l}$

d = deflection of one end of the member relative to the other end, measured in a direction normal to the original position of the member.

l = length of the member in inches.

E = modulus of elasticity of the material.

$K = \frac{I}{l} = \frac{\textit{moment of inertia}}{\text{length}}$ for the member.

It has been proved that the moment at the end A of the member AB, Fig. 1, is given by the equation *

$$M_{AB} = 2EK\ (2\theta_A + \theta_B - 3R) \quad . \quad . \quad . \quad . \quad (1)$$

The moment at the end of a member may be expressed in terms of the changes in the slopes of the ends of the member and the deflection of one end of the member relative to the other end.

If the frame has perfectly rigid connections, and $\frac{I}{l}$ for $A\ B$ and $C\ D$ is represented by K, and $\frac{I}{l}$ for $A\ D$ and $B\ C$ by nK, then, from equation (1)

*See "Wind Stresses in the Steel Frames of Office Buildings," Univ. of Ill. Eng. Exp. Sta., Bul. 80, Equation A, p. 13, 1915.

$$M_{AB} = 2\ E\ K\ \left[\ 2\,\theta_A\ +\ \theta_B - (3R = O)\ \right]\ .\ .\ (2)$$

$$M_{AD} = 2\ E\ nK \left[\ 3\,\theta_A\ +\ \theta_D - 3R\right].\ .\ .\ .\ .\ (3)$$

$$M_{DC} = 2\ E\ K\ \left[\ 2\,\theta_D\ +\ \theta_C - (3R = O)\ \right]\ .\ .\ (4)$$

$$M_{DA} = 2\ E\ nK \left[\ 2\,\theta_D\ +\ \theta_A - 3R\right].\ .\ .\ .\ .\ (5)$$

The point A is in equilibrium; therefore $M_{AB}\ +\ M_{AD} = 0$, and likewise $M_{DA}\ +\ M_{DC}\ =\ 0$. Since the frame is symmetrical about a vertical center line, the sum of the moments at the top and at the bottom of AD is equal to one-half of the shear in the frame multiplied by the height of the frame, that is,

$$M_{AD} + M_{DA} + \frac{Ph}{2} = 0\ .\ .\ .\ .\ .\ .\ .\ .\ .\ .\ .\ (6)$$

Equating the right hand members of equations (2) and (3) gives

$$2\theta_A(1+n) + \theta_B + n\theta_D - 3nR = 0\ .\ .\ .\ .\ .\ .\ (7)$$

Likewise from equations (4) and (5)

$$2\,\theta_D(1+n) + \theta_C + n\,\theta_A - 3nR = 0\ .\ .\ .\ .\ .\ .\ (8)$$

From the conditions assumed, $\theta_B = \theta_A$ and $\theta_C = \theta_D$.

Substituting these values of θ_B and θ_C in equations (7) and (8) gives.

$$\theta_A\,(3+2n) +\ n\theta_D - 3nR = 0\ .\ .\ .\ .\ .\ .\ .\ .\ .\ (9)$$

$$n\theta_A\ +\ \theta_D\,(3\ +\ 2n) - 3nR = 0\ .\ .\ .\ .\ .\ .\ .\ .\ .\ (10)$$

Substituting the values of M_{AD} and M_{DA} from equations (3) and (5) in equation (6) gives

$$\theta_A\ +\theta_B = 2R = -\frac{1}{12}\ \frac{Ph}{E\,nK}\ .\ .\ .\ .\ .\ .\ (11)$$

Solving equations (9), (10), and (11) for θ_A, θ_D, and R gives

$$\theta_A = \frac{1}{24}\ \frac{Ph}{EK}\ .\ .\ .\ .\ .\ .\ .\ .\ .\ .\ .\ .\ (12)$$

$$\theta_D = \frac{1}{24}\ \frac{Ph}{EK}\ .\ .\ .\ .\ .\ .\ .\ .\ .\ .\ .\ .\ (13)$$

$$R = \frac{1}{24}\ \frac{Ph}{EK}\ (1+\frac{1}{n})\ .\ .\ .\ .\ .\ .\ .\ .\ .\ .\ (14)$$

Substituting these values of θ_A, θ_D, and R in equations (2) and (4) gives

$$M_{AB} = 1/4 \ Ph \quad . \quad . \quad . \quad . \quad . \quad . \quad . \quad . \quad . \quad (15)$$

$$M_{DC} = 1/4 \ Ph \quad . \quad . \quad . \quad . \quad . \quad . \quad . \quad . \quad . \quad (16)$$

Assuming that each member maintains a constant section up to the neutral axis of the member which it intersects, equations (14), (15), and (16) give the deflection and the moments in the frame when the connections are perfectly rigid.

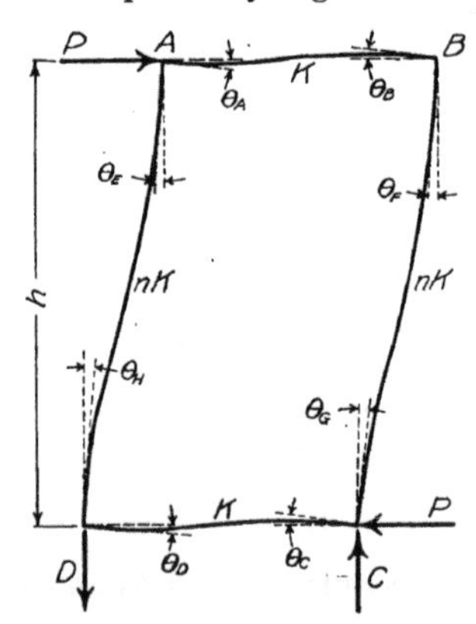

FIG 17. DIAGRAM OF RECTANGULAR FRAME WITH SLIP AT JOINTS

Consider now a rectangular frame having connections which slip. Such a frame is represented by Fig. 17. The changes in the slopes at the ends of the member AB are represented by θ_A at A and θ_B at B; likewise for the member CD the change in slope at C is represented by θ_C and at D by θ_D, for the member AD the change in slope at A is represented by θ_E and at D by θ_H, and for the member BC the change of slope at B is represented by θ_F and at C by θ_G, that is, the angular slip at A equalt $\theta_E - \theta_A$, the angular slip at B equals $\theta_F - \theta_B$, the angular slip at C equals $\theta_G - \theta_C$, and the angular slip at D equals $\theta_H - \theta_D$.

Represent $\frac{I}{l}$ for AB and CD by K, and $\frac{I}{l}$ for AD and BC by nK.

Applying equation (1),* equating the sum of the moments at each of the points A, B, C, and D to zero, and equating the sums of the moments at the ends of the two members AD and BC to zero gives

*The presence of slip does not invalidate equation (1).

$$2EK\ (2\theta_A + \theta_B) + 2EnK\ (2\theta_E + \theta_H - 3R) = 0 \quad . \quad . \quad (17)$$

$$2EK\ (2\theta_B + \theta_A) + 2EnK\ (2\theta_F + \theta_G - 3R) = 0 \quad . \quad . \quad (18)$$

$$2EK\ (2\theta_C + \theta_D) + 2EnK\ (2\theta_G + \theta_F - 3R) = 0 \quad . \quad . \quad (19)$$

$$2EK\ (2\theta_D + \theta_C) + 2EnK\ (2\theta_H + \theta_E - 3R) = 0 \quad . \quad . \quad (20)$$

$$2\ EnK\ (2\theta_E + \theta_H - 3R + 2\theta_H + \theta_E - 3R + 2\theta_F + \theta_G - 3R + 2\theta_G + \theta_F - 3R) + Ph = 0. \quad . \quad . \quad . \quad . \quad . \quad . \quad (21)$$

Letting A represent the quantity, slip at A divided by R, and letting B, C, and D represent the corresponding quantities at the points B, C, and D, respectively, gives

$$A = \frac{\theta_E}{R} - \frac{\theta_A}{R}$$

$$B = \frac{\theta_F}{R} - \frac{\theta_B}{R}$$

$$C = \frac{\theta_G}{R} - \frac{\theta_C}{R}$$

$$D = \frac{\theta_H}{R} - \frac{\theta_D}{R}$$

Substituting the values of θ_E, θ_F, θ_G, and θ_H from these equations in the preceding equations gives

$$\frac{2\theta_A}{R}\ (1+n) + \frac{\theta_B}{R} + \frac{n\theta_D}{R} = n\ (3-2A-D) \quad . \quad . \quad (22)$$

$$\frac{2\theta_B}{R}\ (1+n) + \frac{\theta_A}{R} + \frac{n\theta_C}{R} = n\ (3-2B-D) \quad . \quad . \quad (23)$$

$$\frac{2\theta_B}{R}\ (1+n) + \frac{\theta_D}{R} + \frac{n\theta_B}{R} = n\ (3-2C-B) \quad . \quad . \quad (24)$$

$$\frac{2\theta_D}{R}\ (1+n) + \frac{\theta_C}{R} + \frac{n\theta_A}{R} = n\ (3-2D-A) \quad . \quad . \quad (25)$$

$$2EnKR = \frac{1/3\ Ph}{\frac{\theta_A}{R} + \frac{\theta_B}{R} + \frac{\theta_C}{R} + \frac{\theta_R}{R} + 4 - (A + B + C + D).} \quad (26)$$

These five equations contain four unknown angles and one unknown deflection. Solving these equations and substituting the values for the θ's and R in the expressions for the moments gives

$$M_{AD}=\frac{1}{3}\;\frac{Ph}{4-(A+B+C+D)}$$

$$\left[\frac{n^2(6A+3B-9)+n(16A+5B+4C+5D-30)+(6A+3D-9)}{(n+3)\quad(3n+1)}\right] \quad (27)$$

$$M_{DA}=\frac{1}{3}\;\frac{Ph}{4-(A+B+C+D)}$$

$$\left[\frac{n^2(6D+3C-9)+n(16D+5C+4B+5A-30)+(6D+3A-9)}{(n+3)\quad(3n+1)}\right] \quad (28)$$

$$M_{BC}=\frac{1}{3}\;\frac{Ph}{4-(A+B+C+D)}$$

$$\left[\frac{n^2(6B+3A-9)+n(16B+5A+4D+5C-30)+(6B+3C-9)}{(n+3)\quad(3n+1)}\right] \quad (29)$$

$$M_{CB}=\frac{1}{3}\;\frac{Ph}{4-(A+B+C+D)}$$

$$\left[\frac{n^2(6C+3D-9)+n(16C+5D+4A+5B-30)+(6C+3B-9)}{(n+3)\quad(3n+1)}\right] \quad (30)$$

The values of $\frac{\theta_A}{R}$, $\frac{\theta_B}{R}$, $\frac{\theta_C}{R}$, and $\frac{\theta_D}{R}$, determined from equations (22) to (25) substituted in equation (26) give

$$R=\frac{1}{4}\,(RA+RB+RC+RD)+\frac{1}{6}\,Ph\frac{(n+1)}{EnK} \quad . \quad . \quad (31)$$

In this equation RA, RB, RC, and RD represent the slips in the connections at A, B, C, and D, respectively. If the slips are measured, R can be computed from equation (31). Knowing R and the slip, the values of A, B, C, and D can be computed. Substituting the values of A, B, C, and D in equations (11) to (14), inclusive, the moments in a frame having connections which are not rigid can be determined.

A comparison of the moments given by equations (27) to (30) with the moments in a similar frame having rigid connections, determined by equations (15) and (16), will give the changes in the moments due to the slip in the connections. A comparison of the deflection of the frame, as given by equation (31), with the deflection in a similar frame having rigid connections, as given by equation (14), will give the change in the deflection due to the slip in the connections.

If *A*, *B*, *C*, and *D* of equations (27) to (30), inclusive, are equal to each other, that is, if the slips in all the connections of the rectangular frame represented by Fig. 17 are equal, equations (27) to (30) reduce to the following forms:

$$M_{AD} = 1/4\,Ph \qquad (27a)$$

$$M_{DA} = 1/4\,Ph \qquad (28a)$$

$$M_{BC} = 1/4\,Ph \qquad (29a)$$

$$M_{CB} = 1/4\,Ph \qquad (30a)$$

These moments are the same as the moments given by equations (15) and (16).

That is, if the slips in all the connections of the rectangular frame shown in Fig. 17 are equal, the stresses in the frame are the same as they are in a similar frame having connections which are perfectly rigid.

8. *Graphic Records.*—The quantities measured in the tests are recorded graphically in Figs. 20 to 25. In these diagrams, measured quantities are represented by full lines, and computed quantities by broken lines.

9. *Strength of Test Pieces.*—An engineer is interested in the rigidity of a connection at usual working stresses or at stresses slightly higher. In computing the allowable working loads upon the test pieces the following unit stresses were used:

Axial bending stress . . .	16,000 pounds per square inch.
Shear on rivets	12,000 pounds per square inch.
Bearing on rivets	24,000 pounds per square inch.

The computed working loads are given in Column 3 of Table 2. Methods of computing the working loads on connections of the types used in the test pieces have not been standardized. The methods used in computing the working loads given in Table 2 are as follows:

TABLE 2

LOADS ON TEST PIECES

Test Piece	Load at Failure on One Girder Lb.	Working Load Lb.[1]	1.5 Times Working Load Lb.[2]	Load at Failure Divided by Working Load	Moment on Connection at Failure In. Lb.	Manner of Failure
(1)	(2)	(3)	(4)	(5)	(6)	(7)
*A*1	44,450	17,500	26,250	2.54	5,334,000	Column buckled
*A*2	37,500	18,540	27,800	2.02	4,500,000	Gusset plate buckled
*A*3	10,250	3,030	4,545	3.38	537,000	Lug angle opened
*A*4	13,550	5,000	7,500	2.71	570,000	Column buckled and angle opened
*A*5	19,000	3,400	5,100	5.59	1,860,000	Connection angle opened. Rivet failed in tension
*A*6	11,250	2,750	4,125	4.10	506,000	Connection angles opened

1 Based upon the following unit stresses:
Axial bending stress, 16,000 lb. per sq. in.
Shear on rivets, 12,000 " " " "
Bearing, Rivets, 24,000 " " " "
2 Corresponding to allowable working stresses for wind loads.

*A*1.—The strength of the connection of test piece *A*1 to resist moment is considered to be the moment of a couple composed of two horizontal forces, one force is applied at the centroid of each flange; the magnitude of each force is the working strength in bearing of the eleven rivets connecting each flange of the girder to the gusset plate. The working load on the girder is the load which applied at the outer end of the girder produces the given moment on a vertical section through the outer row of rivets in the gusset plate. The vertical shear is considered to be taken by the vertical splice plates connecting the girder web to the gusset plate.

*A*2.—The working load for test piece *A*2 was determined in the same manner as for test piece *A*1.

*A*3.—The strength of the connection of test piece *A*3 to resist moment is considered to be the moment of a couple composed of two horizontal forces; one force is applied at the top surface of the girder and the other at the bottom surface of the girder. The magnitude of each force is the working strength in shear of the two rivets connecting the lug angle to the girder. The working load on the girder is the load which applied at the outer end of the girder produces the given moment on a vertical section through the rivets connecting the lug angles to the girder.

*A*4.—The strength of the connection of test piece *A*4 to resist moment is determined by the same method as for test piece *A*3. The working load on the girder is the load which applied at the outer end of the girder produces the given moment on a vertical section through the middle rivet connecting the lug angle to the girder.

*A*5.—If the strength of the connection of test piece *A*5 to resist moment is regarded as determined by the strength to resist moment of the rivets attaching the connection angles to the girder, if the outer rivet at each end of the connection angles is considered to be in double shear, if the strength of the intermediate rivets is considered as determined by bearing on the web plate of the girder, reduced one-third for the loose fill, and if the stress in the rivets varies as the distance from the center of gravity of all the rivets, the working load on the girder is the load which applied at the outer end of the girder produces the given moment on a vertical section through the rivets attaching the connection angles to the girder. The vertical stress upon the rivets is neglected.

*A*6.—Consider the strength of the connection of test piece *A*6 to resist moment to be determined by the strength of the rivets attaching the connection angles to the girder web, consider the stress on each rivet due to moment to vary as the distance from that rivet to the center of gravity of all the rivets, and consider the vertical shear to be evenly distributed over all the rivets. The working load upon the girder is the load which applied at the outer end of the girder produces on a vertical section through the center of gravity of the rivets attaching the connection angles to the girder web a moment equal to the resisting moments of the rivets.

The unit stresses which have been used in determining the working loads on the girders as given in Column 3 of Table 2 are the usual allowable working stresses due to dead and to live loads. This type of connections for the test pieces is used largely for building frames in which the bending stresses are due to wind loads. When wind load stresses are combined with dead and live load stresses, the allowable working stresses are usually fifty per cent greater than the allowable working stresses for dead and live loads alone. In discussing the results of the tests, loads one and one-half times the working loads for dead and live loads alone, corresponding to wind load working stresses, are used. These loads are given in Column 4 of Table 2.

10. *Interpretation of Results.*—For the curves of Figs. 20 to 25, the relation of the deviation of the full lines from the dotted lines to the deviation of the dotted lines from the vertical axis, that is, the relation of the slip to the elastic strain, is meaningless inasmuch as the direction of the dotted line, or the magnitude of the elastic strain, is dependent upon the distances of the points A, B, and C from O, Fig. 8. In order to determine the practical significance of the slip in the analysis of the wind stresses in the frame of an office building, the deflection and the stresses in a rectangular frame were computed for the case in which the connections are assumed to be perfectly rigid and for the case in which the connections slip by the amounts determined in the test.

The change in the deflection and the change in the distribution of the moments in a frame due to slip depend upon the $\frac{I}{l}$ of the members of the frame, that is, they depend upon the lengths as well as upon the sections of the members. In the discussion of the effect of slip, a frame fifteen feet high and twenty feet long is considered. This frame corresponds to average practice in office building construction.

The effect of slip upon the distribution of stresses in a rectangular frame depends upon differences in the slips rather than upon the slips themselves. If more specimens had been tested, it is probable that greater differences might have been found in slips than in the slips obtained from the two specimens of each type tested. This fact must be kept in mind in considering the interpretation of the results.

Two methods have been used in interpreting the results of the tests. In Sections 11 to 16, the stresses were determined in a rectangular frame for which the slips in the connections were taken equal to the slips corresponding to the loads given in Column 4 of Table 2, as measured in the tests. In Section 17 the stresses were determined in a rectangular frame for which the slips in two connections were taken equal to the maximum slip corresponding to the loads given in Column 4 of Table 2 as measured, and the slips in the other two connections were taken equal to one-half of the maximum slip. In other words, the differences in the slips were taken equal to one-half of the maximum measured slip.

The interpretation of the results in Sections 11 to 17, inclusive, are therefore based upon arbitrarily fixed conditions, and that fact must be kept in mind in considering the interpretations presented.

11. *Test Piece A*1.—The angular strains for test piece *A*1 are shown graphically in Fig. 20. The strain measured by dial *UN*8 should equal the sum of the strains measured by dials *N*4 and *UN*1. A similar relation exists between the strains measured by dials *LN*8, *N*4, and *LN*1; *US*8, *S*4, and *US*1; and *LS*8, *S*4, and *LS*1. These quantities which should be equal are approximately equal, indicating that the angular strains as measured are at least reasonably accurate. The fact that the observed quantities for *UN*8, *LN*8, *US*8, and *LS*8 are on smooth curves and the fact that the curves are so nearly alike are additional reasons for believing that the angular strains measured by these dials are reasonably accurate. The discussion which follows is based upon the angular strains as measured by dials *UN*8, *LN*8, *US*8, and *LS*8.

In Fig. 20, comparing the full line curves with the broken-line curves, it is apparent that for small loads the angular strain of C and *B* relative to *A* (Fig. 8), as measured and as computed, agree very closely, but that as the load increases, the difference between the measured and the computed strains increases very rapidly.

If the slip corresponds to a load one and one-half times the usual working dead and live loads, the load, as given in Column 4 of Table 2, is 26,250 pounds. Reading from the curves of Fig. 20, the slip measured in radians, that is, the quantities represented by the horizontal distances between the full lines and the broken lines, corresponding to a load of 26,250 pounds are .0010 for *UN*8, .0013 for *LN*8, and .0025 for *US*8 and *LS*8. The properties of the columns and girders are given in Table 1.

A load of 26,250 pounds on the end of a girder, as tested, produces a moment of 3,150,000 inch-pounds on the connection between the column and the girder. With a rectangular frame subjected to a shear as shown in Fig. 1, if the moment on each connection is 3,150,000 inch-pounds, since the total shear on the frame times the height of the frame equals the sum of the moments on all four connections, $Ph = 4 \times 3{,}150{,}000 = 12{,}600{,}000$ inch-pounds.

From the tests the slips in the connections due to a load upon the girder of 26,250 pounds as measured by dials *UN*8, *LN*8, *US*8, and *LS*8 and as given previously are .0010, .0013, .0025, and .0025. These quantities correspond to RA, RB, RC, and RD of equation (31).

From Table 1:

$$K = 8.2$$
$$nK = 3.25$$
$$n = .4$$
$$R = 1/4 \left[.0073 + .0302 \right] = .009375.$$

$d = 180 \times .009375 = 1.688$ inches, horizontal deflection of the top of the frame relative to the bottom.

If all the connections are perfectly rigid,

$$R = 1/4 \times .0302 = .0075, \text{ and}$$
$$d = 180 \times .0075 = 1.35 \text{ inches.}$$

The increase in the deflection of the columns from the vertical due to the slip is therefore $1.688 - 1.350 = .338$ inches, or 25 per cent.

Referring to Equation (31), d, the horizontal deflection of the frame which equals $R \times l$, is made up of two quantities, one quantity contains the slip in the connections and the other quantity contains n and K. The first quantity is the deflection due to slip; the second quantity is the deflection due to the elastic strain of the material. The first quantity is independent of the sections of the members; the second quantity decreases as both n and K increase. The actual deflection of a frame due to slip in the connections is, therefore, independent of the K's of the members, but the ratio of the deflection due to slip in the connections to the deflection due to the elastic strain of the material increases as the K's of the members increase.

If the slips in all the connections are equal, the distribution of the stresses is the same as if the connections were perfectly rigid.* If the slips in the connections are not equal, the effect of the slip upon the distribution of the moments depends upon the location of the connections at which the different slips occur. The three following distributions of the slips have been considered:

Case I

$RA = .0025$, slip at A in radians
$RB = .0025$, slip at B in radians
$RC = .0010$, slip at C in radians
$RD = .0013$, slip at D in radians

* See page 30.

Case II

$RA = .0025$, slip at A in radians
$RB = .0010$, slip at B in radians
$RC = .0025$, slip at C in radians
$RD = .0013$, slip at D in radians

Case III

$RA = .0025$, slip at A in radians
$RB = .0010$, slip at B in radians
$RC = .0013$, slip at C in radians
$RD = .0025$, slip at D in radians

For Case I the large slips occur at the tops of the columns, for Case II one large slip occurs at the top of one column, and the other large slip occurs at the bottom of the other column, and for Case III one large slip occurs at the top of one column, and the other large slip at the bottom of the same column. The moments M_{AD}, M_{DA}, M_{BC}, and M_{CB}, as given by equations (27) to (30), are given in Table 3.

TABLE 3

EFFECT OF SLIP IN CONNECTIONS OF A1 UPON DISTRIBUTION OF MOMENTS IN RECTANGULAR FRAME

CASE I

$K = 8.2$, $n = .4$ (See Table 2)

Moment	Error[1]
$M_{AD} = 3{,}030{,}000$ in. lb.	—3.8 per cent
$M_{DA} = 3{,}240{,}000$ in. lb.	+3.0 per cent
$M_{BC} = 3{,}050{,}000$ in. lb.	—3.0 per cent
$M_{CB} = 3{,}300{,}000$ in. lb.	+4.6 per cent

TABLE 3 (Continued)

CASE II

$K = 8.2$, $n = .4$ (See Table 2)	
Moment	Error[1]
M_{AD} = 3,050,000 in. lb.	—3.2 per cent
M_{DA} = 3,225,000 in. lb.	+2.0 per cent
M_{BC} = 3,275,000 in. lb.	+4.0 per cent
M_{CB} = 3,060,000 in. lb.	—3.0 per cent

CASE III

Moment	Error[1]
M_{AD} = 2,980,000 in. lb.	—5.4 per cent
M_{DA} = 2,978,000 in. lb.	—5.5 per cent
M_{BC} = 3,360,000 in. lb.	+6.8 per cent
M_{CB} = 3,308,000 in. lb.	+5.0 per cent

[1] "Error" in this table means the difference between the moments determined from equations (27) to (30) and the moments based upon the assumptions that the connections are perfectly rigid and that each member maintains a constant section up to the neutral axis of the member which it intersects.

It is apparent from Table 3 that for the frame considered the slip does not materially affect the distribution of the stresses. It remains to determine the effect of the size of the section of the members upon the change in the distribution of the stresses resulting from the slip. It is to be expected that slip in the connections of a rectangular frame will have a greater effect upon the distribution of the stresses if the members of the frame are short and stiff than if they are long and flexible. Consider a frame exactly like the one for which the moments have been determined except that the I's of the columns are changed. That is, n of equations (27) to (30) is assigned different values. The moments in frames having values of n equal to 1 and 2 were computed on the basis that the connections are perfectly rigid, and also upon the basis that the slip is the same as specified for Case I, Case II, and Case III (page 34). The differences in results obtained by the two methods are designated as the errors due to the assumptions that the connections are perfectly rigid and that each member maintains its section up to the axis of the member which it intersects. The errors are represented graphically in Fig. 18.

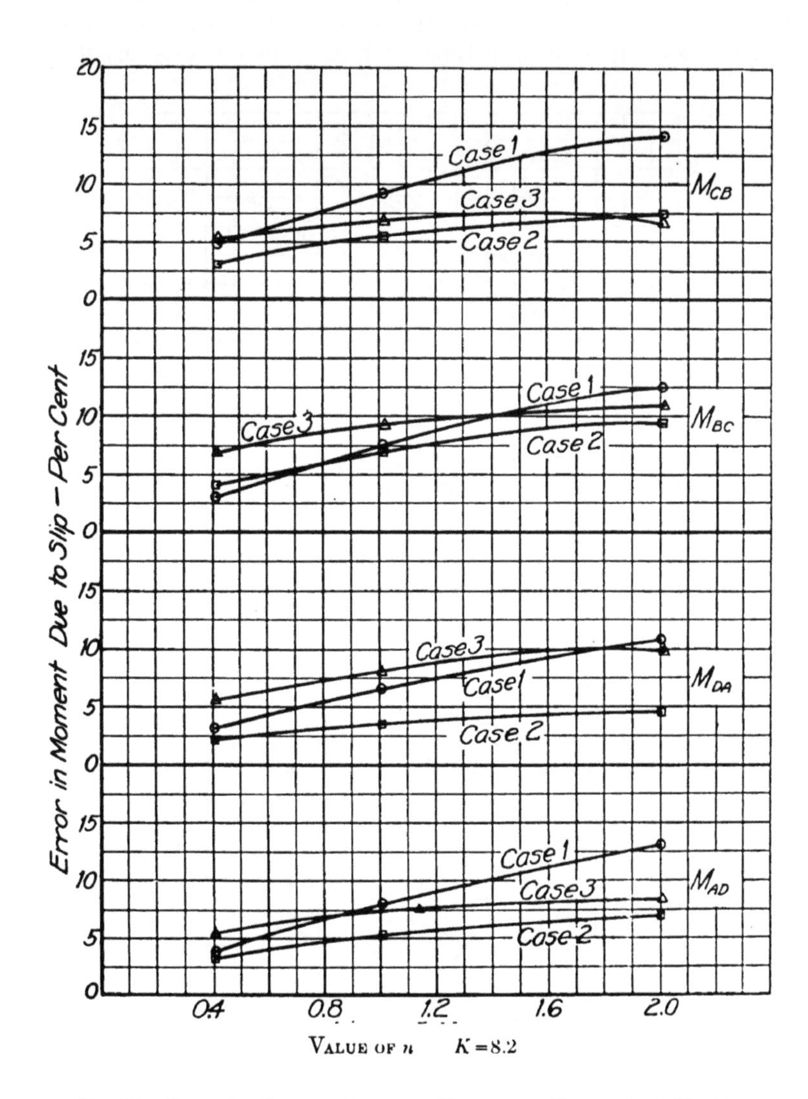

Fig. 18. Error in Computations for Moment in Frame with Varying Values of n

In order to study further the effect of the stiffness of the members of a frame upon the error due to slip in the connections, the moments were determined in frames for which the columns and the

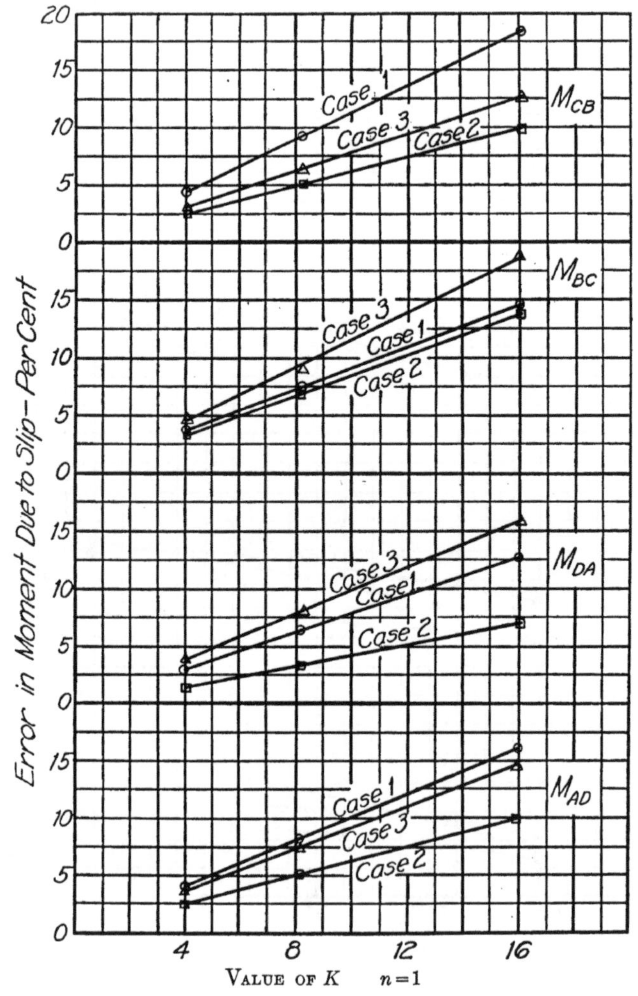

VALUE OF K $n=1$

FIG. 19. ERROR IN COMPUTATIONS FOR MOMENT IN FRAME WITH VARYING VALUES OF K

girders have the same values of K, that is, n is made equal to 1, and for which K for both columns and girders has the values 4, 8, 12, and 16. The slips in the connections and the total moment upon the frame were taken the same as in Case I, II, and III. The magnitude of the errors due to slip is represented in Fig. 19.

It is apparent from Figs. 18 and 19 that in judging the seriousness of the error in the moments in a frame due to slip in the connections, it is necessary to consider the K's of the members. The girder of $A1$ is about 8.5 per cent stronger in moment than the connection. The distances between columns of bents of office-building frames are usually between fifteen feet and twenty-five feet. It, therefore, seems that the value of K equaling 8.2 for the girder of $A1$, corresponding to a distance between columns of twenty feet, is as large as would be used in an office building in conjunction with the connection of test piece $A1$.

The size of a column of an office building is usually determined by the axial load rather than by the bending moment due to the wind load. The relation between the axial stress in a column due to dead and to live load and the bending stress in the column due to the wind load depends upon many features of the building and is distinct for each building.

Taking a position intermediate between the two extremes, a column section consisting of one web plate 14 by 5/8, four flange angles 5 by 4 by 5/8 and two cover plates 14 by 1-1/8, having a moment of inertia of 3006 in.^{4}, was used at a point on the column where the moment on the connection corresponded to a load of 26,250 pounds upon the girder of $A1$ as tested. For this column $n = \dfrac{3006 \times 20}{1966 \times 15} = 2$. With K equal to 8.2 and n equal to 2, the maximum error is in M_{CB} under Case I and is slightly less than 15 per cent.

In the tests, furthermore, the loads were so applied that slip in a connection did not reduce the moment to which the connection was subjected, whereas in engineering structures the moment is automatically transferred from a connection which slips to the more rigid connections. The addition to the moment on the more rigid connections and the reduction of the moment on the less rigid connections tend to make the slip on all the connections more nearly equal than in the case of the specimens tested.

It seems, therefore, that the tests of the two specimens marked $A1$ support the following statement:

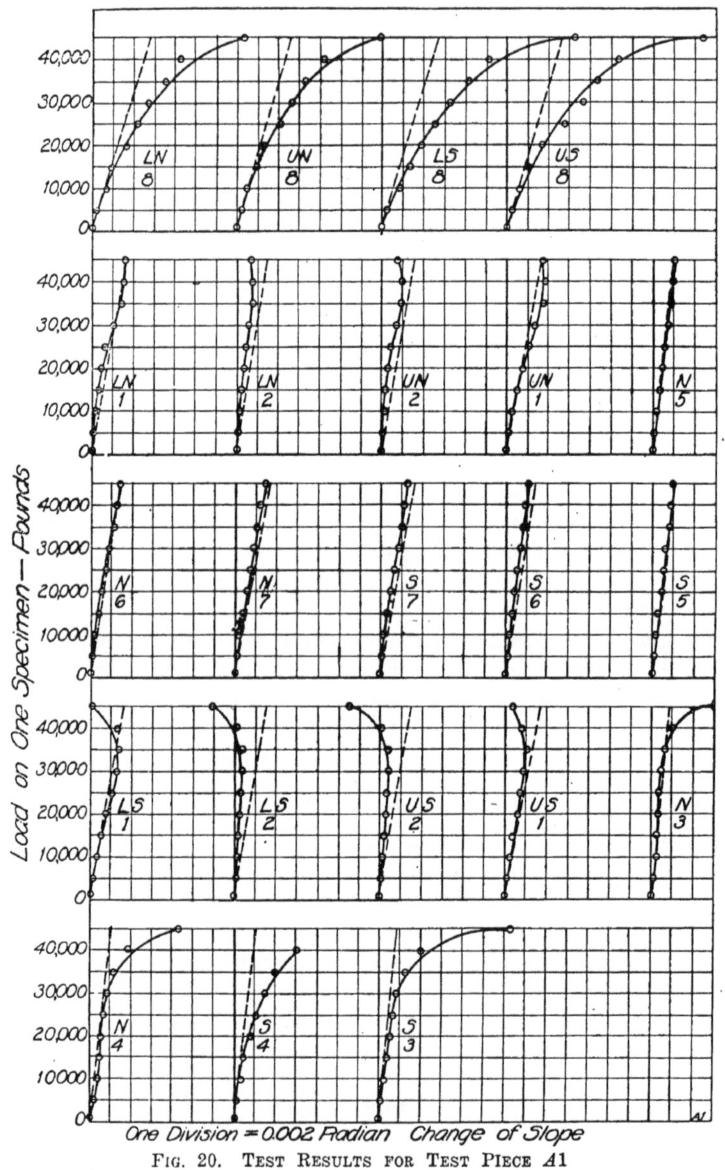

FIG. 20. TEST RESULTS FOR TEST PIECE A1

NOTE.—In Figs. 20 to 25 the location of the measurements is given by a combination of letters and numbers. The letters refer to general location. *U*, upper side ; *L*, lower side; *N*, north end; *S*, south nd; *E*, east side; *W*, west side. The numbers refer to the location of the measurements on the test piece as shown by Figs. 10 to 15, inclusive.

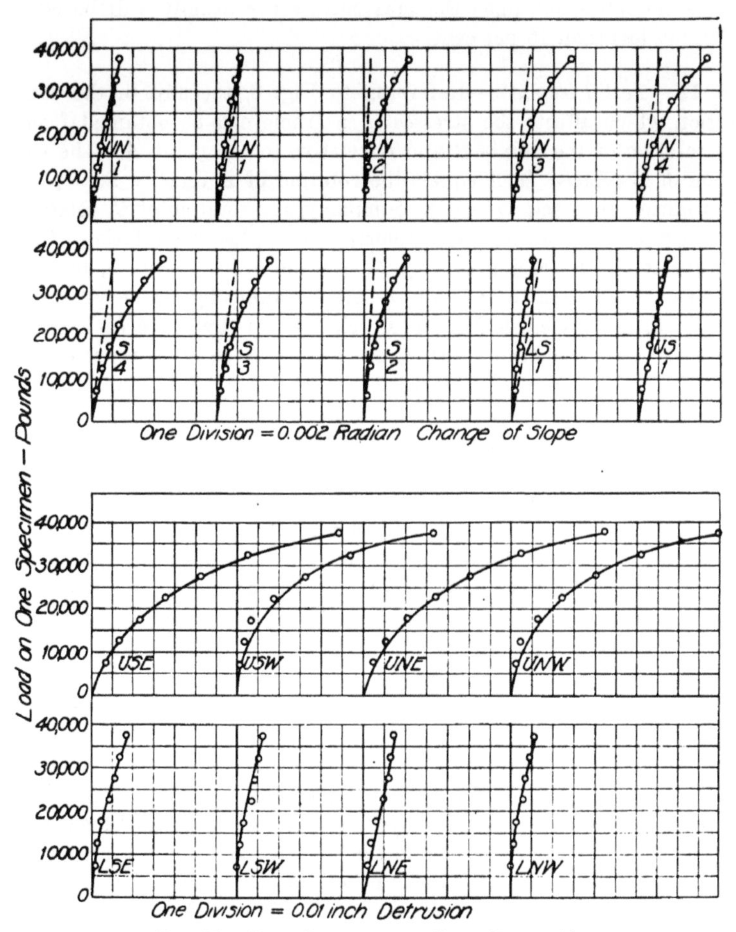

FIG. 21. TEST RESULTS FOR TEST PIECE A2

For rectangular frames having opposite sides alike, the assumptions that the connections are perfectly rigid and that each member maintains its section up to the neutral axis of the member to which it is connected produce errors in the moments which are dependent upon the stiffness of the members. The magnitude of the error depends upon the K's of all the members of the frame. For wind braces for office buildings the maximum error usually will be considerably less than 15 per cent.

12. *Test Piece A2.*—Referring to Fig. 8, the slip in the connection which affects the stresses in a frame can be obtained from the rotation of the points *B* and *C* relative to the point *A*. The rotation of *B* relative to *A* equals the rotation of *B* relative to *O* plus the rotation of *A* relative to *O*. Similarly, the rotation of *C*, relative to *A* equals the rotation of *C* relative to *O* plus the rotation of *A* relative to *O*. Referring to Fig. 21, for the north specimen, the rotation of *B* relative to *O* is represented graphically by curve *UN*1, *A* relative to *O* by curve *N*4, and *C* relative to *O* by curve *LN*1. For the south specimen, the rotation of *B* relative to *O* is represented graphically by curve *US*1, *A* relative to *O* by curve *S*4, and *C* relative to *O* by curve *LS*1. The slip of *B* relative to *A* is represented by the sum of the horizontal distances between the full lines and the broken lines for curves *UN*1 and *N*4 for the north specimen, and by the sum of the corresponding distances for curves *US*1 and *S*4 for the south specimen. Similarly the slip of *C* relative to *A* is represented by the sum of the horizontal distances between the full lines and the broken lines of curves *LN*1 and *N*4 for the north specimen, and by the sum of the corresponding distances for curves *LS*1 and *S*4 for the south specimen.

With a load of 27,800 pounds on the girder (see Column 4, Table 2), the rotation of *B* relative to *A* due to slip, obtained as outlined previously, is .0018 for the north specimen and .002 for the south specimen. The rotation of *C* relative to *A* due to slip is .0013 for the north specimen and .0014 for the south specimen. The differences between the slips for specimen *A*2 are only about 42 per cent as great as the differences in the slips for *A*1, and the errors in the moments based upon the assumption that the connections are perfectly rigid are accordingly less for *A*2 than for *A*1.

The fact that the lower part of the connection is more rigid than the upper part is due to the fact that the lower part is in compression and the upper part is in tension. The lug angles of the connection and the flange angles of the column are more rigid to resist compression than tension. If the frame represented by Fig. 1 has connections like *A*2, the lug angles and flange angles are in compression for connections at *B* and *D* and are in tension for connections at *A* and *C*.

In order to determine the error in the moments in a frame due to slip in the connections of the type used for *A*2, a frame fifteen feet high and twenty feet long with column and girder sections the same as for specimens *A*2 was used. For a moment on each connection corresponding to a load of 27,800 pounds on a girder as the specimens were tested, let the slip at *A* and *C* equal the average slips of the two connections having lug angles in compression and let the slip at *B* and *D* equal the average of the slips of the two connections having the lug angles in tension. That is,

$$RC = RA = \frac{.0018 + .0020}{2} = .0019, \text{ slips at } A \text{ and } C \text{ in radians.}$$

$$RD = RB = \frac{.0013 + .0014}{2} = .00135, \text{ slips at } B \text{ and } D \text{ in radians.}$$

$$Ph = 4 \times 27{,}800 \times 120 = 13{,}350{,}000 \text{ inch-pounds.}$$

From Table 1,

$$K = 8.2 \text{ in.}^3$$
$$nK = 3.25 \text{ in.}^3$$
$$n = .4$$

Substituting these values in equations (27) to (30) gives the moments in the frame. The moments and the errors due to slip are given in Table 4.

TABLE 4

EFFECT OF SLIP IN CONNECTIONS OF *A*2 UPON DISTRIBUTION OF MOMENTS IN RECTANGULAR FRAME

$K = 8.2$, $n = .4$ (See Table 2.)	
Moment	Error[1]
$M_{AD} = 3{,}300{,}000$ in. lb.	—1 per cent
$M_{DA} = 3{,}375{,}000$ in. lb.	+1 per cent
$M_{BC} = 3{,}375{,}000$ in. lb.	+1 per cent
$M_{CB} = 3{,}300{,}000$ in. lb.	—1 per cent

1 "Error" in this table means the difference between the moments determined from equations (27) to (30) and the moments based upon the assumptions that the connections are perfectly rigid and that each member maintains a constant section up to the neutral axis of the member which it intersects.

Comparing Table 4 with Table 3 it is apparent that with $n = .4$ the maximum error in the moments due to slip is much less for *A*2 than for *A*1.

The horizontal deflection of the frame with connections like the connections used for $A2$ as given by equation (31) is 1.728 inches, of which .288 inches or 20 per cent are due to the slip in the connections.

With $n = 2$ and all other quantities the same as before, the moments are as given in Table 5.

TABLE 5

EFFECT OF SLIP IN CONNECTIONS OF $A2$ UPON DISTRIBUTION OF MOMENTS IN RECTANGULAR FRAME

$K = 8.2$, $n = 2.0$ (See Table 2.)	
Moment	Error[1]
$M_{AD} = 3{,}250{,}000$ in. lb.	—2.7 per cent
$M_{DA} = 3{,}430{,}000$ in. lb.	+3 per cent
$M_{BC} = 3{,}430{,}000$ in. lb.	+3 per cent
$M_{CB} = 3{,}250{,}000$ in. lb.	—2.7 per cent

[1] "Error" in this table means the difference between the moments determined from equations (27) to (30) and the moments based upon the assumptions that the connections are perfectly rigid and that each member maintains a constant section up to the neutral axis of the member which it intersects.

Comparing Table 5 with Table 3, with $n = 2.0$, the maximum error in the moments due to slip is 3 per cent for $A2$; whereas the corresponding error due to slip for $A1$ is 9.5 per cent.

Although sufficient tests have not been made to justify a final conclusion, the tests of the two specimens indicate that the type of connection used for specimen $A2$ is more rigid than that used for $A1$. The relation between moment and slip was, moreover, more nearly the same for the two specimens $A2$ than for the two specimens $A1$. Inasmuch as the strain for $A2$ was largely due to the elastic strain of the material, a quantity independent of the workmanship of the assembler, whereas the strain for $A1$ was largely due to slip of rivets, a quantity dependent upon the workmanship of the assembler, it is reasonable to expect the connections of $A2$ to behave more consistently than the connections of $A1$. Since it is differences of slips rather than absolute slips which affect the distribution of moments, slip of the connections $A2$ will cause less error in the analysis of the stresses than slip of the connections $A1$. With the type of connection used for specimen $A2$, for frames of usual proportions, furthermore, the assumptions that the connections are perfectly rigid and that a member maintains its section up to the neutral axis of the member to which it is connected produce a very small error in the moments in the frame.

13. *Test Piece A3.*—Referring to Fig. 22 the rotation of the girder relative to the column is measured by dial N for the north

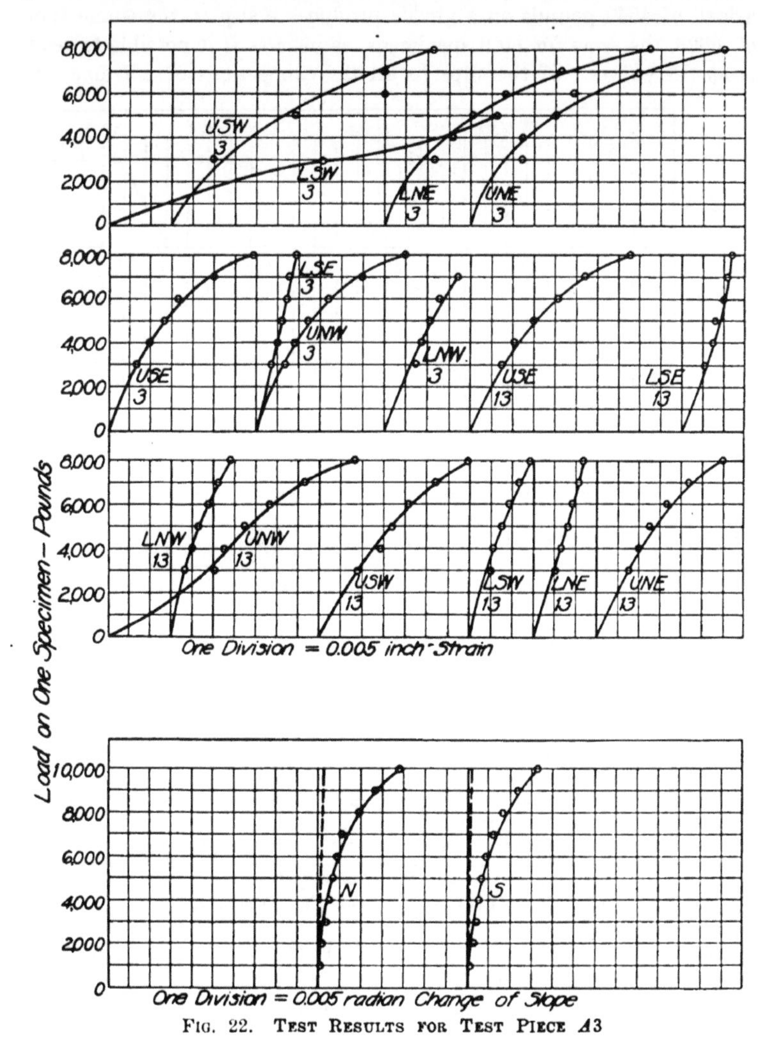

FIG. 22. TEST RESULTS FOR TEST PIECE A3

specimen and by dial S for the south specimen. The slips in the connections are measured by the horizontal distances between the full-line and the broken-line curves.

Referring to Table 2, one and one-half times the working load for usual dead and live load stresses is 4,545 pounds. From Fig. 22, a load of 4545 pounds on a girder produces a slip in the connection of .0028, the same for both specimens, as near as it is possible to read from the curves. If the slips in all the connections of a frame are

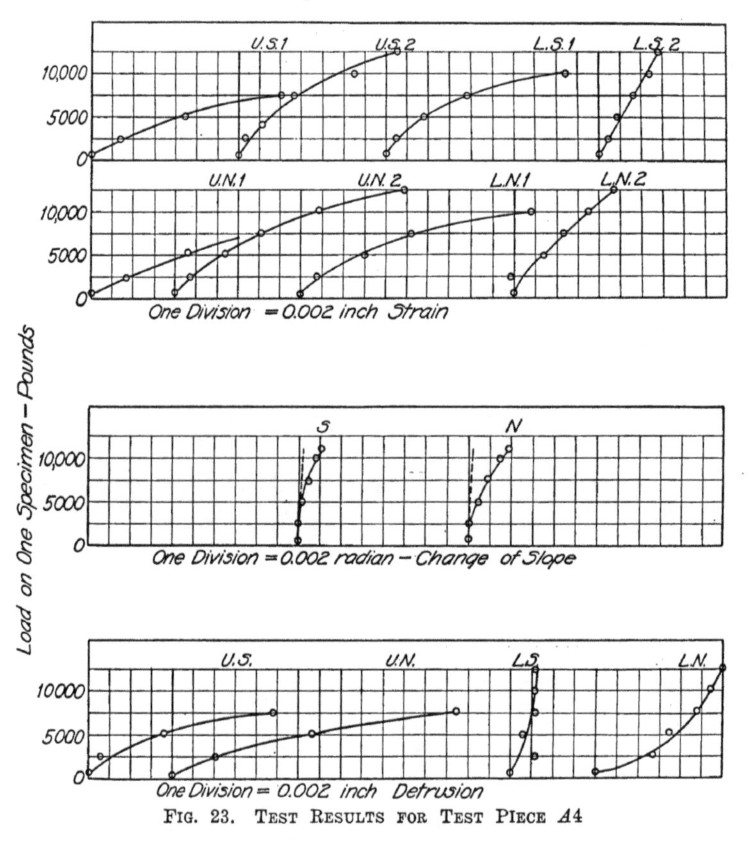

FIG. 23. TEST RESULTS FOR TEST PIECE *A*4

equal, the distribution of the moments is the same as if the connections were perfectly rigid.

Although large equal slips in the connections of a frame do not affect the distribution of the moments, a large slip in the connections does seriously affect the stiffness of the frame. If a frame fifteen feet high and twenty feet long made up of 12 inch — 31 1/2 pound I-beams is joined with connections like the ones used for *A*3, the horizontal deflection of the columns due to horizontal shear on the frame producing a slip of .0028 in the connections, as given by equation (31), is 1.260 inches, of which .672 inches or 53.4 per cent are due to the slip in the connections.

14. *Test Piece A*4.—Referring to Fig. 23, rotation of the girder relative to the column is measured by dial *N* for the north specimen and by dial *S* for the south specimen. The slips in the connections are measured by the horizontal distances between the full-line and the broken-line curves.

Referring to Table 2, one and one-half times the working load for usual dead and live load stresses is 7,500 pounds.

From Fig. 23, a load of 7,500 pounds on a girder produces a slip in the connection of .00144 radians for the north specimen and .00072 radians for the south specimen. In each case the angle given is the slip between *A* and *O*. The slip upon which the distribution of the moments depends is the slip between *A* and *B*, and between *A* and *C*. This slip was not determined for these specimens, and it is therefore impossible to determine the effect of the slip upon the distribution of the moments. Judging from the results of this test, however, for connections which are apparently alike, a given moment produces radically different slips in different connections, and the distribution of the moment is therefore seriously affected by the slip in the connections.

15. *Test Piece A*5.—Referring to Fig. 24, rotation of the girder relative to the column is measured by dials *UN, LN,* and *N*3 for the north specimen, and by dials *US, LS,* and *S*3 for the south specimen. The total slip in the connection for the upper part of the column is the slip between *B* and *O* plus the slip between *A* and *O*. This is represented graphically by the sum of the horizontal distances between the full-line and the broken-line curves for *UN* and *N*3. The total slip in the connection for the lower part of the column is the slip between *C* and *O* plus the slip between *A* and *O*. This is represented graphically by the sum of the horizontal distances between the full-line and broken-line curves *LN* and *N*3.

Referring to Table 2, one and one-half times the working load for usual dead and live load stresses is 5,100 pounds. Referring to Fig. 24, the slips in the connections for a load of 5,100 pounds are

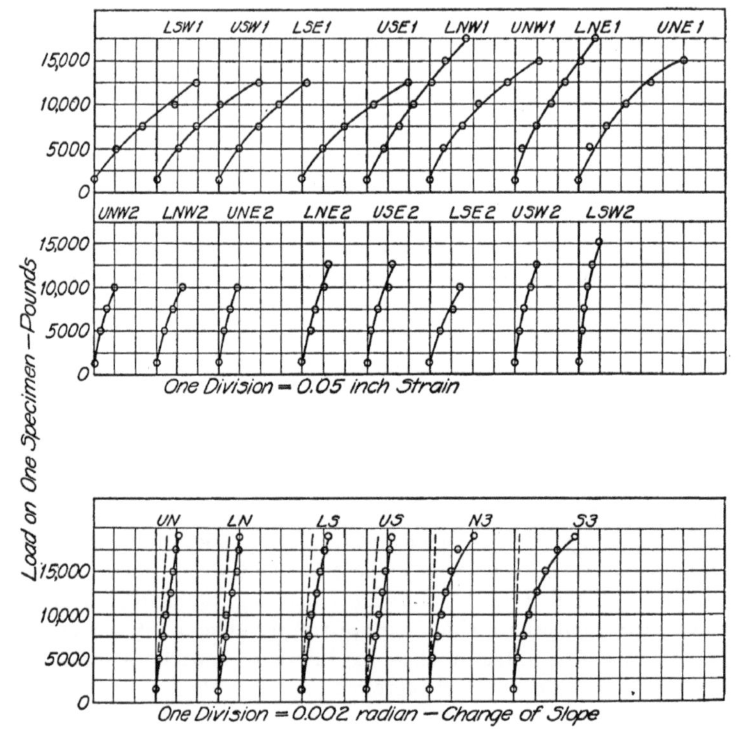

FIG. 24. TEST RESULTS FOR TEST PIECE $A5$

equal for all connections as nearly as can be determined from the curves, but for larger loads the slip in the south specimen is a little greater than in the north specimen. That is, although the connections slip, the slips in the different connections at working loads are so nearly alike that the moment in a rectangular frame is distributed almost exactly the same as it is when all the connections are perfectly rigid.

To determine the effect of the slip in the connections upon the horizontal deflection in a rectangular frame, consider a frame fifteen feet high and twenty feet long having column and girder sections and connections, like the ones used for $A5$. As determined from equation (31) the deflection of the frame when the moment in each connection equals the moment in the connection of $A5$ due to a load of 5,100 pounds is .339 inches of which .122 inches, or 36 per cent, are due to slip in the connections.

16. *Test Piece A6.*—Referring to Fig. 25, the rotation of the

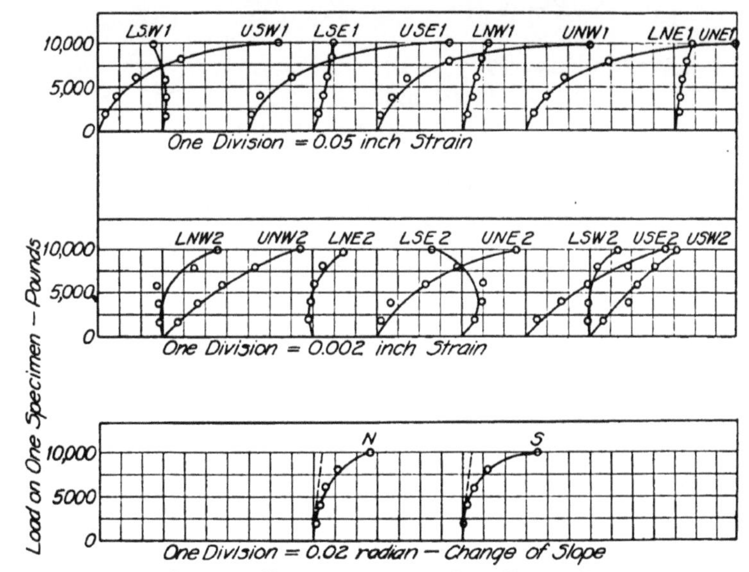

FIG. 25. TEST RESULTS FOR TEST PIECE A6

girder relative to the column is measured by dial N for the north specimen and by dial S for the south specimen. The slip is represented graphically by the horizontal distance between the full-line and the broken-line curves marked N, for the north specimen and by the corresponding distance on the curves marked S, for the south specimen.

From Table 2, one and one-half times the working load for usual dead and live load stresses is 4,125 pounds.

Referring to Fig. 25 for a load of 4,125 pounds on one girder the slip in the connection for the north specimen is .0052 radians, and the slip for the south specimen is .0033 radians.

Let a rectangular frame similar to Fig. 1, fifteen feet high and twenty feet, long be made up of girders and columns having the same section as the corresponding members of specimen A6 and joined with the type of connection used for A6. If the average moment at each connection of the frame is equal to the moment produced in the connection of A6 due to a load of 4,125 pounds on one girder, the quantity $Ph = 743{,}000$ inch-pounds. Consider the slip at A and D to be .0052 radians and at B and C to be .0033 radians.

$K = .9$
$nK = .755$
$n = .86$
Ph $= 743{,}000$ in. lb.
$RA = .0052$ slip at A in radians
$RB = .0033$ slip at B in radians
$RC = .0033$ slip at C in radians
$RD = .0052$ slip at D in radians

The horizontal deflection of the frame as given by equation (31) is 1.211 inches of which .765 inches or 63 per cent are due to slip in the connections.

The moments in the frame, as given by equations (27) to (30), are given in Table 6.

TABLE 6

EFFECT OF SLIP IN CONNECTIONS OF A6 UPON DISTRIBUTION OF MOMENTS IN RECTANGULAR FRAME

$K = .9$, $n = .86$ (see Table 2)	
Moment	Error[1]
$M_{AD} = 150{,}000$ in. lb.	—19.4 per cent
$M_{DA} = 150{,}000$ in. lb.	—19.4 per cent
$M_{BC} = 222{,}000$ in. lb.	+19.4 per cent
$M_{CB} = 222{,}000$ in. lb.	+19.4 per cent

1 "Error" in this table means the difference between the moments determined from equations (27) to (30) and the moments based upon the assumptions that the connections are perfectly rigid and that each member maintains a constant section up to the neutral axis of the member which it intersects.

If the frame is 7.5 feet high and 10 feet long,

$K = 1.8$ in.^{3}
$nK = 1.55$ in.^{3}
$n = .86$
$Ph = 743{,}000$ in. lb.
$RA = .0052$ slip at A in radians
$RB = .0033$ slip at B in radians
$RC = .0033$ slip at C in radians
$RD = .0052$ slip at D in radians

The horizontal deflection of the frame as given by equation (31) is .495 inches of which .382 inches or 77 per cent are due to slip in the connections.

The moments in the frame are given in Table 7.

TABLE 7

EFFECT OF SLIP IN CONNECTIONS OF A6 UPON DISTRIBUTION OF MOMENTS IN RECTANGULAR FRAME

$K = 1.8$, $n = .86$ (see Table 2)	
Moment	Error[1]
$M_{DA} = 113{,}000$ in. lb.	—39.0 per cent
$M_{DA} = 113{,}000$ in. lb.	—39.0 per cent
$M_{BC} = 260{,}000$ in. lb.	+40.0 per cent
$M_{CB} = 260{,}000$ in. lb.	+40.0 per cent

[1] "Error" in this table means the percentage of difference between the moment determined from equations (27) to (30) and the moments based upon the assumptions that the connections are perfectly rigid and that each member maintains a constant section up to the neutral axis of the member which it intersects.

17. *Comparison of Results of Tests of Different Test Pieces.*—Since the effect of slip in the connections upon the distribution of the stresses in a rectangular frame depends upon quantities independent of the type of connection used, in making any comparison of the different types of connections certain quantities must be fixed arbitrarily. It must be clearly borne in mind that the results of the comparison are true only under the conditions specified. Analyses of the effect of slip in the connections upon the stresses in a rectangular frame under certain arbitrary fixed conditions are presented in Sections 10 to 15, inclusive. As a further study, the effect of slip has been determined under the following arbitrarily fixed conditions:

(1) The moment on the connection is the moment due to a load one and one-half times the working load corresponding to usual dead and live load stresses; that is, it is the moment to which the connection would be subjected as a part of wind bracing. This load is given in Column 4 of Table 2 and is repeated in Column 2 of Table 8.

(2) The slips at A and B are taken equal to each other and equal to the maximum slip measured at the load specified. The

slips at *C* and *D* are taken equal to each other and equal to one half of the slips at *A* and *B*. Differences in slips rather than slips themselves affect the distribution of the stresses in the frame. The arbitrarily fixed differences in the slips are used rather than the differences obtained in the tests because it is considered that tests of two specimens cannot be relied upon to bring out the differences in the behavior of connections which are supposed to be identical. It is only fair, however, to point out that for *A*1 the differences in the slips for the two specimens as determined by tests were a little greater than one-half the maximum slip; whereas for *A*2 the differences in the slips were less than one half of the maximum slip as determined by the tests.

(3) The *K*'s of all members of the frame are taken equal to 10, that is, if the moment of inertia of the member is small the member is short, whereas if the moment of inertia of the member is large the member is long, a condition prevalent in structures.

The quantities used in the determination of the moments are given in Columns 2 to 4 of Table 8; the moments in the frames are given in Columns 5 and 6. The errors in the calculated moments resulting from the slip, due to the assumptions that the connections are perfectly rigid and that the member maintains its section up to the neutral axis of the member to which it is connected, are given in Column 7, all in Table 8.

From Column 7, Table 8, it is apparent that under the conditions assumed, the connections for specimens *A*1 and *A*2 can, for the purpose of analyzing stresses in a rectangular frame, be considered perfectly rigid without introducing prohibitive error, but the connections for specimens *A*3, *A*4, *A*5, and *A*6, for the purpose of analyzing stresses, cannot be considered perfectly rigid.

The error due to slip cannot exceed one hundred per cent of the computed moment. Errors greater than one hundred per cent given in Column 8 of Table 8 were obtained because, for the conditions upon which the error due to slip is based, slip in a connection is not considered as reducing the moment which produces the slip. As a matter of fact if a connection slips, the moment to which it is subjected is automatically transferred to other connections. The error in a frame due to slip in the connections is therefore less than the values given in Column 8 of Table 8, the difference between the true error and the values given in Table 8 depending upon the size of the error. The values given in Column 8 of Table 8 are therefore relative rather than actual errors. The actual errors are less than the values given in the table.

TABLE 8

COMPARISON OF EFFECTS OF SLIPS IN CONNECTIONS UPON MOMENTS IN RECTANGULAR FRAMES

$K=10,\ n=1.$

Slip at *A* and *B* is maximum slip as determined by tests at a load corresponding to one and one-half times the usual dead and live working loads. Slip at *C* and *D* is taken as one-half of the slip at *A* and *B*.

Test Piece	1.5 Times Working Load	Slip at A and $B=$ $RA=RB$	Slip at C and $D=$ $RC=RD$	$M_{AD}=M_{BC}$ in Lb.	$M_{DA}=M_{BC}$ in Lb.	Errors Due to Slip	$P\times h$ in Lb.
*A*1	26,250	.0025	.00125	2,870,000	3,434,000	9 per cent	12,600,000
*A*2	27,800	.0020	.0010	3,095,000	3,560,000	7 per cent	13,300,000
*A*3	4,545	.0028	.0014	—88,000	541,000	138 per cent	910,000
*A*4	7,500	.00144	.00072	161,000	483,300	50 per cent	1,290,000
*A*5	5,100	.0046	.0023	—22,700	1,023,000	104 per cent	2,000,000
*A*6	4,125	.0052	.0026	—439,000	805,000	435 per cent	742,000

18. *Conclusions.*—The main object of these tests was to determine whether serious error is introduced into computations for stresses in steel frames by the assumption that the joints are perfectly rigid.

The rigidity of various types of joints has been studied by means of tests, and the error introduced into computations studied by means of mathematical analysis of the action of frames with slip at the joints of a magnitude such as was observed in these tests. The action under a load producing stresses equal to one and one-half times the working stress has been taken as a criterion, and for the joints studied the following conclusions reached:

Connections of the type used for specimens *A*1 and *A*2 are so rigid that for the purpose of analyzing stresses in rectangular frames the connections can be considered as perfectly rigid without introducing serious errors into the results.

The errors due to slip in the connections is less for *A*2 than for *A*1.

Connections of the type used for specimens *A*3, *A*4, *A*5, and *A*6 for the purpose of analyzing stresses cannot be considered perfectly rigid.

LIST OF
PUBLICATIONS OF THE ENGINEERING EXPERIMENT STATION

Bulletin No. 1. Tests of Reinforced Concrete Beams, by Arthur N. Talbot, 1904. *None availab le*

Circular No. 1. High-Speed Tool Steels, by L. P. Breckenridge. 1905. *None available.*

Bulletin No. 2. Tests of High-Speed Tool Steels on Cast Iron, by L. P. Breckenridge and Henry B. Dirks. 1905. *None available.*

Circular No. 2. Drainage of Earth Roads, by Ira O. Baker. 1906. *None available.*

Circular No. 3. Fuel Tests with Illinois Coal (Compiled from tests made by the Technological Branch of the U. S. G. S., at the St. Louis, Mo., Fuel Testing Plant, 1904–1907), by L. P. Breckenridge and Paul Diserens. 1908. *Thirty cents.*

Bulletin No. 3. The Engineering Experiment Station of the University of Illinois, by L. P. Breckenridge. 1906. *None available.*

Bulletin No. 4. Tests of Reinforced Concrete Beams, Series of 1905, by Arthur N. Talbot. 1906. *Forty-five cents.*

Bulletin No. 5. Resistance of Tubes to Collapse, by Albert P. Carman and M. L. Carr. 1906. *None available.*

Bulletin No. 6. Holding Power of Railroad Spikes, by Roy I. Webber, 1906. *None available.*

Bulletin No. 7. Fuel Tests with Illinois Coals, by L. P. Breckenridge, S. W. Parr, and Henry B. Dirks. 1906. *None available.*

Bulletin No. 8. Tests of Concrete: I, Shear; II, Bond, by Arthur N. Talbot. 1906. *None available.*

Bulletin No. 9. An Extension of the Dewey Decimal System of Classification Applied to the Engineering Industries, by L. P. Breckenridge and G. A. Goodenough. 1906. Revised Edition 1912. *Fifty cents.*

Bulletin No. 10. Tests of Concrete and Reinforced Concrete Columns, Series of 1906, by Arthur N. Talbot. 1907. *None available.*

Bulletin No. 11. The Effect of Scale on the Transmission of Heat through Locomotive Boiler Tubes, by Edward C. Schmidt and John M. Snodgrass. 1907. *None available.*

Bulletin No. 12. Tests of Reinforced Concrete T-Beams, Series of 1906, by Arthur N. Talbot. 1907. *None available.*

Bulletin No. 13. An Extension of the Dewey Decimal System of Classification Applied to Architecture and Building, by N. Clifford Ricker. 1907. *None available.*

Bulletin No. 14. Tests of Reinforced Concrete Beams, Series of 1906, by Arthur N. Talbot. 1907. *None available.*

Bulletin No. 15. How to Burn Illinois Coal Without Smoke, by L. P. Breckenridge. 1908. *None available.*

Bulletin No. 16. A Study of Roof Trusses, by N. Clifford Ricker. 1908. None available.

Bulletin No. 17. The Weathering of Coal, by S. W. Parr, N. D. Hamilton, and W. F. Wheeler. 1908. *None available.*

Bulletin No. 18. The Strength of Chain Links, by G. A. Goodenough and L. E. Moore. 1908. *Forty cents.*

Bulletin No. 19. Comparative Tests of Carbon, Metallized Carbon and Tantalum Filament Lamps, by T. H. Amrine. 1908. *None available.*

Bulletin No. 20. Tests of Concrete and Reinforced Concrete Columns, Series of 1907, by Arthur N. Talbot. 1908. *None available.*

Bulletin No. 21. Tests of a Liquid Air Plant, by C. S. Hudson and C. M. Garland. 1908. *Fifteen cents.*

Bulletin No. 22. Tests of Cast-Iron and Reinforced Concrete Culvert Pipe, by Arthur N. Talbot. 1908. *None available.*

Bulletin No. 23. Voids, Settlement and Weight of Crushed Stone, by Ira O. Baker. 1908. *Fifteen cents.*

**Bulletin No. 24.* The Modification of Illinois Coal by Low Temperature Distillation, by S. W. Parr and C. K. Francis. 1908. *Thirty cents.*

Bulletin No. 25. Lighting Country Homes by Private Electric Plants, by T. H. Amrine. 1908 *Twenty cents.*

*A limited number of copies of bulletins starred is available for free distribution.

Bulletin No. 26. High Steam-Pressures in Locomotive Service. A Review of a Report to the Carnegie Institution of Washington, by W. F. M. Goss. 1908. *Twenty-five cents.*

Bulletin No. 27. Tests of Brick Columns and Terra Cotta Block Columns, by Arthur N. Talbot and Duff A. Abrams. 1909. *Twenty-five cents.*

Bulletin No. 28. A Test of Three Large Reinforced Concrete Beams, by Arthur N. Talbot. 1909. *Fifteen cents.*

Bulletin No. 29. Tests of Reinforced Concrete Beams: Resistance to Web Stresses, Series of 1907 and 1908, by Arthur N. Talbot. 1909. *Forty-five cents.*

**Bulletin No. 30.* On the Rate of Formation of Carbon Monoxide in Gas Producers, by J. K. Clement, L. H. Adams, and C. N. Haskins. 1909. *Twenty-five cents.*

**Bulletin No. 31.* Fuel Tests with House-Heating Boilers, by J. M. Snodgrass. 1909. *Fifty-five cents.*

Bulletin No. 32. The Occluded Gases in Coal, by S. W. Parr and Perry Barker. 1909. *Fifteen cents.*

Bulletin No. 33. Tests of Tungsten Lamps, by T H Amrine and A Guell. 1909. *Twenty cents.*

**Bulletin No. 34.* Tests of Two Types of Tile-Roof Furnaces under a Water-Tube Boiler, by J. M. Snodgrass. 1909. *Fifteen cents.*

Bulletin No 35. A Study of Base and Bearing Plates for Columns and Beams, by N. Clifford Ricker. 1909 *Twenty cents*

Bulletin No. 36. The Thermal Conductivity of Fire-Clay at High Temperatures, by J. K. Clement and W. L. Egy. 1909. *Twenty cents.*

Bulletin No. 37. Unit Coal and the Composition of Coal Ash, by S. W. Parr and W. F. Wheeler. 1909. *Thirty-five cents.*

**Bulletin No. 38.* The Weathering of Coal, by S. W. Parr and W. F. Wheeler. 1909. *Twenty-five cents.*

**Bulletin No. 39.* Tests of Washed Grades of Illinois Coal, by C. S. McGovney. 1909. *Seventy-five cents.*

Bulletin No. 40. A Study in Heat Transmission, by J K. Clement and C. M. Garland. 1910 *Ten cents.*

Bulletin No. 41. Tests of Timber Beams, by Arthur N. Talbot. 1910. *Thirty-five cents.*

**Bulletin No. 42.* The Effect of Keyways on the Strength of Shafts, by Herbert F. Moore. 1910. *Ten cents.*

Bulletin No. 43. Freight Train Resistance, by Edward C. Schmidt. 1910. *Seventy-five cents.*

Bulletin No. 44. An Investigation of Built-up Columns Under Load, by Arthur N. Talbot and Herbert F. Moore. 1911. *Thirty-five cents.*

**Bulletin No. 45.* The Strength of Oxyacetylene Welds in Steel, by Herbert L. Whittemore. 1911. *Thirty-five cents.*

**Bulletin No. 46.* The Spontaneous Combustion of Coal, by S. W. Parr and F. W. Kressman. 1911. *Forty-five cents.*

**Bulletin No. 47.* Magnetic Properties of Heusler Alloys, by Edward B. Stephenson, 1911. *Twenty-five cents.*

**Bulletin No. 48.* Resistance to Flow Through Locomotive Water Columns, by Arthur N. Talbot and Melvin L. Enger. 1911. *Forty cents.*

**Bulletin No. 49.* Tests of Nickel-Steel Riveted Joints, by Arthur N. Talbot and Herbert F. Moore. 1911. *Thirty cents.*

**Bulletin No. 50.* Tests of a Suction Gas Producer, by C. M. Garland and A. P. Kratz. 1912. *Fifty cents.*

Bulletin No. 51. Street Lighting, by J. M. Bryant and H. G. Hake. 1912. *Thirty-five cents.*

**Bulletin No. 52.* An Investigation of the Strength of Rolled Zinc, by Herbert F. Moore. 1912. *Fifteen cents.*

**Bulletin No. 53.* Inductance of Coils, by Morgan Brooks and H. M. Turner. 1912. *Forty cents.*

**Bulletin No. 54.* Mechanical Stresses in Transmission Lines, by A. Guell. 1912. *Twenty cents.*

**Bulletin No. 55.* Starting Currents of Transformers, with Special Reference to Transformers with Silicon Steel Cores, by Trygve D. Yensen. 1912. *Twenty cents.*

**Bulletin No. 56.* Tests of Columns: An Investigation of the Value of Concrete as Reinforcement for Structural Steel Columns, by Arthur N. Talbot and Arthur R. Lord. 1912. *Twenty-five cents.*

**Bulletin No. 57.* Superheated Steam in Locomotive Service. A Review of Publication No. 127 of the Carnegie Institution of Washington, by W. F. M. Goss. 1912. *Forty cents.*

*A limited number of copies of bulletins starred is available for free distribution.

**Bulletin No. 58*. A New Analysis of the Cylinder Performance of Reciprocating Engines, by J. Paul Clayton 1912. *Sixty cents.*

**Bulletin No. 59*. The Effect of Cold Weather Upon Train Resistance and Tonnage Rating, by Edward C. Schmidt and F. W. Marquis. 1912. *Twenty cents.*

**Bulletin No. 60*. The Coking of Coal at Low Temperatures, with a Preliminary Study of the By-Products, by S. W. Parr and H. L. Olin. 1912. *Twenty-five cents.*

**Bulletin No. 61*. Characteristics and Limitation of the Series Transformer, by A. R. Anderson and H. R. Woodrow. 1913. *Twenty-five cents.*

Bulletin No. 62. The Electron Theory of Magnetism, by Elmer H. Williams. 1913. *Thirty-five cents.*

Bulletin No. 63. Entropy-Temperature and Transmission Diagrams for Air, by C. R. Richards. 1913. *Twenty-five cents.*

**Bulletin No. 64*. Tests of Reinforced Concrete Buildings Under Load, by Arthur N. Talbot and Willis A. Slater. 1913. *Fifty cents.*

**Bulletin No. 65*. The Steam Consumption of Locomotive Engines from the Indicator Diagrams, by J. Paul Clayton. 1913. *Forty cents.*

Bulletin No. 66. The Properties of Saturated and Superheated Ammonia Vapor, by G. A. Goodenough and William Earl Mosher. 1913. *Fifty cents.*

Bulletin No. 67. Reinforced Concrete Wall Footings and Column Footings, by Arthur N. Talbot. 1913. *Fifty cents.*

**Bulletin No. 68*. The Strength of I-Beams in Flexure, by Herbert F. Moore. 1913. *Twenty cents.*

Bulletin No. 69. Coal Washing in Illinois, by F. C. Lincoln. 1913. *Fifty cents.*

Bulletin No. 70. The Mortar-Making Qualities of Illinois Sands, by C. C. Wiley. 1913. *Twenty cents.*

Bulletin No. 71. Tests of Bond between Concrete and Steel, by Duff A. Abrams. 1913. *One dollar.*

**Bulletin No. 72*. Magnetic and Other Properties of Electrolytic Iron Melted in Vacuo, by Trygve D. Yensen. 1914. *Forty cents.*

Bulletin No. 73. Acoustics of Auditoriums, by F. R. Watson. 1914. *Twenty cents.*

**Bulletin No. 74*. The Tractive Resistance of a 28-Ton Electric Car, by Harold H. Dunn. 1914. *Twenty-five cents.*

Bulletin No. 75. Thermal Properties of Steam, by G. A. Goodenough. 1914. *Thirty-five cents.*

Bulletin No. 76. The Analysis of Coal with Phenol as a Solvent, by S. W. Parr and H. F. Hadley. 1914. *Twenty-five cents.*

**Bulletin No. 77*. The Effect of Boron upon the Magnetic and Other Properties of Electrolytic Iron Melted in Vacuo, by Trygve D. Yensen. 1915. *Ten cents.*

**Bulletin No. 78*. A Study of Boiler Losses, by A. P. Kratz. 1915. *Thirty-five cents.*

**Bulletin No. 79*. The Coking of Coal at Low Temperatures, with Special Reference to the Properties and Composition of the Products, by S. W. Parr and H. L. Olin. 1915. *Twenty-five cents.*

**Bulletin No. 80*. Wind Stresses in the Steel Frames of Office Buildings, by W. M. Wilson and G. A. Maney. 1915. *Fifty cents.*

**Bulletin No. 81*. Influence of Temperature on the Strength of Concrete, by A B. McDaniel. 1915. *Fifteen cents.*

Bulletin No. 82. Laboratory Tests of a Consolidation Locomotive, by E. C. Schmidt, J. M. Snodgrass, and R. B. Keller. 1915. *Sixty-five cents.*

**Bulletin No. 83*. Magnetic and Other Properties of Iron-Silicon Alloys, Melted in Vacuo, by Trygve D. Yensen. 1915. *Thirty-five cents.*

Bulletin No. 84. Tests of Reinforced Concrete Flat Slab Structure, by A. N. Talbot and W. A. Slater. 1916. *Sixty-five cents.*

**Bulletin No. 85*. The Strength and Stiffness of Steel Under Biaxial Loading, by A. T Becker. 1916. *Thirty-five cents.*

**Bulletin No. 86*. The Strength of I-Beams and Girders, by Herbert F. Moore and W. M. Wilson. 1916. *Thirty cents.*

**Bulletin No. 87*. Correction of Echoes in the Auditorium, University of Illinois, by F. R. Watson and J. M. White. 1916. *Fifteen cents.*

Bulletin No. 88. Dry Preparation of Bituminous Coal at Illinois Mines, by E. A. Holbrook. 1916. *Seventy cents.*

---

* A limited number of copies of bulletins starred is available for free distribution

**Bulletin No. 89*. Specific Gravity Studies of Illinois Coal, by Merle L. Nebel. 1916. *Thirty cents.*

**Bulletin No. 90*. Some Graphical Solutions of Electric Railway Problems, by A. M. Buck. 1916. *Twenty cents.*

**Bulletin No. 91*. Subsidence Resulting from Mining, by L. E. Young and H. H. Stoek. 1916. *One dollar.*

**Bulletin No. 92*. The Tractive Resistance on Curves of a 28-Ton Electric Car, by E. C. Schmidt and H. H. Dunn. 1916. *Twenty-five cents.*

**Bulletin No. 93*. A Preliminary Study of the Alloys of Chromium, Copper, and Nickel, by D. F. McFarland and O. E. Harder. 1916. *Thirty cents.*

**Bulletin No. 94*. The Embrittling Action of Sodium Hydroxide on Soft Steel, by S. W. Parr 1917. *Thirty cents.*

**Bulletin No. 95*. Magnetic and Other Properties of Iron-Aluminum Alloys Melted in Vacuo, by T. D. Yensen and W. A. Gatward. 1917. *Twenty-five cents.*

**Bulletin No. 96*. The Effect of Mouthpieces on the Flow of Water Through a Submerged Short Pipe, by Fred B. Seely. 1917. *Twenty-five cents.*

**Bulletin No. 97*. Effects of Storage Upon the Properties of Coal, by S. W. Parr. 1917 *Twenty cents.*

**Bulletin No. 98*. Tests of Oxyacetylene Welded Joints in Steel Plates, by Herbert F. Moore 1917. *Ten cents.*

Circular No. 4. The Economical Purchase and Use of Coal for Heating Homes, with Special Reference to Conditions in Illinois. 1917. *Ten cents.*

**Bulletin No. 99*. The Collapse of Short Thin Tubes, by A. P. Carman. 1917. *Twenty cents*

**Circular No. 5*. The Utilization of Pyrite Occurring in Illinois Bituminous Coal, by E. A. Holbrook. 1917. *Twenty cents.*

**Bulletin No. 100*. Percentage of Extraction of Bituminous Coal with Special Reference to Illinois Conditions, by C. M. Young. 1917.

**Bulletin No. 101*. Comparative Tests of Six Sizes of Illinois Coal on a Mikado Locomotive, by E. C. Schmidt, J. M. Snodgrass, and O. S. Beyer, Jr. 1917. *Fifty cents.*

**Bulletin No. 102*. A Study of the Heat Transmission of Building Materials, by A. C. Willard and L. C. Lichty. 1917.

**Bulletin No. 103*. An Investigation of Twist Drills, by Bruce W. Benedict and W. Penn Lukens. 1917.

**Bulletin No. 104*. Tests to Determine the Rigidity of Riveted Joints of Steel Structures by Wilbur M. Wilson and Herbert F. Moore. 1917. *Twenty-five cents.*

---

*A limited number of copies of bulletins starred is available for free distribution.

THE UNIVERSITY OF ILLINOIS

THE STATE UNIVERSITY

Urbana

EDMUND J. JAMES, Ph. D., LL. D., President

THE UNIVERSITY INCLUDES THE FOLLOWING DEPARTMENTS:

The Graduate School

The College of Liberal Arts and Sciences (Ancient and Modern Languages and Literatures; History, Economics, Political Science, Sociology; Philosophy, Psychology, Education; Mathematics; Astronomy; Geology; Physics; Chemistry; Botany, Zoology, Entomology; Physiology; Art and Design)

The College of Commerce and Business Administration (General Business, Banking, Insurance, Accountancy, Railway Administration, Foreign Commerce; Courses for Commercial Teachers and Commercial and Civic Secretaries)

The College of Engineering (Architecture; Architectural, Ceramic, Civil, Electrical, Mechanical, Mining, Municipal and Sanitary, and Railway Engineering)

The College of Agriculture (Agronomy; Animal Husbandry; Dairy Husbandry; Horticulture and Landscape Gardening; Agricultural Extension; Teachers' Course; Household Science)

The College of Law (three years' course)

The School of Education

The Course in Journalism

The Courses in Chemistry and Chemical Engineering

The School of Railway Engineering and Administration

The School of Music (four years' course)

The School of Library Science (two years' course)

The College of Medicine (in Chicago)

The College of Dentistry (in Chicago)

The School of Pharmacy (in Chicago; Ph. G. and Ph. C. courses)

The Summer Session (eight weeks)

Experiment Stations and Scientific Bureaus: U. S. Agricultural Experiment Station; Engineering Experiment Station; State Laboratory of Natural History; State Entomologist's Office; Biological Experiment Station on Illinois River; State Water Survey; State Geological Survey; U. S. Bureau of Mines Experiment Station.

The library collections contain (December 1, 1917) 411,737 **volumes and** 104,524 **pamphlets.**

For catalogs and information address

THE REGISTRAR
URBANA, ILLINOIS

UNIVERSITY OF ILLINOIS BULLETIN
ISSUED WEEKLY
Vol. XV MAY 13, 1918 No. 37
[Entered as second-class matter Dec. 11, 1912, at the Post Office at Urbana, Ill., under the Act of Aug. 24, 1912.]

HYDRAULIC EXPERIMENTS WITH VALVES, ORIFICES, HOSE, NOZZLES, AND ORIFICE BUCKETS

BY
ARTHUR N. TALBOT, FRED B SEELY,
VIRGIL R FLEMING, MELVIN L. ENGER

BULLETIN No. 105

ENGINEERING EXPERIMENT STATION
PUBLISHED BY THE UNIVERSITY OF ILLINOIS, URBANA

PRICE: THIRTY-FIVE CENTS
EUROPEAN AGENT
CHAPMAN & HALL, LTD., LONDON

THE Engineering Experiment Station was established by act of the Board of Trustees, December 8, 1903. It is the purpose of the Station to carry on investigations along various lines of engineering and to study problems of importance to professional engineers and to the manufacturing, railway, mining, constructional, and industrial interests of the State.

The control of the Engineering Experiment Station is vested in the heads of the several departments of the College of Engineering. These constitute the Station Staff and, with the Director, determine the character of the investigations to be undertaken. The work is carried on under the supervision of the Staff, sometimes by research fellows as graduate work, sometimes by members of the instructional staff of the College of Engineering, but more frequently by investigators belonging to the Station corps.

The results of these investigations are published in the form of bulletins, which record mostly the experiments of the Station's own staff of investigators. There will also be issued from time to time, in the form of circulars, compilations giving the results of the experiments of engineers, industrial works, technical institutions, and governmental testing departments.

The volume and number at the top of the front cover page are merely arbitrary numbers and refer to the general publications of the University of Illinois: *either above the title or below the seal is given the number of the Engineering Experiment Station bulletin or circular which should be used in referring to these publications.*

For copies of bulletins, circulars, or other information address the

ENGINEERING EXPERIMENT STATION,
URBANA, ILLINOIS.

UNIVERSITY OF ILLINOIS
ENGINEERING EXPERIMENT STATION

BULLETIN No. 105 — MAY, 1918

HYDRAULIC EXPERIMENTS WITH VALVES, ORIFICES, HOSE, NOZZLES, AND ORIFICE BUCKETS

PART I

LOSS OF HYDRAULIC HEAD IN SMALL VALVES

BY ARTHUR N. TALBOT
PROFESSOR OF MUNICIPAL AND SANITARY ENGINEERING
IN CHARGE OF THEORETICAL AND APPLIED MECHANICS

AND

FRED B SEELY
ASSISTANT PROFESSOR OF THEORETICAL AND APPLIED MECHANICS

PART II

THE FLOW OF WATER THROUGH SUBMERGED ORIFICES

BY FRED B SEELY
ASSISTANT PROFESSOR OF THEORETICAL AND APPLIED MECHANICS

PART III

FIRE STREAMS FROM SMALL HOSE AND NOZZLES

BY VIRGIL R FLEMING
ASSISTANT PROFESSOR OF APPLIED MECHANICS

PART IV

THE ORIFICE BUCKET FOR MEASURING WATER

BY MELVIN L. ENGER
ASSOCIATE PROFESSOR OF MECHANICS AND HYDRAULICS

ENGINEERING EXPERIMENT STATION
PUBLISHED BY THE UNIVERSITY OF ILLINOIS, URBANA

PREFACE

AS a part of the experimental work conducted in the Hydraulic Laboratory of the University of Illinois a number of problems has been investigated which has not been large enough in scope to warrant publication as separate bulletins. It seems well, however, to put on record the results of such experiments, and this bulletin presents a record of four of these problems. It is believed that the four papers will be found to be of use in various aspects of engineering practice even though the experiments are not exhaustive investigations.

The investigations for the most part have been the outgrowth of experimental work begun by students, largely as thesis work, and carried on over a period of several years.

The variety of conditions under which the flow of water takes place, the possibility of large changes in the state of the flow due apparently to small changes in the form of the passages through which the water flows, and the necessity of persistent effort in subjecting assumptions and analytical deductions to experimental verification, make it desirable to report all hydraulic experimental results which are believed to be reliable.

A part of the experimental results herein reported has appeared in the publication of a technical society. The material, however, has been expanded in this bulletin and will be found in a more convenient form for use.

ARTHUR N. TALBOT
FRED B SEELY
Editors

Part I

LOSS OF HYDRAULIC HEAD IN SMALL VALVES

By ARTHUR N. TALBOT

Professor of Municipal and Sanitary Engineering
in charge of Theoretical and Applied Mechanics

AND

FRED B SEELY

Assistant Professor of Theoretical and Applied Mechanics

CONTENTS

PART I

LOSS OF HYDRAULIC HEAD IN SMALL VALVES

LIST OF FIGURES

LIST OF TABLES

HYDRAULIC EXPERIMENTS

WITH

VALVES, ORIFICES, HOSE, NOZZLES, AND ORIFICE BUCKETS

PART I

LOSS OF HYDRAULIC HEAD IN SMALL VALVES

I. INTRODUCTION

1. *Preliminary.*—Part I of this bulletin presents the results of experiments on the flow of water through 1-in. and 2-in. gate valves, 1-in. and 2-in. globe valves, and 1-in. and 2-in. angle valves. The loss of head caused by each valve, expressed in terms of the velocity head in the pipe, is given for four different ratios of the height of the valve opening to the diameter of the full valve orifice, namely, ¼, ½, ¾, and 1. The coefficients of discharge are also given for the gate valves for each of the four valve openings.

In a long pipe line the total amount of lost head is caused chiefly by pipe friction, the resistance due to a valve being comparatively small except for very small valve openings.

In a variety of cases, however, where valves are used on comparatively short pipe lines as, for example, in hydraulic elevator service, in office buildings, and in special apparatus it is important to know the lost head caused by small valves of different kinds when set at various positions. Very few experimental results have been published on this subject, particularly for globe and angle valves. Any experimental work, furthermore, which helps to indicate the laws governing the flow of water should prove of value. With these facts in mind the results herein recorded have been prepared.

2. *Acknowledgment.*—The experiments herein considered were performed as student thesis work in the Hydraulic Laboratory of the University of Illinois by M. E. THOMAS, class of 1906, under the direction of PROFESSOR ARTHUR N. TALBOT. Unusual care in the experimenting is reflected in the congruity of the data presented in Mr. Thomas' thesis.

II. Apparatus and Method of Experimenting

3. *Valves.*—The valves used were bought in the open market and tested just as received. The passages through the 2-in. globe valve were then modified by the use of plaster of paris to give a more gradual change of section (see Fig. 10), and this valve was tested again. The types or forms of the interiors of the valves and the dimensions of some of the passageways through the valves are shown in Fig. 1. The 1-in. globe valve and the 1-in. angle valve were made by the Western Tube Company. All the other valves were made by the Crane Company.

4. *Method of Experimenting.*—The arrangement of the apparatus is shown in Fig. 2. The test valve was placed in a horizontal pipe to which water was supplied from a standpipe under a static head of about 50 feet. The quantity of water discharged through the valve was regulated by another valve downstream from the test valve. The volume discharged in a certain time was measured in a calibrated pit and the time taken with an ordinary watch from which the rate of discharge was calculated. Three-way gage connections for obtaining the pressure head in the pipe were made at a section one foot upstream and one foot downstream from the valve. Care was taken to avoid having these connections project into the interior of the pipe. It was found by experiment that when any two of the three pressure connections at either section were closed, the same difference of head was registered as when all three connections at either section were open. The three-way connections were used, however, in all the experiments. The difference in the pressure heads at the two sections was measured by a differential mercury gage. A Crosby pressure gage was also attached at each section to serve as a rough check on the differential gage. The lost head due to the pipe friction for the two feet of pipe between the two sections was assumed to be as given in Weston's Tables of Friction of Water in Pipes. This amount of lost head was subtracted from the reading of the differential mercury gage in determining the loss of head caused by the valve.

The loss of head and the corresponding rate of discharge and velocity in the pipe were determined for each of four valve openings for each of the six valves tested. The valve openings used were such that the heights of the openings were one-fourth, one-half, three-fourths, and one times the diameter of the full valve orifice.

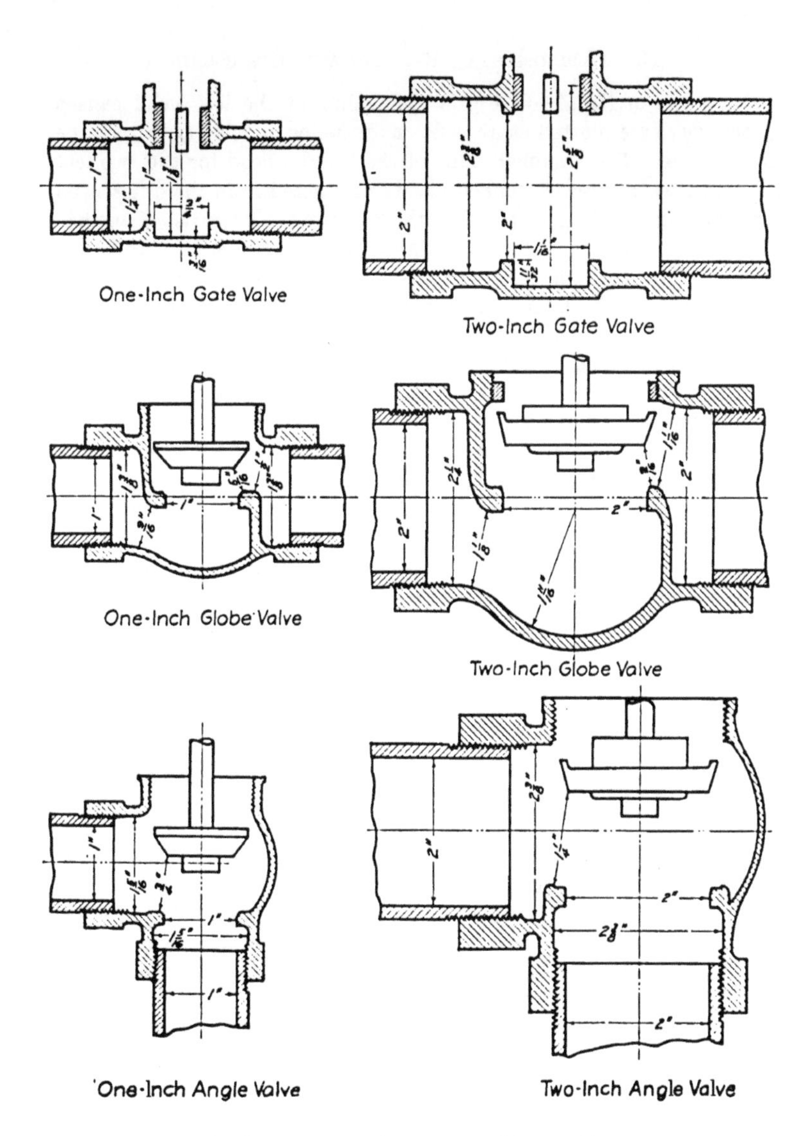

FIG. 1. LONGITUDINAL SECTIONS OF VALVES TESTED

III. Experimental Results and Discussion

5. *Loss of Head.*—In Fig. 3 to 8 values of the lost head caused by the valve are plotted as abscissas and the mean velocity in the pipe as ordinates. The assumed value of the friction head for the two feet of pipe between pressure connections is subtracted from the differential mercury gage reading in plotting the abscissas. There is, of course, some doubt concerning the correct allowance to be made for this pipe friction. The loss of head due to this cause, however, will be relatively small except for the larger valve openings. It will be noted from the curves in Fig. 3 to 8 that the range of velocity in the pipe varied of course with the kind of valve and with the amount of valve opening. The smallest mean velocity in any case was about ¼ ft. per sec., while the maximum mean velocity was about 40 ft. per sec.

The curves in Fig. 3 to 8 give values of the loss of head caused by the valves which vary as the square of the velocity in the pipe, that is, the lost head due to the valve may be expressed in terms of the velocity head in the pipe. This fact is shown very clearly by plotting the values from the curves in Fig. 3 to 8 on logarithmic paper; the curves showing the relation between the lost head, h, and the velocity in the pipe, v, become parallel straight lines with a slope varying but little from two, the slope indicating the exponent in the equation $h=kv^n$. That is, $h=kv^2$ or, when expressed in terms of the velocity head in the pipe, $h=\frac{mv^2}{2g}$ in which m is called the coefficient of loss.

Values of the coefficients of loss for the valves with the various valve openings as obtained from the curves in Fig. 3 to 8 are given in Table 1. These values have been plotted in Fig. 9 as abscissas against the valve openings as ordinates. From these curves and also from Table 1 the resistance to flow caused by the three kinds of valves may be compared at various valve openings. It will be noted that the loss of head varies in a quite different manner with the amount of valve opening for these three kinds of valves, for instance, a comparison of the results for the valves when completely opened shows that a globe valve causes more than twice as much loss of head as the corresponding size of angle valve, while a gate valve causes markedly less loss than either a globe or an angle valve, the velocity in the pipe being the same in the three cases. As the valve is gradually closed, the resistance to flow of the angle valves increases the least (decreasing at first) while the resistance

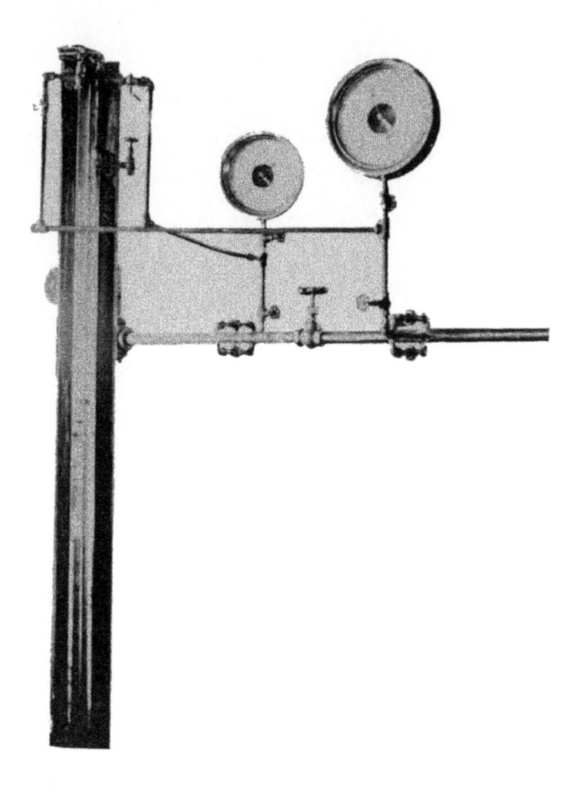

Fig. 2. Arrangement of Apparatus

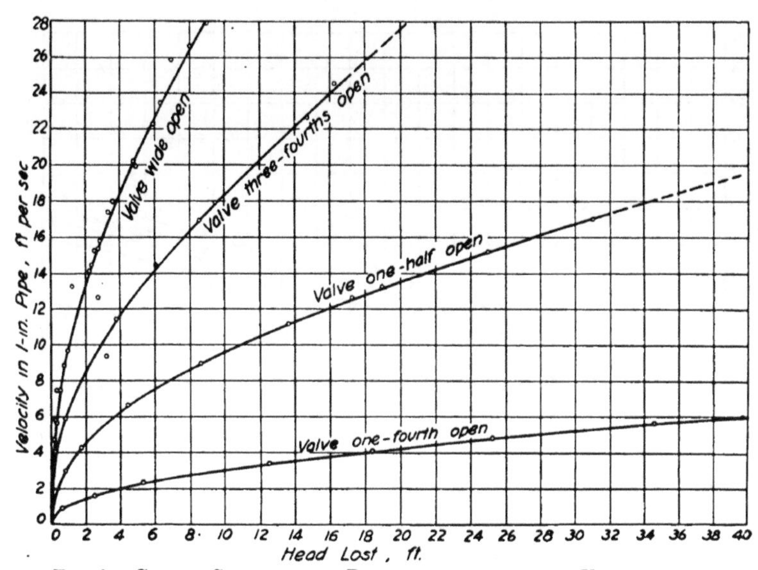

FIG. 3. CURVES SHOWING THE RELATION BETWEEN THE VELOCITY AND HEAD LOST IN 1-INCH GATE VALVE

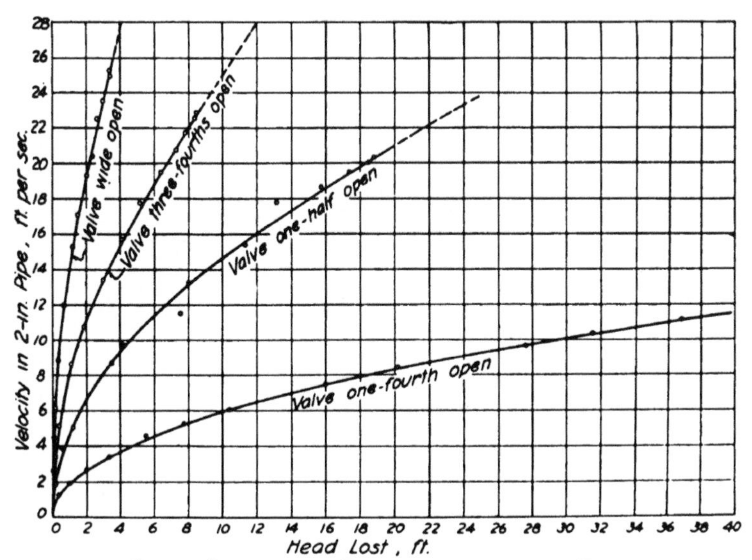

FIG. 4. CURVES SHOWING THE RELATION BETWEEN THE VELOCITY AND HEAD LOST IN 2-INCH GATE VALVE

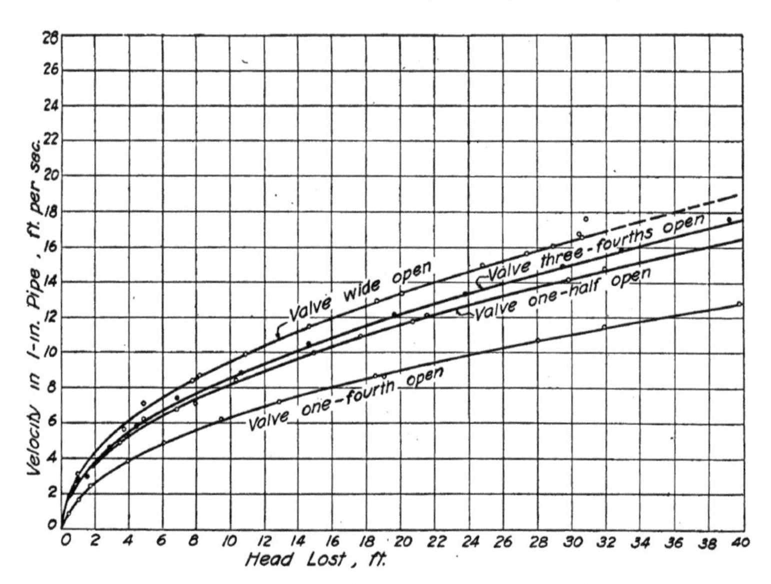

FIG. 5. CURVES SHOWING THE RELATION BETWEEN THE VELOCITY AND HEAD LOST IN 1-INCH GLOBE VALVE

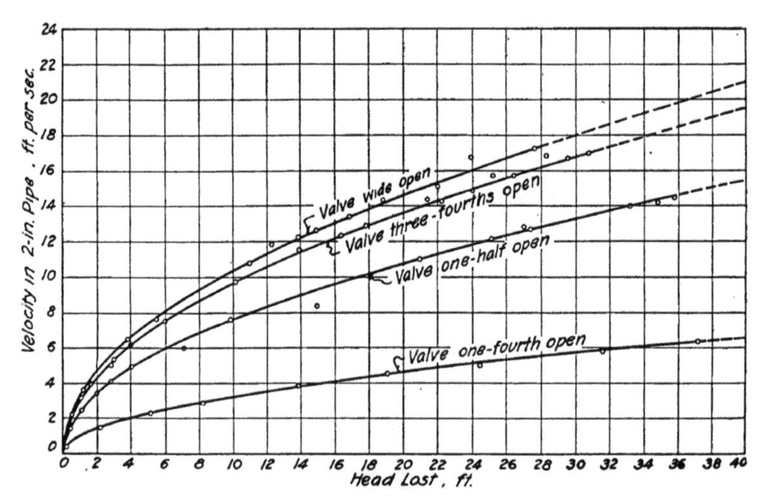

FIG. 6. CURVES SHOWING THE RELATION BETWEEN THE VELOCITY AND HEAD LOST IN 2-INCH GLOBE VALVE

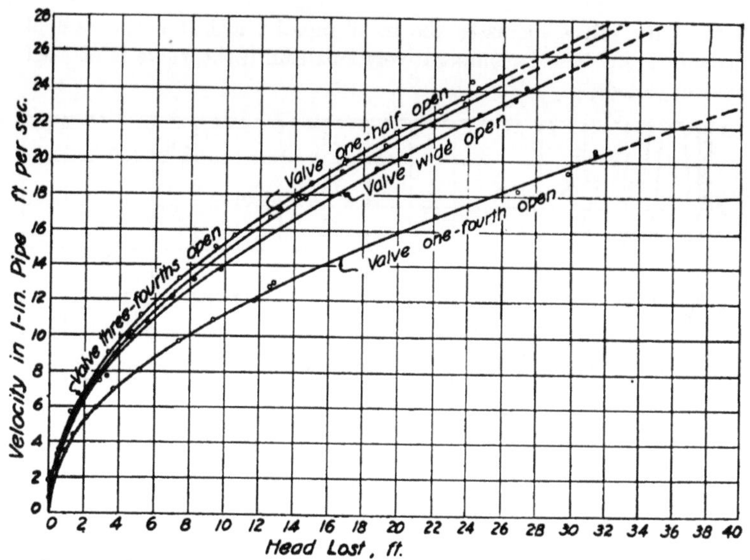

FIG. 7. CURVES SHOWING THE RELATION BETWEEN THE VELOCITY AND HEAD LOST IN 1-INCH ANGLE VALVE

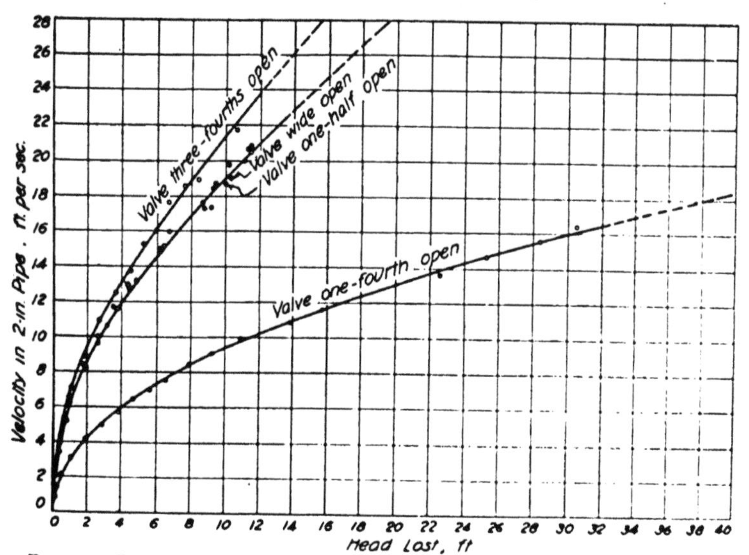

FIG. 8. CURVES SHOWING THE RELATION BETWEEN THE VELOCITY AND HEAD LOST IN 2-INCH ANGLE VALVE

of the gate valves increases the most rapidly, although the rate of increase in any case is comparatively small until the valve is at least one-half closed.

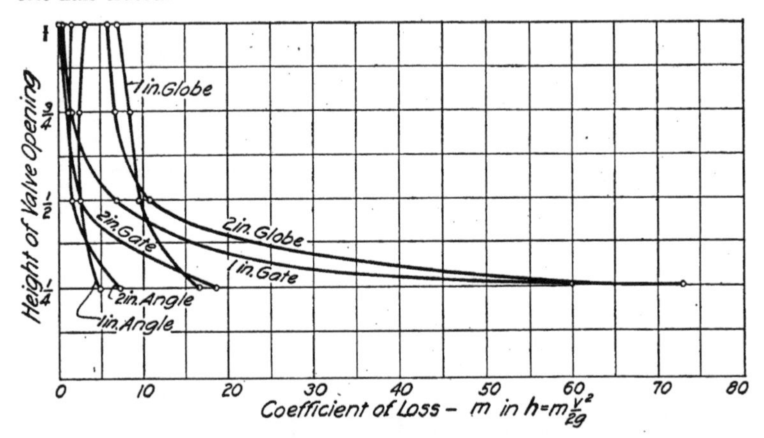

FIG. 9. CURVES SHOWING THE RELATION BETWEEN COEFFICIENTS OF LOSS AND VALVE OPENINGS

Fig. 9 also indicates that the proportions or form or shape of the passageways of the valve of a given type or kind is a very important factor in causing loss of head. This fact is shown by a comparison of the results for the 1-in. globe valve with those for the 2-in. globe valve and also by a comparison of the results of the 1-in. angle valve with those of the 2-in. angle valve. Each of these 1-in. valves was of a somewhat different form from that of the corresponding 2-in. valve as may be seen in Fig. 1. It will be noted from Fig. 9 and Table 1 that

TABLE 1

EXPERIMENTAL VALUES OF COEFFICIENTS OF LOSS

Values of m in $h = \frac{mv^2}{2g}$

Ratio of Height of Valve-Opening to Diameter of Full Valve Opening	Gate Valves		Globe Valves		Angle Valves	
	1-inch Diameter	2-inch Diameter	1-inch Diameter	2-inch Diameter	1-inch Diameter	2-inch Diameter
¼	73.0	18.8	16.6	60.0	5.00	7.3
½	7.0	2.94	9.62	10.9	2.90	1.70
¾	1.84	1.06	8.75	6.84	2.72	1.44
1	0.74	0.35	7.12	6.0	3.23	1.70

for the smaller valve openings the 1-in. globe valve and the 1-in. angle valve cause less resistance to flow than the corresponding 2-in. valves. The difference is especially large in the case of the globe valves. This unexpected result seems to be due chiefly to the better shaped discharge passages (more gradual expansion) as the water makes its exit from the 1-in. globe valve.

Experiments were made on the 2-in. globe valve to see if a more gradual change in sections through the valve would cause less loss of head. This gradual change was made by filling in part of the passageway with plaster of paris, as shown in Fig. 10. This modified valve was then tested with the valve one-half open and wide open, the results for which are shown in Fig. 10. It will be seen that this modification had no

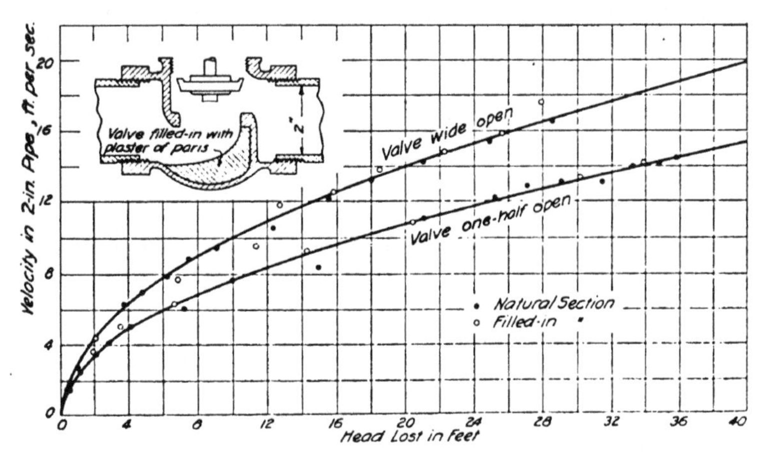

FIG. 10. CURVES SHOWING THE EFFECT OF GRADUAL CHANGE OF SECTION THROUGH 2-INCH GLOBE VALVE

effect on the amount of head lost. This suggests that the lost head in a small globe valve is caused more by the form or shape of the passageway at exit from the valve than by the form of the passages through the valve. Other valve openings and other modifications of the passageways, however, may give better results.

In the case of angle valves the loss of head is not a minimum for the greatest valve opening as is shown in Table 1 and in Fig. 9. For the 2-in. angle valve the lost head is the same when the valve is only one-half open as it is when the valve is wide open, the velocity in the pipe for the two valve openings being the same, that is, the coefficient of loss is the same for these two valve openings. When this valve is three-fourths open, however, the coefficient of loss is about 20 per cent less than when the valve is one-half open or wide open. The 1-in. globe valve caused a smaller amount of lost head when it was one-half and three-fourths open than it did when wide open, the velocity in the pipe being the same for each of the valve settings. The difference, however, between the coefficients of loss for these three valve openings is not large. The reason that the minimum loss of head in the angle valves occurs when the valve is about three-fourths open is probably because at this opening the water can flow through comparatively large openings all around the valve disc meeting with less abrupt changes of directions than when the valve is wide open. In the latter case there is much turbulent action due to the impact of the water against the bottom of the valve. As the valve opening decreases from the three-fourths open position, the greater resistance due to the narrowing passages causes the lost head to increase again.

The assumption is sometimes made that for comparatively small valves of like type or kind the loss of head varies inversely with the diameter of the valve. For the larger valve openings this assumption is probably approximately true, but from the foregoing results and discussion it would seem that at least for globe and angle valves the form or shape of the passages of the valve is a determining factor in the amount of head lost at the smaller valve openings.

6. *Earlier Experiments on Gate Valves.*—Among the first reliable published results on valves were those by Weisbach.* The largest gate valve used by Weisbach was a little less than two inches. Globe and angle valves, at least of modern construction, were not tested. Other experiments on gate valves have been reported by Magruder† on ⅜-in., ½-in., ¾-in., 1-in., and 1½-in. gate valves, by Folwell‡ on a 4-in. gate valve, by Kuichling¶ on a 24-in. gate valve, and by J. Waldo Smith§ on a 30-in. gate valve. In Smith's experiments the 30-in. valve was located in a 42-in. pipe with increaser-shaped or Venturi-shaped approaches, and in Kuichling's experiments the valve was placed in one branch of a Y only a few feet from the section where the Y started to branch. The methods of determining the lost head in the

* Mechanics of Engineering (Coxe's translation).
† Engineering Record, Vol. XL, p. 78, 1899.
‡ Engineering News, Vol. XLVII, p. 302, 1902.
¶ Trans. Am. Soc. Civ. Eng. Vol. XXVI, p. 439, and Vol. XXXIV.
§ Trans. Am. Soc. Civ. Eng. Vol. XXXIV, p. 235 (p. 243), 1895.

various experiments were also different. For these reasons it is obvious that the results of these experiments are not directly comparable.

TABLE 2

VALUES OF THE COEFFICIENT OF LOSS FOR GATE VALVES OF VARIOUS DIAMETERS DUE TO PARTIAL CLOSURE ONLY

Ratio of Height of Opening to Diameter of Full Valve Orifice	Weisbach	Kuichling	Smith	Folwell		This Bulletin	
	2½-inch Diameter	24-inch Diameter	30-inch Diameter	4-inch Diameter		2-inch and 1-inch	
	Parallel Sides	Parallel Sides	Parallel Sides Venturi-Shaped Approaches	Parallel Sides	Wedge Shaped	Parallel Sides	
						2-inch Diameter	1-inch Diameter
0	......	......	......	......	......	......	......
3/100	......	......	950.0	......	......	......	......
1/10	......	......	128.0	......	......	......	......
⅛	98.0	......	90.0	72.3	104	......	......
¼	17.0	22.7	17.0	16.8	20.5	18.45	72.3
⅜	5.5	8.63	7.5	6.19	8.0	7.0[1]	16.0[1]
½	2.1	3.27	3.5	2.58	2.72	2.59	6.26
⅝	0.81	1.09[1]	1.5	1.22	1.5	1.2[1]	2.5[1]
¾	0.26	0.25	0.50	0.55	0.66	0.71	1.10
⅞	0.07	0.019	0.19	0.20	0.16	0.15[1]	0.70[1]
1	0.00	0.00	0.00	0.00	0.00	0.00	0.00

[1] Interpolated from curve.

In Table 2 are given the values of the coefficients of loss as obtained by the various experimenters mentioned previously, as well as the values obtained in the experiments herein reported. These values of the coefficients of loss are those due to partial closure of the valves only, that is, in excess of the loss of head caused by the valve when wide open. Smith's experiments are the only ones in which valve openings less than one-eighth were used. There is considerable chance for error in the results obtained for the very small valve openings, due chiefly to the uncertainty in securing the valve setting desired. Table 2 indicates a rather close agreement in the coefficient of loss for all the gate valves having diameters of 2 in. or greater, and for valve openings of ¼ or perhaps ⅛ and greater. The values for the 1-in. valve show a considerable increase in the lost head over that for valves of 2-in. diameter and greater. It is probable also that there is considerable variation in the smaller valves of any one type and size.

7. *Coefficients of Discharge for Gate Valves.*—In order to determine the rate of discharge through a pipe a partially closed valve has sometimes been used. This requires the values of the coefficients of discharge of the valve for various valve openings since the rate of discharge, q, is found from the expression, $cA\sqrt{2gh}$, in which c is the coefficient of discharge, A the area of the valve opening, and h the difference in pressure heads on the two sides of the valve (lost head), velocity of approach being neglected. The average values of the coefficients of discharge for the 1-in. and 2-in. gate valves as found in the experiments reported in this bulletin are given in Table 3. The value of the

TABLE 3

EXPERIMENTAL VALUES OF THE COEFFICIENTS OF DISCHARGE FOR GATE VALVES

Values of $c = \frac{q}{A\sqrt{2gh}}$

Ratio of Height of Valve-Opening to Diameter of Full Valve-Opening	Coefficient of Discharge		Area of Valve-Opening Square Inch	
	1-inch Valve	2-inch Valve	1-inch Valve	2-inch Valve
¼	.48	.88	.195	.826
½	.67	1.00	.450	1.80
¾	.88	1.12	.660	2.67
1	1.16	1.70	.785	3.14

coefficient varied somewhat with the velocity for any given valve opening. Because of the uncertainty of obtaining the exact valve setting desired and the corresponding uncertainty in the area of the valve opening, the values of the coefficients of discharge given in Table 3 cannot be considered as refined determinations.

It will be noted that the coefficient of discharge increases directly with the valve opening for each of the gate valves for a range of valve openings of ¼ to ¾ or perhaps greater. The more the valve is opened the greater is the velocity of approach toward the valve and since the velocity of approach is not considered in the calculation of the coefficient of discharge the value of the coefficient increases with the valve opening. The coefficient of discharge for the valves used by Kuichling and Smith decreased slightly until the valve was about one-fourth open and then increased rapidly for further openings of the valve. Gibson with a 2½-in. flat disc stop valve found nearly a constant coefficient of discharge of 0.80. These variations in the coefficients of discharge are not surprising considering the wide range of conditions covered by the experiments. They suggest, however, that if gate valves are to be used for determining the rate of discharge in pipes with reasonable accuracy much more experimenting is required, or better, where it is possible, experiments should be performed under service conditions to calibrate the particular valve to be used.

8. *Summary.*—The following brief summary is given as applying to 1-in. and 2-in. valves of the three kinds tested (gate valves, globe valves, and angle valves) with valve settings ranging from one-fourth open to wide open.

(1) The loss of head caused by small valves varies as the square of the velocity in the pipe for all the valve openings; hence the lost head may be expressed as a constant times the velocity head in the pipe, $\left(h=\frac{mv^2}{2g}\right)$.

(2) When wide open a globe valve causes more than twice as much loss of head as an angle valve of the same size, while a gate valve causes much less loss of head than either a globe or an angle valve, the velocity in the pipe being the same in the three cases.

(3) The loss of head for an angle valve is somewhat less when about three-fourths open than when wide open, the velocity in the pipe being the same in each case.

(4) The loss of head for each valve, as the valve is closed from a wide open position, varies comparatively little with the valve opening until the valve is at least one-half closed. As further closure takes place the loss of head of the globe valves and gate valves increases rapidly and is considerably larger than that of the angle valves.

(5) The form or shape of the passageways through a globe or angle valve has a large influence on the loss of head for the small

valve openings. The portion of the passageways in which the form seems of greatest importance is in the exit from the valve rather than in the passageways leading to the valve disc or seat. On account of the influence of the form or shape of the valve no law giving the relation of the lost head to the diameter of the valve can be stated for valve settings less than five-eighths open. For larger valve openings than this, the lost head seems to vary approximately inversely as the diameter.

(6) The use of the lost head through a partially closed valve as a means of determining the flow can be only a very rough method of measurement unless the particular valve to be used is calibrated under service conditions. Even then the difficulty in obtaining the desired valve setting may introduce considerable uncertainty in the results.

PART II

THE FLOW OF WATER THROUGH SUBMERGED ORIFICES

BY FRED B SEELY
ASSISTANT PROFESSOR OF THEORETICAL AND APPLIED MECHANICS

CONTENTS

PART II

THE FLOW OF WATER THROUGH SUBMERGED ORIFICES

LIST OF FIGURES

LIST OF TABLES

PART II

THE FLOW OF WATER THROUGH SUBMERGED ORIFICES

IV. INTRODUCTION

9. *Preliminary.*—Part II of this bulletin presents the results of experiments on submerged sharp-edged orifices of various shapes and sizes discharging under moderately low and under very low heads. The orifices used were of three shapes, circular orifices with diameters from 1 in. to 6 in., square orifices with sides from ½ in. to 5½ in., and rectangular orifices having one side range from ½ in. to 2 in., the other side being 6 in. in each case. The coefficient of discharge is given for each orifice for a velocity range of approximately ½ ft. per sec. to 4 ft. per sec. This range corresponds roughly to a range of head on the orifice of 0.006 ft. to 0.08 ft.

Considerable experimenting has been done on orifices discharging into air, particularly on sharp-edged circular orifices of rather small size although the results are somewhat discordant. Comparatively little experimental work, however, has been carried out on submerged orifices. While the orifice has lost some of its importance as a water measuring device due to the development of other methods, it is, nevertheless, of importance to determine how the rate of discharge is affected by the shape and the size of the orifice and also by the head on the orifice, particularly the effect of very low heads which the submerged orifice makes possible.

The submerged orifice may be of particular importance in cases which require the measurement of water with as small a loss of head as possible as, for example, in determining the discharge from a water turbine when operating under a low head. The decrease in the available head on the turbine made necessary by the proper setting of a weir may be an important factor in the installation.

There is a feeling among some engineers that the importance of the so-called standard orifice (sharp edges, complete contraction without velocity of approach, etc.) has been over-emphasized and that beveled-edged orifices are better adapted at least to conditions where the orifice may be obstructed and the edge soon worn off, as, for example, in measuring the water supplied to water wheels through flume or bulkhead openings. There exist, no doubt, some grounds for this feeling. A sharp edged orifice (an opening in a thin plate), however, is subject to less variation in its construction than a beveled-edged orifice. This fact is of considerable importance where accuracy is essential. It is felt that the submerged orifice, both beveled-edged and sharp-edged, is worthy of more attention than has been accorded it.

10. *Acknowledgment.*—The experimenting was done in the Hydraulic Laboratory of the University of Illinois. Some of the results herein presented have been taken from the theses of W. R. ROBINSON of the class of 1906 and G. D. PHILLIPS of the class of 1907, and some of the results also, particularly at the low heads, have been taken from a second thesis presented by Mr. Robinson in 1909. All the thesis work was conducted under the direction of PROFESSOR ARTHUR N. TALBOT. The careful way in which this preliminary experimenting was done has made the results of the theses of much value. During 1914 and 1915 the writer spent considerable time in checking the results of the theses work and extending certain parts of the investigation.

V. Apparatus and Method of Experimenting

11. *Orifices.*—The orifices used were of three different shapes. Four of the orifices were circular with diameters of 1 in., 2 in., 4 in., and 6 in. Five were square with sides of ½ in., 1 in., 2 in., 4 in., and 5½ in. Three were rectangular with dimensions of ½ in. by 6 in., 1 in. by 6 in., and 2 in. by 6 in. In each case the orifice was formed in a cast iron plate ½ in. thick and 10½ in. in diameter, a sharp edge being formed by beveling at 45 degrees. Except for a few small nicks the edges were sharp and the areas closely true to shape. The dimensions of the orifices were carefully determined (except for the 1-in. circular orifice) by an inside micrometer for dimensions greater than 1 in. and inside screw calipers for dimensions less than 1 in. A list of the orifices used and the areas as determined from the measured dimensions are given in Table 4. The 1-in. circular orifice was broken before

Table 4

List of Orifices Used

Form of Orifice	Nominal Size	Measured Area square feet
Circular	1 in. diam.	not measured
	2 in. diam.	0.0219
	4 in. diam.	0.0883
	6 in. diam.	0.1967
Square	½ in. by ½ in.	0.001735
	1 in. by 1 in.	0.00698
	2 in. by 2 in.	0.0279
	4 in. by 4 in.	0.1109
	5½ in. by 5½ in.	0.2105
Rectangular	½ in. by 6 in.	0.0206
	1 in. by 6 in.	0.0418
	2 in. by 6 in.	0.0838

its dimensions were taken so that the nominal diameter (1 in.) has been used in the calculations. There may be some error, therefore, in the results for this orifice.

12. *Tank Used and Method of Experimenting.*—The same tank was used in all the experiments, the dimensions and general arrangement of which is shown in Fig. 11.* The tank was divided into two compartments by a vertical partition in which the orifice was placed, holding the orifice in a vertical plane.

The water coming from the laboratory standpipe was supplied to the tank through a 6-in. supply pipe and also through a ¾-in. pipe, the latter making possible a finer adjustment in maintaining a constant head. After passing through baffle boards the water flowed through the orifice and finally left the downstream compartment by passing out through small openings in the end of the tank, the flow through which was regulated by placing stoppers in some of the holes. These holes were arranged in two narrow portions in the end of the tank, one near each side of the tank, and the stoppers were arranged so as to give nearly a uniform distribution from each of the two sets of openings. This arrangement, it was found, helped to maintain steady conditions.

The quantity of water discharged was determined by weighing for the small discharges and by measuring in a pit for the larger discharges. The pit was about 6 ft. deep, and 7.995 ft. in diameter. The value for the diameter is the average of a large number of readings of a micrometer attached to a rigid stick. The rise in the pit was determined by a vertical graduated rod which could be read directly to 0.02 ft. and to 0.004 ft. by estimating. A float was attached to the bottom of the rod and a still basin was provided. The water was wasted into another pit through a movable spout until the surface of the water in the measuring pit became fairly still so that an accurate reading of the rod could be taken. A hook gage was used to test the accuracy of the float and rod. At the end of the experiment the water was again wasted in the same manner. A calibrated stop watch gave the time corresponding to the rise in the pit.

The head causing flow through the orifice is the difference in the levels of the water surfaces in the two compartments of the tank. This head was measured in nearly all the experiments by means of hook gages. These gages were read directly to 0.001 ft. and to 0.0005 ft. by estimating. Vertical 2-in. pipes attached toward the bottom of the tank served as still basins for the hook gages. The level of the water in the upstream compartment was determined by the use of one

*A view of the tank is shown in Fig. 5 of Bulletin No. 96 of the Engineering Experiment Station of the University of Illinois.

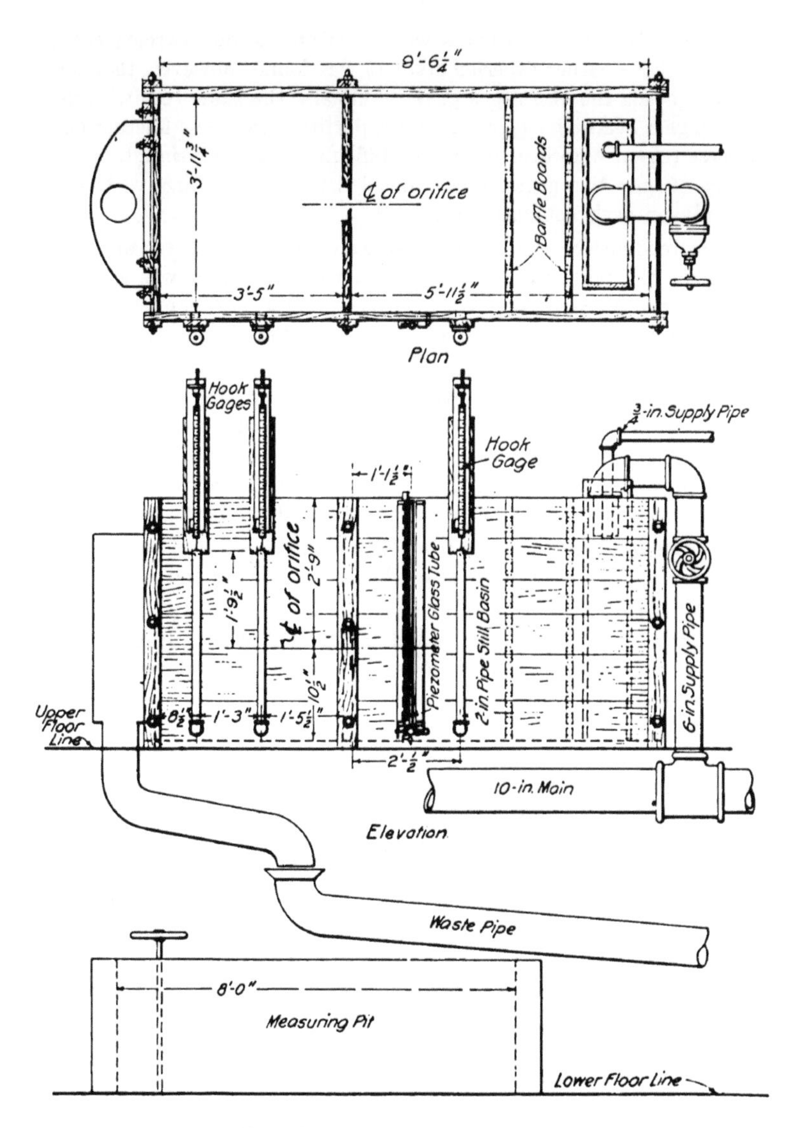

FIG. 11. TANK USED AND ARRANGEMENT OF APPARATUS

hook gage only, while two gages were used on the downstream compartment in the earlier experiments. It was found, however, that for the lower heads the two gages gave practically the same result, while for the higher heads the gage nearer the partition gave less fluctuation. For these reasons and because of less difficulty in getting simultaneous readings of only two gages, it was decided to take readings with one gage only on each compartment.

Zero readings of the hook gages were obtained by reading the gages when the tank was nearly full and when no water was allowed to escape, the levels of the water surfaces in the two compartments then being the same. Zero readings were taken frequently during the experimenting.

For most of the heads above 0.3 ft., the head was measured by two vertical peizometer glasses, one attached near the bottom of each compartment, the difference in readings of which (corrected for zero reading) gave the head to 0.001 ft. These two methods overlapped somewhat so that certain heads were measured by both methods.

Leakage from the tank and from the measuring pit was determined several times during the progress of the experimenting and was found to be negligible.

An experiment or run consisted of the following: A sufficient number of stoppers was removed from the end of the tank to give the desired discharge and the inflow through the 6-in. and ¾-in. pipes was then adjusted until the difference in levels of the water surfaces in the two compartments of the tank became constant. The ¾-in. supply pipe was used to make the final adjustment of the head and to hold the head constant throughout the experiment. After obtaining a constant head, the waste pipe shown in Fig. 11 was pulled from beneath the discharge pipe, thus allowing the water to discharge into the measuring pit until the rise in the pit was sufficient to allow its measurement without appreciable error and also to allow time for an accurate measurement of the head. The head was taken as an average of from two to ten readings of the hook gages, the larger number being necessary with the higher velocities on account of the greater fluctuations of the water levels due to the more turbulent conditions of the water, especially in the downstream compartment. Each experiment was repeated, as a rule, three times, although in some cases as many as eight or ten runs were made.

13. *Method of Calculating the Coefficient of Discharge.*—The head, h, causing flow through the orifice is the difference in the levels of the water surfaces in the two compartments of the tank. The ideal rate of discharge is $A\sqrt{2gh}$ in which A is the area of the orifice in square feet and g is the acceleration due to gravity in feet per second per second; hence the coefficient of discharge, c, is found from,

$$c = \frac{q}{A\sqrt{2gh}}$$

where q is the measured rate of discharge in cubic feet per second, as determined from the measured weight or volume discharged and the corresponding time.

VI. Experimental Results and Discussion

14. *Coefficients of Discharge.*—Fig. 12, 13, and 14 show the experimental values of the coefficients of discharge for the various orifices tested. Each plotted point represents the average of from two to ten experiments at practically the same head. It will be noted that

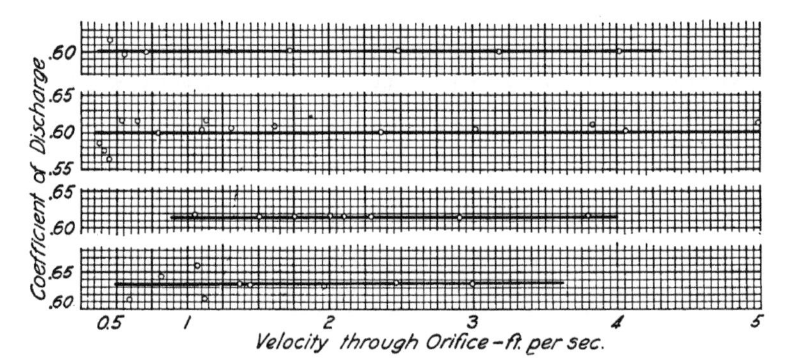

Fig. 12. Diagrams Showing Values of Coefficients of Discharge of Circular Submerged Orifices for Various Velocities

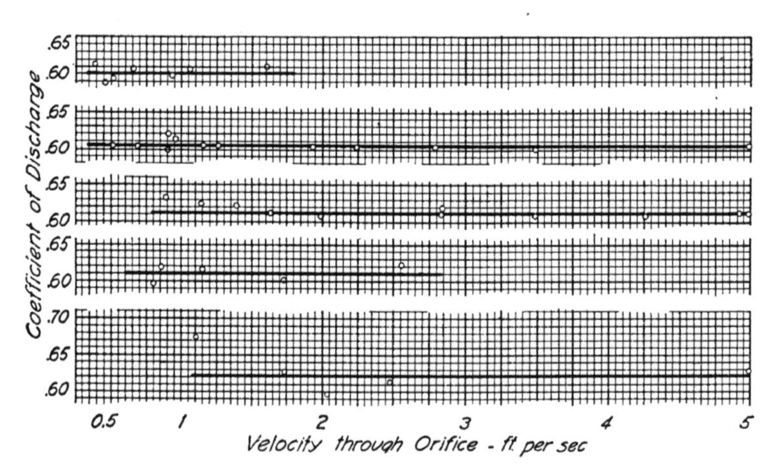

Fig. 13. Diagrams Showing Values of Coefficients of Discharge of Square Submerged Orifices for Various Velocities

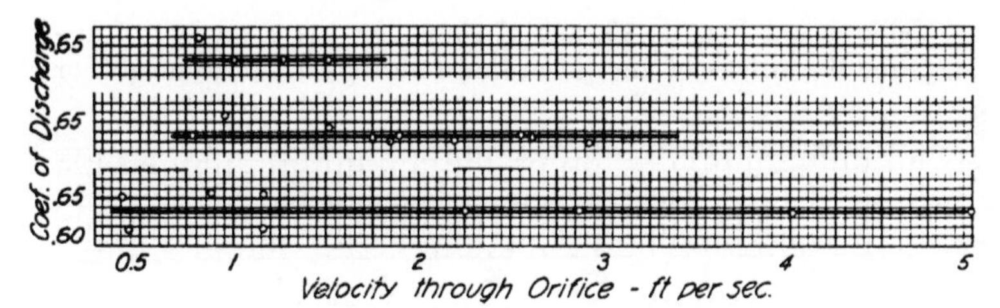

FIG. 14. DIAGRAMS SHOWING VALUES OF COEFFICIENTS OF DISCHARGE OF RECTANGULAR SUBMERGED ORIFICES FOR VARIOUS VELOCITIES

for any given orifice the coefficient is constant for the whole range of velocity used in these experiments which in most of the cases is about ½ ft. per sec. to 4 or 5 ft. per sec. This velocity range corresponds roughly to a range in head of 0.006 to 0.08 ft. and as may be expected the values of the coefficient show the greatest variation at the very low heads.

TABLE 5

VALUES OF COEFFICIENT OF DISCHARGE FOR SUBMERGED ORIFICES FOR VELOCITIES FROM ONE-HALF TO FIVE FEET PER SECOND

Kind of Orifice	Nominal Size	Coefficient of Discharge
Circular	1 in. diameter	0.635[1]
	2 in. diameter	0.615
	4 in. diameter	0.600
	6 in. diameter	0.600
Square	½ in. by ½ in.	0.620
	1 in. by 1 in.	0.610
	2 in. by 2 in.	0.610
	4 in. by 4 in.	0.605
	5½ in. by 5½ in.	0.600
Rectangular	½ in. by 6 in.	0.635
	1 in. by 6 in.	0.635
	2 in. by 6 in.	0.635

[1] Probably somewhat in error since diameter was not measured; nominal diameter used in calculations.

Table 5 and Fig. 15, 16, and 17 show how the coefficient of discharge for the orifices of any given shape varies with the diameter or

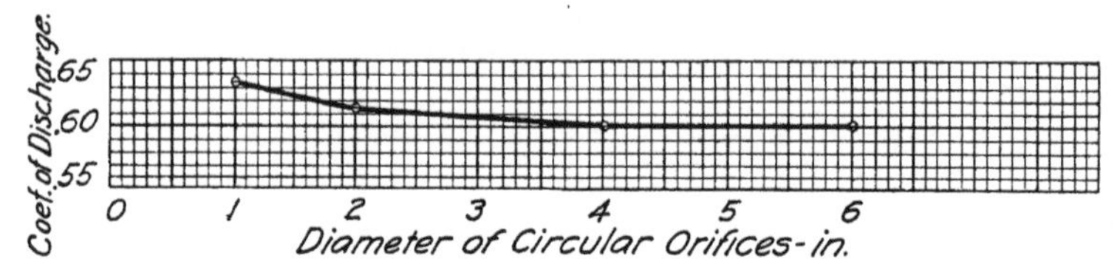

FIG. 15. CURVE SHOWING THE RELATION BETWEEN COEFFICIENT OF DISCHARGE FOR CIRCULAR ORIFICE AND DIAMETER OF ORIFICE

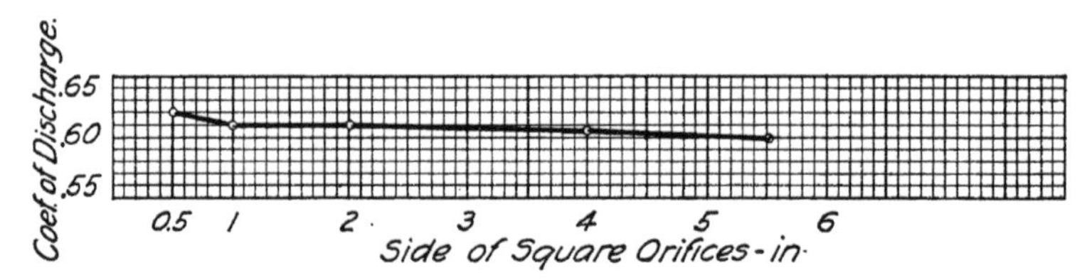

FIG. 16. CURVE SHOWING THE RELATION BETWEEN COEFFICIENT OF DISCHARGE FOR SQUARE ORIFICE AND SIDE OF ORIFICE

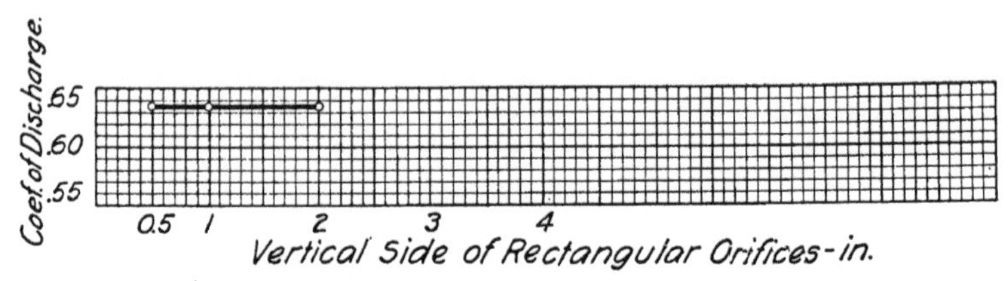

FIG. 17. CURVE SHOWING THE RELATION BETWEEN COEFFICIENT OF DISCHARGE OF RECTANGULAR ORIFICE AND SHORT SIDE OF ORIFICE (OTHER SIDE BEING SIX INCHES IN EACH CASE)

side of the orifice, while from Fig. 18 a comparison may be made between the coefficients of discharge for the different shaped orifices on the basis of their areas. These figures show that the coefficient of discharge for circular and square orifices decrease as the size increases until an area of 8 or 10 square inches is reached after which the coefficient has a constant value of not far from 0.60. This indicates that complete contraction does not take place with the smaller orifices. Because of the uncertainty of the exact diameter there is some doubt, however, concerning the correct value for the 1-in. circular orifice. It will be noted also that the coefficient of discharge for the rectangular orifices

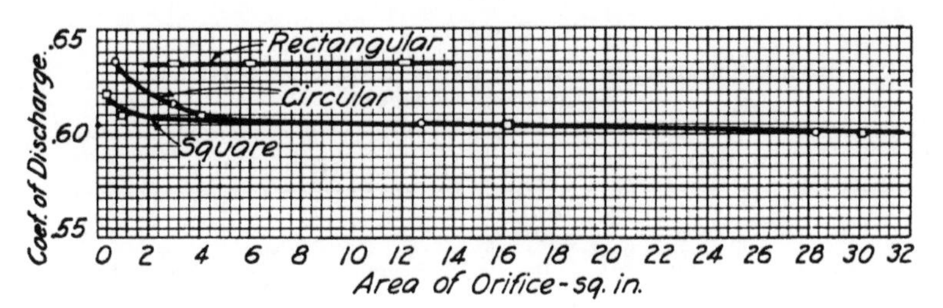

FIG. 18. CURVES SHOWING RELATION BETWEEN COEFFICIENT OF DISCHARGE AND AREA OF ORIFICES

remain constant for the range of areas used in these experiments and that its value is larger than that for circular and square orifices of the same area. Fig. 18 indicates furthermore that as the area of the orifices decreases below 8 sq. in., the coefficient of discharge for circular orifices increases faster than that for square orifices. These observations suggest that the longer side of the rectangular orifices has a controlling influence in determining the rate of discharge for a given head and that the corners of a small square orifice are inefficient in discharging water as compared with the form of a circular orifice of the same area.

15. *Results Obtained by Earlier Experimenters.*—In order to compare the results given in this bulletin with those of earlier investigations and to extend the study to include higher heads and velocities, the results given in Table 6 have been condensed from available published data. It will be noted that the results are not entirely concordant, but considering the different arrangements and methods of measuring the head and the rate of discharge, the results show a very good agreement. The low value of the coefficient of discharge found by Francis is due no doubt to the fact that the rate of discharge was measured over a weir on which the head was rather small. From Table 6 it will be seen that in some of the earlier investigations the coefficient of discharge increased slightly with the head while in others the coefficient decreased, and in still others it showed no systematic change. In all cases the value of the coefficient of discharge is not far from 0.60. The small square orifice (1.2 in. by 1.2 in.) used by Hamilton Smith gave a slightly larger coefficient than the circular orifice with a diameter of 1.2 inches. This result is the reverse of that found in the experiments described in this bulletin. The values also of the coefficient of discharge for circular and square orifices as found by

TABLE 6

RESULTS OBTAINED BY EARLIER EXPERIMENTERS ON SUBMERGED SHARP-EDGED ORIFICES

Circular Orifices

Source	Diameter inches d	Head feet h	Coefficient of Discharge c
Francis	1.22	1.024	.592
		1.324	.592
		1.490	.592
		1.499	.593
		1.514	.591
Hamilton Smith, Jr.	0.6	0.437	.6183
		2.16	.6041
		4.08	.6016
Hamilton Smith, Jr.	1.2	0.250	.6048
		0.648	.6027
		0.985	.6025
		1.51	.6006
		2.00	.6006
		2.58	.5997
		2.99	.5989
		3.57	.5987
		3.97	.5992
Ellis	12.0	2.60	.607
		4.71	.590
		6.41	.606
		8.10	.599
		8.80	.600
		12.09	.600
		14.25	.601
		16.29	.602
		18.66	.599
Balch	12.0	0.145	.5909
		0.469	.5902
		0.851	.5912
		1.254	.5993
		1.612	.5921
		2.012	.5924
		2.421	.5954
		2.949	.5967
		3.410	.6006
		4.015	.6054

Square Orifices

Source	Dimensions inches	Head feet h	Coefficient of Discharge c
Hamilton Smith, Jr.	0.6 by 0.6	0.35	.6201
		2.21	.6092
		4.06	.6068
Hamilton Smith, Jr.	1.2 by 1.2	0.207	.6117
		0.410	.6091
		0.771	.6053
		1.52	.6055
		2.32	.6040
		3.11	.6052
		3.95	.6048
Ellis	12 by 12	2.32	.600
		3.92	.602
		7.99	.606
		11.58	.605
		14.31	.611
		16.22	.606
		18.45	.606
Balch	12 by 12	0.363	.5940
		0.750	.5940
		0.771	.5932
		0.826	.5982
		0.905	.5950
		1.134	.5960
		1.371	.5970
		2.097	.6056
		2.636	.6105
		3.220	.6095
		3.975	.6148
Stewart	48 by 48 (3.72 in. thick)	.05	.626
		.10	.608
		.15	.605
		.20	.605
		.25	.606
		.30	.610

Rectangular Orifices

Source	Dimensions inches	Head feet h	Coefficient of Discharge c
Hamilton Smith, Jr.	0.6 by 3.6	0.614	.6219
		1.63	.6207
		2.77	.6188

Hamilton Smith are slightly less than those herein reported in Table 5. Omitting the values as given by Francis it will be observed that there is very little difference between the coefficients for the small and the large orifices, the value of the coefficient varying only slightly from 0.60. From the results obtained in the present investigation as given in Table 5 and in Fig. 12, 13, and 14, it will be seen that the coefficient varies more with the size of the orifice than is shown by the results of the earlier experiments, as given in Table 6.

It will be observed also that the coefficient of discharge for the rectangular orifice used by Hamilton Smith is somewhat smaller than that herein reported. It may seem that the diverging sides of the orifices used in the experiments reported in this bulletin (orifice plate ½-in. thick) would form a diverging mouthpiece, particularly in the case of the smaller orifices, but experiments* on diverging mouthpieces have shown that a mouthpiece having a total angle of divergence of 90 degrees has very little, if any, effect on the rate of discharge.

16. *Comparison with Discharge into Air.*—The experiments on sharp-edged orifices with discharge into air are more numerous than for submerged discharge. The experiments of Bilton and to a less degree those by Judd and King, and those by Mair and by Ellis indicate that there is a critical head for each circular orifice above which the coefficient remains constant. Bilton concludes that "circular orifices of 2½-in. diameter, and over, under heads of 17 in., and over, have a common coefficient of discharge lying between 0.59 and 0.60 but which is probably about 0.598 (subject to the head being not less than 2 or 3 diameters)." The results of the experiments of Hamilton Smith, as is well known, indicate that the coefficient of discharge gradually decreases as the size of the orifice increases, and also decreases as the head increases until at a head of 100 ft. all orifices, regardless of the size or the shape, have a common coefficient of discharge.

The results of the experiments on submerged orifices herein reported seem to indicate, as previously noted, that orifices having diameters greater than about 2½ in. (or sides, if square) have a common coefficient of discharge which is very close to 0.60. There seems, however, to be no evidence of a critical head since the coefficient remains constant for the whole range of head used, nor is there evidence of a critical head in the results obtained by earlier experimenters on submerged orifices as given in Table 6.

---

*"The Effect of Mouthpieces on the Flow of Water Through a Submerged Short Pipe." Univ. of Ill. Eng. Exp. Sta., Bul. 96, 1917.

From a study of the experimental results on orifices with discharge into air it is believed that the coefficient of discharge for submerged orifices are the same as those for discharge into air for the same heads and sizes and shapes (except for very small heads). It is doubtful if the statement sometimes made, namely, that the coefficient of discharge for submerged orifices is about one per cent less than that for free discharge, is justified.

17. *Summary.*—The following brief summary is given as applying to submerged sharp-edged orifices for velocities from ½ to 5 ft. per sec.

(1) The coefficient of discharge for a circular, a square, or a rectangular submerged orifice does not vary with the velocity.

(2) Circular and square submerged orifices having areas greater than about 10 sq. in. have a common coefficient of discharge varying but little from 0.60.

(3) Rectangular submerged orifices having one side from 3 to 12 times the other side have a constant coefficient of discharge which is larger than that for circular and square orifices of the same size, particularly for the larger areas, at least up to a size of 12 sq. in.

(4) The flow of water through submerged sharp-edged orifices is very nearly the same as that for the same kind of orifices with discharge into air, provided the head is not less than 2 or 3 diameters when the discharge is into air.

PART III

FIRE STREAMS FROM SMALL HOSE AND NOZZLES

By VIRGIL R FLEMING
Assistant Professor of Applied Mechanics

CONTENTS

PART III

FIRE STREAMS FROM SMALL HOSE AND NOZZLES

LIST OF FIGURES

LIST OF TABLES

PART III

FIRE STREAMS FROM SMALL HOSE AND NOZZLES*

VII. INTRODUCTION

18. *Scope of Experiments.*—Part III presents the results of experiments on 1½-in. hose and nozzles. Both rubber-lined hose and unlined linen hose were used. Three sizes of conical nozzles were tested, the diameters of the nozzle openings being 5/16 in., 7/16 in., and ½ in.

The loss of head in the hose due to friction and the corresponding friction factor are given for each hose for a range in velocity from about 4 to 8 ft. per sec. The coefficient of discharge for each nozzle is recorded for a range in pressure at the base of the nozzle from about 10 to 85 lb. per sq. in. The height and the horizontal distance which the jets reached are also recorded. The influence of a cylindrical tip on a nozzle is brought out and some discussion is given concerning the quantity of water required for temporary fire protection for the interior of buildings.

The importance of adequate fire protection has become so well recognized that most buildings, even those of moderate size, are equipped with some sort of fire apparatus for immediate service in case of fire in the interior of the building and until the city fire department arrives. The ordinary water buckets and portable chemical fire extinguishers have in a large measure been supplemented with small fire hose. Few data are available concerning the hydraulics of small fire streams. Many inquiries concerning the discharge from small nozzles and the loss of head in small hose led to the tests which are herein described. The tests were undertaken with the object of acquiring data and putting the results into so workable a form that it would be easy to compute the quantity of water delivered by a nozzle of the size ordinarily used in the fire protection of the interior of buildings or to compute the pressure necessary in the mains to give an effective fire stream from such nozzles, and also to throw some light upon the quantity of water which would be considered sufficient for temporary protection.

* The experiments used in Part III of this Bulletin were reported in the Proceedings of the Fifth Meeting of the Illinois Water Supply Association, p. 170, 1913.

19. *Acknowledgment.*—The experiments here used were conducted at the University of Illinois under the direction of the writer by E. O. KORSMO and A. B. NEININGER of the class of 1911 as thesis work. Much credit is due them for the care and thought given the problem and the thoroughness with which they did their work. The water for the experiments was drawn from the University mains. The experiments for determining the height and the horizontal distance the jets would reach were conducted out of doors. The other experiments were carried on in the Hydraulics Laboratory.

VIII. Apparatus and Method of Experimenting

20. *Hose and Nozzles.*—Rubber-lined cotton hose and unlined linen hose having a nominal diameter of 1½ in. were used, the length of the test section for determining the lost head being 50 ft. in each case. The hose taken was from the racks in the University buildings and is representative of hose of this size commonly in use.

Three 1½-in. conical nozzles having different sizes of openings, as shown in Fig. 19, were tested. The first nozzle had a diameter of

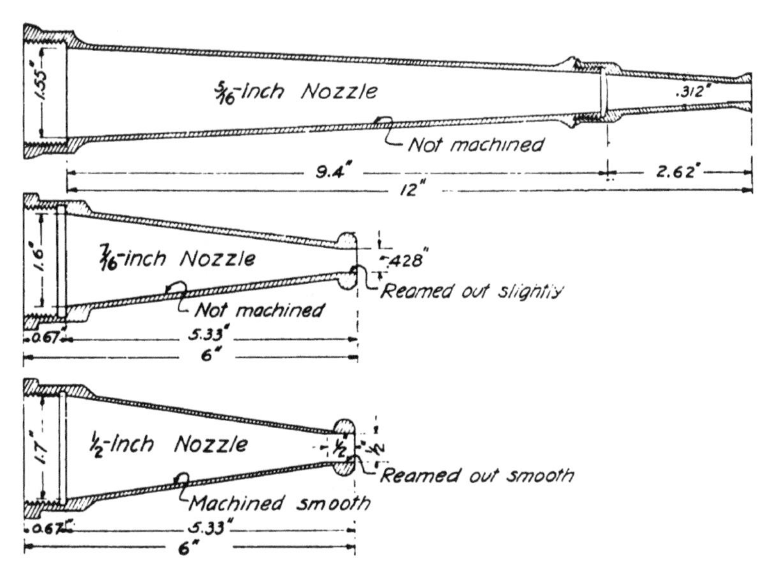

Fig. 19. Longitudinal Sections of Nozzles Tested

5/16 in. The second nozzle had a diameter of 0.428 in., which is approximately 7/16 in., and in compiling the tables, corrections were made so as to apply to a 7/16-in. nozzle. The third nozzle had a diameter of ½ in. The 5/16-in. nozzle was 12 in. long while the other two were only 6 in. long (see Fig. 19). The 5/16-in. and the 7/16-in. nozzles were rough on the interior surfaces, having been left just as they came from the molds, the prints of the sand core being plainly visible. The tips had been smoothed slightly by running a drill through the opening, but the cylindrical portion made by the drill was very short in both cases. The ½-in. nozzle was made from a 7/16-in. nozzle. The entire inner surface was machined smooth and a ½-in. reamer was run through the opening making a cylindrical portion ½ in. long.

21. *Method of Experimenting.*—The loss of head was measured over a length of fifty feet of the hose by means of a differential mercury gage. The average pressure at a section of the hose was obtained with a piezometer connection or coupling of the Freeman type. A cross-section of one of these couplings is shown in Fig. 20. The discharge

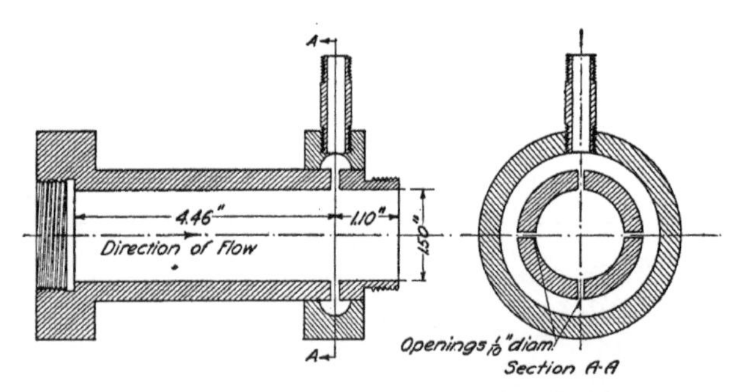

FIG. 20. CROSS-SECTION OF PIEZOMETER COUPLING

through the hose when determining the lost head in the hose was measured with a calibrated nozzle. When determining the coefficient of discharge for the nozzles the discharge was measured by weighing. The pressure at the base of the nozzle was measured with a calibrated pressure gage.

The vertical heights attained by the streams were determined by means of a transit and the horizontal distances reached were found by measuring with a tape from stakes which were driven in the ground at frequent intervals and at known distances from the nozzle.

IX. Experimental Results and Discussion

22. *Results from Freeman's Experiments.*—In 1888 John R. Freeman conducted an extensive series of tests upon 2½-in. fire hose and nozzles.* In general, Freeman arrived at the following conclusions: Smooth conical nozzles give coefficients of discharge as high as any other form of nozzle, the jets reach farther and the streams remain solid for greater distances than for any other form of nozzle of the same size of opening and with the same pressure at the base of the nozzle. For smooth conical nozzles 1⅛ or 1¼ in. in diameter, a coefficient of discharge of 0.977 may be taken with great confidence that it will not be more than one-half of 1 per cent in error. The coefficient will be slightly larger for smaller nozzles. The nozzle makes a very convenient method of measuring water. The friction is but slightly more in smooth rubber-lined hose than in clean iron pipe of the same diameter. The friction in unlined linen hose is about two and one-third times as much as in smooth rubber-lined hose. A hose elongates from 2 per cent to 5 per cent with a pressure of 50 lb. per sq. in. This elongation produces a sinuosity which increases the loss of head about 6 per cent. Care should be exercised that there is no abrupt change of section in the hose couplings and that no washers or gaskets are so left as to impede the flow of water.

It is frequently recommended that a 250 gal. per min. fire stream be used in business districts, while a 175 or a 200 gal. per min. stream may be used in a residential district. These discharges correspond to a nozzle pressure of 40 to 50 lb. per sq. in., and a hydrant pressure of 80 to 110 lb. per sq. in. These values refer to outside service. Table 7 gives data for 2½-in. hose and nozzles for three different sizes of nozzle openings taken from Freeman's results. This table is convenient for making calculations for outside fire protection.

23. *Experimental Data.*—Table 8 gives the more important data of the experiments with hose and nozzles herein reported. Values are given for the pressures at the base of the nozzles, the discharges, the loss of head in the hose, and the vertical and horizontal distances reached by the jets. Other results discussed have been calculated from the data in this table.

---

* "Experiments Relating to the Hydraulics of Fire Streams." Trans. Am. Soc. Civ. Eng., Vol. XXI, p. 304, 1889.

TABLE 7

FREEMAN'S RESULTS FOR 1-IN. 1⅛-IN. AND 1¼-IN. NOZZLES ATTACHED TO 2½-IN. HOSE

1-INCH NOZZLE

Pressure Base of Nozzle	Discharge	Loss of Head in 100 Feet of Hose		Vertical Height of Jet for Good Fire Stream	Horizontal Distance	
		Rubber Lined	Unlined Linen		Jet for Good Fire Stream	Extreme Drops at Level of Nozzle
Lb. per sq. in.	Gallons per minute	Lb. per sq. in.	Lb. per sq. in.	Feet	Feet	Feet
20	132	5	10	35	37	77
30	161	7	15	51	47	109
40	186	10	20	64	55	133
50	208	12	25	73	61	152
60	228	15	30	79	67	167
70	246	17	35	85	72	179

1⅛-INCH NOZZLE

Pressure Base of Nozzle	Discharge	Loss of Head in 100 Feet of Hose		Vertical Height of Jet for Good Fire Stream	Horizontal Distance	
		Rubber Lined	Unlined Linen		Jet for Good Fire Stream	Extreme Drops at Level of Nozzle
Lb. per sq. in.	Gallons per minute	Lb. per sq. in.	Lb. per sq. in.	Feet	Feet	Feet
20	168	8	16	36	38	80
30	206	12	25	52	50	115
40	238	16	33	65	59	142
50	266	20	41	75	66	162
60	291	24	49	83	72	178
70	314	28	57	88	77	191

1¼-INCH NOZZLE

Pressure Base of Nozzle	Discharge	Loss of Head in 100 Feet of Hose		Vertical Height of Jet for Good Fire Stream	Horizontal Distance	
		Rubber Lined	Unlined Linen		Jet for Good Fire Stream	Extreme Drops at Level of Nozzle
Lb. per sq. in.	Gallons per minute	Lb. per sq. in.	Lb. per sq. in.	Feet	Feet	Feet
20	209	12	25	37	40	83
30	256	19	38	53	54	119
40	296	25	51	67	63	148
50	331	31	63	77	70	169
60	363	37	76	85	76	186
70	392	43	88	91	81	200

TABLE 8

RESULTS OF EXPERIMENTS AT UNIVERSITY OF ILLINOIS WITH 5/16-IN., 7/16-IN. AND ½-IN. NOZZLES ATTACHED TO 1½-IN. HOSE

5/16-INCH NOZZLE

Pressure Base of Nozzle	Discharge	Loss of Head in 100 Feet of Hose		Vertical Height of Jet for Good Fire Stream	Horizontal Distance	
		Rubber Lined	Unlined Linen		Jet for Good Fire Stream	Extreme Drops at Level of Nozzle
Lb. per sq. in.	Gallons per minute	Lb. per sq. in.	Lb. per sq. in.	Feet	Feet	Feet
20	12	.7	1.3	28	15	53
30	15	1.1	1.9	32	18	63
40	17	1.5	2.6	34	21	71
50	19	1.8	3.2	35	23	78
60	21	2.2	3.9	36	26	84
70	23	2.6	4.5	37	28	90
80	24	2.9	5.2	38	29	96
90	26	3.3	5.9	39	30	102
100	28	3.7	6.5	40	31	107

$\frac{7}{16}$-INCH NOZZLE

Pressure Base of Nozzle	Discharge	Loss of Head in 100 Feet of Hose		Vertical Height of Jet for Good Fire Stream	Horizontal Distance	
		Rubber Lined	Unlined Linen		Jet for Good Fire Stream	Extreme Drops at Level of Nozzle
Lb. per sq. in.	Gallons per minute	Lb. per sq. in.	Lb. per sq. in.	Feet	Feet	Feet
20	25	2.8	5.1	23	10	45
30	30	4.2	7.7	27	13	54
40	35	5.6	10.2	30	16	63
50	39	7.0	12.8	32	18	70
60	43	8.5	15.3	33	20	77
70	47	9.8	17.8	34	21	84
80	50	11.1	20.3	35	23	94
90	53	12.7	22.9	36	24	99
100	56	14.1	25.5	37	25	106

$\frac{1}{2}$-INCH NOZZLE

Pressure Base of Nozzle	Discharge	Loss of Head in 100 Feet of Hose		Vertical Height of Jet for Good Fire Stream	Horizontal Distance	
		Rubber Lined	Unlined Linen		Jet for Good Fire Stream	Extreme Drops at Level of Nozzle
Lb. per sq. in.	Gallons per minute	Lb. per sq. in.	Lb. per sq. in.	Feet	Feet	Feet
20	33	5.2	9.5	34	15	63
30	40	7.7	14.4	37	20	79
40	46	10.2	18.8	38	25	91
50	52	12.8	23.8	39	30	102
60	57	15.4	28.5	40	33	111
70	61	18.0	32.7	41	37	120
80	65	20.5	38.4	42	40	127
90	69	23.0	42.0	43	43	134
100	73	25.6	47.0	44	46	140

24. *Friction Factors.*—The curves of Fig. 21 show the friction factors for each kind of hose used and for velocities in the hose ranging

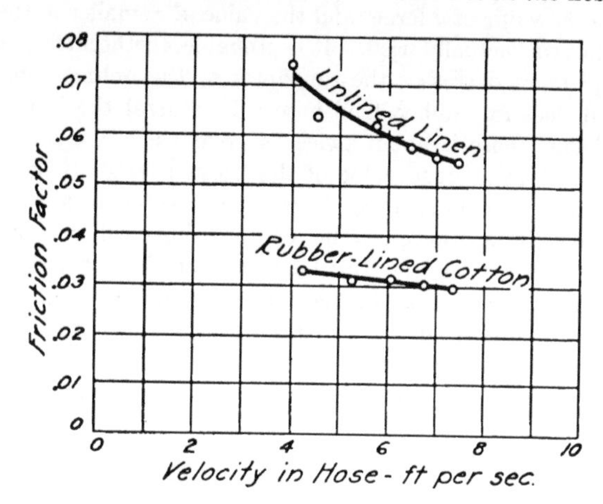

FIG. 21. DIAGRAM SHOWING FRICTION FACTORS IN RUBBER-LINED AND UNLINED HOSE

from 4 to 8 ft. per sec. These curves cover the range of velocities which would be met in ordinary use. The friction factor f is computed from the formula

$$h=\frac{fl}{d}\frac{v^2}{2g}$$

h = head lost in feet of water
l = length of hose in feet
d = diameter of hose in feet
v = velocity of the water in the hose in feet per second
g = acceleration due to gravity in feet per second per second

The loss of head in the rubber-lined hose varies almost directly as the square of the velocity and is about the same as the loss of head in clean iron pipe of the same diameter. The friction factor for the unlined linen hose decreases as the velocity increases, or in other words the loss of head does not vary directly as the square of the velocity, the ratio of the lost head to the square of the velocity being larger for the lower velocities. The reason that the friction factor for unlined linen hose decreased more rapidly with the velocity than does that for rubber-lined cotton hose may be that the diameter of the unlined hose is increased more than that of the rubber-lined cotton hose by the increasing pressures which accompany the increasing velocities. This would make the value of d larger and the value of v smaller in the equation for f than was actually used. It is probable, furthermore, that the increasing pressure decreases the roughness of the unlined linen hose more than it does for rubber-lined hose. In general the lost head in the unlined linen hose is about twice as great as in the rubber-lined cotton hose. If an average value of the friction factor (0.06) is used for the unlined linen hose, no great error will enter into the results under ordinary circumstances. The length of hose will ordinarily not be mose than 100 feet and for this length about 10 lb. per sq. in. will be the maximum loss of head in the unlined linen hose under working conditions with nozzles giving streams up to ½ in. in diameter. An error as large as 10 per cent in the calculation of the loss of head in the hose would affect the nozzle pressure not more than one pound per square inch.

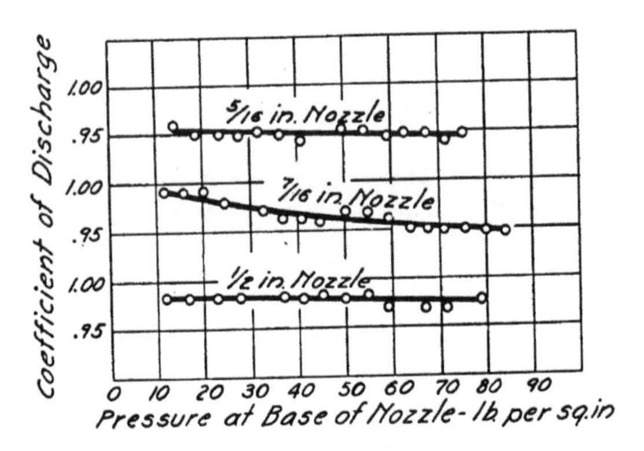

FIG. 22. DIAGRAM SHOWING COEFFICIENTS OF DISCHARGE OF NOZZLES

While the loss of head in the unlined linen hose is about twice as great as the loss of head in the rubber-lined hose, the linen hose has several advantages. It is much lighter to handle, folds up in less space on the wall racks, costs only about 50 to 60 per cent of the cost of rubber-lined hose and, in an ordinary building, its life is much longer.

25. *Coefficients of Discharge.*—The coefficients of discharge for each of the three sizes of nozzles are given in Fig. 22 for pressures

at the base of the nozzle ranging from about 10 to 85 lb. per sq. in. This range of pressures corresponds to a range in the velocity of the issuing jet from a minimum of about 35 ft. per sec. with the ½-in. nozzle to a maximum of about 185 ft. per sec. with the 5⁄16-in. nozzle. The coefficient of discharge is the ratio of the measured discharge to the ideal discharge. The measured discharge was weighed and the volume computed from the weights. The ideal discharge was computed from the formula

$$q = A\sqrt{2gh}$$

q = discharge in cubic feet per second. A = area of the opening of the nozzle in square feet. g = acceleration due to gravity in feet per second per second. h = pressure at the base of the nozzle in feet of water. The velocity of approach to the nozzle was negligible and was, therefore, not considered in the equation for the ideal discharge. The pressure at the base of the nozzles was measured with calibrated pressure gages.

The coefficient of discharge for the 7⁄16-in. and the ½-in. nozzles is nearly constant for all pressures and averages 0.98. The coefficient is slightly lower than 0.98 for the 7⁄16-in. nozzle at the higher pressures. The 5⁄16-in. nozzle gives a coefficient of 0.95. The 5⁄16-in. nozzle is 12 inches long while the other two are only 6 inches long, and this greater length adds somewhat to the friction and lowers the coefficient of discharge for the 5⁄16-in. nozzle.

The 5⁄16-in. and the 7⁄16-in. nozzles were rough on the interior surfaces, having been left just as they came from the molds. The tips had been smoothed slightly by running a drill through the opening, but the cylindrical portion made by the drills was very short in each case and the nozzles gave streams which sprayed badly a short distance away. The ½-in. nozzle was made from a 7⁄16-in. nozzle. The entire inner surface was first machined out in hopes that it would prevent the spraying of the jet, but the nozzle gave a stream which appeared no better than before machining. Then a 15⁄32-in. reamer and finally a ½-in. reamer were run through the opening, each reducing the spraying. The ½-in. reamer made the cylindrical portion of the opening ½-in. long, and the resulting nozzle gave a very good stream. An opening larger than ½-in. could not be made in the nozzle because of the thinness of the walls.

26. *Height and Horizontal Distance of Jets.*—The heights and the horizontal distances reached by the jets from each of the three nozzles

used are given in Table 8. As stated, the vertical heights were measured by means of a transit and the horizontal distances were measured with a tape from stakes which were driven in the ground at frequent known space intervals. The observations were made when a moderate wind was blowing which interfered with the streams considerably. A stream was considered good for the distance in which practically all the water would pass through a circle whose diameter was 18 inches. The value was an arbitrary selection and the streams might be considered by some as effective for greater distances than those given in Table 8. The streams, however, beyond the sections chosen, diverged rapidly and the selection of a circle larger than 18 inches would have added but a few feet to the distances given in Table 8 in any case.

27. *Effect of Cylindrical Tip.*—The tests show clearly the importance of a smooth cylindrical opening at the tip of the nozzle. A comparison of the results of the tests on the 7/16-in. and the 1/2-in. nozzles for vertical heights and horizontal distances of the jets will show this difference. In the case of the 7/16-in. nozzle with a pressure of 30 lb. per sq. in. at the base of the nozzle the vertical height of the jet was 27 ft. as compared with 37 ft. for the 1/2-in. nozzle for the same pressure. Likewise the horizontal distance reached with the 7/16-in. nozzle was 13 ft. as compared with 20 ft. with the 1/2-in. nozzle. Similar comparisons may be made for other pressures at the base of the nozzle. The appearance of the jets showed a much greater difference than the data would indicate. It must be remembered that the two nozzles were alike and gave streams which appeared to be the same before one was reamed out to a larger size.

It will be noted also that in the case of the 5/16-in. nozzle for a pressure of 30 lb. per sq. in. at the base of the nozzle, the vertical and horizontal distances reached by the stream were respectively 32 and 18 ft., which indicate that the improvement in the carrying capacity of the 1/2-in. nozzle over that of the 7/16-in. nozzle was not due to the smoother condition of the interior surface of the 1/2-in. nozzle, but rather to the effect of the cylindrical tip. The condition of the interior surface of the nozzle to within one-half inch of the end does not seem to affect appreciably either the quantity of discharge or the quality of the stream.

It seems important, therefore, that the tip of the nozzle should be reamed out for a distance of at least 1/2 in. in order to obtain a good fire stream. It is probably true also that for nozzles somewhat larger than those used in these experiments the length of the cylindrical portion should be more than ½ in., perhaps equal to the diameter of the issuing stream.

28. *Requirements for Temporary Fire Protection for the Interior of Buildings.*—Small fire hose and nozzles should be used as a temporary protection and brought into play until greater relief is at hand. They must necessarily operate under ordinary working pressures in the mains more often than under fire pressures. With 40 lb. per sq. in. as an average pressure in the mains, there should be, after deducting for losses in the hose and connecting pipes, about 30 lb. per sq. in. at the nozzle. This pressure, of course, would be still further reduced if the nozzle used was at a higher elevation than the main. With a nozzle pressure of 30 lb. per sq. in. the ½-in. nozzle will discharge 40 gal. per min., the 7/16-in. and the 5/16-in. nozzles will discharge 30 and 15 gal. per min., respectively. It is felt that the discharge from the two smaller nozzles is not great enough for effective work. It is true that the pressure at the nozzle for the smaller sizes with a given pressure in the main will be somewhat greater than for the ½-in. nozzle, because of the decreased velocity in the hose which will give a smaller loss of head, but this difference in pressure will not be enough to increase the discharge materially for an ordinary length of hose. The discharge from the 5/16-in. nozzle is too small to be very effective even at higher pressures. The discharge for a pressure of 100 lb. per sq. in. is but 28 gal. per min. It is recommended that ½-in. nozzles be used with 1½-in. hose. For nozzles larger than ½ in., the discharge would become greater and increase the loss of head in the hose to such an extent that there would not be enough nozzle pressure left to produce a stream which would carry a sufficient distance.

With the aid of the tables the discharge for any of the nozzles may be readily computed for any pressure in the mains. If the nozzle is at a higher elevation than the main, subtract from the pressure in the main an amount equal to 0.434 times the difference in elevation in feet between the nozzle and the main. Take a discharge from the table for any pressure at the base of the nozzle for the size of nozzle used, then take the corresponding value of the head lost in the kind of hose used, multiply this value by the length of hose in feet used and divide by 100. The result gives the total loss in the hose for the assumed discharge. If there is any connecting pipe, the loss in it will be the same as the loss in a corresponding length of rubber-lined hose.

Add the losses in the pipe and hose to the pressure at the base of the nozzle for the assumed discharge to obtain the pressure in the main (corrected for the difference in elevation) necessary to produce this discharge. The discharge will vary as the square root of this pressure. Letting q' = the assumed discharge, P' = the pressure in the main (corrected for the difference in elevation) which will produce this discharge, q = the discharge to be determined, P = the actual pressure in the mains and H = difference in elevation between the nozzle and the main in feet gives the relation

$$q = q'\sqrt{\frac{P - 0.434H}{P'}}$$

which gives the required discharge.

To illustrate the use of the formula the following assumptions are made. Pressure in mains, P = 60 lb. per sq. in., 80 ft. of linen hose, 50 ft. of 1½-in. connecting pipe, elevation of nozzle above main 30 ft. and ½-in. nozzle used.

Assume a discharge of 46 gal. per min. and from the table the following values are obtained:

$$\text{Nozzle pressure} = 40$$

$$\text{Loss in hose} = \frac{80 \times 18.8}{100} = 15.0$$

$$\text{Loss in pipe} = \frac{50 \times 10.2}{100} = 5.1$$

$$\text{Total} = P' = 60.1$$

Substituting in the formula

$$q = 40.7 \text{ gal. per min.}$$

The following method may be used to determine the discharge for any size of nozzle for any pressure in the mains. Assume any pressure at the base of the nozzle, h', in feet of water. The discharge for this pressure may be determined by the formula

$$q' = cA\sqrt{2gh'}$$

q' = discharge in cu. ft. per sec.

c = coefficient of discharge and may be taken as 0.98

A = area of opening of nozzle in sq. ft.

$2g$ = 64.4 ft. per sec. per sec.

Determine the velocity in the hose for this discharge from the formula

$$v=\frac{q'}{a}$$

v = velocity in hose in ft. per sec.
a = area of hose in sq. ft.

Determine the loss in the hose from the formula

$$h_2=\frac{fl\,v^2}{d\,2g}$$

h_2 = head lost, in feet
f = friction factor which may be taken as 0.03 for rubber lined hose or 0.06 for unlined linen hose
l = length of hose in feet
d = diameter of hose in feet
v = velocity in hose in ft. per sec.
$2g$ = 64.4 ft. per sec. per sec.

If there is any pipe connecting the hose to the main, the loss for it may be computed by the same formula as for the hose, using 0.03 for the friction factor for 1½-in. pipe. Call this lost head h_3.

The pressure in the main to give the assumed nozzle pressure is

$$H'=h'+h_2+h_3$$

This pressure will be in feet of water. Then using the relation

$$q=q'\sqrt{\frac{H}{H'}}$$

gives the required discharge. If the main is below the nozzle, subtract the difference in elevation in feet from H in the formula.

It is recognized that this method is not strictly accurate since the head does not vary exactly as the square of the discharge, but the results obtained will be close enough for practical use.

29. *Summary.*—The following brief summary is given as applying to small hose and nozzles with velocities in the hose ranging from about 4 to 8 ft. per sec. and with pressures at the base of the nozzle ranging from about 10 to 85 lb. per sq. in.

(1) The friction factor (f in the equation for the lost head, $h=f\,\frac{l\,v^2}{d\,2g}$) for rubber-lined hose varies but little with the velocity in the hose and is nearly the same as for clean iron pipe of the same diameter.

(2) The friction factor for unlined linen hose decreases as the velocity increases. In general the loss of head in unlined linen hose is about twice as great as in rubber-lined hose of the same diameter and for the same velocity.

(3) The nozzle should have a smooth cylindrical tip at least one-half inch long to keep the jet from spraying. A cylindrical tip is a much more important factor in securing a good fire stream than a smooth surface in the interior of the nozzle.

(4) Nozzle openings commonly in use to supply fire streams in the interior of buildings seem too small for adequate temporary fire protection. It is recommended that a nozzle with a ½-in. opening be used with a 1½-in. hose in order to secure a sufficient quantity of water for an effective fire stream.

(5) The coefficient of discharge of a small conical nozzle varies but little with the velocity and is close to 0.98. The value of 0.95 obtained with the 5/16-in. nozzle, which was 12 in. long as compared with 6 in. for the other nozzles tested, indicates, however, that the nozzle should be short to obtain the value of 0.98. A cylindrical tip on the nozzle seems to have little influence on the coefficient of discharge.

Part IV

THE ORIFICE BUCKET FOR MEASURING WATER

By MELVIN L. ENGER
Associate Professor of Mechanics and Hydraulics

CONTENTS

PART IV

THE ORIFICE BUCKET FOR MEASURING WATER

LIST OF FIGURES

PART IV

THE ORIFICE BUCKET FOR MEASURING WATER

X. INTRODUCTION

30. *Purpose.*—The purpose of Part IV is to describe a method of measuring water by means of a simple, portable, and inexpensive device, here called an orifice bucket, and to present experimental data applying thereto for a range of conditions sufficient to indicate that the device is reliable for use in engineering practice. An orifice bucket is a cylindrical vessel into which water to be measured falls vertically and passes out through a number of holes or orifices in the bottom. A vertical glass tube placed just outside the bucket is connected to the sides of the bucket near the bottom, and the height of the water in the tube indicates the head on the orifice.

The orifice bucket was devised for the purpose of measuring the discharge of several artesian wells pumped by means of air lift, the water from each of which discharged into a separate cistern or small reservoir through a vertical pipe. In each case the water left the pipe with considerable blast and momentum. Several possible methods for the measurement of the discharge were considered but were thought to be impracticable for various reasons or inapplicable for the particular case. After some preliminary laboratory experimenting an orifice bucket was devised which served very satisfactorily in determining the discharge from each of the wells. It was at first feared that the water would enter the bucket with such a blast that entrained air would enter the vertical glass tube and cause trouble in determining the height of water in the bucket. There was, however, no trouble from this cause and the fluctuations of the water level in the glass tube offered no serious difficulties.

The orifice bucket has also given satisfaction in tests made to determine yields of well pumps of the reciprocating type. It should give satisfactory results in the field where simplicity of construction and portability are desirable and where extreme accuracy is not of great importance.

31. *Acknowledgment.*—The orifice bucket was developed by the writer through experimental work in the Hydraulic Laboratory of the University of Illinois during 1910 and 1911.* Considerable improvement has been made in the arrangement of certain parts of the bucket by I. W. Fisk, P. S. Biegler, and P. J. Nilsen, of the department of electrical engineering, in connection with tests on electric motor-driven deep-well pumps. Some of the experimental data herein presented were obtained by them, to whom acknowledgment is made.

* A part of the results here presented was published in the Proceedings of the Third Meeting of the Illinois Water Supply Association, p. 87, 1911.

XI. Apparatus and Method of Calibrating

32. *Orifice Bucket.*—Fig. 23 shows the construction and dimensions of one of the first orifice buckets used in the experiments, and Fig. 24 shows the bucket in use. This bucket weighed 23 lb.

As previously stated an orifice bucket is a cylindrical vessel having holes or orifices in its bottom and into which water to be measured

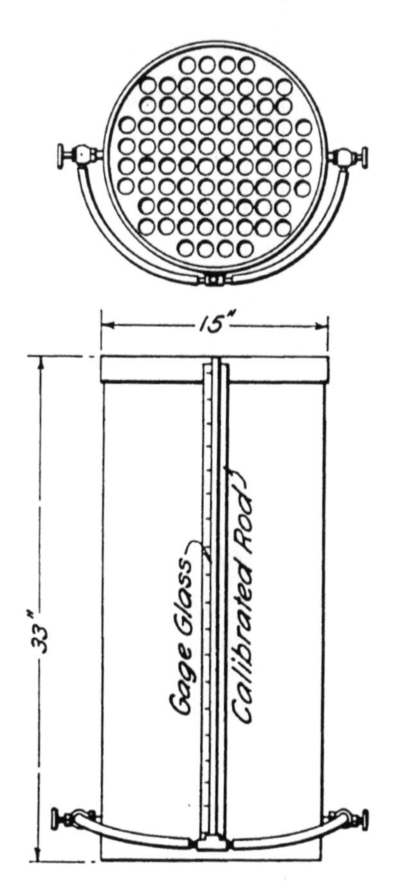

Fig. 23. Fifteen-Inch Orifice Bucket Having Fifty-six Orifices

falls vertically, the head of water on the orifices being indicated by the height of the water in a vertical glass piezometer tube attached near the bottom of the bucket.

Fig. 25 shows the construction of the most elaborate orifice bucket which has been used. It is provided with a short tube checker-work to smooth out the flow of the water on its way to the orifices in the

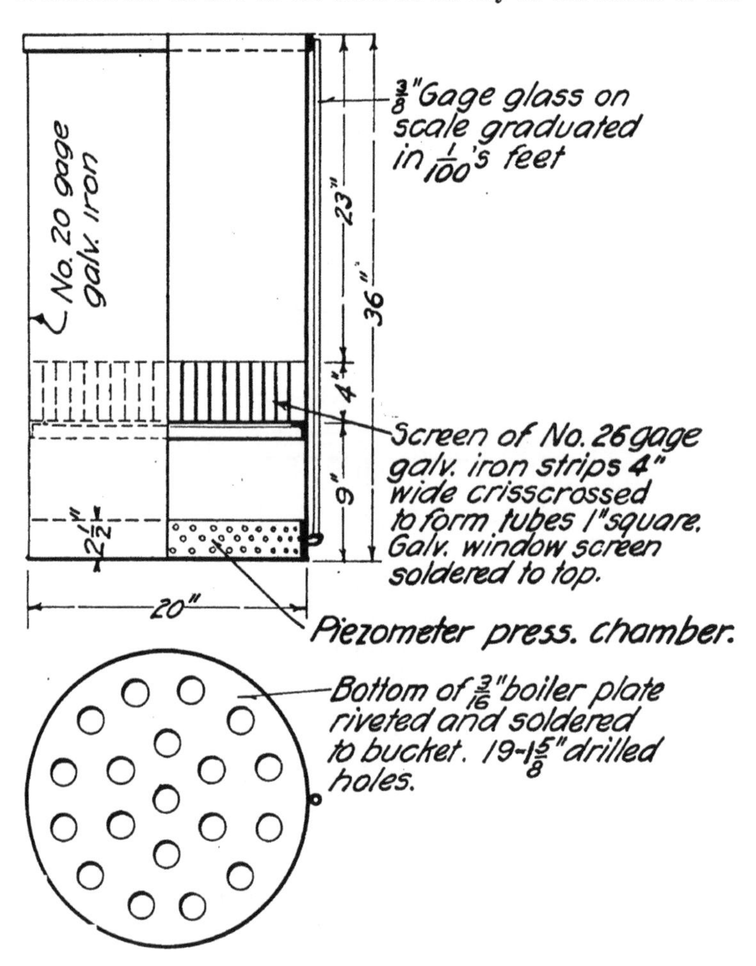

FIG. 25. TWENTY-INCH ORIFICE BUCKET HAVING NINETEEN ORIFICES

Fig. 24. View Showing Orifice Bucket in Use

bottom of the bucket. The vertical glass tube is connected to a piezometer chamber or ring around the base of the bucket, pressure being transmitted to the piezometer ring through a large number of small holes.

The orifice bucket may be adapted for the measurement of water for a considerable range in the discharge by varying the head on the orifices and also by varying the number of holes which are stopped or plugged with corks or wooden stoppers. The range in the capacities of the orifice buckets which have been used is from about 40 to 1000 gal. per min.

33. *Method of Calibrating Orifice Bucket.*—In calibrating the orifice bucket it was hung underneath a vertical pipe as shown in Fig. 24. The quantity of water discharged was measured with a 6-in. Venturi meter in most of the calibration tests although a calibrated measuring pit was used in some of the tests to determine the volume discharged in a given time.

With a given number of holes open, the flow in the orifice bucket was regulated by means of a valve between the Venturi meter and the bucket until the height in the bucket remained constant. The Venturi meter reading and the head on the orifices were then taken. This procedure was repeated for several different heads and for different numbers of orifices open.

The effect of varying the conditions of flow was investigated somewhat. The height of the free fall of the water from the inflow pipe to the orifice bucket was varied; likewise different sizes of pipe were used giving different velocities to the stream entering the bucket. The stream was also allowed to enter near to one side of the bucket instead of at the center. Different groupings of the open orifices, furthermore, were tried, and different methods were employed in attempting to spread or distribute the inflowing stream.

In using the orifice bucket it is necessary to estimate the average head shown in the glass tube because there is some fluctuation. The amount of the fluctuation may be reduced by throttling the valve in the connection of the glass tube to the orifice bucket. If the proper conditions are observed, there should be little trouble from this source. It should be remembered that the rate of discharge is proportional to the square root of the head and that the effect of the error which might occur in the head reading itself is thus reduced in determining the discharge.

XII. Experimental Data and Discussion

34. *Fifteen-inch Orifice Bucket Having Fifty-six Orifices.*—Fig. 26 shows the calibration curves for the 15-inch orifice bucket shown in Fig. 23 and 24. There were fifty-six 1-in. holes in the bottom of the bucket giving a maximum capacity of about 1000 gal. per min. With all the orifices open the rate of discharge was varied from about 600 to 1000 gal. per min. by varying the head from about ¾ ft. to 2 ft. With thirty-two orifices open the discharge had a range of about 300 to 600 gal. per min. by varying the head from about ½ ft. to 2.5 ft. In closing the twenty-four orifices, corks were used of such size that they projected but little into the bucket. It was found that in filling the orifices a symmetrical arrangement gave somewhat steadier action, particularly when the orifices near the circumference were the ones filled. The inflowing stream was discharged from an 8-in. pipe. A

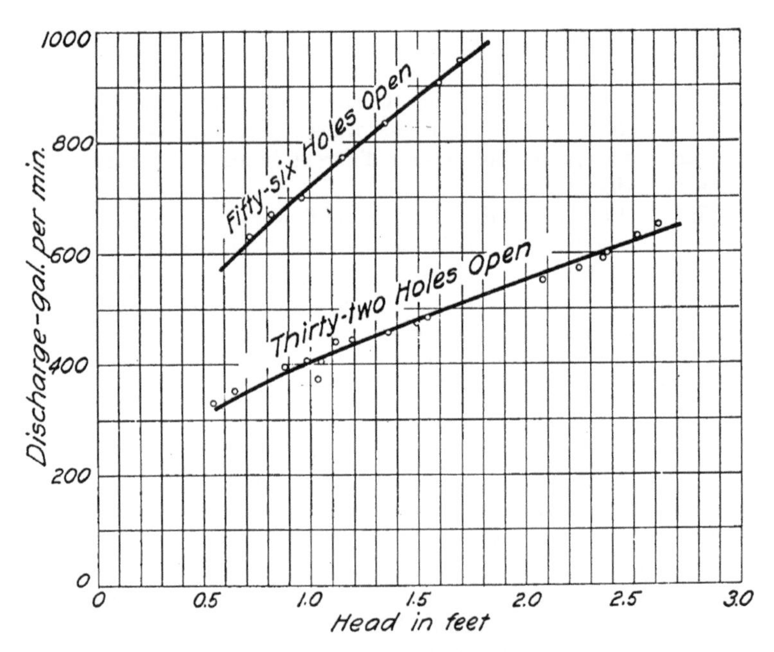

Fig. 26. Calibration Curves for 15-Inch Orifice Bucket Having Fifty-six Orifices

3-in. pipe was also tried but did not give satisfactory results, on account of the high velocity which produced an extremely agitated condition of the water in the bucket. This condition may be overcome, however, by use of a deflector or distributor, such as an open bag or sack attached to the end of the discharge pipe.

The rate of discharge for any other number of open orifices for this bucket may be obtained from the equation

$$q=12.8\, n \sqrt{h}$$

which represents fairly well the relation between the quantity, q, in gal. per min., the number of orifices open, n, and the head in the bucket, h, in ft. The experiments give an average coefficient of discharge for the 1-in. orifices of this bucket of about 0.63.

35. *Fifteen-inch Orifice Bucket Having Only Three Orifices.*—Fig. 27 shows an orifice bucket of the same dimensions as the one just described but with three iron tubes about 1 in. long inserted in a 1-in. wooden bottom. It was provided with two screens through which the water passed on its way to the orifices in the bottom of the bucket.

Fig. 27 also shows the calibration curves for this orifice bucket. It will be noted that the discharge ranges from about 35 to 115 gal. per min. This orifice was constructed and calibrated for immediate use and not for experimental purposes. The calibration curves are of value in indicating the reliability of the orifice bucket under a rather wide range in the details of its construction.

36. *Twenty-inch Orifice Bucket Having Nineteen Orifices.*—An illustration of the most elaborate orifice bucket used in the experiments is shown in Fig. 25, the capacity of which is about 1000 gal. per min. It contains a checkerwork of vertical tubes through which the water flows in passing to the orifices. The gage glass which indicates the head on the orifices is connected to a piezometer ring or chamber around the base of the bucket. The pressure of the water in the bucket is transmitted to the piezometer chamber through a large number of ⅛-in. holes. The bottom of the bucket consists of 3/16-in. boiler plate in which nineteen 1⅝-in. circular holes are drilled.

The calibration curves for this orifice bucket are shown in Fig. 28, for all holes open and for ten holes open. The discharge for any other number of orifices open may be found with a fair degree of accuracy from the equation

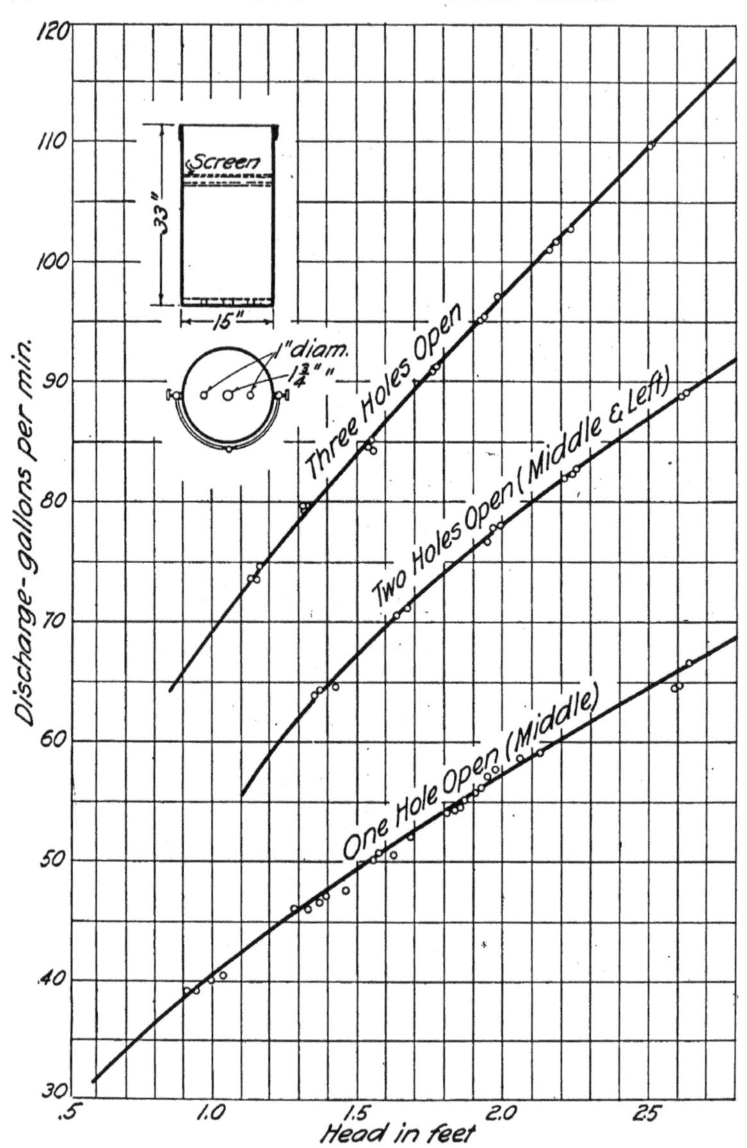

Fig. 27. Calibration Curves for 15-Inch Orifice Bucket Having Only Three Orifices

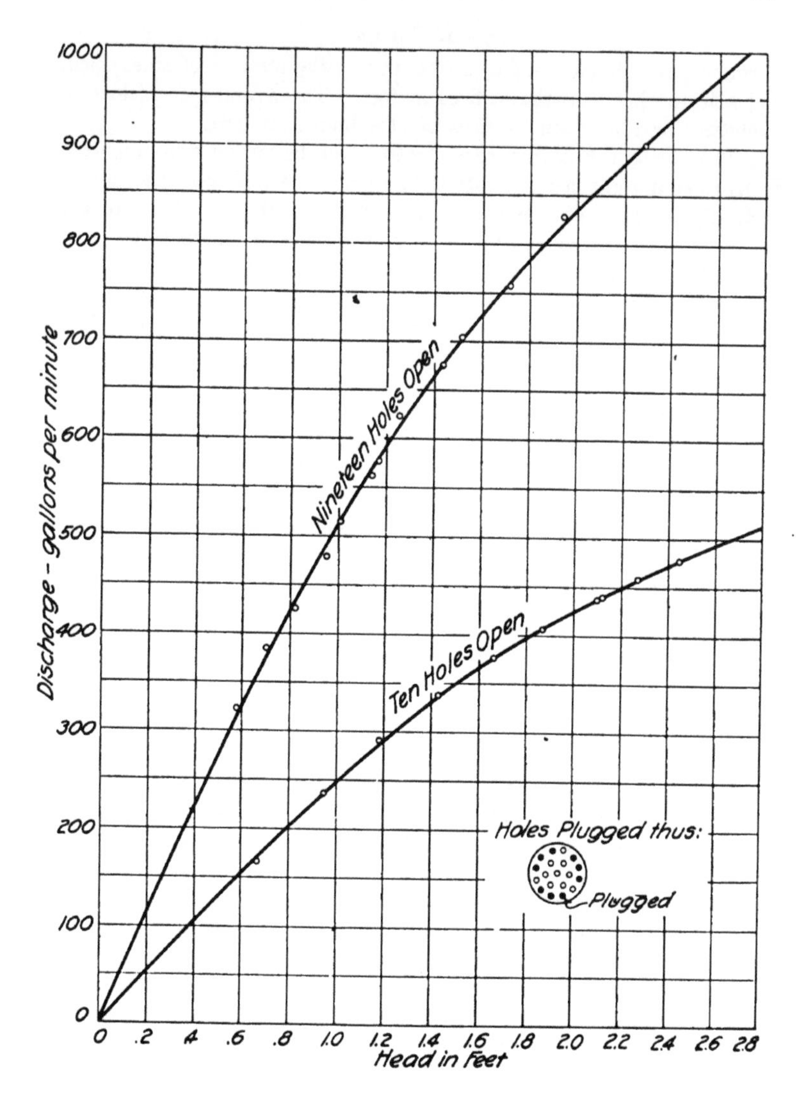

FIG. 28. CALIBRATION CURVES FOR 20-INCH ORIFICE BUCKET HAVING NINETEEN ORIFICES

$$q = 32.7\, n \sqrt{h}$$

in which q is expressed in gal. per min., n is the number of holes open, and h is the head on the orifices in feet. The average coefficient of discharge for the 1⅝-in. orifices of this bucket is 0.61.

The curves in Fig. 28 were obtained when the bucket was supported firmly in an upright position with the stream to be measured discharging vertically in the center of the bucket and with the free fall into the bucket small. The velocity of the inflowing stream, furthermore, was not high (2 or 3 ft. per sec.), thereby causing but little agitation of the water in the bucket. Experiments, however, in which more or less variation from these conditions were allowed indicated that no serious errors resulted.

37. *Conclusions.*—The conditions under which the discharge of water has to be measured are so varied and the purpose or aim in determining the discharge differs so much in different problems that nearly any one of the many common methods of measuring water has a rather restricted field of usefulness, while some methods apply only to very special conditions.

The orifice bucket is designed to meet rather special conditions. It is peculiarly adapted for the measurement of water where a device which is portable (light weight and small size), simple in construction, and low in cost are essential features. The measuring capacity, moreover, covers a considerable range. The orifice bucket is particularly fitted for the measurement of water when the water discharges with considerable blast and momentum from the end of a vertical pipe, in such a manner that the spray covers the entire surface of the water in the bucket, as in the case of air lift-pumping. When so used the orifice bucket gives results which should be correct within 5 per cent if the proper precautions are observed in its use, as is shown by the calibration curves, and correct within 10 per cent for the more unfavorable conditions to be met in the field. The highest accuracy is obtained when the orifice bucket is supported rigidly in an upright position with the center of the discharging stream vertically over the center of the bucket. The free fall of the water should be as small as possible and the velocity of the water as it enters the bucket should not be large, unless the stream is distributed, so as to avoid high local velocities in the bucket. The orifice bucket, however, gives very satisfactory results even when there are considerable deviations from these desirable conditions and renders a service for which other measuring devices may not be adapted.

LIST OF
PUBLICATIONS OF THE ENGINEERING EXPERIMENT STATION

Bulletin No. 1. Tests of Reinforced Concrete Beams, by Arthur N. Talbot, 1904. *None available.*

Circular No. 1. High-Speed Tool Steels, by L. P. Breckenridge. 1905. *None available.*

Bulletin No. 2. Tests of High-Speed Tool Steels on Cast Iron, by L. P. Breckenridge and Henry B. Dirks. 1905. *None available.*

Circular No. 2. Drainage of Earth Roads, by Ira O. Baker. 1906. *None available.*

Circular No. 3. Fuel Tests with Illinois Coal (Compiled from tests made by the Technological Branch of the U. S. G. S., at the St. Louis, Mo., Fuel Testing Plant, 1904–1907), by L. P. Breckenridge and Paul Diserens. 1908. *Thirty cents.*

Bulletin No. 3. The Engineering Experiment Station of the University of Illinois, by L. P. Breckenridge. 1906. *None available.*

Bulletin No. 4. Tests of Reinforced Concrete Beams, Series of 1905, by Arthur N. Talbot. 1906. *Forty-five cents.*

Bulletin No. 5. Resistance of Tubes to Collapse, by Albert P. Carman and M. L. Carr. 1906. *None available.*

Bulletin No. 6. Holding Power of Railroad Spikes, by Roy I. Webber. 1906. *None available.*

Bulletin No. 7. Fuel Tests with Illinois Coals, by L. P. Breckenridge, S. W. Parr, and Henry B. Dirks. 1906. *None available.*

Bulletin No. 8. Tests of Concrete: I, Shear; II, Bond, by Arthur N. Talbot. 1906. *None available.*

Bulletin No. 9. An Extension of the Dewey Decimal System of Classification Applied to the Engineering Industries, by L. P. Breckenridge and G. A. Goodenough. 1906. Revised Edition 1912. *Fifty cents.*

Bulletin No. 10. Tests of Concrete and Reinforced Concrete Columns, Series of 1906, by Arthur N. Talbot. 1907. *None available.*

Bulletin No. 11. The Effect of Scale on the Transmission of Heat through Locomotive Boiler Tubes, by Edward C. Schmidt and John M. Snodgrass. 1907. *None available.*

Bulletin No. 12. Tests of Reinforced Concrete T-Beams, Series of 1906, by Arthur N. Talbot. 1907. *None available.*

Bulletin No. 13. An Extension of the Dewey Decimal System of Classification Applied to Architecture and Building, by N. Clifford Ricker. 1907. *None available.*

Bulletin No. 14. Tests of Reinforced Concrete Beams, Series of 1906, by Arthur N. Talbot. 1907. *None available.*

Bulletin No. 15. How to Burn Illinois Coal Without Smoke, by L. P. Breckenridge. 1908. *None available.*

Bulletin No. 16. A Study of Roof Trusses, by N. Clifford Ricker. 1908. *None available.*

Bulletin No. 17. The Weathering of Coal, by S. W. Parr, N. D. Hamilton, and W. F. Wheeler. 1908. *None available.*

Bulletin No. 18. The Strength of Chain Links, by G. A. Goodenough and L. E. Moore. 1908. *Forty cents.*

Bulletin No. 19. Comparative Tests of Carbon, Metallized Carbon and Tantalum Filament Lamps, by T. H. Amrine. 1908. *None available.*

Bulletin No. 20. Tests of Concrete and Reinforced Concrete Columns, Series of 1907, by Arthur N. Talbot. 1908. *None available.*

Bulletin No. 21. Tests of a Liquid Air Plant, by C. S. Hudson and C. M. Garland. 1908. *Fifteen cents.*

Bulletin No. 22. Tests of Cast-Iron and Reinforced Concrete Culvert Pipe, by Arthur N. Talbot. 1908. *None available.*

Bulletin No. 23. Voids, Settlement and Weight of Crushed Stone, by Ira O. Baker. 1908. *Fifteen cents.*

**Bulletin No. 24.* The Modification of Illinois Coal by Low Temperature Distillation, by S. W. Parr and C. K. Francis. 1908. *Thirty cents.*

Bulletin No. 26. Lighting Country Homes by Private Electric Plants, by T. H. Amrine. 1908. *Twenty cents.*

---

*A limited number of copies of bulletins starred is available for free distribution.

Bulletin No. 26. High Steam-Pressure in Locomotive Service. A Review of a Report to the Carnegie Institution of Washington, by W. F. M. Goss. 1908. *Twenty-five cents.*

Bulletin No. 27. Tests of Brick Columns and Terra Cotta Block Columns, by Arthur N. Talbot and Duff A. Abrams. 1909. *Twenty-five cents.*

Bulletin No. 28. A Test of Three Large Reinforced Concrete Beams, by Arthur N. Talbot. 1909. *Fifteen cents.*

Bulletin No. 29. Tests of Reinforced Concrete Beams: Resistance to Web Stresses, Series of 1907 and 1908, by Arthur N. Talbot. 1909. *Forty-five cents.*

**Bulletin No. 30.* On the Rate of Formation of Carbon Monoxide in Gas Producers, by J. K. Clement, L. H. Adams, and C. N. Haskins. 1909. *Twenty-five cents.*

**Bulletin No. 31.* Tests with House-Heating Boilers, by J. M. Snodgrass. 1909. *Fifty-five cents.*

Bulletin No. 32. The Occluded Gases in Coal, by S. W. Parr and Perry Barker. 1909. *Fifteen cents.*

Bulletin No. 33. Tests of Tungsten Lamps, by T. H. Amrine and A. Guell. 1909. *Twenty cents.*

**Bulletin No. 34.* Tests of Two Types of Tile-Roof Furnaces under a Water-Tube Boiler, by J. M. Snodgrass. 1909. *Fifteen cents.*

Bulletin No. 35. A Study of Base and Bearing Plates for Columns and Beams, by N. Clifford Ricker. 1909. *Twenty cents.*

Bulletin No. 36. The Thermal Conductivity of Fire-Clay at High Temperatures, by J. K. Clement and W. L. Egy. 1909. *Twenty cents.*

Bulletin No. 37. Unit Coal and the Composition of Coal Ash, by S. W. Parr and W. F. Wheeler. 1909. *Thirty-five cents.*

**Bulletin No. 38.* The Weathering of Coal, by S. W. Parr and W. F. Wheeler. 1909. *Twenty-five cents.*

**Bulletin No. 39.* Tests of Washed Grades of Illinois Coal, by C. S. McGovney. 1909. *Seventy-five cents.*

Bulletin No. 40. A Study in Heat Transmission, by J. K. Clement and C. M. Garland. 1910 *Ten cents.*

Bulletin No. 41. Tests of Timber Beams, by Arthur N. Talbot. 1910. *Thirty-five cents.*

**Bulletin No. 42.* The Effect of Keyways on the Strength of Shafts, by Herbert F. Moore. 1910. *Ten cents.*

Bulletin No. 43. Freight Train Resistance, by Edward C. Schmidt. 1910. *Seventy-five cents.*

Bulletin No. 44. An Investigation of Built-up Columns Under Load, by Arthur N. Talbot and Herbert F. Moore. 1911. *Thirty-five cents.*

**Bulletin No. 45.* The Strength of Oxyacetylene Welds in Steel, by Herbert L. Whittemore. 1911. *Thirty-five cents.*

**Bulletin No. 46.* The Spontaneous Combustion of Coal, by S. W. Parr and F. W. Kressman. 1911. *Forty-five cents.*

**Bulletin No. 47.* Magnetic Properties of Heusler Alloys, by Edward B. Stephenson. 1911. *Twenty-five cents.*

**Bulletin No. 48.* Resistance to Flow Through Locomotive Water Columns, by Arthur N. Talbot and Melvin L. Enger. 1911. *Forty cents.*

**Bulletin No. 49.* Tests of Nickel-Steel Riveted Joints, by Arthur N. Talbot and Herbert F. Moore. 1911. *Thirty cents.*

**Bulletin No. 50.* Tests of a Suction Gas Producer, by C. M. Garland and A. P. Kratz. 1912. *Fifty cents.*

Bulletin No. 51. Street Lighting, by J. M. Bryant and H. G. Hake. 1912. *Thirty-five cents.*

**Bulletin No. 52.* An Investigation of the Strength of Rolled Zinc, by Herbert F. Moore. 1912. *Fifteen cents.*

**Bulletin No. 53.* Inductance of Coils, by Morgan Brooks and H. M. Turner. 1912. *Forty cents.*

**Bulletin No. 54.* Mechanical Stresses in Transmission Lines, by A. Guell. 1912. *Twenty cents.*

**Bulletin No. 55.* Starting Currents of Transformers, with Special Reference to Transformers with Silicon Steel Cores, by Trygve D. Yensen. 1912. *Twenty cents.*

**Bulletin No. 56.* Tests of Columns: An Investigation of the Value of Concrete as Reinforcement for Structural Steel Columns, by Arthur N. Talbot and Arthur R. Lord. 1912. *Twenty-five cents.*

**Bulletin No. 57.* Superheated Steam in Locomotive Service. A Review of Publication No. 127 of the Carnegie Institution of Washington, by W. F. M. Goss. 1912. *Forty cents.*

*A limited number of copies of bulletins starred is available for free distribution.

**Bulletin No. 58.* A New Analysis of the Cylinder Performance of Reciprocating Engines, by J. Paul Clayton. 1912. *Sixty cents.*

**Bulletin No. 59.* The Effect of Cold Weather Upon Train Resistance and Tonnage Rating, by Edward C. Schmidt and F. W. Marquis. 1912. *Twenty cents.*

**Bulletin No. 60.* The Coking of Coal at Low Temperatures, with a Preliminary Study of the By-Products, by S. W. Parr and H. L. Olin. 1912. *Twenty-five cents.*

**Bulletin No. 61.* Characteristics and Limitation of the Series Transformer, by A. R. Anderson and H. R. Woodrow. 1913. *Twenty-five cents.*

Bulletin No. 62. The Electron Theory of Magnetism, by Elmer H. Williams. 1913. *Thirty-five cents.*

Bulletin No. 63. Entropy-Temperature and Transmission Diagrams for Air, by C. R. Richards. 1913. *Twenty-five cents.*

**Bulletin No. 64.* Tests of Reinforced Concrete Buildings Under Load, by Arthur N. Talbot and Willis A. Slater. 1913. *Fifty cents.*

**Bulletin No. 65.* The Steam Consumption of Locomotive Engines from the Indicator Diagrams, by J. Paul Clayton. 1913. *Forty cents.*

Bulletin No. 66. The Properties of Saturated and Superheated Ammonia Vapor, by G. A. Goodenough and William Earl Mosher. 1913. *Fifty cents.*

Bulletin No. 67. Reinforced Concrete Wall Footings and Column Footings, by Arthur N. Talbot. 1913. *Fifty cents.*

Bulletin No. 68. The Strength of I-Beams in Flexure, by Herbert F. Moore. 1913. *Twenty cents.*

Bulletin No. 69. Coal Washing in Illinois, by F. C. Lincoln. 1913. *Fifty cents.*

Bulletin No. 70. The Mortar-Making Qualities of Illinois Sands, by C. C. Wiley. 1913. *Twenty cents.*

Bulletin No. 71. Tests of Bond between Concrete and Steel, by Duff A. Abrams. 1914. *One dollar.*

**Bulletin No. 72.* Magnetic and Other Properties of Electrolytic Iron Melted in Vacuo, by Trygve D. Yensen. 1914. *Forty cents.*

Bulletin No. 73. Acoustics of Auditoriums, by F. R. Watson. 1914. *Twenty cents.*

**Bulletin No. 74.* The Tractive Resistance of a 28-Ton Electric Car, by Harold H. Dunn. 1914. *Twenty-five cents.*

Bulletin No. 75. Thermal Properties of Steam, by G. A. Goodenough. 1914. *Thirty-five cents.*

Bulletin No. 76. The Analysis of Coal with Phenol as a Solvent, by S. W. Parr and H. F. Hadley. 1914. *Twenty-five cents.*

**Bulletin No. 77.* The Effect of Boron upon the Magnetic and Other Properties of Electrolytic Iron Melted in Vacuo, by Trygve D. Yensen. 1915. *Ten cents.*

**Bulletin No. 78.* A Study of Boiler Losses, by A. P. Kratz. 1915. *Thirty-five cents.*

**Bulletin No. 79.* The Coking of Coal at Low Temperatures, with Special Reference to the Properties and Composition of the Products, by S. W. Parr and H. L. Olin. 1915. *Twenty-five cents.*

**Bulletin No. 80.* Wind Stresses in the Steel Frames of Office Buildings, by W. M. Wilson and G. A. Maney. 1915. *Fifty cents.*

**Bulletin No. 81.* Influence of Temperature on the Strength of Concrete, by A. B. McDaniel. 1915. *Fifteen cents.*

Bulletin No. 82. Laboratory Tests of a Consolidation Locomotive, by E. C. Schmidt, J. M. Snodgrass, and R. B. Keller. 1915. *Sixty-five cents.*

**Bulletin No. 83.* Magnetic and Other Properties of Iron Silicon Alloys. Melted in Vacuo, by Trygve D. Yensen. 1915. *Thirty-five cents.*

Bulletin No. 84. Tests of Reinforced Concrete Flat Slab Structure, by A. N. Talbot and W. A. Slater. 1916. *Sixty-five cents.*

**Bulletin No. 85.* The Strength and Stiffness of Steel Under Biaxial Loading, by A. T. Becker. 1916. *Thirty-five cents.*

**Bulletin No. 86.* The Strength of I-Beams and Girders, by Herbert F. Moore and W. M. Wilson 1916. *Thirty cents.*

**Bulletin No. 87.* Correction of Echoes in the Auditorium, University of Illinois, by F. R. Watson and J. M. White. 1916. *Fifteen cents.*

Bulletin No. 88. Dry Preparation of Bituminous Coal at Illinois Mines, by E. A. Holbrook. 1916. *Seventy cents.*

*A limited number of copies of bulletins starred is available for free distribution.

**Bulletin No. 89.* Specific Gravity Studies of Illinois Coal, by Merle L. Nebel. 1916. *Thirty cents.*

**Bulletin No. 90.* Some Graphical Solutions of Electric Railway Problems, by A. M. Buck. 1916. *Twenty cents.*

**Bulletin No. 91.* Subsidence Resulting from Mining, by L. E. Young and H. H. Stoek. 1916. *None available.*

**Bulletin No. 92.* The Tractive Resistance on Curves of a 28-Ton Electric Car, by E. C. Schmidt and H. H. Dunn. 1916. *Twenty-five cents.*

**Bulletin No. 93.* A Preliminary Study of the Alloys of Chromium, Copper, and Nickel, by D. F. McFarland and O. E. Harder. 1916. *Thirty cents.*

**Bulletin No. 94.* The Embrittling Action of Sodium Hydroxide on Soft Steel, by S. W. Parr. 1917. *Thirty cents.*

**Bulletin No. 95.* Magnetic and Other Properties of Iron-Aluminum Alloys Melted in Vacuo, by T. D. Yensen and W. A. Gatward. 1917. *Twenty-five cents.*

**Bulletin No. 96.* The Effect of Mouthpieces on the Flow of Water Through a Submerged Short Pipe, by Fred B. Seely. 1917. *Twenty-five cents.*

**Bulletin No. 97.* Effects of Storage Upon the Properties of Coal, by S. W. Parr. 1917 *Twenty cents.*

**Bulletin No. 98.* Tests of Oxyacetylene Welded Joints in Steel Plates, by Herbert F. Moore. 1917. *Ten cents.*

Circular No. 4. The Economical Purchase and Use of Coal for Heating Homes, with Special Reference to Conditions in Illinois. 1917. *Ten cents.*

**Bulletin No. 99.* The Collapse of Short Thin Tubes, by A. P. Carman. 1917. *Twenty cents.*

**Circular No. 5.* The Utilization of Pyrite Occurring in Illinois Bituminous Coal, by E. A. Holbrook. 1917.

**Bulletin No. 100.* Percentage of Extraction of Bituminous Coal with Special Reference to Illinois Conditions, by C. M. Young. 1917.

**Bulletin No. 101.* Comparative Tests of Six Sizes of Illinois Coal on a Mikado Locomotive, by E. C. Schmidt, J. M. Snodgrass, and O. S. Beyer, Jr. 1917. *Fifty cents.*

**Bulletin No. 102.* A Study of the Heat Transmission of Building Materials, by A. C. Willard and L. C. Lichty. 1917.

**Bulletin No. 103.* An Investigation of Twist Drills, by B. Benedict and W. P. Lukens. 1917. *Sixty cents.*

**Bulletin No. 104.* Tests to Determine the Rigidity of Riveted Joints of Steel Structures, by W. M. Wilson and H. F. Moore. 1917. *Twenty-five cents.*

Circular No. 6. The Storage of Bituminous Coal, by H. H. Stoek. 1918. *Forty cents.*

Circular No. 7. Fuel Economy in the Operation of Hand Fired Power Plants. 1918. *Twenty-five cents.*

**Bulletin No. 105.* Hydraulic Experiments with Valves, Orifices, Hoze, Nozzles, and Orifice Buckets, by Arthur N. Talbot, Fred B Seely, Virgil R Fleming and Melvin L. Enger. 1918. *Thirty-five cents.*

---

*A limited number of copies of bulletins starred is available for free distribution.

THE UNIVERSITY OF ILLINOIS

THE STATE UNIVERSITY

Urbana

EDMUND J. JAMES, Ph. D., LL. D., President

THE UNIVERSITY INCLUDES THE FOLLOWING DEPARTMENTS:

The Graduate School

The College of Liberal Arts and Sciences (Ancient and Modern Languages and Literatures; History, Economics, Political Science, Sociology; Philosophy, Psychology, Education; Mathematics; Astronomy; Geology; Physics; Chemistry; Botany, Zoology, Entomology; Physiology; Art and Design)

The College of Commerce and Business Administration (General Business, Banking, Insurance, Accountancy, Railway Administration, Foreign Commerce; Courses for Commercial Teachers and Commercial and Civic Secretaries)

The College of Engineering (Architecture; Architectural, Ceramic, Civil, Electrical, Mechanical, Mining, Municipal and Sanitary, and Railway Engineering)

The College of Agriculture (Agronomy; Animal Husbandry; Dairy Husbandry; Horticulture and Landscape Gardening; Agricultural Extension; Teachers' Course; Household Science)

The College of Law (three years' course)

The School of Education

The Course in Journalism

The Courses in Chemistry and Chemical Engineering

The School of Railway Engineering and Administration

The School of Music (four years' course)

The School of Library Science (two years' course)

The College of Medicine (in Chicago)

The College of Dentistry (in Chicago)

The School of Pharmacy (in Chicago; Ph. G. and Ph. C. courses)

The Summer Session (eight weeks)

Experiment Stations and Scientific Bureaus: U. S. Agricultural Experiment Station; Engineering Experiment Station; State Laboratory of Natural History; State Entomologist's Office; Biological Experiment Station on Illinois River; State Water Survey; State Geological Survey; U. S. Bureau of Mines Experiment Station.

The library collections contain (December 1, 1917) 411,737 volumes and 104,524 pamphlets.

For catalogs and information address

THE REGISTRAR
URBANA, ILLINOIS

106

UNIVERSITY OF ILLINOIS BULLETIN
ISSUED WEEKLY
Vol. XV May 27, 1918 No. 39
[Entered as second-class matter Dec. 11, 1912, at the Post Office at Urbana, Ill., under the Act of Aug. 24, 1912.]

TEST OF A FLAT SLAB FLOOR OF THE WESTERN NEWSPAPER UNION BUILDING

BY
ARTHUR N. TALBOT
AND
HARRISON F. GONNERMAN

BULLETIN No. 106

ENGINEERING EXPERIMENT STATION
PUBLISHED BY THE UNIVERSITY OF ILLINOIS, URBANA

PRICE: TWENTY CENTS
EUROPEAN AGENT
CHAPMAN & HALL, LTD., LONDON

THE Engineering Experiment Station was established by act of the Board of Trustees, December 8, 1903. It is the purpose of the Station to carry on investigations along various lines of engineering and to study problems of importance to professional engineers and to the manufacturing, railway, mining, constructional, and industrial interests of the State.

The control of the Engineering Experiment Station is vested in the heads of the several departments of the College of Engineering. These constitute the Station Staff and, with the Director, determine the character of the investigations to be undertaken. The work is carried on under the supervision of the Staff, sometimes by research fellows as graduate work, sometimes by members of the instructional staff of the College of Engineering, but more frequently by investigators belonging to the Station corps.

The results of these investigations are published in the form of bulletins, which record mostly the experiments of the Station's own staff of investigators. There will also be issued from time to time, in the form of circulars, compilations giving the results of the experiments of engineers, industrial works, technical institutions, and governmental testing departments.

The volume and number at the top of the front cover page are merely arbitrary numbers and refer to the general publications of the University of Illinois: *either above the title or below the seal is given the number of the Engineering Experiment Station bulletin or circular which should be used in referring to these publications.*

For copies of bulletins, circulars, or other information address the

ENGINEERING EXPERIMENT STATION,
URBANA, ILLINOIS.

UNIVERSITY OF ILLINOIS
ENGINEERING EXPERIMENT STATION

BULLETIN No. 106 MAY, 1918

TEST OF A FLAT SLAB FLOOR OF THE WESTERN NEWSPAPER UNION BUILDING

BY
ARTHUR N. TALBOT
PROFESSOR OF MUNICIPAL AND SANITARY ENGINEERING
AND IN CHARGE OF THEORETICAL AND APPLIED MECHANICS
AND
HARRISON F. GONNERMAN
RESEARCH ASSOCIATE IN THEORETICAL AND APPLIED MECHANICS

ENGINEERING EXPERIMENT STATION
PUBLISHED BY THE UNIVERSITY OF ILLINOIS, URBANA

CONTENTS

LIST OF FIGURES

LIST OF TABLES

TEST OF A FLAT SLAB FLOOR OF THE WESTERN NEWS-PAPER UNION BUILDING

1. *Preliminary.*—This bulletin gives the results of the test made on a four-way reinforced concrete flat slab floor of the Western Newspaper Union Building in Chicago in August and September, 1917. A load of 913 lb. per sq. ft. was applied over four panels. The building, which was nine years old at the time of the test, was to be torn down to clear the site for the new Union Passenger Station; the opportunity was utilized to apply a test load much greater in proportion to the design load than had been used in previous tests of buildings. The test was carried far enough to give stresses in the reinforcing bars and concrete markedly higher than have been obtained in other building tests. The information on the action of the slab in its various parts given by the strain measurements has an important bearing on the design of the flat slab structure.

2. *Acknowledgment.*—The test was made as investigative work of the Engineering Experiment Station The testing work was done under the direct supervision of Mr. Gonnerman. He and Mr. N. E. Ensign, Associate in Theoretical and Applied Mechanics, acted as observers. The results have been reduced and prepared for publication as a bulletin of the Engineering Experiment Station.

Acknowledgment of valuable aid received in carrying out the test is made. The Portland Cement Association furnished the labor for preparing for the test and for hauling the loading material and loading and unloading the floor. The Universal Portland Cement Company assisted in making arrangements and gave assistance on the test. The pig iron used for loading material was lent by the Illinois Steel Company. The freight on the pig iron was borne jointly by the Pennsylvania Railroad and the Portland Cement Association. Opportunity to use the building for the purpose of the test was given by the Chicago Union Station Company; the test was made at the suggestion of A. J. Hammond, Principal Assistant Engineer. The Condron Company provided an assistant for tracing and checking the transfer of the loading material.

3. *The Building.*—The Western Newspaper Union Building was an eight-story reinforced concrete structure located at Clinton and Adams Streets, Chicago. The building was erected in the spring of 1909 by the George Hinchcliff Company, contractors, according to plans furnished by S. N. Crowen, architect, and Ritter and Mott, engineers. It had been in use by a printing company until 1916. The floor tested had been occupied by printing presses. Fig. 1 is a view of the building at the time of the test; the wrecking of the building had begun.

Two types of floor construction were used in the building. The first five floors were slab and girder type; the sixth, seventh, and eighth floors were Turner mushroom flat slab type (four-way reinforcement). The floors of the building were divided into panels 17 ft. 5½ in. by 19 ft. 4½ in. The test was made on the sixth floor. This floor was designed for a live load of 250 lb. per sq. ft. and was nominally 8½ in. thick. A considerable variation in thickness was found, the measured thickness over the test area ranging from 7.5 to 9.8 in. Fig. 3 gives the thickness of the floor at a number of places as determined by readings with an engineer's level. In general, the thickness was greater away from the columns than in the vicinity of the columns. The interior columns were octagonal in form, 24 in. in short diameter below the floor tested and 21 in. in short diameter above it. The inside diameters of the hooping of the columns on the fifth and sixth floors are given on the plans as 21 in. and 18 in., respectively. The column capitals were pyramidal; the short diameter at the top of the capital was 54 in. The building plans called for 15 ⅝-in. round bars in each of the four bands of reinforcement in the floor slab and indicated that over most of the columns in the test area there were laps in certain bands. After the test was made, the floor was broken into and the location and extent of all laps and the position of reinforcing bars with respect to the surfaces of the slab were found. Fig. 4 shows the arrangement of the reinforcement found over the test area, including the position of the laps. In several places the arrangement of reinforcement differs from that given in the building plans. In three places in rectangular bands the reinforcement for positive moment was double that given on the plans (30 bars instead of 15). The lapping of bars at columns was generally greater than that indicated on the plans. At column 15 three bands were lapped; at columns 14, 16, 21, 26, 27, and 28, two bands; and at columns 22 and 23, one band. In most cases, the length of lap and its position were such that the extra

FIG. 1. VIEW OF WESTERN NEWSPAPER UNION BUILDING AT TIME OF TEST

FIG. 2. VIEW SHOWING FULL LOAD ON FLOOR

metal was effective in regions of greatest moment. In some cases the laps were poorly arranged, as at columns 15, 16, and 27. No reason is apparent for the lapping of bars between columns. There was no reinforcement for negative moment in the region midway between columns. The eight 1¼-in. column rods were bent out into the slab, and two circumferential ring rods (circles of 5 ft. 6 in. and 8 ft. 6 in. diameters) rested on these and supported the lower layers of reinforcing bars. The measurement of position of bars with respect to

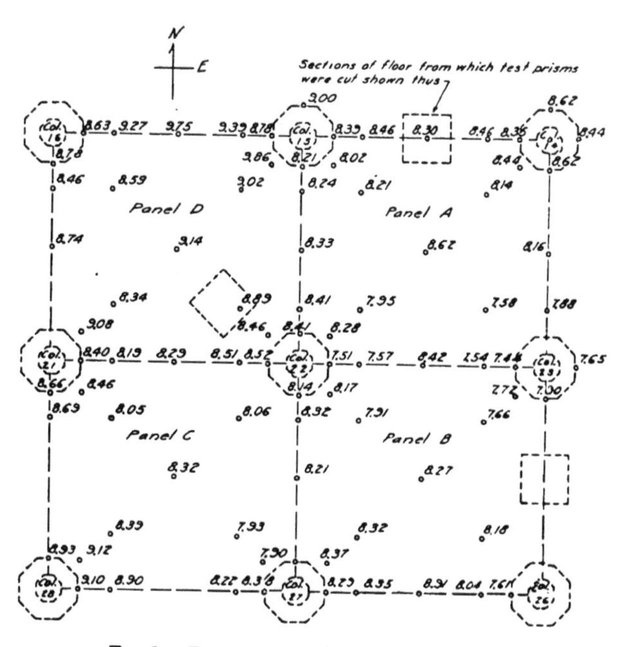

FIG. 3. THICKNESS OF FLOOR IN TEST AREA

the surface of the slab showed considerable variation at the several columns. The method of lapping was not always the same; in some cases the laps of a given band were not at the same level—at columns 15 and 22, for example, there were five layers of ⅝-in. bars besides the circumferential and bent-out column bars. At the columns the distance of the centers of the bars of the top layer from the upper surface of the slab varied from 0.90 in. to 2.00 in., and that of the lower layer from 3.60 to 4.00 in. At points between columns the centers of the bars of the rectangular bands were from 2.30 in one case to 4.20 in. above the lower surface of the slab. In the region of the center of the panel the centers of the lower layer of diagonal bars were from 1.20 to 2.20 in. above the lower surface of the slab.

The variation in amount of reinforcement available at the different sections of the slab due to the diversity in amount and position

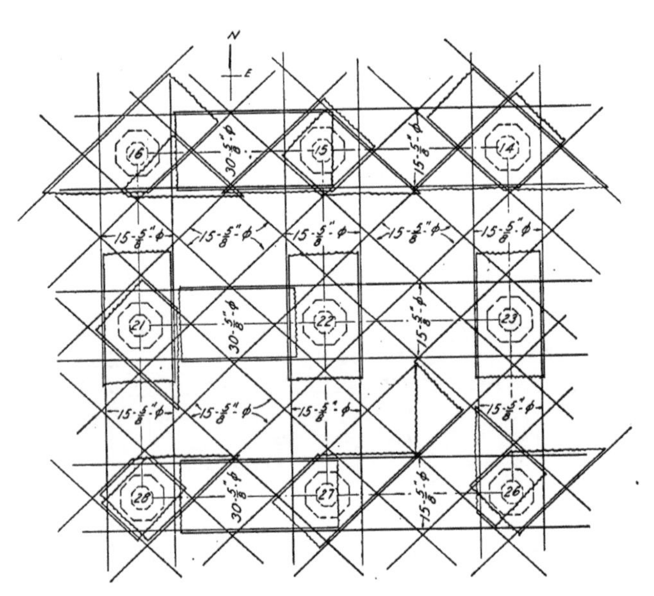

FIG. 4. ARRANGEMENT OF REINFORCEMENT IN TEST AREA

of laps, and the variation in depth of reinforcing bar and thickness of slab, as well as variations in the quality and stiffness of the concrete in different parts of the loaded area, may be expected to cause some lack of uniformity in the stresses and deflections at points similarly located on the test area.

The concrete in the slab was 1-2-4 mix; in the columns 1-1-2 mix was used. The coarse aggregate was gravel. At the time of the test the building was eight years old. Pieces of the concrete were cut out from the floor and sawed into test prisms approximately 5 in. square and 16 in. long; one face of the prism was coincident with the upper surface of the slab. The concrete was taken from parts of the slab (indicated in Fig. 3) which had not been highly stressed and which was relatively free from reinforcing bars and cracks. The results of the tests are given in Table 1. The strength of these test prisms

TABLE 1

COMPRESSION TESTS OF CONCRETE PRISMS

Specimen	From Panel	Length inches	Section inches	Loaded Area sq. in.	Maximum Applied Load pounds	Unit Compressive Strength lb. per sq. in.	Modulus of Elasticity lb. per sq. in.
A1	A	16.9	4.6 by 4.8	22.1	71 000	3210	4 500 000
A2	A	16.8	4.7 by 4.7	22.1	119 300	5400	5 100 000
A3	A	16.2	4.9 by 3 9	19.1	104 400	5460	4 800 000
					Average	4690	4 800 000
B1	B	18.0	6.5 by 4.5	29.3	93 400	3190	4 200 000
B2	B	15 0	4.7 by 4.7	22.1	79 100	3580	3 500 000
					Average	3385	3 850 000
D1	D	16.0	4.5 by 3.0	13.5	48 600	3600	4 600 000
D2	D	16.5	5.1 by 2.5	12.8	42 800	3340	4 500 000
D3	D	10.0	4.9 by 3.1	15.2	60 600	3990	....
D4	D	9.5	5.0 by 3.1	15.5	59 700	3850	....
D5	D	8.0	4.5 by 3.2	14.4	48 200	3350	....
					Average	3626	4 550 000

ranged from 3190 lb. per sq. in. in panel B to 5460 lb. per sq. in. in panel A, and the initial modulus of elasticity from 3 500 000 to 5 100-000 lb. per sq. in. When the floor was broken up after the test, a noticeable difference was found in the quality of the concrete in the four test panels. The concrete in panels A and D appeared much stronger and harder than that in panels B and C. That in panel D was very hard.

Steel coupons were cut from reinforcing bars at different places in the tested floor. The results of the tension tests of these bars are given in Table 2. The bars gave an average yield point by drop of beam of 63 600 lb. per sq. in. and an average ultimate strength of 101 300 lb. per sq. in.

4. *The Test.*—The method of testing was similar to that used in previous buildings tests, as described in Bulletin No. 64 of the University of Illinois Engineering Experiment Station, "Tests of Reinforced Concrete Buildings Under Load." The loading material was pig

TABLE 2

TENSION TESTS OF REINFORCING BARS

Specimen No.	Average Diameter inches	Yield Point lb. per sq. in.	Ultimate Strength lb. per sq. in.	Elongation in 8 Inches per cent	Reduction of Area per cent
1	.602	66 800	103 500	17.2	37.6
2	.603	59 900	96 200	17.8	34.0
3	.619	69 100	115 600	17.8	33.4
4	.625	55 400	86 700	21.5	48.0
5	.626	64 000	100 000	16.2	39.4
6	.625	66 500	115 000	14.6	45.3
7	.625	64 900	101 600	15.1	45.3
8	.615	60 000	99 000	18.6	45.3
9	.621	65 700	102 700	16.9	41.7
Average		63 600	101 300	17.2	41.1

iron. The pig iron was hauled from the freight yards to the building in auto trucks. The net weight of each truck load of iron was obtained by weighing on a certified scale before it was hauled to the building. At the building the pig iron was loaded on hand trucks, hoisted to the test floor by means of an electric freight elevator, and then placed on the test area by hand. A record was kept of the number of truck loads placed on each panel and the total weight on each panel was obtained from the truck weights.

The load was applied over the four interior panels of the sixth floor. Fig. 5 gives the location of the panels tested. The load on each panel was divided into quarters by means of aisles 6 to 8 in. wide extending at right angles to each other along the center lines of the panels and along the boundaries between panels. The space occupied by the aisles and by the boxes built on the floor to afford access to the gage lines amounted to 17 per cent of the panel areas. The final load on the slab was 913 lb. per sq. ft., a total load of 308 400 lb. per panel. Fig. 2 shows the full load in place.

Gage lines were prepared in advance of the test—103 on the reinforcing bars and 75 on the concrete. Fig. 6 and 7 show the location of the gage lines on the upper and lower sides of the slab. To insure reliability of initial readings, three sets of strain gage readings (and more on many of the gage lines) were taken before the load was applied. Strain gage readings were taken at loads of 234, 425, 637, 855, and 913 lb. per sq. ft. of panel area. In each case except for the first load two complete sets of strain readings were taken on the reinforcing bars and the concrete, and sometimes more. The deflection of the slab was measured at 20 points. The location of the deflection points are shown in Fig. 8. The appearance of cracks was also noted. Readings of deformation and deflection at the more important points were taken from time to time as the load was being placed on the floor.

Readings were also taken three days after the maximum load had

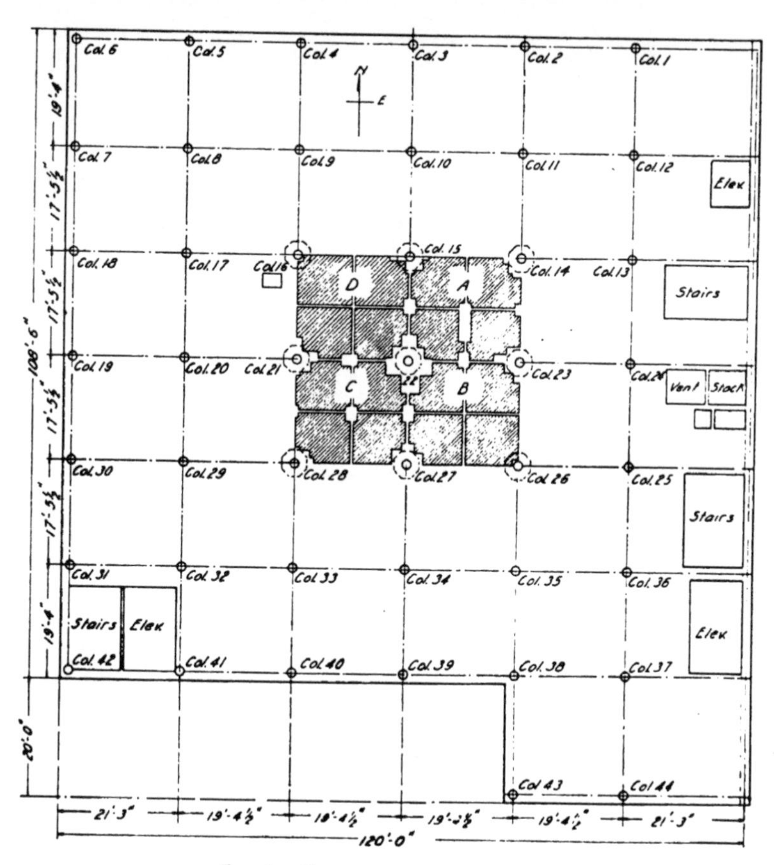

FIG. 5. LOCATION OF TEST PANELS

been placed on the floor in order to get information on the time effect of the load upon deformations and deflections. After the removal of the load readings were taken to find the amount of recovery in the floor.

The test area was chosen where there would be the least effect of floor openings and where the building plans showed laps in only the rectangular bands over the column in the center of four panels. In a building test the considerable time required to reduce the data of the

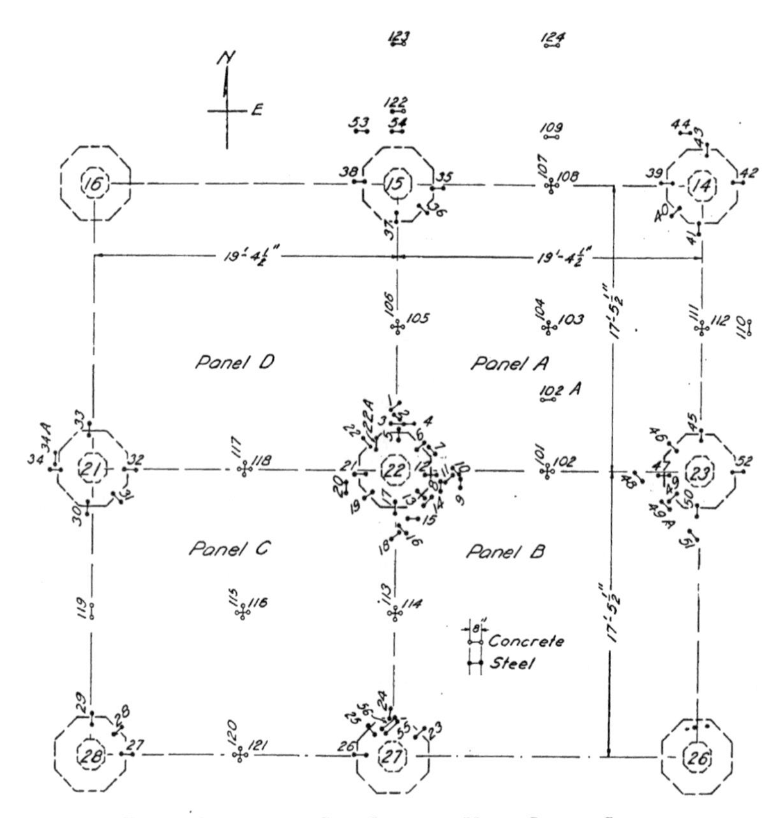

Fig. 6. Location of Gage Lines on Upper Side of Slab

readings into form for analysis renders it necessary to restrict the number of gage lines and so their distribution over the test area becomes a matter of importance. The gage lines in this test were placed with a view of getting some information on (1) the amount and distribution of the stresses in the reinforcement along sections through

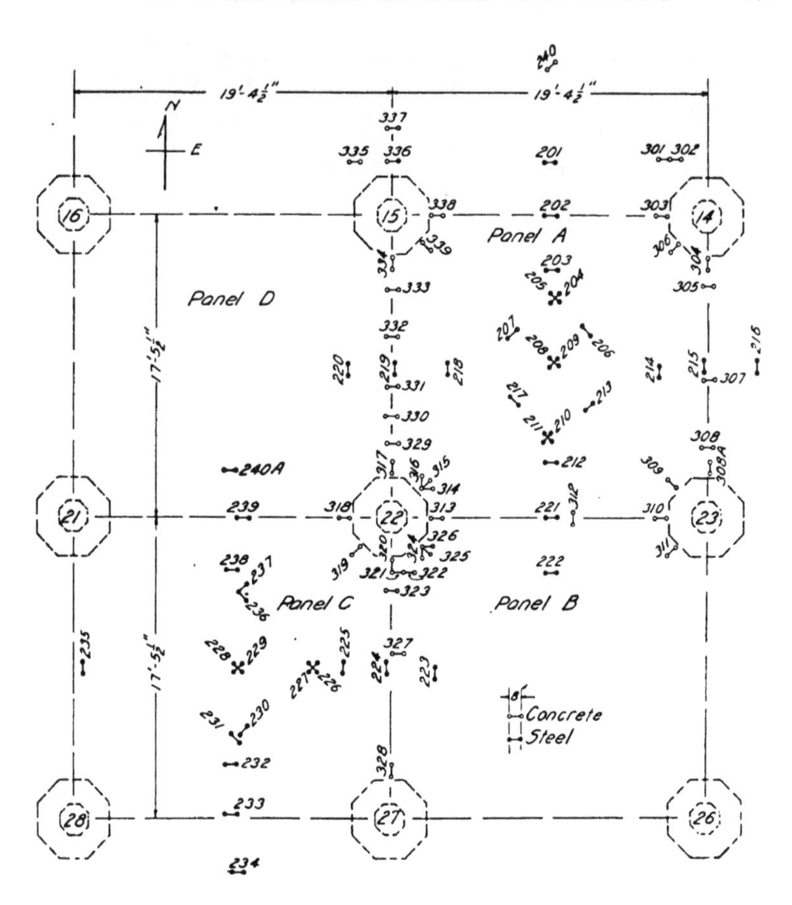

FIG. 7. LOCATION OF GAGE LINES ON LOWER SIDE OF SLAB

the panel centers and through the panel edges, and at other points, (2) the strain in the concrete at the more important points in the slab, (3) the moment of resistance accounted for by the stresses in the reinforcing bars which cross sections through the panel centers and through the panel edges, and (4) something on the stresses in columns at the edges of the loaded area due to bending under the applied load.

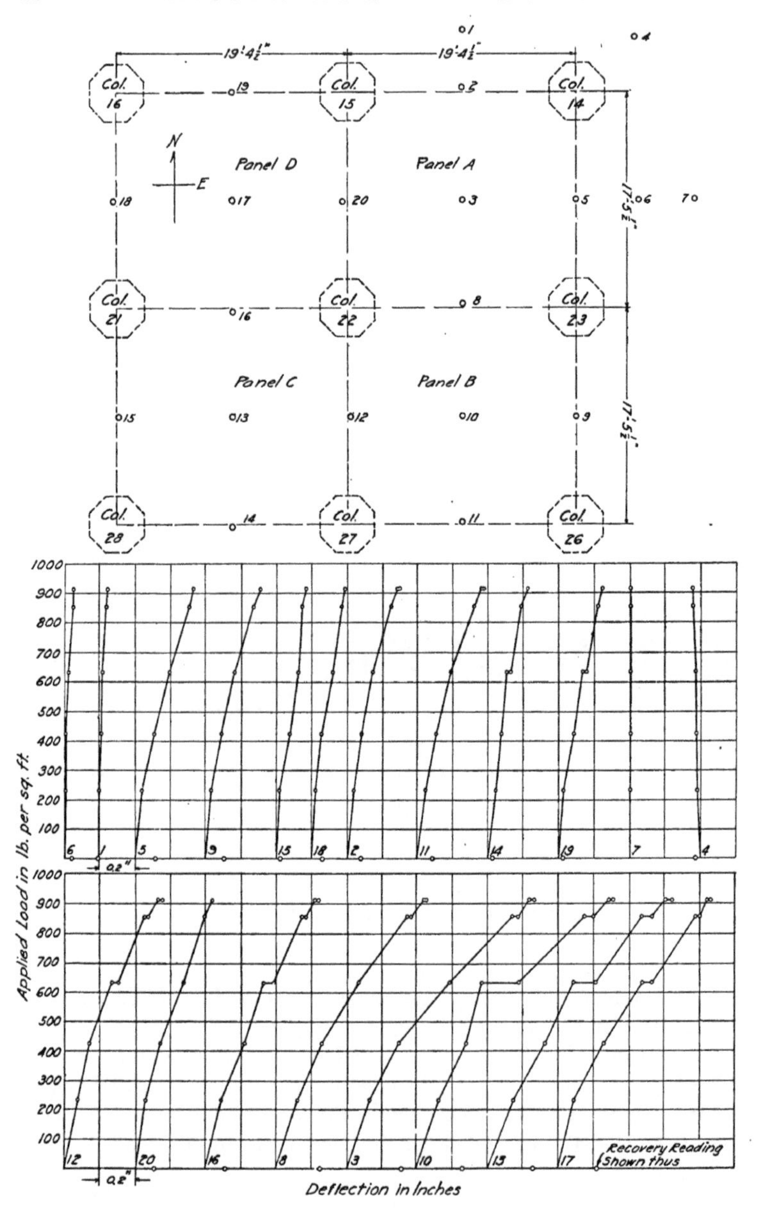

FIG. 8. LOAD-DEFLECTION DIAGRAMS AND LOCATION OF DEFLECTION POINTS

The gage lines in panels A and C with few exceptions were laid out in duplicate in order that readings obtained on the gage lines in the one panel might serve as a check on readings obtained at corresponding gage lines in the other, and also to find out whether there was similarity of action in the two panels.

5. *Deflection of Slab.*—The deflections of the slab at the several deflection points are plotted in Fig. 8, the diagrams for points similarly located being grouped together. The recovery of deflection one day after the load was removed is indicated by the points plotted at the bottom of the diagram. The second point plotted for the load of 637 lb. per sq. ft. is the deflection 16 hours after the last of the load was applied; the second point for the load of 855 lb. per sq. ft., 37 hours after; the second point for the load of 913 lb. per sq. ft., 66 hours after. Where no second point is plotted, the change in deflection was negligible.

The deflections at the centers of panels A, B, C, and D under the load of 913 lb. per sq. ft. were 1.06, 1.12, 1.04, and 0.87 in., respectively. It may be noted in this connection that panels C and D had a greater amount of reinforcement than A and B and that panel D was thicker. The concrete of panels A and D was of unusually good quality.

Of the deflections at the middle of the inner edges of the loaded panels, that at point 8 (Fig. 8) was considerably greater than that at point 16, and that at point 12 was more than at point 20. The differences are explainable by differences in quality of concrete and in amount and arrangement of reinforcement.

At the outer edges of the loaded area, the deflections at points 5 and 9 were considerably greater than at points 15 and 18; and the deflections at points 2 and 11 were greater than at points 14 and 19—the same circumstances explain these differences in deflections.

It may be noted that points 1 and 6 (Fig. 8) distant one-quarter of the panel lengths from the panel edges gave a measurable deflection. Point 7 (center of adjacent panel) remained stationary and point 4 showed an upward movement.

6. *Cracks in Slab.*—It was noted before the load was applied that there were numerous checks in the upper surface of the slab in the regions around the column and along the panel edges. Most of these were evidently surface checks; others were tension cracks formed under previous loads. The latter opened upon the application of load.

Fig. 9 gives the location of the more important cracks on the upper side of the slab which either opened or formed under the test load. These cracks were all open cracks—much more marked than hair cracks. Under the load of 913 lb. per sq. ft. they ranged in width from 0.02 to 0.06 in. These cracks show the regions of high tensile stress in the top of the slab. The main cracks at the columns were generally at or near the edges of the column capital; they branched out to join the cracks extending along the panel edges between the columns. The cracks at the capitals of the columns bordering the loaded area were

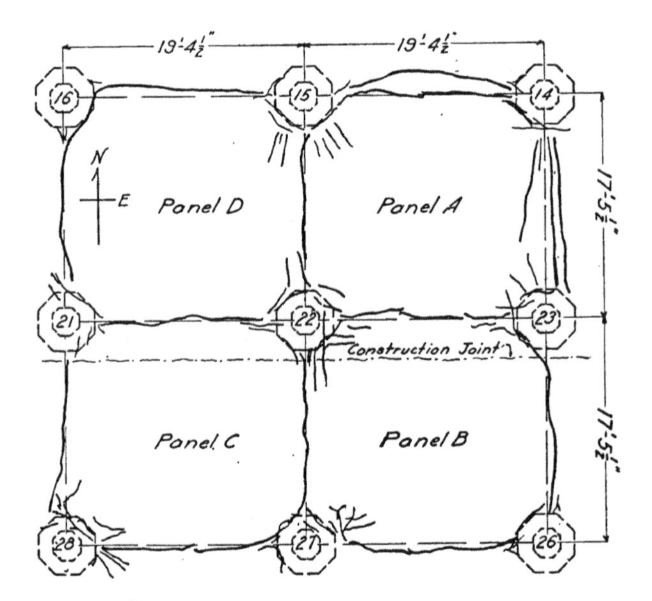

FIG. 9. LOCATION OF MAIN CRACKS IN UPPER SIDE OF SLAB

fully as wide as those at the capital of the central column. The cracks along the panel edges were generally as pronounced as those around the capitals.

Fig. 10 shows the location of the more important cracks on the lower side of the slab in panels A and C, the panels in which the principal tension gage lines on the lower side of the floor were placed. In panels B and D, in addition to those noted in the figure, there were cracks over the panels in positions similar to those noted in panels A and C. In panel D the cracks were not so wide nor so numerous as in the other panels, even at the maximum load. Panel B had larger and more numerous cracks than the other panels. Most of the cracks on the lower side of the slab were found in bands extending in rectangular directions. In a small area at the panel centers cracks extended in diagonal directions. No cracks or checks had been noted on the lower side of the slab before loading was begun. The first hair cracks here were observed at a load of 234 lb. per sq. ft. At a load of 425 lb. per sq. ft., the cracks were fairly well defined and for higher

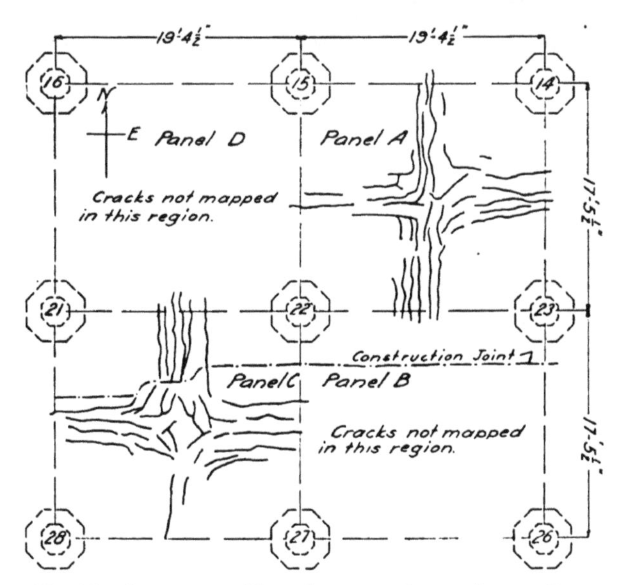

FIG. 10. LOCATION OF MAIN CRACKS IN LOWER SIDE OF SLAB

loads they gradually opened up and extended. The main cracks on the lower side of the slab did not open so much as the main cracks on the upper side of the slab. The construction joint shown in Fig. 9 and 10 opened appreciably under the application of load.

Upon the removal of the load the cracks closed, giving the slab an appearance similar to that which it had before the load was applied.

7. *Load-strain Diagrams.*—In Figs. 11 to 13 the load-strain diagrams are given for gage lines on the upper and lower sides of the

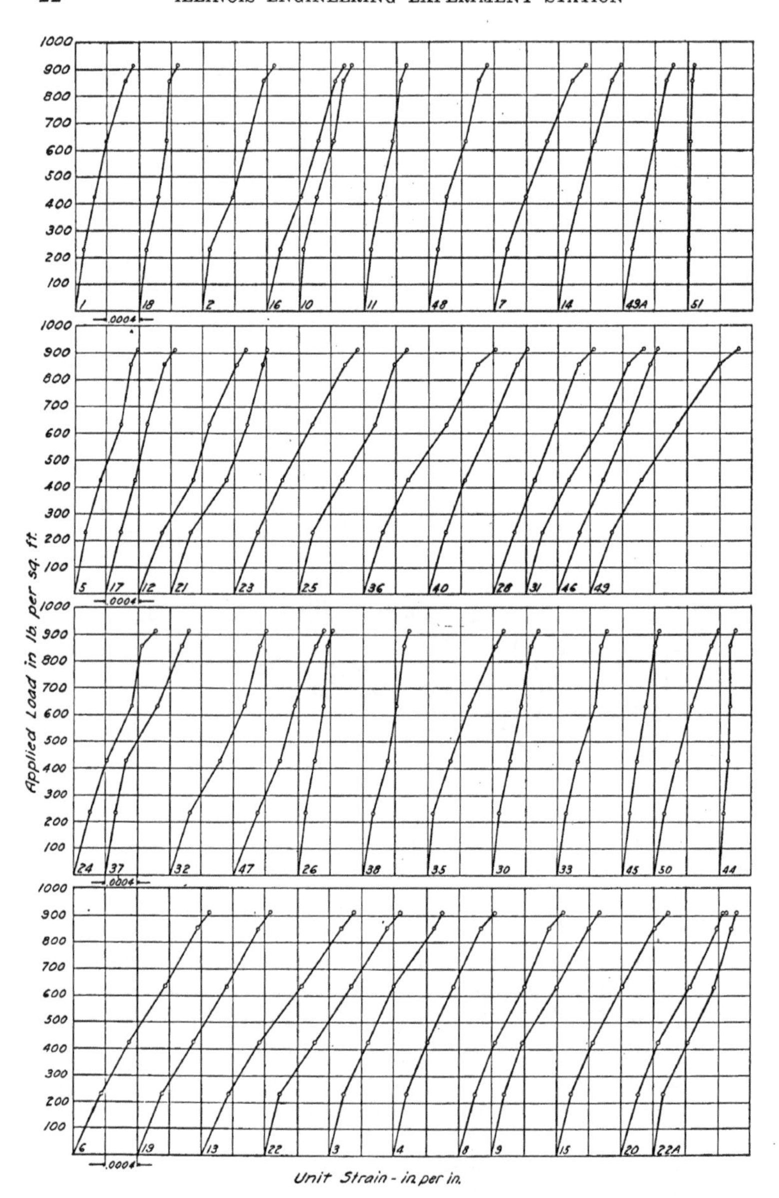

FIG. 11. LOAD-STRAIN DIAGRAMS FOR GAGE LINES ON REINFORCING BARS ON UPPER SIDE OF SLAB

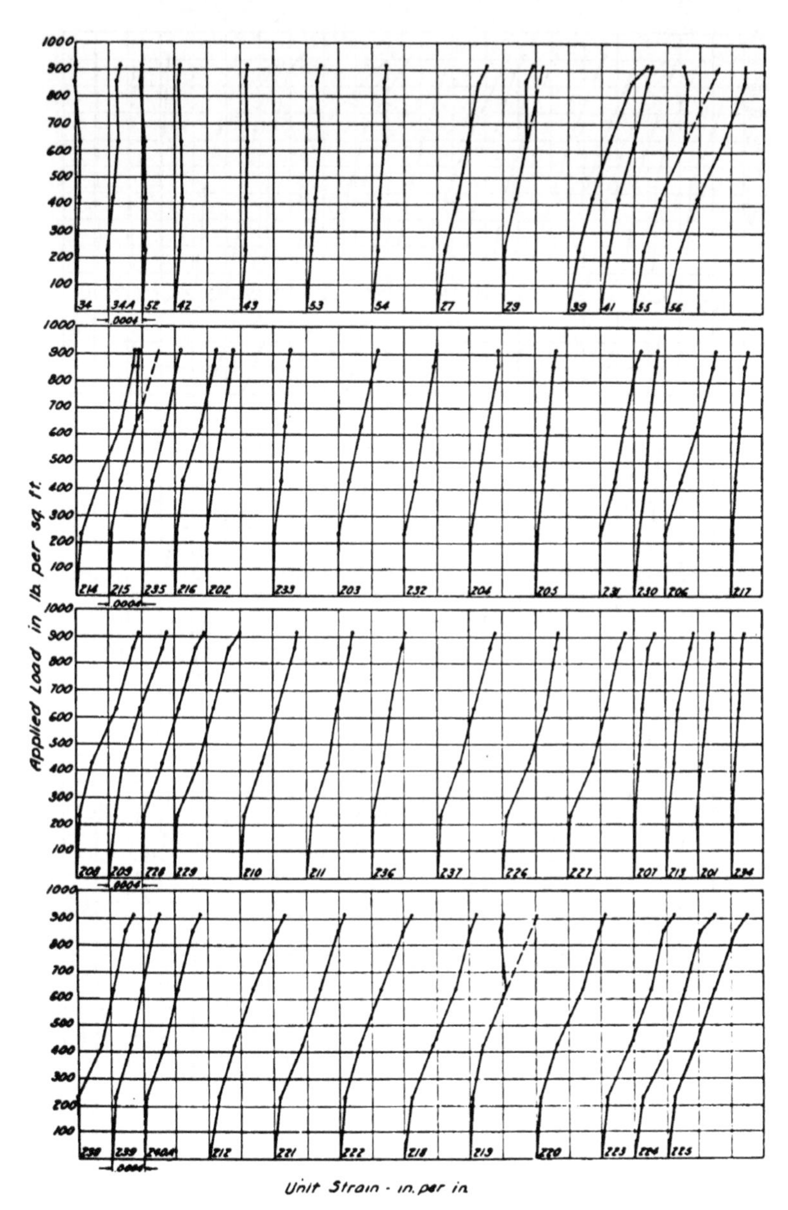

FIG. 12. LOAD-STRAIN DIAGRAMS FOR GAGE LINES ON REINFORCING BARS ON UPPER

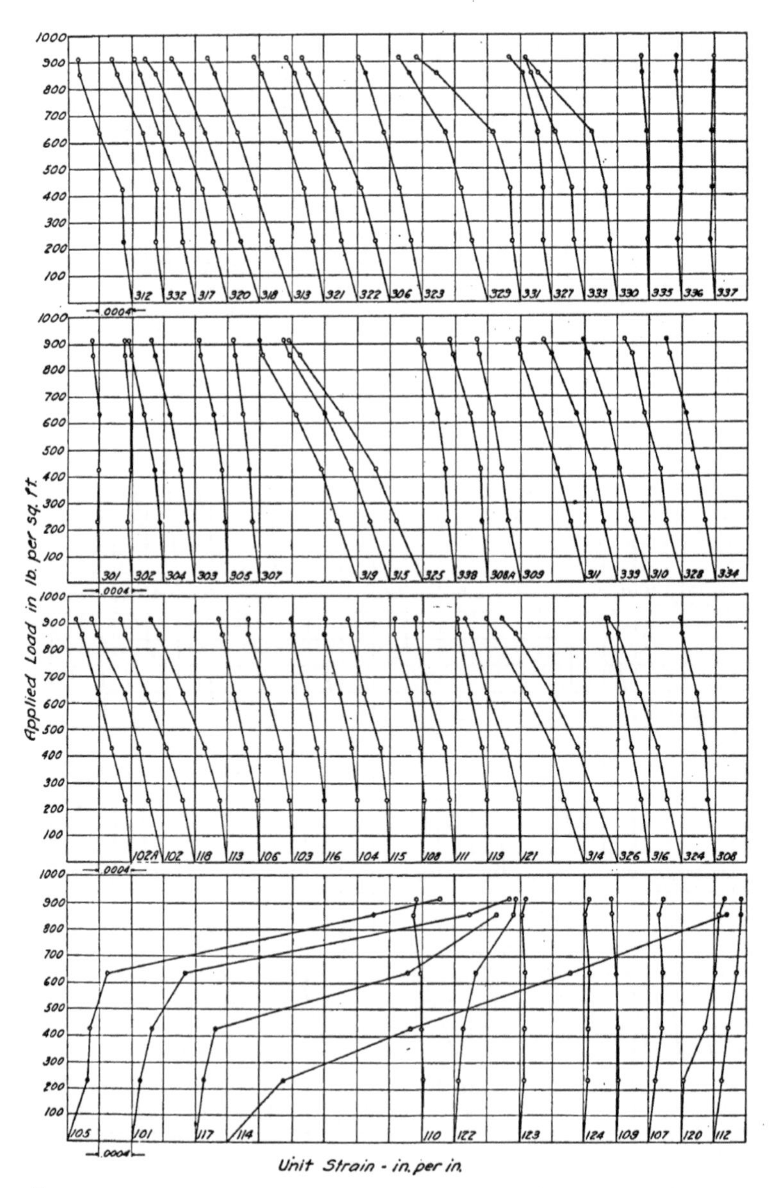

FIG. 13. LOAD-STRAIN DIAGRAMS FOR GAGE LINES IN CONCRETE ON UPPER AND LOWER SIDES OF SLAB

slab. In these diagrams tensile strains are plotted to the right and compressive strains to the left of the axis. Gage lines having similar positions on the floor have been grouped. Averages of the several readings were used in computing the strains. It will be noted that for gage lines on bars in the upper side of the slab the diagrams show deformations of considerable amount at the load of 234 lb. per sq. ft; the diagrams for gage lines on bars on lower side show markedly smaller deformations. For loads greater than 234 lb. per sq. ft. the deformations in bars on the lower side of the slab increased fairly uniformly with increase of load; it has already been noted that the first cracks on the lower side appeared at this load. The diagrams for gage lines on the concrete on both upper and lower sides of the slab show deformations of considerable amount at the first load and a fairly uniform increase in deformation for higher loads.

8. *Stresses in Reinforcing Bars in Upper Side of Slab.*—The stresses in the reinforcing bars at the several loads may be computed from the strains given in Fig. 11 and 12. For convenience of comparison the stresses corresponding to the strains for the maximum load (913 lb. per sq. ft.) are given in Fig. 14. The values shown at points around the columns are in bars at the upper side of the slab; the others are in bars at the lower side.

For bars on the upper side of the slab the greatest stresses were found at gage lines located on diagonal bars at the edge of the column capitals. At column 22 (the central column of the loaded area) the stresses in diagonal bars over the edge of the column capital ranged from 49 200 to 57 300 lb. per sq. in. The stresses in diagonal bars at column 22 at gage lines located some distance from the column capital ranged from 15 000 to 34 500 lb. per sq. in. The stresses in diagonal bars at the columns bordering the loaded area ranged from 36 600 to 54 900 lb. per sq. in. In general, the stresses in the diagonal bars at the columns bordering the loaded area were nearly as great as those observed at corresponding gage lines at the central column. The stresses given do not include the stress due to the load of the slab itself; allowing for this, it is apparent that the yield point of the steel was not reached in even the most highly stressed bars.

The stresses in the east and west rectangular band at column 22 averaged about 41 400 lb. per sq. in. The stresses in the north and south rectangular band at column 22 ranged from 23 700 lb. per sq. in.

at the edge of the column capital to 40 500 lb. per sq. in. at gage lines near the edge of the band on either side of the column. The bars in this band were lapped at column 22, the bars ending about 80 inches north and south of the column center. It will be noted that at gage lines near the edges of the rectangular bands the stresses were greater

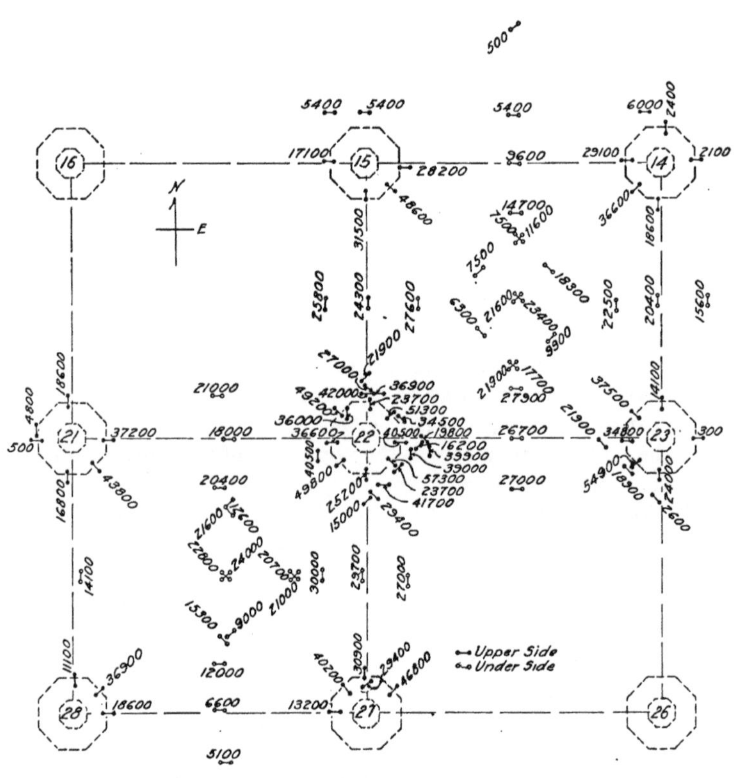

FIG. 14. STRESSES IN REINFORCING BARS IN POUNDS PER SQUARE INCH AT LOAD OF 913 POUNDS PER SQUARE FOOT

than in bars at the middle of the band. In bars outside the loaded area and near the edge of a rectangular band stresses were found as great as 6000 lb. per sq. in. It is evident that portions of the slab outside the loaded area contributed measurably to the resistance developed in the slab.

9. *Stresses in Reinforcing Bars in Lower Side of Slab.*—On the lower side of the slab the greatest stresses were found in the bars of the rectangular bands (Fig. 14). At the maximum load stresses from 24 300 to 30 000 lb. per sq. in. were observed in rectangular bands within the loaded area. In the one which had 30 bars instead of the usual 15, stresses from 18 000 to 21 000 lb. per sq. in. were found.

The stresses in the rectangular bands at the edge of the loaded area varied from 5100 lb. per sq. in. in a bar outside the loaded area to 22 500 lb. per sq. in. in a bar inside the loaded area. At one gage line outside the loaded area and near the edge of the band a stress of 15 600 lb. per sq. in. was found.

The stresses in the diagonal bands in the region of the center of the panels were smaller than those in bars of the rectangular bands in the region between columns. The stresses in diagonal bars at gage lines away from rectangular sections which pass through panel centers ranged from 6300 to 18 300 lb. per sq. in.

The effect of position of reinforcing bars with respect to surface of slab on the stress developed is discussed in another place.

10. *Strains in Concrete at Upper Surface of Slab.*—The unit-strains in the concrete at the several loads are plotted in Fig. 13. For convenience of comparison, the unit-strains at the maximum load are recorded in Fig. 15. The values for points around the columns are for gage lines on the lower side of the slab; a few gage lines which cross the panel edges between columns are also on the lower side. The remaining gage lines are on the upper side of the slab.

On the upper surface of the slab the greatest compressive strains were found at gage lines along the inner panel edges midway between columns. Strains from 0.00089 to 0.00097 in. per in. were observed at these gage lines. Assuming a modulus of elasticity of concrete of 4 000 000 lb. per sq. in. and a straight-line stress-strain relation the stresses in the concrete corresponding to these deformations would be 3560 and 3880 lb. per sq. in. Strains as great as 0.00054 in. per in. (corresponding stress in the concrete on the assumption just given, 2160 lb. per sq. in.) were found at gage lines at the panel centers.

11. *Strains in Concrete at Lower Surface of Slab.*—The greatest strains in concrete on the lower surface of the slab were found close to the edge of the capital of column 22. The strains at this column at

the diagonal gage lines (see Fig. 15) ranged from 0.0012 to 0.0016 in. per in. at the maximum load. These strains are as great as the strains which were found at failure in the tests of the concrete prisms cut from the slab; they represent the range in deformation at the ultimate load usually found in compression tests of concrete. Spalling or chipping

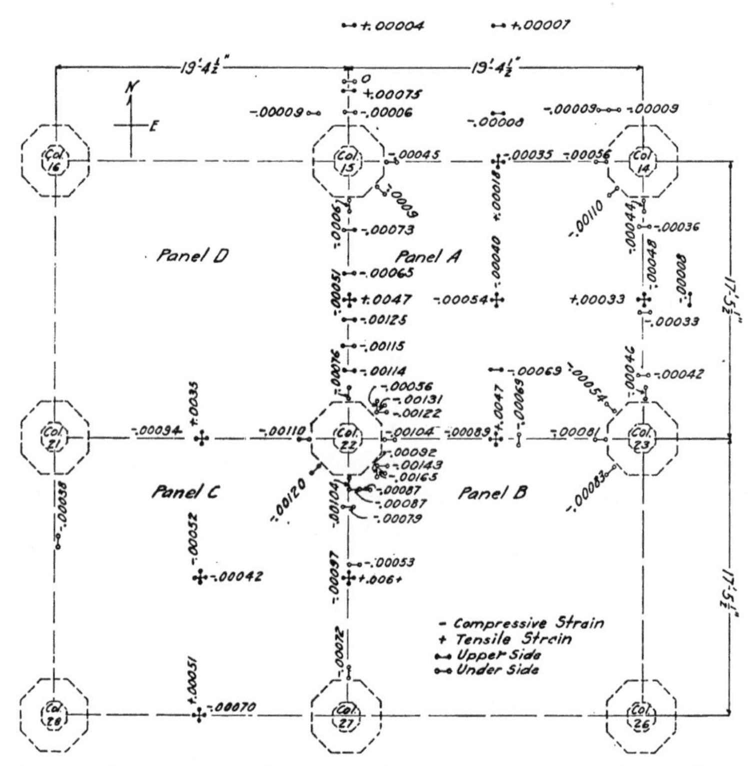

FIG. 15. UNIT-STRAINS IN CONCRETE AT LOAD OF 913 POUNDS PER SQUARE FOOT

of the concrete surface was plainly visible near the edge of the capital of column 22 in panel A. At rectangular gage lines near the capital of this column a strain in the concrete of 0.0014 in. per in. was observed. These high deformations indicate that the concrete near the capital of column 22 was highly stressed and that at certain gage lines it was stressed to its ultimate strength, the action of the surrounding concrete preventing its complete failure.

Near the edges of the capitals of columns bordering the loaded area strains from 0.00054 to 0.0011 in. per in. were observed at diagonal gage lines, and strains from 0.00044 to 0.00081 in. per in. at rectangular gage lines. With a modulus of elasticity of concrete of 4 000 000 lb. per sq. in. the stresses corresponding to these strains would range from 1760 to 4400 lb. per sq. in.

At the gage lines crossing the inner panel edges at a section of negative moment between columns 15 and 22, compressive strains about one-half those found near the column capital were observed at the lower loads, even though there was no tension reinforcement in the upper side of the slab in this region. For the highest load when the concrete at the capital had begun to crush there was a relatively great increase in the strains in this middle region, a value as great as 0.0012 in. per in. being found.

It should be noted that there were compressive strains of some amount in regions of negative moment outside the loaded area.

12. *Influence of Position of Bar.*—The stresses in Fig. 14 are given without reference to the position of the bar with respect to the surface of the slab and without reference to its position in the band. It may be expected that the stresses in the several layers of bars will vary. The average depths of the layers of bars at the edge of the capital of column 22 ranged from 0.90 in. below the upper surface of the slab for the upper layer to 3.65 inches for the lower layer. At this column the layers of bars in the order of their position with respect to the upper surface of the slab were as follows: (1) north and south rectangular bars (one lap), (2) diagonal bars running to northwest, (3) east and west rectangular bars, (4) diagonal bars running northeast, (5) north and south rectangular bars (second lap), (6) circumferential ring bars, (7) radial column bars. (See Fig. 20.)

The strains in steel and concrete at gage lines near the capital of column 22 for the load of 913 lb. per sq. ft. have been plotted in Fig. 16 to show the position of the neutral axis for the various layers of bars on which readings of deformation were taken. The strains found at gage lines 5 and 17 are markedly smaller than those found at gage lines 8, 9, 20, and 22A which were placed on bars of the same layer. Gage line 22A, like gage lines 5 and 17, was placed over the edge of the capital and it will be noted that the deformation found at this gage line is more in harmony with the deformations found at other gage lines on bars of the same layer than are the deformations at 5 and 17.

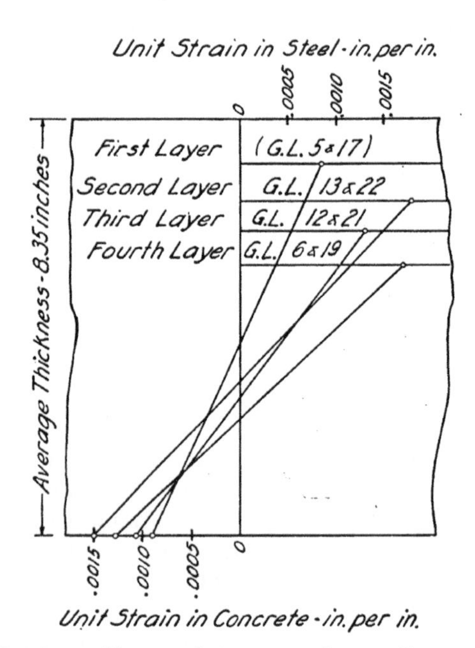

FIG. 16. POSITION OF NEUTRAL AXIS FOR THE SEVERAL LAYERS OF BARS AT COLUMN 22

No reason for the markedly smaller value in gage lines 5 and 17 is apparent. For the gage lines other than 5 and 17 the position of the neutral axis with respect to the under side of slab ranged from 0.43 to 0.46 of the effective depth of the several layers. The value of j (which represents the ratio of the distance between the bar and the centroid of compressive stresses to the distance between the bar and the face of the slab) for these bars is, therefore, about 0.85. The value of j found in a similar way for gage lines located at columns bordering the loaded area was 0.87. It will be noted in Fig. 16 that the strain in the lower layer of diagonal bars is nearly as great as that in the upper layer of diagonal bars. The strain in the layer of rectangular bars between the two layers of diagonal bars was less than that found in the diagonal bars.

No measurements of strain were made on the rings nor on the column bars bent out radially into the slab. These bars were placed low in the slab and near where the neutral axis may be expected to be. It is probable that they were stressed somewhat. As has been stated elsewhere these bars were not taken into account in the calculations of resisting moment.

For the diagonal bars in the central area of the panels, the stresses in the bars of the lower layers were generally considerably greater than those in the upper layer. In the one case where the lower bar shows considerably less stress, the gage lines were near the end of lapped bars.

The bars in the rectangular bands at sections of positive moment were farther from the under surface of the slab than were the bars of the diagonal bands, but the stresses in the former were greater than in the latter except where laps occurred. It should be noted that the bars at the middle of the rectangular bands were farther from the under surface of the slab than were the outer bars. Differences in the magnitude of stresses in bars in similar places at sections of positive moment are partly accounted for by differences in the position of the bars; for example, the bar at gage line 236 was 2.7 in. from the lower surface and the one at 237 was 1.15 in. from the lower surface, while the stresses were 12 600 and 21 600 lb. per sq. in. respectively. Similarly, the bars for gage lines 219, 221, and 239 are higher in the slab than the corresponding bars at the edges of the band.

Gage lines 2, 7 and 10 were placed on the same diagonal bar. Similarly, gage lines 11, 14, and 18 were on another diagonal bar. The bar having gage lines 2, 7, and 10 was about 1.95 in. below the upper surface of the slab and the other bar about 3.15 in. below the upper surface. The stresses at gage lines 2, 7, and 10 were 27 000, 34 500 and 19 800 lb. per sq. in., respectively; the stresses at gage lines 11, 14, and 18 were 16 200, 23 700, 15 000 lb. per sq. in., respectively. It will be seen that the stresses at the gage lines opposite the column capital (7 and 14) were greater than those observed at the other gage lines.

13. *Resisting Moment Accounted for by Stresses in Reinforcing Bars.*—It will be of interest to find the magnitude of the resisting moments developed in sections of the slab and to compare the values

with the bending moment due to the external forces. As the part played by the tensile stresses in the concrete is unknown and uncertain, only that part of the resisting moment found by using the measured stress in the reinforcing bars can be considered. The resisting moment based on the tensile stresses in these bars may not be expected to account fully for the bending moment. The two sections considered will be (1) a section across the panels midway between columns (AB, Fig.

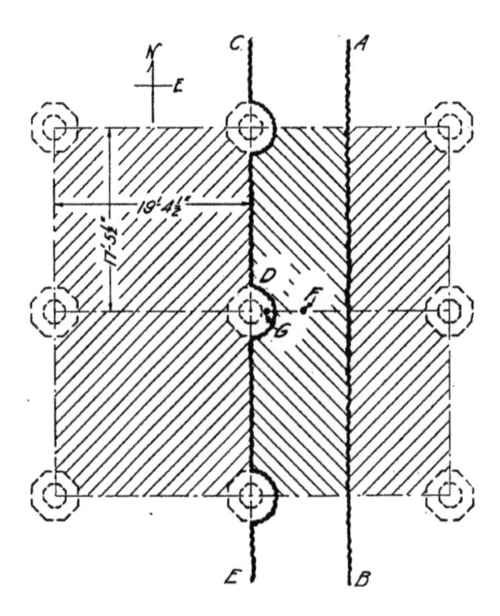

FIG. 17. SECTIONS OF POSITIVE MOMENT AND NEGATIVE MOMENT CONSIDERED IN THE CALCULATIONS

17) and (2) one along an edge of the panels parallel to the first section but skirting the part of the periphery of the column capitals at the corners of the panels (CDE, Fig. 17). It will be noted that the edges of the area here considered are along lines of zero shear, except around the column capitals. The external forces acting on the row of half panels are the load on the two half panels and the reaction or external shear at the three column capitals. The moment of the couple formed by these two external forces will be resisted by the numeri-

cal sum of the resisting moments developed in the two sections AB, and CDE. The moment of the internal stresses at the section of the panel midway between columns is referred to as the positive resisting moment and that at the edge of the panel as the negative resisting moment. The actual distribution of these resisting moments along the sections need not be considered in making the desired comparison.

The point of application of the resultant of the load on the two half panels is shown at F; that of the resultant of the reaction of the three supports at G. The moment of the couple formed by these two resultants is the bending moment due to the external forces and is the moment to be considered. Analysis shows that, for a uniformly distributed load, and round columns, the value of this bending moment for a load two panels wide is given quite closely by the equation

$$M = \frac{2}{8} W l_1 \left[1 - \frac{2}{3} \frac{c}{l_1} \right]^2 \quad \text{. . . . . (1)}$$

for sections at right angles to the long way of the panel, and

$$M = \frac{2}{8} W l_2 \left[1 - \frac{2}{3} \frac{c}{l_2} \right]^2 \quad \text{. . . . . (2)}$$

for sections at right angles to the short way of the panel where

M = bending moment for a width of two panels
W = load on one panel
l_1 = long side of an oblong panel measured from center to center of column
l_2 = short side of an oblong panel measured from center to center of column
c = diameter of column capital.

With the load of 913 lb. per sq. ft. over four panels, the bending moment for a width of two panels, as obtained by equations (1) and (2), is 12 820 000 lb.-in. for the long way of the panels (resisted at north and south sections) and 11 060 000 lb.-in. for the short way (resisted at east and west sections).

The positive moment (the resisting moment at the section across the panels midway between columns) and the negative moment (the resisting moment at the section at the edge of the panels) together must resist this bending moment. With a condition of ends of slab for which the tangent to the curve of flexure at the edges of the panels remains horizontal when the load is applied (usually termed fixed ends), the condition assumed in the analysis, the positive moment will be found by analysis to be one-third of the total resisting moment and the negative moment two-thirds, provided the slab is uniformly stiff throughout. If the tangents at the panel edges deflect somewhat, the positive moment will be greater than one-third and the negative moment less than two-thirds. In the comparisons to be made it will be assumed that the proportions given by analysis are one-third for the positive moment and two-thirds for the negative moment. As stated in "14, Bending of Columns," calculations of the resisting moment developed at sections of the slab at the boundaries of the loaded area as accounted for by the observed stresses in the reinforcing bars, the results of analytical determinations of the moments at sections at the edges and at the middle of the loaded area, and the amount of flexure in the columns all go to show that there is little inaccuracy in this assumption in the case under consideration. The analytical value of the positive resisting moment will be termed the analytical positive moment and that of the negative resisting moment the analytical negative moment. For the north and south sections the magnitude of these analytical moments becomes 4 270 000 and 8 550 000 lb.-in. respectively, and for the east and west sections 3 700-000 and 7 400 000 lb.-in. respectively.

The sections used in obtaining the resisting moment accounted for by the stresses observed in the reinforcing bars are shown in Fig. 18. QKC-BJT is the east and west section of positive moment used, and MNOP the east and west section of negative moment; similarly, IJO-NKL is the north and south section of positive moment, and ABCD and EFGH the north and south sections of negative moment. Sections of negative moment in the north and south direction are taken on two sides of the column capitals, because the available reinforcement differed in the two sections. The sections of positive moment were taken as shown because they cross the greatest number of gage lines.

In the calculation of the resisting moments developed, lapped bars were considered as contributing to the resisting moment whereever the bars extended beyond the section a sufficient distance to insure adequate anchorage with respect to the magnitude of the stress developed in the bar. Many of the laps were made in such a way that the additional section may not be expected to contribute much to the resisting moment. As measurable stresses were found in bars near the edge of the band outside the loaded area, they were taken into account in the calculations. The ring rods around the columns and the column bars which were bent out into the slab were not included in the reinforcement, for no measurements of strain were made on these bars. For the diagonal bars the component of the stress was taken in a direction at right angles to the direction of the panel edge. The average of the stresses at the principal critical gage lines was generally used. For the bands at the edges of the loaded area the stresses were considered to vary over the band from gage line to gage

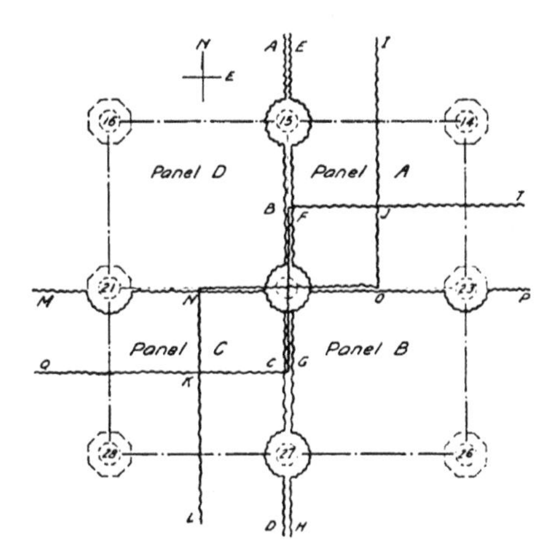

FIG. 18. SECTIONS OF POSITIVE AND NEGATIVE MOMENT CONSIDERED IN THE CALCULATION OF THE RESISTING MOMENTS ACCOUNTED FOR BY STRESSES IN THE REINFORCING BARS

line. Some judgment had to be used in determining the stress variation as well as the availability of lapped bars, but it is believed that the assumptions made give results well below the actual resistance developed in the slab. It should be noted that the maximum stresses observed in bars of negative moment were 15 per cent greater than the average of the stresses used in computing the resisting moment accounted for by these stresses, not counting bars in bands near the edges of the loaded area, and similarly the maximum stresses in bars at sections of positive moment were 25 per cent greater than the average of the stresses used in computing the positive moment. The measured position of the bars with reference to the face of the slab and the measured thicknesses of the slab were used. The position of the neutral axis was determined from the strains measured in the reinforcing bars and in the concrete at the face of the slab; knowing this distance, the values of *jd*, the distance from the bar to the center of gravity of the compressive stresses, were computed in the usual way. The value of *jd* was generally about 0.87 of the distance from the center of gravity of the bars to the face of the slab.

TABLE 3

RESISTING MOMENTS ACCOUNTED FOR BY STRESSES OBSERVED IN REINFORCING BARS

(Moments are given in pound inches)

NORTH AND SOUTH SECTIONS

Negative Moment		Positive Moment	
Width AB Fig. 18 Width BC Width CD	2 760 000 2 720 000 1 690 000	Width IJ Width JO–NK Width KL	560 000 1 570 000 750 000
Total	7 170 000	Total	2 880 000
Width EF Width FG Width GH	2 950 000 2 890 000 2 030 000	....	....
Total	7 870 000	....	....
Average total negative moment	7 520 000	....	....

EAST AND WEST SECTIONS

Negative Moment		Positive Moment	
Width MN Width NO Width OP	2 190 000 2 950 000 1 680 000	Width QK Width KG–FJ Width JT	600 000 1 340 000 730 000
Total	6 820 000	Total	2 670 000

Table 3 gives the resisting moments accounted for by the stresses observed in the reinforcing bars as calculated by the method described. A comparison of these moments with the values of the analytical moments already given indicates that about 88 per cent of the analytical negative moment is accounted for by the stresses in the reinforcing bars in the case of the north and south section and 92 per cent in the case of the east and west section. For positive moments 68 per cent of the analytical positive moment is accounted for in the north and south section and 72 per cent in the east and west section. The average percentages for the sections in the two directions are 90 for the negative moment and 70 for the positive moment.

The reinforcing bars are not the only source of tensile resistance in the slab; tensile strength of concrete assists. It is evident that at sections of positive moment tension in the concrete may make up a considerable part of the resistance offered by the slab, and even at sections of negative moment it may have an effect. This tension will be especially influential in regions away from cracks. At points of the sections which are outside the loaded area and where the stresses in the bars are small it may be expected to form a not inconsiderable part of the total resisting moment developed. Of course, if all panels were loaded, the effect of tension in the concrete would be much less, since its greatest proportional effect is outside the loaded area.

In making comparisons, it must be kept in mind that the unit-strain computed from the measurements gives the average strain over the gage length and that at a crack the stress in the bar will be more than the average stress over the gage length.

Measurements made in beam tests in various laboratories generally show that up to loads near the ultimate load the measured stress in the reinforcing bars is less than that necessary to account for the full bending moment due to the applied load. At the lower stresses the deficiency is considerable; at stresses near the yield point of the steel it may not be much, and the measured stresses may even be larger than necessary to account for the bending moment. The effect is particularly noticeable in concrete of high quality.

On the whole the analytical values of the moments are closely approached, as closely as may be expected in tests of this kind. The negative moment of course is most fully accounted for. The positive moment is not wholly accounted for; but as the section of positive moment involves relatively low stresses the effect of the tensile resistance of the concrete on the measured stresses may be considerable, especially in the part of the slab outside the loaded area. It must not be overlooked, also, that the stress in the bars at the cracks will be larger than the average stress over the gage length.

It should be noted that in the case of negative moments the magnitudes of the moments found for the part of the sections adjacent to the central column are smaller than those for the corresponding parts of the sections adjacent to the outer columns. In the case of column 15, part of this difference is due to the additional amount of reinforcement furnished by the extra laps at this place, and it should be noted too that the larger number of laps at the outer columns makes this portion of the slab stiffer than the portion around the central column. In general, however, most of it appears to be due to the stresses developed in slab and reinforcing bars outside the loaded area, a part of the slab which has not usually been considered to contribute to the resisting moment. In the sections of positive moment, the magnitudes of the resisting moments found are less proportionally at the ends of the sections than at the middle, a condition which may be due to the greater proportional effect of the tensile resistance developed in the concrete in these regions.

The foregoing comparisons have been made on the basis of the full analytical value of the bending moment and by considering one-third of it as positive moment and two-thirds as negative moment. The Joint Committee on Concrete and Reinforced Concrete recommended for the sum of the positive and negative moments a value which is about 85 per cent of the analytical value heretofore used and recommended that the distribution be three-eighths positive moment and five-eighths negative moment. It may be of interest to note that the sum of the positive and negative moments accounted for by the measured stresses in the reinforcing bars has nearly the same value as the sum of the moments recommended by the Joint Committee. The negative moments so accounted for are 13 per cent higher than the value recommended by the Joint Committee, and the positive moments are about 73 per cent of the committee's values. In judging of the results, it should be remembered that at the section of positive moments tensile stresses in the concrete have a considerable influence at the stage of the loading indicated by the stresses in the bars in the loaded area and outside of it. The reference to the tensile resistance of the concrete as contributing to the resisting moment of the slab in the test should not be taken to mean that it will be effective as a resistance when the ultimate load is reached. It is apparent, also, that requirements less than those of the Joint Committee will not provide for all the moment developed by the bars of the slab.

In making a comparison with methods used in design it should be borne in mind that the principal observed maximum stresses were from 15 to 25 per cent greater than the average of the observed stresses which were used in computing the resisting moments accounted for by the stresses in the bars; in designing, a uniform stress over the section is assumed.

14. *Bending of Columns.*—A few gage lines were placed on columns bordering the loaded area. In Fig. 19 the location of these is shown, together with the load-strain diagrams. Although for some

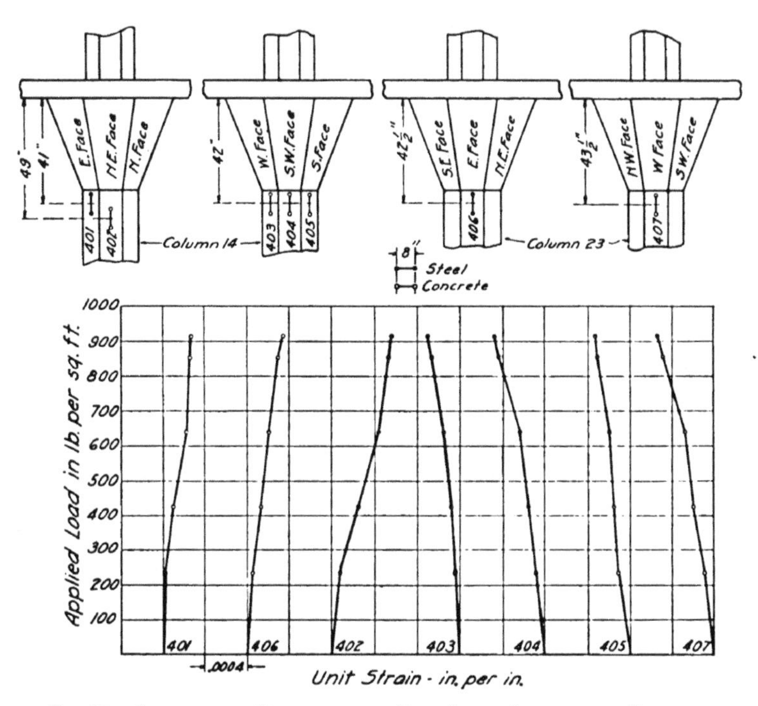

FIG. 19. LOAD-STRAIN DIAGRAMS FOR GAGE LINES LOCATED ON COLUMNS

of the gage lines the strains show as tensile strains, it is probable that for the gage lines given as in tension the compressive strain in the column due to the weight of the floors above was sufficient to overcome the tensile strain measured by the instrument.

The direction of the bending was in the direction which would be expected from the condition of loading. For gage line 406 on a column reinforcing bar the deformation at the load of 913 lb. per sq. ft. corresponded to a flexural tensile stress of 10 500 lb. per sq. in. and for gage line 407 on the concrete on the opposite face of the column the strain was 0.00054 in. per in. corresponding to a stress of 2160 lb. per sq. in. with a modulus assumed as 4 000 000 lb. per sq. in. Similarly at gage line 401 on a column reinforcing bar at the east face of column 14 a stress of 8100 lb. per sq. in. was found, and for gage line 403 on the concrete on the opposite face of the column the compressive stress was 1200 lb. per sq. in. based on the same modulus of elasticity of concrete. The highest compressive strain on column 14 was found at gage line 404 on the southwest face of the column. Gage line 402 was placed across an opening which had been cut into the concrete of the column in an unsuccessful effort to expose column bars at this point. A fine crack formed across this gage line. It is seen that the deformation at this gage line is large.

Fine cracks formed on the tension side of the columns at or near the juncture of the column shaft and capital. At column 15 a crack was found about half way up the column capital. All the cracks noted were fine cracks—much smaller than those observed on the lower surface of the slab.

The amount of flexure in the columns was apparently not sufficient to give more than a slight reduction in the proportion of resisting moment carried by the slab at the sections at the edges of the loaded area, nor more than a slight increase in the moment carried at the sections through the middle of the loaded area. This view is borne out by calculations of the resisting moment developed at sections of the slab at the boundaries of the loaded area, as accounted for by the observed stresses in the reinforcing bars, and also by the results of analytical determinations of the moments at sections at the edges and the middle of the loaded area.

15. *Time Effect of Load and Recovery upon Removal of Load.*—As has been stated, two or more sets of strain gage and deflection readings were generally taken after an increment of load had been applied,—the first set immediately after the last of the load had been placed and another set 12 hours thereafter. A third set was taken 66 hours after the load of 913 lb. per sq. ft. had been applied.

The effect of time on the strains in the steel and concrete was very small; the reading in a few of the gage lines increased slightly after 12 hours, but generally not enough to allow two points to be plotted on Fig. 11 to 13 and not more than may be considered to be within the error of observation. This was true for both steel and concrete even in the most highly stressed places at the maximum load after a period of 66 hours had elapsed.

The deflection readings were affected but little through the 12-hour period until a load of 637 lb. per sq. ft. was reached; Fig. 8 shows that at this load a marked change in deflection occurred at the centers of panels B, C, and D in the 12 hours' time. At the maximum load the increase in deflection through the 66-hour period was small.

Four days were consumed in removing the load. Four hours after the last of the load was removed, readings were taken on the gage lines and deflection points, and twenty hours later another set was taken. The cracks in both upper and lower surfaces of the slab had closed so that they were not easily traced. The recovery in deflection is given by the plotted points at the bottom of the diagrams in Fig. 8. The recovery at the centers of the panels was about 75 per cent; at other points it was generally greater. The recovery in strains in steel and concrete was also large. Even where high stresses and open cracks had been observed the residual strain in the steel was not more than that to be expected from the lack of interlocking of particles at the cracks. In the region of high compressive stress in the concrete the recovery at the principal gage lines was 75 per cent or more; in the reinforcing bars the recovery averaged about 80 per cent. In general the action of the slab was that of concrete of high quality.

16. *General Comments.*—Although there were differences in the stresses and deflections found at corresponding points in the four loaded panels, these differences can generally be accounted for by (1) difference in the arrangement and number of the reinforcing bars, (2) differences in position of the bars with respect to the surfaces of the slab, and (3) differences in strength and stiffness of concrete in the four panels. The large amount of steel in the region of the column capitals bordering the loaded area (where there were two or three bands lapped) added greatly to the stiffness of the slab in this

region. The effect of this added stiffness was to cause higher relative stresses to be developed at the columns bordering the loaded area, and relatively large negative resisting moments were developed at columns where lapped bars were numerous. Another effect of the lapping of bars at columns and in certain of the rectangular bands was to decrease the deflection of the slab in the panels affected by the laps. The influence of the position of the reinforcing bars has already been discussed. Higher stresses in reinforcing bars and greater deflections in the slab were found in the panels for which the compression tests showed weaker and less stiff concrete. Cracks, also, were wider and more numerous in these panels. It is apparent that the tensile resistance of the concrete at sections of positive moment, particularly near the outer edges of the loaded panels, contributed to the resistance of the slab even at the maximum load.

With reference to the design of the slab, it may be noted that the slab was strongly reinforced, though the reinforcing bars were not distributed to the best advantage and the laps were not placed so as to be fully effective. Taking the total reinforcement found over the loaded area and immediately outside, and including such lapped bars as had sufficient anchorage beyond critical sections to be effective (there were many lapped bars which were not counted as effective), and considering the thinness of the slab, the amount of reinforcement for negative moment was on the average as much as that required for the negative moments recommended by the Joint Committee on Concrete and Reinforced Concrete. The amount of reinforcement available for positive moment was on the average more than 50 per cent greater than that required for the positive moments recommended by the same committee. The distribution of the reinforcing bars was, however, quite different from that recommended by this committee. It may be noted also that the amount of reinforcement was greater than that required by most building regulations.

Although the nominal thickness of the slab (8½ in.) was less than that required by building regulations, the slab fulfilled the common requirements for compressive and shearing stresses in concrete of the high quality shown in the tests of prisms taken from the slab. The provisions of the Joint Committee on Concrete and Reinforced Concrete for bending moments and working stresses in concrete of 3000 lb. per sq. in. strength give a thickness about the same as the designed thickness of the slab.

In view of the large number of lapped bars not shown on the building plans, the use of high-carbon steel instead of the mild steel specified, and the unexpectedly high strength of the concrete, it is not strange that the floor carried a much higher load than was anticipated when the test was begun.

17. *Wrecking of the Floor.*—As soon as the load was removed, portions of the floor in the four panels of the loaded area were broken into to uncover the reinforcing bars at important sections. Measurements of the position and location of the reinforcing bars were made for use in calculation. The thickness of the floor was measured at a number of points to check the values obtained from level readings. Photographs were taken to give a record of the position and location of reinforcing bars and their laps; Fig. 21 is a sample. As the wrecking of the floor by the contractor progressed, further measurements of the position of the reinforcing bars and their laps, including those in the bands outside of the loaded area, were taken.

The wrecking of the building by the contractor was an interesting operation. Fires were first built around the columns on the floor below the one to be wrecked; the effect on the concrete at the base of the column after several hours application of heat was to crack and loosen the concrete shell and expose the reinforcement. To assist in cracking the concrete and separating it from the steel, in many cases water was thrown over the columns after they were well heated. A heavy iron pear-shaped weight (about 1600 pounds) was dropped on the floor immediately over the column capital close to where the column of the story above had been. After the column capital had been sheared and shattered by this operation, the portion of the floor surrounding the column and that directly between columns was broken up with the weight and with sledge hammers. After the concrete of sections of the floor had been removed in this way, the reinforcing bars were cut with the oxyacetylene flame. Many of the bars were taken out in good condition. The process was continued until the entire floor was wrecked. The longitudinal reinforcing bars in the column were then cut near the floor below and the columns were pulled down on the floor and broken up with the heavy weight and hammers.

18. *Summary.*—The following deductions have been drawn in the foregoing presentation of the results of the test:

(1) The tests of samples of the concrete from the slab, as well as the hardness and toughness of the concrete observed in breaking up the slab, indicate that the concrete was of unusually good quality and that it had high strength and stiffness. The action of the slab under load was that to be expected with high grade, well-seasoned concrete. The effect of time on the stresses in steel and concrete and on the deflection of the slab under a sustained load was slight, even over a period of 66 hours under the maximum load of 913 lb. per sq. ft.,—conditions which would not exist at an early age of concrete. Upon removal of load, the recovery in deflection at the centers of the panels was about 75 per cent of that under load, and at other points generally more; the recovery in strains in steel and concrete was as large.

(2) The position of the important cracks on both upper and lower sides of the slab may be expected to indicate the region of high tensile stresses in the reinforcing bars; it is also an indication of the general action of the slab in flexure. The cracks on the upper side at the load of 913 lb. per sq. ft. opened to a width of 0.02 to 0.06 in.; those on the lower side were not so wide. Upon removal of the load the cracks closed, leaving the surfaces of the slab with the appearance which they had before the load was applied.

(3) For reinforcing bars in the upper side of the slab in the regions of negative moment, the stresses in bars of diagonal bands were greater than those in bars of rectangular bands, a stress of 57 300 lb. per sq. in. being observed in a diagonal bar and one of 42 000 lb. per sq. in. in a rectangular bar at the maximum load. Stresses were found in both rectangular and diagonal bars at the columns bordering the loaded area nearly as great as those at corresponding points at the central column. Stresses of some magnitude were found in bars outside the loaded area. The stresses given do not include the stress due to the load of the slab itself.

(4) For reinforcing bars in the lower side of the slab in the regions of positive moment, the stresses in bars of rectangular bands were greater than those in bars of diagonal bands even though the former were farther above the lower surface of the slab than the latter; in the one apparent exception, the presence

Fig. 20. View Showing Arrangement of Reinforcement at Column 22

Fig. 21. View of Slab after Reinforcing Bars were Exposed

of laps doubled the usual number of bars. At the maximum load stresses of 24 000 to 30 000 lb. per sq. in. were observed in bars of rectangular bands and of 20 000 to 24 000 lb. per sq. in. in bars of diagonal bands. A stress of 15 600 lb. per sq. in. was observed in a bar outside the loaded area at the edge of a rectangular band.

(5) On the upper surface of the slab the greatest compressive strains were found at gage lines along the inner panel edges midway between columns, ranging from 0.0009 to nearly 0.001 in. per in. at the maximum load; at the centers of the panels values about half as great were found. On the lower surface the greatest strains were found at the middle column; these ranged from 0.0012 to 0.0016 in. per in., values which are as great as the strains found at failure in the tests of the concrete prisms cut from the slab and as great as are ordinarily found in compression tests of concrete at the ultimate load. In some places there was chipping and spalling of the concrete. It is evident that the action of the surrounding concrete assisted in preventing failure. At gage lines crossing the inner panel edges high compressive strains also were found, even though there was no tension reinforcement in the upper side of the slab in this region. It may be noted that the maximum compressive strains on the upper surface of the slab were one-half to three-fifths those found on the lower surface. The intensity of the strains at various points along sections of positive moment and negative moment will give some measure of the distribution of intensity of moments along those sections.

(6) The observed stresses in the reinforcing bars accounted for about 90 per cent of the analytical negative moment and about 70 per cent of the analytical positive moment, as given by the methods used. It should be noted that the observed stresses used are average stresses over the gage length, and the stress at a crack may be expected to be greater than the average over the gage length. It seems probable that tensile resistance of the concrete contributed to the resistance of the slab, particularly in the sections of positive moment and in regions near the edges of the loaded area. A similar influence of the tensile resistance of concrete, when the stresses in the steel are well below its yield point, has been observed in numerous beam tests. That the tensile re-

sistance of the concrete contributed to the resisting moment of the slab in the test should not be taken to mean that it will be effective in resisting moment when the ultimate load is reached. It may be noted also that the sum of the positive and negative moments accounted for by the measured stresses in the reinforcing bars has almost the same value as the sum of the positive and negative moments recommended by the Joint Committee on Concrete and Reinforced Concrete, and that the negative moments so accounted for are about 113 per cent and the positive moments about 73 per cent of the moments recommended by this committee. In making a comparison with methods used in designing it should be borne in mind that the principal maximum stresses were from 15 to 25 per cent greater than the average stresses which were used in computing the resisting moments accounted for by the stresses in the bars; in designing, a uniform stress over the section is assumed.

Although the arrangement of bars was not as recommended, the amount of reinforcement for negative moment, considering all available bars over the area used, was as much as that required for the negative moments recommended by the Joint Committee on Concrete and Reinforced Concrete, even though the slab was thinner than recommended for ordinary concrete. The amount of reinforcement for positive moment was more than 50 per cent greater than that required for the positive moments recommended by the committee. Although the nominal thickness of the slab was less than that required by building regulations, it fulfilled the provisions of the committee for bending moments and working stresses for concrete of a test strength of 3000 pounds per square inch.

(7) The action of the floor slab under test should give added confidence in the suitability and reliability of the flat slab as a load-carrying structure.

LIST OF
PUBLICATIONS OF THE ENGINEERING EXPERIMENT STATION

Bulletin No. 1. Tests of Reinforced Concrete Beams, by Arthur N. Talbot. 1904. *None available.*

Circular No. 1. High-Speed Tool Steels, by L. P. Breckenridge. 1905. *None available.*

Bulletin No. 2. Tests of High-Speed Tool Steels on Cast Iron, by L. P. Breckenridge and Henry B. Dirks. 1905. *None available.*

Circular No. 2. Drainage of Earth Roads, by Ira O. Baker. 1906. *None available.*

Circular No. 3. Fuel Tests with Illinois Coal (Compiled from tests made by the Technological Branch of the U. S. G. S., at the St. Louis, Mo., Fuel Testing Plant, 1904–1907), by L. P. Breckenridge and Paul Diserens. 1908. *Thirty cents.*

Bulletin No. 3. The Engineering Experiment Station of the University of Illinois, by L. P. Breckenridge. 1906. *None available.*

Bulletin No. 4. Tests of Reinforced Concrete Beams, Series of 1905, by Arthur N. Talbot 1906. *Forty-five cents.*

Bulletin No. 5. Resistance of Tubes to Collapse, by Albert P. Carman and M. L. Carr. 1906. *None available.*

Bulletin No. 6. Holding Power of Railroad Spikes, by Roy I. Webber. 1906. *None available.*

Bulletin No. 7. Fuel Tests with Illinois Coals, by L. P. Breckenridge, S. W. Parr, and Henry B. Dirks. 1906. *None available.*

Bulletin No. 8. Tests of Concrete: I, Shear; II, Bond, by Arthur N. Talbot. 1906. *None available.*

Bulletin No. 9. An Extension of the Dewey Decimal System of Classification Applied to the Engineering Industries, by L. P. Breckenridge and G. A. Goodenough. 1906. Revised Edition 1912. *Fifty cents.*

Bulletin No. 10. Tests of Concrete and Reinforced Concrete Columns, Series of 1906, by Arthur N. Talbot. 1907. *None available.*

Bulletin No. 11. The Effect of Scale on the Transmission of Heat through Locomotive Boiler Tubes, by Edward C. Schmidt and John M. Snodgrass. 1907. *None available.*

Bulletin No. 12. Tests of Reinforced Concrete T-Beams, Series of 1906, by Arthur N. Talbot. 1907. *None available.*

Bulletin No. 13. An Extension of the Dewey Decimal System of Classification Applied to Architecture and Building, by N. Clifford Ricker. 1907. *None available.*

Bulletin No. 14. Tests of Reinforced Concrete Beams, Series of 1906, by Arthur N. Talbot. 1907. *None available.*

Bulletin No. 15. How to Burn Illinois Coal Without Smoke, by L. P. Breckenridge. 1908 *None available.*

Bulletin No. 16. A Study of Roof Trusses, by N. Clifford Ricker. 1908. *None available.*

Bulletin No. 17. The Weathering of Coal, by S. W. Parr, N. D. Hamilton, and W. F. Wheeler. 1908. *None available.*

Bulletin No. 18. The Strength of Chain Links, by G. A. Goodenough and L. E. Moore. 1908. *Forty cents.*

Bulletin No. 19. Comparative Tests of Carbon, Metallized Carbon and Tantalum Filament Lamps, by T. H. Amrine. 1908. *None available.*

Bulletin No. 20. Tests of Concrete and Reinforced Concrete Columns, Series of 1907, by Arthur N. Talbot. 1908. *None available.*

Bulletin No. 21. Tests of a Liquid Air Plant, by C. S. Hudson and C. M. Garland. 1908. *Fifteen cents.*

Bulletin No. 22. Tests of Cast-Iron and Reinforced Concrete Culvert Pipe, by Arthur N. Talbot. 1908. *None available.*

Bulletin No. 23. Voids, Settlement and Weight of Crushed Stone, by Ira O. Baker. 1908. *Fifteen cents.*

**Bulletin No. 24.* The Modification of Illinois Coal by Low Temperature Distillation, by S. W. Parr and C. K. Francis. 1908. *Thirty cents.*

Bulletin No. 25. Lighting Country Homes by Private Electric Plants, by T. H. Amrine. 1908 *Twenty cents.*

*A limited number of copies of bulletins starred is available for free distribution.

Bulletin No. 26. High Steam-Pressures in Locomotive Service. A Review of a Report to the Carnegie Institution of Washington, by W. F. M. Goss. 1908. *Twenty-five cents.*

Bulletin No. 27. Tests of Brick Columns and Terra Cotta Block Columns, by Arthur N. Talbot and Duff A. Abrams. 1909. *Twenty-five cents.*

Bulletin No. 28. A Test of Three Large Reinforced Concrete Beams, by Arthur N. Talbot. 1909. *Fifteen cents.*

Bulletin No. 29. Tests of Reinforced Concrete Beams: Resistance to Web Stresses, Series of 1907 and 1908, by Arthur N. Talbot. 1909. *Forty-five cents.*

**Bulletin No. 30.* On the Rate of Formation of Carbon Monoxide in Gas Producers, by J. K. Clement, L. H. Adams, and C. N. Haskins. 1909. *Twenty-five cents.*

**Bulletin No. 31.* Tests with House-Heating Boilers, by J. M. Snodgrass. 1909. *Fifty-five cents.*

Bulletin No. 32. The Occluded Gases in Coal, by S. W. Parr and Perry Barker. 1909. *Fifteen cents.*

Bulletin No. 33. Tests of Tungsten Lamps, by T. H. Amrine and A. Guell. 1909. *Twenty cents.*

**Bulletin No. 34.* Tests of Two Types of Tile-Roof Furnaces under a Water-Tube Boiler, by J. M. Snodgrass. 1909. *Fifteen cents.*

Bulletin No. 35. A Study of Base and Bearing Plates for Columns and Beams, by N. Clifford Ricker. 1909. *Twenty cents.*

Bulletin No. 36. The Thermal Conductivity of Fire-Clay at High Temperatures, by J. K. Clement and W. L. Egy. 1909. *Twenty cents.*

Bulletin No. 37. Unit Coal and the Composition of Coal Ash, by S. W. Parr and W. F. Wheeler. 1909. *Thirty-five cents.*

**Bulletin No. 38.* The Weathering of Coal, by S. W. Parr and W. F. Wheeler. 1909. *Twenty-five cents.*

**Bulletin No. 39.* Tests of Washed Grades of Illinois Coal, by C. S. McGovney. 1909. *Seventy-five cents.*

Bulletin No. 40. A Study in Heat Transmission, by J. K. Clement and C. M. Garland. 1910. *Ten cents.*

Bulletin No. 41. Tests of Timber Beams, by Arthur N. Talbot. 1910. *Thirty-five cents.*

**Bulletin No. 42.* The Effect of Keyways on the Strength of Shafts, by Herbert F. Moore. 1910. *Ten cents.*

Bulletin No. 43. Freight Train Resistance, by Edward C. Schmidt. 1910. *Seventy-five cents.*

Bulletin No. 44. An Investigation of Built-up Columns Under Load, by Arthur N. Talbot and Herbert F. Moore. 1911. *Thirty-five cents.*

**Bulletin No. 45.* The Strength of Oxyacetylene Welds in Steel, by Herbert L. Whittemore. 1911. *Thirty-five cents.*

**Bulletin No. 46.* The Spontaneous Combustion of Coal, by S. W. Parr and F. W. Kressman. 1911. *Forty-five cents.*

**Bulletin No. 47.* Magnetic Properties of Heusler Alloys, by Edward B. Stephenson. 1911. *Twenty-five cents.*

**Bulletin No. 48.* Resistance to Flow Through Locomotive Water Columns, by Arthur N. Talbot and Melvin L. Enger. 1911. *Forty cents.*

**Bulletin No. 49.* Tests of Nickel-Steel Riveted Joints, by Arthur N. Talbot and Herbert F. Moore. 1911. *Thirty cents.*

**Bulletin No. 50.* Tests of a Suction Gas Producer, by C. M. Garland and A. P. Kratz. 1912. *Fifty cents.*

Bulletin No. 51. Street Lighting, by J. M. Bryant and H. G. Hake. 1912. *Thirty-five cents.*

**Bulletin No. 52.* An Investigation of the Strength of Rolled Zinc, by Herbert F. Moore. 1912. *Fifteen cents.*

**Bulletin No. 53.* Inductance of Coils, by Morgan Brooks and H. M. Turner. 1912. *Forty cents.*

**Bulletin No. 54.* Mechanical Stresses in Transmission Lines, by A. Guell. 1912. *Twenty cents.*

**Bulletin No. 55.* Starting Currents of Transformers, with Special Reference to Transformers with Silicon Steel Cores, by Trygve D. Yensen. 1912. *Twenty cents.*

**Bulletin No. 56.* Tests of Columns: An Investigation of the Value of Concrete as Reinforcement for Structural Steel Columns, by Arthur N. Talbot and Arthur R. Lord. 1912. *Twenty-five cents.*

**Bulletin No. 57.* Superheated Steam in Locomotive Service. A Review of Publication No. 127 of the Carnegie Institution of Washington, by W. F. M. Goss. 1912. *Forty cents.*

---

*A limited number of copies of bulletins starred is available for free distribution.

**Bulletin No. 58.* A New Analysis of the Cylinder Performance of Reciprocating Engines, by J. Paul Clayton. 1912. *Sixty cents.*

**Bulletin No. 59.* The Effect of Cold Weather Upon Train Resistance and Tonnage Rating, by Edward C. Schmidt and F. W. Marquis. 1912. *Twenty cents.*

**Bulletin No. 60.* The Coking of Coal at Low Temperatures, with a Preliminary Study of the By-Products, by S. W. Parr and H. L. Olin. 1912. *Twenty-five cents.*

**Bulletin No. 61.* Characteristics and Limitation of the Series Transformer, by A. R. Anderson and H. R. Woodrow. 1913. *Twenty-five cents.*

Bulletin No. 62. The Electron Theory of Magnetism, by Elmer H. Williams. 1913. *Thirty-five cents.*

Bulletin No. 63. Entropy-Temperature and Transmission Diagrams for Air, by C. R. Richards. 1913. *Twenty-five cents.*

**Bulletin No. 64.* Tests of Reinforced Concrete Buildings Under Load, by Arthur N. Talbot and Willis A. Slater. 1913. *Fifty cents.*

**Bulletin No. 65.* The Steam Consumption of Locomotive Engines from the Indicator Diagrams, by J. Paul Clayton. 1913. *Forty cents.*

Bulletin No. 66. The Properties of Saturated and Superheated Ammonia Vapor, by G. A. Goodenough and William Earl Mosher. 1913. *Fifty cents.*

Bulletin No. 67. Reinforced Concrete Wall Footings and Column Footings, by Arthur N. Talbot. 1913. *Fifty cents.*

Bulletin No. 68. The Strength of I-Beams in Flexure, by Herbert F. Moore. 1913. *Twenty cents.*

Bulletin No. 69. Coal Washing in Illinois, by F. C. Lincoln. 1913. *Fifty cents.*

Bulletin No. 70. The Mortar-Making Qualities of Illinois Sands, by C. C. Wiley. 1913. *Twenty cents.*

Bulletin No. 71. Tests of Bond between Concrete and Steel, by Duff A. Abrams. 1913. *One dollar.*

**Bulletin No. 72.* Magnetic and Other Properties of Electrolytic Iron Melted in Vacuo, by Trygve D. Yensen. 1914. *Forty cents.*

Bulletin No. 73. Acoustics of Auditoriums, by F. R. Watson. 1914. *Twenty cents.*

**Bulletin No. 74.* The Tractive Resistance of a 28-Ton Electric Car, by Harold H. Dunn. 1914. *Twenty-five cents.*

Bulletin No. 75. Thermal Properties of Steam, by G. A. Goodenough. 1914. *Thirty-five cents.*

Bulletin No. 76. The Analysis of Coal with Phenol as a Solvent, by S. W. Parr and H. F. Hadley. 1914. *Twenty-five cents.*

**Bulletin No. 77.* The Effect of Boron upon the Magnetic and Other Properties of Electrolytic Iron Melted in Vacuo, by Trygve D. Yensen. 1915. *Ten cents.*

**Bulletin No. 78.* A Study of Boiler Losses, by A. P. Kratz. 1915. *Thirty-five cents.*

**Bulletin No. 79.* The Coking of Coal at Low Temperatures, with Special Reference to the Properties and Composition of the Products, by S. W. Parr and H. L. Olin. 1915. *Twenty-five cents.*

**Bulletin No. 80.* Wind Stresses in the Steel Frames of Office Buildings, by W. M. Wilson and G. A. Maney. 1915. *Fifty cents.*

**Bulletin No. 81.* Influence of Temperature on the Strength of Concrete, by A. B. McDaniel. 1915. *Fifteen cents.*

Bulletin No. 82. Laboratory Tests of a Consolidation Locomotive, by E. C. Schmidt, J. M. Snodgrass, and R. B. Keller. 1915. *Sixty-five cents.*

**Bulletin No. 83.* Magnetic and Other Properties of Iron-Silicon Alloys. Melted in Vacuo, by Trygve D. Yensen. 1915. *Thirty-five cents.*

Bulletin No. 84. Tests of Reinforced Concrete Flat Slab Structure, by Arthur N. Talbot and W. A. Slater. 1916. *Sixty-five cents.*

**Bulletin No. 85.* The Strength and Stiffness of Steel Under Biaxial Loading, by A. T. Becker 1916. *Thirty-five cents.*

**Bulletin No. 86.* The Strength of I-Beams and Girders, by Herbert F. Moore and W. M. Wilson. 1916. *Thirty cents.*

**Bulletin No. 87.* Correction of Echoes in the Auditorium, University of Illinois, by F. R. Watson and J. M. White. 1916. *Fifteen cents.*

Bulletin No. 88. Dry Preparation of Bituminous Coal at Illinois Mines, by E. A. Holbrook. 1916. *Seventy cents.*

---

*A limited number of copies of bulletins starred is available for free distribution.

**Bulletin No. 89.* Specific Gravity Studies of Illinois Coal, by Merle L. Nebel. 1916. *Thirty cents.*

**Bulletin No. 90.* Some Graphical Solutions of Electric Railway Problems, by A. M. Buck. 1916. *Twenty cents.*

Bulletin No. 91. Subsidence Resulting from Mining, by L. E. Young and H. H. Stoek. 1916. *None available.*

**Bulletin No. 92.* The Tractive Resistance on Curves of a 28-Ton Electric Car, by E. C. Schmidt and H. H. Dunn. 1916. *Twenty-five cents.*

**Bulletin No. 93.* A Preliminary Study of the Alloys of Chromium, Copper, and Nickel, by D. F. McFarland and O. E. Harder. 1916. *Thirty cents.*

**Bulletin No. 94.* The Embrittling Action of Sodium Hydroxide on Soft Steel, by S. W. Parr. 1917. *Thirty cents.*

**Bulletin No. 95.* Magnetic and Other Properties of Iron-Aluminum Alloys Melted in Vacuo, by T. D. Yensen and W. A. Gatward. 1917. *Twenty-five cents.*

**Bulletin No. 96.* The Effect of Mouthpieces on the Flow of Water Through a Submerged Short Pipe, by Fred B. Seely. 1917. *Twenty-five cents.*

**Bulletin No. 97.* Effects of Storage Upon the Properties of Coal, by S. W. Parr. 1917. *Twenty cents.*

**Bulletin No. 98.* Tests of Oxyacetylene Welded Joints in Steel Plates, by Herbert F. Moore. 1917. *Ten cents.*

Circular No. 4. The Economical Purchase and Use of Coal for Heating Homes, with Special Reference to Conditions in Illinois. 1917. *Ten cents.*

**Bulletin No. 99.* The Collapse of Short Thin Tubes, by A. P. Carman. 1917. *Twenty cents.*

**Circular No. 5.* The Utilization of Pyrite Occurring in Illinois Bituminous Coal, by E. A. Holbrook. 1917. *Twenty cents.*

**Bulletin No. 100.* Percentage of Extraction of Bituminous Coal with Special Reference to Illinois Conditions, by C. M. Young. 1917.

**Bulletin No. 101.* Comparative Tests of Six Sizes of Illinois Coal on a Mikado Locomotive, by E. C. Schmidt, J. M. Snodgrass, and O. S. Beyer, Jr. 1917. *Fifty cents.*

**Bulletin No. 102.* A Study of the Heat Transmission of Building Materials, by A. C. Willard and L. C. Lichty. 1917. *Twenty-five cents.*

**Bulletin No. 103.* An Investigation of Twist Drills, by Bruce W. Benedict and W. P. Lukens. 1917. *Sixty cents.*

**Bulletin No. 104.* Tests to Determine the Rigidity of Riveted Joints of Steel Structures, by W. M. Wilson and H. F. Moore. 1917. *Twenty-five cents.*

Circular No. 6. The Storage of Bituminous Coal, by H. H. Stoek. 1918. *Forty cents.*

Circular No. 7. Fuel Economy in the Operation of Hand Fired Power Plants. 1918. *Twenty cents.*

**Bulletin No. 105.* Hydraulic Experiments with Valves, Orifices, Hoze, Nozzles, and Orifice Buckets, by Arthur N. Talbot, F. B. Seely, V. R. Fleming and M. L. Enger. 1918. *Thirty-five cents.*

**Bulletin No. 106.* Test of a Flat Slab Floor of the Western Newspaper Union Building, by Arthur N. Talbot and Harrison F. Gonnerman. 1918. *Twenty cents.*

---

*A limited number of copies of bulletins starred is available for free distribution.

THE UNIVERSITY OF ILLINOIS

THE STATE UNIVERSITY

Urbana

EDMUND J. JAMES, Ph. D., LL. D., *President*

THE UNIVERSITY INCLUDES THE FOLLOWING DEPARTMENTS:

The Graduate School

The College of Liberal Arts and Sciences (Ancient and Modern Languages and Literatures; History, Economics, Political Science, Sociology; Philosophy, Psychology, Education; Mathematics; Astronomy; Geology; Physics; Chemistry; Botany, Zoology, Entomology; Physiology; Art and Design)

The College of Commerce and Business Administration (General Business, Banking, Insurance, Accountancy, Railway Administration, Foreign Commerce; Courses for Commercial Teachers and Commercial and Civic Secretaries)

The College of Engineering (Architecture; Architectural, Ceramic, Civil, Electrical, Mechanical, Mining, Municipal and Sanitary, and Railway Engineering)

The College of Agriculture (Agronomy; Animal Husbandry; Dairy Husbandry; Horticulture and Landscape Gardening; Agricultural Extension; Teachers' Course; Household Science)

The College of Law (three years' course)

The School of Education

The Course in Journalism

The Courses in Chemistry and Chemical Engineering

The School of Railway Engineering and Administration

The School of Music (four years' course)

The School of Library Science (two years' course)

The College of Medicine (in Chicago)

The College of Dentistry (in Chicago)

The School of Pharmacy (in Chicago; Ph. G. and Ph. C. courses)

The Summer Session (eight weeks)

Experiment Stations and Scientific Bureaus: U. S. Agricultural Experiment Station; Engineering Experiment Station; State Laboratory of Natural History; State Entomologist's Office; Biological Experiment Station on Illinois River; State Water Survey; State Geological Survey; U. S. Bureau of Mines Experiment Station.

The library collections contain (December 1, 1917) 411,737 volumes and 104,524 pamphlets.

For catalogs and information address

THE REGISTRAR
URBANA, ILLINOIS

UNIVERSITY OF ILLINOIS BULLETIN
Issued Weekly
Vol. XVI October 21, 1918 No. 8
[Entered as second-class matter December 11, 1912, at the post office at Urbana, Illinois, under the Act of August 24, 1912. Acceptance for mailing at the special rate of postage provided for in section 1103 Act of October 3, 1917, authorized July 31, 1918]

ANALYSIS AND TESTS OF RIGIDLY CONNECTED REINFORCED CONCRETE FRAMES

BY
MIKISHI ABE

BULLETIN No. 107
ENGINEERING EXPERIMENT STATION
Published by the University of Illinois, Urbana

Price: Fifty Cents
European Agent
Chapman & Hall, Ltd., London

THE Engineering Experiment Station was established by act of the Board of Trustees, December 8, 1903. It is the purpose of the Station to carry on investigations along various lines of engineering and to study problems of importance to professional engineers and to the manufacturing, railway, mining, constructional, and industrial interests of the State.

The control of the Engineering Experiment Station is vested in the heads of the several departments of the College of Engineering. These constitute the Station Staff and, with the Director, determine the character of the investigations to be undertaken. The work is carried on under the supervision of the Staff, sometimes by research fellows as graduate work, sometimes by members of the instructional staff of the College of Engineering, but more frequently by investigators belonging to the Station corps.

The results of these investigations are published in the form of bulletins, which record mostly the experiments of the Station's own staff of investigators. There will also be issued from time to time, in the form of circulars, compilations giving the results of the experiments of engineers, industrial works, technical institutions, and governmental testing departments.

The volume and number at the top of the front cover page are merely arbitrary numbers and refer to the general publications of the University of Illinois: *either above the title or below the seal is given the number of the Engineering Experiment Station bulletin or circular which should be used in referring to these publications.*

For copies of bulletins, circulars, or other information address the

ENGINEERING EXPERIMENT STATION,
URBANA, ILLINOIS.

UNIVERSITY OF ILLINOIS
ENGINEERING EXPERIMENT STATION

BULLETIN No. 107 OCTOBER, 1918

ANALYSIS AND TESTS OF RIGIDLY CONNECTED REINFORCED CONCRETE FRAMES

BY
MIKISHI ABE
FORMERLY STUDENT IN THE GRADUATE SCHOOL OF THE
UNIVERSITY OF ILLINOIS

ENGINEERING EXPERIMENT STATION
PUBLISHED BY THE UNIVERSITY OF ILLINOIS, URBANA

CONTENTS

III. TESTS OF RIGIDLY CONNECTED REINFORCED CONCRETE FRAMES

LIST OF FIGURES

LIST OF TABLES

ANALYSIS AND TESTS OF RIGIDLY CONNECTED REINFORCED CONCRETE FRAMES*

I. Introduction

1. *Preliminary.*—The rigidly connected frame has frequent applications in reinforced concrete construction. It is made use of in engineering structures in a variety of forms. With a wider knowledge of the analysis of rigidly connected frames of reinforced concrete it is believed that the frame as an element in design will become more important and will be utilized in various ways in gaining economies and in securing effective designs.

Notwithstanding the importance, the exact determination of stresses as they actually occur in a rigidly connected reinforced concrete frame is not usually attempted. In most cases the determination requires considerable time and labor to work out the formulas for statically indeterminate quantities which depend upon the number and the fixity of the supports. There are few publications in English which treat this subject systematically, and formulas for practical use are not generally available. In some quarters there may still be a lack of acceptance of the principle of continuity and the effect of rigidity of joints as applied to reinforced concrete construction.

In this bulletin formulas for moments and other indeterminate quantities for several types of indeterminate structures are presented. For the derivation of the formulas the method which involves the use of the principle of least work was chosen. To test the applicability and reliability of the methods of analysis test frames designed according to the formulas found by the analyses were fabricated and the deformations produced in various parts of the members under the action of the loads measured, and a study and comparison of the results of analysis

*This bulletin presents a *résumé* of the analytical treatment of rigidly connected reinforced concrete frames and the principal results of the experimental work on reinforced concrete frames which were given in the thesis of Mikishi Abe as presented in partial fulfillment of the requirements for the degree of Doctor of Philosophy in Engineering in the Graduate School of the University of Illinois, June, 1914. The work of condensation and reformulation necessary to bring the text within the space available for the bulletin has been done by members of the department of Theoretical and Applied Mechanics. The complete thesis is on file in the library of the University of Illinois. The original data of the tests and more detailed work of the analyses are on file at the Laboratory of Applied Mechanics of the University of Illinois.

It may be helpful to note that Dr. Abe's name is pronounced in two syllables; thus, Ah'be.

and test have been made. The general rigidity at the places where the members are joined has been investigated. The work is presented in the thought that it will be helpful in bringing into wider use the principles applicable to the design of rigidly connected reinforced concrete constructions.

2. *The Use and the Advantages of the Rigidly Connected Reinforced Concrete Frame.*—Since about 1905 reinforced concrete frame construction has been extensively used in continental Europe. Many examples can be found in the German texts and magazines. In England also frame constructions of reinforced concrete viaducts and other structures have been built in recent years. There is also a tendency in America to use reinforced concrete frames for buildings and bridges.

The field of the application of rigid frames is almost unlimited, for most reinforced concrete structures are composed of elements of rigid frames. It covers such constructions as buildings, bridge structures, trestles and viaducts, culverts and sewers, subway construction, retaining walls, and reservoirs and water tanks. In these structures rigid connections are used between members and in many or most of them the bending moments are statically indeterminate.

It is clear that every building construction of reinforced concrete may be considered as a rigidly connected frame, for columns, girders, beams and slabs are all rigidly connected with each other, even though the effect of this condition is not fully considered in the design. In continental European countries it is most common to use frames in building constructions, such as roofs, balconies, towers, and the building as a whole. In the design the requirements and the advantages of the frame are taken into account.

Bridge structures are in the field of the rigid frame. Arches, beam and bent construction, and most bridge structures can be designed as frames on a rigid analytical basis. In highway bridges, for example, a spandrel-braced arch is frequently used. In such a case columns are rigidly connected to the arch ribs and to the superstructure, and therefore the design should be made as a rigidly connected frame. The designing of trestles and viaducts as a frame will secure safety and at the same time obtain the best proportioning of parts.

Box culverts and the box type of construction for subways give sections which are examples of the rigidly connected frame and which may not be rationally designed without a sufficient knowledge of rigid frames.

In water tanks and reservoirs of a rectangular or a polygonal form, the unknown negative bending moment due to a rigid connection of wall to wall or base to wall will exist at each corner. These moments are modified by the relative thickness of walls and the other dimensions of the structure. A knowledge of the rigid frame will suggest the proper method of solution.

It can thus be seen that most monolithic construction falls within the field of the rigid frame. A study of the rigid frame will assist in developing judgment for use in the design of such construction.

In building and structural design, insufficient attention is often given to the bending of columns caused by the rigidity of connections. The bending moment for a beam is frequently taken as an assumed fraction of Pl (where P is the load and l is the span) while bending moments at the ends and in the columns are disregarded entirely, thus leaving the structure inadequate or making one part stronger at the expense of the other.

The reinforced concrete frame is advantageous in that material can be saved and a much better result obtained from the theoretical and structural point of view. In ordinary concrete building construction the element of rigidity is usually not fully taken advantage of. With the concrete frame construction, however, the rigidity of the connection of the members may be used. The rigid frame is capable of exact design, and therefore the economical distribution of materials can be realized.

One reason why some engineers hesitate to use concrete frames extensively is that they hardly believe in the continuity of the parts of the structure and doubt the effect of the rigidity of the connection. The question is also naturally raised if the formulas deduced from the elastic work of deformation of a non-homogeneous material like reinforced concrete will hold good for such composite members with fair agreement; furthermore the secondary stresses may act to modify the results. Under actual conditions, as is well known, the fundamental assumptions which underlie the static considerations can seldom be more than partially fulfilled even under carefully prepared specifications and well executed designs. These things must be considered before coming to a conclusion as to the reliability of rigidly connected reinforced concrete frames. It is evident that reinforced concrete frames will be reliable if there is perfect continuity or complete rigidity of joint, close agreement between theory and experiment, and a small effect of stresses of a secondary character. It is not known that an experimental study of this subject has before been made. It may be expected then that careful experiments and investigation will give information which will help to settle these questions.

3. *Scope of Investigation and Acknowledgment.*—In this bulletin formulas for several types of statically indeterminate structures which have been deduced by the use of the principle of least work are given. For vertical load the following cases have been analyzed: (1) single story, single span; (2) single story, three spans; (3) trestle bent with tie, single span; (4) building frame with several stories and several spans; and (5) bridge trestle. For horizontal load the following cases have been analyzed: (1) single story, single span; (2) octagonal reservoir or tank; and (3) rectangular reservoir or tank.

In order to put to practical test the reliability of these formulas for reinforced concrete structures, eight test frames designed according to the formulas found by the analyses were made, and the deformations produced in the various parts of the members by the series of test loads were measured. In the design of the frames requirements not touched upon by the analyses referred to were provided for in a practical way. The analyses and the results of the tests have been subjected to critical study and discussion. The specimens were made in November and December, 1913, and January, 1914, and were tested in January, February, and March, 1914. In making these tests the purpose was to obtain experimental information along the following lines which have a bearing on the design of rigidly connected reinforced concrete frames:

(1) The amount and the distribution of stresses in the reinforcement and in the concrete
(2) The continuity of the composing members of a frame
(3) The location of sections of critical stress
(4) The reliability of a reinforced concrete frame
(5) The applicability of the theoretical formulas in the design of frames.

The experimental work was done as a research problem of the Engineering Experiment Station of the University of Illinois.

The work was under the charge of PROFESSOR ARTHUR N. TALBOT, to whom, as well as to other members of the staff, acknowledgment is due for valuable suggestions and aid.

The limits of space set for this bulletin will not permit publication of even a small part of the details of the derivation of the formulas found nor of the observations, calculations, and other data of the tests. Instead the plan has been followed generally of giving the formulas found from the analyses without the details of the derivation and, in the case of the experimental work, of showing graphically the main stresses that were observed at the principal loads, and of not including details of the data. The original and the reduced data and more detailed work of the analyses are on file at the Laboratory of Applied Mechanics of the University of Illinois.

II. The Analysis of Rigidly Connected Frames

4. *Notation.*—The following notation is used generally throughout the bulletin:

A = area of cross-section of a member. Numerical suffixes are used for individual members when a frame is composed of members of different sizes.
A is also used as a coefficient to represent certain algebraic expressions.
a = distance from the left corner or axis of a frame to the point of application of a concentrated load on a top beam.
b = distance from the right corner or axis of a frame to the point of application of a concentrated load on a top beam.
E = modulus of elasticity (considered as constant) of the material.
H = horizontal reaction acting at the end of a column.
I = moment of inertia in general.
h = total vertical height of frame.
l = total length of horizontal span of frame.
s = length of an inclined member.
m = ratio of moment of inertia of horizontal member to that of vertical member.
$n = \frac{h}{l}$ = ratio of height of frame to length of span.
M = bending moment in general.
N = normal force or stress on a section (total internal force normal to the section).
P = a concentrated load.
p = intensity of a uniformly distributed load.
V = vertical reaction.
f_s = unit stress in steel in tension.
f_s' = unit stress in steel in compression.
f_c = unit stress in concrete in compression.

In the diagrams representing the forms of the frames analyzed, the ends of members are indicated by lower case letters and the symbols for the properties, forces, and moments use these letters as subscripts to indicate the members to which they apply as well as the point of application. The method of use will be made clear by the following examples and by reference to Fig. 6:

H_a = horizontal reaction at a.
I_{ab} = moment of inertia of member ab.
h_{bc} = vertical height of bc.
l_{bc} = length of horizontal projection of bc.
M_{bc} = bending moment at any point in bc.

5. *Statically Determinate and Indeterminate Systems and Number of Statically Indeterminates.*—A force is said to be statically determinate when its direction and magnitude and its point of application are known from the conditions of static equilibrium. The conditions of static equilibrium for any number of forces in a plane, as is generally well known, are three, that is to say,

(1) $Y=0$, or the algebraic sum of all vertical forces acting on a body is equal to zero.
(2) $X=0$, or the algebraic sum of all horizontal forces acting on a body is equal to zero.
(3) $M=0$, or the algebraic sum of the moments of all forces is equal to zero.

The loads to which structures may be subjected are always given. The other external forces are the reactions due to the loads. The reactions are exerted by the supports of the structure, and in order that they may be determined from the statical conditions the total number of unknowns must not exceed three. The ordinary trusses without redundant members are always statically determinate if a frictionless pin is used at each joint and if in the determination the effect of the longitudinal deformation of members on the stresses is neglected. If a case in which two members meet at a joint is considered as shown in Fig. 1, two unknown forces exist at the joint, the vertical

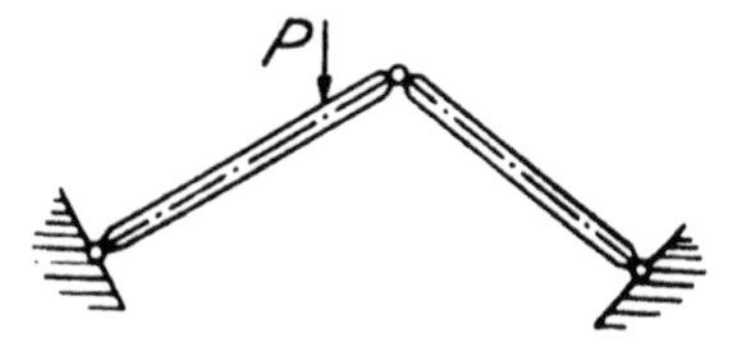

FIG. 1. SIMPLE HINGED FRAME WITH CONCENTRATED LOAD

and horizontal forces, and therefore the total number of unknown forces due to the external force P is six. But each member will give three statical conditions as stated before, and therefore this is a statically determinate system.

In studying the behavior of statically determinate and statically indeterminate systems, the conception of the connection of members by means of joint bars is a convenience. In Fig. 2 (a) the two members are not connected. They are entirely free to move horizontally and

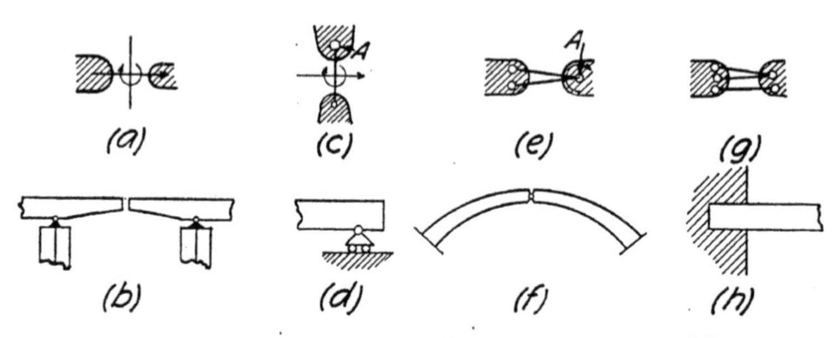

FIG. 2. ILLUSTRATIONS OF DEGREE OF INDETERMINATENESS BY MEANS OF JOINT BARS

vertically and also are free in rotation; accordingly it may be called the arrangement having three freedoms in motion. An example of this arrangement is a touching joint between the free ends of cantilever beams as shown in Fig. 2 (b). In Fig. 2 (c) two members are connected by a single bar, and they are prevented from moving vertically, but are free to move horizontally and to rotate about a point A. This may be called the arrangement having two freedoms in motion. An example of this arrangement is the frictionless roller end of a cantilever as shown in Fig. 2 (d). In Fig. 2 (e) two members are connected by two connecting bars and have only one freedom in motion, that is, the rotation of a member about A, the intersecting point of two bars. The crown hinge of an arch, Fig. 2 (f), is an example. In Fig. 2 (g), two members are connected by three joint-bars, and may be called a rigid connection which allows no freedom of motion. The restrained end of a cantilever beam, Fig. 2 (h), is an example of this arrangement.

To make a rigid joint it is necessary to have three joint bars at each connection between members, and $3S$ conditions of equilibrium must be set up to determine $3S$ unknowns when the structure is composed of S members. If the structure is rigidly connected to the ground, more conditions than $3S$ are required. Let a be the number of joint-bars needed to connect one member to another, and b be the number of joint bars needed to connect the member to the ground; then from the existing $3S$ conditions the following relation is necessary to make the structure statically determinate,

$$a+b=3S$$

When the members in the structure are all rigidly connected to each other, $a+b$ always exceeds $3S$ and therefore the case becomes a statically indeterminate system in which

$$a+b-3S=m$$

where m represents the number of the statically indeterminate forces. Such a system is called m-fold statically indeterminate, and m additional equations of condition are necessary to determine these unknowns.

Fig. 3 gives a few examples.

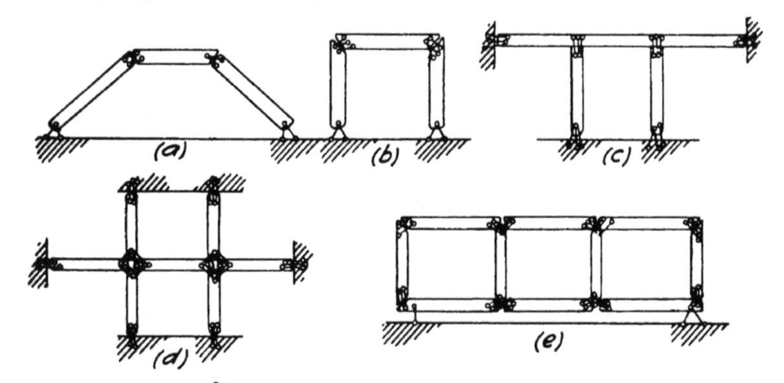

FIG. 3. TYPES OF FRAMES OF VARIOUS DEGREES OF INDETERMINATENESS

Case a. $a+b-3S=5+4-9=0$, therefore statically determinate.

Case b. $a+b-3S=6+4-9=1$, 1-fold statically indeterminate.

Case c. $a+b-3S=6+12-9=9$, 9-fold statically indeterminate.

Case d. $a+b-3S=18+18-21=15$, 15-fold statically indeterminate.

Case e. $a+b-3S=36+3-30=9$, 9-fold statically indeterminate.

In general the change in length of a member due to the direct stresses in the member will have an effect on the magnitude of the stresses developed. However, as is shown in a later paragraph, this effect is very slight in all ordinary forms of construction and has been neglected in stating the method of determining the degree of indeterminateness. Consequently in any member in which the change in length due to the direct stress has an appreciable effect on the stress developed, the criterion stated does not apply. Fig. 4 illustrates such a case.

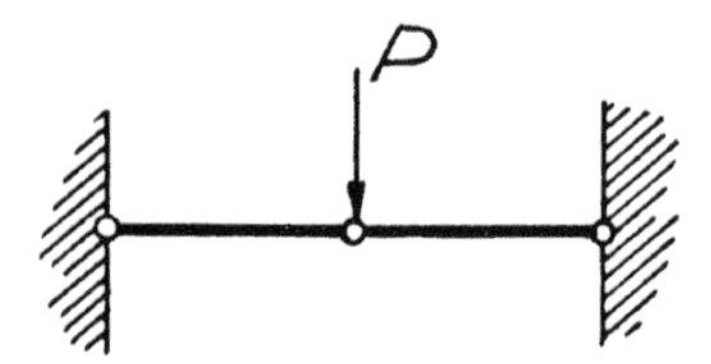

FIG. 4. TYPE OF STRUCTURE IN WHICH EFFECT OF NORMAL FORCE IS GREAT

6. *Principle of Least Work.*—By a law of nature, the principle of least work, the resisting forces will develop no more energy than the minimum which is necessary to maintain equilibrium with the external forces; or in other words, the external forces are so adjusted, among themselves, as to develop internal forces in the structure which will make the total internal work of resistance of the internal forces a minimum. When forces act upon an elastic system in which the deformations are proportional to the stresses, the principle of least work may be applied to determine the statically indeterminate forces.

The principle of least work has been known for a hundred years, but the first complete announcement of this theorem was given by Castigliano in 1879. Professor Cain expresses the principle in the following words:*

"The elastic forces experienced between the molecules after deformation correspond to a minimum of the work of deformation of the system, expressed as a function of certain stresses, taken with respect to these stresses successively, regarded as independent during the differentiation."

Professor Hiroi† translates Castigliano's expression of the principle of least work in the following words: "The partial derivatives

*See Trans. Amer. Soc. Civ. Eng., Vol. XXIV, p. 291.
†See Statically Indeterminate Stresses, by I. Hiroi.

of the work of resistance with respect to statically indeterminate forces which are so chosen that the forces themselves perform no work are equal to zero."

The total internal work may be subdivided into the parts due to bending moment, normal stress, and shearing stress.

The total work due to the bending moment M in a member will be

$$w_1=\int\frac{M^2dx}{2EI}$$

For the total internal work due to a total normal stress* N on a section

$$w_2=\int\frac{N^2dx}{2EA}$$

If shearing stress S is uniform over the cross-section, the expression for the internal work due to shearing stress is

$$w=\int\frac{S^2dx}{2GA}$$

where G expresses the shearing modulus of elasticity of a material. Since, however, the shearing stress is not uniform over the cross-sections, the expression for the internal work due to shear is modified, and

$$w_3=\int\frac{KS^2dx}{2GA}$$

where K is a known factor for a specified form of the cross-section. Therefore for the total work of resistance,

$$W=w_1+w_2+w_3=\frac{1}{2}\int\frac{M^2dx}{EI}+\frac{1}{2}\int\frac{N^2dx}{EA}+\frac{1}{2}\int\frac{KS^2dx}{GA}$$

Suppose that there are n statically indeterminate forces P_1, P_2,..... P_n in an elastic system. According to the theorem of Castigliano

$$\frac{\partial W}{\partial P_1}=\int\frac{M}{EI}\,\frac{\partial M}{\partial P_1}dx+\int\frac{N}{EA}\,\frac{\partial N}{\partial P_1}dx+\int\frac{KS}{GA}\,\frac{\partial S}{\partial P_1}dx=0$$

$$\frac{\partial W}{\partial P_2}=\int\frac{M}{EI}\,\frac{\partial M}{\partial P_2}dx+\int\frac{N}{EA}\,\frac{\partial N}{\partial P_2}dx+\int\frac{KS}{GA}\,\frac{\partial S}{\partial P_2}dx=0$$

. . . .

$$\frac{\partial W}{\partial P_n}=\int\frac{M}{EI}\,\frac{\partial M}{\partial P_n}dx+\int\frac{N}{EA}\,\frac{\partial N}{\partial P_n}dx+\int\frac{KS}{GA}\,\frac{\partial S}{\partial P_n}dx=0$$

*Normal to a cross-section of the member.

These furnish as many equations of condition as there are unknown quantities. The solution of these equations will give the exact formulas for statically indeterminate quantities.

7. *The Effect of Work of Normal Force on the Magnitude of Statically Indeterminate Forces.*—In the general equation

$$\int \frac{M}{EI}\,\frac{\partial M}{\partial P}dx + \int \frac{N}{EA}\,\frac{\partial N}{\partial P}dx + \int \frac{KS}{GA}\,\frac{\partial S}{\partial P}dx = 0$$

the second term will disappear when $\frac{\partial N}{\partial P}$ is equal to zero or, in other words, when the normal force* does not contain any of the statically indeterminate quantities.

When a frame is fixed at its column ends, the vertical and horizontal reactions and the bending moment at the fixed column ends are statically indeterminate. Generally the normal force in any member contains these reactions as factors. But if a frame having a single span is symmetrical in form and in the manner of loading, the vertical reactions become statically determinate, and therefore the horizontal reaction is the only indeterminate term which enters into the expression of the normal force. If, at the same time, the columns are vertical, the horizontal reactions do not affect the normal forces in the columns, and in the horizontal members only are the normal forces affected by a statically indeterminate force, namely, the horizontal reaction.

From these statements it will be seen that the form of frame which will be largely affected by the normal force is that having a sloped column under a vertical load.

The frame shown in Fig. 5 is used to illustrate the method of

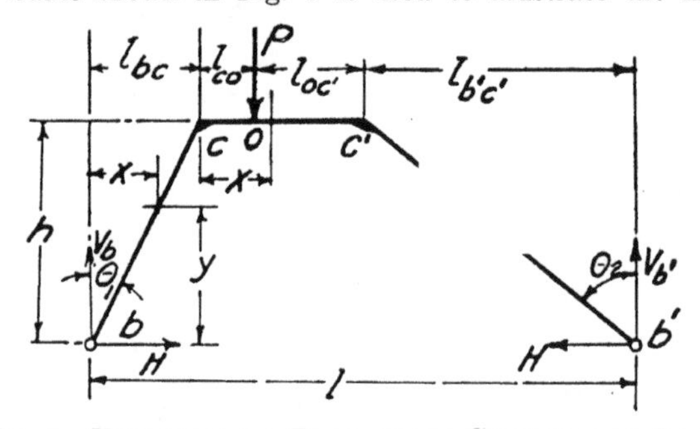

FIG. 5. UNSYMMETRICAL FRAME UNDER CONCENTRATED LOAD

*Normal to a cross-section of the member.

TABLE 1

ELEMENTS USED IN CONDITIONAL EQUATIONS FOR SOLUTION OF AN INDETERMINATE STRUCTURE

Member	Moment of Inertia	Limits of Integration	Bending Moment in the Member	Normal Forces in the Member	$\frac{\partial M}{\partial H}$	$\frac{\partial N}{\partial H}$
bc	I_{bc}	From zero to h	$V_b x - Hy = V_b y \tan\theta_1 - Hy$	$V_b \cos\theta_1 + H \sin\theta_1$	$-y$	$\sin\theta_1$
co	$I_{oc'}$	From zero to l_{co}	$V_b(l_{bc}+x) - Hh$	H	$-h$	$+1$
oc'	$I_{cc'}$	From l_{co} to $l_{oc'}$	$V_b(l_{bc}+x) - Hh - P(x - l_{co})$	H	$-h$	$+1$
$b'c'$	$I_{b'c'}$	From zero to h	$V_{b'} x - Hy = V_{b'} y \tan\theta_2 - Hy$	$V_{b'} \cos\theta_2 + H \sin\theta_2$	$-y$	$\sin\theta_2$

analysis and to bring out the effect of the work of the direct normal force on the magnitude of the statically indeterminate forces. In this case H is the only statically indeterminate force.

Taking the moment of all forces about b'

$$V_b\, l - P(l_{oc'} + l_{b'c'}) = 0 \text{ or } V_b = \frac{l_{b'c'} + l_{oc'}}{l} P = KP$$

and

$$V_{b'} = P - KP \text{ or } V_{b'} = (1-K)P \quad .$$

K being used as a general coefficient.

In general the internal work due to shearing stress may be neglected, and the general equation becomes

$$\int \frac{M}{EI} \frac{\partial M}{\partial H} dx + \int \frac{N}{EA} \frac{\partial N}{\partial H} dx = 0$$

All necessary elements in forming this equation are arranged in Table 1.

Inserting these values in the general equation gives the following expression, in which it is assumed that E and I are constant:

$$\frac{1}{EI_{bc}} \int_0^h (V_b \tan\theta_1 y - Hy)(-y) \sec\theta_1 dy + \frac{1}{EI_{co'}} \int_0^{l_{co}} (V_b l_{bc} + V_b x - Hh)(-h) dx$$

$$+ \frac{1}{EI_{cc'}} \int_{l_{co}}^{l_{cc'}} \left\{ V_b l_{bc} + V_b x - Hh - P(x - l_{co}) \right\} (-h) dx$$

$$+ \frac{1}{EI_{b'c'}} \int_0^h (V_{b'} \tan\theta_2 y - Hy)(-y) \sec\theta_2 dy$$

$$+ \frac{1}{EA_{bc}} \int_0^h (V_b \cos\theta_1 + H \sin\theta_1) \sin\theta_1 \sec\theta_1 dy + \frac{1}{EA_{co'}} \left\{ \int_0^{l_{co}} H dx + \int_{l_{co}}^{l_{cc'}} H dx \right\}$$

$$+ \frac{1}{EA_{b'c'}} \int_0^h \left\{ V_{b'} \cos\theta_2 + H \sin\theta_2 \right\} \sin\theta_2 \sec\theta_2 dy = 0$$

Integrating and simplifying this equation gives the following general expression for the statically indeterminate force H, in which the work of the normal force is fully counted:

$$= \frac{K\left[\frac{l_{bc}S_{bc}}{3I_{bc}}+\frac{2l_{bc}l_{cc'}+l^2_{cc'}}{2I_{cc'}}-\frac{l_{b'c'}S_{b'c'}}{3I_{b'c'}}-\frac{sin\,\theta_1}{A_{bc}}+\frac{sin\,\theta_2}{A_{b'c'}}\right]-\left[\frac{(l_{cc'}-l_{oc})^2}{2I_{cc'}}-\frac{S_{b'c'}l_{b'c'}}{3I_{b'c'}}+\frac{sin\,\theta_2}{A_{b'c'}}\right]}{h\left[\frac{S_{bc}}{3I_{bc}}+\frac{l_{cc'}}{I_{cc'}}+\frac{S_{b'c'}}{3I_{b'c'}}\right]+\left[\frac{sin\,\theta_1\,tan\,\theta_1}{A_{bc}}+\frac{sin\,\theta_2\,tan\,\theta_2}{A_{b'c'}}+\frac{l_{oc'}}{hA_{cc'}}\right]}P$$

where $K = \frac{l_{b'c'}+l_{oc'}}{l}$

In this formula, the terms which contain A_{bc}, $A_{cc'}$, and $A_{b'c'}$ enter because of taking into consideration the work of the normal forces represented by the second term of the general equation

$$\int\frac{M}{EI}\frac{\partial M}{\partial H}dx+\int\frac{N}{EA}\frac{\partial N}{\partial H}dx=0$$

If the effect of the normal force is neglected, then

$$H = \frac{K\left[\frac{l_{bc}S_{bc}}{3I_{bc}}+\frac{2l_{bc}l_{cc'}+l^2_{cc'}}{2I_{cc'}}-\frac{l_{b'c'}S_{b'c'}}{3I_{b'c'}}\right]-\left[\frac{(l_{cc'}-l_{oc})^2}{2I_{cc'}}-\frac{S_{b'c'}l_{b'c'}}{3I_{b'c'}}\right]}{h\left[\frac{S_{bc}}{3I_{bc}}+\frac{l_{cc'}}{I_{cc'}}+\frac{S_{b'c'}}{3I_{b'c'}}\right]}P$$

In most cases a value of θ greater than 30 degrees will not be used, because an increase in θ rapidly increases the horizontal reaction at the end of the column.

Assume a case in which $l_{bc}=l_{cc'}=l_{b'c'}=h=120$ inches. $\theta_1=\theta_2=$ 45 degrees, $A_{bc}=A_{cc'}=A_{b'c'}=10$ by 12 inches. $I_{bc}=I_{cc'}=I_{b'c'}=1{,}000$ in.^{4} $K=\frac{1}{2}$

When the effect of the work of the direct force is considered,

$$H=0.5637P$$

If the effect of the work of the direct force is neglected,

$$H=0.5643P$$

The difference $0.0006P$ ($0.0011H$) is inconsiderable.

In the foregoing example it is seen that the final formula is very much complicated by taking the direct force into consideration and that the effect of the internal work of all the direct stresses on the final value for statically indeterminate stresses is inconsiderable when compared with that of the bending moment. The work of the normal force is therefore disregarded in making the analyses of the frames treated in this bulletin.

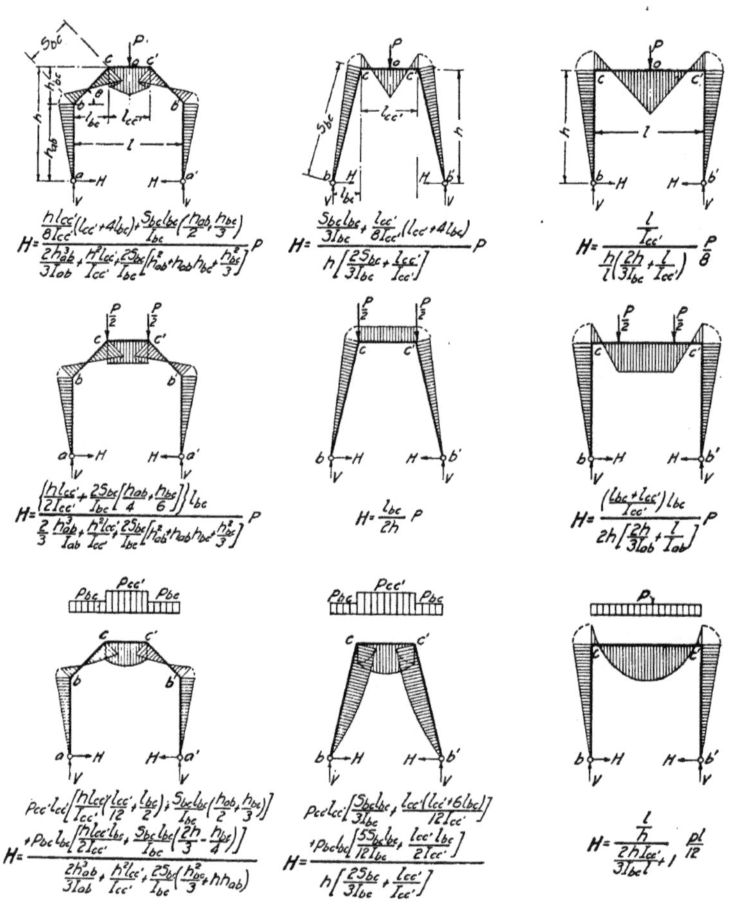

FIG. 6. SIMPLE FRAMES UNDER VERTICAL LOAD; LOWER ENDS OF COLUMNS HINGED

Attention is called to the fact that even though the effect of the internal work of the direct stresses may be neglected in determining the reactions the direct stresses themselves can not be neglected when calculating the total stress in any member. The direct stresses may be added algebraically after the statically indeterminate stresses are found.

The effect of the work of the deformation due to shear is generally so insignificant when compared with that due to the bending that it may be entirely neglected without sensible error in the calculation of the internal work.

8. *Simple Frames under Vertical Load.*—In Fig. 6 are given forms of a type of simple frame. The form at the left of the figure is the general form of the inverted U-frame, and the others are special cases of this frame. The column connections at the base are hinged and all other joints are rigid. Three forms of vertical loading are presented—a single concentrated load, two concentrated loads, and a uniform load. For this frame the statically indeterminate force is the horizontal reaction H. Formulas for H, derived by analysis using the principle of least work, are given in the figure for the three forms of frame and for the three loadings.

Knowing the horizontal reaction H, the bending moment and the forces at any section of the frame may be determined by the ordinary analytical method. For example, the bending moment at the middle of the top beam for the frames with two concentrated loads shown in Fig. 6 is equal to the algebraic sum of the moment of the horizontal reaction H with a moment arm equal to the height of the frame, the moment of the vertical reaction V about the section considered, and the moment of a load $\frac{P}{2}$ about the section.

As a specimen application of the method of using the principle of least work, the general solution of the frame and loading shown in the upper left-hand corner of Fig. 6 is given. The statically indeterminate force is H. The effect of normal forces being neglected, the general equation of condition is

$$\int \frac{M}{EI} \frac{\partial M}{\partial H} dx = 0$$

All quantities necessary in forming the conditional equation for this case are arranged in Table 2.

TABLE 2

ELEMENTS USED IN CONDITIONAL EQUATIONS FOR SOLUTION OF INVERTED U-FRAME

Member		I	M	$\frac{\partial M}{\partial H}$
ab	$a'b'$	I_{ab}	$-Hy$	$-y$
bc	$b'c'$	I_{bc}	$\frac{Px}{2} - H(h_{ab} + x \tan\theta)$	$-(h_{ab} + x \tan\theta)$
co	co'	$I_{cc'}$	$\frac{Px}{2} - Hh$	$-h$

Substituting the proper values in the general equation of condition

$$\frac{2}{EI_{ab}}\int_0^{h_{ab}} Hy^2dy - \frac{2\sec\theta}{EI_{bc}}\int_0^{l_{bc}}\left[\frac{Px}{2} - H(h_{ab} + x\tan\theta)\right](h_{ab} + x\tan\theta)\,dx$$

$$-\frac{2h}{EI_{cc'}}\int_{l_{bc}}^{l_{bc}+\frac{l_{cc'}}{2}}\left(\frac{Px}{2} - Hh\right)dx = 0 \quad . \quad . \quad . \quad . \quad . \quad . \quad . \quad (1)$$

Integrating equation (1)

$$\frac{2}{EI_{ab}}\left[\frac{Hy^3}{3}\right]_0^h - \frac{2\sec\theta}{EI_{bc}}\left[\frac{Px^2h_{ab}}{4} + \frac{Px^3}{6}\tan\theta - H(h_{ab}^2x + h_{ab}x^2\tan\theta + \frac{x^3}{3}\tan^2\theta)\right]_0^{l_{bc}}$$

$$-\frac{2h}{EI_{cc'}}\left[\frac{Px^2}{4} - Hhx\right]_{l_{bc}}^{l_{bc}+\frac{l_{cc'}}{2}} = 0 \quad . \quad . \quad . \quad . \quad . \quad . \quad . \quad . \quad (2)$$

Substituting the limits in equation (2)

$$\frac{2Hh_{ab}^3}{3I_{ab}} - \frac{2\sec\theta}{I_{bc}}\left[\frac{Pl_{bc}^2h_{ab}}{4} + \frac{Pl_{bc}^3}{6}\tan\theta - H(h_{ab}^2l_{bc} + h_{ab}l_{bc}^2\tan\theta + \frac{l_{bc}^3}{3}\tan^2\theta)\right]$$

$$-\frac{2h}{I_{cc'}}\left[\frac{P}{4}\left(l_{bc}l_{cc'} + \frac{l_{cc'}^2}{4}\right) - Hh\frac{l_{cc'}}{2}\right] = 0 \quad . \quad . \quad . \quad . \quad . \quad . \quad . \quad . \quad (3)$$

Collecting the terms involving H and those involving P

$$\frac{2h_{ab}^3H}{3I_{ab}}+\frac{2\sec\theta}{I_{bc}}\left(h_{ab}^2l_{bc}+h_{ab}l_{bc}^2\tan\theta+\frac{l_{bc}^3}{3}\tan^2\theta\right)H+\frac{2h^2l_{cc'}H}{2I_{cc'}}$$

$$=\frac{2\sec\theta\ P}{I_{bc}}\left[\frac{l_{bc}^2h_{ab}}{4}+\frac{l_{bc}^3}{6}\tan\theta\right]+\frac{2h}{I_{cc'}}\frac{P}{4}\left(l_{bc}\ l_{cc'}+\frac{l_{cc'}^2}{4}\right)\quad . \quad . \quad . \quad (4)$$

Solving for H,

$$H=\frac{\dfrac{2\sec\theta}{I_{bc}}\left[\dfrac{l_{bc}^2h_{ab}}{4}+\dfrac{l_{bc}^3}{6}\tan\theta\right]+\dfrac{2h}{4I_{cc'}}\left(l_{bc}\ l_{cc'}+\dfrac{l_{cc'}^2}{4}\right)}{\dfrac{2h_{ab}^3}{3I_{ab}}+\dfrac{2\sec\theta}{I_{bc}}\left(h_{ab}^2l_{bc}+h_{ab}l_{bc}^2\tan\theta+\dfrac{l_{bc}^3}{3}\tan^2\theta\right)+\dfrac{h^2l_{cc'}}{I_{cc'}}}P\quad . \quad . \quad . \quad (5)$$

Substituting $l_{bc}\sec\theta=S_{bc}$ and $h_{bc}=l_{bc}\tan\theta$

$$H=\frac{\dfrac{hl_{cc'}}{8I_{cc'}}\left(4l_{bc}+l_{cc'}\right)+\dfrac{S_{bc}\ l_{bc}}{I_{bc}}\left[\dfrac{h_{ab}}{2}+\dfrac{h_{bc}}{3}\right]}{\dfrac{2h_{ab}^3}{3I_{ab}}+\dfrac{h^2l_{cc'}}{I_{cc'}}+\dfrac{2S_{bc}}{I_{bc}}\left(h_{ab}^2+h_{ab}h_{bc}+\dfrac{h_{bc}^2}{3}\right)}P\quad . \quad . \quad . \quad . \quad . \quad . \quad . \quad . \quad (6)$$

Fig. 7 gives sketches of the inverted U-frame having the lower ends of the columns fixed. Equations of the statically indeterminates, the horizontal reaction, and the bending moment at the lower end of the column, as determined from analyses, are given in the figure.

Knowing these indeterminates, the moments, and the forces at any section of the frame may be determined by ordinary analysis. Thus the moment at the middle of the top beam for the frame in the upper left-hand corner of Fig. 7 is equal to the algebraic sum of the bending moment at the lower end of the column, M_a, the moment of the horizontal reaction H with a moment arm equal to the height of the frame, the moment of the vertical reaction V about the section considered, and the moment of one load $\frac{P}{2}$ about the section. Fig. 7 indicates the manner in which the moment varies along the members composing the frame.

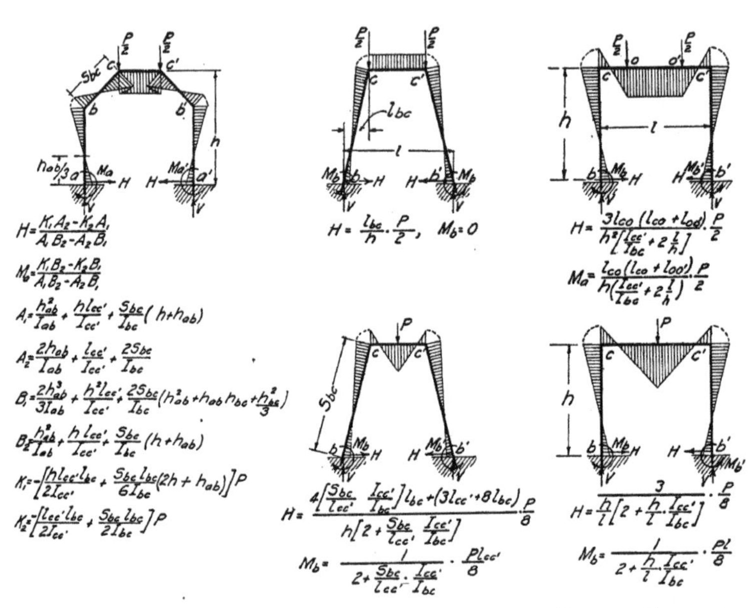

FIG. 7. SIMPLE FRAMES UNDER VERTICAL LOAD; LOWER ENDS OF COLUMNS FIXED

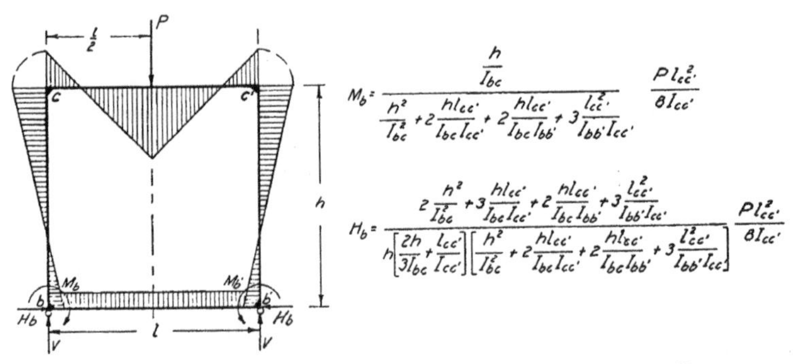

FIG. 8. RECTANGULAR FRAME WITH RIGIDLY CONNECTED TIE AT BASE OF COLUMNS

Fig. 8 shows a rectangular frame in which all the joints are rigid, the horizontal cross tie at the bottom of the frame having rigid connection to the columns. The equations of the indeterminates M_b and H_b

for this frame are given in the figure. The figure indicates the manner in which the moment varies along the members.

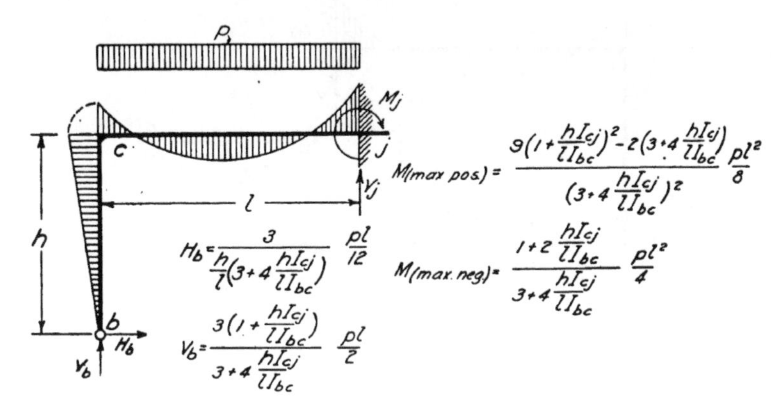

FIG. 9. L-FRAME WITH COLUMN HINGED AT SUPPORT

Fig. 9 shows an unsymmetrical frame, here termed an L-frame, under uniform load and with the column hinged at the support. For this frame the horizontal and vertical reactions H_b and V_b at the column end are given in the figure.

The maximum positive bending moment in this frame occurs at the distance x from c where

$$x = \frac{3\left(1+\frac{hI_{cj}}{lI_{bc}}\right)}{3+\frac{4hI_{cj}}{lI_{bc}}}\ \frac{l}{2}$$

The value of the maximum positive bending moment is given in the figure. The maximum bending moment is the negative moment at the wall, which is also given in the figure.

Fig. 10 shows an L-frame in which the column is fixed at the support. On account of the fixity of the column there are three statically indeterminates for this frame, H_b, V_b, and M_b or M_j. The values of these are given in the figure.

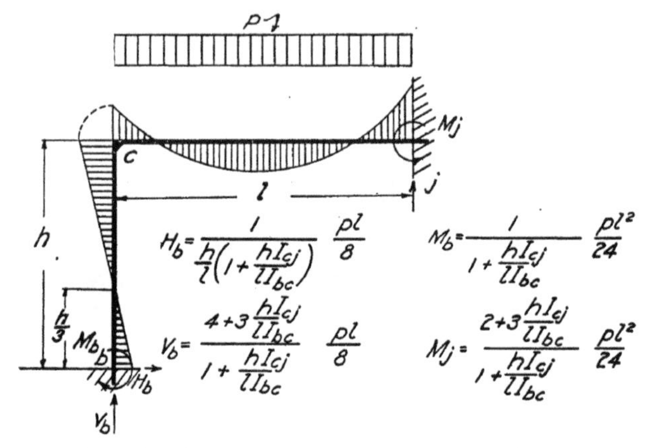

FIG. 10. L-FRAME WITH COLUMN FIXED AT SUPPORT

The formulas for the maximum positive moment in the beam and the distance of the section of maximum positive moment from c are

$$M_{max\ pos} = \frac{3\left(1+\frac{3hI_{cj}}{4lI_{bc}}\right)^2 - 2\left(1+\frac{hI_{cj}}{lI_{bc}}\right)}{\left(1+\frac{hI_{cj}}{lI_{bc}}\right)^2} \frac{pl^2}{24}$$

$$x = \frac{1+\frac{3hI_{cj}}{4lI_{bc}}}{1+\frac{hI_{cj}}{lI_{bc}}} \frac{l}{2}$$

The manner in which the bending moment varies along the members composing the frames is indicated in Fig. 9 and 10.

9. *Single Story Construction with Three Spans.*—In the design of a beam-and-girder or a flat-slab construction of a single story, many engineers do not take the effect of the bending of columns on the moments in other portions of the structure into consideration. Authors also have tried to analyze the stress distribution without taking this bending into account. Obviously a bending in the columns will allow an increased bending moment at the center of the span of a girder or slab loaded unsymmetrically with respect to the column, and stresses in the slab will be modified by variations in the ratio of the moment of inertia of the girder or slab to that of the column and of the column

height to the span length. Tests have shown that columns may be subjected to severe bending, and it will be seen that this will be more important for a single story structure than for others.

In actual cases, there may be twenty or more spans in succession, with different span lengths and different cross-sections of members, and consequently an exact analysis is hardly possible with any assumption.

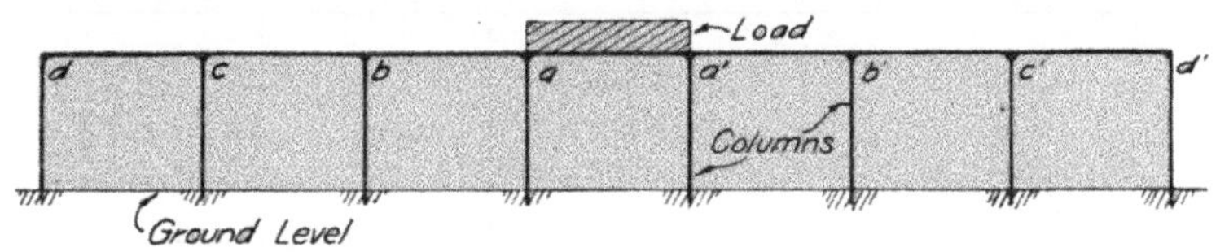

FIG. 11. CONTINUOUS SINGLE STORY SPANS; ONE PANEL LOADED

If panel aa' in Fig. 11 is loaded, the bending moments in slab bc, cd, $b'c'$, and $c'd'$ are so small as to be negligible in actual cases. It may be assumed, therefore, that the end condition of slabs or beams ab and $a'b'$ will, perhaps, be between the hinged and the fixed state at b and b', the degree of fixity depending upon the ratio of moments of inertia of the column and the slab at that joint. Formulas will, therefore, be given for both conditions.

Fig. 12 shows the manner in which the bending moment varies along the members composing the frame for four combinations of end

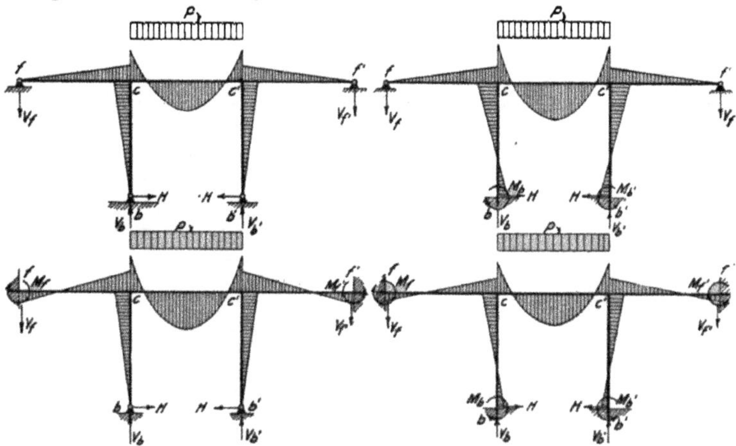

FIG. 12. SINGLE STORY THREE-SPAN FRAME, HAVING MIDDLE SPAN UNIFORMLY LOADED

conditions of beams and columns outside the loaded panel. The formulas which have been derived for these four cases are given in Table 3.

TABLE 3

FORMULAS FOR REACTIONS, BENDING MOMENTS, AND POINTS OF INFLECTION FOR SINGLE STORY THREE-SPAN FRAMES

	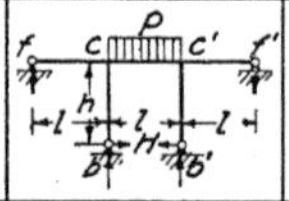	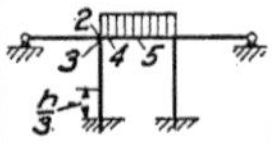	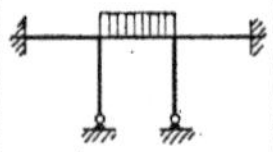	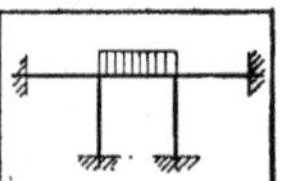
H Hor. reaction at col. ends	$\frac{1}{4n(3+5mn)}pl$	$\frac{1}{2n(4+5mn)}pl$	$\frac{1}{12n(1+2mn)}pl$	$\frac{1}{4n(2+3mn)}pl$
V_f Vert. reaction at beam ends	$\frac{mn}{4(3+5mn)}pl$	$\frac{mn}{4(4+5mn)}pl$	$\frac{mn}{6(1+2mn)}pl$	$\frac{mn}{4(2+3mn)}pl$
V_b Vert. reaction at col. ends	$\frac{6+11mn}{4(3+5mn)}pl$	$\frac{8+11mn}{4(4+5mn)}pl$	$\frac{3+7mn}{6(1+2mn)}pl$	$\frac{4+7mn}{4(2+3mn)}pl$
M_f Moment at beam ends	0	0	$\frac{mn}{(1+2mn)}\frac{pl^2}{18}$	$\frac{mn}{(2+3mn)}\frac{pl^2}{12}$
M_b Moment at column ends	0	$\frac{1}{(4+5mn)}\frac{pl^2}{6}$	0	$\frac{1}{(2+3mn)}\frac{pl^2}{12}$
M_2 Moment at point 2	$\frac{-mn}{(3+5mn)}\frac{pl^2}{4}$	$\frac{-mn}{(4+5mn)}\frac{pl^2}{4}$	$\frac{-mn}{(1+2mn)}\frac{pl^2}{9}$	$\frac{-mn}{(2+3mn)}\frac{pl^2}{6}$
M_3 Moment at point 3	$\frac{-1}{(3+5mn)}\frac{pl^2}{4}$	$\frac{-1}{(4+5mn)}\frac{pl^2}{3}$	$\frac{-1}{(1+2mn)}\frac{pl^2}{12}$	$\frac{-1}{(2+3mn)}\frac{pl^2}{6}$
M_4 Moment at point 4	$\frac{-(1+mn)}{(3+5mn)}\frac{pl^2}{4}$	$\frac{-(4+3mn)}{(4+5mn)}\frac{pl^2}{12}$	$\frac{-(3+4mn)}{3(1+2mn)}\frac{pl^2}{12}$	$\frac{-(1+mn)}{(2+3mn)}\frac{pl^2}{6}$
M_5 Moment at point 5	$\frac{(1+3mn)}{(3+5mn)}\frac{pl^2}{8}$	$\frac{(4+9mn)}{(4+5mn)}\frac{pl^2}{24}$	$\frac{(3+10mn)}{3(1+2mn)}\frac{pl^2}{24}$	$\frac{(2+5mn)}{(2+3mn)}\frac{pl^2}{24}$
Height of point of inflection in col.	0	$\frac{h}{3}$	0	$\frac{h}{3}$
Distance from column to point of inflect. in cent. span	$\frac{l}{2}\left[1\pm\sqrt{1-\frac{2(1+mn)}{(3+5mn)}}\right]$	$\frac{l}{2}\left[1\pm\sqrt{1-\frac{2(4+3mn)}{3(4+5mn)}}\right]$	$\frac{l}{2}\left[1\pm\sqrt{1-\frac{2(3+4mn)}{9(1+2mn)}}\right]$	$\frac{l}{2}\left[1\pm\sqrt{1-\frac{4(1+mn)}{3(2+3mn)}}\right]$
	$m=\frac{I_{cc'}}{I_{bc}}$	$n=\frac{h}{l}$		

TABLE 4

COEFFICIENTS OF BENDING MOMENT FOR SINGLE STORY THREE-SPAN FRAME

α = coefficient of pl^2 for bending moment at end of middle span.
β = coefficient of pl^2 for bending moment at top of column.

	Coefficient	Values of m $m=\frac{I_{cc'}}{I_{bc}}$	Values of n $n=\frac{h}{l}$					
			0.20	0.50	0.75	1.00	1.25	1.50
Extreme Ends of Columns and Beams Hinged	$\alpha=\frac{1+mn}{(3+5mn)4}$	0.5	.0785	.0735	.0705	.0682	.0664	.0648
		1.0	.0750	.0682	.0648	.0625	.0608	.0595
		1.5	.0722	.0648	.0615	.0595	.0582	.0570
		2.0	.0700	.0625	.0595	.0577	.0564	.0555
		2.5	.0682	.0608	.0582	.0564	.0554	.0546
		3.0	.0667	.0595	.0570	.0556	.0546	.0540
	$\beta=\frac{1}{(3+5mn)4}$	0.5	.0714	.0588	.0513	.0455	.0408	.0370
		1.0	.0625	.0455	.0370	.0312	.0270	.0238
		1.5	.0556	.0370	.0290	.0238	.0202	.0175
		2.0	.0500	.0312	.0238	.0192	.0161	.0139
		2.5	.0455	.0270	.0202	.0161	.0134	.0115
		3.0	.0417	.0238	.0175	.0139	.0115	.0098
Ext. Ends of Beams Hinged Lower Ends of Columns Fixed	$\alpha=\frac{4+3mn}{(4+5mn)12}$	0.5	.0796	.0754	.0728	.0706	.0687	.0672
		1.0	.0767	.0706	.0672	.0647	.0630	.0616
		1.5	.0743	.0672	.0637	.0616	.0600	.0587
		2.0	.0723	.0648	.0616	.0595	.0581	.0572
		2.5	.0706	.0630	.0600	.0581	.0568	.0559
		3.0	.0690	.0616	.0588	.0571	.0559	.0550
	$\beta=\frac{1}{(4+5mn)3}$	0.5	.0741	.0635	.0575	.0513	.0468	.0430
		1.0	.0667	.0513	.0430	.0371	.0326	.0290
		1.5	.0607	.0430	.0346	.0290	.0250	.0219
		2.0	.0555	.0371	.0290	.0238	.0202	.0175
		2.5	.0513	.0326	.0250	.0202	.0170	.0147
		3.0	.0477	.0290	.0219	.0175	.0147	.0126
Extreme Ends of Columns and Beams Fixed	$\alpha=\frac{1+mn}{(2+3mn)6}$	0.5	.0798	.0757	.0732	.0714	.0700	.0686
		1.0	.0774	.0714	.0686	.0667	.0652	.0640
		1.5	.0745	.0686	.0658	.0640	.0628	.0619
		2.0	.0728	.0667	.0640	.0624	.0614	.0606
		2.5	.0714	.0652	.0626	.0614	.0604	.0598
		3.0	.0702	.0640	.0619	.0606	.0598	.0591
	$\beta=\frac{1}{(2+3mn)6}$	0.5	.0695	.0556	.0476	.0416	.0370	.0333
		1.0	.0595	.0416	.0333	.0278	.0238	.0208
		1.5	.0521	.0333	.0256	.0208	.0175	.0151
		2.0	.0463	.0278	.0208	.0167	.0139	.0119
		2.5	.0417	.0238	.0175	.0139	.0115	.0098
		3.0	.0379	.0208	.0151	.0119	.0098	.0077

In Table 4 are given values of the bending moment coefficients for both hinged and fixed ends at top of column and at end of middle span for six ratios of moments of inertia.

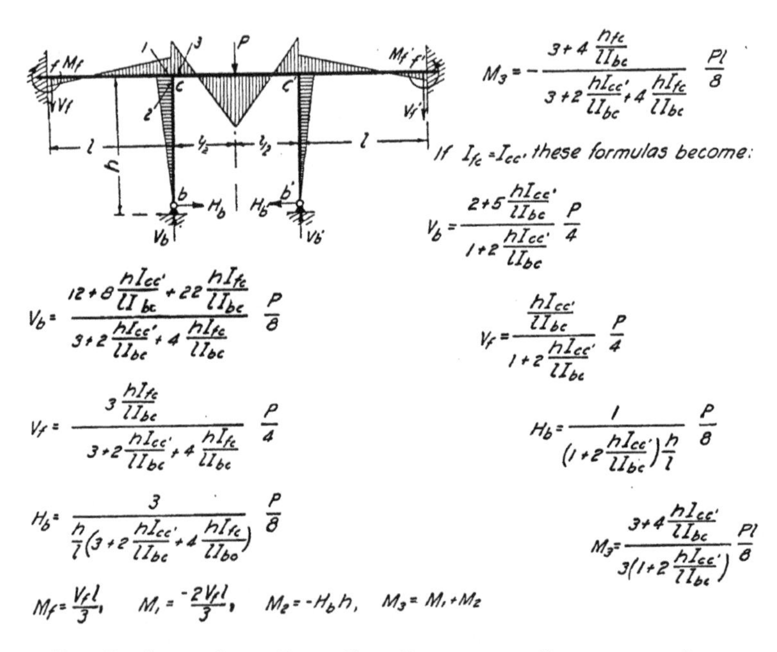

FIG. 13. SINGLE STORY THREE-SPAN FRAME UNDER CONCENTRATED LOAD

For a three-span frame with concentrated load at the center of the middle panel, Fig. 13 shows the manner in which the moment varies along the several members composing the frame.

Assuming symmetry about the vertical center line of this frame the formulas for the horizontal and vertical reactions are given in Fig. 13, and also formulas for bending moments at top of column and at end of middle span.

10. *Trestle Bent with Tie.*—The frame shown in Fig. 14, frequently termed the A-frame, may be used in trestle construction. Formulas for $M_{cc'}$, H_b, and H_c are given in the figure.

Knowing the values of $M_{cc'}$, H_c, and H_b, the stresses at any section of the frame may be computed. When $\theta = 0$, $I_{bc} = 0$, and l_{bc} approaches 0, making $h_{cd} = h$ (that is when $k = 1$), this frame becomes the same as that shown in Fig. 8 and the formulas for $M_{cc'}$ and H_c reduce to the same form as those given in Fig. 8.

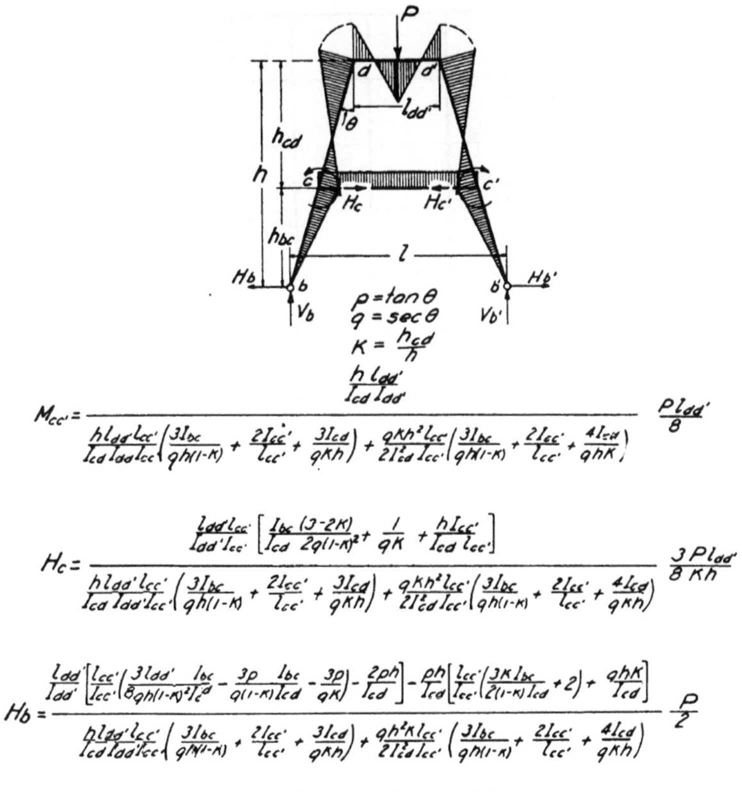

FIG. 14. A-FRAME SHOWING FORM OF MOMENT CURVE

11. *Building Construction with Several Stories and Spans.*—In the actual construction of buildings of reinforced concrete, it is common to use a continuous slab for floors supported by a number of columns. A loading which produces serious bending in columns is, of course, an eccentric arrangement, such as is shown in Fig. 15. The moments of inertia of columns are sometimes smaller than those of slabs. Accordingly, the bending moment in the floor slab $fcc'f'$ is greatly modified by the flexure of columns bc, $b'c'$, cd, and $c'd'$.

In present practice, frequently little attention is paid to this point, and columns are assumed to be rigid enough to resist the bending. This assumption may be approximately true for the lower stories,

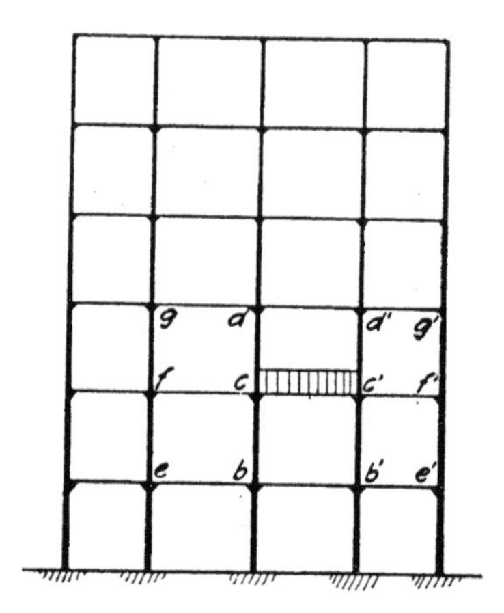

FIG. 15. CROSS-SECTION OF BUILDING FRAME WITH SINGLE PANEL LOADED

where the columns have large diameters, but it is not true for the upper stories, where the cross-section of columns is usually small, and serious bending stress may exist in the column due to eccentric loading. An exact analysis is hardly possible because there are many unknown conditions entering into the solution. From a practical standpoint, it is easily understood that the bending moment in floor slabs, *gdd'g'* and *ebb'e'*, (see Fig. 15) due to the load on the floor *cc'* is so small as to be inconsiderable if the floors are of moderate thickness. That is, the columns *cb*, *c'b'*, *cd*, and *c'd'* are practically fixed at *b*, *b'*, *d*, and *d'*, respectively. If the floor slabs are not thick enough to keep the

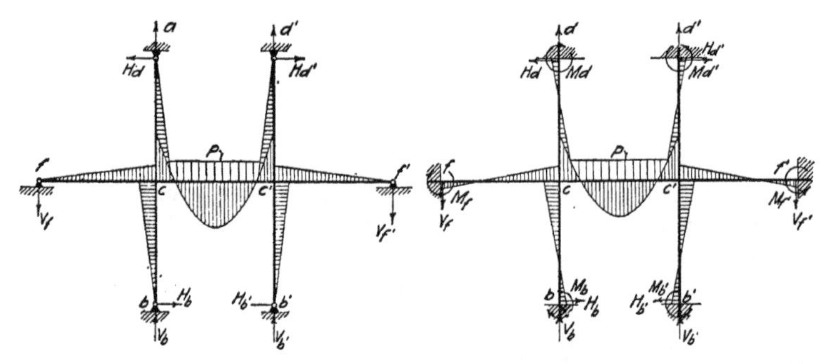

FIG. 16. TWO-STORY THREE-SPAN FRAME HAVING MIDDLE SPAN UNIFORMLY LOADED

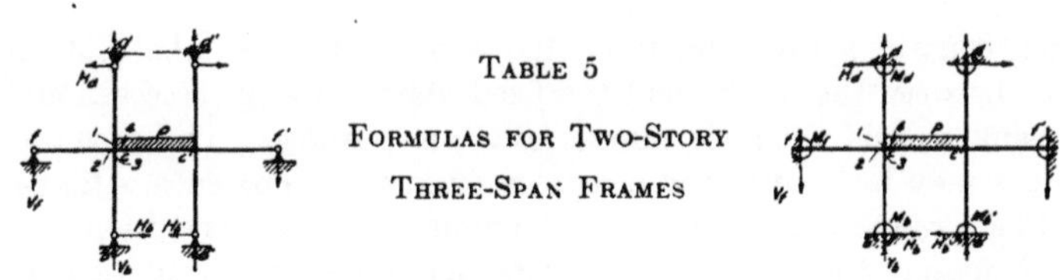

TABLE 5

FORMULAS FOR TWO-STORY THREE-SPAN FRAMES

ENDS HINGED

	GENERAL CASE (a)	SPECIAL CASE (b) $l_{cc'}=l_{fc}=l.\ I_{cc'}=I_{fc}$	SPECIAL CASE (c) $h_{cd}=h_{bc}=l_{fc}=l_{cc'}=l$, $I_{cc'}=I_{bc}=I_{fc}=I_{cd}$
H_b = Horizontal reaction at b	$\frac{1}{\frac{h_{bc}}{l_{cc'}}\left(3+3\frac{h_{bc}I_{cd}}{h_{cd}I_{bc}}+3\frac{h_{bc}I_{fc}}{l_{fc}I_{bc}}+2\frac{h_{bc}I_{cc'}}{l_{cc'}I_{bc}}\right)}\ \frac{pl_{cc'}}{4}$	$\frac{1}{\frac{h_{bc}}{l}\left(3+3\frac{h_{bc}I_{cd}}{h_{cd}I_{bc}}+5\frac{h_{bc}I_{cc'}}{lI_{bc}}\right)}\ \frac{pl}{4}$	$\frac{pl}{44}$
H_d = Horizontal reaction at d	$\frac{1}{\frac{h_{cd}}{l_{cc'}}\left(3+3\frac{h_{cd}I_{bc}}{h_{bc}I_{cd}}+3\frac{h_{cd}I_{fc}}{l_{fc}I_{cd}}+2\frac{h_{cd}I_{cc'}}{l_{cc'}I_{cd}}\right)}\ \frac{pl_{cc'}}{4}$	$\frac{1}{\frac{h_{cd}}{l}\left(3+3\frac{h_{cd}I_{bc}}{h_{bc}I_{cd}}+5\frac{h_{cd}I_{cc'}}{lI_{cd}}\right)}\ \frac{pl}{4}$	$\frac{pl}{44}$
V_f = Vertical reaction at f	$\frac{1}{\frac{l_{fc}}{l_{cc'}}\left(3+3\frac{l_{fc}I_{cd}}{h_{cd}I_{fc}}+3\frac{l_{fc}I_{bc}}{h_{bc}I_{fc}}+2\frac{l_{fc}I_{cc'}}{l_{cc'}I_{fc}}\right)}\ \frac{pl_{cc'}}{4}$	$\frac{1}{5+3\frac{l\,I_{cd}}{h_{cd}I_{cc'}}+3\frac{l\,I_{bc}}{h_{bc}I_{cc'}}}\ \frac{pl}{4}$	$\frac{pl}{44}$
V_b = Vertical reaction at b	$\frac{6+\frac{l_{cc'}}{l_{fc}}+6\frac{l_{fc}I_{cd}}{h_{cd}I_{fc}}+6\frac{l_{fc}I_{bc}}{h_{bc}I_{fc}}+4\frac{l_{fc}I_{cc'}}{l_{cc'}I_{fc}}}{3+3\frac{l_{fc}I_{cd}}{h_{cd}I_{fc}}+3\frac{l_{fc}I_{bc}}{h_{bc}I_{fc}}+2\frac{l_{fc}I_{cc'}}{l_{cc'}I_{fc}}}\ \frac{pl_{cc'}}{8}$	$\frac{11+6\frac{l}{I_{cc'}}\left(\frac{I_{cd}}{h_{cd}}+\frac{I_{bc}}{h_{bc}}\right)}{5+3\frac{l}{I_{cc'}}\left(\frac{I_{cd}}{h_{cd}}+\frac{I_{bc}}{h_{bc}}\right)}\ \frac{pl}{8}$	$23\frac{pl}{88}$
M_1 = Bending moment at 1	$\frac{1}{3+3\frac{h_{cd}I_{bc}}{h_{bc}I_{cd}}+3\frac{h_{cd}I_{fc}}{l_{fc}I_{cd}}+2\frac{h_{cd}I_{cc'}}{l_{cc'}I_{cd}}}\ \frac{pl^2_{cc'}}{4}$	$\frac{1}{3+3\frac{h_{cd}I_{bc}}{h_{bc}I_{cd}}+5\frac{h_{cd}I_{cc'}}{lI_{cd}}}\ \frac{pl^2}{4}$	$\frac{pl^2}{44}$
M_2 = Bending moment at 2	$\frac{1}{3+3\frac{l_{fc}I_{cd}}{h_{cd}I_{fc}}+3\frac{l_{fc}I_{bc}}{h_{bc}I_{fc}}+2\frac{l_{fc}I_{cc'}}{l_{cc'}I_{fc}}}\ \frac{pl^2_{cc'}}{4}$	$\frac{1}{5+3\frac{l}{I_{cc'}}\left(\frac{I_{cd}}{h_{cd}}+\frac{I_{bc}}{h_{bc}}\right)}\ \frac{pl^2}{4}$	$\frac{pl^2}{44}$
M_3 = Bending moment at 3	$\frac{1}{3+3\frac{h_{bc}I_{cd}}{h_{cd}I_{bc}}+3\frac{h_{bc}I_{fc}}{l_{fc}I_{bc}}+2\frac{h_{bc}I_{cc'}}{l_{cc'}I_{bc}}}\ \frac{pl^2_{cc'}}{4}$	$\frac{1}{3+3\frac{h_{bc}I_{cd}}{h_{cd}I_{bc}}+5\frac{h_{bc}I_{cc'}}{lI_{bc}}}\ \frac{pl^2}{4}$	$\frac{pl^2}{44}$
M_4 = Bending moment at 4	$M_1+M_2+M_3$	$M_1+M_2+M_3$	$3\frac{pl^2}{44}$

ENDS FIXED

	GENERAL CASE (a)	SPECIAL CASE (b) $l_{cc'}=l_{fc}=l.\ I_{cc'}=I_{fc}$	SPECIAL CASE (c) $h_{cd}=h_{bc}=l_{fc}=l_{cc'}=l$, $I_{cc'}=I_{fc}=I_{bc}=I_{cd}$
M_b = Moment at col ends b & b'	$\frac{1}{2+2\frac{h_{bc}I_{fc}}{l_{fc}I_{bc}}+2\frac{h_{bc}I_{cd}}{h_{cd}I_{bc}}+\frac{h_{bc}I_{cc'}}{l_{cc'}I_{bc}}}\ \frac{pl^2_{cc'}}{12}$	$\frac{1}{2+2\frac{h_{bc}I_{cd}}{h_{cd}I_{bc}}+3\frac{h_{bc}I_{cc'}}{lI_{bc}}}\ \frac{pl^2}{12}$	$\frac{pl^2}{84}$
M_d = Moment at col ends d & d'	$\frac{1}{2+2\frac{h_{cd}I_{fc}}{l_{fc}I_{cd}}+2\frac{h_{cd}I_{bc}}{h_{bc}I_{cd}}+\frac{h_{cd}I_{cc'}}{l_{cc'}I_{cd}}}\ \frac{pl^2_{cc'}}{12}$	$\frac{1}{2+2\frac{h_{cd}I_{bc}}{h_{bc}I_{cd}}+3\frac{h_{cd}I_{cc'}}{lI_{cd}}}\ \frac{pl^2}{12}$	$\frac{pl^2}{84}$
M_f = Moment at beam ends f & f'	$\frac{1}{2+2\frac{l_{fc}I_{bc}}{h_{bc}I_{fc}}+2\frac{l_{fc}I_{cd}}{h_{cd}I_{fc}}+\frac{l_{fc}I_{cc'}}{l_{cc'}I_{fc}}}\ \frac{pl^2_{cc'}}{12}$	$\frac{1}{2\frac{l}{I_{cc'}}\left(\frac{I_{bc}}{h_{bc}}+\frac{I_{cd}}{h_{cd}}\right)+3}\ \frac{pl^2}{12}$	$\frac{pl^2}{84}$
H_b = Horizontal reaction at b & b'	$\frac{1}{\frac{h_{bc}}{l_{cc'}}\left(2+2\frac{h_{bc}I_{fc}}{l_{fc}I_{bc}}+2\frac{h_{bc}I_{cd}}{h_{cd}I_{bc}}+\frac{h_{bc}I_{cc'}}{l_{cc'}I_{bc}}\right)}\ \frac{pl^2_{cc'}}{4}$	$\frac{1}{\frac{h_{bc}}{l}\left(2+2\frac{h_{bc}I_{cd}}{h_{cd}I_{bc}}+3\frac{h_{bc}I_{cc'}}{lI_{bc}}\right)}\ \frac{pl}{4}$	$\frac{pl}{28}$
H_d = Horizontal reaction at d & d'	$\frac{1}{\frac{h_{cd}}{l_{cc'}}\left(2+2\frac{h_{cd}I_{fc}}{l_{fc}I_{cd}}+2\frac{h_{cd}I_{bc}}{h_{bc}I_{cd}}+\frac{h_{cd}I_{cc'}}{l_{cc'}I_{cd}}\right)}\ \frac{pl^2_{cc'}}{4}$	$\frac{1}{\frac{h_{cd}}{l}\left(2+2\frac{h_{cd}I_{bc}}{h_{bc}I_{cd}}+3\frac{h_{cd}I_{cc'}}{lI_{cd}}\right)}\ \frac{pl}{4}$	$\frac{pl}{28}$
V_f = Vertical reaction at f & f'	$\frac{1}{\frac{l_{fc}}{l_{cc'}}\left(2+2\frac{l_{fc}I_{bc}}{h_{bc}I_{fc}}+2\frac{l_{fc}I_{cd}}{h_{cd}I_{fc}}+\frac{l_{fc}I_{cc'}}{l_{cc'}I_{fc}}\right)}\ \frac{pl_{cc'}}{4}$	$\frac{1}{3+2\frac{l}{I_{cc'}}\left(\frac{I_{bc}}{h_{bc}}+\frac{I_{cd}}{h_{cd}}\right)}\ \frac{pl}{4}$	$\frac{pl}{28}$
V_b = Vertical reaction at b & b'	$\frac{4+\frac{l_{cc'}}{l_{fc}}+4\frac{l_{fc}I_{bc}}{h_{bc}I_{fc}}+4\frac{l_{fc}I_{cd}}{h_{cd}I_{fc}}+2\frac{l_{fc}I_{cc'}}{l_{cc'}I_{fc}}}{2+2\frac{l_{fc}I_{bc}}{h_{bc}I_{fc}}+2\frac{l_{fc}I_{cd}}{h_{cd}I_{fc}}+\frac{l_{fc}I_{cc'}}{l_{cc'}I_{fc}}}\ \frac{pl_{cc'}}{8}$	$\frac{7+4\frac{l}{I_{cc'}}\left(\frac{I_{bc}}{h_{bc}}+\frac{I_{cd}}{h_{cd}}\right)}{2\frac{l}{I_{cc'}}\left(\frac{I_{bc}}{h_{bc}}+\frac{I_{cd}}{h_{cd}}\right)+3}\ \frac{pl}{8}$	$15\frac{pl}{56}$
M_1 Moment at 1	$2M_d$	$2M_d$	$\frac{pl^2}{42}$
M_2 " " 2	$2M_f$	$2M_f$	$\frac{pl^2}{42}$
M_3 " " 3	$2M_b$	$2M_b$	$\frac{pl^2}{42}$
M_4 " " 4	$2(M_b+M_f+M_d)$	$2(M_b+M_f+M_d)$	$\frac{3}{42}pl^2$

column ends in a fixed condition, the end condition of the columns will be between the hinged and the fixed state. Using these assumptions an analysis which is almost exact is possible. The resulting formulas may be used in the design of buildings. For such a frame, Fig. 16 shows the manner in which the moment varies along the members composing the frame for two combinations of end conditions of beams and columns. The formulas for the horizontal and vertical reactions and the bending moments are given in Table 5 for ends of columns and beams hinged and for ends of columns and beams fixed, the end spans being equal in both cases. Formulas are also given for equal spans and equal moments of inertia of the beams and for equal spans, story heights, and moments of inertia of beams and columns. Fig. 17 gives numerical values of the coefficients of pl^2 for the case in which ends of beams and columns are fixed, α being the coefficient of the bending moment at the top of the lower column and β the coefficient at the foot of the upper column. In Table 6 are given values

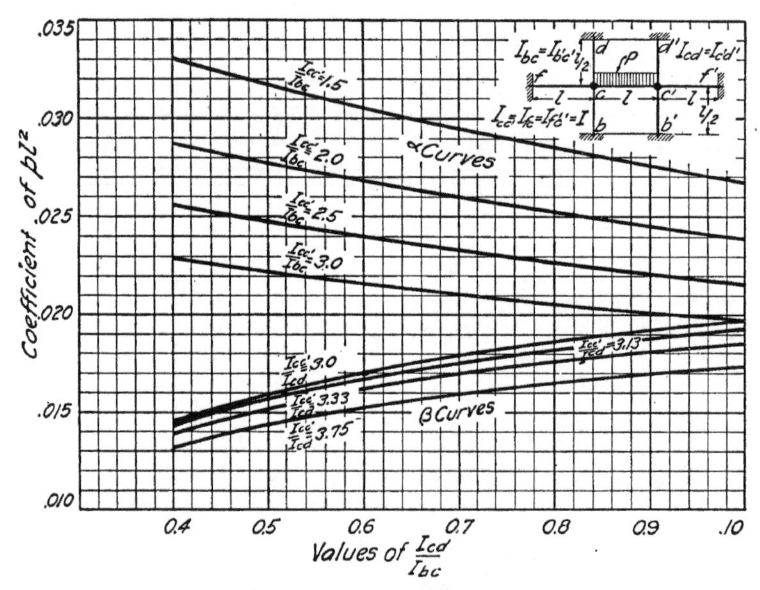

FIG. 17. COEFFICIENT OF BENDING MOMENTS FOR TOP OF LOWER COLUMN AND FOOT OF UPPER COLUMN FOR TWO-STORY THREE-SPAN FRAME HAVING ALL EXTERNAL CONNECTIONS FIXED

TABLE 6

COEFFICIENTS OF BENDING MOMENT FOR TWO-STORY THREE-SPAN FRAME WITH ENDS OF COLUMNS AND BEAMS FIXED

α = coefficient of pl^2 for bending moment at top of lower column.
β = coefficient of pl^2 for bending moment at foot of upper column.

	$\frac{I_{fc}}{I_{bc}}$	$\frac{I_{cd}}{I_{bc}}$			
		0.4	0.6	0.8	1.0
α	1.5	.0330	.0306	.0285	.0267
	2.0	.0287	.0269	.0253	.0238
	2.5	.0255	.0240	.0227	.0215
	3.0	.0228	.0216	.0206	.0196
	$\frac{I_{fc}}{I_{cd}}$				
β	3.75	.0132	.0153	.0165	.0173
	3.33	.0139	.0162	.0175	.0185
	3.13	.0143	.0167	.0181	.0192
	3.00	.0145	.0170	.0185	.0196

of the bending moment coefficients α and β for four ratios of the moments of inertia. The frame with hinged ends has nine statically indeterminate quantities, while the frame with fixed ends has fifteen statically indeterminates, but the condition of symmetrical loading shown greatly reduces the number of these quantities. In the analyses it has been assumed that the vertical reactions at b and d (also at b' and d') are the same. This assumption may not be the real condition in actual cases, but no effect is produced on bending moments by it.

12. *Frame with Three Spans.*—In bridge or trestle construction across a wide stream or valley, several spans may be built continuously as a monolith. Because of the necessity of providing expansion joints, the number of spans thus connected is frequently limited to three.

Rigidly connected frames with three spans, equal or unequal, may advantageously be used for bridges of moderate spans.

No analytical formulas for such frames have, to the writer's knowledge, been published. Fig. 18 and 19 give the formulas for a three-span frame under various conditions of load. Fig. 20 shows the manner in which the moment varies along the members composing the frames. Table 7 gives values of the bending moment coefficients

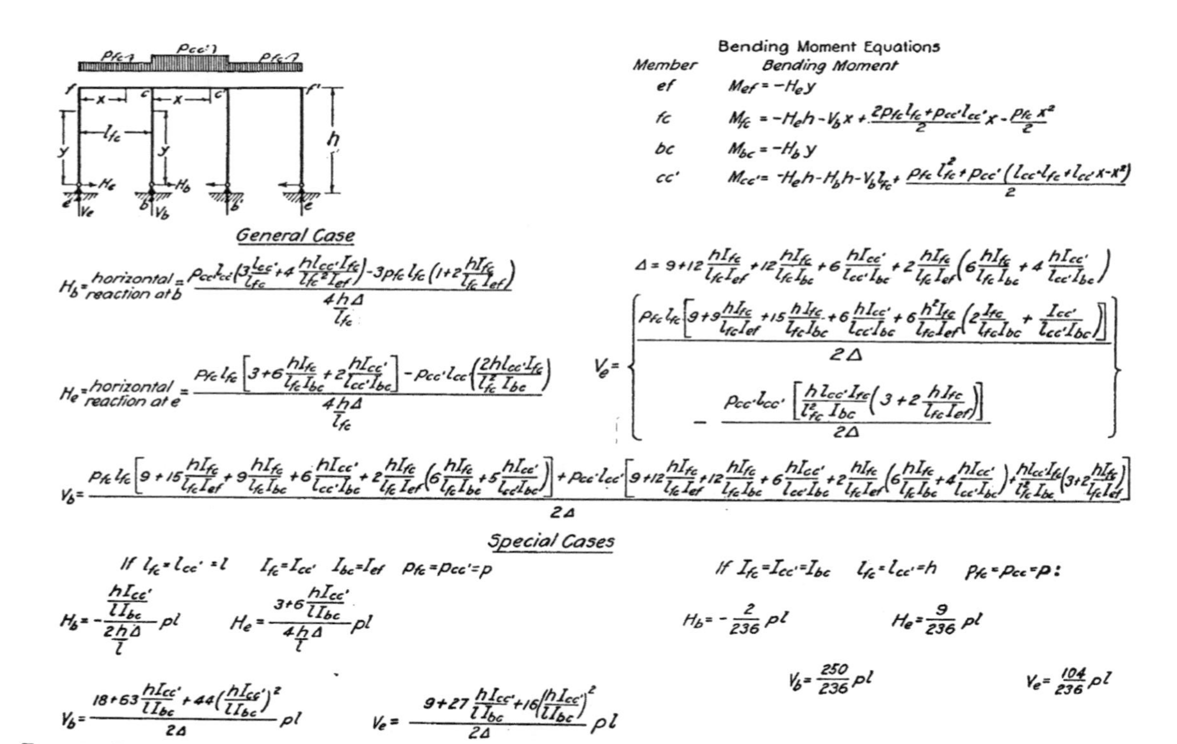

FIG. 18. FORMULAS FOR SINGLE STORY THREE-SPAN FRAME WITH FOUR COLUMNS HAVING COLUMNS HINGED AT LOWER END

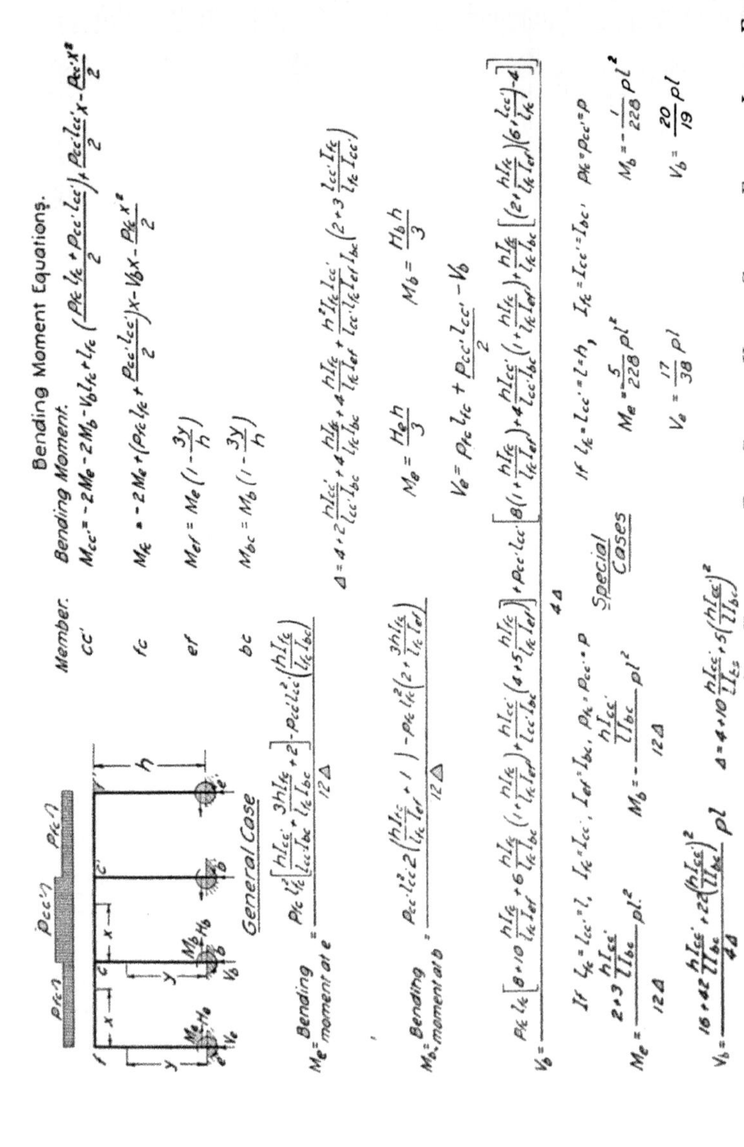

19. Formulas for Single Story Three-Span Frame with Four Columns Fixed at Lower End

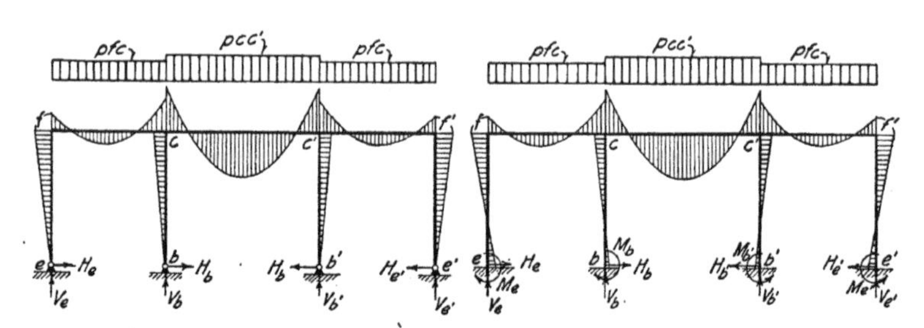

FIG. 20. SINGLE STORY THREE-SPAN FRAME WITH FOUR COLUMNS HAVING MIDDLE SPAN UNIFORMLY LOADED

TABLE 7

COEFFICIENTS OF BENDING MOMENT FOR SINGLE STORY THREE-SPAN FRAME HAVING FOUR COLUMNS

α = coefficient of pl^2 for bending moment at end of middle span.
β = coefficient of pl^2 for bending moment at top of intermediate column.

		$\frac{I_{cc}}{I_{bc}}$	$\frac{h}{l}$				
			0.50	0.75	1.00	1.25	1.50
Lower Ends of Columns Hinged	α	0.5	.0739	.0711	.0689	.0672	.0658
		1.0	.0689	.0658	.0634	.0620	.0605
		1.5	.0656	.0625	.0605	.0591	.0580
		2.0	.0634	.0605	.0587	.0573	.0565
		3.0	.0607	.0579	.0565	.0550	.0547
	β	0.5	.0563	.0488	.0431	.0387	.0351
		1.0	.0431	.0351	.0297	.0257	.0227
		1.5	.0351	.0276	.0227	.0194	.0169
		2.0	.0297	.0227	.0185	.0155	.0134
		3.0	.0228	.0169	.0134	.0112	.0096
Lower Ends of Columns Fixed	α	0.5	.0758	.0732	.0711	.0694	.0680
		1.0	.0711	.0680	.0658	.0640	.0626
		1.5	.0680	.0648	.0626	.0610	.0597
		2.0	.0658	.0626	.0605	.0592	.0580
		3.0	.0626	.0598	.0580	.0567	.0559
	β	0.5	.0610	.0542	.0488	.0444	.0408
		1.0	.0488	.0408	.0352	.0308	.0276
		1.5	.0408	.0330	.0276	.0238	.0210
		2.0	.0352	.0276	.0228	.0194	.0168
		3.0	.0276	.0210	.0168	.0142	.0122

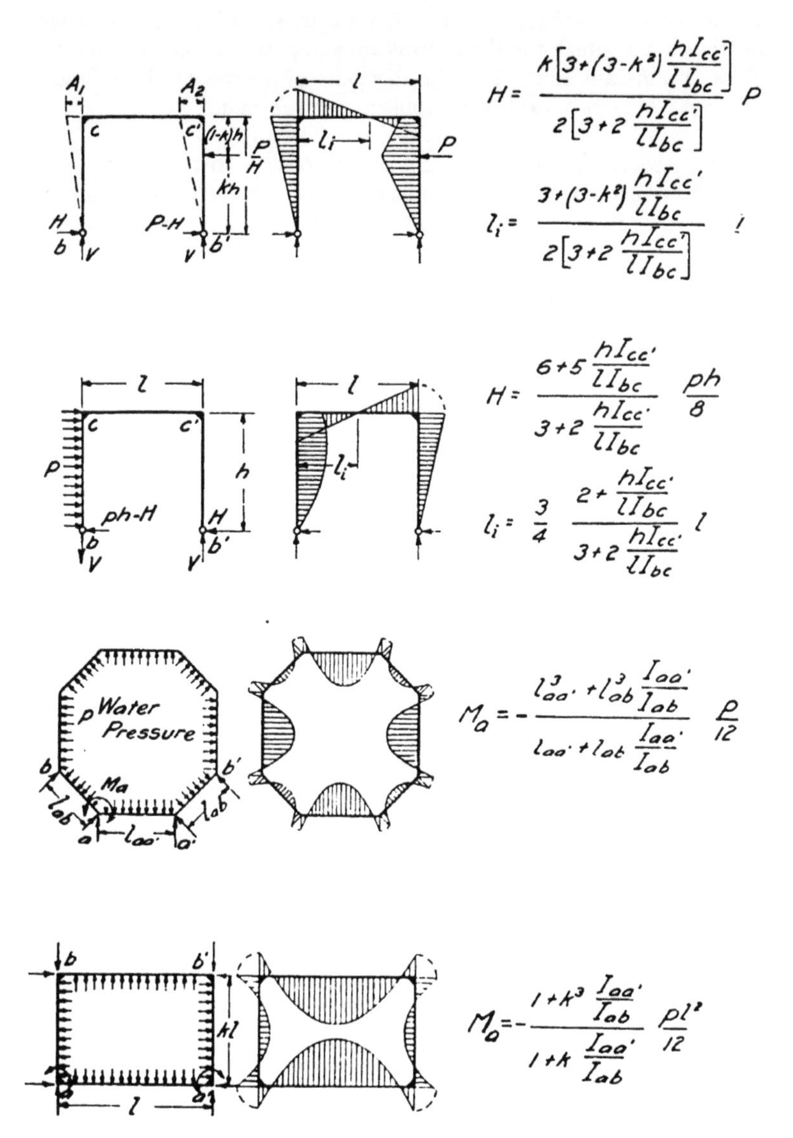

Fig. 21. Formulas for Rectangular Frame under Horizontal Load and for Octagonal and Rectangular Reservoirs

at the top of the middle columns and at the end of the middle span for the case in which the three spans are equal, the moments of inertia of the beams are equal, and the moments of inertia of the columns are equal, a uniform load being applied over the middle span.

13. *Square Frame under Horizontal Load.*—Hitherto only the cases in which the load was applied vertically have been discussed. It is frequently necessary to solve for the statically indeterminate stresses due to a horizontal force, such as a wind pressure or the braking

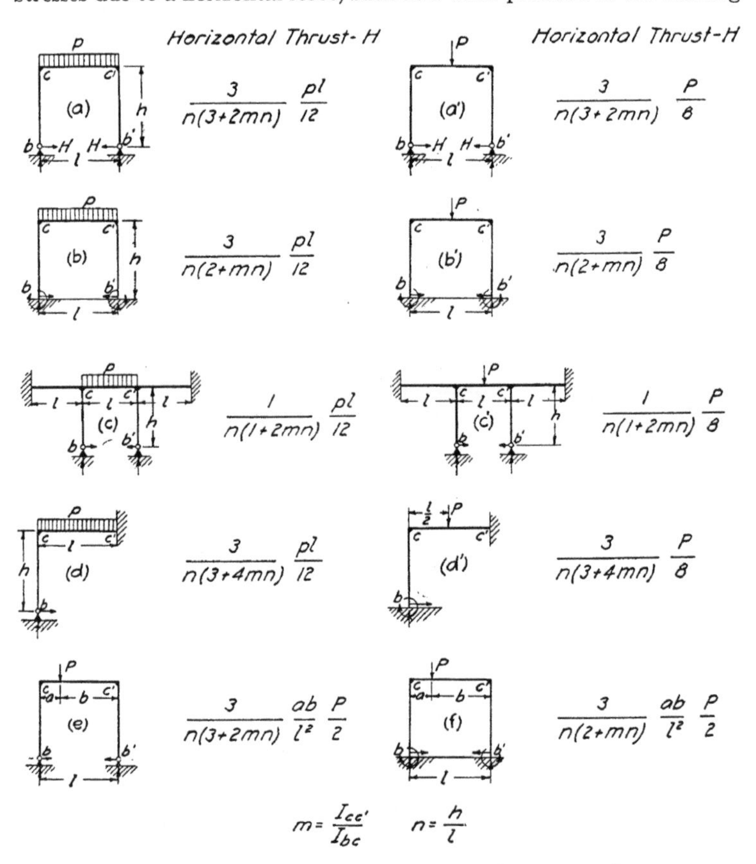

FIG. 22. HORIZONTAL REACTIONS FOR DISTRIBUTED LOADS AND CONCENTRATED LOADS

force of a locomotive. The method of determination of the statically unknowns is the same as that used for frames with vertical loads. A few cases have been taken as illustrations and the resulting equations are given in Fig. 21.

Another application is the water tank or reservoir subject to the static pressure of water, such as may be found in filter plants. Fig. 21 gives two examples of framed constructions of this character. It is seen that for a square tank having the same walls on the four sides the negative moment from the formula for rectangular tank becomes $\frac{1}{12}pl^2$ and the positive moment $\frac{1}{24}pl^2$, as is known from other sources.

14. *The Nature of the Resulting Formulas; Relation between Horizontal Reactions in Frame under Uniform Load and under Concentrated Load.*—It is interesting and important to note from the results of the foregoing analysis that there is a fixed relation between the horizontal reactions in the symmetrical frame under distributed loads and those in the same frame under concentrated loads. To show this relation a few cases have been selected as illustrative. These are shown in Fig. 22.

It has been stated previously that there is also a fixed relation between the horizontal thrust and the bending moment at the fixed column or beam ends, and the bending moment can be expressed in terms of the horizontal thrust. The bending moment at any section of a frame is a function of the horizontal thrust. Therefore, it may be stated that the statically indeterminate stresses in the *symmetrical* frame have a fixed relation under distributed and concentrated loads.

From the foregoing illustrations it will be seen that the horizontal thrusts at the column ends due to uniform and centrally concentrated loads may be expressed in the following forms:

Uniform load, $$H = K\frac{pl^2}{12}$$

Centrally concentrated load, $$H = K\frac{Pl}{8}$$

The coefficient K is the same in both cases, but varies with the form of frame. The formula for the horizontal thrust in the frame under concentrated load may be written directly if the formula for thrust in the frame under uniform load is known. An analysis of the statically indeterminate forces for a given case should first be made to find the form of the function.

The bending moment at the end of the span in these frames is:

For Case a, $$M=\frac{3}{3+2mn}\,\frac{pl^2}{12}=K_1\frac{pl^2}{12}$$

For Case a′, $$M=\frac{3}{3+2mn}\,\frac{Pl}{8}=K_1\frac{Pl}{8}$$

For Case b, $$M=\frac{2}{2+mn}\,\frac{pl^2}{12}=K_2\frac{pl^2}{12}$$

For Case b′, $$M=\frac{2}{2+mn}\,\frac{Pl}{8}=K_2\frac{Pl}{8}$$

It is known that when a beam is perfectly fixed at its ends the negative bending moments due to a distributed load and a centrally concentrated load are $\frac{pl^2}{12}$ and $\frac{Pl}{8}$, respectively. It is seen, therefore, that for these cases the bending moment at the end of the beam is obtained from the value of the end bending moment of a fixed beam by multiplying by K, a coefficient which depends upon the form of the frame, but is independent of whether the load is applied uniformly or is concentrated at the center of the span.

Returning to the nature of the formulas for the horizontal thrust at the lower column end of a frame, it is further seen from Fig. 22 that the given constant relation between the values of the horizontal thrusts for a frame under a distributed load and under a concentrated load still holds for the case in which a frame is subjected to a non-symmetrical load. These simple frames are sufficient to illustrate the general relation. It appears, therefore, that for the same frame the coefficient K remains constant and independent of the method of loading. This statement can easily be extended to the case of multiple concentrated loads, for then the horizontal thrust is the sum of the horizontal thrusts due to the individual concentrated loads. It will be found that this statement applies also to the non-symmetrical frames of Cases *d* and *d′*.

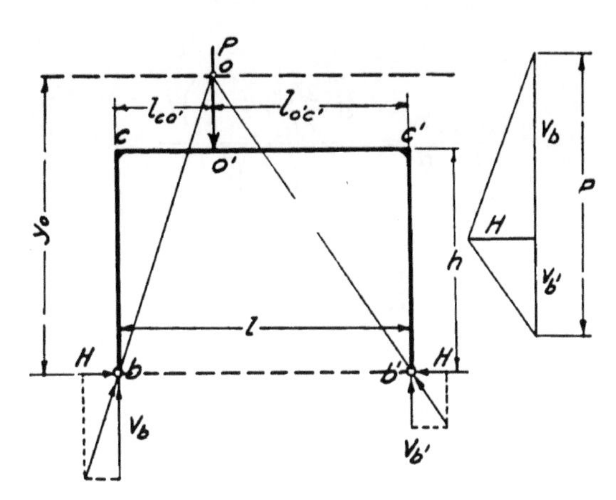

Fig. 23. Locus of Intersection of Reaction Lines in Single Span Single Story Frame under Single Concentrated Load

For a concentrated load, it will be of interest to find the locus of y_o, the point of intersection of the lines of action of the reactions with the line of action of the load (see Fig. 23). In a complicated form of frame, there are, of course, many statically indeterminate quantities, but H is an important one. The remaining statically indeterminates have always the same factor in the denominator as H. Therefore, it is very interesting to know the form of the expression for H. It is evident that H (Fig. 23) is a function of l, h, I, and P in a given case.

Since the moments at b and b' (Fig. 23) are zero, the equilibrium polygon for the load P must pass through these points. Taking the moments of H and V_b about the point o

$$V_b l_{co'} - H y_o = 0 \qquad y_o = \frac{V_b l_{co'}}{H}$$

H and V_b are known in this case when P and $l_{co'}$ are given, and

$$H = \frac{\left(\frac{l}{h}\right)^2}{2\left(\frac{2I_{cc'}}{3I_{bc}} + \frac{l}{h}\right)} \cdot \frac{l_{co'} l_{o'c'}}{l^2} P$$

$$V_b = \frac{l_{o'c'}}{l} P$$

Therefore

$$y_o = \frac{l_{co'} l_{o'c'}}{l} \frac{P}{H} = \frac{2\left(\frac{2I_{cc'}}{3I_{bc}} + \frac{l}{h}\right)l}{\left(\frac{l}{h}\right)^2}.$$

or

$$y_o = 2h\left(1 + \frac{2hI_{cc'}}{3lI_{bc}}\right)$$

This equation is entirely free from $l_{co'}$, $l_{o'c'}$, and P; therefore, y_o is a constant quantity for a given frame and is not changed by the change of the point of application of a load P. Accordingly the locus of the point o is a straight line parallel to bb'.

In the case in which $\frac{I_{cc'}}{I_{bc}} = 1.0$ and $\frac{h}{l} = 1.0$, $y_o = \frac{10h}{3}$. For $\frac{I_{cc'}}{I_{bc}} = 2.0$ and $\frac{h}{l} = 1.0$, $y_o = \frac{14h}{3}$.

The equation for y_o permits the determination of the position of loads which gives the maximum reaction and stress in any member. The same method may be extended to any case, if it is remembered that when a column is fixed at its end the point of application of the reaction deviates from the neutral line of the column by $\frac{M_b}{V_b}$, where M_b is the end moment and V_b is the vertical reaction at that point.

15. *Effect of Variation in Moment of Inertia and Relative Height of Frame on Bending Moment in Horizontal Member.*—Fig. 24, 25, 26, 27, and 28 give bending moment coefficients for the beam of the central span for several cases of a three-span frame in which the spans are equal, the moments of inertia of the three beams are equal, and the moments of inertia of the columns are equal. The effect on the bending moment caused by variation in the relative values of the moments of inertia of members and in heights of frames is shown in these figures.

For a general comparison it is only necessary to consider the bending moment at the center of the span for load applied eccentrically with respect to the columns, since the effect on moments at other places

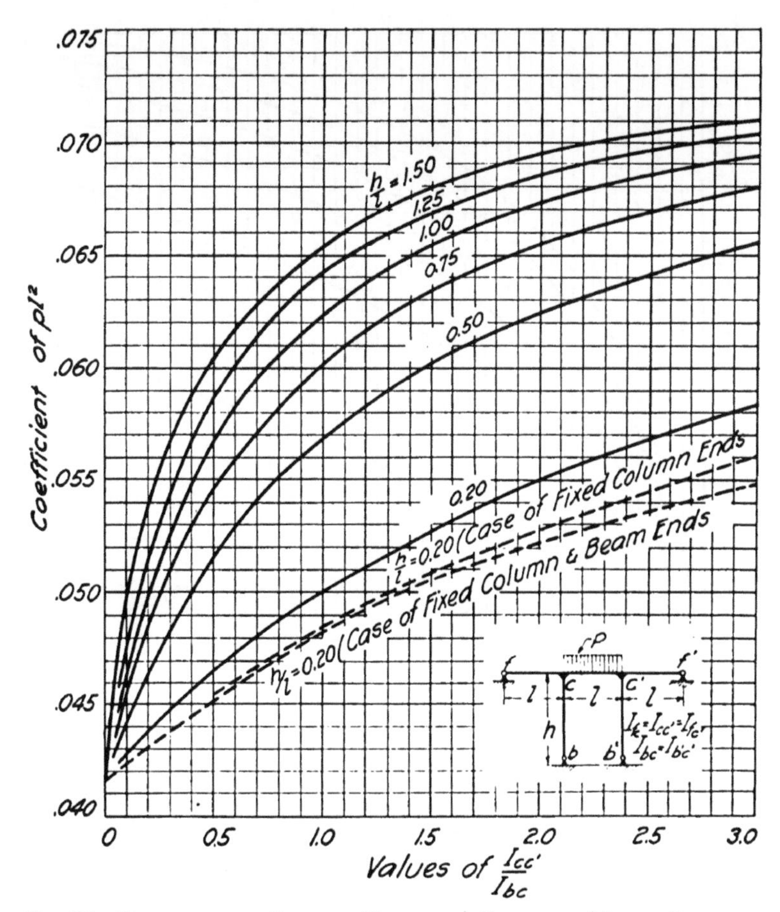

FIG. 24. COEFFICIENT OF BENDING MOMENT AT CENTER OF MIDDLE SPAN FOR SINGLE STORY THREE-SPAN FRAME HAVING EXTREME ENDS OF COLUMNS AND BEAMS HINGED

and on thrusts will be similar. From the general nature of the curves shown the following conclusions are drawn:

(1) The bending moment is increased rapidly as the value of $\frac{I_{cc'}}{I_{bc}}$ increases from 0 to 1.5, but beyond that range the increase in bending moment is comparatively small.

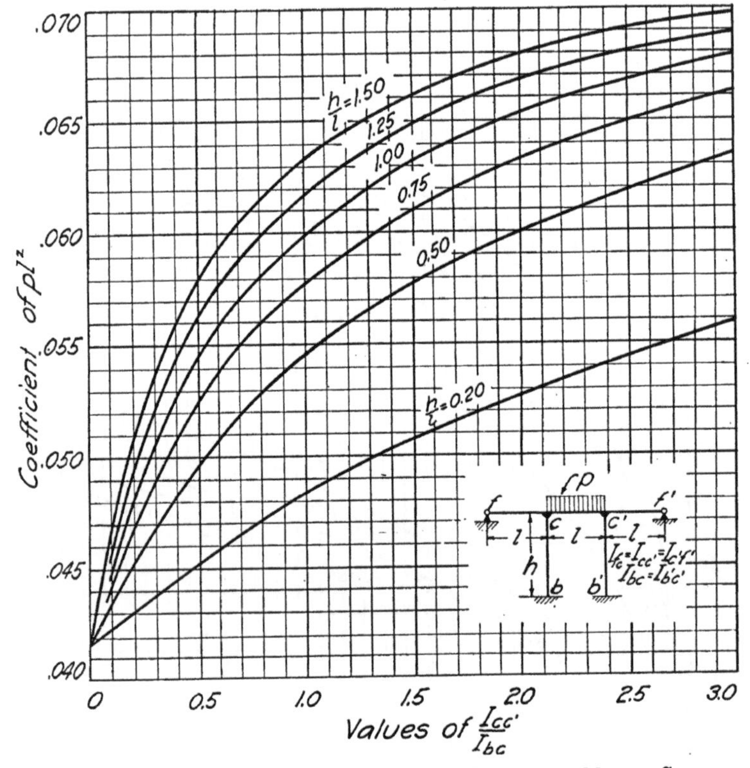

Fig. 25. Coefficient of Bending Moment at Center of Middle Span for Single Story Three-Span Frame Having Extreme Ends of Beams Hinged and Column Ends Fixed

(2) An increase in the height of the frame has an effect of the same nature on the bending moment as an increase in the ratio $\frac{I_{cc'}}{I_{bc}}$.

(3) The variation in coefficient of bending moment is wider in the frame hinged at ends of columns and beams than in the case of fixed ends.

(4) By the fixing of ends of columns and end beams the coefficient of positive bending moment is slightly decreased from that for hinged ends.

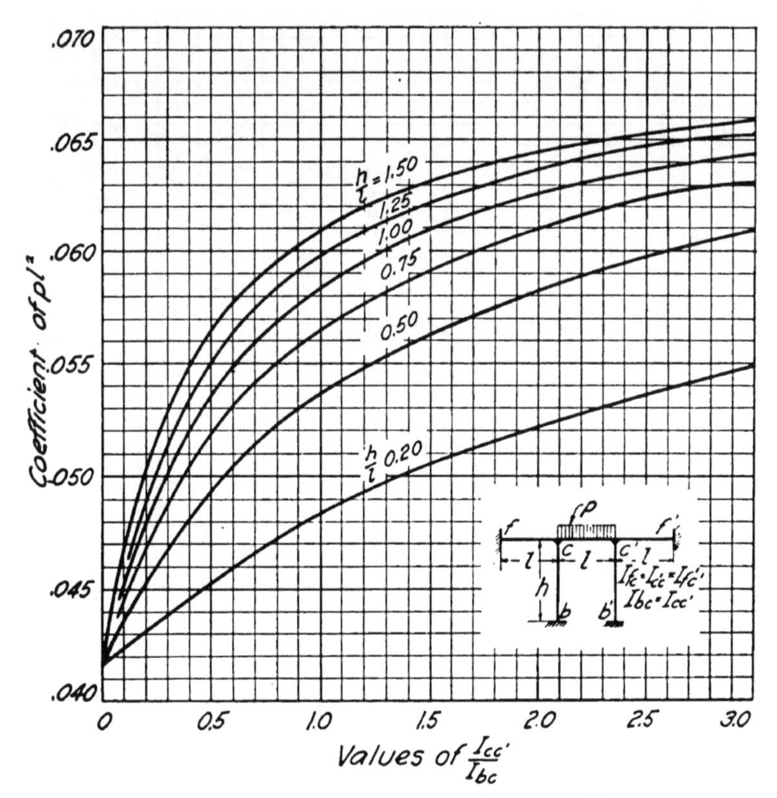

FIG. 26. COEFFICIENT OF BENDING MOMENT AT CENTER OF MIDDLE SPAN FOR SINGLE STORY THREE-SPAN FRAME HAVING ALL EXTERNAL CONNECTIONS FIXED

(5) In most common cases of panels under uniform load where the ratio $\frac{h}{l}$ is not far from 1.0 and $\frac{I_{cc'}}{I_{bc}}$ varies from 1.5 to 3.0, the bending moment at the center of the loaded span (case of equal spans) varies from about $\frac{1}{16}pl^2$ to about $\frac{1}{14}pl^2$, and may be conveniently assumed as $\frac{1}{15}pl^2$.

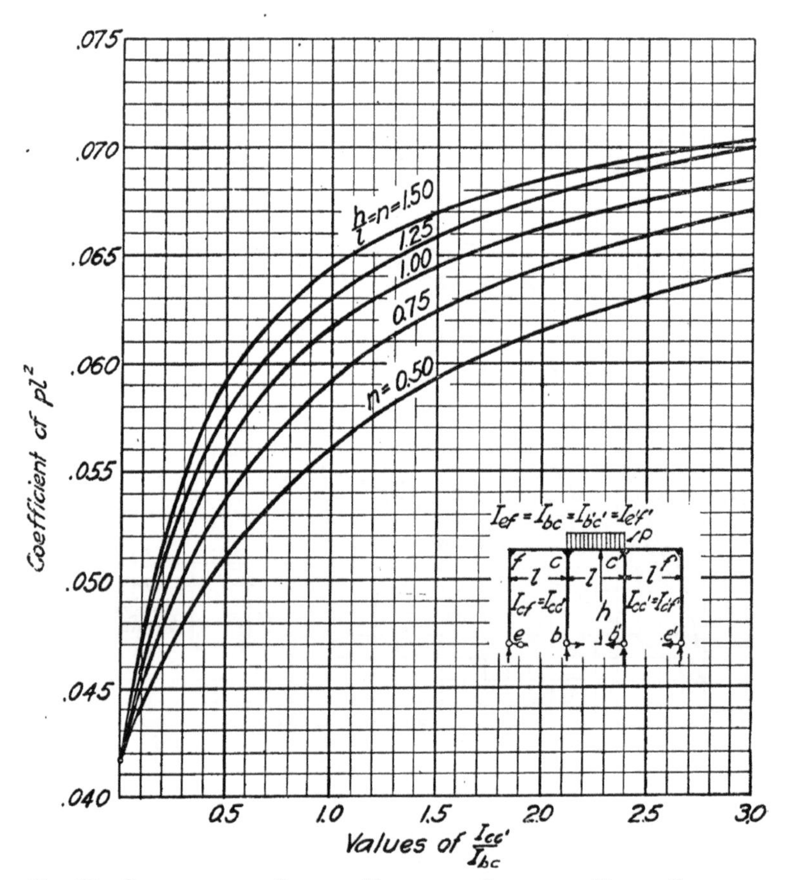

Fig. 27. Coefficient of Bending Moment at Center of Middle Span for Single Story Three-Span Frame Having Lower Ends of Columns Hinged

16. *Effect of Variation in Moment of Inertia on Bending Moment in Vertical Member.*—The variation in bending moments in column ends due to the variation in properties of the members for several cases of a three-span frame is shown in Fig. 17 and 29 and in Tables 4, 6, and 7. In Table 6 the three spans are taken as equal and the story height is taken equal to half the span.

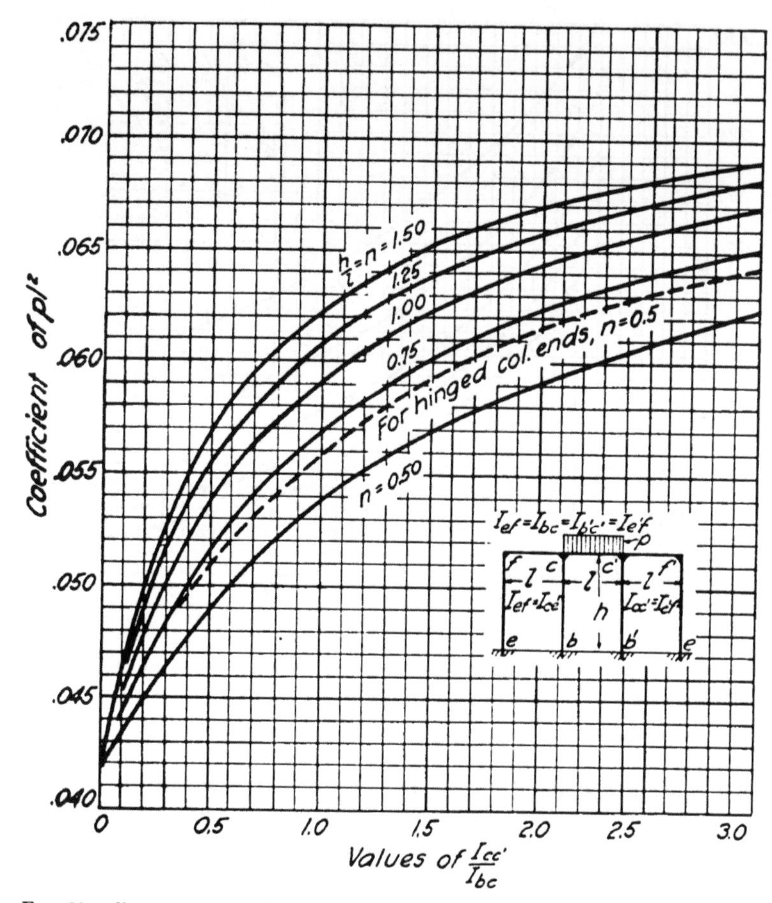

FIG. 28. COEFFICIENT OF BENDING MOMENT AT CENTER OF MIDDLE SPAN FOR SINGLE STORY THREE-SPAN FRAME HAVING LOWER ENDS OF COLUMNS FIXED

Values of coefficients of pl^2 for various values of $\frac{I_{cc'}}{I_{bc}}$ and $\frac{I_{cd}}{I_{bc}}$ are plotted in Fig. 17. It is seen from the diagram that for structures having the relations between span lengths and moments of inertia assumed in Table 6, higher bending stress will exist at the top of the

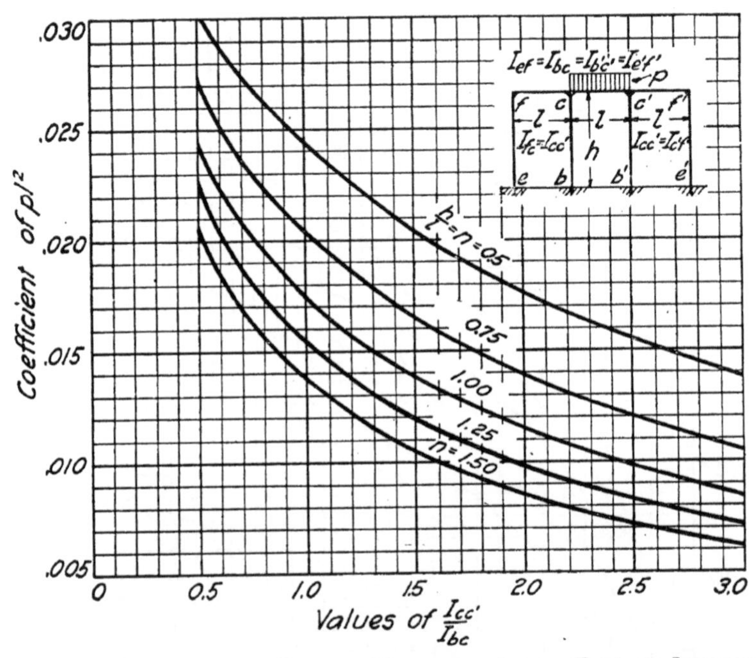

Fig. 29. Coefficient of Bending Moment at Lower Ends of Interior Columns for Single Story Three-Span Frame Having Lower Ends of Columns Fixed

lower column than at the foot of the upper column and that the variations in moments of inertia assumed cause less variation in the moment in the upper columns than in the lower columns.

III. Tests on Rigidly Connected Reinforced Concrete Frames

17. *Test Specimens.*—Five types of frames were selected for the tests. The cross-section of the composing members varied from 8 by 8 in. to 8½ by 17⅜ in. The length of span of the frames was 6 ft. on centers except Frame No. 8, which had three spans of 4 ft. 8 in. The height of the frames varied from about 5 ft. to about 10 ft. The size and disposition of the reinforcing bars and the dimensions of the frames are shown in Fig. 30 to 34. Data of the frames are given in Table 8.

Care was taken in designing the test specimens to secure continuity of connected members and to obtain such proportions between moments of inertia and spans as would result in high bending stresses in the columns and beams at nearly the same time. The ends of the steel reinforcing bars were bent into hooks in the specimens having columns fixed at the ends. Bars continuous from one end to another were used for all frames. The radius of bends of the main rods was about 5 in. Several bars were welded and these welds were located at points where the bending moment was very small.

In the frames with stirrups, U-shaped or double U-shaped stirrups were used. They passed under the longitudinal bars and extended to the top of the beam. The size and spacing of the stirrups are given in Table 8.

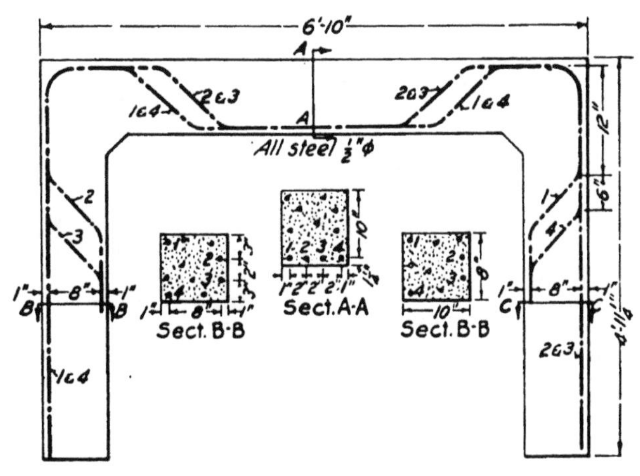

Fig. 30. Size and Distribution of Reinforcing Bars in Frames 1 and 6

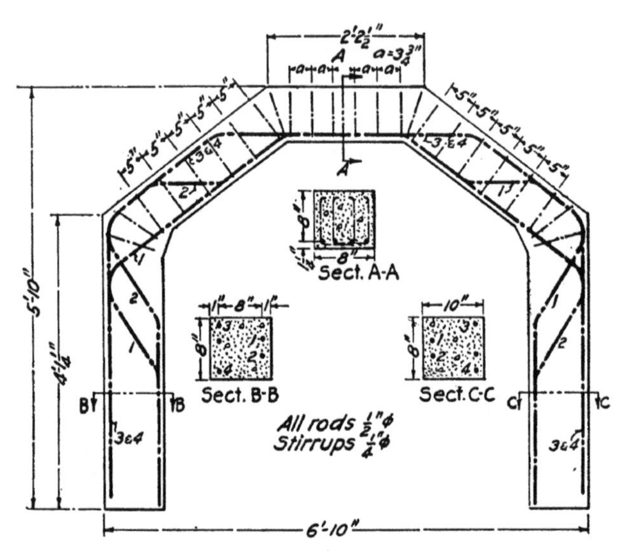

FIG. 31. SIZE AND DISTRIBUTION OF REINFORCING BARS IN FRAMES 2 AND 4

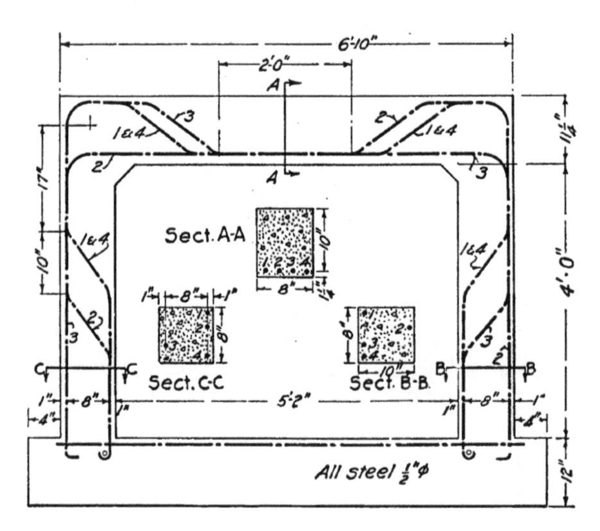

FIG. 32. SIZE AND DISTRIBUTION OF REINFORCING BARS IN FRAMES 3 AND 7

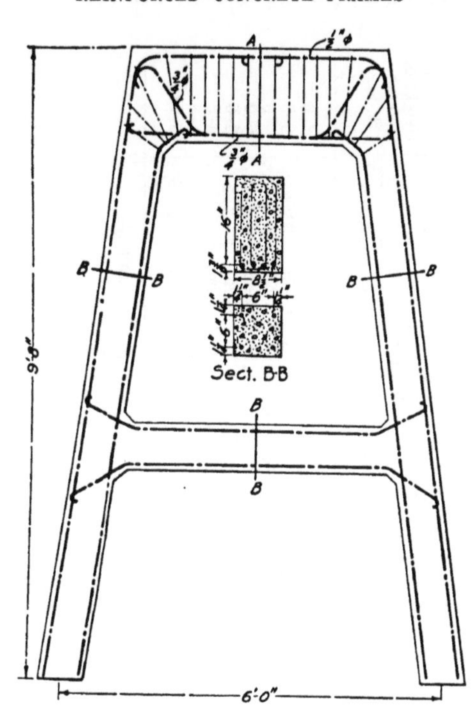

FIG. 33. SIZE AND DISTRIBUTION OF REINFORCING BARS IN FRAME 5

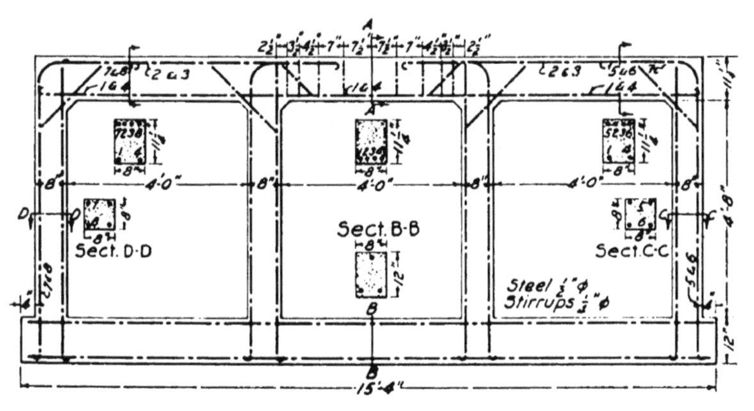

FIG. 34. SIZE AND DISTRIBUTION OF REINFORCING BARS IN FRAME 8

TABLE 8

DATA OF TEST FRAMES

Frame No.	Span Length c. to c.		Cross-Section		Longitudinal Reinforcement				Stirrups		Nominal Height of Frame ft.-in.
					Column (top)		Beam (center)				
	At Top ft.-in.	At Bottom ft.-in.	Column in. x in.	Beam in. x in.	Description	Per cent	Description	Per cent	Description	Spacing	
1	6–0	6–0	8 x10	8 x11¼	3–½ in.	0.82	4–½ in.	0.98	None	....	4–6
2	6–0	6–0	8 x10	8 x 9¼	3–½ in.	0.82	4–½ in.	1.23	¼–in.	3¾&5[1]	6–0
3	6–0	6–0	8 x10	8 x11¼	4–½ in.	1.09	4–½ in.	0.98	None	....	4–6
4	6–0	6–0	8 x10	8 x 9¼	4–½ in.	0.82	4–½ in.	1.23	¼–in.	3¾&5[1]	6–0
5	3–6	6–0	8½x 8½	8½x17⅜	4–½ in.	1.28	4–¾ in.	1.30	¼–in.	2½	9–0
6	6–0	6–0	8 x10	8 x11¼	3–½ in.	0.82	4–½ in.	0.98	None	....	4–6
7	6–0	6–0	8 x10	8 x11¼	4–½ in.	1.09	4–½ in.	0.98	None	....	4–6
8	4–8[2]	4–8[2]	8 x 8	8 x11¼	4–½ in.	1.40	4–½ in.	0.98	¼–in.	From 3½ to 7½[3]	5–2¼

[1] Double loop.
[2] Three equal spans.
[3] Only used in the middle span.

All reinforcement of plain round bars.
All concrete of 1–2–4 mix.

18. *Materials.*—The materials used in making the test frames were similar to those ordinarily used in reinforced concrete construction. The sand, stone, and cement were taken from the stock of the Laboratory of Applied Mechanics.

A good quality of crushed limestone ordered to pass over a ¼-inch sieve and through a 1-inch sieve was used. The sand was of good quality, hard, sharp, well-graded, and generally clean.

The reinforcing bars were plain round rods of open hearth mild steel. Test pieces were taken from the test frames after the test. Table 9 gives the results of the tension tests of the steel.

TABLE 9

TENSION TESTS OF REINFORCING STEEL

Nominal Diameter inches	Yield Point lb. per sq. in.	Ultimate Strength lb. per sq. in.	Per Cent Elongation in 8 in.
½	36 200	55 100	26.3
½	36 200	54 200	26.9
½	36 900	54 700	28.7
½	36 700	54 600	26.9
½	37 700	54 200	25.0
½	37 400	55 900	30.0
Average	36 850	54 783	27.3

Universal Portland cement was used for all specimens. Standard briquettes of neat cement gave an average tensile strength of 575 lb. per sq. in. at 7 days and 670 lb. per sq. in. at 28 days, and standard briquettes of 1–3 mortar 207 lb. per sq. in. at 7 days and 303 lb. per sq. in. at 28 days. Briquettes of 1–3 mortar made with the sand used in the concrete gave a strength of 279 lb. per sq. in. at 7 days and 353 lb. per sq. in. at 28 days. Tests with the Vicat needle indicated that initial set occurred in 3 hours and 15 minutes and final set in 6 hours.

Men skilled in this kind of work were employed in making the concrete. Care was taken in measuring, mixing, and tamping to secure concrete as nearly uniform as possible. All the concrete was made in the proportions, 1 part cement, 2 parts sand, and 4 parts stone, by volume. The mixing was done with a concrete mixing machine.

The results of compression tests on 6-in. cubes made from the concrete used in the frames are given in Table 10. Tests were made

TABLE 10

COMPRESSION TESTS OF CONCRETE CUBES AND CYLINDERS

Frame No.	Age at Test days	Maximum Load lb. per sq. in.		Frame No.	Age at Test days	Maximum Load lb. per sq. in.	
		6 in. Cube	8x16 in. Cyl.			6 in. Cube	8x16 in. Cyl.
1	64	1780	1150	5	61	3070	2670
1	64	1750		5	61	3100	
1	64	1680		5	61	2580	
Average		1740		Average		2920	
2	62	2210	1850	6	62	2605	2310
2	62	2250		6	62	2445	
2	62	2540		6	62	2510	
Average		2330		Average		2520	
3	73	2860	2050	7	60	2140	1970
3	73	2820		7	60	2390	
3	73	2840		7	60	2220	
Average		2840		Average		2250	
4	66	2600	1910	8	63	3288	3060
4	66	2580		8	63	3900	
4	66	2570		8	63	3653	
Average		2580		Average		3614	

on one 8 by 16-in. cylinder for each frame, and the axial deformation was measured to give a means of judging of the modulus of elasticity of the concrete used in the frames. Fig. 35 gives the stress-deformation diagrams for these cylinders. Table 10 gives the compressive strength of the cylinders.

19. *Making and Storage of Test Frames.*—It had been hoped to make the frames in a vertical position similar to that in practice, but because of the difficulty and added expense in doing this, all the frames were built directly on the concrete floor of the laboratory in a horizontal position with a strip of building paper beneath the forms.

The forms were generally removed after seven days, and the frames were lifted from the horizontal position after thirty days and were kept in a vertical position in the laboratory where they were made until the day they were tested. They were dampened every morning for two weeks after making to prevent too rapid drying, and were dampened occasionally after that time. The temperature of the room ranged from 55 to 70 degrees F.

20. *Testing.*—To develop high stresses in the beam and in the columns nearly at the same time one-third point loadings were used for many of the frames. In Frame 5 the centrally concentrated load was used to develop as high a flexural stress in the columns as possible.

In Frame 8 in order to see the effect of the eccentric load on the adjacent spans and at the same time to produce high bending stresses in the middle beam and in the central columns, a uniform load on the middle span was selected, the load being applied through a number of spiral springs.

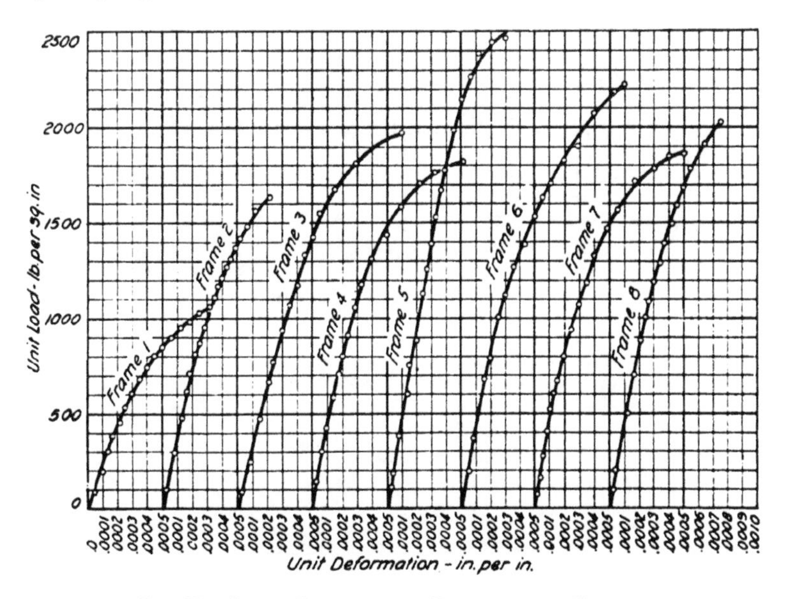

Fig. 35. Stress-Deformation Diagrams for Cylinders

The positions of the loads for the different frames are shown in Fig. 36 to 43. The specimens were tested in the 600,000-lb. Riehle testing machine in the Laboratory of Applied Mechanics of the University of Illinois. Deflections were read on some of the frames. The deformations of the steel and of the concrete were measured at the various parts of the frames for each load applied.

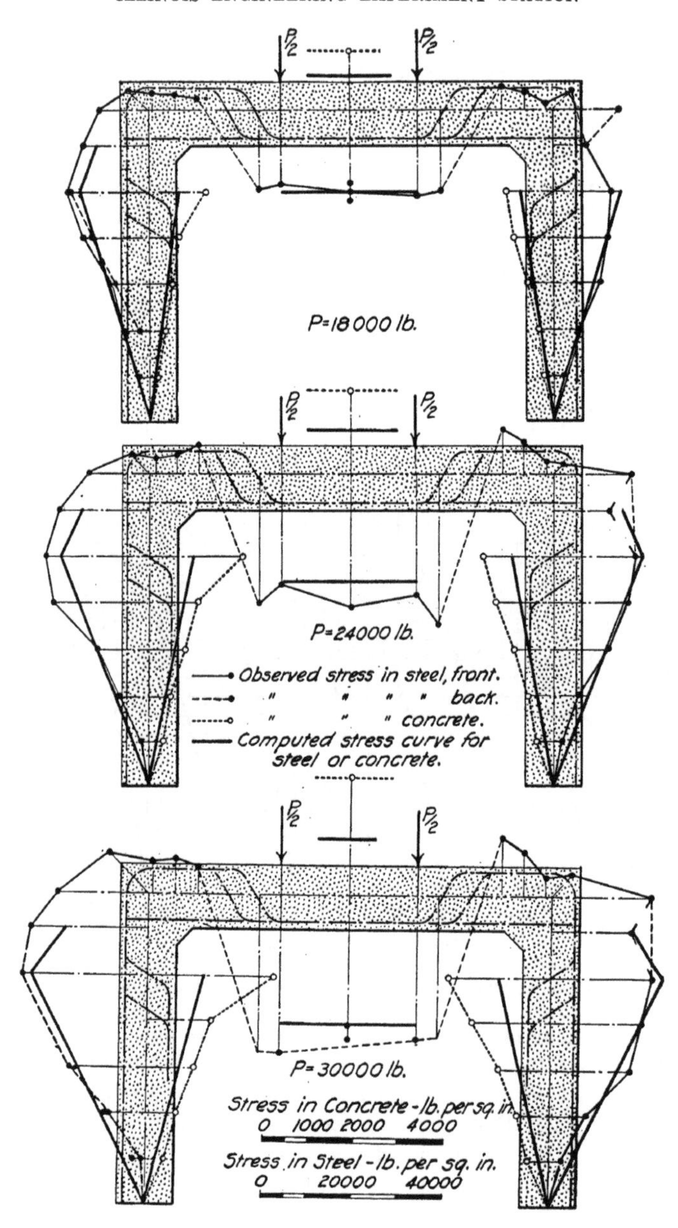

FIG. 36. OBSERVED AND COMPUTED STRESSES IN FRAME 1

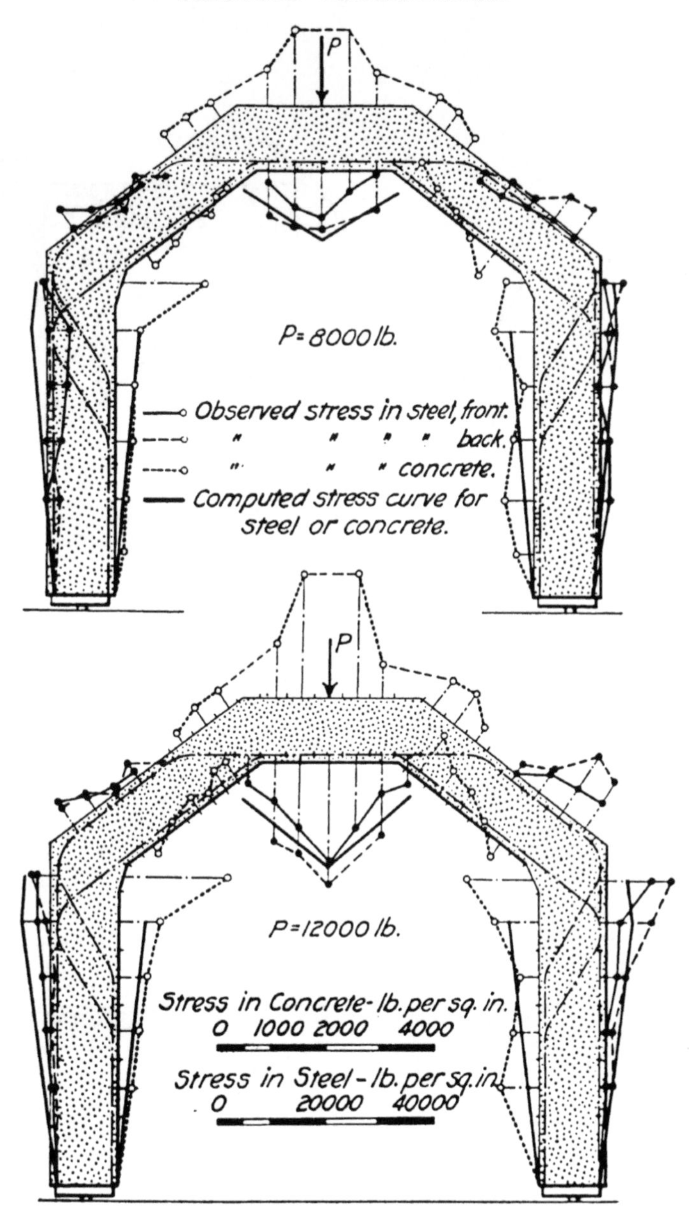

FIG. 37. OBSERVED AND COMPUTED STRESSES IN FRAME 2

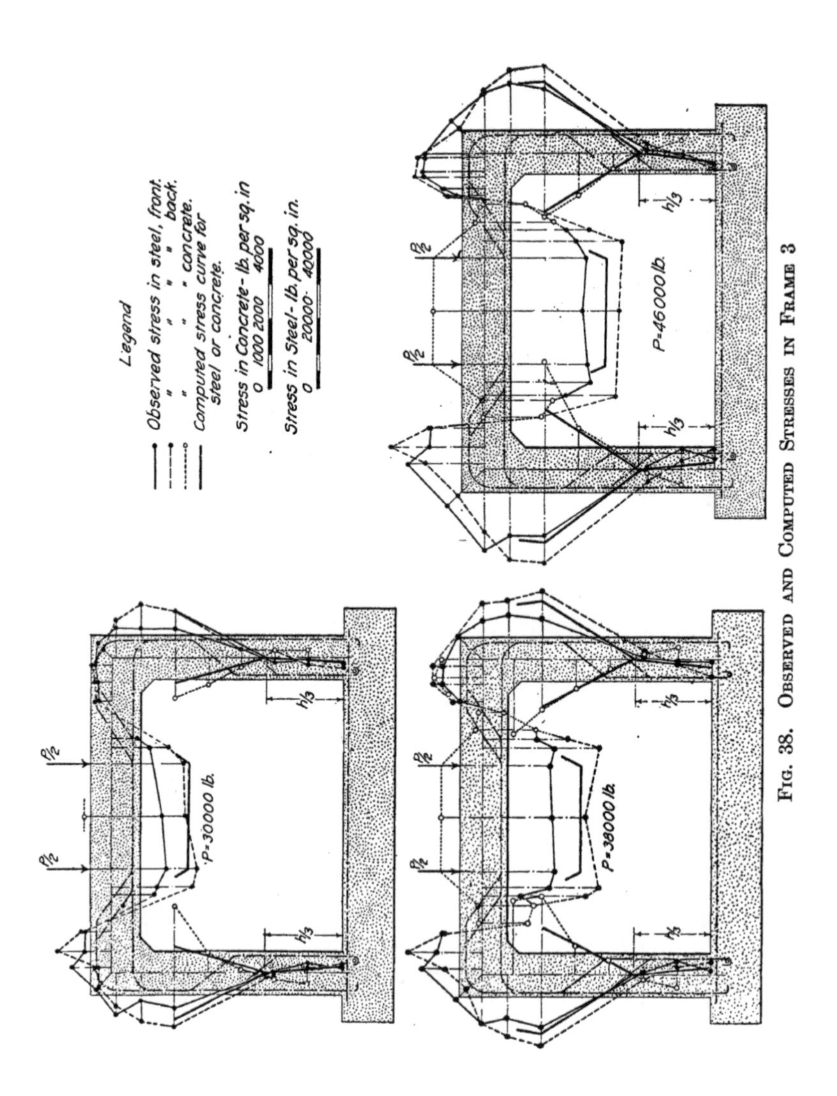

FIG. 38. OBSERVED AND COMPUTED STRESSES IN FRAME 3

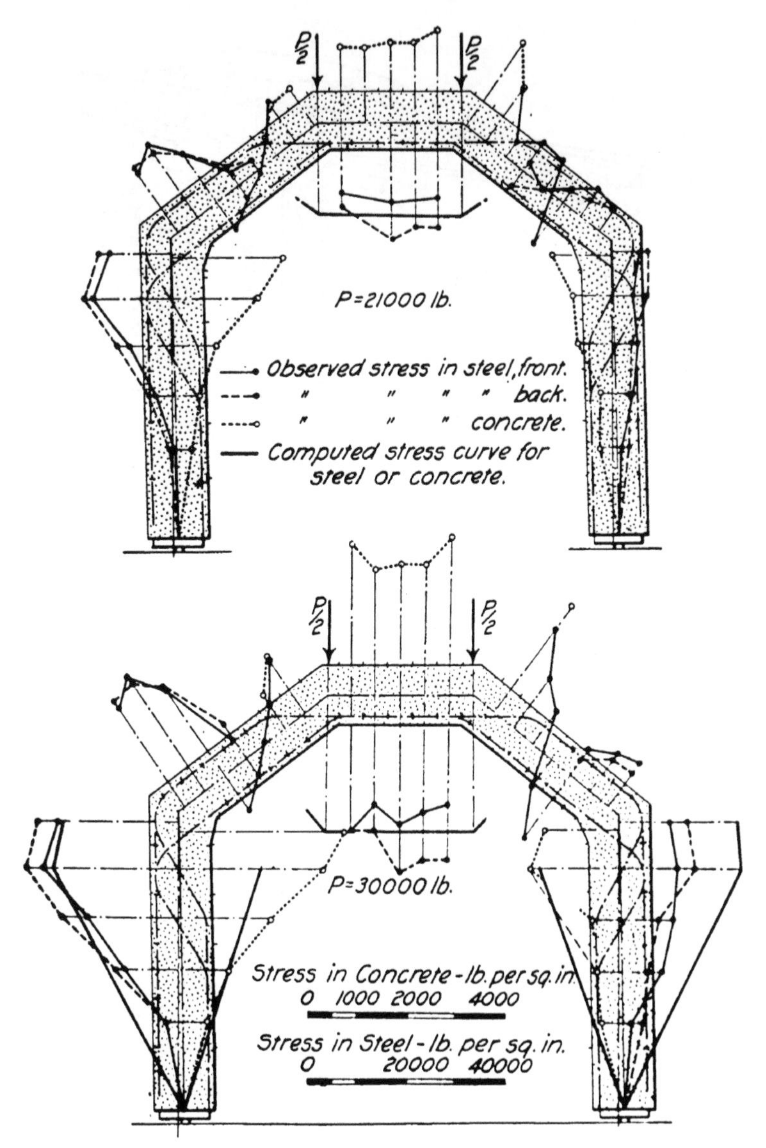

FIG. 39. OBSERVED AND COMPUTED STRESSES IN FRAME 4

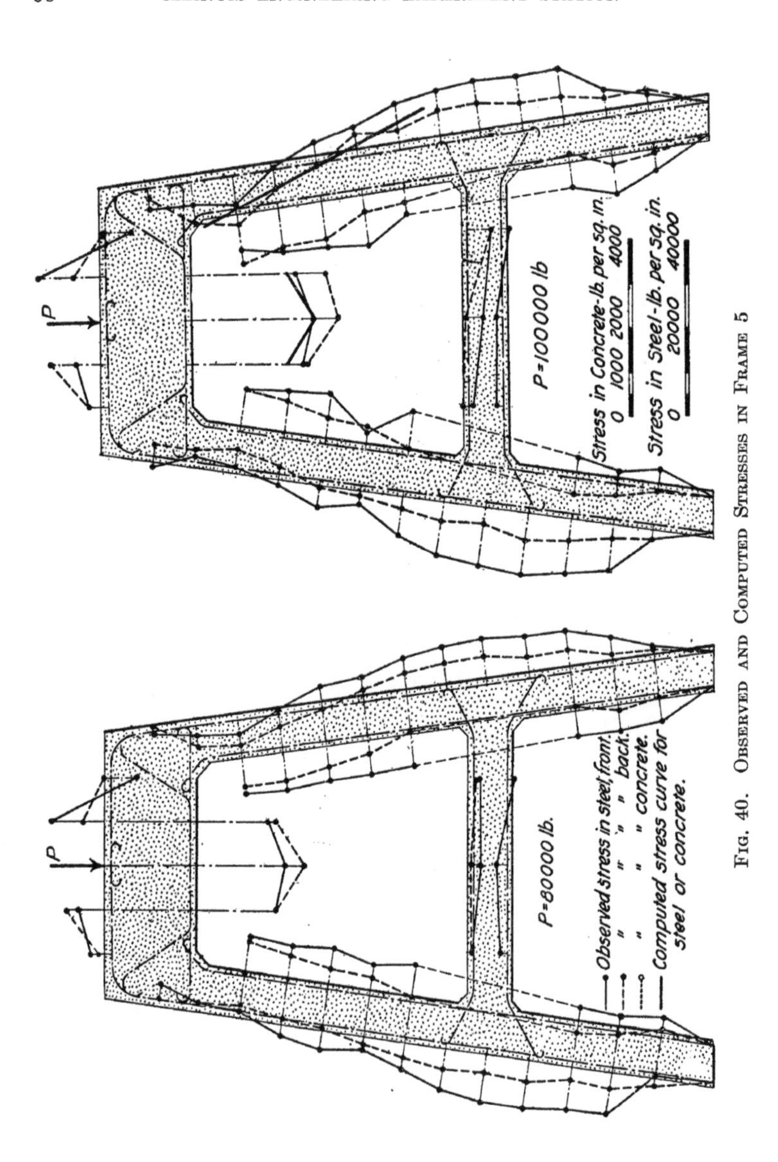

FIG. 40. OBSERVED AND COMPUTED STRESSES IN FRAME 5

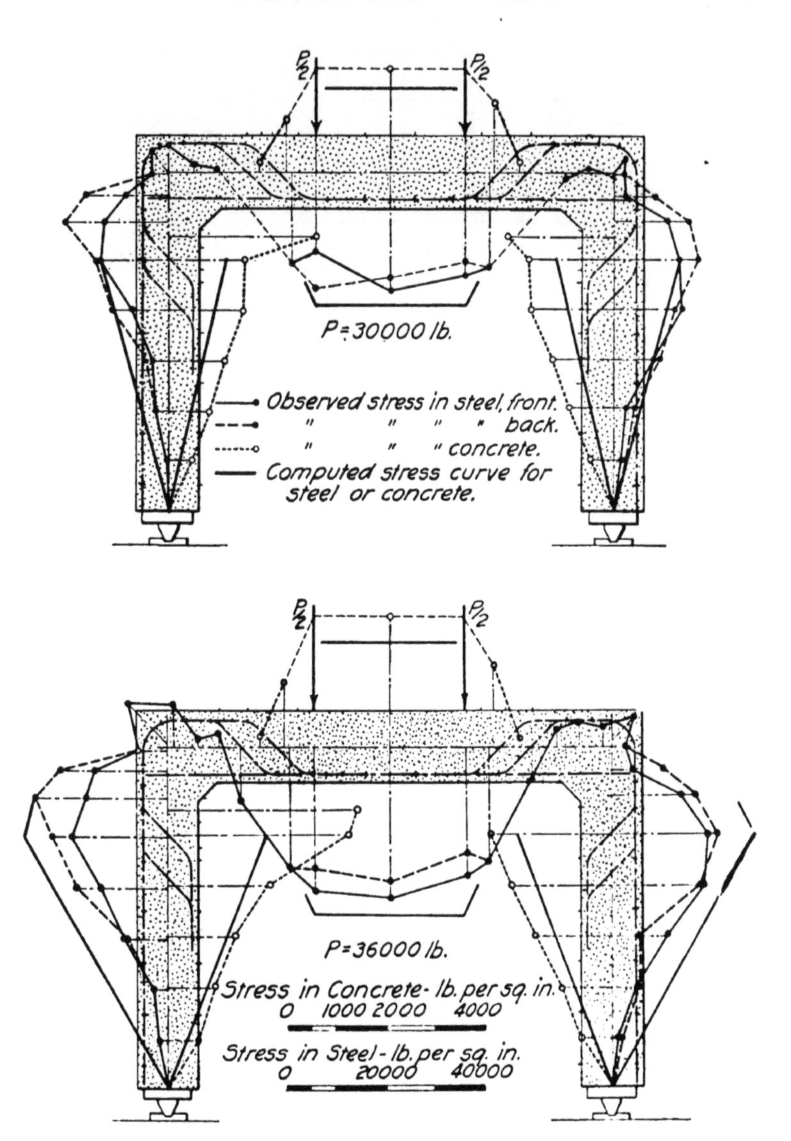

FIG. 41. OBSERVED AND COMPUTED STRESSES IN FRAME 6

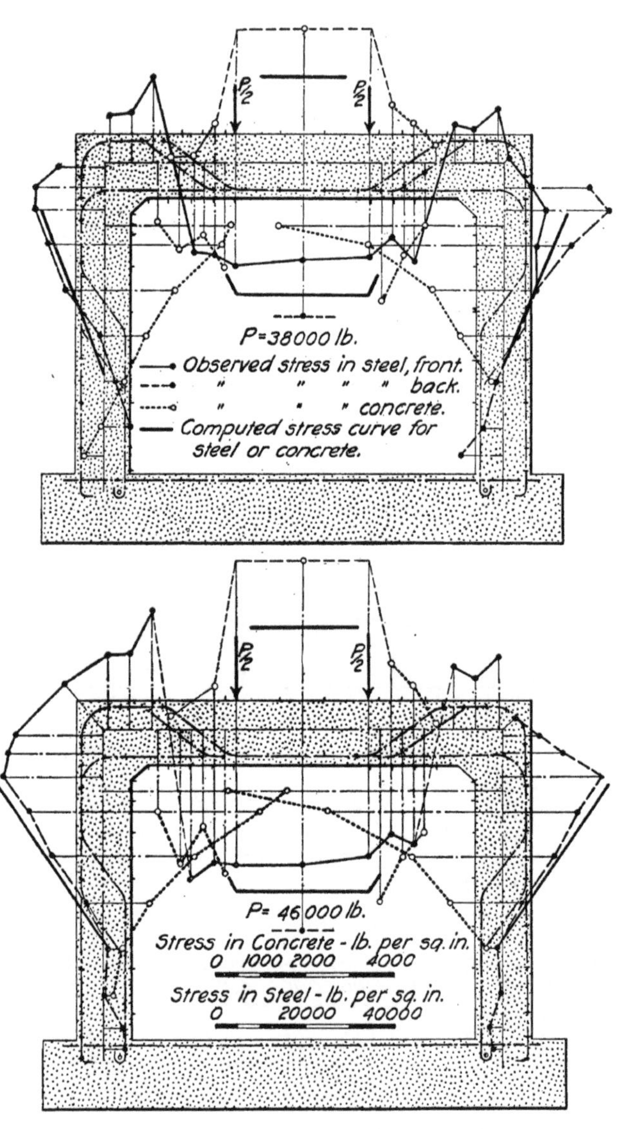

FIG. 42. OBSERVED AND COMPUTED STRESSES IN FRAME 7

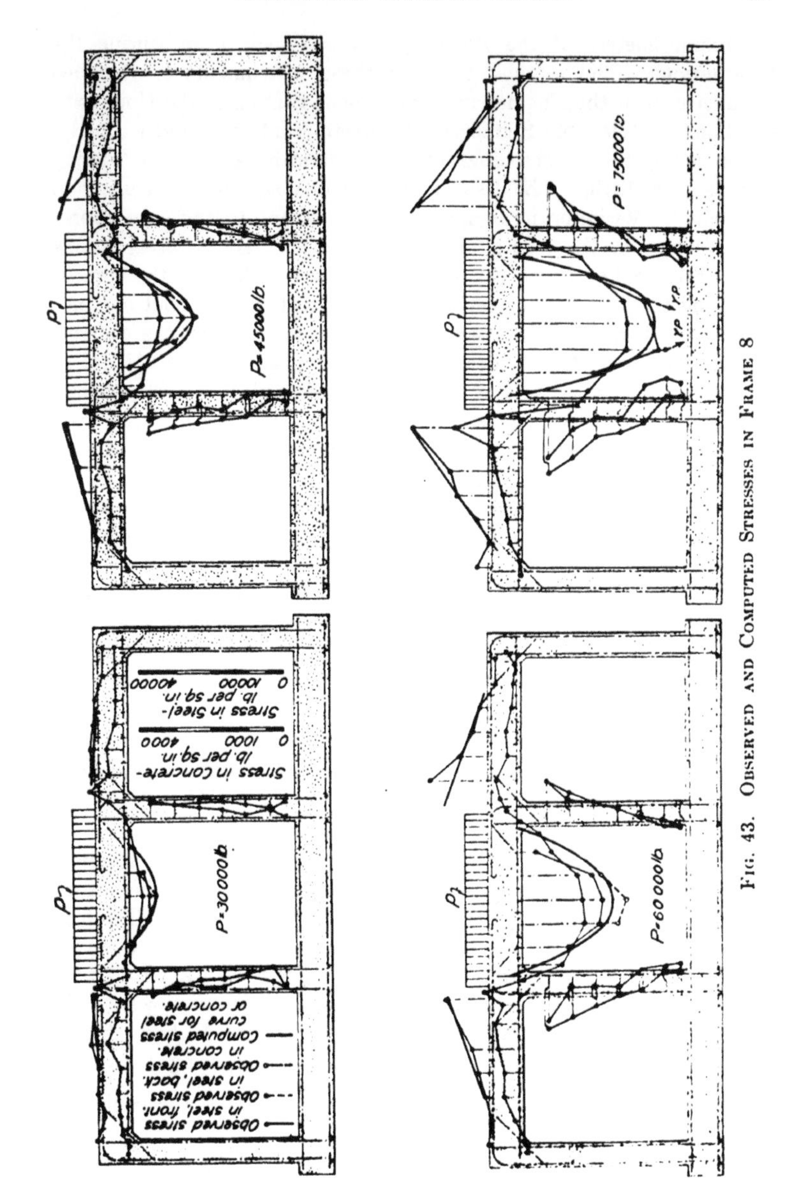

Fig. 43. Observed and Computed Stresses in Frame 8

Extensometers of the Berry type were used in measuring the deformations. The method of using these instruments is described in Bulletin 64 of the Engineering Experiment Station of the University of Illinois, "Tests of Reinforced Concrete Buildings under Load," and in a paper in *Proceedings* of the American Society for Testing Materials for 1913, "The Use of the Strain Gage in the Testing of Materials." Variation in temperature is sufficient to cause an appreciable change in the length of the instrument. Hence observations on an unstressed standard bar of invar steel were taken for the purpose of making temperature corrections. Small steel plugs, about one inch long, were set in plaster of paris in the concrete, where the concrete deformations were to be measured. Small gage holes, 0.055 in. in diameter, were drilled in the reinforcing bars and in the steel plugs. Two sets of initial readings were taken before the application of load. A complete set of observations of the deformations was taken at each increment of load. In reducing the strain gage readings to stress, temperature corrections were made. These were based on an assumed linear variation of length with time between successive readings on the standard bar.

The smallest number of gage lines on any frame was 75 on Frame 1; the greatest number, 163, on Frame 8. The gage lengths used were 2 in., 4 in., and 8 in. The average deformation over the gage length was used.

The faces of the test frames were whitewashed to enable the appearance of cracks and their growth to be more easily observed. The extent of the cracks at the several loads was marked on the specimen during the test.

21. *Explanation of Tables and Diagrams.*—The loads given in the various tables and figures are the loads applied by the testing machine and do not include the weight of the frame itself. The load at first crack is the load noted when the first fine crack was observed during the test. The ultimate load is the highest load applied to the specimen just before the load carried began to decrease slowly. The maximum tensile and compressive stresses are the highest stresses observed at the points specified. The vertical shearing stress was calculated with the ordinary formula $v = \frac{V}{bjd}$, where v represents the vertical shearing unit-stress in the concrete, V the total vertical shear at the end of the beam, b the breadth of the beam, and jd the distance

from the center of the steel to the center of the compressive stresses in the concrete. The bond stress in the beam was computed by means of the formula $u = \frac{V}{mojd}$, where u is the bond stress per unit of area of the surface of the reinforcing steel, m the number of reinforcing bars, and o the circumference or periphery of one reinforcing bar. The values of jd were selected with reference to the amount of reinforcing steel and the modulus of elasticity of the concrete.

Loads are given in pounds, unit-stresses in pounds per square inch, and moments in inch-pounds.

TABLE 11

VALUES OF MODULUS OF ELASTICITY OF CONCRETE USED IN STRESS COMPUTATIONS

Frame	E	Frame	E
1	2 100 000	5	3 000 000
2	3 600 000	6	3 900 000
3	2 070 000	7	3 700 000
4	3 600 000	8	3 300 000

The so-called observed stresses have been obtained from the observed deformations by using a modulus of elasticity of 30,000,000 lb. per sq. in. for the steel, and for the concrete the values given in Table 11.

Table 12 contains general data of the tests of the frames.

Table 13 gives computed stresses at the three points in each frame which are shown in Fig. 44, calculated by means of the formulas given in the preceding pages, the values being expressed in terms of the load applied to the frame. The values in columns marked I were obtained on the assumption that the concrete has full tensile strength; those in columns marked II on the assumption of no tensile strength. The division of the direct stress between concrete and steel was computed by the usual formulas for reinforced concrete columns.

Other explanations of tables and diagrams are made elsewhere.

22. *Phenomena of Frame Tests.*—As may be expected in reinforced concrete flexural members, the tensile stresses in the steel were very small at low loads. Undoubtedly this effect was largely due to the ability of the concrete to carry tensile stress. As soon as the con-

TABLE 12

GENERAL DATA OF TESTS OF FRAMES

Frame No.	Age days	Method of Loading	Load at First Crack pounds	Maximum Load at which Deformation was measured pounds	Ultimate Load at Failure pounds	Maximum Stress pounds per square inch							
						Tensile Stress in Steel				Compressive Stress in Concrete		Vertical Shearing Stress in Top Beam	Bond Stress in Top Beam
						Center of Top Beam	Ends of Top Beam	Bent up Bar in Beam	In Column	Center of Top Beam	In Column		
1	63	[illegible]ed Loads at Third Point	12 000	36 000	40 500	36 000	16 000	21 700	35 200	3 320[1]	3 130[1]	266	333
2	62	[illegible]ed Load at Center	8 000	14 000	....	32 900	28 900	......	12 300	3 840[1]	1 570	124	159
3	63	Concentrated Loads at Third Point	21 000	46 000	61 000	39 500	20 300	15 700	23 800	1 500	3 050	326	415
4	63	Concentrated Loads at Third Point	10 000	30 000	50 000	25 300	21 800	......	27 600	2 590	2 920	267	340

TABLE 12—(CONTINUED)

GENERAL DATA OF TESTS OF FRAMES

Frame No.	Age days	Method of Loading	Load at First Crack pounds	Maximum Load at which Deformation was measured pounds	Ultimate Load at Failure pounds	Maximum Stress pounds per square inch							
						Tensile Stress in Steel				Compressive Stress in Concrete		Vertical Shearing Stress in Top Beam	Bond Stress in Top Beam
						Center of Top Beam	Ends of Top Beam	Bent up Bar in Beam	In Column	Center of Top Beam	In Column		
5	61	Concentrated Load at Center	40 000	100 000	146 000	36 400	5 700	......	11 800	....	12 200[2]	425	383
6	58	Concentrated Lds at Third Point	18 000	36 000	46 000	29 800	7 400	10 300	27 000	2 750	3 840[1]	253	323
7	62	Concentrated Loads at Third Point	21 000	46 000	61 000	44 400	25 500	18 100	30 200	3 700[1]	4 130[1]	323	413
8	61	Uniform Load over Middle Span	45 000	60 000[3]	134 000	30 900[1]	13 200	......	14 700	2 540[4]	14 700[2]	425	540

[1] Not reliable, elastic limit exceeded.
[2] Steel stress in compression side of column.
[3] Not maximum load.
[4] Stress at point one inch distant from extreme fiber.

TABLE 13

COMPUTED STRESSES FOR TEST FRAMES IN TERMS OF TOTAL APPLIED LOAD

No. of Frame	H	Point	M inch unit	p	Direct Stress		I			II		
					Steel	Concrete	k	f_s	f_c	k	f_s	f_c
1	0.082 P	A	—3.43 P	.0068	.070 P	.0049 P				0.35	0.82 P	.09 P
		B	—4.04 P	.0109	.070 P	.0049 P				0.41	0.60 P	.01 P
		C	+7.43 P	.0098	.012 P	.0009 P				0.41	1.09 P	.054 P
2	0.121 P	A	—4.84 P	.0082	.049 P	.0059 P	0.58	0.15 P	.041 P	0.31	0.97 P	.09 P
		B	+5.28 P	.0123	.015 P	.0018 P				0.36	0.93 P	.07 P
		C	+9.78 P	.0123	.015 P	.0018 P				0.36	1.75 P	.23 P
3	0.145 P	A	+1.37 P	p .0054 p' = .0027	.053 P	.0059 P	0.52	0.032 P	.016 P			
		B	—3.02 P	.0109	.052 P	.0058 P	0.54	0.11 P	.029 P	0.35	0.44 P	.036 P
		C	+5.40 P	.0098	.015 P	.0017 P	0.60	0.17 P	.032 P	0.35	0.76 P	.045 P
4	0.100 P	A	—4.30 P	.0082	.049 P	.0059 P	0.58	0.15 P	.041 P	0.31	0.85 P	.04 P
		B	+4.90 P	.0123	.013 P	.0015 P	0.61	0.20 P	.034 P	0.36	0.87 P	.03 P
		C	+4.90 P	.0123	.013 P	.0015 P				0.36	0.87 P	.03 P
5		A	0.23 P	p=p'=.0064	.060 P		0.50	0.047 P	.074 P[1]			
		A_1	0.10 P	p=p'=.0064	.060 P		0.50	0.055 P	.067 P[1]			
		B	0.62 P	p=p'=.0064	.060 P		0.50	0.024 P	.097 P[1]	0.242	0.19 P	.066 P
		C	8.40 P	.013	.003 P		0.57	0.150 P		0.42	0.30 P	
6	0.077 P	A	—3.42 P	.0068	.040 P	.0052 P	0.58	0.10 P	.030 P	0.28	0.81 P	.047 P
		B	—4.04 P	.0109	.040 P	.0052 P	0.59	0.12 P	.033 P	0.34	0.60 P	.046 P
		C	+7.48 P	.0098	.007 P	.0009 P	0.59	0.22 P	.042 P	0.33	1.06 P	.065 P

TABLE 13—(CONTINUED)

COMPUTED STRESSES FOR TEST FRAMES IN TERMS OF TOTAL APPLIED LOAD

No. of Frame	H	Point	M inch unit	p	Direct Stress		I			II		
					Steel	Concrete	k	f_s	f_c	k	f_s	f_c
7	0.145 P	A	2.03 P	p = .0054 p′ = .0027	.048 P	.0059 P	0.52	f_s = 0.028P f'_s = 0.178P	.024 P			
		B	—3.02 P	.0109	.047 P	.0058 P	0.54	0.13 P	.029 P	0.35	0.43 P	.038 P
		C	+5.40 P	.0098	.014 P	.0017 P				0.35	0.76 P	.048 P
8	I: H_a .009 P; II: .010 P	A		.0098	Neglected		0.33	0.46 P	.028 P	0.33	0.51 P	.032 P
	I: H_b .003 P; II: .023 P	A_1		p = .0073 p′ = .0049	Neglected		0.58	0.47 P	.007 P	0.28	0.24 P[2] 0.34 P[3]	.0102 P[2] .0146 P[3]
	I: V_a .047 P; II: .039 P	B		p = p′ = .0062	.070 P		0.50	0.166 P	.216 P[1]			
	I: V_b .547 P; II: .530 P	C		p = p′ = .0049	Neglected		0.50	0.06 P	.06 P[1]	0.24	0.31 P	.047 P[1]

[1] Compressive stress in steel, f'_s.

[2] Same conditions as I except central beam is considered as broken in tension.

[3] Same conditions as I except beams and intermediate columns are considered as broken in tension.

crete on the tension side of the member was sufficiently stretched, a vertical tension crack formed on the beam underneath the load and then a crack formed at the side near the juncture of the column and the beam, in most cases. After the formation of these cracks, the tension in these parts was taken mainly by the reinforcing bars. As the loads were increased the cracks developed and new cracks appeared on the tension side between the points of application of the load on the beam, and horizontal cracks formed at regular intervals in the columns.

The tensile stress due to the negative bending moment within the space occupied by the juncture of the beam and the column was small, and in these places tension cracks did not form in many frames until high loads were applied. The bent-up bars in the beam came into action as soon as tension cracks formed in their vicinity, and in the bent-up portions tensile stresses as high as 22,000 pounds per square inch were developed in several instances. The tensile stresses in the steel at the fixed ends of the columns were rather low. The tensile strength of the concrete in this part apparently reduced the tensile stress in the steel.

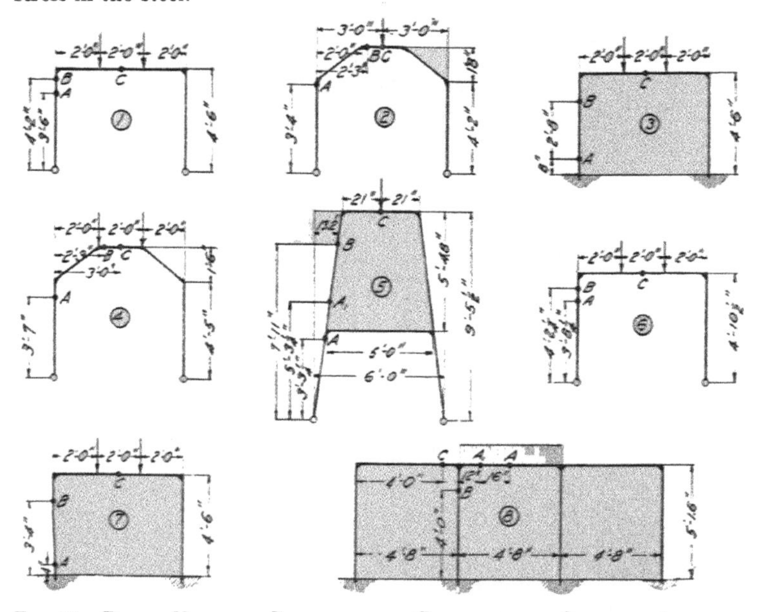

FIG. 44. POINTS USED FOR COMPARISON OF COMPUTED AND OBSERVED STRESSES

High compressive stresses were developed in the concrete in the upper portion of the columns below the intersection with the beam, and the maximum compression was observed along the sharp corner at the juncture of the beam and the column, as might be expected. This is due to the curved beam action at the rigid joint. In each frame the maximum load was higher than the load expected.

Views of the frames which show the location of cracks are given in Fig. 49 to 57.

The general phenomena of the tests of the individual frames are given in the following brief notes.

Frame 1—Square Frame with Columns Hinged at Lower End.—Nominal span length was 6 feet. Total height was 5 ft. 2 in. Frame was loaded at the one-third points of the span of the horizontal member. Fig. 49 shows the frame in the testing machine. The location of the cracks is shown in Fig. 51. At the 12,000-lb. load the first fine crack in the beam appeared directly under a load point and extended from the bottom of the member vertically 2 in. to the level of the reinforcement. At the same load the first noticeable cracks appeared in the columns, one in the outside edge of the column on a level with the bottom surface of the beam and one at 2 ft. 5 in. from the bottom of each column end. No crack appeared in the top side of the beam until the load was increased to 36,000 lb. At that load cracks appeared 8 in. from the top corner of the frame and extended vertically downward.

The frame carried 40,500 lb. and the load was held for a few minutes and then dropped very slowly. The cracks were well distributed in the tension zone of the frame and no crack due to diagonal tension was formed. The frame failed by tension in the reinforcement of the beam.

Frame 2—Inverted U-frame with Columns Hinged at Lower End.—Nominal span length and total height were each 6 ft. Load was applied at the center of the span of the horizontal member. The location of cracks is shown in Fig. 52. At 8,000 lb. two cracks appeared 2 in. on each side of the center of the top beam and extended upward 2 in. and 3 in., respectively. At 12,000 lb. these cracks had extended vertically 6 in. from the bottom surface of the beam, and a new crack appeared just inside the right-hand corner 10 inches from the center of the beam and extended diagonally toward the load point. At the same load four cracks appeared at the outer shoulders. At 14,000 lb.

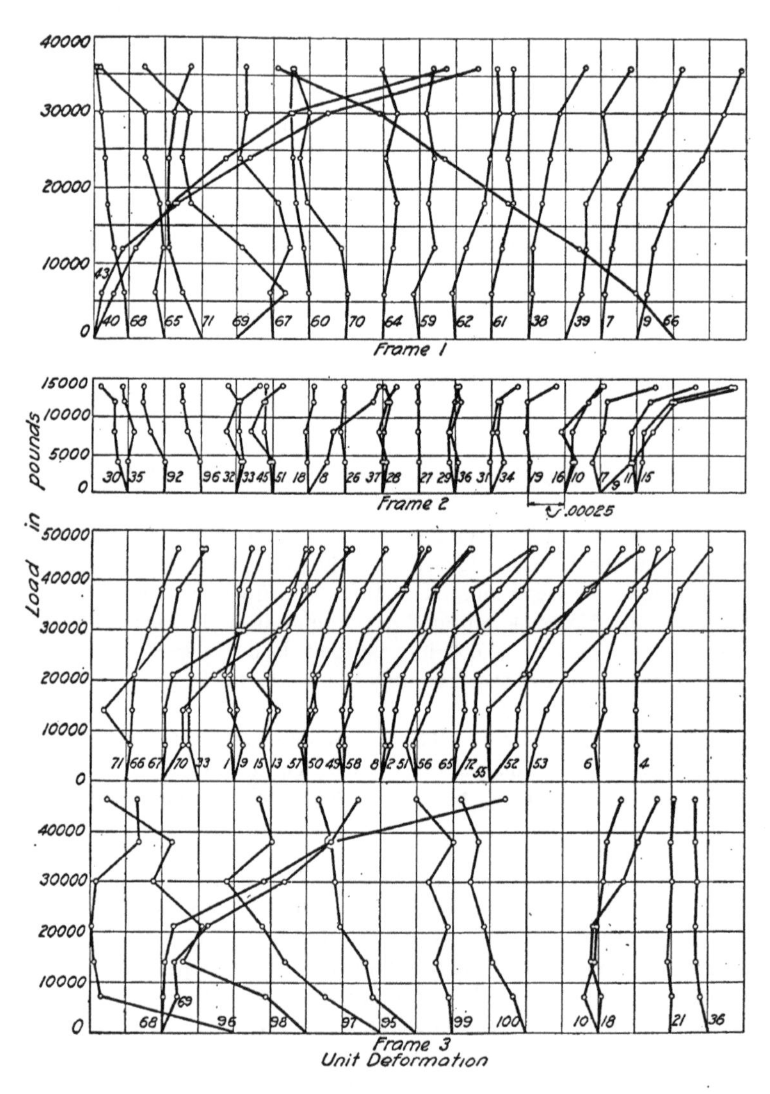

Fig. 45. Load-Deformation Diagrams for Additional Gage Lines, Frames 1, 2, 3

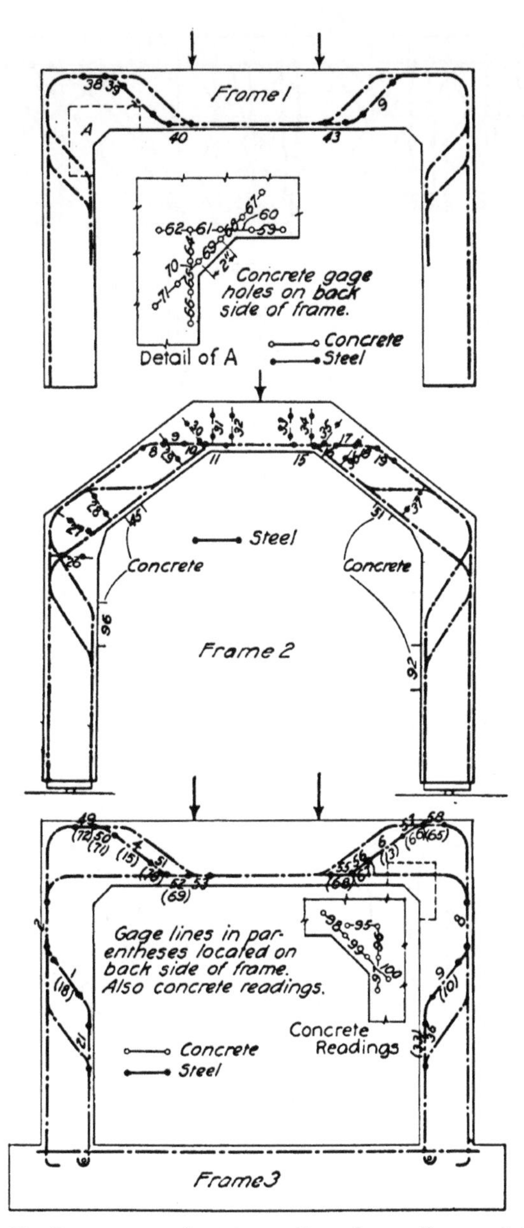

FIG. 46. LOCATION OF ADDITIONAL GAGE LINES, FRAMES 1, 2, 3

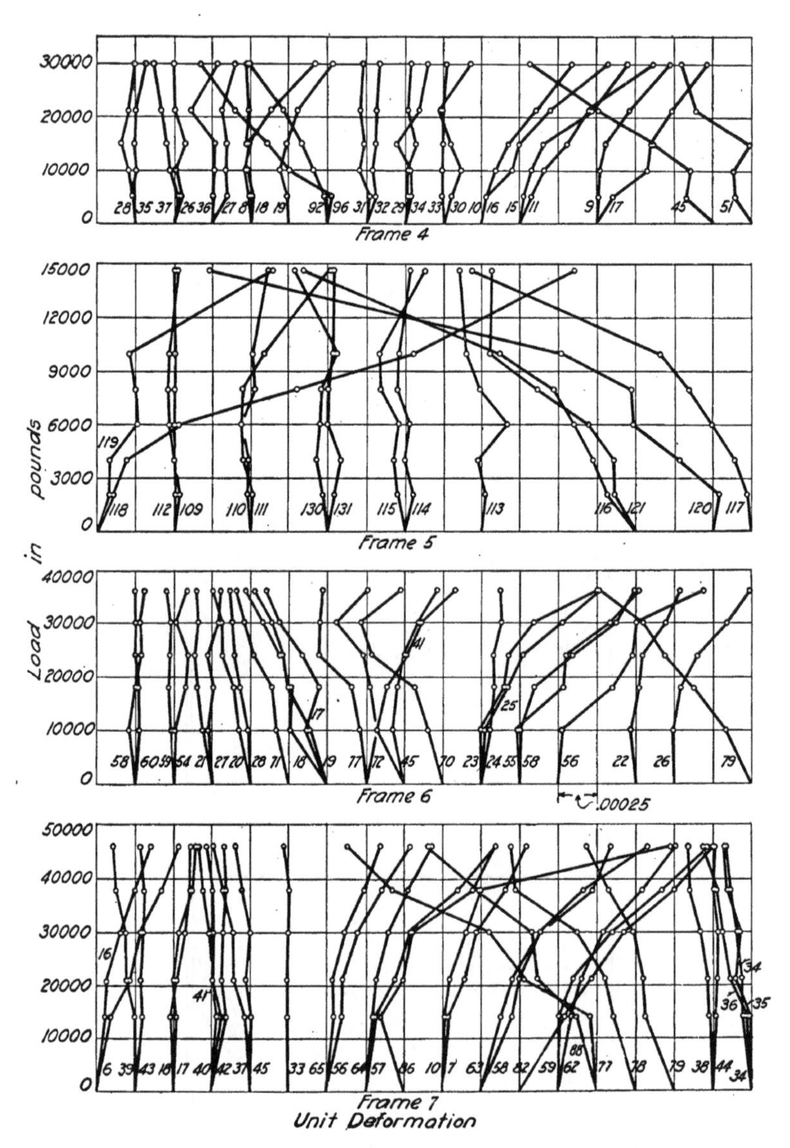

FIG. 47. LOAD-DEFORMATION DIAGRAMS FOR ADDITIONAL GAGE LINES, FRAMES 4, 5, 6, 7

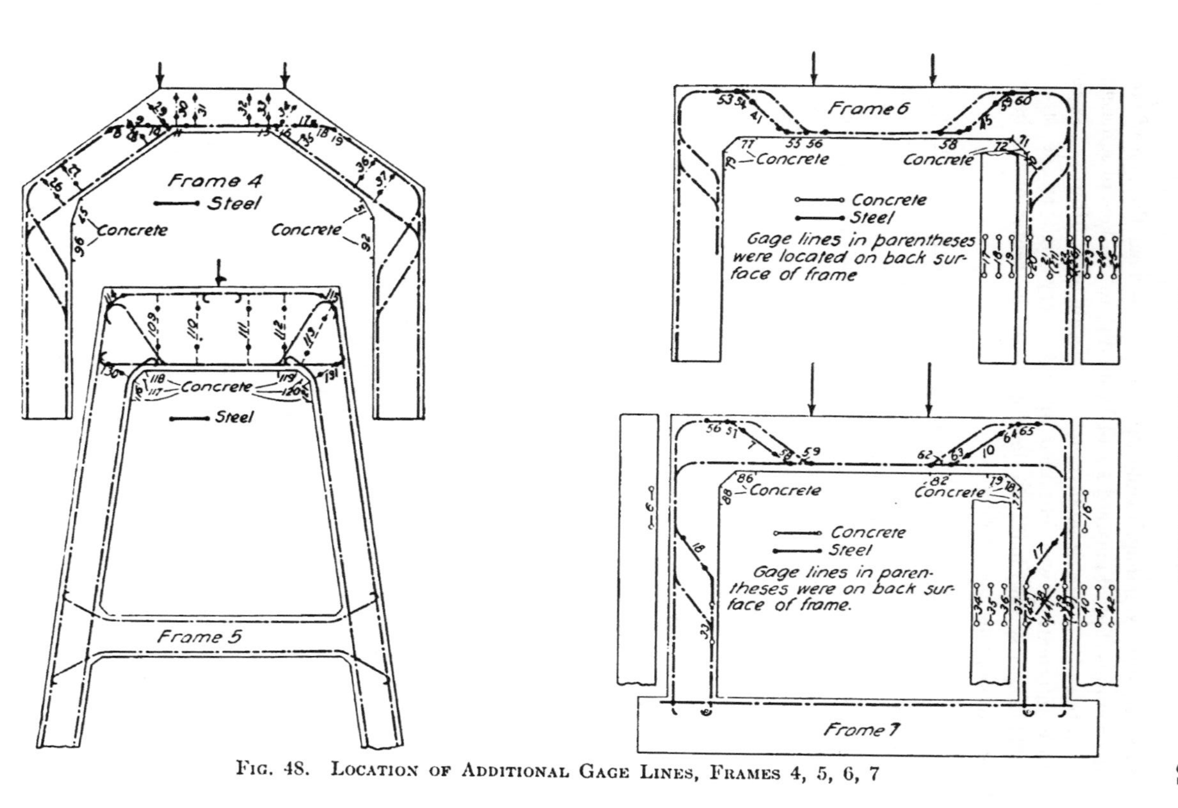

FIG. 48. LOCATION OF ADDITIONAL GAGE LINES, FRAMES 4, 5, 6, 7

the cracks were extending. Unfortunately at this load the foot of one of the columns slipped outward about ¼ in. due to the lack of sufficient friction to resist the horizontal thrust at the support. However, satisfactory information was obtained because very high tensile stress (32,900 lb. per sq. in.) had been developed at the center of the beam before the slipping occurred.

Frame 3—Square Frame with Columns Fixed at Lower End.—Nominal span length was 6 ft. Total height of frame from fixed column end was 4 ft. 11 in. Loads were applied at one-third points of span of horizontal member. No noticeable cracks appeared until 21,000 lb. had been applied, when three cracks appeared at the bottom between loads and several cracks appeared in both columns. At 30,000 lb. new cracks appeared in the beam and columns, and one crack appeared at the top of the beam. The location of the cracks is shown in Fig. 53. The cracks in the upper part of the columns were located within 14 in. downward from the extended line of the bottom face of the beam. No crack was observed at the fixed ends of the columns. The frame carried 60,000 lb. and the load was held for a few minutes, then dropped very slowly, and there appeared to be no danger of sudden failure. No diagonal tension crack appeared in the beam, and the frame failed by tension in the longitudinal steel of the beam.

Frame 4—Inverted U-frame with Columns Hinged at Lower End.—Nominal span length was 6 ft. Total height of frame was 6 ft. 3 in. from hinged end of columns. Loads were applied at one-third points. At 10,000 lb. the first noticeable crack appeared at the left-hand inside top corner, and extended 2½ in. upward. Cracks are shown in Fig. 54. Accidentally the frame was built slightly out of form, the columns being out of plumb 1¼ in. in the height of 4 ft., and more stress was thrown to the left-hand column than to the other. The distribution of the cracks shows this clearly. The frame, however, carried a comparatively high load (50,000 lb.). Failure was by tension in the steel in the horizontal beam and at the rigid joint between the columns and the sloped beam.

Frame 5—Trestle Bent with Tie—(A-frame).—Span length center to center at the supported column ends was 6 feet. Total height from the base to the top of the frame was 10 ft. 1½ in. Load was applied at the center of the top beam. The cross-section of the top beam

Fig. 49. View of Frame 1 in Testing Machine

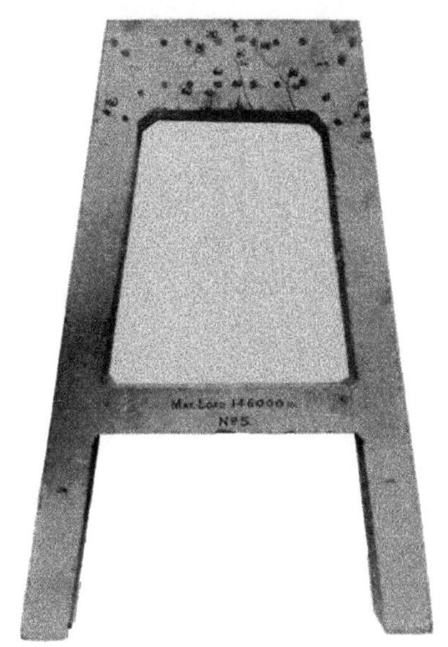

Fig. 50. View of Frame 5 after Test

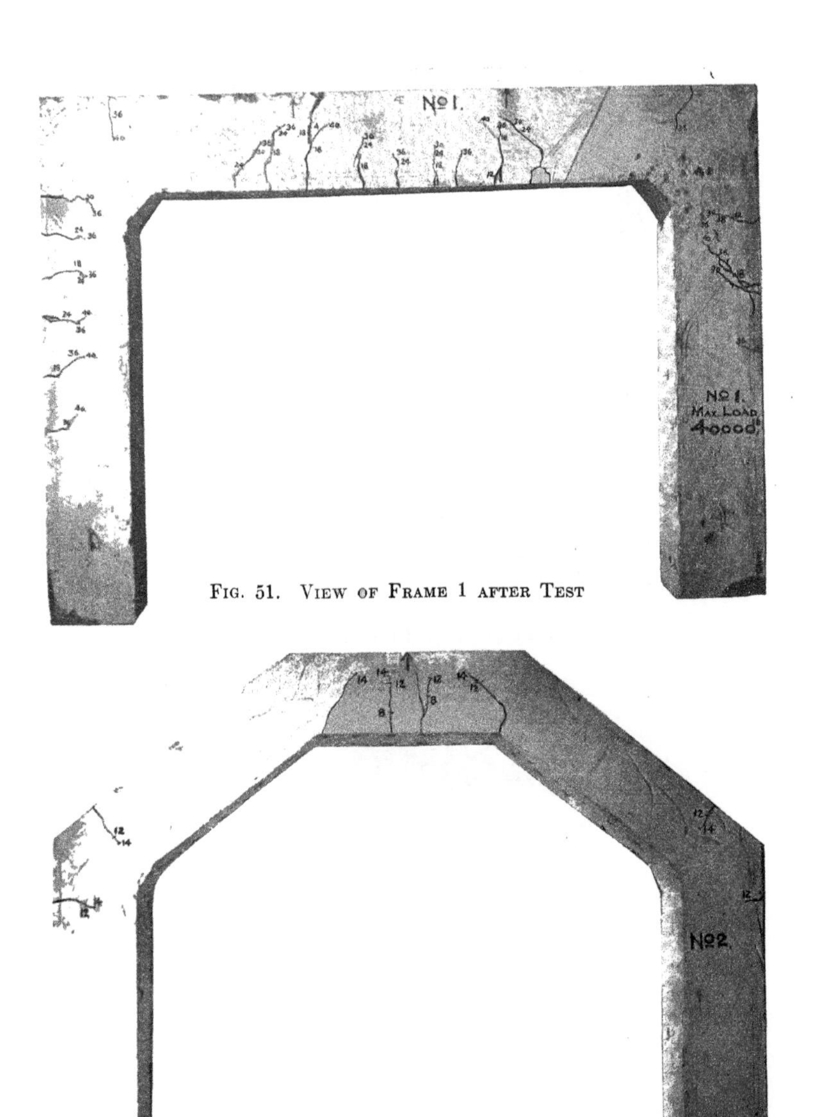

FIG. 51. VIEW OF FRAME 1 AFTER TEST

FIG. 52. VIEW OF FRAME 2 AFTER TEST

Fig. 53. View of Frame 3 after Test

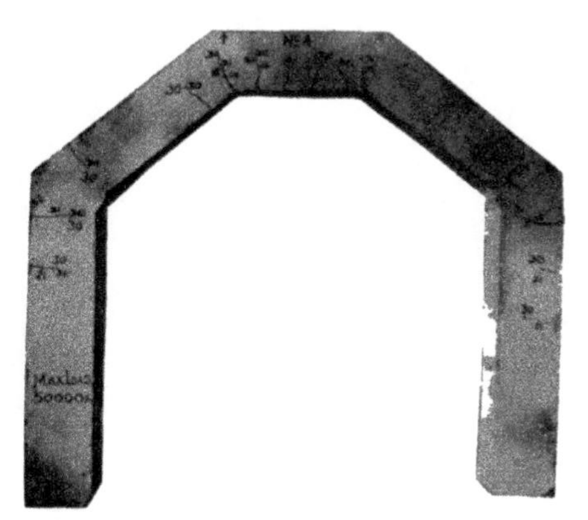

Fig. 54. View of Frame 4 after Test

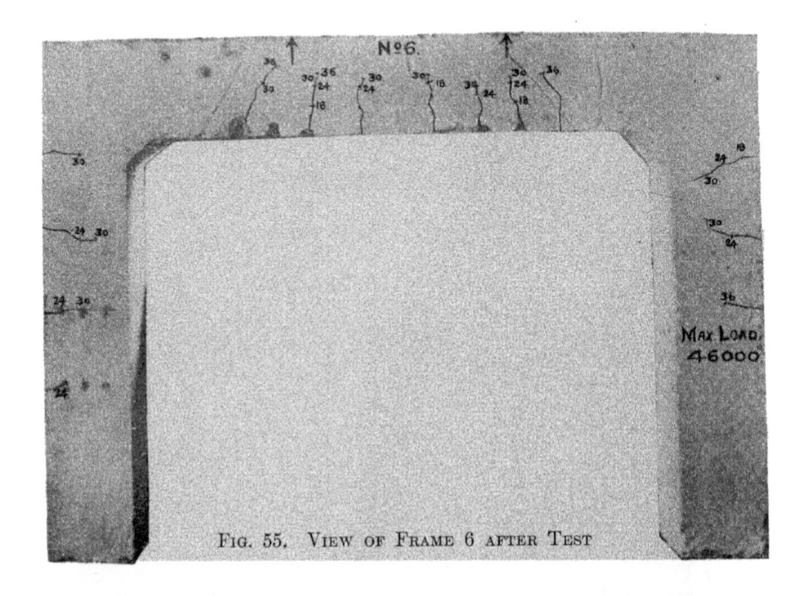

Fig. 55. View of Frame 6 after Test

Fig. 56. View of Frame 7 after Test

was 8½ by 16 in. and the column section 8½ by 8½ in. At 40,000 lb. the first two noticeable cracks appeared. These were under the load points of the beam. For location of cracks see Fig. 50. The outer cracks of the beam finally extended diagonally nearly to the load point. At 100,000 lb. the first cracks appeared in the column at the outside near its juncture with the beam. At 140,000 lb. load a crack suddenly occurred at the right-hand rigid joint between the tie and the column with a breaking sound. The frame carried 146,000 lb. and the load gradually dropped. The maximum load was controlled by the failure of the top beam which failed by tension in the reinforcement.

Frame 6—Square Frame with Columns Hinged at Lower Ends. Same as Frame 1.—Load was applied at one-third points of span of horizontal member. Fig. 55, a view of the frame after the test, shows the appearance of the cracks. At 18,000 lb. four cracks appeared, two of them under the load points, one near the center of the beam, and one at the upper part of the right-hand column. At 24,000 lb. the cracks extended further and additional cracks appeared in the beam and columns at regular intervals. At 30,000 and 36,000 lb. new cracks appeared in the beam where the longitudinal bars were bent up and these cracks ran diagonally almost to the load points. No crack appeared on the top side of the beam. The frame carried 46,000 lb. and the load then dropped slowly. Failure was by tension in the longitudinal steel in the beam.

Frame 7—Square Frame with Columns Fixed at Lower Ends. Same as Frame 3.—Load was applied at one-third points of the span of the horizontal member. The location of the cracks is shown in Fig. 56. The first noticeable cracks appeared at 21,000 lb., three in the beam and three in the columns.

At 30,000 lb., the cracks had extended further and two cracks due to the negative bending moment appeared at the ends 8 in. from the outside face of the columns. At the same load three cracks formed in the bottom half of the beam. As the load increased, the crack located on the outside of the left-hand load point extended diagonally almost to the load point, and the cracks at both ends of the beam extended vertically downward nearly to the bottom side of the beam. The ultimate load carried by the frame was 61,000 lb. At this load sudden failure took place at both inside corners of the lower ends of the columns and the cracks extended horizontally and vertically almost through the concrete base and almost through the columns. This fact shows that considerable positive bending was developed there. The frame also failed by tension in the longitudinal steel of the top beam.

Frame 8—Frame with Three Spans.—Span lengths were 4 ft. 8 in. center to center. Total height of frame was 6 ft. 7¼ in. The base of the frame was 15 ft. 4 in. in length, while the length of the base of the testing machine was 10 ft. 6 in. Consequently the ends of the frame projected beyond the base of the testing machine upon which the frame was bedded in plaster of paris. To observe the upward deflection of the ends of the frame under test an Ames dial was attached at each end of the frame. The movements of both ends were observed as the load increased. The maximum movements (upward) were observed at 60,000 lb. and the amounts were 1/263 in. at the east end and 1/300 in. at the west end. Therefore the steel stress in the beam of the side span may have been slightly modified by the movement, but the structure as a whole probably was not appreciably affected. Uniform load was applied to the upper horizontal member of the middle span. Fig. 57 shows the appearance of the frame after the test with the location of the cracks. No cracks were observed until the load had reached 45,000 lb., when three cracks appeared in the middle span, and one in each outer span on the top side of the horizontal member near the intermediate columns. The former are due to the positive bending moment and the latter are due to the negative moment. These cracks were located symmetrically and they extended vertically about 6 in. At 60,000 lb. they extended deeper. The frame was subjected to the load of 60,000 lb. over 20 hours, but the fall in the applied load was only 300 lb.

At 75,000 lb. several new cracks appeared at both ends of the middle span and also in the upper part of the intermediate columns. At this load the reinforcement at the bottom of the middle beam was stressed in tension beyond the elastic limit of the steel. The frame, however, carried an increasing load in good condition and the highest load was 134,000 lb. At this load the crack at the center of the middle span had opened considerably and the steel at this place had scaled, indicating failure by tension in the steel. At the same time the concrete at the top of the intermediate columns had crushed. Also the concrete base was cracked at the bottom end of the right-hand intermediate column. It is noted that the stresses in the outside columns were very low, even at the maximum load.

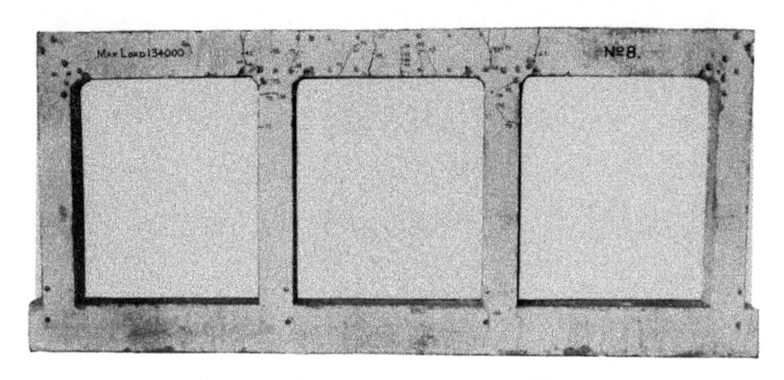

FIG. 57. VIEW OF FRAME 8 AFTER TEST

23. *Conditions Considered in the Comparison of Tests with Analysis.*— In the comparison of test results with analysis, the following modifying conditions have been taken into consideration:

(1) The quality of the concrete was not uniform over the cross-section of the member. The frames were made in a horizontal position on the floor of the laboratory, and the concrete on the rear side of the frame (the bottom side for the position in which the frame was made) seems to have been richer than that on the front side. The concrete on the rear side was stiffer and stronger than that on the front side and the distribution of steel stresses seem to have been modified by this fact. There was more stress in the steel at the rear side of the member than in the steel at the front side. It appears reasonable to take the average value of the observed stresses in the part in question for the purpose of comparison.

(2) The steel stresses are greatly modified by the presence of tension in the concrete for the low loads. Therefore in comparing the computed stresses with the observed stresses, two cases must be considered, one in which the concrete is considered to take tension and the other in which the concrete is considered to be broken in tension.

(3) The cross-sections of the test frames were designedly made larger in proportion to the span than would commonly be used in practice. In most of the test pieces, the column width occupied nearly one-seventh of the nominal span (distance center to center of the columns). In addition to this the corner at the juncture of the beam and the column was provided with a fillet. Under these conditions the bending moment at the center of the beam will be less than that calculated on the basis of the nominal span length—the difference being greater than that occurring with the dimensions found in practice. In the computations for the horizontal reactions the nominal span and height of the frames (distances center to center) were used. In finding the numerical values of the bending moment in the beam, and also of those in columns having fixed ends, the horizontal reactions computed as described previously were used, but when the equations involved further use of span lengths the nominal span length was replaced by the clear length of the span.

(4) The design of Frames 1 to 7 was such as to cause high stresses in columns and beams at about the same time. In Frame 8 the presence

of the members of unloaded panels caused the moments in the columns to be much smaller than in the loaded beam. The base also offered so much restraint as to give the column ends almost a fixed condition. Consequently loads which would cause high stresses in the beams would not produce cracks even in the intermediate columns although the moment there would be greater than in the outer columns. It will be seen then that for Frame 8 there are two cases to be considered in comparing the experimental results with the analyses, one in which tensile strength in the concrete is considered in all members and the other in which the beams and the intermediate columns are cracked on the tension side. The bending moment in the outside column is very small, and there is no chance for tension cracks. The moments in the intermediate columns and in the beams of the side spans are also small except at the extreme end, and only one crack appeared in this member. Therefore in the calculation of the moment of inertia of the cross-section for the second case it is not correct to neglect entirely the tensile strength of the concrete in these two members. A probable value for the moment of inertia of these members will be an average between that obtained by using the full cross-section and a section which neglects the part outside the tension rods. This assumption was made in the numerical computation of the moments and stresses for Frame 8.

In making the computations of stress in the concrete of the frames a constant modulus of elasticity was used, that is, a straight line stress-deformation relation was assumed. The selection of a proper value of the modulus of elasticity was somewhat dependent upon a knowledge of the qualities of the concrete of the various frames and a comparison of the behavior of the frames with that of the corresponding control cylinders. Naturally the modulus of elasticity used was generally less than the initial modulus. The values of modulus of elasticity of the concrete used in the computations of stress are given in Table 11.

24. *Comparison of Test Results with Analyses.*—The observed stresses, i. e., those obtained from the observed deformations by the process already described, have been plotted in Fig. 36 to 43. The light full line represents stress in the steel at the nearer side of the specimen; the dashed line the stress in the farther side of the specimen. The dotted line represents stress in the concrete, usually at the median plane. In general, the stresses have been plotted from the central longitudinal axis of the member in which they were observed. Because of the possibility of confusion resulting from this method an exception

was made in Frames 2, 5, and 8, where the stresses have been plotted from the line showing the location of the reinforcing bar. Tension in steel and compression in concrete when measured at or near the inner surface of a member have been plotted in, that is, toward the central part of the frame as a whole; tension in steel and compression in concrete when measured at or near the outer surface of a member have been plotted out. In only a few cases have tension in concrete and compression in steel been plotted. Both have been plotted out if measured on the inner surface of a member and in if measured on the outer surface. With very few exceptions, consequently, steel stresses represent tension regardless of whether in the diagrams they appear as positive or as negative with reference to the coördinate axes. Plotted concrete stresses, likewise, represent compression. The point at which the stress line crosses the coördinate axis represents the position of the point of inflection and not a change in the sign of the stress. The heavy full line represents the computed stresses for both steel and concrete which have been calculated for various points by means of the formulas given in preceding pages.

In Fig. 45 and 47 are given load-deformation curves for gage lines not represented in the diagrams of Fig. 36 to 43. The location of the gage lines is shown in Fig. 46 and 48.

In Table 14 are given values of both observed and computed stresses at three points for several loads for all the frames. Where I precedes the computed stress the calculation considered the tensile strength of the concrete; where II precedes it and where neither I nor II is given, the computed stress is based on the assumption of no tensile strength in the concrete.

A study of the tables and diagrams seems to justify the following statements:

(1) The experimental and the computed values of steel stress at the center of the loaded top beam are in fair agreement for each kind of frame tested except in a few instances in which the load was comparatively low or extremely high.

(2) The experimental and the computed values of the steel stress in the columns are also in fair agreement, but the maximum difference between experimental and theoretical values is higher than in the beams, owing to the fact that the direct stress is not equally distributed over the cross-section of the column.

TABLE 14

COMPUTED AND OBSERVED STRESSES IN FRAMES

Computed Values are given in Roman Type; Observed Stresses in Italics; f_s is Unit Stress in Steel; f_c is Unit Stress in Concrete.

Frame and Load	A		B		C	
	f_s	f_c	f_s	f_c	f_s	f_c
Frame 1						
18 000	14 800	710	10 800	.....	16 900	840
	16 100	*1 140*	*8 900*	.....	*17 900*	*1 240*
30 000	24 600	1 180	18 000	.....	28 200	1 400
	25 300	*2 560*	*17 400*	.....	*28 600*	*2 500*
36 000	29 500	1 420	21 600	.....	33 800	.....
	31 900	.....	*21 300*	.....	*36 000*	.....
Frame 2						
12 000 I	1 800	490	......	.....	......	.....
II	11 600	710	11 200	800	21 000	.....
	5 000	*690*	*11 600*	*730*	*21 100*	.....
Frame 3						
30 000 I	600	480	......	.....	......	.....
II	.....	.....	13 200	1 080	22 800	1 350
	1 400	*250*	*11 900*	*1 520*	*18 600*	*820*
38 000 I	900	610	......	.....	......	.....
II	.....	.....	16 700	1 370	28 900	1 710
	2 000	*340*	*16 700*	*1 840*	*25 700*	*1 190*
46 000 I	1 100	740	......	.....	......	.....
II	.....	.....	20 200	1 660	35 000	2 070
	2 800	*400*	*22 300*	*1 850*	*34 300*	*1 500*
Frame 4						
10 000 I	1 500	410	2 000	340	......	.....
II	8 500	540	8 700	630	8 700	630
	3 400	*590*	*3 500*	*680*	*7 200*	*730*
21 000 I	3 200	860	......	.....	......	.....
II	17 800	1 130	18 300	1 320	18 300	1 320
	11 700	*1 550*	*16 300*	*1 670*	*19 300*	*1 540*
30 000	25 500	1 620	26 100	1 890	26 100	1 890
	21 400	*2 350*	*23 700*	.....	*29 900*	*2 550*
Frame 5						
40 000 I	— 1 900	— 3 000[1]	— 900	— 3 900[1]	.....	.....
II	.....	......	......	......	12 000	.....
	— 2 000	*— 3 700*[1]	*+ 1 300*	*— 2 200*[1]	*10 100*	.....
80 000 I	— 3 700	— 5 900[1]	— 1 900	— 7 800[1]	......	.....
II	.....	......	+15 200	— 5 300[1]	24 000	.....
	— 2 400	*— 7 300*[1]	*+ 4 500*	*— 8 500*[1]	*24 800*	.....
100 000 I	— 4 700	— 7 400[1]	— 2 400	— 9 700[1]	......	.....
II	.....	......	+19 000	— 6 600[1]	30 000	.....
	— 2 000	*— 9 400*[1]	*+ 9 300*	*—10 900*[1]	*33 500*	.....

TABLE 14—(CONTINUED)

COMPUTED AND OBSERVED STRESSES IN FRAMES

Computed Values are given in Roman Type; Observed Stresses in Italics; f_s is Unit Stress in Steel; f_c is Unit Stress in Concrete.

Frame and Load	A		B		C	
	f_s	f_c	f_s	f_c	f_s	f_c
Frame 6						
18 000 I	1 800	540	2 200	600	4 000	760
I	.....	.....	......	.....	19 600[2]	1 170[2]
II	.....	.....	10 800	830	9 900	840
	3 700	*900*	*6 900*	*1 090*	*9 300*	*1 080*
30 000 I	3 000	900	......	.....	31 800[2]	1 950[2]
II	24 300	1 320	18 000	1 380	26 500	1 680
	14 800	*1 670*	*17 700*	*2 430*	*21 900*	*2 050*
36 000 I	.....	.....	......	.....	38 200[2]	2 340[2]
II	29 100	1 930	21 600	1 660	32 900	2 020
	23 600	*2 460*	*22 100*	.....	*28 100*	*2 750*
Frame 7						
21 000 I	900	3 700[1]	2 700	610	......	.....
II	.....	.....	9 000	800	16 000	1 000
	— *100*	*3 400*[1]	*4 900*	*1 350*	*13 100*	*1 530*
38 000 I	1 600	6 800[1]	4 900	1 100	......	.....
II	.....	.....	16 300	1 440	28 900	1 830
	400	*6 000*[1]	*12 100*	....[3]	*27 500*	*2 900*
46 000 I	1 900	8 200[1]	......	.....	......	.....
II	.....	.....	19 800	1 750	35 000	2 210
	700	*7 700*[1]	*16 500*	....[3]	*37 000*	....[3]
Frame 8						
30 000 I	8 900	850	5 000	6 500[1]	1 800	1 800[1]
	7 000	*1 490*	*1 300*	*5 100*[1]	*2 100*	*3 000*[1]
60 000 I	27 700[2]	1 700[2]	9 900	13 000[1]	3 600	3 600[1]
II	......	.....	.....	......	18 300	2 800[1]
	24 500	*2 540*	*7 200*	*13 100*[1]	*12 800*	*6 200*[1]
75 000 I	......	.....	12 400	16 200[1]	......	.....
II	38 500	2 360	......	......	23 000	3 500[1]
	37 500	*2 750*	*11 500*	*17 700*[1]	*19 100*	*6 700*[1]

[1] Compressive stress in steel.
[2] Concrete in beam under load considered as broken in tension.
[3] Very high stress.

(3) The observed compressive stresses in the concrete at the low loads which developed a unit-stress up to about 800 lb. per sq. in. agree reasonably well with the computed values though in most of these cases the observed concrete stresses were somewhat higher than the computed stresses. In some instances the discrepancies ran up to 50 per cent, and for higher stresses the discrepancies were frequently even greater. Undoubtedly these differences are partly due to the fact that the modulus of elasticity used does not represent correctly the modulus of elasticity of the concrete in the frames. Other matters difficult of explanation probably cause further discrepancies.

25. *Effect of End Condition of Column on Results.*—The secondary stresses which would be expected as a result of the friction in the bearings at the free ends of the columns in Frames 1, 2, 4, and 6 seem to have been very small and may be neglected without appreciable error.

The concrete bases used for Frames 3, 7, and 8 to secure the fixity of the column ends were, of course, not entirely rigid, and a slight bending in the base due to a load may be expected to have an influence on the bending in the other members. Deformation readings at the middle point of the base were taken at each increase in the load. The results of these observations showed practically no bending stress for all loads except the ultimate load.

26. *Distribution of Stress over the Cross-Section.*—In the observations it was found that stress in the steel on bars near a front corner of a member differed from that on bars near a back corner of the member, the front of the member being the top side of the frame as poured and the back side being the bottom. In Table 15 are given stresses in bars at front and back at two places on the frame for one load generally near the maximum. It is seen that generally the stress in a bar near the front of the member (top of the member as poured) is less than that in a bar near the back (bottom of the member as poured). To investigate the distribution further, special measurements were made in the columns of Frames 6 and 7, the gage lines being placed where bending was not sufficiently large to produce tension cracks in the concrete. The gage lines were located on the four faces of the column, and the observations were made at each load. The front outer corner developed the lowest tensile stress and the front inner corner the highest compressive stress. The back outer corner developed the highest tensile

TABLE 15

STRESSES IN BARS AT FRONT AND BACK OF MEMBERS OF TEST FRAMES

Frame No.	At Center of Span				At Upper Part of Column			
	Observed Stress in Steel				Observed Stress in Steel			
	Back Side	Front Side	Difference	Per cent	Back Side	Front Side	Difference	Per cent
1	22 200	27 100	— 4 900	22.3	28 600	35 200	— 6 600	23.1
2	32 900	27 400	+ 5 500	16.7	11 300	9 500	+ 1 800	15.9
3	39 500	29 000	+10 500	26.6	25 500	18 900	+ 6 600	25.9
4	34 400	24 000	+10 400	30.2	31 000	27 600	+ 3 400	11.0
5	36 400	30 600	+ 5 800	16.4	35 600	22 700	+12 900	36.2
6	29 800	26 400	+ 3 400	11.4	23 500	19 300	+ 4 200	17.9
7	33 500	21 500	+12 500	35.8	22 300	10 500	+11 800	52.9
8	30 900	18 400	+12 500	40.4	21 400	15 200	+ 6 200	29.0

stress and the back inner corner the lowest compressive stress. The distribution was not much altered by the increase of load.

The observations indicate a lateral bending and twisting of the frame. It seems probable that the main source of the difference in stresses from front to back was the lack of homogeneity of the concrete, that at the bottom of the member as poured being stronger and stiffer than that at the top. A difference in stiffness would at least partially account for the phenomena.

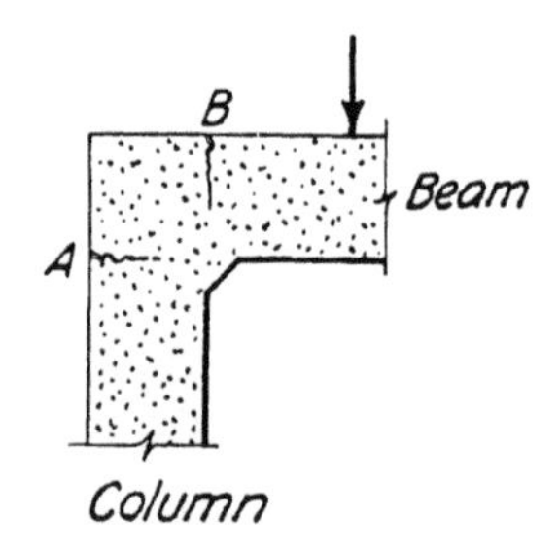

FIG. 58. CONNECTION OF BEAM AND COLUMN SHOWING TYPICAL LOCATION OF FIRST CRACKS

27. *Position of Point of Inflection in Columns.*—The position of the point of inflection in a member of a structure which is subject to flexure is an important element for use in designing the frame. To determine from observed deformations the position of the point of inflection for a member, it is necessary to separate the deformation into that caused by direct stress and that due to flexure of the member. The deformation in the columns of Frames 3 and 7 were thus separated, a straight line stress-deformation relation being assumed, and the position of the point of inflection found. The position of the point of inflection in the columns of these frames changed very little during the progress of the loading. For these frames the point of inflection was found to be almost exactly at one-third the height of the column, as is indicated by the analysis.

28. *Continuity of the Composing Members of a Frame.*—In the tests of the frames there was no sign of discontinuity of members whatever. It is apparent from the action of the frames and from the stresses

observed that the stresses and therefore the moments were well transmitted from member to member by the connection. From the results it is felt that there is every reason to have confidence in the rigidity of connections in frames that are properly designed.

In the frames free to turn at the lower column ends there was a tendency for a crack to form near the juncture at A (Fig. 58) at a lower load than that at which a crack appeared at B. In the frames with rigid connection at the lower column ends, the crack at B appeared at nearly the same time as that at A.

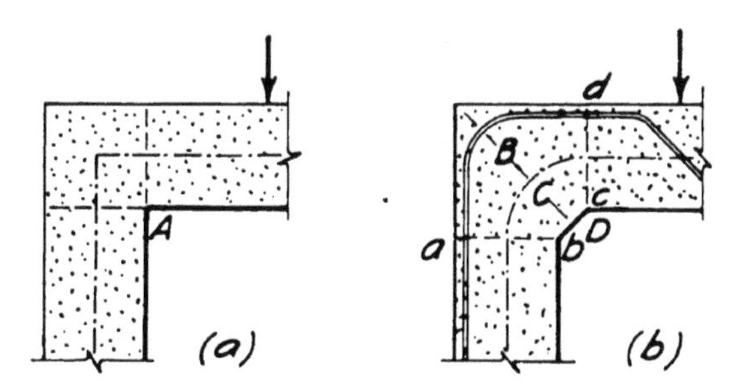

FIG. 59. RECTANGULAR JOINT WITH AND WITHOUT FILLET

29. *Stresses at Corners.*—In the design of a frame a square corner such as that shown at A in Fig. 59 (a) should be avoided for all connections, for it is well known that theoretically in resisting bending the material at the corner would develop excessively high stresses. It is therefore common to design such corners with fillets as shown in Fig. 59 (b). No attempt will be made to compute the stresses at the fillets. The observed deformations at gage lines in the neighborhood of the fillets and within the space occupied by the intersection of the two members are plotted in Fig. 45 and 47. These deformations are of interest. Some of these values have been converted into concrete stresses by the use of the moduli of elasticity of the concrete already assumed and are given in Table 16.

TABLE 16

OBSERVED STRESSES AT SHARP CORNERS IN FRAMES 1, 3 and 6

Stress in lb. per sq. in.

Load lb.	Frame 1 Gage Line 69	Frame 3 Gage Line 81	Frame 6 Gage Line 71
7 000	....	300	....
10 000	....	....	310
12 000	80	....	....
14 000	....	720	....
18 000	670	....	450
21 000	....	760	....
24 000	810	....	900
30 000	....	1 460	1 100
36 000	1 410	....	1 320
38 000	....	2 120	....

30. *Conclusions and General Comments.*—Some of the conclusions which may be drawn from the tests and the discussion are as follows:

(1) Considering the errors involved in the measurement of the deformations and in the determination of the modulus of elasticity of the concrete, as well as those due to assumptions with reference to the distribution of stresses across the section and over the gage length, the results presented indicate a fair agreement between analyses and tests and justify the conclusion that the formulas given in the bulletin for statically indeterminate stresses as applied to reinforced concrete structures will give values for stresses in the members well within the limit of accuracy required in design.

(2) The elastic action of the frames under external load and the manner of stress distribution along the members of the frame agree fairly well with the analyses given.

(3) The location of the point of inflection in the members of the frames under load agrees closely with the location found by analyses.

(4) If a frame is carefully designed and well reinforced, there need be no anxiety as to the rigidity of a joint. Effective continuity of members has been found in the tests.

(5) No sudden failure took place in the frames tested. The increase in the deflection was uniform, indicating as great reliability for reinforced concrete frames as for steel structures.

(6) The load at which the first fine crack appears near the juncture of members is increased by fixing the lower column ends of a frame. This is obviously due to the increase in horizontal thrust at the lower column end over that developed when the lower end is free to turn.

(7) At sharp inside corners, high compressive stresses were developed in the concrete due to so-called curved beam action and in several cases local failure occurred by the crushing of the concrete at these corners under high loads.

(8) A slight deviation of the axis of vertical members from a vertical line, that is to say, a slight "out-of-form" of the vertical columns, produced an appreciable variation in the stress distribution in the frame.

(9) Owing to the existence of a horizontal thrust (which varies from $\frac{1}{8}P$ to $\frac{1}{18}P$ in most common cases of simple frames) at the ends of a vertical or inclined member, it is advisable to incline the member slightly toward the direction of the reaction at the end. Such arrangement will greatly reduce the bending stress in the member. If this arrangement is not practicable, a slight increase in the top width of a vertical member and a slight decrease in its bottom width, brought about by inclining the inner surface and making the outer surface vertical will add materially to the rigidity of a frame without a proportional increase in the amount of material used.

(10) For a frame having an inclined column, it may be possible to select the form of frame in such a way that the column will take no bending stress throughout its length.

(11) Due attention should be paid to the rigid joint of a tie member to insúre the stiff connection with a main member. A marked tendency to cause a sudden breaking of such a joint accompanied an increase of bending moment in a main member.

(12) The use of a footing rigidly connected to the lower end of a vertical member is advisable, for it will reduce the bending moment at the juncture of the vertical and inclined members. A frame having such a footing is solvable analytically, since it approaches the case halfway between that of the hinged end and that of the fixed end of the vertical member, provided the foundation is sufficiently unyielding. A little consideration is needed to provide proper reinforcement at the juncture of the column and the footing.

(13) The formulas derived by analysis may be applied to a variety of forms of frames and are of wide applicability.

LIST OF
PUBLICATIONS OF THE ENGINEERING EXPERIMENT STATION

Bulletin No. 1. Tests of Reinforced Concrete Beams, by Arthur N. Talbot. 1904. *None available.*

Circular No. 1. High-Speed Tool Steels, by L. P. Breckenridge. 1905. *None available.*

Bulletin No. 2. Tests of High-Speed Tool Steels on Cast Iron, by L. P. Breckenridge and Henry B. Dirks. 1905. *None available.*

Circular No. 2. Drainage of Earth Roads, by Ira O. Baker. 1906. *None available.*

Circular No. 3. Fuel Tests with Illinois Coal (Compiled from tests made by the Technological Branch of the U. S. G. S., at the St. Louis, Mo., Fuel Testing Plant, 1904–1907), by L. P. Breckenridge and Paul Diserens. 1908. *Thirty cents.*

Bulletin No. 3. The Engineering Experiment Station of the University of Illinois, by L. P. Breckenridge. 1906. *None available.*

Bulletin No. 4. Tests of Reinforced Concrete Beams, Series of 1905, by Arthur N. Talbot. 1906. *Forty-five cents.*

Bulletin No. 5. Resistance of Tubes to Collapse, by Albert P. Carman and M. L. Carr. 1906. *None available.*

Bulletin No. 6. Holding Power of Railroad Spikes, by Roy I. Webber. 1906. *None available.*

Bulletin No. 7. Fuel Tests with Illinois Coals, by L. P. Breckenridge, S. W. Parr, and Henry B. Dirks. 1906. *None available.*

Bulletin No. 8. Tests of Concrete: I, Shear; II, Bond, by Arthur N. Talbot. 1906. *None available.*

Bulletin No. 9. An Extension of the Dewey Decimal System of Classification Applied to the Engineering Industries, by L. P. Breckenridge and G. A. Goodenough. 1906. Revised Edition 1912. *Fifty cents.*

Bulletin No. 10. Tests of Concrete and Reinforced Concrete Columns, Series of 1906, by Arthur N. Talbot. 1907. *None available.*

Bulletin No. 11. The Effect of Scale on the Transmission of Heat through Locomotive Boiler Tubes, by Edward C. Schmidt and John M. Snodgrass. 1907. *None available.*

Bulletin No. 12. Tests of Reinforced Concrete T-Beams, Series of 1906, by Arthur N. Talbot. 1907. *None available.*

Bulletin No. 13. An Extension of the Dewey Decimal System of Classification Applied to Architecture and Building, by N. Clifford Ricker. 1907. *None available.*

Bulletin No. 14. Tests of Reinforced Concrete Beams, Series of 1906, by Arthur N. Talbot. 1907. *None available.*

Bulletin No. 15. How to Burn Illinois Coal without Smoke, by L. P. Breckenridge. 1908. *None available.*

Bulletin No. 16. A Study of Roof Trusses, by N. Clifford Ricker. 1908. *None available.*

Bulletin No. 17. The Weathering of Coal, by S W. Parr, N. D. Hamilton, and W. F. Wheeler. 1908. *None available.*

Bulletin No. 18. The Strength of Chain Links, by G. A. Goodenough and L. E. Moore. 1908. *Forty cents.*

Bulletin No. 19. Comparative Tests of Carbon, Metallized Carbon and Tantalum Filament Lamps, by T. H. Amrine. 1908. *None available.*

Bulletin No. 20. Tests of Concrete and Reinforced Concrete Columns, Series of 1907, by Arthur N. Talbot. 1908. *None available.*

Bulletin No. 21. Tests of a Liquid Air Plant, by C. S. Hudson and C. M. Garland. 1908. *Fifteen cents.*

Bulletin No. 22. Tests of Cast-Iron and Reinforced Concrete Culvert Pipe, by Arthur N. Talbot 1908. *None available.*

Bulletin No. 23. Voids, Settlement and Weight of Crushed Stone, by Ira O. Baker. 1908. *Fifteen cents.*

**Bulletin No. 24.* The Modification of Illinois Coal by Low Temperature Distillation, by S. W. Parr and C. K. Francis. 1908. *Thirty cents.*

Bulletin No. 25. Lighting Country Homes by Private Electric Plants, by T. H. Amrine. 1908 *Twenty cents.*

*A limited number of copies of bulletins starred is available for free distribution.

Bulletin No. 26. High Steam-Pressure in Locomotive Service. A Review of a Report to the Carnegie Institution of Washington, by W. F. M. Goss. 1908. *Twenty-five cents.*

Bulletin No. 27. Tests of Brick Columns and Terra Cotta Block Columns, by Arthur N. Talbot and Duff A. Abrams. 1909. *Twenty-five cents.*

Bulletin No. 28. A Test of Three Large Reinforced Concrete Beams, by Arthur N. Talbot. 1909. *Fifteen cents.*

Bulletin No. 29. Tests of Reinforced Concrete Beams: Resistance to Web Stresses, Series of 1907 and 1908, by Arthur N. Talbot. 1909. *Forty-five cents.*

**Bulletin No. 30.* On the Rate of Formation of Carbon Monoxide in Gas Producers, by J. K. Clement, L. H. Adams, and C. N. Haskins. 1909. *Twenty-five cents.*

**Bulletin No. 31.* Tests with House-Heating Boilers, by J. M. Snodgrass. 1909. *Fifty-five cents.*

Bulletin No. 32. The Occluded Gases in Coal, by S. W. Parr and Perry Barker. 1909. *Fifteen cents.*

Bulletin No. 33. Tests of Tungsten Lamps, by T. H. Amrine and A. Guell. 1909. *Twenty cents.*

**Bulletin No. 34.* Tests of Two Types of Tile-Roof Furnaces under a Water-Tube Boiler, by J. M. Snodgrass. 1909. *Fifteen cents.*

Bulletin No. 35. A Study of Base and Bearing Plates for Columns and Beams, by N. Clifford Ricker. 1909. *Twenty cents.*

Bulletin No. 36. The Thermal Conductivity of Fire-Clay at High Temperatures, by J. K. Clement and W. L. Egy. 1909. *Twenty cents.*

Bulletin No. 37. Unit Coal and the Composition of Coal Ash, by S. W. Parr and W. F. Wheeler. 1909. *Thirty-five cents.*

Bulletin No. 38. The Weathering of Coal, by S. W. Parr and W. F. Wheeler. 1909. *Twenty-five cents.*

**Bulletin No. 39.* Tests of Washed Grades of Illinois Coal, by C. S. McGovney. 1909. *Seventy-five cents.*

Bulletin No. 40. A Study in Heat Transmission, by J. K. Clement and C. M. Garland. 1910. *Ten cents.*

Bulletin No. 41. Tests of Timber Beams, by Arthur N. Talbot. 1910. *Thirty-five cents.*

**Bulletin No. 42.* The Effect of Keyways on the Strength of Shafts, by Herbert F. Moore. 1910. *Ten cents.*

Bulletin No 43. Freight Train Resistance, by Edward C. Schmidt. 1910. *Seventy-five cents.*

Bulletin No. 44. An Investigation of Built-up Columns under Load, by Arthur N. Talbot and Herbert F. Moore. 1911. *Thirty-five cents.*

**Bulletin No. 45.* The Strength of Oxyacetylene Welds in Steel, by Herbert L. Whittemore. 1911. *Thirty-five cents.*

**Bulletin No. 46.* The Spontaneous Combustion of Coal, by S. W. Parr and F. W. Kressman. 1911. *Forty-five cents.*

**Bulletin No. 47.* Magnetic Properties of Heusler Alloys, by Edward B. Stephenson. 1911. *Twenty-five cents.*

**Bulletin No. 48.* Resistance to Flow through Locomotive Water Columns, by Arthur N. Talbot and Melvin L. Enger. 1911. *Forty cents.*

**Bulletin No. 49.* Tests of Nickel-Steel Riveted Joints, by Arthur N. Talbot and Herbert F. Moore. 1911. *Thirty cents.*

**Bulletin No. 50.* Tests of a Suction Gas Producer, by C. M. Garland and A. P. Kratz. 1912. *Fifty cents.*

Bulletin No. 51. Street Lighting, by J. M. Bryant and H. G. Hake. 1912. *Thirty-five cents.*

**Bulletin No. 52.* An Investigation of the Strength of Rolled Zinc, by Herbert F. Moore. 1912. *Fifteen cents.*

**Bulletin No. 53.* Inductance of Coils, by Morgan Brooks and H. M. Turner. 1912. *Forty cents.*

**Bulletin No. 54.* Mechanical Stresses in Transmission Lines, by A. Guell. 1912. *Twenty cents.*

**Bulletin No. 55.* Starting Currents of Transformers, with Special Reference to Transformers with Silicon Steel Cores, by Trygve D. Yensen. 1912. *Twenty cents.*

**Bulletin No. 56.* Tests of Columns: An Investigation of the Value of Concrete as Reinforcement for Structural Steel Columns, by Arthur N. Talbot and Arthur R. Lord. 1912. *Twenty-five cents.*

**Bulletin No. 57.* Superheated Steam in Locomotive Service. A Review of Publication No. 127 of the Carnegie Institution of Washington, by W. F. M. Goss. 1912. *Forty cents.*

*A limited number of copies of bulletins starred is available for free distribution.

**Bulletin No. 58.* A New Analysis of the Cylinder Performance of Reciprocating Engines, by J. Paul Clayton. 1912. *Sixty cents.*

**Bulletin No. 59.* The Effect of Cold Weather upon Train Resistance and Tonnage Rating, by Edward C. Schmidt and F. W. Marquis. 1912. *Twenty cents.*

**Bulletin No. 60.* The Coking of Coal at Low Temperatures, with a Preliminary Study of the By-Products, by S. W. Parr and H. L. Olin. 1912. *Twenty-five cents.*

**Bulletin No. 61.* Characteristics and Limitation of the Series Transformer, by A. R. Anderson and H. R. Woodrow. 1913. *Twenty-five cents.*

Bulletin No. 62. The Electron Theory of Magnetism, by Elmer H. Williams. 1913. *Thirty-five cents.*

Bulletin No. 63. Entropy-Temperature and Transmission Diagrams for Air, by C. R. Richards. 1913. *Twenty-five cents.*

**Bulletin No. 64.* Tests of Reinforced Concrete Buildings under Load, by Arthur N. Talbot and Willis A. Slater. 1913. *Fifty cents.*

**Bulletin No. 65.* The Steam Consumption of Locomotive Engines from the Indicator Diagrams, by J. Paul Clayton. 1913. *Forty cents.*

Bulletin No. 66. The Properties of Saturated and Superheated Ammonia Vapor, by G. A. Goodenough and William Earl Mosher. 1913. *Fifty cents.*

Bulletin No. 67. Reinforced Concrete Wall Footings and Column Footings, by Arthur N. Talbot. 1913. *Fifty cents.*

Bulletin No. 68. The Strength of I-Beams in Flexure, by Herbert F. Moore. 1913. *Twenty cents.*

Bulletin No. 69. Coal Washing in Illinois, by F. C. Lincoln. 1913. *Fifty cents.*

Bulletin No. 70. The Mortar-Making Qualities of Illinois Sands, by C. C. Wiley. 1913. *Twenty cents.*

Bulletin No. 71. Tests of Bond between Concrete and Steel, by Duff A. Abrams. 1914. *One dollar.*

**Bulletin No. 72.* Magnetic and Other Properties of Electrolytic Iron Melted in Vacuo, by Trygve D. Yensen. 1914. *Forty cents.*

Bulletin No. 73. Acoustics of Auditoriums, by F. R. Watson. 1914. *Twenty cents.*

**Bulletin No. 74.* The Tractive Resistance of a 28-Ton Electric Car, by Harold H. Dunn. 1914. *Twenty-five cents.*

Bulletin No. 75. Thermal Properties of Steam, by G. A. Goodenough. 1914. *Thirty-five cents.*

Bulletin No. 76. The Analysis of Coal with Phenol as a Solvent, by S. W. Parr and H. F. Hadley. 1914. *Twenty-five cents.*

**Bulletin No. 77.* The Effect of Boron upon the Magnetic and Other Properties of Electrolytic Iron Melted in Vacuo, by Trygve D. Yensen. 1915. *Ten cents.*

**Bulletin No. 78.* A Study of Boiler Losses, by A. P. Kratz. 1915. *Thirty-five cents.*

**Bulletin No. 79.* The Coking of Coal at Low Temperatures, with Special Reference to the Properties and Composition of the Products, by S. W. Parr and H. L. Olin. 1915. *Twenty-five cents.*

**Bulletin No. 80.* Wind Stresses in the Steel Frames of Office Buildings, by W. M. Wilson and G. A. Maney. 1915. *Fifty cents.*

**Bulletin No. 81.* Influence of Temperature on the Strength of Concrete, by A. B. McDaniel. 1915. *Fifteen cents.*

Bulletin No. 82. Laboratory Tests of a Consolidation Locomotive, by E. C. Schmidt, J. M. Snodgrass, and R. B. Keller. 1915. *Sixty-five cents.*

**Bulletin No. 83.* Magnetic and Other Properties of Iron Silicon Alloys. Melted in Vacuo, by Trygve D. Yensen. 1915. *Thirty-five cents.*

Bulletin No. 84. Tests of Reinforced Concrete Flat Slab Structure, by Arthur N. Talbot and W. A. Slater. 1916. *Sixty-five cents.*

**Bulletin No. 85.* The Strength and Stiffness of Steel under Biaxial Loading, by A. T. Becker. 1916. *Thirty-five cents.*

**Bulletin No. 86.* The Strength of I-Beams and Girders, by Herbert F. Moore and W. M. Wilson. 1916. *Thirty cents.*

**Bulletin No. 87.* Correction of Echoes in the Auditorium, University of Illinois, by F. R. Watson and J. M. White. 1916. *Fifteen cents.*

Bulletin No. 88. Dry Preparation of Bituminous Coal at Illinois Mines, by E. A. Holbrook. 1916. *Seventy cents.*

---

*A limited number of copies of bulletins starred is available for free distribution.

**Bulletin No. 89.* Specific Gravity Studies of Illinois Coal, by Merle L. Nebel. 1916. *Thirty cents.*

**Bulletin No. 90.* Some Graphical Solutions of Electric Railway Problems, by A. M. Buck. 1916. *Twenty cents.*

**Bulletin No. 91.* Subsidence Resulting from Mining, by L. E. Young and H. H. Stoek. 1916. *None available.*

**Bulletin No. 92.* The Tractive Resistance on Curves of a 28-Ton Electric Car, by E. C. Schmidt and H. H. Dunn. 1916. *Twenty-five cents.*

**Bulletin No. 93.* A Preliminary Study of the Alloys of Chromium, Copper, and Nickel, by D. F. McFarland and O. E. Harder. 1916. *Thirty cents.*

**Bulletin No. 94.* The Embrittling Action of Sodium Hydroxide on Soft Steel, by S. W. Parr. 1917. *Thirty cents.*

**Bulletin No. 95.* Magnetic and Other Properties of Iron-Aluminum Alloys Melted in Vacuo, by T. D. Yensen and W. A. Gatward. 1917. *Twenty-five cents.*

**Bulletin No. 96.* The Effect of Mouthpieces on the Flow of Water Through a Submerged Short Pipe, by Fred B Seely, 1917. *Twenty-five cents.*

**Bulletin No. 97.* Effects of Storage Upon the Properties of Coal, by S. W. Parr. 1917. *Twenty cents.*

**Bulletin No. 98.* Tests of Oxyacetylene Welded Joints in Steel Plates, by Herbert F. Moore. 1917. *Ten cents.*

Circular No. 4. The Economical Purchase and Use of Coal for Heating Homes, with Special Reference to Conditions in Illinois. 1917. *Ten cents.*

**Bulletin No. 99.* The Collapse of Short Thin Tubes, by A. P. Carman. 1917. *Twenty cents.*

**Circular No. 5.* The Utilization of Pyrite Occurring in Illinois Bituminous Coal, by E. A. Holbrook. 1917. *Twenty cents.*

**Bulletin No. 100.* Percentage of Extraction of Bituminous Coal with Special Reference to Illinois Conditions, by C. M. Young. 1917.

**Bulletin No. 101.* Comparative Tests of Six Sizes of Illinois Coal on a Mikado Locomotive, by E. C. Schmidt, J. M. Snodgrass, and O. S. Beyer, Jr. 1917. *Fifty cents.*

**Bulletin No. 102.* A Study of the Heat Transmission of Building Materials, by A. C. Willard and L. C. Lichty. 1917. *Twenty-five cents.*

**Bulletin No. 103.* An Investigation of Twist Drills, by B. Benedict and W. P. Lukens. 1917. *Sixty cents.*

**Bulletin No. 104.* Tests to Determine the Rigidity of Riveted Joints of Steel Structures by W. M. Wilson and H. F. Moore. 1917. *Twenty-five cents.*

Circular No. 6. The Storage of Bituminous Coal, by H. H. Stoek. 1918. *Forty cents.*

Circular No 7. Fuel Economy in the Operation of Hand Fired Power Plants. 1918. *Twenty cents.*

**Bulletin No. 105.* Hydraulic Experiments with Valves, Orifices, Hose, Nozzles, and Orifice Buckets, by Arthur N. Talbot, Fred B Seely, Virgil R. Fleming and Melvin L. Enger. 1918. *Thirty-five cents.*

**Bulletin No. 106.* Test of a Flat Slab Floor of the Western Newspaper Union Building, by Arthur N. Talbot and Harrison F. Gonnerman. 1918. *Twenty cents.*

Circular No. 8. The Economical Use of Coal in Railway Locomotives. 1918. *Twenty cents.*

**Bulletin No. 107.* Analysis and Tests of Rigidly Connected Reinforced Concrete Frames, by Mikishi Abe. 1918. *Fifty cents.*

---

*A limited number of copies of bulletins starred is available for free distribution.

THE UNIVERSITY OF ILLINOIS

THE STATE UNIVERSITY

Urbana

EDMUND J. JAMES, Ph. D., LL. D., President

THE UNIVERSITY INCLUDES THE FOLLOWING DEPARTMENTS:

The Graduate School

The College of Liberal Arts and Sciences (Ancient and Modern Languages and Literatures; History, Economics, Political Science, Sociology; Philosophy, Psychology, Education; Mathematics; Astronomy; Geology; Physics; Chemistry; Botany, Zoology, Entomology; Physiology; Art and Design)

The College of Commerce and Business Administration (General Business, Banking, Insurance, Accountancy, Railway Administration, Foreign Commerce; Courses for Commercial Teachers and Commercial and Civic Secretaries)

The College of Engineering (Architecture; Architectural, Ceramic, Civil, Electrical, Mechanical, Mining, Municipal and Sanitary, and Railway Engineering)

The College of Agriculture (Agronomy; Animal Husbandry; Dairy Husbandry; Horticulture and Landscape Gardening; Agricultural Extension; Teachers' Course; Household Science)

The College of Law (three years' course)

The School of Education

The Course in Journalism

The Courses in Chemistry and Chemical Engineering

The School of Railway Engineering and Administration

The School of Music (four years' course)

The School of Library Science (two years' course)

The College of Medicine (in Chicago)

The College of Dentistry (in Chicago)

The School of Pharmacy (in Chicago; Ph. G. and Ph. C. courses)

The Summer Session (eight weeks)

Experiment Stations and Scientific Bureaus: U. S. Agricultural Experiment Station; Engineering Experiment Station; State Laboratory of Natural History; State Entomologist's Office; Biological Experiment Station on Illinois River; State Water Survey; State Geological Survey; U. S. Bureau of Mines Experiment Station.

The library collections contain (December 1, 1917) 411,737 volumes and 104,524 pamphlets.

For catalogs and information address

THE REGISTRAR
URBANA, ILLINOIS

UNIVERSITY OF ILLINOIS BULLETIN
ISSUED WEEKLY
Vol. XVI   NOVEMBER 4, 1918   No. 10
[Entered as second-class matter December 11, 1912, at the post office at Urbana, Illinois, under the Act of August 24, 1912. Acceptance for mailing at the special rate of postage provided for in section 1103, Act of October 3, 1917, authorized July 31, 1918.]

ANALYSIS OF STATICALLY INDETERMINATE STRUCTURES BY THE SLOPE DEFLECTION METHOD

BY
W. M. WILSON
F. E. RICHART AND CAMILLO WEISS

BULLETIN No. 108

ENGINEERING EXPERIMENT STATION

PUBLISHED BY THE UNIVERSITY OF ILLINOIS, URBANA

PRICE: ONE DOLLAR
EUROPEAN AGENT
CHAPMAN & HALL, LTD., LONDON

THE Engineering Experiment Station was established by act of the Board of Trustees, December 8, 1903. It is the purpose of the Station to carry on investigations along various lines of engineering and to study problems of importance to professional engineers and to the manufacturing, railway, mining, constructional, and industrial interests of the State.

The control of the Engineering Experiment Station is vested in the heads of the several departments of the College of Engineering. These constitute the Station Staff and, with the Director, determine the character of the investigations to be undertaken. The work is carried on under the supervision of the Staff, sometimes by research fellows as graduate work, sometimes by members of the instructional staff of the College of Engineering, but more frequently by investigators belonging to the Station corps.

The results of these investigations are published in the form of bulletins, which record mostly the experiments of the Station's own staff of investigators. There will also be issued from time to time, in the form of circulars, compilations giving the results of the experiments of engineers, industrial works, technical institutions, and governmental testing departments.

The volume and number at the top of the front cover page are merely arbitrary numbers and refer to the general publications of the University of Illinois: *either above the title or below the seal is given the number of the Engineering Experiment Station bulletin or circular which should be used in referring to these publications.*

For copies of bulletins, circulars, or other information address the

ENGINEERING EXPERIMENT STATION,
URBANA, ILLINOIS.

UNIVERSITY OF ILLINOIS
ENGINEERING EXPERIMENT STATION

BULLETIN NO. 108 NOVEMBER, 1918

ANALYSIS OF STATICALLY INDETERMINATE STRUCTURES BY THE SLOPE DEFLECTION METHOD

BY

W. M. WILSON
ASSISTANT PROFESSOR OF CIVIL ENGINEERING

F. E. RICHART
INSTRUCTOR IN THEORETICAL AND APPLIED MECHANICS

AND

CAMILLO WEISS
INSTRUCTOR IN STRUCTURAL ENGINEERING

ENGINEERING EXPERIMENT STATION
PUBLISHED BY THE UNIVERSITY OF ILLINOIS, URBANA

CONTENTS

ANALYSIS OF STATICALLY INDETERMINATE STRUCTURES BY THE SLOPE DEFLECTION METHOD

I. Preliminary

1. *Object and Scope of Investigation.*—Frames composed of rectangular elements must in general be designed with stiff connections between the members at the joints, in order that loads may be carried. These connections must be capable of transferring not only direct axial tensile and compressive forces, but also bending moments. It follows that frames made up of rectangular elements are usually statically indeterminate; that is, the stresses in them can be found only by taking into account the relative stiffness and deformations of the various members. The common use of rectangular frames in engineering structures makes it highly desirable that the most convenient methods of analyzing their stresses should be developed. The stresses in a number of such rectangular frames have been analyzed by the writers. This bulletin describes the methods used and presents the formulas derived.

The bulletin is divided into two parts: the first part is devoted to the derivation of fundamental equations; in the second part, methods and equations are derived for use in determining moments, stresses, and deflections for a variety of typical structures.

2. *Acknowledgments.*—The investigation here reported was made under the auspices of the Department of Civil Engineering of which Dr. F. H. Newell is the head. A portion of the work was done in 1915 in connection with the development of a thesis in partial fulfillment of the requirements for the degree of Master of Science in Civil Engineering by F. E. Richart. Many of the analyses have been checked by W. L. Parish, graduate student in Architectural Engineering, and Yi Liu, graduate student in Civil Engineering, to whom the authors gratefully acknowledge their indebtedness.

PART I

DERIVATION OF FUNDAMENTAL EQUATIONS

II. Propositions upon which Fundamental Equations Are Based

3. *Statement of Propositions.*—The fundamental equations used in these investigations are derived from the principal propositions of the moment-area method.* These may be expressed as follows:

(1) *When a member is subjected to flexure, the difference in the slope of the elastic curve between any two points is equal in magnitude to the area of the* $\frac{M}{EI}$ *diagram for the portion of the member between the two points.*

(2) *When a member is subjected to flexure, the distance of any point Q on the elastic curve, measured normal to initial position of member, from a tangent drawn to the elastic curve at any other point P is equal in magnitude to the first or statical moment of the area of the* $\frac{M}{EI}$ *diagram between the two points, about the point Q.*

The $\frac{M}{EI}$ diagram is a graph in which the ordinate at any point is obtained by dividing the resisting moment, M, by the product of modulus of elasticity of material, E, and the moment of inertia of the section, I, at that point. If E and I are constant, the diagram will be similar in shape to the moment diagram for the member.

4. *Proof of Propositions.*—The line AB, Fig. 1, represents the elastic curve of a member in flexure. Consider the elementary length, ds, of the member shown in Fig. 2. The angle between radii at the ends of ds will be denoted by $d\theta$. The linear deformation of a fibre at a distance c from the neutral surface is $cd\theta$, and the unit deformation of the same fibre is $\frac{cd\theta}{ds}$. From the well known flexure formula the

*The principles of the moment-area method were given in an article by O. Mohr, Beitraege zur Theorie der Holz-und Eisenkonstruktionen, Zeitschrift des Arch.-und Ing. Ver. zu Hannover, 1868, p. 19. About the same time the method was presented by C. E. Greene in lectures at the University of Michigan. Several modern textbooks on mechanics give the method; see, for instance, Strength of Materials, by J. E. Boyd, Second Ed., 1917.

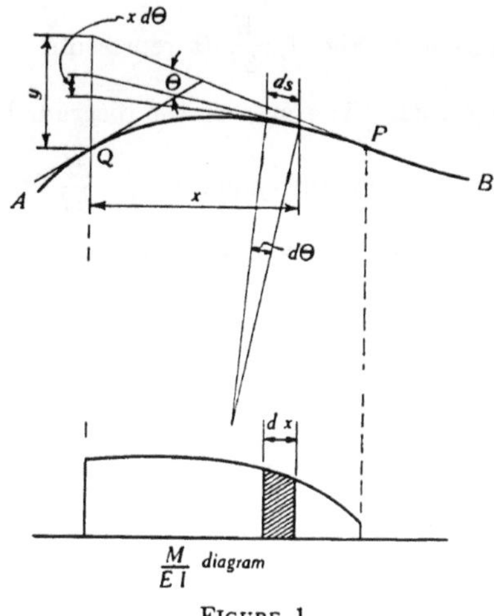

FIGURE 1

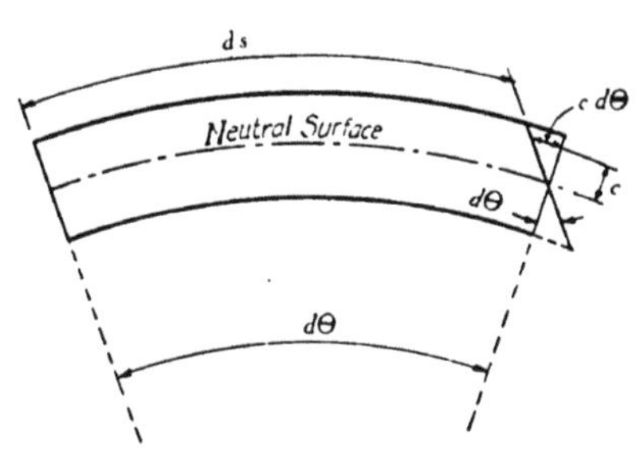

FIGURE 2

accompanying unit stress in the fibre is $S=\frac{Mc}{I}$, in which M is the resisting moment and I moment of inertia of section.

Since the modulus of elasticity is the ratio of unit stress to unit deformation, E is equal to $\frac{Mc}{I}$ divided by $\frac{cd\theta}{ds}$. Hence $d\theta=\frac{M}{EI}ds$. Since in a well designed beam, the curvature and slope are small, dx may be substituted for ds without material error, and $d\theta=\frac{M}{EI}dx$.

In the $\frac{M}{EI}$ diagram of Fig. 1, $\frac{M}{EI}dx$ represents the area of the diagram for the length dx. The area of the diagram between points Q and P on the elastic curve is then equal to $\int_Q^P \frac{M}{EI}dx$. But the difference of slope of the tangent to the elastic curve is also represented by

$$\theta=\int d\theta=\int_Q^P \frac{M}{EI}dx \qquad (1)$$

Hence proposition (1) of the preceding section is proved. It may be noted that if M is taken as the resisting moment acting on the portion of the member to the left of any section, by applying the conventions of section 5 the area of the $\frac{M}{EI}$ diagram is positive; also the direction of integration from Q to P is positive, and the difference in slope θ is positive. Other terms involved may be considered as scalar quantities. These conventions apply to any case, as, for instance, difference in slope from P to Q is negative, since the direction of integration is negative.

In Fig. 1 the tangents at the extremities of the element of the elastic curve, ds, are extended until they intersect the vertical line through the point Q. The intercept on this vertical line between the two consecutive tangents is $xd\theta$. The total vertical distance, y, of Q from the tangent drawn at P is the algebraic sum of all the intercepts between tangents for the portion of the curve between Q and P; that is, $y=\int_Q^P xd\theta$. Substituting the value of $d\theta$ found previously,

$$y=\int_Q^P \frac{M}{EI}x\,dx \qquad (2)$$

In the $\frac{M}{EI}$ diagram of Fig. 1, $\frac{M}{EI}dx$ represents the area of the diagram for the length dx, and $\frac{M}{EI}dx$ times x represents the moment of this area about the point Q. The moment of the entire area of the $\frac{M}{EI}$ diagram between points Q and P about the point Q may now be expressed by $\int_{Q}^{P} \frac{M}{EI} x dx$. Since this expression is identical with the right-hand member of equation (2), proposition (2) of the preceding section is proved. The conventions of section 5 apply here as explained in the proof of proposition (1).

III. Derivation of Fundamental Equations

5. *Conventional Signs.*—The signs of the quantities used in the equations in this bulletin are determined by the following conventional rules:

When the tangent to the elastic curve of a member has been turned through a clockwise direction, measured *from* its initial position, the change in slope, or the angular deformation, is positive.

When the line joining the ends of a member is rotated, the movement of one end of the member relative to the other, measured perpendicular to the initial position of member is called a deflection, and is so used throughout the following discussion. The deflection is positive when such rotation is in a clockwise direction from the initial position of member.

The resisting moment or moment of the internal stresses on a section is positive when the internal or resisting couple acts in a clockwise direction upon *the portion of the member considered.* According to this rule the portion of the member considered must always be specified, and will be indicated by the subscripts used with the moments. For example, if C is a point on a member between the ends A and B, M_{CA} is equal to $-M_{CB}$.

The moment of an external force or couple is positive if it tends to cause a clockwise rotation.

6. *Derivation of Equations for Moments at Ends of Members in Flexure—Member Restrained at the Ends with No Intermediate Loads.*—The line AB in Fig. 3 represents the elastic curve of a member which is not acted upon by any external forces or couples except at the ends. The resisting moment at A is represented by M_{AB} and at B by M_{BA}. The change in the slope of the elastic curve at A from its initial position is represented by θ_A, and that at B by θ_B. The deflection of A from its original position A' is d. The distance of B from the tangent drawn to the curve at A is equal to $(d-l\theta_A)$.

From proposition (2), section 3, $(d-l\theta_A)$ may be expressed as the statical moment of the $\frac{M}{EI}$ diagram for member AB about the end B.

The quantities E and I will here be considered as constant throughout the length AB. If M represents the resisting moment on the portion of member to the left of a section, M is equal to $-M_{AB}$ at A, and to $+M_{BA}$ at B. The $\frac{M}{EI}$ diagram of Fig. 3 can best be treated as the

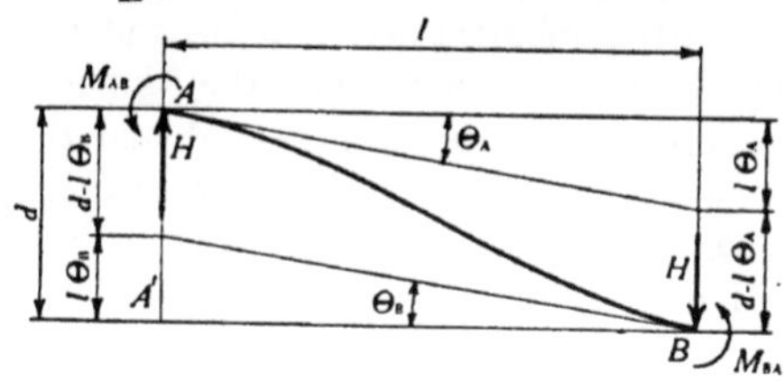

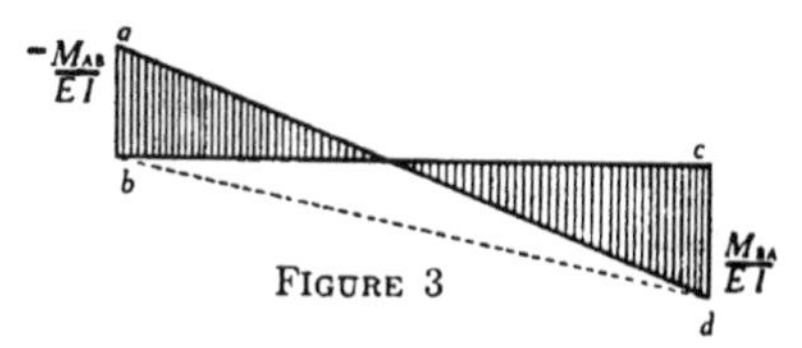

FIGURE 3

algebraic sum of the two triangles *bad* and *bcd*. Hence the statical moment of the $\frac{M}{EI}$ diagram about B is equal to the area of triangle *bad* times the distance to its centroid, $\frac{2}{3}l$, plus the area of triangle *bcd* times the distance to its centroid, $\frac{1}{3}l$. This gives

$$d-l\theta_A=\frac{-M_{AB}l^2}{3EI}+\frac{M_{BA}l^2}{6EI} \quad . \quad . \quad . \quad (3)$$

From proposition (1), section 3, $\theta_B-\theta_A$ is equal to the area of the $\frac{M}{EI}$ diagram for member AB, or the algebraic sum of areas *bad* and *bcd*. This gives

$$\theta_B-\theta_A=\frac{-M_{AB}l}{2EI}+\frac{M_{BA}l}{2EI} \quad . \quad . \quad . \quad (4)$$

Combining equations (3) and (4) to eliminate M_{BA}, letting $\frac{I}{l}=K$ and $\frac{d}{l}=R$, gives

$$M_{AB}=2EK(2\theta_A+\theta_B-3R) \quad . \quad . \quad . \quad (5)$$

Similarly combining equations (3) and (4) to eliminate M_{AB} gives

$$M_{BA}=2EK(2\theta_B+\theta_A-3R) \quad . \quad . \quad . \quad (6)$$

Since the signs of all quantities in equations (3) and (4) are independent of the sense of the quantities themselves it follows that equations (3) and (4) are general; and they give the sense as well as magnitude of the moments, no matter what the senses of θ_A, θ_B, and R may be, provided the method of determining signs given in section 5 is followed. As before noted, M_{AB} is the resisting moment acting at the end A of the member AB. The moment which AB exerts upon the support at A is equal in magnitude but opposite in sense to M_{AB}. A and B are not necessarily supports of a member but may be any two points along the length of a member, provided there is no intermediate load on the member between them.

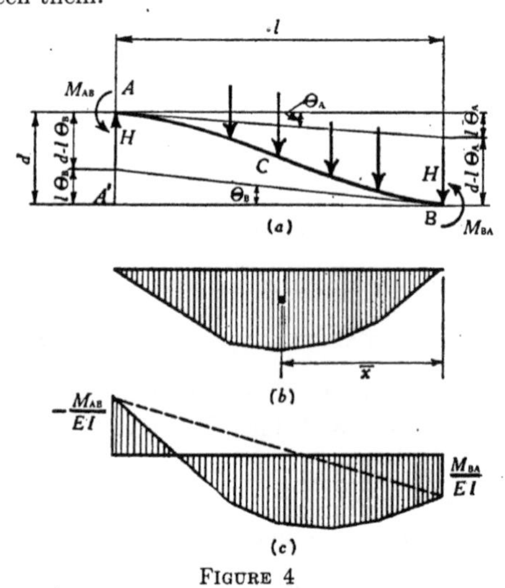

FIGURE 4

Equations (5) and (6) are fundamental equations.* They may be expressed as follows:—The moment at the end of any member carrying no intermediate loads is equal to $2EK$ times the quantity: Twice the change in slope at the near end plus the change in slope at the far end minus three times the ratio of deflection to length. E is the modulus

*The slope-deflection equations for a member acted upon only by forces and couples at the ends were deduced by Manderla in 1878. See Annual Report of the Technische Hochschule, Munich, 1879, and Allgemeine Bauzeitung, 1880. The use of these equations has been developed by several writers, among whom are:

Mohr, Otto, "Abhandlungen aus dem Gebiete der Technischen Mechanik," Second Ed., 1914.

Kunz, F. C. "Secondary Stresses," Engineering News, Vol. 66, p. 397, Oct. 5, 1911.

Wilson and Maney, "Wind Stresses in the Steel Frames of Office Buildings," Univ. of Ill. Eng. Exp. Sta., Bul. 80, 1915.

of elasticity of the material, and K is the ratio of moment of inertia to length of member.

7. *Derivation of Equations for Moments at Ends of Members in Flexure—Member Restrained at the Ends with Any System of Intermediate Loads.*—The line AB, Fig. 4-a, represents the elastic curve of the member of Fig. 3, but acted upon by a system of intermediate loads. The moments, slopes, and deflections at A and B are similar to those of Fig. 3. The $\frac{M}{EI}$ diagram, however, is affected by the intermediate loads. The quantity EI will again be considered constant. From well known principles of mechanics, the $\frac{M}{EI}$ diagram of Fig. 4-c may be obtained by superimposing the $\frac{M}{EI}$ diagram for a simple beam under the same intermediate loads (see Fig. 4-b) upon the $\frac{M}{EI}$ diagram of Fig. 3. This is merely the algebraic addition of the different moments at any section, just as in an algebraic analysis the moment at the end of a girder is combined with the moment of the shear at the end and of the external loads about the given section. Denote the area of the simple beam diagram of Fig. 4-b by F, and the distance of its centroid from B by $\bar{x}$. Then, using the propositions of section 3 as before, the statical moment of the $\frac{M}{EI}$ diagram about B is equal to $(d-\theta_A l)$.

$$(d-\theta_A l)=\frac{-M_{AB}l}{3EI}+\frac{M_{BA}l}{6EI}-\frac{F\bar{x}}{EI} \quad . \quad . \quad . \quad (7)$$

The area of the $\frac{M}{EI}$ diagram is equal to $\theta_B-\theta_A$.

$$(\theta_B-\theta_A)=\frac{-M_{AB}l}{2EI}+\frac{M_{BA}l}{2EI}-\frac{F}{EI} \quad . \quad . \quad . \quad (8)$$

Combining equations (7) and (8) to eliminate M_{BA}, letting $\frac{I}{l}=K$ and $\frac{d}{l}=R$, gives

$$M_{AB}=2EK\,(2\theta_A+\theta_B-3R)\;-\;\frac{2F}{l^2}(3\bar{x}-l) \quad . \quad . \quad . \quad (9)$$

Similarly, combining equations (7) and (8) to eliminate M_{AB} gives

$$M_{BA}=2EK\,(2\theta_B+\theta_A-3R)\;+\;\frac{2F}{l^2}\,(2l-3\bar{x}) \quad . \quad . \quad . \quad (10)$$

It is seen that equations (9) and (10) are identical with equations (5) and (6) except that they contain an additional term in the right-hand members of the equations. This additional term is independent of the slopes and deflections of the member, and depends solely upon the intermediate loads. Further significance is given to this term if the slopes and deflections are made equal to zero, as is true in a fixed beam with supports on same level. The last term then becomes the resisting moment acting on the end of the fixed beam. Hence it is seen that in general *the resisting moment at the end of a member with any system of intermediate loads can be expressed as the algebraic sum of the resisting moment at the end of a member with no intermediate loads, as given by equations* (5) *and* (6), *and the resisting moment at the end of a fixed beam with an equal span and carrying the same system of intermediate loads.*

If the resisting moment at the end of a fixed beam with supports on same level be expressed by C, with subscripts similar to those used for moments, M, equations (9) and (10) may be written in the following general form

$$M_{AB}=2EK(2\theta_A+\theta_B-3R)-C_{AB} \qquad (11)$$

$$M_{BA}=2EK(2\theta_B+\theta_A-3R)+C_{BA} \qquad (12)$$

These are the general *slope deflection* equations which apply to any condition of loading and restraint.

The sign of the constant C may be determined as follows: *In a fixed beam the sign of the resisting moment at the end of a member is opposite to that of the moment of external loads.* For instance, in Fig. 4 the moment of external loads about the end A is clockwise, so the resisting moment C_{AB} is counter clockwise or negative; and since the moment of the loads is counter clockwise about B, C_{BA} is clockwise or

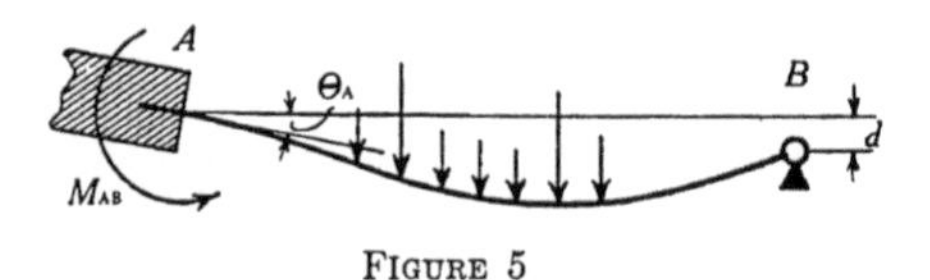

FIGURE 5

positive. If the loads were upward instead of downward, the signs of C_{AB} and C_{BA} would be reversed. With signs thus treated, C becomes merely a numerical, or scalar, quantity.

It has been noted that equations (11) and (12) apply to any condition of restraint of the ends of a member. Fig. 5 shows a member restrained at A and hinged to the support at B, so that the resisting moment at B is zero. Equations (11) and (12) may be written:

$$M_{AB}=2EK(2\theta_A+\theta_B-3R)-C_{AB}$$
$$0=M_{BA}=2EK(2\theta_B+\theta_A-3R)+C_{BA}$$

Combining these two equations to eliminate θ_B gives

$$M_{AB}=EK(3\theta_A-3R)-(C_{AB}+\frac{C_{BA}}{2}) \quad . \quad . \quad . \quad . \quad . \quad . \quad . \quad . \quad (13)$$

If the beam is fixed at A and hinged at B, with the supports on the same level, θ_A and R in equation (13) are zero, and therefore the term $-(C_{AB}+\frac{C_{BA}}{2})$ represents the resisting moment at the end A, and can be readily calculated for any given loading.

By similar reasoning, when a beam is restrained at the end B and hinged to support at A, it is found that

$$M_{BA}=EK(3\theta_B-3R)+(C_{BA}+\frac{C_{AB}}{2}) \quad . \quad . \quad . \quad . \quad . \quad . \quad . \quad . \quad (14)$$

For more convenient reference let the quantity $(C_{AB}+\frac{C_{BA}}{2})$ be denoted by H_{AB}, and the quantity $(C_{BA}+\frac{C_{AB}}{2})$ by H_{BA}.

Equations (13) and (14) then take the general form

$$M_{AB}=EK(3\theta_A-3R)-H_{AB} \quad . \quad . \quad . \quad . \quad . \quad . \quad . \quad . \quad . \quad . \quad . \quad (15)$$

$$M_{BA}=EK(3\theta_B-3R)+H_{BA} \quad . \quad . \quad . \quad . \quad . \quad . \quad . \quad . \quad . \quad . \quad . \quad (16)$$

The term H represents the resisting moment at the fixed end of a beam which is fixed at one end and hinged to the support at the other, with supports at same level. The sign of H is determined in the same way as the sign of C in equations (11) and (12). That is, the sign of H is always opposite to the sign of the moment of the external loads about the fixed end of the member. If the external loads act upward instead of downward, the values of H in equations (15) and (16) must be reversed.

Equations (11) and (12) are the general equations for the ends of a member in flexure. Equations (15) and (16) are special forms of equations (11) and (12), applicable to members having one end hinged. For convenience in reference these four equations are given in Table 1 where they are denoted as equations (A), (B), (C), and (D), respectively.

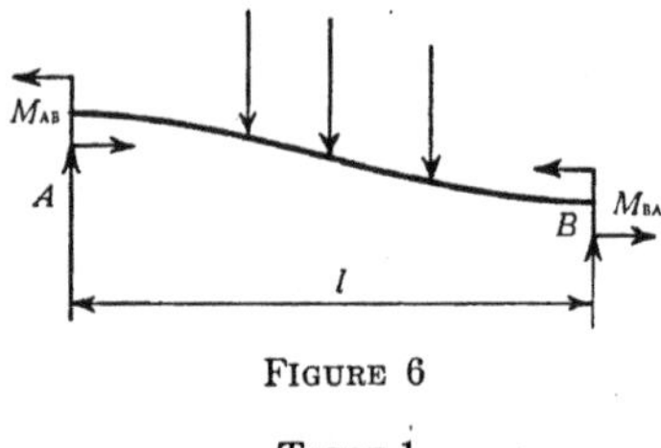

FIGURE 6

TABLE 1

GENERAL EQUATIONS FOR THE MOMENTS AT THE ENDS OF A MEMBER AB IN FIG. 6

$$M_{AB} = 2EK(2\theta_A + \theta_B - 3R) \mp C_{AB} \quad \text{(A)}$$

$$M_{BA} = 2EK(2\theta_B + \theta_A - 3R) \pm C_{BA} \quad \text{(B)}$$

If end B is hinged,

$$M_{AB} = EK(3\theta_A - 3R) \mp H_{AB} \quad \text{(C)}$$

If end A is hinged,

$$M_{BA} = EK(3\theta_B - 3R) \pm H_{BA} \quad \text{(D)}$$

NOTE.—The signs of the quantities used in these equations are determined by the following rules:

θ is positive (+) when the tangent to the elastic curve is turned in a clockwise direction.

R is positive (+) when the member is deflected in a clockwise direction.

The moment of the internal stresses on a section is positive (+) when the internal couple acts in a clockwise direction upon the *portion of the member considered.*

If the moment of the external forces on the member about the end at which the moment is to be determined is positive (+), the sign before the constant is minus (—); if the moment of the external forces about the end at which the moment is to be determined is negative (—), the sign before the constant is plus (+). With the forces acting downward as shown in the sketch, for the moment at A, C_{AB} and H_{AB} are preceded by a minus (—) sign, but for the moment at B, C_{BA} and H_{BA} are preceded by a plus (+) sign.

8. *Derivation of Equations for Moments at Ends of Members in Flexure—Member Restrained at the Ends, with Special Cases of* Loading.— One method of determining the quantities C and H in equations (A), (B), (C), and (D) of Table 1 has been explained. To illustrate the method, some special cases will be considered here.

Fig. 7 shows a member restrained at the ends with a concentrated load at a distance a from A, and a distance b from B. In the simple beam moment diagram, the maximum ordinate is $\frac{-Pab}{l}$, the area F is $\frac{-Pab}{2}$, and the distance $\bar{x}$ of the centroid of the area F from B is $\frac{1}{3}(l+b)$.

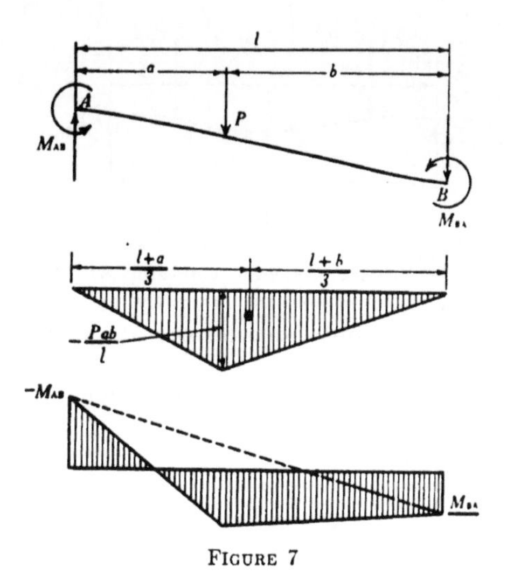

FIGURE 7

Hence putting these values in the last terms of equations (9) and (10) gives

$$-\frac{2F}{l^2}(3\bar{x}-l)=\frac{-Pab^2}{l^2}=-C_{AB} \quad . \quad . \quad . \quad . \quad . \quad . \quad . \quad . \quad . \quad . \quad (17)$$

and

$$\frac{2F}{l^2}(2l-3\bar{x})=\frac{+Pa^2b}{l^2}=+C_{BA} \quad . \quad . \quad . \quad . \quad . \quad . \quad . \quad . \quad . \quad . \quad (18)$$

If the member had been hinged instead of being restrained at B, the value of H_{AB} could have been found from the last term of equation (13), in which

$$-(C_{AB}+\frac{C_{BA}}{2})=\frac{-Pab}{2l^2}(l+b)=-H_{AB} \quad . \quad . \quad . \quad . \quad . \quad . \quad . \quad (19)$$

Similarly, if the member had been hinged at A and restrained at B, the value of H_{BA} could have been found from the last term of equation (14) in which

$$+(C_{BA}+\frac{C_{AB}}{2})=\frac{Pab(l+a)}{2l^2}=H_{BA} \quad . \quad . \quad . \quad . \quad . \quad . \quad . \quad . \quad . \quad (20)$$

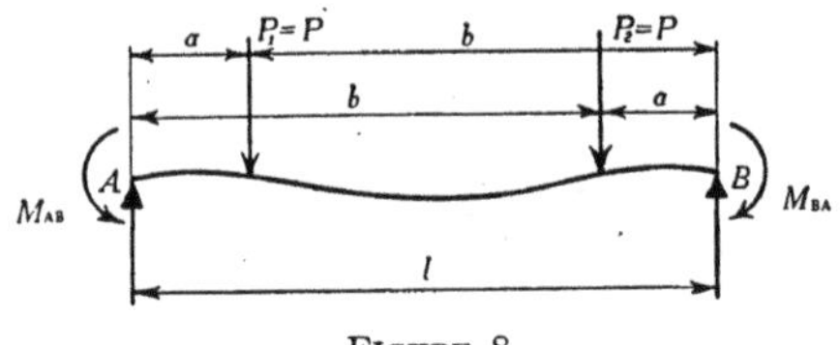

FIGURE 8

As another common case, consider a loading which is symmetrical about the middle of the member, as shown by Fig. 8. It is obvious that the centroid of the simple beam moment diagram will be at the middle of the member, so that $\bar{x}=\frac{l}{2}$. Substituting this in the last term of equations (9) and (10),

$$-\frac{2F}{l^2}(3\bar{x}-l)=-\frac{F}{l}=-C_{AB} \quad . \quad . \quad . \quad . \quad . \quad . \quad . \quad . \quad . \quad . \quad (21)$$

and

$$+\frac{2F}{l^2}(2l-3\bar{x})=+\frac{F}{l}=+C_{BA} \quad . \quad . \quad . \quad . \quad . \quad . \quad . \quad . \quad . \quad . \quad (22)$$

Similarly, for a member having the end A hinged, the last term of equation (13) gives

$$-(C_{AB}+\frac{C_{BA}}{2})=-\frac{3}{2}\frac{F}{l}=-H_{AB} \quad . \quad . \quad . \quad . \quad . \quad . \quad . \quad . \quad . \quad (23)$$

For a member having the end B hinged, the last term of equation gives

$$+(C_{BA}+\frac{C_{AB}}{2})=+\frac{3}{2}\frac{F}{l}=+H_{BA} \quad . \quad . \quad . \quad . \quad . \quad . \quad . \quad . \quad (24)$$

A geometrical meaning is attached to the term $\frac{F}{l}$ since it represents the average ordinate of the moment diagram for a simple beam under the given loading.

From these illustrations it is seen that values of C and H may be found by the use of equations (9), (10), (13), and (14). Values are also given for the more common cases of loading in text books on strength of materials, but when so determined, the sign must be fixed in accordance with the rules of section 5.

Another method for determining C and H may be readily applied to any kind of loading. For a member carrying a single concentrated load P, as shown in Fig. 7, the value of C_{AB} is $\frac{Pab^2}{l^2}$, and the value of C_{BA} is $\frac{Pa^2b}{l^2}$, as given in equations (17) and (18). If there are several concentrated loads on the member, by summation $C_{AB}=\sum\frac{Pab^2}{l^2}$, and $C_{BA}=\sum\frac{Pa^2b}{l^2}$.

If there is a distributed load on the member the same method may be used, by performing an integration in place of the summation. Let w be the unit loading on an element of length dx, which is at a distance x from the left end, and a distance $l-x$ from the right end of the member. In the expression $\frac{Pab^2}{l^2}$, replace P by wdx, a by x, and b by $l-x$, whence $C_{AB}=\int\frac{wx(l-x)^2dx}{l^2}$. Similarly, $C_{BA}=\int\frac{wx^2(l-x)dx}{l^2}$. The limits of the definite integral are fixed by the length of the member under load.

If the unit load w is not constant, its variation may be expressed in terms of x, and the general value for the total load on a length dx thus found substituted for P in the given expression for a single concentrated load, after which the integration may be performed as just indicated.

Values of C and H for different systems of loads are given for reference in Tables 2 and 3.

TABLE 2

VALUES OF CONSTANTS C AND H FOR DIFFERENT SYSTEMS OF LOADS TO BE USED IN THE EQUATIONS OF TABLE 1

No.	Condition of Loading	C_{AB}	C_{BA}	H_{AB}	H_{BA}
1	No intermediate external loads.	0	0	0	0
2	Single concentrated load at any point.	$\frac{Pab^2}{l^2}$	$\frac{Pba^2}{l^2}$	$\frac{Pab}{2l^2}(l+b)$	$\frac{Pab}{2l^2}(l+a)$
3	Any number of concentrated loads.	$\frac{1}{l^2}\sum Pab^2$	$\frac{1}{l^2}\sum Pba^2$	$\frac{1}{2l^2}\sum Pab\ (l+b)$	$\frac{1}{2l^2}\sum Pab(l+a)$
4	Any distributed load over any portion of the member.	$\frac{1}{l^2}\int_b^d yx^2(l-x)dx$	$\frac{1}{l^2}\int_c^a yx(l-x)^2dx$	$\frac{1}{2l^2}\int_b^d yx(l^2-x^2)dx$	$\frac{1}{2l^2}\int_c^a yx(l-x)(2l-x)dx$
5	Uniform load over any portion of the member.	$\frac{w}{12l^2}\left[d^3(4l-3d)-b^3(4l-3b)\right]$	$\frac{w}{12l^2}\left[a^3(4l-3a)-c^3(4l-3c)\right]$	$\frac{w}{8l^2}(d^2-b^2)(2l^2-b^2-d^2)$	$\frac{w}{8l^2}(a^2-c^2)(2l^2-a^2-c^2)$

	oad rtion mber at one end.				
7	LOAD W, ml, u, A, B, l Uniformly varying load over entire member.	$\frac{l^2}{60}(5u+3ml)$	$\frac{l^2}{60}(5u+2ml)$	$\frac{l^2}{120}(15u+8ml)$	$\frac{l^2}{120}(15u+7ml)$
8	TOTAL LOAD W, A, B, l Load varying uniformly from zero at one end to a maximum at the other end.	$\frac{Wl}{10}$	$\frac{Wl}{15}$	$\frac{2}{15}Wl$	$\frac{7}{60}Wl$
9	a, TOTAL LOAD W, A, B, l Load varying uniformly from zero at any point to a maximum at one end.	$\frac{Wa}{30l^2}(3a^2-10al+10l^2)$	$\frac{Wa^2}{30l^2}(5l-3a)$	$\frac{Wa}{60l^2}(3a^2-15al+20l^2)$	$\frac{Wa}{60l^2}(10l^2-3a^2)$
10	ANY LOAD SYMMETRICAL ABOUT CENTER LINE. A, B, $\frac{l}{2}$, l, $\frac{l}{2}$ Any load symmetrical about center of member.	$\frac{F}{l}$	$\frac{F}{l}$	$\frac{3}{2}\cdot\frac{F}{l}$	$\frac{3}{2}\cdot\frac{F}{l}$

F is the area of the moment diagram of a simple beam having the same length and carrying the same load as the member in question.

Values of $\frac{F}{l}$ for different loads are given in Table 3.

TABLE 3

VALUES OF CONSTANTS C AND H FOR DIFFERENT SYSTEMS OF LOADS TO BE USED IN THE EQUATIONS OF TABLE 1

All Loads Symmetrical about Center of Member

No.	Condition of Loading / Moment Diagram	$C_{AB}=C_{BA}=\frac{F}{l}$	$H_{AB}=H_{BA}=\frac{3}{2}\frac{F}{l}$
1	Single load at the center.	$\frac{1}{8}Pl$	$\frac{3}{16}Pl$
2	Two equal loads.	$\frac{Pa}{l}(l-a)$	$\frac{3}{2}\frac{Pa}{l}(l-a)$
3	Equal loads at the third points.	$\frac{2}{9}Pl$	$\frac{1}{3}Pl$
4	Equal loads at the quarter points.	$\frac{5}{16}Pl$	$\frac{15}{32}Pl$
5	Uniform load over entire span	$\frac{1}{12}Wl$	$\frac{1}{8}Wl$

TABLE 3—CONTINUED

No.	Condition of Loading	Moment Diagram	$C_{AB} = C_{BA} = \frac{F}{l}$	$H_{AB} = H_{BA} = \frac{3}{2}\frac{F}{l}$
6	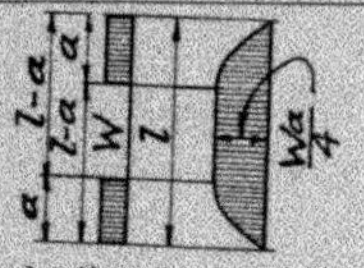 Equal uniform loads at the ends.	$\frac{Wa}{4}$	$\frac{Wa}{12l}(3l-2a)$	$\frac{Wa}{8l}(3l-2a)$
7	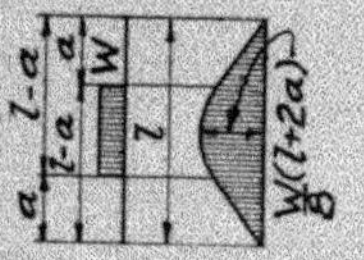 Uniform load at the center.	$\frac{W}{8}(l+2a)$	$\frac{W}{12l}(l^2+2al-2a^2)$	$\frac{W}{8l}(l^2+2al-2a^2)$
8	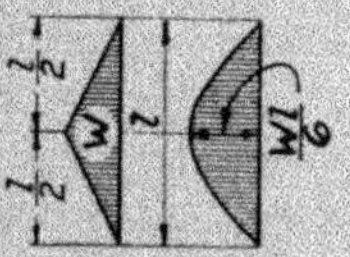 Load increasing uniformly from zero at the ends.	$\frac{Wl}{6}$	$\frac{5}{48}Wl$	$\frac{5}{32}Wl$
9	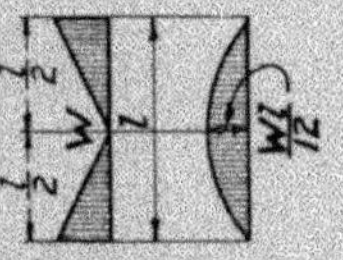 Load increasing uniformly from zero at the center.	$\frac{Wl}{12}$	$\frac{1}{16}Wl$	$\frac{3}{32}Wl$
10	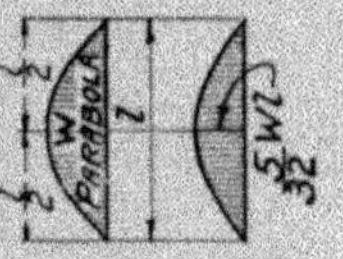 Load varying as the ordinates of a parabola.	$\frac{5}{32}Wl$	$\frac{1}{10}Wl$	$\frac{3}{20}Wl$

PART II

DETERMINATION OF STRESSES IN STATICALLY INDETERMINATE STRUCTURES

9. *Assumptions upon which the Analyses are Based.*—The analyses in this bulletin are based upon the following assumptions:

(1) That the connections are perfectly rigid.
(2) That the length of a member is not changed by axial stress.
(3) That the shearing deformation is zero.

Recent tests by Abe* show that the first assumption is approximately true for reinforced concrete frames, and tests by Wilson and Moore† show that this assumption is also approximately true for certain types of riveted connections of steel frames.

The error due to assumptions (2) and (3) depends upon the geometrical properties of the frame, but for frames of usual proportions the error is not large. These assumptions are discussed in detail in sections 67 and 68. The error due to slip in connections is discussed in section 69.

10. *Notation.*—The following notation has been used:

a = distance from end A of a member to a load.

b = distance from end B of a member to a load.

d = deflection of one end of a member with respect to the other end, measured perpendicular to initial position of member.

e = eccentricity of load.

h = vertical height of a structure.

k = error in resisting moment due to neglect of shearing strain.

l = length of a member.

m = change in the rate of loading in a unit distance.

n = ratio of K of top member to K of left-hand column for a four-sided frame.

p = ratio of K of top member to K of bottom member for a four-sided frame.

---

*Abe, Mikishi, "Analysis and Tests of Rigidly Connected Reinforced Concrete Frames," Univ. of Ill. Eng. Exp. Sta., Bul. 107, 1918.

†Wilson, W. M., and Moore, H. F., "Tests to Determine the Rigidity of Riveted Joints of Steel Structures," Univ. of Ill. Eng. Exp. Sta., Bul 104, 1917.

q = ratio of the length of the left-hand column to the length of the right-hand column of a two-legged bent.

s = ratio of K of top member to K of right-hand column for a four-sided frame.

u = load per unit of length (variable).

w = uniformly distributed load per unit of length.

A = area of section of a member.

C_{AB} = resisting moment at end A of a member AB fixed at both ends and having both ends at the same level.

E = modulus of elasticity in tension and compression.

F = area of the moment diagram of a simple beam.

G = modulus of elasticity in shear.

H = reaction.

H_{AB} = resisting moment at end A of a member AB fixed at A and hinged at B and having both ends at the same level.

I = moment of inertia of section of a member.

K = ratio of moment of inertia of section to length of a member.

M = moment of an external couple.

M_A = statical moment of external forces about point A.

M_{AB} = resisting moment acting at the end A of a member AB.

M_{BA} = resisting moment acting at the end B of a member AB.

N = restraint factor, depending on manner in which the ends of a member are held.

P = concentrated load.

$R = \frac{d}{l}$ = ratio of the deflection of one end of a member (with respect to the other end) to the length of the member.

S = shear.

W = total distributed load on a member.

$\alpha = n^2+2pn+2n+3p$, for a symmetrical four-sided frame.

$\beta = 6n+p+1$, for a symmetrical four-sided frame.

$\Delta = 22(pns+ps+ns+np)+2(p^2s+ps^2+pn^2+p^2n+s^2+s+n^2+n)+6(n^2s+ns^2+p^2+p)$, for a rectangular frame.

$\Delta_o = 2[ns(4+3q+4q^2)+(s^2+s)+q^2(n^2+n)+3(q^2sn^2+s^2n)]$, for a two-legged rectangular bent with unequal legs.

$\Delta_o = 2(3ns^2+11ns+s^2+s+3n^2s+n^2+n)$, for a two-legged rectangular bent.

θ = change in the slope of the tangent to the elastic curve of a member.

IV. GIRDERS HAVING RESTRAINED ENDS

11. *Moments at the Ends of a Girder Having Fixed Ends—Both Supports on the Same Level.*—If a girder is fixed at the ends and if both supports are on the same level, θ_A, θ_B, and R of equations (A) and (B), Table 1, equal zero. This being the case, $M_{AB} = \mp C_{AB}$ and $M_{BA} = \pm C_{BA}$. Values of C_{AB} and C_{BA} for different systems of loads are given in Table 2.

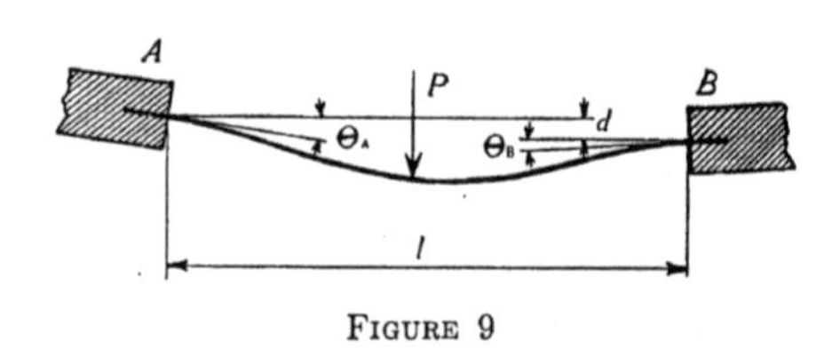

FIGURE 9

12. *Moments at the End of a Girder Having One End Fixed and the Other End Hinged, Both Supports on the Same Level.*—If a girder is fixed at one end, θ for that end equals zero. Likewise if both supports are on the same level, $R=0$. This being the case, the moment at the fixed end, as given by equations (C) and (D), Table 1, is $\mp H_{AB}$ or $\pm H_{BA}$. Values of H_{AB} and H_{BA} for different systems of loads are given in Table 2.

13. *Moments at the Ends of a Girder Having Ends Restrained but not Fixed.*—Fig. 9 represents a girder restrained at A and B. P represents the resultant of any system of forces on AB. The change in slope at A is θ_A and at B is θ_B. The deflection of B relative to A is d $R = \frac{d}{l}$.

Applying equations (A) and (B), Table 1, gives

$$M_{AB} = 2EK(2\theta_A + \theta_B - 3R) - C_{AB} \quad . \quad . \quad . \quad . \quad (25)$$

$$M_{BA} = 2EK(2\theta_B + \theta_A - 3R) + C_{BA} \quad . \quad . \quad . \quad . \quad (26)$$

In order to determine M_{AB} and M_{BA}, θ_A, θ_B, and R must be known. As shown in Fig. 9, θ_A and R are positive and θ_B is negative. If P

had been upward instead of downward, C_{AB} would have been preceded by a plus (+) sign and C_{BA} by a minus (−) sign. Values of C_{AB} and C_{BA} for different systems of loads to be substituted in equations (25) and (26) are given in Table 2.

14. *Moment at End of a Girder Having One End Hinged and the Other End Restrained but not Fixed.*—Fig. 10 represents a girder hinged

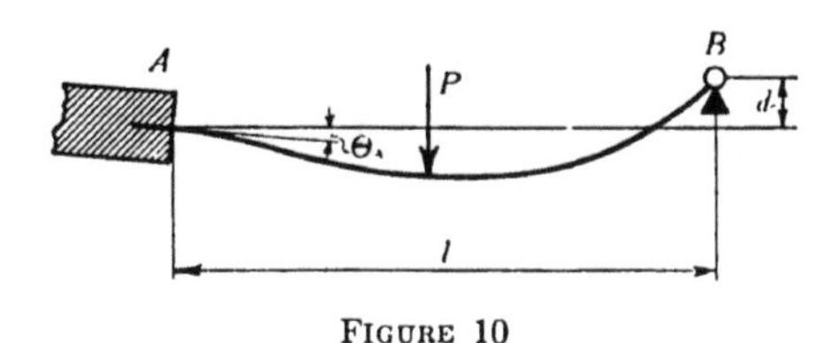

FIGURE 10

at B and restrained but not fixed at A. P represents the resultant of any system of forces on AB. The change in slope at A is θ_A, and the deflection of B relative to A is d. $R=\frac{d}{l}$.

Applying equation (C) of Table 1 gives

$$M_{AB}=EK(3\theta_A-3R)-H_{AB} \quad . \; . \; . \; . \; . \; . \; . \; . \; . \; . \quad (27)$$

As shown in Fig. 10, θ_A is positive (+) and R is negative (−). If P had been upward, H_{AB} would have been preceded by a plus (+) sign.

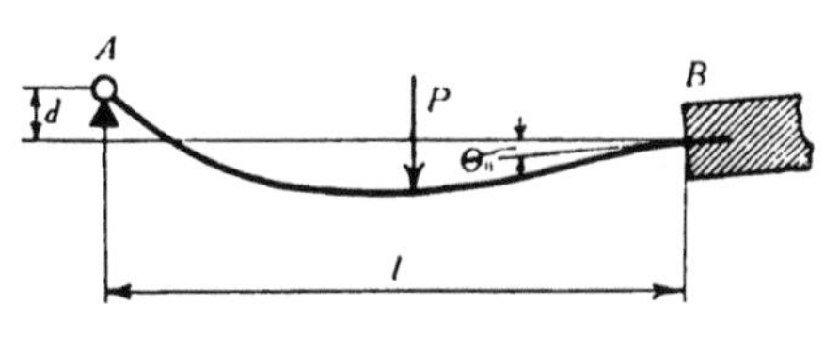

FIGURE 11

For the girder represented by Fig. 11

$$M_{BA}=EK(3\theta_B-3R)+H_{BA} \quad . \; . \; . \; . \; . \; . \; . \; . \; . \; . \quad (28)$$

in which R is positive (+) and θ_B is negative (−).

Values of H_{AB} and H_{BA} for different systems of loads to be used in equations (27) and (28) are given in Table 2.

V. Continuous Girders

15. *Girder Continuous over Any Number of Supports and Carrying Any System of Vertical Loads. Supports all on the Same Level. General Equation of Three Moments—Two Intermediate Spans.*—Although the

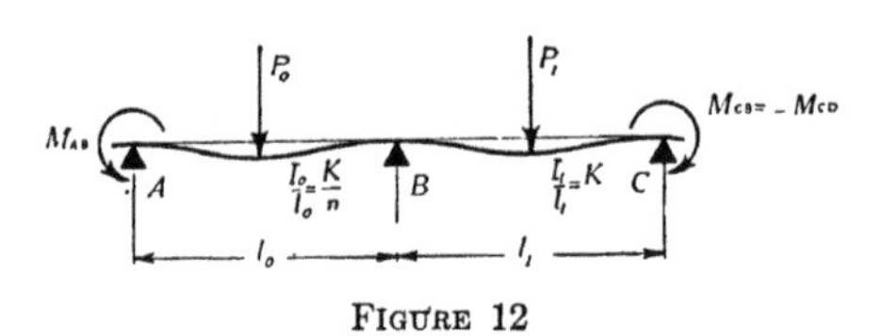

FIGURE 12

results of this section are included in the following section, detailed procedure is given here to show how the slope-deflection equations are to be applied to a continuous girder. Fig. 12 represents two intermediate spans of a continuous girder extending over a number of spans. All supports are on the same level. P_o represents the resultant of the forces on AB, and P_1 represents the resultant of the forces on BC. $\frac{I}{l}$ for span BC is K. $\frac{I}{l}$ for span AB is $\frac{K}{n}$. Since all the supports are on the same level, $R=0$ for all spans.

Applying the equations of Table 1 gives

$$M_{AB}=\frac{2EK}{n}(2\theta_A+\theta_B)-C_{AB} \quad (29)$$

$$M_{BA}=\frac{2EK}{n}(2\theta_B+\theta_A)+C_{BA} \quad (30)$$

$$M_{BC}=2EK(2\theta_B+\theta_C)-C_{BC} \quad (31)$$

$$M_{CB}=2EK(2\theta_C+\theta_B)+C_{CB} \quad (32)$$

$$M_{BA}+M_{BC}=0 \quad (33)$$

Substituting the value of θ_A from equation (29) in equation (30) gives

$$2M_{BA}-M_{AB}=\frac{6EK}{n}\theta_B+2C_{BA}+C_{AB} \quad (34)$$

Substituting the value of θ_C from equation (31) in equation (32) gives

$$M_{CB}-2M_{BC}=-6EK\theta_B+C_{CB}+2C_{BC} \quad (35)$$

Substituting the value of θ_B from equation (35) in equation (34), and substituting $-M_{BC}$ for M_{BA} gives

$$nM_{AB}+2M_{BC}(n+1)+M_{CD}=-[n(2C_{BA}+C_{AB})+(C_{CB}+2C_{BC})] \quad (36)$$

In determining the values of C and H given in Table 3, it was found that

$$H_{AB}=\frac{2C_{AB}+C_{BA}}{2} \text{ and } H_{BA}=\frac{2C_{BA}+C_{AB}}{2}$$

Equation (36) can, therefore, be written in the form

$$nM_{AB}+2M_{BC}(n+1)+M_{CD}=-2[nH_{BA}+H_{BC}] \quad . \quad . \quad . \quad . \quad (37)$$

This is the general form of the well-known "Equation of Three Moments."* It may be applied to a continuous girder having all supports on the same level, no matter what the type of loading to which the girder is subjected. As applied to two adjacent spans, $K=\frac{I}{l}$ for the right-hand span and $\frac{K}{n}=\frac{I}{l}$ for the left-hand span; that is, $n=\frac{I}{l}$ for the right-hand span divided by $\frac{I}{l}$ for the left-hand span. Values of H for different types of loading are given in Table 2.

16. *Girder Continuous over Any Number of Supports and Carrying Any System of Vertical Loads. Supports All on the Same Level. General Equation of Three Moments—Two Adjacent Spans at One End. End of Girder Hinged.*—Fig. 13 represents the two spans at the left-hand end of a continuous girder. All supports are on the same level. P_0 represents the resultant of the loads on AB, and P_1 represents the

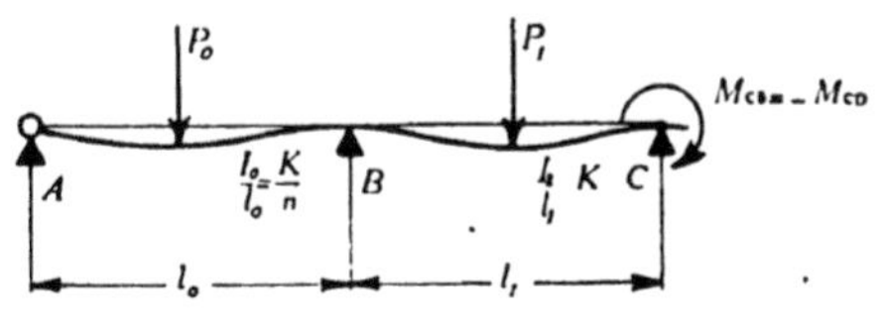

FIGURE 13

*The Equation of Three Moments was first deduced for a girder carrying uniform loads by Clapeyron, in 1857, and was published in Comptes Rendus des Séances de l'Académie des Sciences, Paris, Vol. 45, p. 1076. It has been extended and generalized for other loadings by Bresse, Cours de Mécanique Appliquée, Paris, 1862; Winkler, Die Lehre von der Elasticitat und Festigkeit, Prague, 1867, and others

resultant of the loads on BC. The girder is hinged at A. Equation (37), having been derived for the general case, is applicable. As the girder is hinged at A, $M_{AB}=0$. Equation (37), therefore, takes the form

$$2M_{BC}(n+1)+M_{CD}=-2[nH_{BA}+H_{BC}] \quad . \quad . \quad . \quad . \quad . \quad . \quad (38)$$

for two adjacent spans at the left-hand end of the girder when the left-hand end is hinged. Likewise,

$$nM_{AB}+2M_{BC}(n+1)=-2\,[n\,H_{BA}+H_{BC}] \quad . \quad . \quad . \quad . \quad . \quad . \quad (39)$$

for two adjacent spans at the right-hand end of the girder when the right-hand end is hinged.

17. *Girder Continuous over Any Number of Supports and Carrying Any System of Vertical Loads. Supports All on the Same Level. General Equation of Three Moments—Two Adjacent Spans at One End. End of Girder Restrained.*—Fig. 14 represents the two spans at the left-

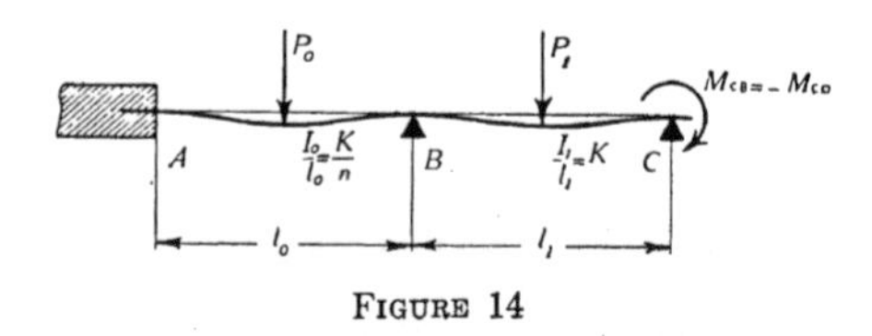

FIGURE 14

hand end of a continuous girder. All supports are on the same level. P_o represents the resultant of the loads on AB, and P_1 represents the resultant of the loads on BC. The girder is restrained at A.

The values of the moments depend upon the restraint of the point A, and therefore the moments cannot be determined unless either the moment at A or the slope of the elastic curve of the girder at A is known. If the moment at A is known, equation (37) is applicable. If the slope at A is known, θ_A is a known quantity.

Substituting the value θ_B from equation (29) in equation (30) gives

$$2n\,M_{AB}=6EK\theta_A-n\,M_{BC}-2n\,H_{AB} \quad . \quad . \quad . \quad . \quad . \quad . \quad . \quad . \quad (40)$$

Substituting M_{AB} from equation (40) in equation (37) gives

$$M_{BC}(4+3n)+2M_{CD}=2n\,H_{AB}-4(nH_{BA}+H_{BC})-6EK\,\theta_A \quad . \quad (41)$$

Equation (41) is applicable to the two adjacent spans at the left-hand end of a continuous girder restrained at the left-hand end.

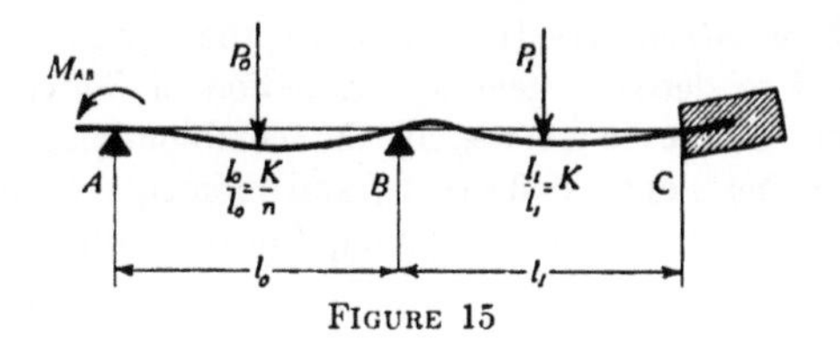

FIGURE 15

Fig. 15 represents the two spans at the right-hand end of a continuous girder. All supports are on the same level. P_o represents the resultant of the loads on AB, and P_1 represents the resultant of the loads on BC. The girder is restrained at C.

From equation (37)

$$2nM_{AB}+4M_{BC}(n+1)-2M_{CB}=-4(nH_{BA}+H_{BC}) \quad . \quad . \quad . \quad (42)$$

Applying the equations of Table 1 gives

$$M_{BC}=2EK(2\theta_B+\theta_C)-C_{BC} \quad . \quad . \quad . \quad (43)$$

$$M_{CB}=2EK(2\theta_C+\theta_B)+C_{CB} \quad . \quad . \quad . \quad (44)$$

Eliminating θ_B from equations (43) and (44) gives

$$-2M_{CB}=-M_{BC}-6EK\theta_C-2H_{CB} \quad . \quad . \quad . \quad (45)$$

Substituting the value of M_{CB} from equation (45) in equation (42) gives

$$2nM_{AB}+M_{BC}(4n+3)=2H_{CB}-4[nH_{BA}+H_{BC}]+6EK\theta_C \quad . \quad (46)$$

Equation (46) is applicable to the two adjacent spans at the right-hand end of a continuous girder.

18. *Girder Continuous over Any Number of Supports and Carrying Any System of Vertical Loads. Supports All on the Same Level. General Equation of Three Moments—Girder Fixed at the Ends.*—If the girder is fixed at the ends, θ_A of equation (41) and θ_C of equation (46) equal zero. Equations (41) and (46) then take the form

$$M_{BC}(4+3n)+2M_{CD}=2nH_{AB}-4[nH_{BA}+H_{BC}] \quad . \quad . \quad . \quad (47)$$

$$2nM_{AB}+M_{BC}(4n+3)=2H_{CB}-4[nH_{BA}+H_{BC}] \quad . \quad . \quad . \quad (48)$$

Equation (47) is applicable to the two adjacent spans at the left-hand end of a continuous girder fixed at the left-hand end, and equation (48) is applicable to the two adjacent spans at the right-hand end of a continuous girder fixed at the right-hand end. Values of H for different systems of loads are given in Table 2.

19. *Girder Continuous over Three Supports and Carrying Any System of Vertical Loads. Supports on Different Levels. General*

Equation of Three Moments.—In section 15, the equations of Table 1 have been used to derive a general "Equation of Three Moments." The derivation is seen to be merely the combination of four linear equations involving slopes, deflections, and moments at the supports for each span into an equation involving the same quantities for any two adjacent spans.

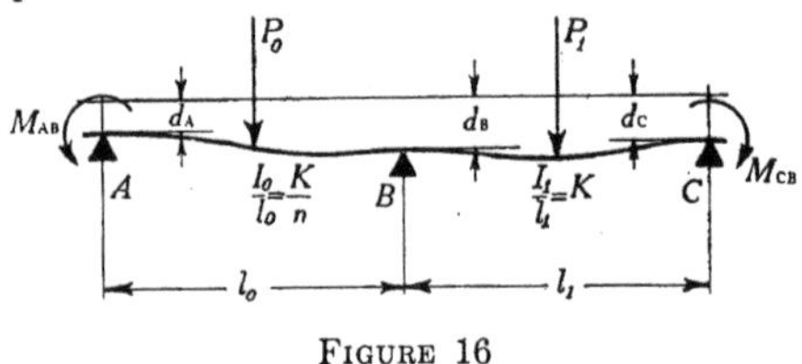

FIGURE 16

Fig. 16 represents two adjacent intermediate spans of a girder continuous over a number of supports. Some of the supports have settled. The vertical distances of the points A, B, and C from a horizontal base line are d_A, d_B, and d_C. P_o represents the resultant of all loads on AB, and P_1 represents the resultant of all loads on BC.

The moments in the girder may be considered as made up of two parts, one part due to the loads represented by P_o and P_1 when the supports are on the same level, and the other part due to the settlement of the supports, it being considered that the girder remains in contact with all supports.

Applying the equations of Table 1 gives

$$M_{AB}=\frac{2EK}{n}(2\theta_A+\theta_B-3R_o)-C_{AB} \quad (49)$$

$$M_{BA}=\frac{2EK}{n}(2\theta_B+\theta_A-3R_o)+C_{BA} \quad (50)$$

$$M_{BC}=2EK(2\theta_B+\theta_C-3R_1)-C_{BC} \quad (51)$$

$$M_{CB}=2EK(2\theta_C+\theta_B-3R_1)+C_{CB} \quad (52)$$

$$M_{BA}+M_{BC}=0 \quad (53)$$

in which $\frac{I}{l}$ for the right-hand span equals K and $\frac{I}{l}$ for the left-hand span equals $\frac{K}{n}$.

From Fig. 16 $R_o=\frac{d_B-d_A}{l_o}$ and $R_1=\frac{d_C-d_B}{l_1}$.

Equations (49) to (53) are identical with equations (29) to (33) of section 15, except for the additional terms R_o and R_1. Hence the method of solving for the moments is the same as in section 15, and will not be repeated here in detail. The general equation of three moments for a continuous girder carrying any system of vertical loads and with supports on different levels, together with some special forms of the equation, are given in Table 4.*

20. *Girder Continuous over Three Supports and Carrying Any System of Vertical Loads. Supports on Different Levels. Various Conditions of Restraint of Ends.*—Table 4, section 19, gives the general equation of three moments which is to be applied here. Various cases of restraint of the ends of girder and the effect upon the quantities in the general equation will be considered. It is seen that if the end of the girder is hinged, the resisting moment there is zero. If it is fixed, the slope at that point is zero. If the end is restrained, or partially fixed, the moments at the other supports can be found, if either the slope or the moment at the end is known. Modifications of the equations of Table 4, for these various cases of restraint, are given in Table 5.

21. *Girder Continuous over Four Supports and Carrying Any System of Vertical Loads. Supports on Different Levels. Various Conditions of Restraint of Ends.*—Table 4, section 19, gives the general equation of three moments which may be applied here. The method of using the equation of three moments is to apply it successively to each pair of adjacent spans in the continuous girder, thus deriving one less equation than the number of spans. If the conditions of restraint at the ends of the continuous girder are known, these equations may be solved simultaneously for the unknown moment at each support.

When the end of the girder is hinged, the moment there is zero. When the end is fixed, the slope there is zero. When the end is restrained, if either the slope or the moment at the end is known, the moments at the other supports may be found.

The equations of Table 4 have been applied to the girder described in this section, and the values of moments obtained for various cases of restraint of ends are given in Table 6.

*Johnson, Bryan, and Turneaure, "The Theory and Practice of Modern Framed Structures," Part II, p. 19, Ninth Ed. 1911, give the general Equation of Three Moments for a continuous girder with supports on different levels and carrying concentrated loads. The pocketbook, Die Huette, and Lanza, Applied Mechanics, have similar equations.

A. Ostenfeld, Teknisk Statik, Vol. 2, Second Ed., Copenhagen, 1913, pp. 98-162, gives a comprehensive treatment of continuous girders, including the case of varying moment of inertia from point to point, and of girder resting upon elastic supports. Both analytical and graphical methods are presented.

TABLE 4

CONTINUOUS GIRDERS

Equations of Three Moments.

Supports on Different Levels.[1] $K=\frac{I}{l}$ for right-hand span.

Any System of Vertical Loads.[2] $\frac{K}{n}=\frac{I}{l}$ for left-hand span.

No.	Portion of Girder Considered	Equations of Three Moments
(a)	Fig. 16 Intermediate spans	$nM_{AB}+2M_{BC}(n+1)+M_{CD}=\frac{6EK}{l_o l_1}\Big[l_1(d_B-d_A)-l_o(d_C-d_B)\Big]-2\Big[H_{BC}+nH_{BA}\Big]$
(b)	Fig. 17 Two adjacent spans at left-hand end. End of girder hinged.	$2M_{BC}(n+1)+M_{CD}=\frac{6EK}{l_o l_1}\Big[l_1(d_B-d_A)-l_o(d_C-d_B)\Big]-2\Big[nH_{BA}+H_{BC}\Big]$
(c)	Fig. 18 Two adjacent spans at right-hand end. End of girder hinged.	$nM_{AB}+2M_{BC}(n+1)=\frac{6EK}{l_o l_1}\Big[l_1(d_B-d_A)-l_o(d_C-d_B)\Big]-2\Big[nH_{BA}+H_{BC}\Big]$
(d)	Fig. 19 Two adjacent spans at left-hand end. End of girder restrained.	If M_{AB} is known $2M_{BC}(n+1)+M_{CD}=\frac{6EK}{l_o l_1}\Big[l_1(d_B-d_A)-l_o(d_C-d_B)\Big]-2\Big[H_{BC}+nH_{BA}\Big]-nM_{AB}$ If θ_A is known $M_{BC}(4+3n)+2M_{CD}=\frac{6EK}{l_o l_1}\Big[3l_1(d_B-d_A)-2l_o(d_C-d_B)\Big]+2nH_{AB}-4\Big[H_{BC}+nH_{BA}\Big]-6EK\,\theta_A$ If the girder is fixed at A, $\theta_A=0$
(e)	Fig. 20 Two adjacent spans at right-hand end. End of girder restrained.	If M_{CB} is known $nM_{AB}+2M_{BC}(n+1)=\frac{6EK}{l_o l_1}\Big[l_1(d_B-d_A)-l_o(d_C-d_B)\Big]-2\Big[H_{BC}+nH_{BA}\Big]+M_{CB}$ If θ_C is known $2nM_{AB}+M_{BC}(4n+3)=\frac{6EK}{l_o l_1}\Big[2l_1(d_B-d_A)-3l_o(d_C-d_B)\Big]-4\Big[H_{BC}+nH_{BA}\Big]+2H_{CB}+6EK\,\theta_C$ If the girder is fixed at C, $\theta_C=0$

[1] If there is no settlement of supports, let all values of d in these equations equal zero.

[2] If there are no loads on girder except at supports, let all values of H in these equations equal zero.

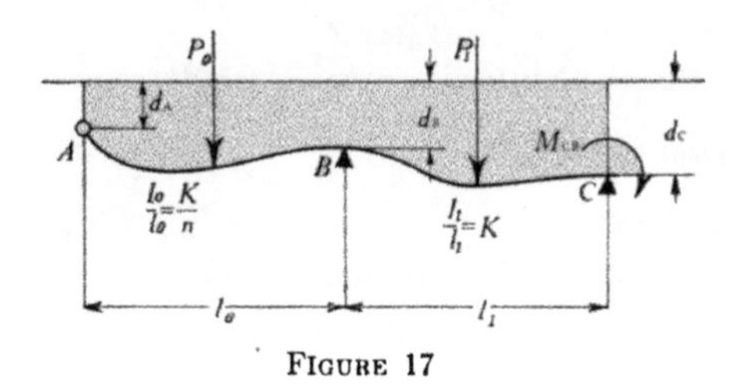

FIGURE 17

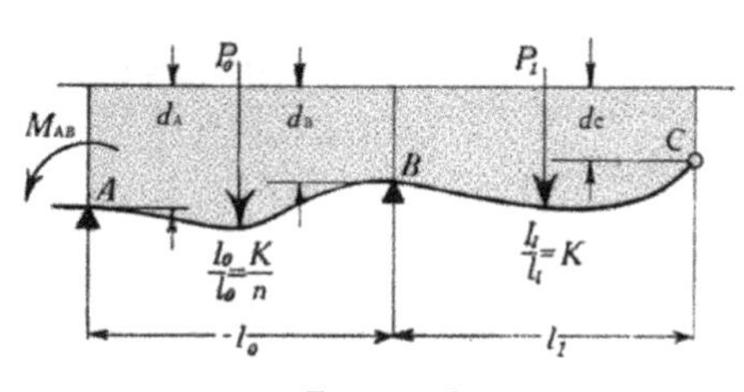

FIGURE 18

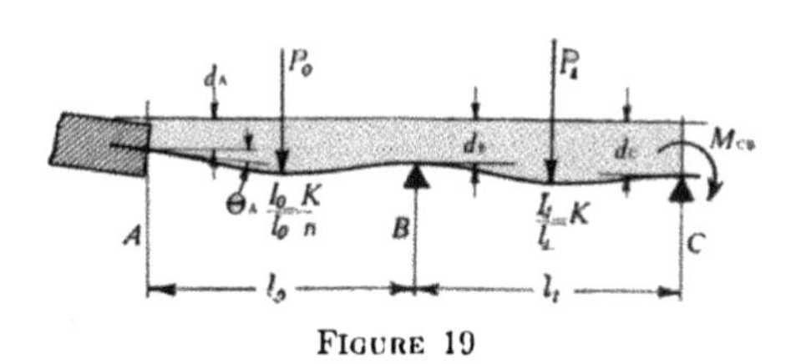

FIGURE 19

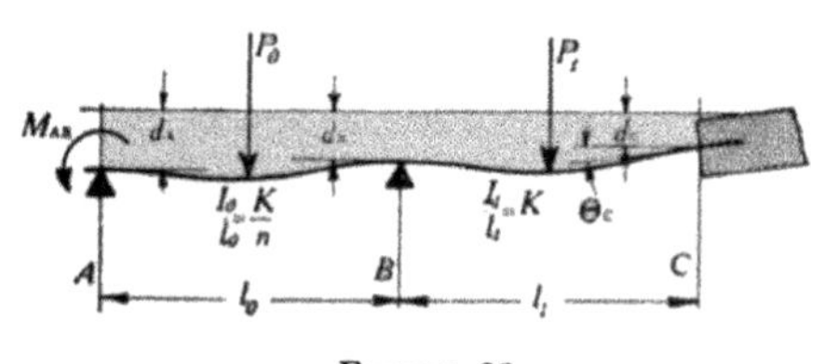

FIGURE 20

TABLE 5

GIRDER CONTINUOUS OVER THREE SUPPORTS

Supports on Different Levels.[1] $K=\frac{I}{l}$ for right-hand span.

Any System of Vertical Loads.[2] $\frac{K}{n}=\frac{I}{l}$ for left-hand span.

(a)	Fig. 21 Ends of girder hinged	General Case $M_{BC}=\frac{6EK}{2l_o l_1(n+1)}\left[l_1(d_B-d_A)-l_o(d_C-d_B)\right]-\frac{2}{n+1}\left[nH_{BA}+H_{BC}\right]$ Spans Identical Except for Loads $M_{BC}=\frac{3EK}{2l^2}\left[l(2d_B-d_A-d_C)\right]-\left[H_{BA}+H_{BC}\right]$
(b)	Fig. 22 Ends of girder restrained	General Case θ_A and M_{CB} known $M_{BC}=\frac{1}{3n+4}\left\{\frac{6EK}{l_o l_1}\left[3l_1(d_B-d_A)-2l_o(d_C-d_B)\right]+2nH_{AB}-4(H_{BC}+nH_{BA})-6EK\theta_A+2M_{CB}\right\}$ $M_{AB}=\frac{3EK}{n}\left(\theta_A-\frac{d_B-d_A}{l_o}\right)-\frac{M_{BC}}{2}-H_{AB}$ θ_C and M_{AB} known $M_{BC}=\frac{1}{4n+3}\left\{\frac{6EK}{l_o l_1}\left[2l_1(d_B-d_A)-3l_o(d_C-d_B)\right]-4(H_{BC}+nH_{BA})+2H_{CB}+6EK\theta_C-2nM_{AB}\right\}$ $M_{CB}=\frac{M_{BC}}{2}+3EK\left[\theta_C-\frac{d_C-d_B}{l_1}\right]+H_{CB}$ θ_A and θ_C known $M_{BC}=\frac{1}{n+1}\left\{\frac{6EK}{l_o l_1}\left[l_1(d_B-d_A)-l_o(d_C-d_B)\right]-nC_{BA}-C_{BC}-2EK(\theta_A-\theta_C)\right\}$ $M_{AB}=\frac{3EK}{n}\left[\theta_A-\frac{d_B-d_A}{l_o}\right]-\frac{M_{BC}}{2}-H_{AB}$ $M_{CB}=3EK\left[\theta_C-\frac{d_C-d_B}{l_1}\right]+\frac{M_{BC}}{2}+H_{CB}$

[1] If there is no settlement of supports, let all values of d in these equations equal zero.

[2] If there are no loads on girder except at supports, let all values of H and C in these equations equal zero.

If an end is fixed, θ for that end is zero.

If an end is hinged, M for that end is zero.

TABLE 5—CONTINUED

(b) | Fig. 22 | Ends of girder restrained

General Case—Continued

M_{AB} and M_{CB} known

$$M_{BC}=\frac{1}{2(n+1)}(M_{CB}-nM_{AB})+\frac{3EK}{l_o l_1(n+1)}\Big[l_1(d_B-d_A)-l_o(d_C-d_B)\Big]-\frac{nH_{BA}+H_{BC}}{n+1}$$

Spans Identical Except for Load

θ_A and M_{CB} known

$$M_{BC}=\frac{1}{7}\left\{\frac{6EK}{l}\Big[3(d_B-d_A)-2(d_C-d_B)\Big]+2H_{AB}-4(H_{BC}+H_{BA})-6EK\theta_A+2M_{CB}\right\}$$

$$M_{AB}=3EK\left[\theta_A-\frac{d_B-d_A}{l}\right]-\frac{M_{BC}}{2}-H_{AB}$$

θ_C and M_{AB} known

$$M_{BC}=\frac{1}{7}\left\{\frac{6EK}{l}\Big[2(d_B-d_A)-3(d_C-d_B)\Big]-4(H_{BC}+H_{BA})+2H_{CB}+6EK\theta_C-2M_{AB}\right\}$$

$$M_{CB}=\frac{M_{BC}}{2}+3EK\left[\theta_C-\frac{d_C-d_B}{l}\right]+H_{CB}$$

θ_A and θ_C known

$$M_{BC}=\frac{1}{2}\left\{\frac{6EK}{l}\Big[2d_B-d_C-d_A\Big]-C_{BA}-C_{BC}-2EK(\theta_A-\theta_C)\right\}$$

$$M_{AB}=3EK\left[\theta_A-\frac{d_B-d_A}{l}\right]-\frac{M_{BC}}{2}-H_{AB}$$

$$M_{CB}=3EK\left[\theta_C\quad\frac{d_C-d_B}{l}\right]+\frac{M_{BC}}{2}+H_{CB}$$

M_{AB} and M_{CB} known

$$M_{BC}=\frac{1}{4}\Big[M_{CB}-M_{AB}\Big]+\frac{3EK}{2l}\Big[2d_B-d_A-d_C\Big]-\frac{H_{BA}+H_{BC}}{2}$$

TABLE 6

GIRDER CONTINUOUS OVER FOUR SUPPORTS

Supports on Different Levels.[1] Any System of Vertical Loads.[2]

$\frac{I}{l}$ for span $AB=\frac{K_1}{n_o}$. $\frac{I}{l}$ for span $BC=\frac{K_2}{n_1}=K_1$. $\frac{I}{l}$ for span $CD=K_2$.

(a)	Fig. 23 Ends of girder hinged	General Case $M_{BC}=\frac{6EK_1\left[2(n_1+1)l_1l_2(d_B-d_A)-(3n_1+2)l_ol_2(d_C-d_B)+n_1l_ol_1(d_D-d_C)\right]}{l_ol_1l_2\left[4(n_o+1)(n_1+1)-n_1\right]}-\left[\frac{4(n_1+1)(n_oH_{BA}+H_{BC})-2(n_1H_{CB}+H_{CD})}{4(n_o+1)(n_1+1)-n_1}\right]$ $M_{CD}=\frac{6EK_2\left[(2n_o+3)l_ol_2(d_C-d_B)-2(n_o+1)l_ol_1(d_D-d_C)-l_1l_2(d_B-d_A)\right]}{l_ol_1l_2\left[4(n_o+1)(n_1+1)-n_1\right]}-\left[\frac{4(n_o+1)(n_1H_{CB}+H_{CD})-2n_1(n_oH_{BA}+H_{BC})}{4(n_o+1)(n_1+1)-n_1}\right]$ All Spans Identical Except for Loads $M_{BC}=\frac{2EK}{5l}(9d_B-6d_C+d_D-4d_A)-\frac{2}{15}\left[4(H_{BA}+H_{BC})-(H_{CB}+H_{CD})\right]$ $M_{CD}=\frac{2EK}{5l}(9d_C-6d_B-4d_D+d_A)-\frac{2}{15}\left[4(H_{CB}+H_{CD})-(H_{BA}+H_{BC})\right]$

[1] If there is no settlement of supports, let all values of d in these equations equal zero.

[2] If there are no loads on the girder except at supports, let all values of H in these equations equal zero.

If an end is fixed, θ for that end is zero. If an end is hinged, M for that end is zero.

TABLE 6—CONTINUED

(b)	Fig. 24 Ends of girder restrained	General Case

General Case

θ_A and M_{DC} known

$$M_{BC}=\frac{6EK_1\left[3(n_1+1)l_1l_2(d_B-d_A)-(3n_1+2)l_ol_2(d_C-d_B)+n_1l_ol_1(d_D-d_C)\right]}{l_ol_1l_2\left[3(n_o+1)(n_1+1)+1\right]}-\frac{2\left[(n_1+1)(2n_oH_{BA}-n_oH_{AB}+2H_{BC}+3EK_1\theta_A)-(n_1H_{CB}+H_{CD})\right]+M_{DC}}{3(n_o+1)(n_1+1)+1}$$

$$M_{CD}=\frac{3EK_2\left[3(n_o+2)l_ol_2(d_C-d_B)-(3n_o+4)l_ol_1(d_D-d_C)-3l_1l_2(d_B-d_A)\right]}{l_ol_1l_2\left[3(n_o+1)(n_1+1)+1\right]}+\frac{2n_1(2n_oH_{BA}-n_oH_{AB}+2H_{BC}+3EK_1\theta_A)-(3n_o+4)(2n_1H_{CB}+2H_{CD}-M_{DC})}{2\left[3(n_o+1)(n_1+1)+1\right]}$$

$$M_{AB}=-\frac{M_{BC}}{2}-\frac{H_{AB}+3EK_1}{n_ol_o}\left[l_o\theta_A-(d_B-d_A)\right]$$

θ_D and M_{AB} known

$$M_{BC}=\frac{3EK_1\left[(4n_1+3)l_1l_2(d_B-d_A)-3(2n_1+1)l_ol_2(d_C-d_B)+3n_1l_ol_1(d_D-d_C)\right]}{l_ol_1l_2\left[3(n_o+1)(n_1+1)+n_on_1\right]}-\left[\frac{(4n_1+3)(2n_oH_{BA}+2H_{BC}+n_oM_{AB})+2(H_{DC}-2H_{CD}-2n_1H_{CB}+3EK_2\theta_D)}{2\left[3(n_o+1)(n_1+1)+n_on_1\right]}\right]$$

TABLE 6 CONTINUED

GIRDER CONTINUOUS OVER FOUR SUPPORTS

Supports on Different Levels. Any System of Vertical Loads.

$\frac{I}{l}$ for span $AB = \frac{K_1}{n_o}$. $\frac{I}{l}$ for span $BC = \frac{K_2}{n_1} = K_1$. $\frac{I}{l}$ for span $CD = K_2$

(b) Fig. 24 Ends of girder restrained (Continued)

General Case—Continued

$$M_{CD} = \frac{6EK_2\left[(2n_o+3)l_ol_2(d_C-d_B)-3(n_o+1)l_ol_1(d_D-d_C)-l_1l_2(d_B-d_A)\right]}{l_ol_1l_2\left[3(n_o+1)(n_1+1)+n_on_1\right]}$$

$$+\frac{2(n_o+1)(H_{DC}-2H_{CD}-2n_1H_{CB}+3EK_2\theta_D)+n_1(2n_oH_{BA}+2H_{BC}+n_oM_{AB})}{3(n_o+1)(n_1+1)+n_on_1}$$

$$M_{DC} = \frac{M_{CD}}{2}+H_{DC}+3EK_2\left[\theta_D-\frac{(d_D-d_C)}{l_2}\right]$$

θ_A and θ_D known

$$M_{BC} = \frac{6EK_1\left[(4n_1+3)\,l_1l_2(d_B-d_A)-2(2n_1+1)l_ol_2(d_C-d_B)+2n_1l_ol_1(d_D-d_C)\right]}{l_ol_1l_2\left[4(n_o+1)(n_1+1)-n_o\right]}$$

$$-\frac{2\left\{(4n_1+3)(2n_oH_{BA}-n_oH_{AB}+2H_{BC})+2(H_{DC}-2H_{CD}-2n_1H_{CB})+3EK_1\left[(4n_1+3)\theta_A+2n_1\theta_D\right]\right\}}{3\left[4(n_o+1)(n_1+1)-n_o\right]}$$

$$M_{CD} = \frac{6EK_2\left[2(n_o+2)l_ol_2(d_C-d_B)-(3n_o+4)l_ol_1(d_D-d_C)-2l_1l_2(d_B-d_A)\right]}{l_ol_1l_2\left[4(n_o+1)(n_1+1)-n_o\right]}$$

$$+\frac{2\left\{(3n_o+4)(H_{DC}-2H_{CD}-2n_1H_{CB})+2n_1(2n_oH_{BA}-n_oH_{AB}+2H_{BC})+3EK_2\left[(3n_o+4)\theta_D+2\theta_A\right]\right\}}{3\left[4(n_o+1)(n_1+1)-n_o\right]}$$

TABLE 6—CONTINUED

(b)	Fig 24 Ends of girder restrained (Continued)	$M_{AB} = -\frac{M_{BC}}{2} - H_{AB} + \frac{3EK_1}{n_o}\left[\theta_A - \frac{(d_B - d_A)}{l_o}\right]$ $M_{DC} = \frac{M_{CD}}{2} + H_{DC} + 3EK_2\left[\theta_D - \frac{(d_D - d_C)}{l_2}\right]$ M_{AB} and M_{DC} known $M_{BC} = \frac{6EK_1\left[2(n_1+1)l_1l_2(d_B-d_A) - (3n_1+2)l_ol_2(d_C-d_B) + n_1l_ol_1(d_D-d_C)\right]}{l_ol_1l_2\left[4(n_o+1)(n_1+1) - n_1\right]}$ $-\left[\frac{4(n_1+1)(n_oH_{BA}+H_{BC}) - 2(n_1H_{CB}+H_{CD}) + 2n_o(n_1+1)M_{AB} + M_{DC}}{4(n_o+1)(n_1+1) - n_1}\right]$ $M_{CD} = \frac{6EK_2\left[(2n_o+3)l_ol_2(d_C-d_B) - 2(n_o+1)l_ol_1(d_D-d_C) - l_1l_2(d_B-d_A)\right]}{l_ol_1l_2\left[4(n_o+1)(n_1+1) - n_o\right]}$ $-\left[\frac{4(n_o+1)(n_1H_{CB}+H_{CD}) - 2n_1(n_oH_{BA}+H_{BC}) - 2(n_o+1)M_{DC} - n_on_1M_{AB}}{4(n_o+1)(n_1+1) - n_1}\right]$

TABLE 6—CONTINUED

GIRDER CONTINUOUS OVER FOUR SUPPORTS

Supports on Different Levels. Any System of Vertical Loads.

$\frac{I}{l}$ for span $AB = \frac{K_1}{n_o}$ $\qquad$ $\frac{I}{l}$ for span $BC = \frac{K_2}{n_1} = K_1.$ $\qquad$ $\frac{I}{l}$ for span $CD = K_2.$

(b)	Fig. 24 Ends of girder restrained (Continued)	General Case—Continued
		All Spans Identical Except for Loads
		θ_A and M_{DC} known
		$M_{BC} = \frac{6EK(11d_B - 6d_A - 6d_C + d_D) - 2l\left[2(2H_{BA} - H_{AB} + 2H_{BC} + 3EK\theta_A) - (H_{CB} + H_{CD})\right] - lM_{DC}}{13l}$
		$M_{CD} = \frac{3EK(16d_C - 7d_D - 12d_B + 3d_A) + l\left[(2H_{BA} - H_{AB} + 2H_{BC} + 3EK\theta_A) - \frac{7}{2}(2H_{CB} + 2H_{CD} - M_{DC})\right]}{13l}$
		$M_{AB} = -\frac{M_{BC}}{2} - H_{AB} + \frac{3EK}{l}\left[l\theta_A - d_B + d_A\right]$
		θ_D and M_{AB} known
		$M_{BC} = \frac{3EK(16d_B - 7d_A - 12d_C + 3d_D) - l\left[\frac{7}{2}(2H_{BA} + 2H_{BC} + M_{AB}) + H_{DC} - 2H_{CD} - 2H_{CB} + 3EK\theta_D\right]}{13l}$
		$M_{CD} = \frac{6EK(11d_C - 6d_D - 6d_B + d_A) + l\left[4(H_{DC} - 2H_{CD} - 2H_{CB} + 3EK\theta_D) + 2H_{BA} + 2H_{BC} + M_{AB}\right]}{13l}$
		$M_{DC} = \frac{M_{CD}}{2} + H_{DC} + 3EK\left[\theta_D - \frac{(d_D - d_C)}{l}\right]$

TABLE 6—CONTINUED

(b)	**Fig. 24 Ends of girder restrained (Continued)**	θ_A and θ_D known $M_{BC} = \frac{2EK}{5l}(13d_B - 8d_C - 7d_A + 2d_D) - \frac{2}{45}\Big[7(2H_{BA} - H_{AB} + 2H_{BC}) + 2(H_{DC} - 2H_{CD} - 2H_{CB})$ $+ 3EK(7\theta_A + 2\theta_D)\Big]$ $M_{CD} = \frac{2EK}{5l}(13d_C - 8d_B - 7d_D + 2d_A) + \frac{2}{45}\Big[7(H_{DC} - 2H_{CD} - 2H_{CB}) + 2(-H_{AB} + 2H_{BA} + 2H_{BC})$ $+ 3EK(7\theta_D + 2\theta_A)\Big]$ $M_{AB} = -\frac{1}{2}M_{BC} - H_{AB} + 3EK\left[\theta_A - \frac{(d_B - d_A)}{l}\right]$ $M_{DC} = \frac{1}{2}M_{CD} + H_{DC} + 3EK\left[\theta_D - \frac{(d_D - d_C)}{l}\right]$ M_{AB} and M_{DC} known $M_{BC} = \frac{2EK}{5l}(9d_B - 6d_C - 4d_A + d_D) - \frac{1}{15}\left[8(H_{BA} + H_{BC}) - 2(H_{CB} + H_{CD}) + 4M_{AB} + M_{DC}\right]$ $M_{CD} = \frac{2EK}{5l}(9d_C - 6d_B - 4d_D + d_A) - \frac{1}{15}\left[8(H_{CB} + H_{CD}) - 2(H_{BA} + H_{BC}) - 4M_{DC} - M_{AB}\right]$

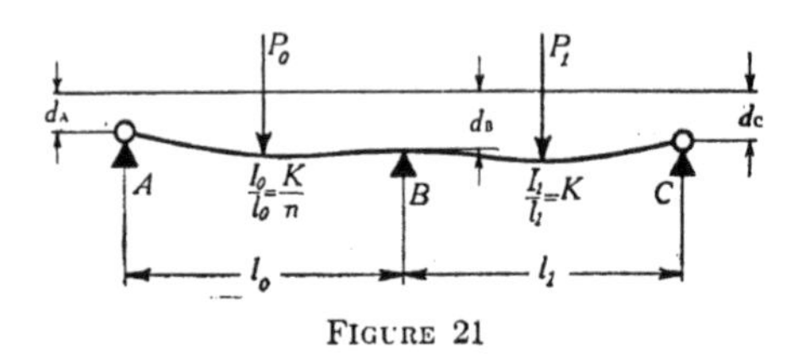

FIGURE 21

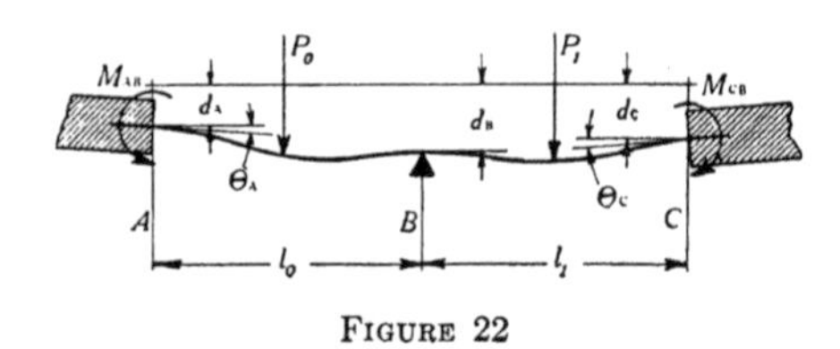

FIGURE 22

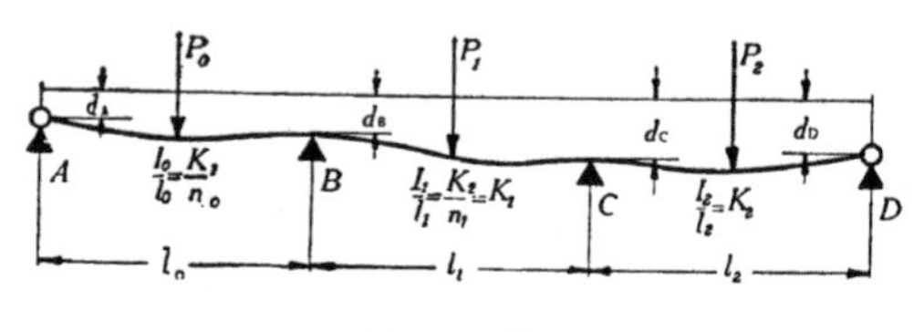

FIGURE 23

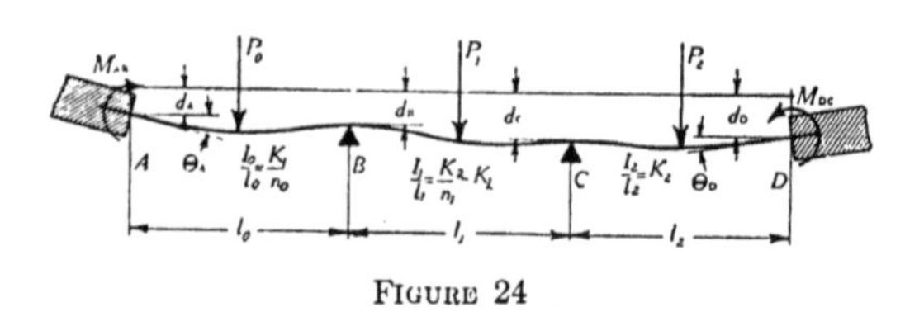

FIGURE 24

VI. Two-legged Rectangular Bent. Legs of Equal Length

22. *Two-legged Rectangular Bent. Concentrated Horizontal Force at the Top—Legs Hinged at the Bases.*—Fig. 25 represents a two-legged rectangular bent having legs hinged at the bases. A single force P is applied at the top of the bent.

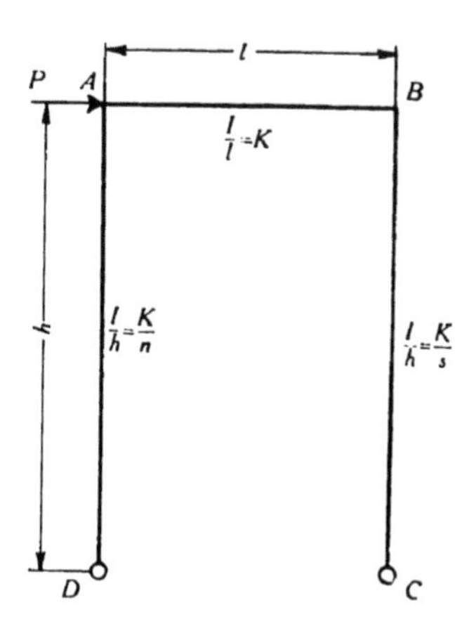

Figure 25

Applying the equations of Table 1

$$M_{AB}=2EK(2\theta_A+\theta_B) \quad . \quad . \quad . \quad . \quad . \quad . \quad . \quad . \qquad (54)$$

$$M_{AD}=\frac{EK}{n}(3\theta_A-3R) \quad . \quad . \quad . \quad . \quad . \quad . \quad . \qquad (55)$$

$$M_{BA}=2EK(2\theta_B+\theta_A) \quad . \quad . \quad . \quad . \quad . \quad . \quad . \quad . \qquad (56)$$

$$M_{BC}=\frac{EK}{s}(3\theta_B-3R) \quad . \quad . \quad . \quad . \quad . \quad . \quad . \qquad (57)$$

$$M_{AD}+M_{BC}+Ph=0 \quad . \quad . \quad . \quad . \quad . \quad . \quad . \quad . \qquad (58)$$

In these equations, from the definition of resisting moment, $M_{AB}=-M_{AD}$ and $-M_{BA}=+M_{BC}$. Hence M_{AB} and M_{BC} may be considered as the two unknown moments to be found. Elimination may be done as follows:

Subtracting equation 56 from equation 54 gives

$$M_{AB}+M_{BC}=2EK(\theta_A-\theta_B) \qquad (59)$$

Subtracting s times equation 57 from n times equation 55 gives

$$-nM_{AB}-sM_{BC}=EK(3\theta_A-3\theta_B) \qquad (60)$$

Combining equations 59 and 60 to eliminate the quantity $(\theta_A-\theta_B)$ gives

$$M_{AB}(3+2n)+M_{BC}(3+2s)=0 \qquad (61)$$

Combining equation 61 with equation 58 gives

$$M_{AB}=\frac{Ph}{2}\left(\frac{3+2s}{3+n+s}\right) \qquad (62)$$

$$M_{BC}=-\frac{Ph}{2}\left(\frac{3+2n}{3+n+s}\right) \qquad (63)$$

If $n=s$, that is, if the sections of AD and BC have the same moments of inertia, equations 62 and 63 take the form

$$M_{BC}=-M_{AB}=-\frac{Ph}{2} \qquad (64)$$

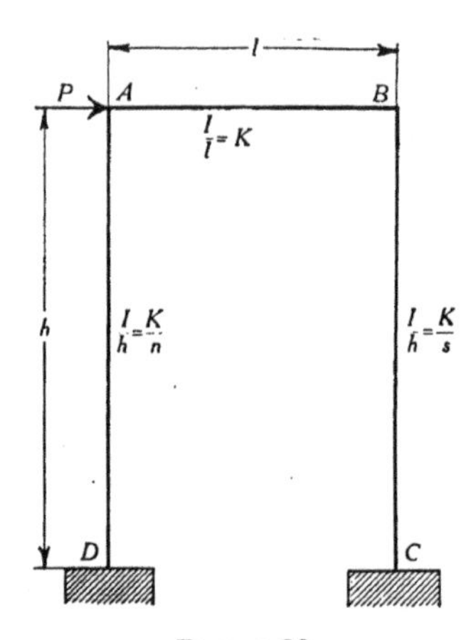

FIGURE 26

23. *Two-legged Rectangular Bent. Concentrated Horizontal Force at the Top—Legs Fixed at the Bases.*—Fig. 26 represents a two-legged rectangular bent having legs fixed at the bases. A single force P is applied at the top of the bent.

Applying equation (A), Table 1,

$$M_{AB}=2EK(2\theta_A+\theta_B) \qquad (65)$$

$$M_{AD}=\frac{2EK}{n}(2\theta_A-3R) \qquad (66)$$

$$M_{DA}=\frac{2EK}{n}(\theta_A-3R) \qquad (67)$$

$$M_{BA}=2EK(2\theta_B+\theta_A) \qquad (68)$$

$$M_{BC}=\frac{2EK}{s}(2\theta_B-3R) \qquad (69)$$

$$M_{CB}=\frac{2EK}{s}(\theta_B-3R) \qquad (70)$$

$$M_{AD}+M_{DA}+M_{BC}+M_{CB}+Ph=0 \qquad (71)$$

$$M_{AB}+M_{AD}=0 \qquad (72)$$

$$M_{BA}+M_{BC}=0 \qquad (73)$$

Substituting the values of M_{AB} and M_{AD} from equations 65 and 66 in equation 72 and simplifying gives

$$\frac{2\theta_A}{R}(1+n)+\frac{n\theta_B}{R}=3 \qquad (74)$$

Substituting the values of M_{BA} and M_{BC} from equations 68 and 69 in equation 73 and simplifying gives

$$\frac{2\theta_B}{R}(1+s)+\frac{s\theta_A}{R}=3. \qquad (75)$$

Substituting the values for the moments in equation 71 and simplifying gives

$$\frac{s\theta_A}{R}+\frac{n\theta_B}{R}=2(s+n)-\frac{n\,s\,Ph}{6EK}\cdot\frac{1}{R} \qquad (76)$$

Solving equation 74 for $\frac{\theta_B}{R}$ gives

$$\frac{\theta_B}{R}=\frac{3}{n}-\frac{2\,\theta_A}{R}\left(\frac{1+n}{n}\right) \qquad (77)$$

Substituting the value of $\frac{\theta_B}{R}$ from equation 77 in equation 75 gives

$$\frac{\theta_A}{R}=\frac{3(2s+2-n)}{3ns+4n+4s+4} \qquad (78)$$

Substituting the value of $\frac{\theta_A}{R}$ from equation 78 in equation 77 and simplifying gives

$$\frac{\theta_B}{R} = \frac{3(2n+2-s)}{3ns+4n+4s+4} \quad \text{. . . . .} \quad (79)$$

Substituting the values of $\frac{\theta_A}{R}$ and $\frac{\theta_B}{R}$ from equation 78 and equation 79 in equation 76 gives

$$\frac{3s(2s+2-n)}{3ns+4n+4s+4} + \frac{3n(2n+2-s)}{3ns+4n+4s+4} = 2(s+n) - \frac{ns\,Ph}{6EK} \cdot \frac{1}{R}$$

Solving this equation for $\frac{1}{R}$ gives

$$\frac{1}{R} = \frac{6EK}{n\,s\,Ph} \; \frac{2(3ns^2+11ns+s^2+s+3n^2s+n^2+n)}{3ns+4n+4s+4} \quad \text{. .} \quad (80)$$

Substituting the value of $\frac{1}{R}$ from equation 80 in equation 79 and solving for θ_B gives

$$\theta_B = \frac{Ph}{12EK} \; \frac{3ns(2n+2-s)}{3ns^2+11ns+s^2+s+3n^2s+n^2+n} \quad \text{. . . . .} \quad (81)$$

Substituting the value of $\frac{1}{R}$ from equation 80 in equation 78 and solving for θ_A gives

$$\theta_A = \frac{Ph}{12EK} \; \frac{3ns(2s+2-n)}{3ns^2+11ns+s^2+s+3n^2s+n^2+n} \quad \text{. . . . .} \quad (82)$$

Substituting the values of θ_B and θ_A from equations 81 and 82 in equation 65 gives

$$M_{AB} = \frac{Ph}{2} \; \frac{3ns(s+2)}{3ns^2+11ns+s^2+s+3n^2s+n^2+n} \quad \text{. . . .} \quad (83)$$

Substituting the values of θ_B and θ_A from equations 81 and 82 in equation 68 and substituting $-M_{BC}$ for M_{BA} gives

$$M_{BC} = -\frac{Ph}{2} \; \frac{3ns(n+2)}{3ns^2+11ns+s^2+s+3n^2s+n^2+n} \quad \text{. . . .} \quad (84)$$

From equation 67

$$M_{DA}=\frac{2EKR}{n}\left[\frac{\theta_A}{R}-3\right]. \qquad (85)$$

Substituting the value of R from equation 80 and of $\frac{\theta_A}{R}$ from equation 78 gives

$$M_{DA}=-\frac{Ph}{2}\,\frac{s(2s+2+5n+3ns)}{3ns^2+11ns+s^2+s+3n^2s+n^2+n} \qquad (86)$$

From equation 70

$$M_{CB}=\frac{2EKR}{s}\left(\frac{\theta_B}{R}-3\right)$$

Substituting the value of R from equation 80 and of $\frac{\theta_B}{R}$ from equation 79 gives

$$M_{CB}=-\frac{Ph}{2}\,\frac{n(2n+2+5s+3ns)}{(3ns^2+11ns+s^2+s+3n^2s+n^2+n)} \qquad (87)$$

Letting Δ_o represent $2(3ns^2+11ns+s^2+s+3n^2s+n^2+n)$, the equation for the moments in the frame are

$$M_{AD}=-\frac{Ph}{\Delta_o}3ns(s+2) \qquad (88)$$

$$M_{BC}=-\frac{Ph}{\Delta_o}3ns(n+2) \qquad (89)$$

$$M_{CB}=-\frac{Ph}{\Delta_o}n(2n+2+5s+3ns) \qquad (90)$$

$$M_{DA}=-\frac{Ph}{\Delta_o}s(2s+2+5n+3ns) \qquad (91)$$

If $n=s$, that is, if the section of AD has the same moment of inertia as the section of BC, equations 88 to 91 take the form

$$M_{AD}=M_{BC}=-\frac{Ph}{2}\,\frac{3n}{6n+1} \qquad (92)$$

$$M_{CB}=M_{DA}=-\frac{Ph}{2}\,\frac{3n+1}{6n+1} \qquad (93)$$

If $n=s=1$, equations 92 and 93 take the form

$$M_{AD}=M_{BC}=-\frac{3}{14}Ph \quad (94)$$

$$M_{CB}=M_{DA}=-\frac{4}{14}Ph \quad (95)$$

24. *Two-legged Rectangular Bent. Any System of Horizontal Loads on One Leg. Legs Hinged at the Bases.*—Fig. 27 represents a

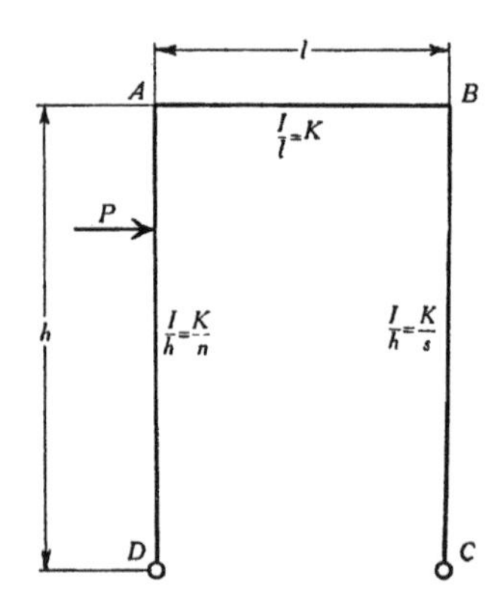

FIGURE 27

two-legged rectangular bent with any horizontal load on the leg AD. The legs are hinged at D and C. P represents the resultant of any system of horizontal forces acting on AD.

If M_D represents the moment of P about D,

$$M_{AD}+M_{BC}+M_D=0$$

Applying the equations of Table 1 gives

$$M_{AB}=2EK(2\theta_A+\theta_B) \quad (96)$$

$$M_{AD}=\frac{EK}{n}(3\theta_A-3R)+H_{AD} \quad (97)$$

$$M_{BA}=2EK(2\theta_B+\theta_A) \quad (98)$$

$$M_{BC}=\frac{EK}{s}(3\theta_B-3R) \quad (99)$$

$$M_{AD}+M_{BC}+M_D=0 \quad (100)$$

Equations 96 to 100 are almost identical in form to equations 54 to 58 of section 22. Hence the solution for the moments by eliminating values of θ and R, as in section 22, gives

$$M_{AB}=\frac{1}{2}\ \frac{M_D(2s+3)-2n\ H_{AD}}{n+s+3} \quad . \quad . \quad . \quad . \quad . \quad . \quad . \quad . \quad (101)$$

$$M_{BC}=-\frac{1}{2}\ \frac{M_D(2n+3)+2n\ H_{AD}}{n+s+3} \quad . \quad . \quad . \quad . \quad . \quad . \quad . \quad (102)$$

If $n=s$, that is, if the sections of AD and BC have the same moments of inertia, equations 102 and 101 take the form

$$M_{BC}=-\frac{1}{2}\left[M_D+\frac{2n\ H_{AD}}{2n+3}\right] \quad . \quad . \quad . \quad . \quad . \quad . \quad . \quad . \quad (103)$$

$$M_{AB}=\frac{1}{2}\left[M_D-\frac{2n\ H_{AD}}{2n+3}\right] \quad . \quad . \quad . \quad . \quad . \quad . \quad . \quad . \quad . \quad (104)$$

If $n=s=1$, equations 103 and 104 take the form

$$M_{BC}=-\frac{1}{10}\left[5M_D+2H_{AD}\right] \quad . \quad . \quad . \quad . \quad . \quad . \quad . \quad . \quad . \quad (105)$$

$$M_{AB}=\frac{1}{10}\left[5M_D-2H_{AD}\right] \quad . \quad . \quad . \quad . \quad . \quad . \quad . \quad . \quad . \quad (106)$$

If both legs of the bent are loaded and if the bent and loads are symmetrical about a vertical center line

$$M_{AB}=M_{BC}=-\left[\frac{2n}{2n+3}\right]H_{AD} \quad . \quad . \quad . \quad . \quad . \quad . \quad . \quad . \quad (107)$$

Values of H_{AD} to be used in equations 101 to 107 are given in Table 2.

25. *Two-legged Rectangular Bent. Any System of Horizontal Loads on One Leg—Legs Fixed at the Bases.*—Fig. 28 represents a two-

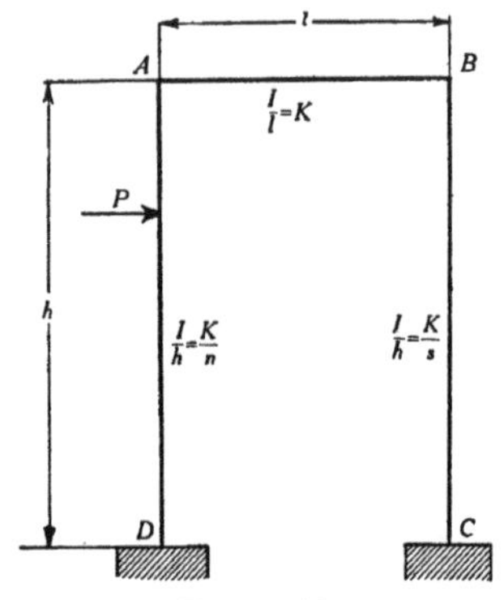

FIGURE 28

legged rectangular bent with any horizontal load on the leg AD. The legs are fixed at D and C. P represents the resultant of all the forces on AD.

If M_D represents the moment of P about D,

$$M_{AD}+M_{DA}+M_{BC}+M_{CB}+M_D=0$$

Applying the equations of Table 1 gives

$$M_{AB}=2EK(2\theta_A+\theta_B) \quad . \quad . \quad . \quad (108)$$

$$n\,M_{AD}=2EK(2\theta_A-3R)+n\,C_{AD} \quad . \quad . \quad . \quad (109)$$

$$n\,M_{DA}=2EK(\theta_A-3R)-n\,C_{DA} \quad . \quad . \quad . \quad (110)$$

$$M_{BA}=2EK(2\theta_B+\theta_A) \quad . \quad . \quad . \quad (111)$$

$$s\,M_{BC}=2EK(2\theta_B-3R) \quad . \quad . \quad . \quad (112)$$

$$s\,M_{CB}=2EK(\theta_B-3R) \quad . \quad . \quad . \quad (113)$$

$$M_{AD}+M_{DA}+M_{BC}+M_{CB}+M_D=0 \quad . \quad . \quad . \quad (114)$$

The equations 108 to 114 are similar to equations 65 to 73 of section 23. The method of solving for the four unknown moments will be done in a different way from that of section 23. The equations will first be combined to eliminate θ and R, then the resulting equations solved simultaneously for the moments.

Adding equation 112 and two times equation 110 and subtracting equation 109 and two times equation 113 to eliminate θ_A, θ_B, and R gives

$$n\,M_{AB}+s\,M_{BC}-2s\,M_{CB}+2n\,M_{DA}=-n(C_{AD}+2C_{DA}) \quad . \quad (115)$$

Adding equation 109 and two times equation 112 and subtracting equations 110 and 111 and two times equation 113 gives

$$-n\,M_{AB}+(2s+1)M_{BC}-2s\,M_{CB}-n\,M_{DA}=n(C_{AD}+C_{DA}) \quad (116)$$

Adding equations 108, 110 and 112, and subtracting equations 109, 111 and 113 gives

$$(n+1)M_{AB}+(s+1)M_{BC}-s\,M_{CB}+n\,M_{DA}= -n(C_{AD}+C_{DA}) \quad (117)$$

Equations 114 to 117 are rewritten in Table 7. In this table the unknown moments are written at the heads of the columns and the coefficients are written below.

TABLE 7

EQUATIONS FOR THE TWO-LEGGED RECTANGULAR BENT OF FIG. 28

No. of Equation	Left-hand Member of Equation				Right-hand Member of Equation	How Equation was Obtained
	M_{AB}	M_{BC}	M_{CB}	M_{DA}		
114	-1	1	1	1	$-M_D$	114
115	n	s	$-2s$	$2n$	$-n(C_{AD}+2C_{DA})$	$-(109)+(112)+2(110)$ $-2(113)$
116	$-n$	$2s+1$	$-2s$	$-n$	$n(C_{AD}+C_{DA})$	$-(111)+2(112)-2(113)$ $-(110)+(109)$
117	$n+1$	$s+1$	$-s$	n	$-n(C_{AD}+C_{DA})$	$(108)-(109)+(110)$ $-(111)+(112)-(113)$

Solving these equations simultaneously and letting
$\Delta_o=2(11sn+3sn^2+3s^2n+s^2+s+n^2+n)$ gives

$$M_{AD}=-\frac{n}{\Delta_o}\left[3s\,(s+2)\,(M_D-C_{DA})-C_{AD}\,(6ns+3s^2+5s+2n)\right] \quad (118)$$

$$M_{BC}=-\frac{n}{\Delta_o}\left[3s(n+2)(M_D-C_{DA})+C_{AD}(3ns-n-4s)\right] \quad (119)$$

$$M_{CB}=-\frac{n}{\Delta_o}\left[(3ns+2n+5s+2)\,(M_D-C_{DA})+C_{AD}\,(3ns+3n-3s-1)\right] \quad (120)$$

$$M_{DA} = -\frac{1}{\Delta_o}\Big[s\,(3ns + 2s + 5n + 2)\ (M_D - C_{DA})$$

$$+ n\, C_{AD}(3s^2 + 12s + 1) \Big] - C_{DA} \quad . . . . (121)$$

If $n = s$ equations 118 and 121 take the form

$$M_{AD} = -\frac{n}{2}\left[\frac{3(M_D - C_{DA})}{6n+1} - C_{AD}\left(\frac{1}{n+2} + \frac{3}{6n+1}\right)\right] \quad . \ (122)$$

$$M_{BC} = -\frac{n}{2}\left[\frac{3(M_D - C_{DA})}{6n+1} + C_{AD}\left(\frac{1}{n+2} - \frac{3}{6n+1}\right)\right] \quad . \ (123)$$

$$M_{CB} = -\frac{1}{2}\left[\frac{(3n+1)\ (M_D - C_{DA})}{6n+1} - C_{AD}\left(\frac{1}{n+2} - \frac{3n}{6n+1}\right)\right]$$
$$. . . . . (124)$$

$$M_{DA} = -\frac{1}{2}\left[\frac{(3n+1)\ (M_D - C_{DA})}{6n+1} + C_{AD}\left(\frac{1}{n+2} + \frac{3n}{6n+1}\right)\right]$$
$$- C_{DA} \quad . . . . (125)$$

If both legs of the bent are loaded and if the bent and loads are symmetrical about a vertical center line

$$M_{AB} = M_{BC} = -\frac{n}{n+2}\, C_{AD} \quad . . . . (126)$$

$$M_{CB} = -M_{DA} = \frac{1}{n+2}\, C_{AD} + C_{DA} \quad . . . . (127)$$

Values of C_{AD} and C_{DA} to be used in equations 118 to 127 are given in Table 2.

26. *Two-legged Rectangular Bent. Any System of Vertical Loads on the Top—Legs Hinged at the Bases.*—Fig. 29 represents a two-legged rectangular bent having any system of vertical loads on AB. The legs are hinged at D and C. P represents the resultant of the loads on AB.

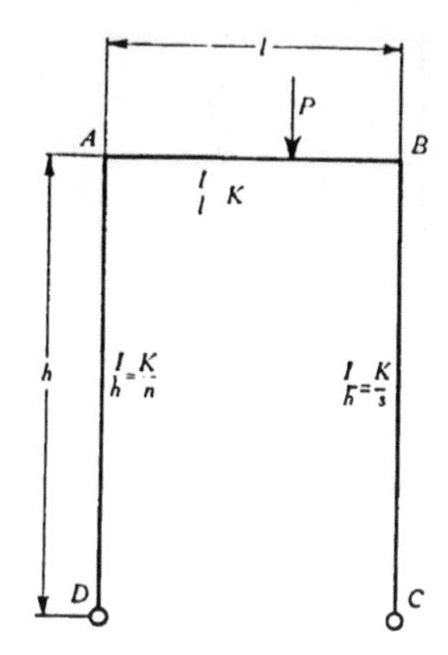

FIGURE 29

Applying the equations of Table 1 gives

$$M_{AB}=2EK(2\theta_A+\theta_B)-C_{AB} \quad \text{. . . . .} \quad (128)$$

$$M_{AD}=\frac{EK}{n}(3\theta_A-3R) \quad \text{. . . . .} \quad (129)$$

$$M_{BA}=2EK(2\theta_B+\theta_A)+C_{BA} \quad \text{. . . . .} \quad (130)$$

$$M_{BC}=\frac{EK}{s}(3\theta_B-3R) \quad \text{. . . . .} \quad (131)$$

$$M_{AD}+M_{BC}=0 \quad \text{. . . . .} \quad (132)$$

Eliminating values of θ and R as in section 22 gives

$$M_{BC}=-\frac{3}{2}\left[\frac{C_{AB}+C_{BA}}{n+s+3}\right] \quad \text{. . . . .} \quad (133)$$

$$M_{AD}=\frac{3}{2}\left[\frac{C_{AB}+C_{BA}}{n+s+3}\right]=-M_{BC} \quad \text{. . . . . . .} \quad (134)$$

Values of C_{BA} and C_{AB} to be used in equations 133 and 134 are given in Table 2.

27. *Two-legged Rectangular Bent. Any System of Vertical Loads on the Top—Legs Fixed at the Bases.*—Fig. 30 represents a two-legged rectangular bent with any system of vertical loads on AB. The legs are fixed at C and D. P represents the resultant of all the loads on AB.

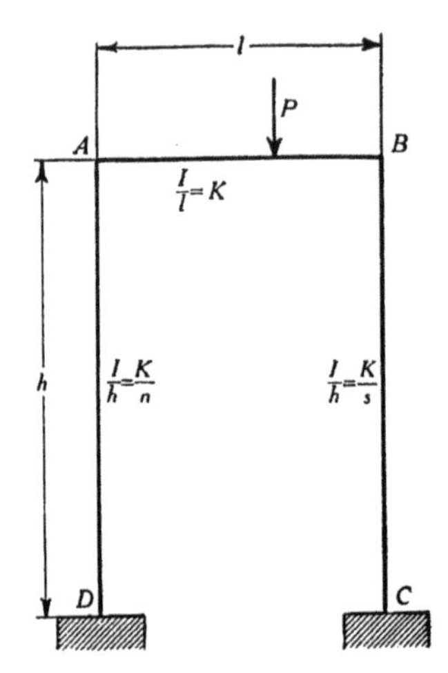

FIGURE 30

Applying the equations of Table 1 gives

$$nM_{DA}=2EK(\theta_A-3R) \quad (135)$$

$$nM_{AB}=-2EK(2\theta_A-3R) \quad (136)$$

$$M_{AB}=2EK(2\theta_A+\theta_B)-C_{AB} \quad (137)$$

$$M_{BC}=-2EK(2\theta_B+\theta_A)-C_{BA} \quad (138)$$

$$sM_{BC}=2EK(2\theta_B-3R) \quad (139)$$

$$sM_{CB}=2EK(\theta_B-3R) \quad (140)$$

$$-M_{AB}+M_{BC}+M_{CB}+M_{DA}=0 \quad (141)$$

Combining these equations as indicated in the table to eliminate θ_A, θ_B, and R gives the equations in Table 8.

Solving these equations simultaneously gives

$$M_{AB}=-\frac{1}{\Delta_o}\left\{C_{BA}(10ns+s^2)+C_{AB}(11ns+2s^2+2s+2n)\right\} \quad (146)$$

$$M_{BC}=-\frac{1}{\Delta_o}\left\{C_{BA}(11ns+2n^2+2n+2s)+C_{AB}(10ns+n^2)\right\} \quad (147)$$

TABLE 8

EQUATIONS FOR THE TWO-LEGGED RECTANGULAR BENT REPRESENTED BY FIG. 30

No. of Equation	Left-hand Member of Equation				Right-hand Member of Equation	How Equation Was Obtained
	M_{AB}	M_{BC}	M_{CB}	M_{DA}		
142	-1	1	1	1	0	(141)
143	n	s	$-2s$	$2n$	0	(136)+(139) +2[(135)−(140)]
144	$-n$	$2s+1$	$-2s$	$-n$	$-C_{BA}$	(138)+2[(139)−(140)] −[(135)+(136)]
145	$n+1$	$s+1$	$-s$	n	$-(C_{AB}+C_{BA})$	(135)+(136)+(137)+(138) +(139)−(140)

$$M_{CB}=-\frac{1}{\Delta_o}\left\{C_{BA}(7ns-2n^2-2n+s)+C_{AB}(8ns-n^2+3n)\right\} \quad (148)$$

$$M_{DA}=\frac{1}{\Delta_o}\left\{C_{BA}(8ns-s^2+3s)+C_{AB}(7ns-2s^2-2s+n)\right\} \quad (149)$$

in which $\Delta_o=2(11sn+3sn^2+3s^2n+s^2+s+n^2+n)$

If the load is symmetrical, that is, if $C_{BA}=C_{AB}=\frac{F}{l}$, equations 146 to 149 take the form

$$M_{AB}=-\frac{F}{l\Delta_o}(21ns+3s^2+2s+2n) \quad (150)$$

$$M_{BC}=-\frac{F}{l\Delta_o}(21ns+3n^2+2n+2s) \quad (151)$$

$$M_{CB}=-\frac{F}{l\Delta_o}(15ns-3n^2+n+s) \quad (152)$$

$$M_{DA}=\frac{F}{l\Delta_o}(15ns-3s^2+s+n) \quad (153)$$

If the bent is symmetrical about a vertical center line, that is, if $n=s$, equations 146 to 149 take the form

$$M_{AB}=-\frac{1}{2}\left\{C_{BA}\left[\frac{2}{n+2}-\frac{1}{6n+1}\right]+C_{AB}\left[\frac{2}{n+2}+\frac{1}{6n+1}\right]\right\} \quad (154)$$

$$M_{BC}=-\frac{1}{2}\left\{C_{BA}\left[\frac{2}{n+2}+\frac{1}{6n+1}\right]+C_{AB}\left[\frac{2}{n+2}-\frac{1}{6n+1}\right]\right\} \quad (155)$$

$$M_{CB}=-\frac{1}{2}\left\{C_{BA}\left[\frac{1}{n+2}-\frac{1}{6n+1}\right]+C_{AB}\left[\frac{1}{n+2}+\frac{1}{6n+1}\right]\right\} \quad . \quad (156)$$

$$M_{DA}=\frac{1}{2}\left\{C_{BA}\left[\frac{1}{n+2}+\frac{1}{6n+1}\right]+C_{AB}\left[\frac{1}{n+2}-\frac{1}{6n+1}\right]\right\} \quad . \quad (157)$$

If the bent and the loading are symmetrical about a vertical center line, that is, if $n=s$ and $C_{BA}=C_{AB}=\frac{F}{l}$, equations 146 to 149 take the form

$$M_{AB}=M_{BC}=-\frac{2}{(n+2)}\frac{F}{l} \quad . \quad . \quad . \quad (158)$$

$$M_{CB}=-M_{DA}=-\frac{1}{n+2}\frac{F}{l} \quad . \quad . \quad . \quad (159)$$

28. *Two-legged Rectangular Bent. External Moment at One Corner—Legs Hinged at the Bases.*—Fig. 31 represents a two-legged

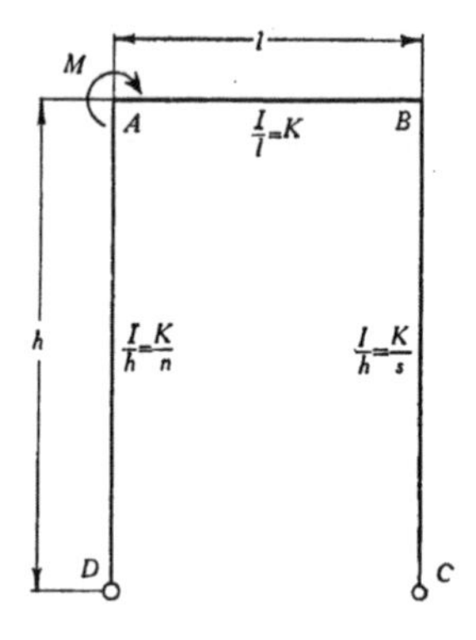

FIGURE 31

rectangular bent having legs hinged at the bases. An external couple whose moment is represented by M is applied at A.

Applying the equations of Table 1 to this bent gives four equations which are identical with the first four equations of section 22. Also, for equilibrium at A,

$$M_{AD}+M_{AB}=M \quad . \quad . \quad . \quad (160)$$

Solving these five equations for the moments gives

$$M_{AB}=M\left[1-\frac{3}{2}\ \frac{1}{(n+s+3)}\right]. \qquad (161)$$

For equilibrium the horizontal reactions at C and D must be equal and opposite. Therefore, $M_{AD}=-M_{BC}$. Combining equations 160 and 161 gives

$$M_{AD}=-M_{BC}=\frac{3}{2}\left(\frac{M}{n+s+3}\right) \qquad (162)$$

If the bent is symmetrical about a vertical center line $n=s$. Equations 161 and 162 then reduce to the form

$$M_{AB}=M\left[1-\frac{3}{2}\ \frac{1}{(2n+3)}\right]. \qquad (163)$$

$$M_{AD}=-M_{BC}=\frac{3}{2}\left(\frac{M}{2n+3}\right) \qquad (164)$$

If $n=s=1$

$$M_{AB}=\frac{7}{10}M \qquad (165)$$

$$M_{AD}=-M_{BC}=\frac{3}{10}M \qquad (166)$$

29. *Two-legged Rectangular Bent. External Moment at One Corner—Legs Fixed at the Bases.*—Fig. 32 represents a two-legged rectangular bent having legs fixed at the bases. An external couple whose moment is represented by M is applied at A.

Applying the equations of Table 1 to this bent gives six equations which are identical with the first six equations of section 23. For equilibrium of the entire bent,

$$-M_{AB}+M_{DA}+M_{BC}+M_{CB}=-M \qquad (167)$$

Also,

$$M_{AB}+M_{AD}=M \qquad (168)$$

$$M_{BA}+M_{BC}=0 \qquad (169)$$

Solving these equations for the moments gives

$$M_{AB}=-\frac{M}{\Delta_o}(11ns+2s^2+2s+2n)+M \quad . \quad . \quad . \quad (170)$$

$$M_{BC}=-\frac{nM}{\Delta_o}(10s+n) \quad . \quad . \quad . \quad (171)$$

$$M_{CB}=-\frac{nM}{\Delta_o}(8s-n+3) \quad . \quad . \quad . \quad (172)$$

$$M_{DA}=\frac{M}{\Delta_o}(7ns-2s^2-2s+n) \quad . \quad . \quad . \quad (173)$$

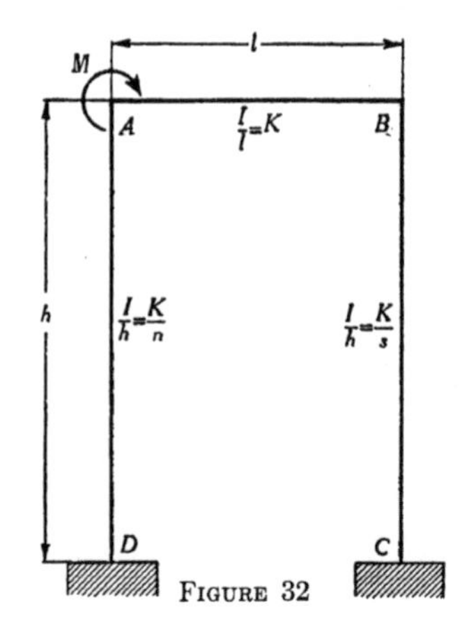

FIGURE 32

$$M_{AD}=\frac{M}{\Delta_o}(11ns+2s^2+2s+2n) \quad . \quad . \quad . \quad (174)$$

in which

$$\Delta_o=22sn+2(s^2+s+n^2+n)+6(sn^2+s^2n)$$

If the bent is symmetrical about a vertical center line $n=s$. Equations 170, 171, 172, 173, and 174 then take the form

$$M_{AB}=\frac{M}{2}\left[1+\frac{n}{n+2}-\frac{1}{6n+1}\right] \quad . \quad . \quad . \quad (175)$$

$$M_{BC}=\frac{M}{2}\left[\frac{n}{n+2}-\frac{6n}{6n+1}\right] \quad . \quad . \quad . \quad (176)$$

$$M_{CB}=-\frac{M}{2}\left[\frac{1}{n+2}+\frac{1}{6n+1}\right] \quad . \quad . \quad . \quad (177)$$

$$M_{DA}=\frac{M}{2}\left[\frac{1}{n+2}-\frac{1}{6n+1}\right] \quad (178)$$

$$M_{AD}=\frac{M}{2}\left[1-\frac{n}{n+2}+\frac{1}{6n+1}\right] \quad (179)$$

If $n=s=1$, equations 170, 171, 172, 173, and 174 take the form

$$M_{AB}=\frac{25}{42}M \quad (180)$$

$$M_{BC}=-\frac{11}{42}M \quad (181)$$

$$M_{CB}=-\frac{10}{42}M \quad (182)$$

$$M_{DA}=\frac{4}{42}M \quad (183)$$

$$M_{AD}=\frac{17}{42}M \quad (184)$$

30. *Two-legged Rectangular Bent. Settlement of Foundations—Legs Hinged at the Bases.*—Fig. 33 represents a two-legged rectangular bent. The legs of the bent are hinged at D and C. The unstrained position of the bent is represented by the broken line $DA'B'C'$. Due to settlement of the foundations the point C' has moved to C. The motion of C can be considered as made up of two parts, as follows: first, without being strained the bent rotates about D until C' is at C'', a point on CD; secondly, the bent is strained by applying a force at C'' acting along DC which moves C'' to C; that is, no matter what motion of C relative to D takes place, the stress in the bent depends only upon the change in the distance of C from D.

The change in the distance from D to C is represented by d. d is made up of two parts, d_1 due to the deflection of AD, and d_2 due to the deflection of BC.

Applying the equations of Table 1 gives

$$M_{AB}=-\frac{EK}{n}(3\theta_A-3R_1) \quad (185)$$

$$M_{AB}=2EK(2\theta_A+\theta_B) \quad (186)$$

$$M_{BC} = -2EK(2\theta_B + \theta_A) \quad (187)$$

$$M_{BC} = \frac{EK}{s}(3\theta_B - 3R_1) + \frac{3EK}{s}\frac{d}{h} \quad (188)$$

$$M_{AB} - M_{BC} = 0 \quad (189)$$

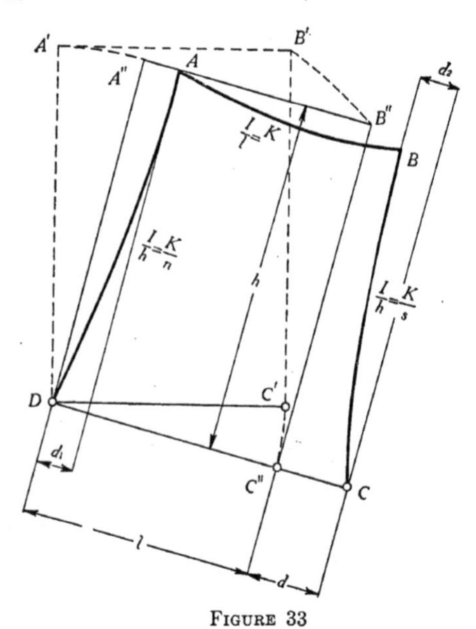

FIGURE 33

Solving these equations for the moments gives

$$M_{AB} = M_{BC} = \frac{d}{h}\frac{3EK}{(n+s+3)} \quad (190)$$

If the bent is symmetrical about a vertical center line $n=s$, and

$$M_{AB} = M_{BC} = \frac{d}{h}\frac{3EK}{2n+3} \quad (191)$$

If $n=s=1$

$$M_{AB} = M_{BC} = \frac{3}{5}\frac{d}{h}EK \quad (192)$$

d in equations 190, 191, and 192 represents the increase in the distance from D to C.

31. *Two-legged Rectangular Bent. Settlement of Foundations—Legs Restrained at the Bases.*—Fig. 34 represents a two-legged reetangular bent. The legs of the bent are restrained at C and D. The unstrained position of the bent is represented by the broken line $DA'B'C'$. Due to settlement of the foundations C' has moved to C.

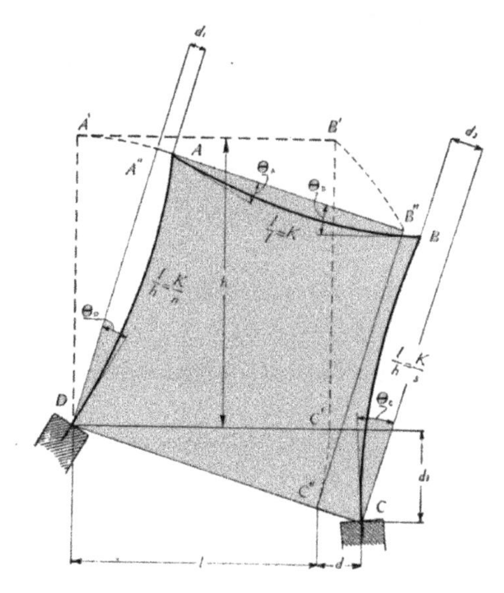

FIGURE 34

The foundations, moreover, have tipped so that whereas the tangents to the neutral axes of the columns at the bases were originally vertical now they are inclined. The motion of the bent can be considered as made up of two parts, as follows: first, without being strained the bent rotates about D until C' is at C'', a point on DC. At the same time the supports rotate so that the tangents to the neutral axes of the columns at the bases are normal to DC. These motions produce no stress in the bent; secondly, C'' moves to C, the support D rotates through the angle θ_D, and the support C rotates through the angle θ_C; that is, no matter what motion of C relative to D takes place, the stress in the bent depends only upon the change in the distance of C from D and upon the rotations of the supports at D and C *relative to the line DC* (not relative to DC').

Applying the equations of Table 1 gives

$$M_{AB} = -\frac{2EK}{n}(2\theta_A + \theta_D - 3R_1) \quad \ldots \ldots \ldots \quad (193)$$

$$M_{AB} = 2EK(2\theta_A + \theta_B) \quad \ldots \ldots \ldots \quad (194)$$

$$M_{BC} = -2EK(2\theta_B + \theta_A) \quad \ldots \ldots \ldots \quad (195)$$

$$M_{BC} = \frac{2EK}{s}(2\theta_B + \theta_C - 3R_2) \quad \ldots \ldots \ldots \quad (196)$$

$$M_{CB} = \frac{2EK}{s}(2\theta_C + \theta_B - 3R_2) \quad \ldots \ldots \ldots \quad (197)$$

$$M_{DA} = \frac{2EK}{n}(2\theta_D + \theta_A - 3R_1) \quad \ldots \ldots \ldots \quad (198)$$

For the columns to be in equilibrium

$$M_{DA} - M_{AB} + M_{BC} + M_{CB} = 0 \quad \ldots \ldots \ldots \quad (199)$$

Combining these equations and solving for the moments gives

$$M_{AB} = \frac{2EK}{\Delta_o}\left[3(6ns + 2n - s)\frac{d}{h} + (9ns + 8n - s)\,\theta_C - (6ns + 2n - 7s - 3s^2)\,\theta_D\right] \quad \ldots \ldots \ldots \quad (200)$$

$$M_{BC} = \frac{2EK}{\Delta_o}\left[3(6ns - n + 2s)\frac{d}{h} + (6ns - 7n + 2s - 3n^2)\,\theta_C - (9ns - n + 8s)\,\theta_D\right] \quad \ldots \ldots \ldots \quad (201)$$

$$M_{CB} = \frac{2EK}{\Delta_o}\left[3(6ns + 5n + 2s + 1)\frac{d}{h} + (12ns + 22n + 4s + 3 + 3n^2)\,\theta_C - (9ns + 7n + 7s + 3)\,\theta_D\right] \quad \ldots \ldots \ldots \quad (202)$$

$$M_{DA} = -\frac{2EK}{\Delta_o}\left[3(6ns + 2n + 5s + 1)\frac{d}{h} + (9ns + 7n + 7s + 3)\,\theta_C - (12ns + 4n + 22s + 3 + 3s^2)\,\theta_D\right] \quad \ldots \ldots \ldots \quad (203)$$

in which

$$\Delta_o = 22ns + 2(s^2 + s + n^2 + n) + 6(n^2 s + s^2 n)$$

If the bent is symmetrical about a vertical center line $n=s$. Equations 200, 201, 202, and 203 then take the form

$$M_{AB}=EK\left[\frac{d}{h}\frac{3}{n+2}+\frac{\theta_C-\theta_D}{n+2}+\frac{3(\theta_C+\theta_D)}{6n+1}\right] \quad . \quad . \quad . \quad . \quad . \quad (204)$$

$$M_{BC}=EK\left[\frac{d}{h}\frac{3}{n+2}+\frac{\theta_C-\theta_D}{n+2}-\frac{3(\theta_C+\theta_D)}{6n+1}\right] \quad . \quad . \quad . \quad . \quad . \quad (205)$$

$$M_{CB}= EK\left[\frac{d}{h}\frac{3(n+1)}{n(n+2)}+\frac{2n+3}{n(n+2)}(\theta_C-\theta_D)+\frac{3(\theta_C+\theta_D)}{6n+1}\right] \quad . \quad (206)$$

$$M_{DA}=-EK\left[\frac{d}{h}\frac{3(n+1)}{n(n+2)}+\frac{2n+3}{n(n+2)}(\theta_C-\theta_D)-\frac{3(\theta_C+\theta_D)}{6n+1}\right] \quad (207)$$

If $n=s=1$

$$M_{AB}=\frac{EK}{21}\left[21\frac{d}{h}+16\theta_C+2\theta_D\right] \quad . \quad . \quad . \quad . \quad . \quad . \quad . \quad . \quad . \quad (208)$$

$$M_{BC}=\frac{EK}{21}\left[21\frac{d}{h}-2\theta_C-16\theta_D\right] \quad . \quad . \quad . \quad . \quad . \quad . \quad . \quad . \quad . \quad (209)$$

$$M_{CB}=\frac{EK}{21}\left[42\frac{d}{h}+44\theta_C-26\theta_D\right] \quad . \quad . \quad . \quad . \quad . \quad . \quad . \quad . \quad . \quad (210)$$

$$M_{DA}=-\frac{EK}{21}\left[42\frac{d}{h}+26\theta_C-44\theta_D\right] \quad . \quad . \quad . \quad . \quad . \quad . \quad . \quad . \quad (211)$$

In all these equations θ is measured from a line normal to the line CD.

If the foundations settle without tipping it is more convenient to measure θ from a line normal to the original position DC'. It is then necessary to consider the vertical settlement d_3.

Proceeding as before, if the tangents to the neutral axes of the columns at their bases remain vertical, if the bases are separated horizontally by an amount represented by d, and if C settles vertically by an amount represented by d_3, it can be proved that

$$M_{AB}=\frac{2EK}{\Delta_o}\left[3(6ns+2n-s)\frac{d}{h}-3(ns+2n+2s+s^2)\frac{d_3}{l}\right] \quad (212)$$

$$M_{BC}=\frac{2EK}{\Delta_o}\left[3(6ns-n+2s)\frac{d}{h}+3(ns+2n+2s+s^2)\frac{d_3}{l}\right] \quad (213)$$

$$M_{CB}=\frac{2EK}{\Delta_o}\left[3(6ns+5n+2s+1)\frac{d}{h}-3(ns+5n-s+n^2)\frac{d_3}{l}\right] \quad (214)$$

$$M_{DA}=-\frac{2EK}{\Delta_o}\left[3(6ns+2n+5s+1)\frac{d}{h}+3(ns-n+5s+s^2)\frac{d_3}{l}\right] \quad (215)$$

If the bent is symmetrical about a vertical center line, $n=s$. Equations 212, 213, 214, and 215 then reduce to the form

$$M_{AB}=EK\left[\frac{d}{h}\frac{3}{n+2}-\frac{d_3}{l}\frac{6}{6n+1}\right] \quad (216)$$

$$M_{BC}=EK\left[\frac{d}{h}\frac{3}{n+2}+\frac{d_3}{l}\frac{6}{6n+1}\right] \quad (217)$$

$$M_{CB}=EK\left[\frac{d}{h}\frac{3(n+1)}{n(n+2)}-\frac{d_3}{l}\frac{6}{6n+1}\right] \quad (218)$$

$$M_{DA}=-EK\left[\frac{d}{h}\frac{3(n+1)}{n(n+2)}+\frac{d_3}{l}\frac{6}{6n+1}\right] \quad (219)$$

If $n=s=1$

$$M_{AB}=EK\left[\frac{d}{h}-\frac{6}{7}\frac{d_3}{l}\right] \quad (220)$$

$$M_{BC}=EK\left[\frac{d}{h}+\frac{6}{7}\frac{d_3}{l}\right] \quad (221)$$

$$M_{CB}=EK\left[\frac{2d}{h}-\frac{6}{7}\frac{d_3}{l}\right] \quad (222)$$

$$M_{DA}=-EK\left[\frac{2d}{h}+\frac{6}{7}\frac{d_3}{l}\right] \quad (223)$$

VII. TWO-LEGGED RECTANGULAR BENT. ONE LEG LONGER THAN THE OTHER

32. *Two-legged Rectangular Bent. One Leg Longer than the Other. Concentrated Horizontal Load at Top of Bent—Legs Hinged at the Bases.*—Fig. 35 represents a two-legged rectangular bent having one leg longer than the other. The legs are hinged at C and D. Let q equal the ratio of the length AD to the length BC.

The horizontal deflections of A and B are equal and are represented by d. R_{AD} is $\frac{d}{h}$ and R_{BC} is $\frac{d}{h}q$. If R_{AD} is represented by R, $R_{BC}=q\ R$.

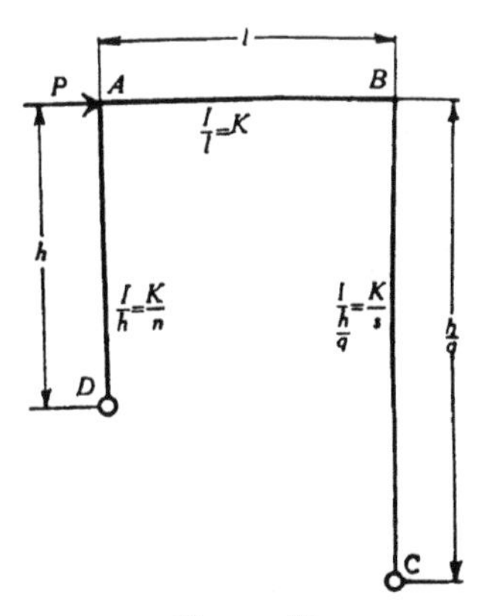

FIGURE 35

Applying the equations of Table 1 gives

$$M_{AB}=2EK(2\theta_A+\theta_B) \qquad (224)$$

$$M_{AD}=\frac{EK}{n}(3\theta_A-3R) \qquad (225)$$

$$M_{BA}=2EK(2\theta_B+\theta_A) \qquad (226)$$

$$M_{BC}=\frac{EK}{s}(3\theta_B-3\ q\ R) \qquad (227)$$

$$M_{AD}+q\ M_{BC}+Ph=0 \qquad (228)$$

Substituting the value of θ_B from equation 224 in equation 226 and substituting $-M_{BC}$ for M_{BA} gives

$$\theta_A = \frac{1}{6EK}(M_{BC}+2M_{AB}) \quad \text{. . . . . (229)}$$

Substituting the value of R from equation 225 and the value of θ_B from equation 224 in equation 227 and substituting $-M_{AB}$ for M_{AD} gives

$$\theta_A = \frac{1}{2EK(6+3q)}\left[M_{AB}(3-2nq)-2s\,M_{BC}\right] \quad \text{. . . . . . (230)}$$

Equating the right-hand members of equations 229 and 230 gives

$$M_{BC}(2+q+2s)+M_{AB}(1+2nq+2q)=0 \quad \text{. . . . . . . (231)}$$

Substituting $-M_{AB}$ for M_{AD} in equation 228 and eliminating M_{AB} from equations 228 and 231 gives

$$M_{BC} = -Ph\left[\frac{2nq+2q+1}{q(2nq+2q+1)+(2s+2+q)}\right] \quad \text{. . . . . . . (232)}$$

$$M_{AB} = Ph\left[\frac{2s+2+q}{q(2nq+2q+1)+(2s+2+q)}\right] \quad \text{. . . . . . . (233)}$$

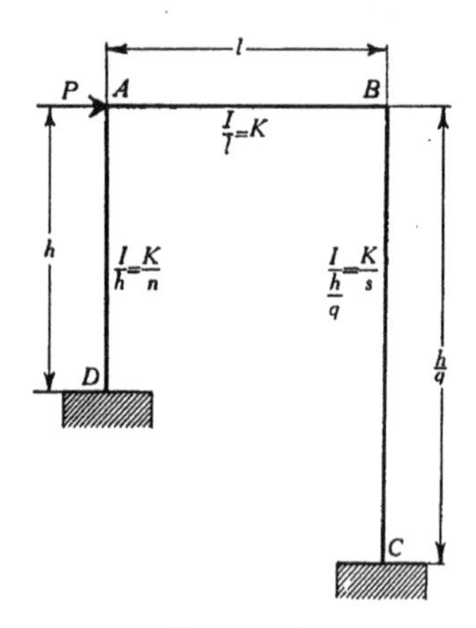

FIGURE 36

33. *Two-legged Rectangular Bent. One Leg Longer than the Other. Concentrated Horizontal Load at Top of Bent—Legs Fixed at the Bases.*—

Fig. 36 represents a two-legged rectangular bent having one leg longer than the other. The legs are fixed at C and D.

The horizontal deflections of A and B are equal and are represented by d. R_{AD} is $\frac{d}{h}$ and R_{BC} is $\frac{d}{h}\,q$. If R_{AD} is represented by R, $R_{BC}=qR$

Applying the equations of Table 1 gives

$$M_{AB}=2EK(2\theta_A+\theta_B) \qquad (234)$$

$$M_{AD}=\frac{2EK}{n}(2\theta_A-3R) \qquad (235)$$

$$M_{DA}=\frac{2EK}{n}(\theta_A-3R) \qquad (236)$$

$$M_{BA}=2EK(2\theta_B+\theta_A) \qquad (237)$$

$$M_{BC}=\frac{2EK}{s}(2\theta_B-3qR) \qquad (238)$$

$$M_{CB}=\frac{2EK}{s}(\theta_B-3qR) \qquad (239)$$

$$M_{AD}+M_{DA}+qM_{BC}+qM_{CB}+Ph=0 \qquad (240)$$

$$M_{AB}+M_{AD}=0 \qquad (241)$$

$$M_{BA}+M_{BC}=0 \qquad (242)$$

Substituting the values of M_{AB} and M_{AD} from equations 234 and 235 in equation 241 gives

$$2\left(\frac{\theta_A}{R}\right)(1+n)+n\left(\frac{\theta_B}{R}\right)=3 \qquad (243)$$

Substituting the values of M_{BA} and M_{BC} from equations 237 and 238 in equation 242 gives

$$s\left(\frac{\theta_A}{R}\right)+2\left(\frac{\theta_B}{R}\right)(s+1)=3q \qquad (244)$$

Substituting the values for the moments in equation 240 gives

$$s\left(\frac{\theta_A}{R}\right)+nq\left(\frac{\theta_B}{R}\right)=2(s+nq^2)-\frac{1}{R}\,\frac{nsPh}{6EK} \qquad (245)$$

Solving equation 243 for $\frac{\theta_B}{R}$ gives

$$\frac{\theta_B}{R}=\frac{3}{n}-\frac{2\theta_A}{R}\left[\frac{1+n}{n}\right] \qquad (246)$$

Substituting the value of $\frac{\theta_B}{R}$ from equation 246 in equation 244 gives

$$\frac{\theta_A}{R}=\frac{3(2s+2-nq)}{3ns+4n+4s+4} \qquad (247)$$

Substituting the value of $\frac{\theta_A}{R}$ from equation 247 in equation 246 gives

$$\frac{\theta_B}{R}=\frac{3(2nq+2q-s)}{3ns+4n+4s+4} \qquad (248)$$

Substituting the values of $\frac{\theta_A}{R}$ and $\frac{\theta_B}{R}$ from equations 247 and 248 in equation 245 and solving for $\frac{1}{R}$ and letting Δ_o represent

$$2(3ns^2+4ns+s^2+s+3nsq+3n^2sq^2+n^2q^2+nq^2+4nsq^2)$$

$$\frac{1}{R}=\frac{6EK}{nsPh}\left(\frac{\Delta_o}{3ns+4n+4s+4}\right) \qquad (249)$$

Substituting the value of $\frac{1}{R}$ from equation 249 in equations 247 and 248 gives

$$\theta_A=\frac{nsPh}{6EK}\;\frac{3(2s+2-nq)}{\Delta_o} \qquad (250)$$

$$\theta_B=\frac{nsPh}{6EK}\;\frac{3(2nq+2q-s)}{\Delta_o} \qquad (251)$$

Substituting the values of θ_A and θ_B from equations 250 and 251 In equations 234 and 237 gives

$$M_{AB}=\frac{ns\,Ph(3s+4+2q)}{\Delta_o} \qquad (252)$$

$$M_{BA}=\frac{n\,s\,Ph(3n\,q+4q+2)}{\Delta_o} \qquad (253)$$

Substituting the values of θ_A and R in equations 236 and 239 gives

$$M_{DA}=-\frac{s\,Ph(3sn+4n+qn+2s+2)}{\Delta_o} \qquad (254)$$

$$M_{CB}=-\frac{n\,Ph(3nsq+4sq+2nq+s+2q)}{\Delta_o} \qquad (255)$$

34. *Two-legged Rectangular Bent. One Leg Longer than the Other. Any Horizontal Load on Vertical Leg—Legs Hinged at the Bases.*—Fig. 37 represents a two-legged bent having one leg longer than the other. P represents the resultant of any system of horizontal forces applied to the leg AD. The legs of the bent are hinged at D and C.

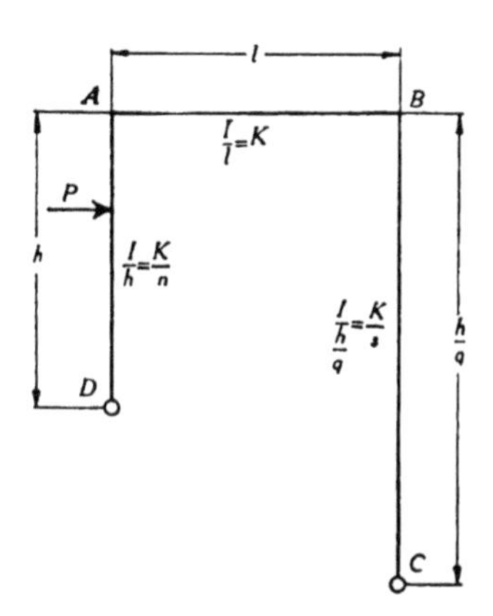

FIGURE 37

Applying the equations of Table 1 gives

$$n\,M_{AB}=-EK(3\theta_A-3R)-nH_{AD} \qquad (256)$$

$$M_{AB}=2EK(2\theta_A+\theta_B) \qquad (257)$$

$$M_{BC}=-2EK(2\theta_B+\theta_A) \qquad (258)$$

$$s\,M_{BC}=EK(3\theta_B-3q\,R) \qquad (259)$$

The forces acting on the members AD and BC are shown in Fig. 38. M_D represents the moment of P about D.

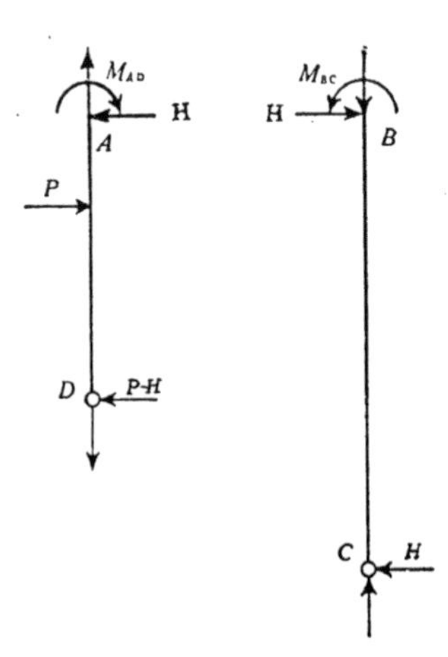

FIGURE 38

Equating to zero the moment of the external forces and reactions, about the point D, gives

$$M_D - M_{AB} + M_{BC} + H\left(\frac{h}{q} - h\right) = 0$$

Equating to zero the moments about the point B gives

$$M_{BC} + H\frac{h}{q} = 0$$

Combining these equations gives

$$M_{AB} - q\,M_{BC} = M_D \quad . \quad . \quad . \quad . \quad . \quad . \quad . \quad . \quad . \quad . \quad . \quad . \quad . \quad (260)$$

Combining equations 256 to 259 as follows

$$-\frac{1}{3}\left\{\left[2(256) - 3(258) - 4(259)\right]\left[2+q\right] - \left[4(256) + 3(\right.\right.$$

$$\left.\left. - 2(259)\right]\left[1+2q\right]\right\}$$

(the numerals in the parentheses are the equation numbers) gives

$$M_{AB}(2qn+2q+1) + M_{BC}(2s+2+q) = -2qn\,H_{AD} \quad . \quad . \quad . \quad . \quad (261)$$

Substituting the value of M_{BC} from equation 260 in equation 261 gives

$$M_{AB}=\frac{1}{2}\,\frac{M_D(2s+2+q)-2q^2n\,H_{AD}}{q^2n+s+1+q+q^2} \quad (262)$$

Substituting the value of M_{AB} in equation 261 gives

$$M_{BC}=-\frac{1}{2}\,\frac{M_D(2qn+2q+1)+2qn\,H_{AD}}{q^2n+s+1+q+q^2} \quad (263)$$

If $n=s$, equations 262 and 263 take the form

$$M_{AB}=\frac{1}{2}\,\frac{M_D(2n+2+q)-2q^2n\,H_{AD}}{q^2n+n+1+q+q^2} \quad (264)$$

$$M_{BC}=-\frac{1}{2}\,\frac{M_D(2qn+2q+1)+2qn\,H_{AD}}{q^2n+n+1+q+q^2} \quad (265)$$

If $n=s=1$, equations 262 and 263 take the form

$$M_{AB}=\frac{1}{2}\,\frac{M_D(4+q)-2q^2\,H_{AD}}{2+q+2q^2} \quad (266)$$

$$M_{BC}=-\frac{1}{2}\,\frac{M_D(1+4q)+2q\,H_{AD}}{2+q+2q^2} \quad (267)$$

Values of H_{AD} for different systems of loadings to be used in equations 262 to 267 are given in Table 2.

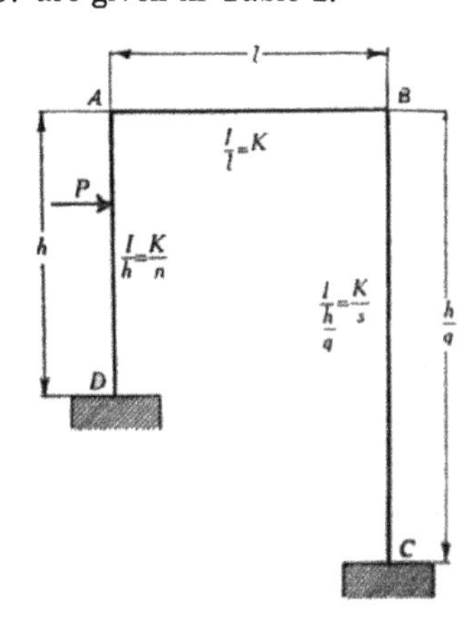

FIGURE 39

35. *Two-legged Rectangular Bent. One Leg Longer than the Other. Any Horizontal Load on Vertical Leg—Legs Fixed at the Bases.*—Fig. 39 represents a two-legged bent having legs of unequal length. P represents the resultant of any system of horizontal forces applied to the leg AD. The legs of the bent are fixed at D and C.

Applying the equations of Table 1 gives

$$n\,M_{AB}=-2EK(2\theta_A-3R)-nC_{AD} \quad (268)$$

$$M_{AB}=2EK(2\theta_A+\theta_B) \quad (269)$$

$$M_{BC}=-2EK(2\theta_B+\theta_A) \quad (270)$$

$$s\,M_{BC}=2EK(2\theta_B-3qR) \quad (271)$$

$$s\,M_{CB}=2EK(\theta_B-3qR) \quad (272)$$

$$n\,M_{DA}=2EK(\theta_A-3R)-nC_{DA} \quad (273)$$

The forces acting on the members AD and BC are shown in Fig. 40.

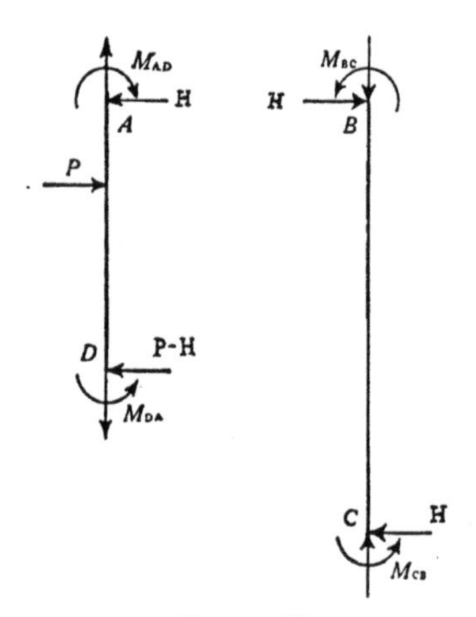

FIGURE 40

M_D represents the moment of P about D. Equating to zero the moment of the external forces and reactions about the point D gives

$$M_D-M_{AB}+M_{BC}-M_{CD}+M_{DA}+H\left(\frac{h}{q}-h\right)=0$$

Also

$$M_{BC}-M_{CD}+H\,\frac{h}{q}=0$$

TABLE 9

EQUATIONS FOR THE MOMENTS IN THE RECTANGULAR BENT REPRESENTED BY FIG. 39

No. of Equation	Left-hand Member of Equation				Right-hand Member of Equation	How Equation Was Obtained
	M_{AB}	M_{BC}	M_{CB}	M_{DA}		
274	-1	q	$+q$	1	$-M_D$	(274)
275	$1+2q-n(2+q)$	$2(s+1)\ (2+q)$	$-s(1+2q)$	$-2n(1+2q)$	$n\ [C_{AD}(2+q)+C_{DA}(2+4q)]$	$[2+q][2(270)-(268)+2(271)] + [1+2q][(269)-2(273) -(272)]$
276	1	$2+3s$	$-3s$	0	0	$(269)+2(270)+3(271) -3(272)$
277	$2(1+n)\ (1+2q)$	$(2+q)-s(1+2q)$	$2s(2+q)$	$n(2+q)$	$-n\ [C_{AD}(2+4q)+C_{DA}(2+q)]$	$[2+q]\ [(270)+2(272) +(273)] + [1+2q] [2(269)-(271)+2(268)]$

Combining these equations gives

$$M_{DA}-M_{AB}+q\,M_{BC}-q\,M_{CD}=-M_D \quad . \quad . \quad . \quad . \quad . \quad . \quad . \quad . \quad (274)$$

Combining the equations as indicated gives the equations of Table 9.

Solving these equations simultaneously gives

$$M_{AB}=\frac{n}{\Delta_o}\left\{(M_D-C_{DA})\,s\,(3s+4+2q)-C_{AD}(6q^2ns+3s^2+4s+qs+2q^2n)\right\} \quad . \quad . \quad . \quad (278)$$

$$M_{BC}=-\frac{n}{\Delta_o}\left\{(M_D-C_{DA})\,s\,(3qn+2+4q)+C_{AD}\,(3qns-2s-2qs-q^2n)\right\} \quad . \quad . \quad . \quad (279)$$

$$M_{CB}=-\frac{n}{\Delta_o}\left\{(M_D-C_{DA})(3qns+s+4qs+2qn+2q)+C_{AD}(3qns-s-2qs+2qn+q^2n-q)\right\} \quad . \quad . \quad . \quad (280)$$

$$M_{DA}=-\frac{1}{\Delta_o}\left\{(M_D-C_{DA})\,s\,(3ns+4n+qn+2s+2)+C_{AD}n\,(3s^2+4s(1+q+q^2)+q^2)\right\}-C_{DA} \quad . \quad . \quad . \quad (281)$$

$\Delta_o=2\left\{ns(4+3q+4q^2)+(s^2+s)+q^2(n^2+n)+3(q^2sn^2+s^2n)\right\}$ for equations 278 to 281 inclusive.

If $n=s$, equations 278 to 281 take the form

$$M_{AB}=\frac{n^2}{\Delta_o}\left\{(M_D-C_{DA})\,(3n+4+2q)-C_{AD}(6q^2n+3n+4+q+2q^2)\right\} \quad . \quad . \quad . \quad (282)$$

$$M_{BC}=-\frac{n^2}{\Delta_o}\left\{(M_D-C_{DA})(3qn+2+4q)+C_{AD}(3qn-2-2q-q^2)\right\} \quad (283)$$

$$M_{CB}=-\frac{n}{\Delta_o}\left\{(M_D-C_{DA})(3qn^2+n+6qn+2q)+C_{AD}(3qn^2-n+q^2n-q)\right\} \quad . \quad . \quad . \quad (284)$$

$$M_{DA}=-\frac{n}{\Delta_o}\Big\{(M_D-C_{DA})\ (3n^2+6n+qn+2)+C_{AD}[3n^2+4n\ (1+q$$

$$+q^2)+q^2]\Big\}-C_{DA} \quad . \quad . \quad . \quad . \quad . \quad . \quad . \quad . \quad . \quad . \quad . \quad . \quad (285)$$

$\Delta_o=2n\Big\{3n^2(1+q^2)+n(5+3q+5q^2)+1+q^2\Big\}$ for equations 282 to 285 inclusive.

If $n=s=1$, equations 282 to 285 take the form

$$M_{AB}=\frac{1}{\Delta_o}\Big\{(M_D-C_{DA})\ (7+2q)-C_{AD}(8q^2+q+7)\Big\} \quad . \quad . \quad . \quad (286)$$

$$M_{BC}=-\frac{1}{\Delta_o}\Big\{(M_D-C_{DA})\ (7q+2)+C_{AD}(q-2-q^2)\Big\} \quad . \quad . \quad (287)$$

$$M_{CB}=-\frac{1}{\Delta_o}\Big\{(M_D-C_{DA})\ (11q+1)+C_{AD}(2q-1+q^2)\Big\} \quad . \quad . \quad (288)$$

$$M_{DA}=-\frac{1}{\Delta_o}\Big\{(M_D-C_{DA})\ (11+q)+C_{AD}(7+4q+5q^2)\Big\}-C_{DA} \quad (289)$$

$\Delta_o=6(3+q+3q^2)$ for equations 286 to 289 inclusive.
Values of C_{AD} and C_{DA} are given in Table 2.

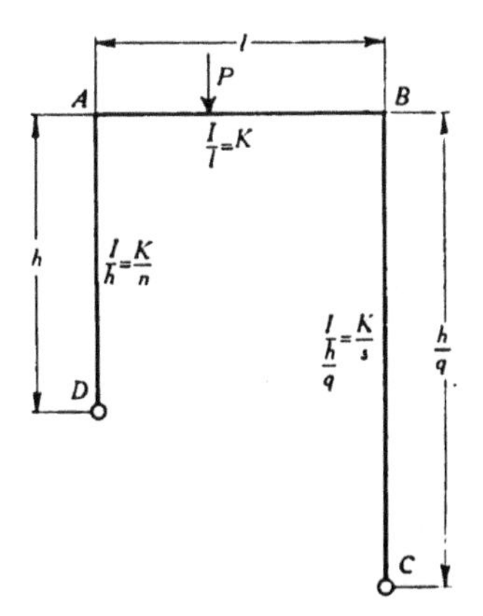

FIGURE 41

36. *Two-legged Rectangular Bent. One Leg Longer than the Other. Any System of Vertical Loads on Top of Bent—Legs Hinged at the Bases.*—Fig. 41 represents a two-legged bent having one leg longer

than the other. P represents the resultant of any system of vertical loads applied to AB. The legs of the bent are hinged at C and D.

Applying the equations of Table 1 gives

$$n\,M_{AB} = -EK(3\theta_A - 3R) \qquad (290)$$

$$M_{AB} = 2EK(2\theta_A + \theta_B) - C_{AB} \qquad (291)$$

$$M_{BC} = -2EK(2\theta_B + \theta_A) - C_{BA} \qquad (292)$$

$$s\,M_{BC} = EK(3\theta_B - 3qR) \qquad (293)$$

Since the sum of the shears in the two legs equals zero,

$$-M_{AB} + q\,M_{BC} = 0 \qquad (294)$$

Combining equations 290 to 293 as follows

$$-\left[\frac{2+q}{3}\right]\left[2(290) - 3(292) - 4(293)\right] + \left[\frac{1+2q}{3}\right]\left[4(290) + 3(291) - 2(293)\right]$$

(the numerals in the parentheses are the equation numbers) gives

$$M_{AB}(2qn + 2q + 1) + M_{BC}(2s + 2 + q) = -[C_{BA}(2+q) + C_{AB}(1+2q)] \qquad (295)$$

Substituting M_{BC} from equation 295 in equation 294 gives

$$M_{AB} = -\frac{q}{2}\,\frac{C_{BA}(2+q) + C_{AB}(1+2q)}{q^2n + s + 1 + q + q^2} \qquad (296)$$

Substituting M_{AB} from equation 296 in equation 294 gives

$$M_{BC} = -\frac{1}{2}\,\frac{C_{BA}(2+q) + C_{AB}(1+2q)}{q^2n + s + 1 + q + q^2} \qquad (297)$$

If the load is symmetrical about the center of $AB, C_{BA} = C_{AB} = \frac{F}{l}$ and equations 296 and 297 take the form

$$M_{AB} = -\frac{3}{2}\,q\,\frac{F}{l}\,\frac{1+q}{q^2n + s + 1 + q + q^2} \qquad (298)$$

$$M_{BC} = -\frac{3}{2}\,\frac{F}{l}\,\frac{1+q}{q^2n + s + 1 + q + q^2} \qquad (299)$$

Values of C_{AB}, C_{BA}, and $\frac{F}{l}$ are given in Tables 2 and 3.

37. *Two-legged Rectangular Bent. One Leg Longer than the Other. Any System of Vertical Loads on Top of Bent—Legs Fixed at the Bases.*— Fig. 42 represents a two-legged bent having legs of unequal length.

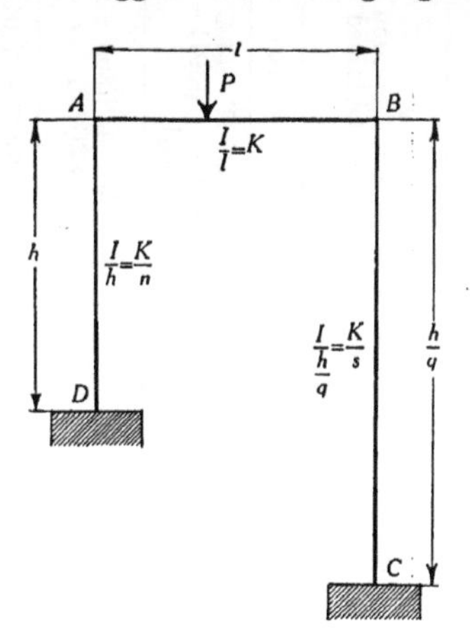

FIGURE 42

P represents the resultant of any system of vertical loads on AB. The legs are fixed at D and C.

Applying the equations of Table 1 gives

$$nM_{AB} = -2EK(2\theta_A - 3R) \quad (300)$$

$$M_{AB} = 2EK(2\theta_A + \theta_B) - C_{AB} \quad (301)$$

$$M_{BC} = -2EK(2\theta_B + \theta_A) - C_{BA} \quad (302)$$

$$sM_{BC} = 2EK(2\theta_B - 3qR) \quad (303)$$

$$sM_{CB} = 2EK(\theta_B - 3qR) \quad (304)$$

$$nM_{DA} = 2EK(\theta_A - 3R) \quad (305)$$

Since the sum of the shears in the two legs equals zero,

$$-M_{AB} + qM_{BC} + qM_{CB} + M_{DA} = 0 \quad (306)$$

Combining equations 300 to 306 as indicated gives the equations of Table 10.

TABLE 10

EQUATIONS FOR THE MOMENTS IN THE RECTANGULAR BENT REPRESENTED BY FIG. 42

No. of Equation	Left-hand Member of Equation				Right-hand Member of Equation	How Equation Was Obtained
	M_{AB}	M_{BC}	M_{CB}	M_{DA}		
306	-1	q	q	1	0	(306)
307	$1+2q-n(2+q)$	$2(s+1)(2+q)$	$-s(1+2q)$	$-2n(1+2q)$	$-[2C_{BA}(2+q)+C_{AB}(1+2q)]$	$[2+q][2(302)-(300)+(2303)]+[1+2q][(301)-2(305)-(304)]$
308	1	$2+3s$	$-3s$	0	$-[2C_{BA}+C_{AB}]$	$(301)+3[(303)-(304)]+2(302)$
309	$+2(1+n)(1+2q)$	$(2+q)-s(1+2q)$	$2s(2+q)$	$n(2+q)$	$-[C_{BA}(2+q)+2C_{AB}(1+2q)]$	$[2+q][(302)+2(304)+(305)]+[1+2q][2(301)-(303)+2(300)]$

Solving these equations simultaneously gives

$$M_{AB} = -\frac{1}{\Delta_o}\Big\{C_{BA}[2qns(3+2q)+s^2]+C_{AB}[qns(3+8q)$$
$$+2s^2+2s+2q^2n]\Big\} \quad . \quad . \quad . \quad . \quad . \quad . \quad . \quad . \quad . \quad . \quad . \quad . \quad (310)$$

$$M_{BC} = -\frac{1}{\Delta_o}\Big\{C_{BA}[ns(8+3q)+2q^2n^2+2q^2n+2s]$$
$$+C_{AB}[2ns(2+3q)+q^2n^2]\Big\} \quad . \quad . \quad . \quad . \quad . \quad . \quad . \quad . \quad . \quad (311)$$

$$M_{CB} = -\frac{1}{\Delta_o}\Big\{C_{BA}[ns(4+3q)-2q^2n^2-2q^2n+s]$$
$$+C_{AB}[2ns(1+3q)-q^2n^2+3qn]\Big\} \quad . \quad . \quad . \quad . \quad . \quad . \quad . \quad (312)$$

$$M_{DA} = \frac{1}{\Delta_o}\Big\{C_{BA}[2qns(3+q)-s^2+3qs]$$
$$+C_{AB}[qns(3+4q)-2s^2-2s+q^2n]\Big\} \quad . \quad . \quad . \quad . \quad . \quad . \quad (313)$$

in which $\Delta_o = 2[ns(4+3q+4q^2)+q^2n(3ns+n+1)+s(3ns+s+1)]$

If the load is symmetrical about the center of $A\ B$, $C_{BA} = C_{AB} = \frac{F}{l}$, and equations 310 to 313 take the form

$$M_{AB} = -\frac{1}{\Delta_o}\frac{F}{l}\ [3qns(3+4q)+3s^2+2s+2q^2n] \quad . \quad . \quad . \quad . \quad . \quad (314)$$

$$M_{BC} = -\frac{1}{\Delta_o}\frac{F}{l}\ [3ns(4+3q)+3q^2n^2+2q^2n+2s] \quad . \quad . \quad . \quad (315)$$

$$M_{CB} = -\frac{1}{\Delta_o}\frac{F}{l}\ [3ns(2+3q)-3q^2n^2+qn(3-2q)+s] \quad . \quad . \quad (316)$$

$$M_{DA} = \frac{1}{\Delta_o}\frac{F}{l}\ [3qns(3+2q)-3s^2+s(3q-2)+q^2n] \quad . \quad . \quad . \quad (317)$$

If $n = s$ equations 310 to 313 take the form

$$M_{AB} = -\frac{n}{\Delta_o}\Big\{C_{BA}n(1+6q+4q^2)+C_{AB}[n(2+3q+8q^2)$$
$$+2+2q^2]\Big\} \quad . \quad . \quad . \quad . \quad . \quad . \quad . \quad . \quad . \quad . \quad . \quad . \quad . \quad . \quad (318)$$

$$M_{BC} = -\frac{n}{\Delta_o}\left\{C_{BA}\left[n(8+3q+2q^2)+2+2q^2\right] + C_{AB}n(4+6q+q^2)\right\} \quad (319)$$

$$M_{CB} = -\frac{n}{\Delta_o}\left\{C_{BA}\left[n(4+3q-2q^2)+1-2q^2\right] + C_{AB}[n(2+6q-q^2)+3q]\right\} \quad (320)$$

$$M_{DA} = \frac{n}{\Delta_o}\left\{C_{BA}\left[n(-1+6q+2q^2)+3q\right] + C_{AB}\left[n(-2+3q+4q^2)-2+q^2\right]\right\} \quad (321)$$

in which $\Delta_o = 2n[3n^2(1+q^2)+n(5+3q+5q^2)+(1+q^2)]$

If $n=s$ and $C_{BA}=C_{AB}=\frac{F}{l}$ equations 310 to 313 take the form

$$M_{AB} = -\frac{n}{\Delta_o}\frac{F}{l}\left[3n(1+3q+4q^2)+2+2q^2\right] \quad (322)$$

$$M_{BC} = -\frac{n}{\Delta_o}\frac{F}{l}\left[3n(4+3q+q^2)+2+2q^2\right] \quad (323)$$

$$M_{CB} = -\frac{n}{\Delta_o}\frac{F}{l}\left[3n(2+3q-q^2)+1+3q-2q^2\right] \quad (324)$$

$$M_{DA} = \frac{n}{\Delta_o}\frac{F}{l}\left[3n(-1+3q+2q^2)-2+3q+q^2\right] \quad (325)$$

Values of C_{BA}, C_{AB}, and $\frac{F}{l}$ are given in Tables 2 and 3.

38. *Two-legged Rectangular Bent. One Leg Longer than the Other. External Moment at One Corner—Legs Hinged at the Bases.*— Fig. 43 represents a two-legged rectangular bent having one leg longer than the other. An external couple with moment M is applied at A. The legs are hinged at D and C.

Applying the equations of Table 1 gives

$$M_{AB} = -\frac{EK}{n}(3\theta_A - 3R) + M \quad (326)$$

$$M_{AB} = 2EK(2\theta_A + \theta_B) \quad (327)$$

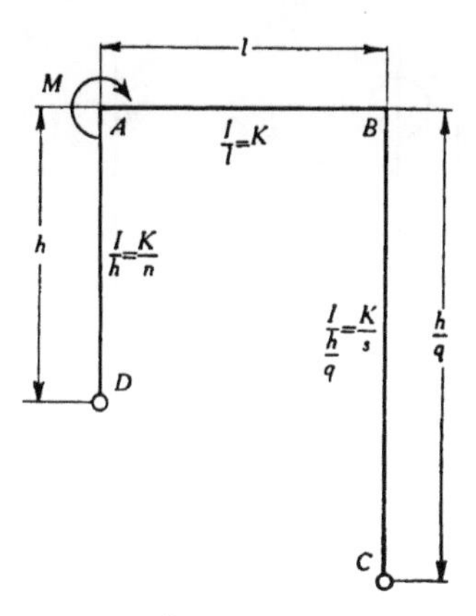

FIGURE 43

$$M_{BC}=-2EK(2\theta_B+\theta_A) \quad . \quad . \quad . \quad . \quad . \quad . \quad . \quad . \quad . \quad . \quad . \quad (328)$$

$$M_{BC}=\frac{EK}{s}(3\theta_B-3qR) \quad . \quad . \quad . \quad . \quad . \quad . \quad . \quad . \quad . \quad . \quad . \quad (329)$$

$$M_{AD}+qM_{BC}=0$$
$$M-M_{AB}+qM_{BC}=0$$
$$M_{AB}-qM_{BC}=M \quad . \quad . \quad . \quad . \quad . \quad . \quad . \quad . \quad . \quad . \quad . \quad . \quad (330)$$

Combining these equations and solving for the moments gives

$$M_{AB}=\frac{M}{2}\,\frac{2s+2+q+2q^2n}{q^2n+s+1+q+q^2} \quad . \quad . \quad . \quad . \quad . \quad . \quad . \quad . \quad . \quad (331)$$

$$M_{AD}=\frac{M}{2}\,\frac{q(1+2q)}{q^2n+s+1+q+q^2} \quad . \quad . \quad . \quad . \quad . \quad . \quad . \quad . \quad . \quad (332)$$

$$M_{BC}=-\frac{M}{2}\,\frac{1+2q}{q^2n+s+1+q+q^2} \quad . \quad . \quad . \quad . \quad . \quad . \quad . \quad . \quad . \quad . \quad (333)$$

39. *Two-legged Rectangular Bent. One Leg Longer than the Other. External Moment at One Corner—Legs Fixed at the Bases.*— Fig. 44 represents a two-legged rectangular bent having one leg longer than the other. An external couple with moment M is applied at A. The legs are fixed at C and D.

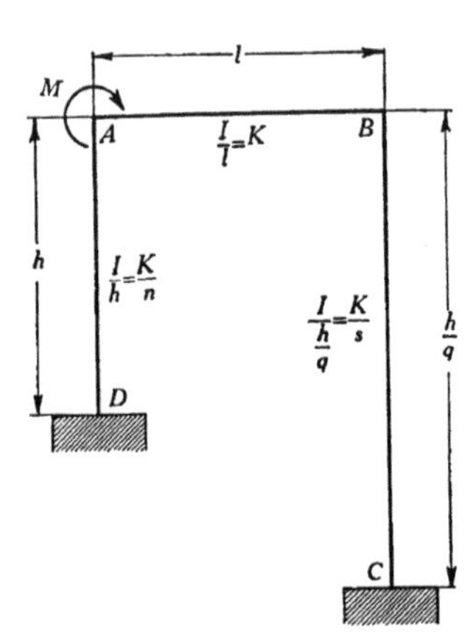

FIGURE 44

Applying the equations of Table 1 gives

$$M_{AB} = -\frac{2EK}{n}(2\theta_A - 3R) + M \quad . \; . \; . \; . \; . \; . \; . \; . \; . \quad (334)$$

$$M_{AB} = 2EK(2\theta_A + \theta_B) \quad . \; . \; . \; . \; . \; . \; . \; . \; . \; . \; . \; . \quad (335)$$

$$M_{BC} = -2EK(2\theta_B + \theta_A) \quad . \; . \; . \; . \; . \; . \; . \; . \; . \; . \; . \; . \quad (336)$$

$$M_{BC} = \frac{2EK}{s}(2\theta_B - 3qR) \quad . \; . \; . \; . \; . \; . \; . \; . \; . \; . \quad (337)$$

$$M_{CB} = \frac{2EK}{s}(\theta_B - 3qR) \quad . \; . \; . \; . \; . \; . \; . \; . \; . \; . \; . \quad (338)$$

$$M_{DA} = \frac{2EK}{n}(\theta_A - 3R) \quad . \; . \; . \; . \; . \; . \; . \; . \; . \; . \; . \quad (339)$$

For AD and BC to be in equilibrium

$$M_{DA} - M_{AB} + q\,M_{BC} + q\,M_{CB} = -M \quad . \; . \; . \; . \; . \; . \; . \quad (340)$$

Combining these equations and solving for the moments gives

$$M_{AB} = +\frac{Mn}{\Delta_o}(6s^2 + 8s + 3qs + 6q^2ns + 2q^2n) \quad . \; . \; . \; . \; . \quad (341)$$

$$M_{BC} = -\frac{Mn}{\Delta_o}(4s + 6qs + q^2n) \quad . \; . \; . \; . \; . \; . \; . \; . \; . \quad (342)$$

$$M_{CB} = -\frac{Mn}{\Delta_o}(2s + 6qs + 3q - q^2n) \quad . \; . \; . \; . \; . \; . \; . \; . \quad (343)$$

$$M_{DA} = +\frac{M}{\Delta_o}(3qns - 2s^2 - 2s + 4q^2ns + q^2n) \quad . \quad . \quad . \quad . \quad . \quad . \quad (344)$$

$$M_{AD} = +\frac{M}{\Delta_o}(3qns + 8q^2ns + 2s^2 + 2s + 2q^2n)$$

in which

$$\Delta_o = 2\left[ns(4 + 3q + 4q^2) + s^2 + s + q^2n(n+1) + 3ns(q^2n + s)\right]$$

40. *Two-legged Rectangular Bent. One Leg Longer than the Other. Settlement of Foundations—Legs Hinged at the Bases.*—Fig. 45 represents a two-legged rectangular bent having one leg longer than the other. The legs of the bent are hinged at D and C. The unstrained position of the bent is represented by the broken line $DA'B'C'$. Due

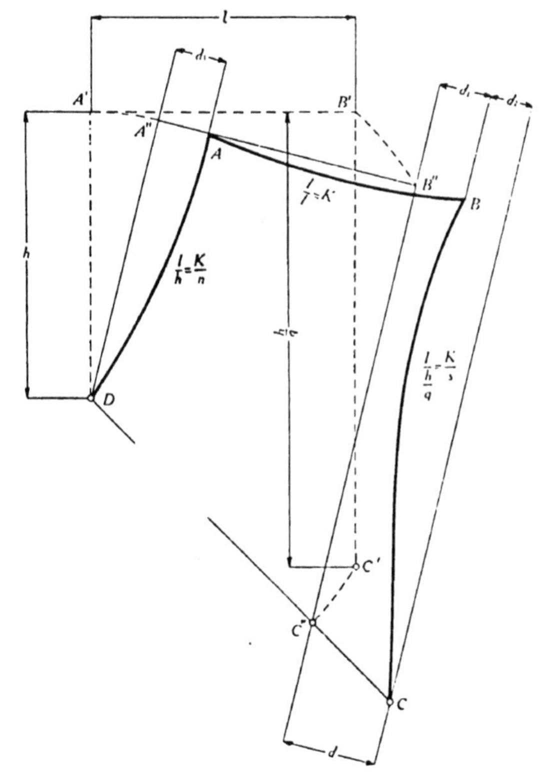

FIGURE 45

to settlement of the foundations the point C' has moved to C. The motion of C can be considered as made up of two parts, as follows: first, without being strained the bent rotates about D until C' is at C'', a point on DC; secondly, the bent is strained by applying a force at C'' which moves C'' to C. It is apparent that no matter what motion of C relative to D takes place, the stress in the bent depends only upon the change in the distance of C from D.

The change in the distance from D to C is proportional to d. d is made up of two parts, d_1 due to the deflection of AD, and d_2 due to the deflection of BC.

Applying the equations of Table 1 gives

$$M_{AB}=-\frac{EK}{n}(3\theta_A-3R_1) \qquad (345)$$

$$M_{AB}=2EK(2\theta_A+\theta_B) \qquad (346)$$

$$M_{BC}=-2EK(2\theta_B+\theta_A) \qquad (347)$$

$$M_{BC}=\frac{EK}{s}(3\theta_B-3qR_1+\frac{3qd}{h}) \qquad (348)$$

For the columns to be in equilibrium

$$M_{AB}-q\,M_{BC}=0 \qquad (349)$$

Combining these equations and solving for the moments gives

$$M_{AB}=\frac{qd}{h}\,\frac{3qEK}{q^2n+s+1+q+q^2} \qquad (350)$$

$$M_{BC}=\frac{qd}{h}\,\frac{3EK}{q^2n+s+1+q+q^2} \qquad (351)$$

41. *Two-legged Rectangular Bent. One Leg Longer than the Other. Settlement of Foundations—Legs Fixed at the Bases.*—Fig. 46 represents a two-legged rectangular bent having one leg longer than the other. The legs of the bent are fixed at D and C. The unstrained position of the bent is represented by the broken line $DA'B'C'$. Due to settlement of the foundations the point C' has moved to C and the tangents to the elastic curves of the legs at their bases, originally vertical, have been rotated to the positions shown. This motion may be considered as made up of three parts, as follows: first, without being strained the

bent rotates about D until C' is at C'', a point on DC; secondly, the bent is strained by applying a force at C'' which moves C'' to C; thirdly, couples are applied at D and C, still further straining the bent so that the tangents to the elastic curves at the bases of the legs make angles of θ_D and θ_C respectively with lines normal to DC. It is apparent that,

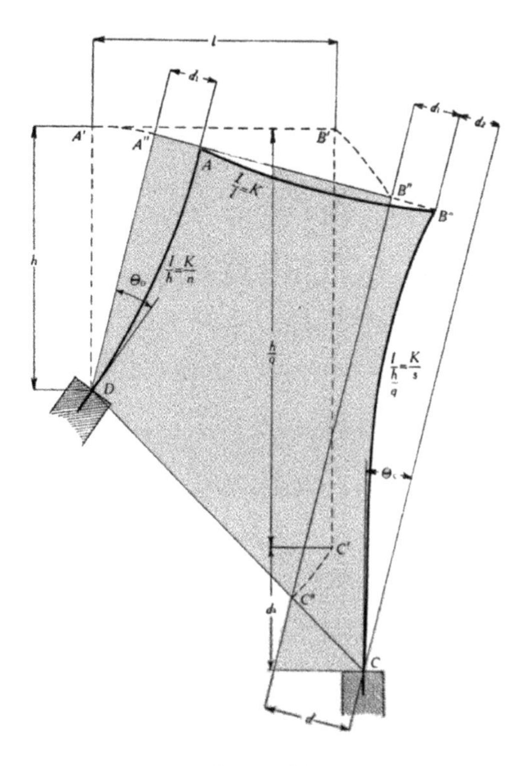

FIGURE 46

no matter what motion of C relative to D takes place, the stress in the bent depends only upon the change in the distance from C to D and upon the angles θ_D and θ_C.

The change in the distance from D to C is proportional to d. d is made up of two parts, d_1 due to the deflection of AD, and d_2 due to the deflection of BC.

Applying the equations of Table 1 gives

$$M_{AB} = -\frac{2EK}{n}(2\theta_A + \theta_D - 3R_1) \quad \ldots \quad (352)$$

$$M_{AB} = 2EK(2\theta_A + \theta_B) \quad \ldots \quad (353)$$

$$M_{BC} = -2EK(2\theta_B + \theta_A) \quad \ldots \quad (354)$$

$$M_{BC} = \frac{2EK}{s}(2\theta_B + \theta_C - 3R_2) \quad \ldots \quad (355)$$

$$M_{CB} = \frac{2EK}{s}(2\theta_C + \theta_B - 3R_2) \quad \ldots \quad (356)$$

$$M_{DA} = \frac{2EK}{n}(2\theta_D + \theta_A - 3R_1) \quad \ldots \quad (357)$$

For the legs of the bent to be in equilibrium

$$M_{DA} - M_{AB} + qM_{BC} + qM_{CB} = 0 \quad \ldots \quad (358)$$

Combining these equations and solving for the moments gives

$$M_{AB} = \frac{2EK}{\Delta_o}\left\{3\frac{qd}{h}[6qns + 2qn - s] + \theta_C[9qns + 2qn(3+q) - s] - \theta_D[6q^2ns + 2q^2n - s(4+3q) - 3s^2]\right\} \quad \ldots \quad (359)$$

$$M_{BC} = \frac{2EK}{\Delta_o}\left\{3\frac{qd}{h}[6ns - qn + 2s] + \theta_C[6ns - qn(3+4q) + 2s - 3q^2n^2] - \theta_D[9qns - q^2n + 2s(1+3q)]\right\} \quad \ldots \quad (360)$$

$$M_{CB} = \frac{2EK}{\Delta_o}\left\{3\frac{qd}{h}[6ns + n(4+q) + 2s + 1] + \theta_C[12ns + 2n(6 + 3q + 2q^2) + 4s + 3q^2n^2 + 3] - \theta_D[9qns + qn(6+q) + s(1+6q) + 3q]\right\} \quad \ldots \quad (361)$$

$$M_{DA} = -\frac{2EK}{\Delta_o}\left\{3\frac{qd}{h}[6qns + 2qn + s(1+4q) + q] + \theta_C[9qns + qn(6+q) + s(1+6q) + 3q] - \theta_D[12q^2ns + 4q^2n + 2s(2 + 3q + 6q^2) + 3s^2 + 3q^2]\right\} \quad \ldots \quad (362)$$

in which

$$\Delta_o = 2\ [ns(4+3q+4q^2)+q^2n(3ns+n+1)+s(3ns+s+1)]$$

It is to be noted that θ_C and θ_D are measured, not from the original direction of AD and BC, but from a line normal to AB.

If the tangents to the elastic curves of the legs at their bases remain vertical the moments can be expressed in terms of the vertical settlement of C relative to D. The θ's are then measured from the original direction of AD and BC and equal zero. The displacement d is measured in a horizontal direction. Proceeding as before the moments are found to be as follows:

$$M_{AB} = \frac{2EK}{\Delta_o}\left\{\frac{3qd}{h}[6qns+2qn-s] - \frac{3}{(4+5q)}\frac{d_3}{l}[q^2ns(13-4q)\right.$$
$$\left. +2qn(2+6q+q^2)+2s(9+q-q^2)+s^2(13-4q)]\right\} \quad . \quad . \quad . \quad (363)$$

$$M_{BC} = \frac{2EK}{\Delta_o}\left\{\frac{3qd}{h}[6ns-qn+2s] + \frac{3}{(4+5q)}\frac{d_3}{l}[ns(4+5q)\right.$$
$$\left. +2qn(1+5q+3q^2)+2s(5+6q-2q^2)+q^2n^2(4+5q)]\right\} \quad . \quad . \quad (364)$$

$$M_{CB} = \frac{2EK}{\Delta_o}\left\{\frac{3qd}{h}[6ns+n(4+q)+2s+1] - \frac{3}{(4+5q)}\frac{d_3}{l}[ns(4+5q)\right.$$
$$+n(8+16q+15q^2+6q^3)-s(3+10q-4q^2)+q^2n^2(4+5q)$$
$$\left. -2(1-2q+q^2)]\right\} \quad . \quad . \quad . \quad . \quad . \quad . \quad . \quad . \quad . \quad . \quad . \quad (365)$$

$$M_{DA} = -\frac{2EK}{\Delta_o}\left\{\frac{3qd}{h}[6qns+2qn+s(1+4q)+q] - \frac{3}{(4+5q)}\frac{d_3}{l}[q^2ns\right.$$
$$(13-4q)-qn(4+2q+3q^2)+s(18+15q+20q^2-8q^3)+s^2(13-4q)$$
$$\left. +2q(1-2q+q^2)]\right\} \quad . \quad . \quad . \quad . \quad . \quad . \quad . \quad . \quad . \quad . \quad . \quad (366)$$

VIII. Two-legged Trapezoidal Bents. Bents and Loading Symmetrical about Vertical Center Line

42. *Two-legged Trapezoidal Bent. Bent and Load Symmetrical about Vertical Center Line. Vertical Load on Top—Legs Hinged at the Bases.*—Fig. 47 represents a two-legged trapezoidal bent. The

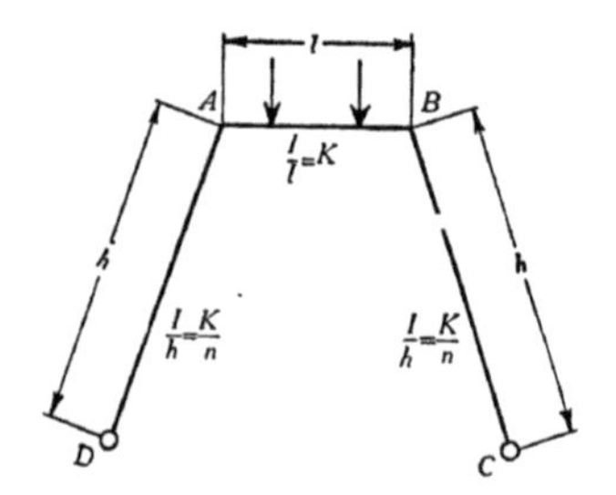

FIGURE 47

bent and the load are symmetrical about a vertical center line. The legs are hinged at C and D.

From symmetry $\theta_A = -\theta_B$. Since the deformation due to shearing and axial stresses may be neglected, the points A and B do not move, and R is zero for all members. By applying the equations of Table 1 five equations are obtained which are identical with equations 128 to 132 of section 26; hence it is seen that the moments in the members of this frame are independent of the angle of inclination of the legs, and this is true of any trapezoidal frame in which loading and frame are symmetrical about a vertical center line. The direct stress does vary with the angle of inclination, and may be found from the equations of statics when the moments are known.

The moments as found in section 26 when applied to this case in which bent and loading are symmetrical about a vertical center line are

$$M_{AD} = -M_{BC} = \left(\frac{3}{3+2n}\right)\frac{F}{l} \quad \ldots \quad (367)$$

If $n=1$

$$M_{AD} = -M_{BC} = \frac{3}{5}\frac{F}{l} \quad \ldots \quad (368)$$

Values of $\frac{F}{l}$ are given in Table 3.

43. *Two-legged Trapezoidal Bent. Bent and Load Symmetrical about Vertical Center Line. Vertical Load on Top—Legs Fixed at the Bases.*—Fig. 48 represents a two-legged trapezoidal bent. The bent

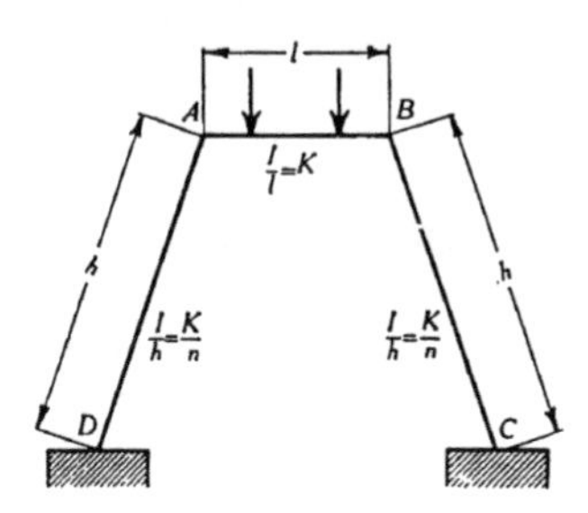

FIGURE 48

and loads are symmetrical about a vertical center line. The legs are fixed at C and D. From the preceding paragraph it is seen that equations 158 and 159 of section 27 apply to this case.

$$M_{AD} = -M_{BC} = \left(\frac{2}{n+2}\right)\frac{F}{l} \qquad (369)$$

$$M_{DA} = -M_{CB} = \left(\frac{1}{n+2}\right)\frac{F}{l} \qquad (370)$$

If $n=1$

$$M_{AD} = -M_{BC} = \frac{2}{3}\frac{F}{l} \qquad (371)$$

$$M_{DA} = -M_{CB} = \frac{1}{3}\frac{F}{l} \qquad (372)$$

Values of $\frac{F}{l}$ for different loads are given in Table 3.

44. *Two-legged Trapezoidal Bent. Bent and Loads Symmetrical about Vertical Center Line. Loads Normal to Legs of Bent—Legs Hinged at the Bases.*—Fig. 49 represents a trapezoidal bent having loads normal

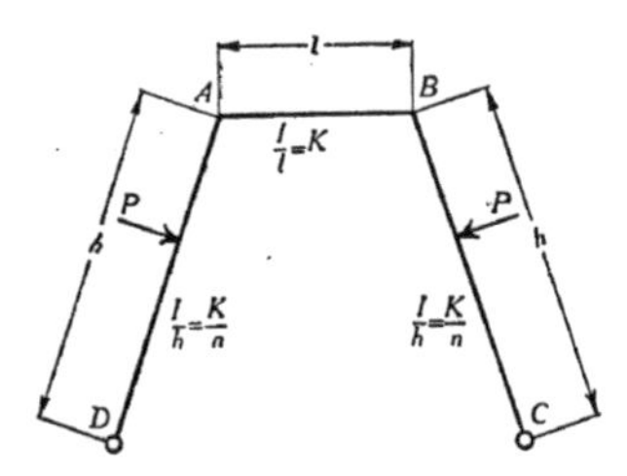

FIGURE 49

to the sides AD and BC. The bent and the loads are symmetrical about a vertical center line. P represents the resultant of the loads on AD, and likewise on BC. The legs are hinged at D and C. As in sections 42 and 43 the moments are independent of the angle of inclination of the legs. Hence equation 107 of section 24 applies.

$$M_{AD} = -M_{BC} = \left(\frac{2n}{2n+3}\right) H_{AD} \quad . \quad . \quad . \quad . \quad . \quad . \quad . \quad . \quad . \quad (373)$$

If $n=1$

$$M_{AD} = \frac{2}{5} H_{AD} \quad . \quad . \quad . \quad . \quad . \quad . \quad . \quad . \quad . \quad . \quad . \quad . \quad . \quad . \quad (374)$$

Values of H_{AD} are given in Table 2.

45. *Two-legged Trapezoidal Bent. Bent and Loads Symmetrical about Vertical Center Line. Loads Normal to Legs of Bent—Legs Fixed at the Bases.*—Fig. 50 represents a trapezoidal bent with loads and members similar to those of Fig. 49, except that the legs are fixed at D and C. Equations 126 and 127 of section 25 apply here.

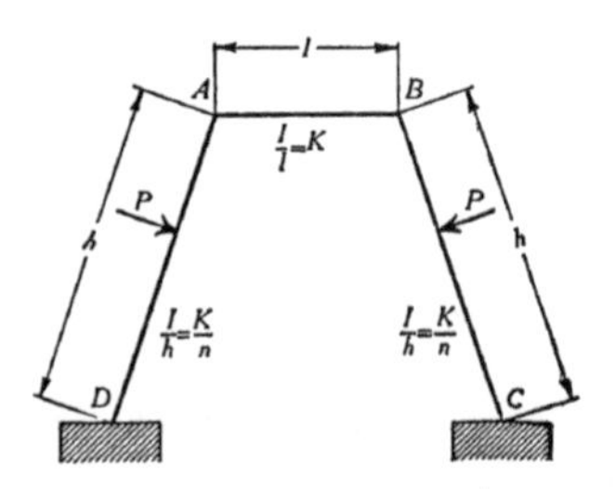

FIGURE 50

$$M_{AD} = -M_{BC} = \left(\frac{n}{n+2}\right) C_{AD} \quad . \quad . \quad . \quad . \quad . \quad . \quad . \quad . \quad . \quad . \quad (375)$$

$$M_{DA} = -M_{CB} = -\left(\frac{C_{AD}}{n+2} + C_{DA}\right) \quad . \quad . \quad . \quad . \quad . \quad . \quad . \quad . \quad (376)$$

If $n=1$

$$M_{AD} = -M_{BC} = \frac{1}{3} C_{AD} \quad . \quad . \quad . \quad . \quad . \quad . \quad . \quad . \quad . \quad . \quad . \quad . \quad (377)$$

$$M_{DA} = -M_{CB} = -\left(\frac{1}{3} C_{AD} + C_{DA}\right) . \quad . \quad . \quad . \quad . \quad . \quad . \quad . \quad . \quad (378)$$

Values of C are given in Table 2.

46. *Two-legged Trapezoidal Bent. Bent and Loads Symmetrical about Vertical Center Line. External Moments at Upper Corners of Bent—Legs Hinged at the Bases.*—Fig. 51 represents a two-legged trape-

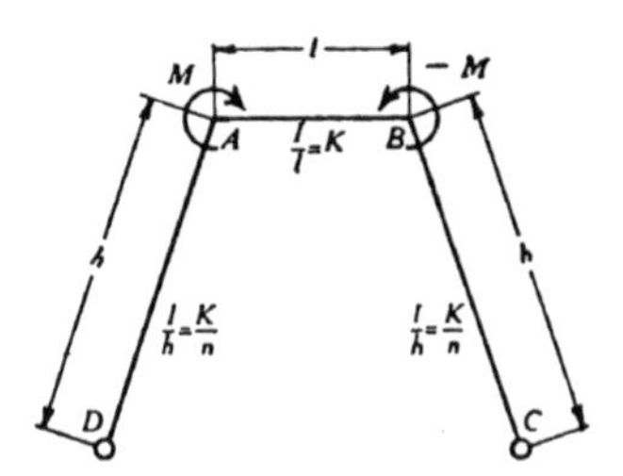

FIGURE 51

zoidal bent having couples acting at A and B. The bent and the couples are symmetrical about a vertical center line. The legs are hinged at C and D.

As in sections 42 and 43 because of the symmetry of loads and bent, the moments are independent of the angle of inclination of the legs. Equations 163 and 164 of section 28 are modified to apply here by the algebraic addition of moments due to the two couples.

$$M_{AD} = \left(\frac{3M}{2n+3}\right) \quad . \quad . \quad . \quad . \quad . \quad . \quad . \quad . \quad . \quad . \quad . \quad . \quad . \quad . \quad (379)$$

$$M_{AB} = \left(\frac{2nM}{2n+3}\right) \quad . \quad . \quad . \quad . \quad . \quad . \quad . \quad . \quad . \quad . \quad . \quad . \quad . \quad . \quad (380)$$

If $n=1$

$$M_{AD}=\frac{3}{5}M \qquad (381)$$

$$M_{AB}=\frac{2}{5}M \qquad (382)$$

47. *Two-legged Trapezoidal Bent. Bent and Loads Symmetrical about Vertical Center Line. External Moments at Upper Corners of Bent—Legs Fixed at the Bases.*—Fig. 52 represents a two-legged trape-

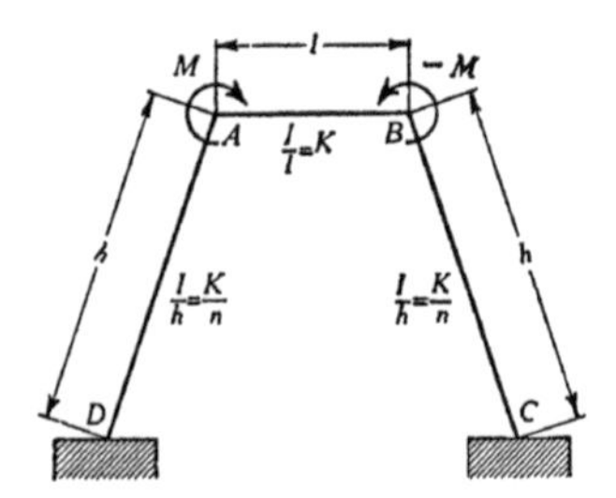

FIGURE 52

zoidal bent having couples acting at A and B. The bent and the couples are symmetrical about a vertical center line. The legs are fixed at C and D.

Since from symmetry of bent and loading, the moments are independent of the angle of inclination of the legs, equations 175 to 179 of section 29 apply here.

$$M_{AD}=-M_{BC}=\frac{2M}{n+2} \qquad (383)$$

$$M_{AB}=-M_{BA}=\frac{nM}{n+2} \qquad (384)$$

$$M_{DA}=-M_{CB}=\frac{M}{n+2} \qquad (385)$$

If $n=1$

$$M_{AD}=-M_{BC}=\frac{2}{3}M \qquad (386)$$

$$M_{AB}=-M_{BA}=\frac{1}{3}M \qquad (387)$$

$$M_{DA}=-M_{CB}=\frac{1}{3}M \qquad (388)$$

IX. Rectangular Frames

48. *Rectangular Frame. Horizontal Force at the Top.*—Fig. 53 represents a rectangular frame having a horizontal force P applied at

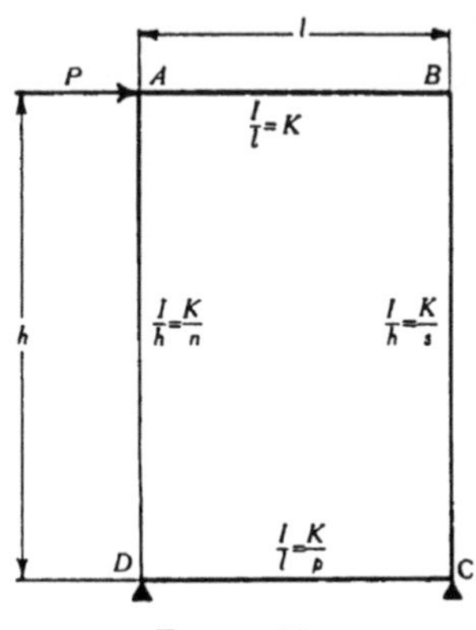

Figure 53

the top. The value of R for members AB and DC is zero, and R for AD equals R for BC.

Applying equation (A), Table 1, gives

$$nM_{DA}=2EK\ (2\theta_D+\theta_A-3R) \quad (389)$$

$$nM_{AB}=-2EK\ (2\theta_A+\theta_D-3R) \quad (390)$$

$$M_{AB}=2EK(2\theta_A+\theta_B) \quad (391)$$

$$M_{BC}=-2EK(2\theta_B+\theta_A) \quad (392)$$

$$sM_{BC}=2EK\ (2\theta_B+\theta_C-3R) \quad (393)$$

$$sM_{CD}=-2EK\ (2\theta_C+\theta_B-3R) \quad (394)$$

$$pM_{CD}=2EK\ (2\theta_C+\theta_D) \quad (395)$$

$$pM_{DA}=-2EK(2\theta_D+\theta_C) \quad (396)$$

Considering AB and DC removed and equating the sum of the moments acting at the tops and bottoms of AD and BC to zero gives

$$M_{AD}+M_{DA}+M_{BC}+M_{CB}+Ph=0$$

Substituting $-M_{AB}$ for M_{AD} and $-M_{CD}$ for M_{CB} gives

$$-M_{AB}+M_{BC}-M_{CD}+M_{DA}=-Ph \quad . \quad . \quad . \quad . \quad . \quad . \quad . \quad . \quad (397)$$

Combining equations 389 to 397 as indicated in Table 11 gives equations 397 to 400 of Table 11.

Solving these equations simultaneously and letting Δ represent the common denominator gives

$$\Delta=22(spn+sp+sn+np)+2(sp^2+s^2p+n^2p+p^2n+s^2+s+n^2+n)$$
$$+6(sn^2+s^2n+p^2+p) \text{ and}$$

$$M_{AB}=\frac{Ph}{\Delta}\left(3s^2n+5nps+2s^2p+2sp^2+6ns+6pn+5ps+3p^2\right) \quad (401)$$

$$M_{BC}=-\frac{Ph}{\Delta}\left(3n^2s+5nps+2n^2p+2np^2+6ns+6ps+5pn+3p^2\right)$$
$$. \quad . \quad . \quad . \quad . \quad . \quad . \quad . \quad . \quad . \quad . \quad . \quad . \quad . \quad . \quad (402)$$

$$M_{CD}=\frac{Ph}{\Delta}\left(3n^2s+6nps+5ns+6ps+5pn+2n+3p+2n^2\right) \quad . \quad (403)$$

$$M_{DA}=-\frac{Ph}{\Delta}\left(3ns^2+6nps+5ns+6pn+5ps+2s+3p+2s^2\right) \quad (404)$$

If the frame is symmetrical about the vertical center line, that is, if AB and BC, Fig. 53, have the same section, $n=s$, and equations 401 to 404 take the form

$$M_{AB}=-M_{BC}=\frac{Ph}{2}\,\frac{3n+p}{\beta} \quad . \quad . \quad . \quad . \quad . \quad . \quad . \quad . \quad . \quad . \quad (405)$$

$$M_{CD}=-M_{DA}=\frac{Ph}{2}\,\frac{3n+1}{\beta} \quad . \quad . \quad . \quad . \quad . \quad . \quad . \quad . \quad . \quad . \quad (406)$$

in which $\beta=6n+p+1$.

TABLE 11

EQUATIONS FOR RECTANGULAR FRAME WITH HORIZONTAL FORCE AT THE TOP

No. of Equation	Left-hand Member of Equation				Right-hand Member of Equation	Method of Obtaining Equation
	M_{AB}	M_{BC}	M_{CD}	M_{DA}		
397	-1	$+1$	-1	$+1$	$-Ph$	Equation (397)
398	n	s	$2s+3p$	$2n+3p$	0	(390) + (393) + 2[(389) + (394)] + 3[(395) + (396)]
399	$-n$	$2s+1$	$2s+p$	$-n$	0	(392) + (395) + 2[(393) + (394)] − (389) − (390)
400	$n+1$	$s+1$	$s+p$	$n+p$	0	(389) + (390) + (391) + (392) + (393) + (394) + (395) + (396)

If the frame is symmetrical about the vertical center line and if the top and the bottom of the frame are alike, that is, if $n=s$ and $p=1$,

$$M_{AB}=-M_{BC}=\frac{Ph}{4} \quad \ldots \ldots \ldots \ldots \quad (407)$$

$$M_{CD}=-M_{DA}=\frac{Ph}{4} \quad \ldots \ldots \ldots \ldots \quad (408)$$

49. *Rectangular Frame. Any System of Horizontal Forces on One Vertical Side.*—Fig. 54 represents a rectangular frame subjected

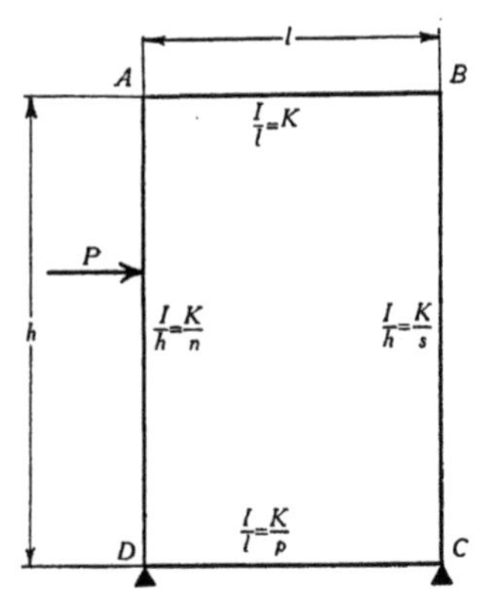

FIGURE 54

to any system of horizontal forces on the side AD.

Let M_D represent the moment of the external forces about D.

Applying the equations of Table 1 gives

$$nM_{DA}=2EK(2\theta_D+\theta_A-3R)-nC_{DA} \quad \ldots \ldots \ldots \quad (409)$$

$$nM_{AB}=-2EK(2\theta_A+\theta_D-3R)-nC_{AD} \quad \ldots \ldots \ldots \quad (410)$$

$$-M_{AB}+M_{BC}-M_{CD}+M_{DA}=-M_D \quad \ldots \ldots \ldots \quad (411)$$

Six other equations which are identical with equations 391 to 396 of section 48 may be written. The values of θ and R in these nine equations are identical with those in equations 389 to 397 of section 48, and hence the combination of equations to eliminate these two quantities is made in the manner indicated in the last column of Table 11. The equations thus obtained are given in Table 12.

TABLE 12

EQUATIONS FOR THE MOMENTS IN THE RECTANGULAR FRAME REPRESENTED BY FIG. 54

No. of Equation	Left-hand Member of Equation				Right-hand Member of Equation
	M_{AB}	M_{BC}	M_{CD}	M_{DA}	
411	-1	1	-1	1	$-M_D$
412	n	s	$2s+3p$	$2n+3p$	$-n(C_{AD}+2C_{DA})$
413	$-n$	$2s+1$	$2s+p$	$-n$	$n(C_{AD}+C_{DA})$
414	$n+1$	$s+1$	$s+p$	$n+p$	$-n(C_{AD}+C_{DA})$

Solving these equations simultaneously gives

$$M_{AB}=\frac{1}{\Delta}\Big\{M_D(3s^2n+5nps+2s^2p+2sp^2+6ns+6pn+5ps+3p^2)$$
$$-C_{AD}n(6sn+2pn+3s^2+17ps+2n+5s+11p+2p^2)$$
$$-C_{DA}n(3s^2+12ps+6s+10p+p^2)\Big\} \quad . \quad . \quad . \quad . \quad . \quad . \quad (415)$$

$$M_{BC}=\frac{1}{\Delta}\Big\{-M_D(3sn^2+2n^2p+5nps+2np^2+6ns+5pn+6ps+3p^2)$$
$$-C_{AD}n(3ns+2pn+5ps-n-7p-4s+2p^2)$$
$$+C_{DA}n(3ns+3pn-3ps+8p+6s-p^2)\Big\} \quad . \quad . \quad . \quad . \quad . \quad (416)$$

$$M_{CD}=\frac{1}{\Delta}\Big\{M_D(3sn^2+2n^2+6nps+5pn+5sn+6ps+2n+3p)$$
$$+C_{AD}n(3ns+6ps+3n+8p-3s-1)$$
$$-C_{DA}n(3ns-4ps+2n-7p-pn+5s+2)\Big\} \quad . \quad . \quad . \quad . \quad (417)$$

$$M_{DA}=\frac{1}{\Delta}\Big\{-M_D(5ns+6nps+3ns^2+2s^2+5ps+6pn+2s+3p)$$
$$-C_{AD}n(3s^2+6ps+12s+10p+1)$$
$$-C_{DA}n(3s^2+5ps+17s+11p+6ns+2pn+2n+2)\Big\} \quad . \quad (418)$$

in which

$$\Delta=22(spn+sp+sn+np)+2(sp^2+s^2p+n^2p+p^2n+s^2+s+n^2+n)$$
$$+6(sn^2+s^2n+p^2+p)$$

If $n=s$, equations 415 to 418 take the form

$$M_{AB}=-\frac{1}{2}\left\{C_{AD}n\left[\frac{n+2p}{\alpha}+\frac{3}{\beta}\right]+C_{DA}n\left[\frac{p}{\alpha}+\frac{3}{\beta}\right]-M_D\frac{3n+p}{\beta}\right\} \quad (419)$$

$$M_{BC}=-\frac{1}{2}\left\{C_{AD}n\left[\frac{n+2p}{\alpha}-\frac{3}{\beta}\right]+C_{DA}n\left[\frac{p}{\alpha}-\frac{3}{\beta}\right]+M_D\frac{3n+p}{\beta}\right\} \quad (420)$$

$$M_{CD}=-\frac{1}{2}\left\{C_{AD}n\left[\frac{1}{\alpha}-\frac{3}{\beta}\right]+C_{DA}n\left[\frac{n+2}{\alpha}-\frac{3}{\beta}\right]-M_D\frac{3n+1}{\beta}\right\} \quad (421)$$

$$M_{DA}=-\frac{1}{2}\left\{C_{AD}n\left[\frac{1}{\alpha}+\frac{3}{\beta}\right]+C_{DA}n\left[\frac{n+2}{\alpha}+\frac{3}{\beta}\right]+M_D\frac{3n+1}{\beta}\right\} \quad (422)$$

in which

$$\alpha=n^2+2pn+2n+3p$$
$$\beta=6n+p+1$$

If $n=s$ and $p=1$ equations 419 to 422 take the form

$$M_{AB}=-\frac{1}{2}\left\{C_{AD}n\left[\frac{n+2}{\alpha}+\frac{3}{\beta}\right]+C_{DA}n\left[\frac{1}{\alpha}+\frac{3}{\beta}\right]-\frac{M_D}{2}\right\} \quad (423)$$

$$M_{BC}=-\frac{1}{2}\left\{C_{AD}n\left[\frac{n+2}{\alpha}-\frac{3}{\beta}\right]+C_{DA}n\left[\frac{1}{\alpha}-\frac{3}{\beta}\right]+\frac{M_D}{2}\right\} \quad (424)$$

$$M_{CD}=-\frac{1}{2}\left\{C_{AD}n\left[\frac{1}{\alpha}-\frac{3}{\beta}\right]+C_{DA}n\left[\frac{n+2}{\alpha}-\frac{3}{\beta}\right]-\frac{M_D}{2}\right\} \quad (425)$$

$$M_{DA}=-\frac{1}{2}\left\{C_{AD}n\left[\frac{1}{\alpha}+\frac{3}{\beta}\right]+C_{DA}n\left[\frac{n+2}{\alpha}+\frac{3}{\beta}\right]+\frac{M_D}{2}\right\} \quad (426)$$

in which

$$\alpha=(n+3)(n+1)$$
$$\beta=2(3n+1)$$

If $n=s=p=1$, equations 423 to 426 take the form

$$M_{AB}=-\frac{1}{8}\left\{3C_{AD}+2C_{DA}-2M_D\right\} \quad . \quad . \quad . \quad . \quad (427)$$

$$M_{BC}=-\frac{1}{8}\left\{-C_{DA}+2M_D\right\} \quad . \quad . \quad . \quad . \quad (428)$$

$$M_{CD}=\frac{1}{8}\left\{C_{AD}+2M_D\right\} \quad . \quad . \quad . \quad . \quad (429)$$

$$M_{DA}=-\frac{1}{8}\left\{2C_{AD}+3C_{DA}+2M_D\right\} \quad . \quad . \quad . \quad . \quad (430)$$

If both vertical sides are loaded and if the frame and the loads are symmetrical about a vertical center line

$$M_{AB}=M_{BC}=-\frac{n}{\alpha}\left[C_{AD}(n+2p)+C_{DA}p\right] \quad . \quad . \quad . \quad . \quad (431)$$

$$M_{CD}=M_{DA}=-\frac{n}{\alpha}\left[C_{AD}+C_{DA}(n+2)\right] \quad . \quad . \quad . \quad . \quad (432)$$

50. *Rectangular Frame. Any System of Vertical Forces on the Top.*—Fig. 55 represents a rectangular frame subjected to any system of vertical forces on the top member AB.

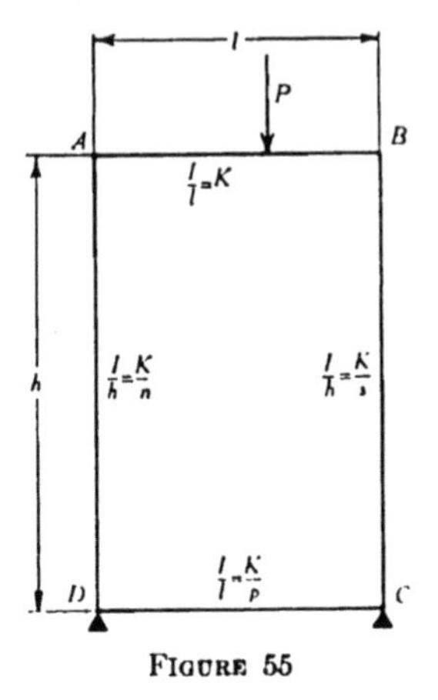

FIGURE 55

Applying the equations of Table 1 gives

$$M_{AB}=2EK(2\theta_A+\theta_B)-C_{AB} \quad (433)$$

$$M_{BC}=-2EK(2\theta_B+\theta_A)-C_{BA} \quad (434)$$

$$-M_{AB}+M_{DA}+M_{BC}-M_{CD}=0 \quad (435)$$

Six other equations which are identical with equations 389, 390, and 393 to 396 of section 48 may be written. Combining these nine equations to eliminate θ and R, as indicated in the last column of Table 11, gives the equations of Table 13.

TABLE 13

EQUATIONS FOR THE MOMENTS IN THE RECTANGULAR FRAME REPRESENTED BY FIG. 55

No. of Equation	Left-hand Member of Equation				Right-hand Member of Equation
	M_{AB}	M_{BC}	M_{CD}	M_{DA}	
435	-1	1	-1	1	0
436	n	s	$2s+3p$	$2n+3p$	0
437	$-n$	$2s+1$	$2s+p$	$-n$	$-C_{BA}$
438	$n+1$	$s+1$	$s+p$	$n+p$	$-(C_{BA}+C_{AB})$

Solving these equations simultaneously gives

$$M_{AB}=-\frac{1}{\Delta}\Big\{C_{BA}(10ns+s^2+12ps+6pn+3p^2)$$
$$+C_{AB}(11ns+2s^2+2s+2n+17ps+5pn+3p^2+6p)\Big\} \quad (439)$$

$$M_{BC}=-\frac{1}{\Delta}\Big\{C_{BA}(11ns+2n^2+2n+2s+17pn+5ps+3p^2+6p)$$
$$+C_{AB}(10ns+n^2+12pn+6ps+3p^2)\Big\} \quad (440)$$

$$M_{CD}=\frac{1}{\Delta}\Big\{C_{BA}(7ns-2n^2-2n+s-5pn+4ps-3p)$$
$$+C_{AB}(8ns-n^2+3n-3pn+6ps+3p)\Big\} \quad (441)$$

$$M_{DA}=\frac{1}{\Delta}\Big\{C_{BA}(8ns-s^2+3s-3ps+6pn+3p)$$

$$+C_{AB}(7ns-2s^2-2s+n-5ps+4pn-3p)\Big\} \quad . \quad . \quad . \quad . \quad . \quad (442)$$

in which

$$\Delta=22(spn+sp+sn+np)+2(sp^2+s^2p+n^2p+np^2+s^2+s+n^2+n)$$
$$+6(sn^2+s^2n+p^2+p)$$

If the load is symmetrical about the center of AB, that is, if $C_{AB}=C_{BA}$, equations 439 to 442 take the form

$$M_{AB}=-\frac{C_{AB}}{\Delta}\Big[21ns+3s^2+2s+2n+29ps+11pn+6p+6p^2\Big] \quad (443)$$

$$M_{BC}=-\frac{C_{AB}}{\Delta}\Big[\ 21ns+3n^2+2n+2s+29pn+11ps+6p+6p^2\ \Big] \quad (444)$$

$$M_{CD}=\frac{C_{AB}}{\Delta}\Big[\ 15ns-3n^2+n+s-8pn+10ps\ \Big] \quad . \quad . \quad . \quad . \quad . \quad (445)$$

$$M_{DA}=\frac{C_{AB}}{\Delta}\Big[\ 15ns-3s^2+s+n-8ps+10pn\ \Big] \quad . \quad . \quad . \quad . \quad (446)$$

If $n=s$ equations 439 to 442 take the form

$$M_{AB}=-\frac{1}{2}\Big\{C_{BA}\Big[\frac{2n+3p}{\alpha}-\frac{1}{\beta}\Big]+C_{DA}\Big[\frac{2n+3p}{\alpha}+\frac{1}{\beta}\Big]\Big\} \quad . \quad (447)$$

$$M_{BC}=-\frac{1}{2}\Big\{C_{BA}\Big[\frac{2n+3p}{\alpha}+\frac{1}{\beta}\Big]+C_{DA}\Big[\frac{2n+3p}{\alpha}-\frac{1}{\beta}\Big]\Big\} \quad . \quad (448)$$

$$M_{CD}=\frac{1}{2}\Big\{C_{BA}\Big[\frac{n}{\alpha}-\frac{1}{\beta}\Big]+C_{DA}\Big[\frac{n}{\alpha}+\frac{1}{\beta}\Big]\Big\} \quad . \quad . \quad . \quad . \quad (449)$$

$$M_{DA}=\frac{1}{2}\Big\{C_{BA}\Big[\frac{n}{\alpha}+\frac{1}{\beta}\Big]+C_{DA}\Big[\frac{n}{\alpha}-\frac{1}{\beta}\Big]\Big\} \quad . \quad . \quad . \quad . \quad (450)$$

in which $\alpha=n^2+2pn+2n+3p$

$\beta=6n+p+1$

If $n=s$ and $C_{AB}=C_{BA}$

$$M_{AB}=M_{BC}=-C_{AB}\frac{2n+3p}{\alpha} \qquad (451)$$

$$M_{CD}=M_{DA}=C_{AB}\frac{n}{\alpha} \qquad (452)$$

If $n=s$ and $p=1$

$$M_{AB}=-\frac{1}{2}\left\{C_{BA}\left[\frac{2n+3}{\alpha}-\frac{1}{\beta}\right]+C_{AB}\left[\frac{2n+3}{\alpha}+\frac{1}{\beta}\right]\right\} \qquad (453)$$

$$M_{BC}=-\frac{1}{2}\left\{C_{BA}\left[\frac{2n+3}{\alpha}+\frac{1}{\beta}\right]+C_{AB}\left[\frac{2n+3}{\alpha}-\frac{1}{\beta}\right]\right. \qquad (454)$$

$$M_{CD}=\frac{1}{2}\left\{C_{BA}\left[\frac{n}{\alpha}-\frac{1}{\beta}\right]+C_{AB}\left[\frac{n}{\alpha}+\frac{1}{\beta}\right]\right\} \qquad (455)$$

$$M_{DA}=\frac{1}{2}\left\{C_{BA}\left[\frac{n}{\alpha}+\frac{1}{\beta}\right]+C_{AB}\left[\frac{n}{\alpha}-\frac{1}{\beta}\right]\right\} \qquad (456)$$

in which

$$\alpha=(n+1)(n+3)$$
$$\beta=2(3n+1)$$

If $n=s$, $p=1$, and $C_{AB}=C_{BA}$

$$M_{AB}=M_{BC}=-C_{AB}\frac{2n+3}{(n+1)(n+3)} \qquad (457)$$

$$M_{CD}=M_{DA}=C_{AB}\frac{n}{(n+1)(n+3)} \qquad (458)$$

If $n=s=p=1$

$$M_{AB}=-\frac{1}{8}\left[2C_{BA}+3C_{AB}\right] \qquad (459)$$

$$M_{BC}=-\frac{1}{8}\left[3C_{BA}+2C_{AB}\right] \qquad (460)$$

$$M_{CD}=\frac{1}{8}\,C_{AB} \quad \ldots \ldots \quad (461)$$

$$M_{DA}=\frac{1}{8}\,C_{BA} \quad \ldots \ldots \quad (462)$$

If $n=s=p=1$, and $C_{AB}=C_{BA}$

$$M_{AB}=M_{BC}=-\frac{5}{8}\,C_{AB} \quad \ldots \ldots \quad (463)$$

$$M_{CD}=M_{DA}=\frac{1}{8}\,C_{AB} \quad \ldots \ldots \quad (464)$$

Values of C_{AB} and C_{BA} to be used in equations 439 to 464 are given in Table 2.

51. *Rectangular Frame. External Moment at Upper Corner.*—Fig. 56 represents a rectangular frame subjected to an external moment M at the upper left-hand corner.

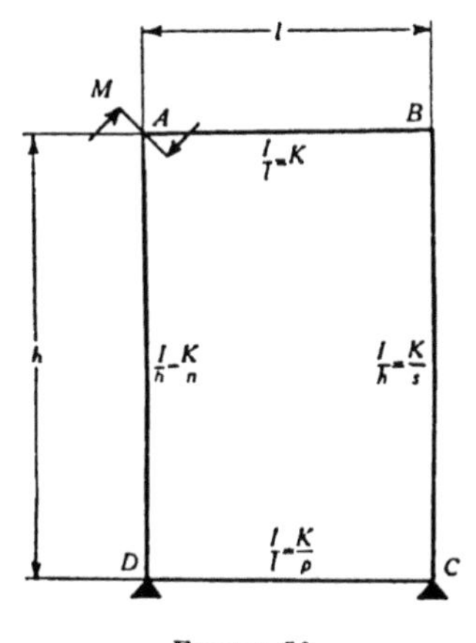

FIGURE 56

For equilibrium at A

$M-M_{AB}-M_{AD}=0$ or

$M_{AB}=M-M_{AD}$

Likewise

$M_{BC}=-M_{BA}$

$M_{CD}=-M_{CB}$

$M_{DA}=-M_{DC}$

$$nM_{AB}=-2EK(2\theta_A+\theta_D-3R)+Mn \quad . \quad . \quad . \quad . \quad . \quad . \quad . \quad (465)$$

$$-M_{AB}+M_{DA}+M_{BC}-M_{CD}=-M \quad . \quad . \quad . \quad . \quad . \quad . \quad . \quad (466)$$

Equations 389 and 391 to 396 of section 48 apply to this case. Eliminating values of θ and R, as indicated in Table 11, gives the equations of Table 14.

TABLE 14

EQUATIONS FOR THE MOMENTS IN THE RECTANGULAR FRAME REPRESENTED BY FIG. 56

No. of Equation	Left-hand Member of Equation				Right-hand Member of Equation
	M_{AB}	M_{BC}	M_{CD}	M_{DA}	
466	-1	1	-1	1	$-M$
467	n	s	$2s+3p$	$2n+3p$	$+Mn$
468	$-n$	$2s+1$	$2s+p$	$-n$	$-Mn$
469	$n+1$	$s+1$	$s+p$	$n+p$	$+Mn$

Solving these equations simultaneously gives

$$M_{AB}=-\frac{M}{\Delta}\left\{11ns+2s^2+2s+2n+17ps+5pn+6p+3p^2\right\}+M \quad (470)$$

$$M_{BC}=-\frac{M}{\Delta}\left\{n^2+10ns+12pn+6ps+3p^2\right\} \quad . \quad . \quad . \quad . \quad . \quad (471)$$

$$M_{CD}=+\frac{M}{\Delta}\left\{3p-n^2-3pn+8ns+6ps+3n\right\} \quad . \quad . \quad . \quad . \quad (472)$$

$$M_{DA}=-\frac{M}{\Delta}\left\{-7ns+2s^2+5ps-4pn+2s+3p-n\right\} \quad . \quad . \quad (473)$$

$$M_{AD}=+\frac{M}{\Delta}\left\{11ns+2s^2+2s+2n+17ps+5pn+6p+3p^2\right\} \quad . \quad (474)$$

in which

$$\Delta=22(pns+sp+sn+np)+2(sp^2+s^2p+n^2p+p^2n+s^2+s+n^2+n)$$
$$+6(sn^2+s^2n+p^2+p)$$

If $n=s$, equations 470 to 474 take the form

$$M_{AB}=\frac{M}{2}\left\{1+\frac{n(n+2p)}{\alpha}-\frac{1}{\beta}\right\} \quad (475)$$

$$M_{BC}=\frac{M}{2}\left\{\frac{n(n+2p)}{\alpha}-\frac{6n+p}{\beta}\right\} \quad (476)$$

$$M_{CD}=\frac{M}{2}\left\{\frac{n}{\alpha}+\frac{1}{\beta}\right\} \quad (477)$$

$$M_{DA}=\frac{M}{2}\left\{\frac{n}{\alpha}-\frac{1}{\beta}\right\} \quad (478)$$

$$M_{AD}=\frac{M}{2}\left\{1-\frac{n(n+2p)}{\alpha}+\frac{1}{\beta}\right\} \quad (479)$$

in which

$$\alpha=n^2+2pn+2n+3p$$
$$\beta=6n+p+1$$

If $n=s$ and $p=1$, equations 475 to 479 take the form

$$M_{AB}=+\frac{M}{2}\left\{1+\frac{n(n+2)}{(n+1)\,(n+3)}-\frac{1}{6n+2}\right\} \quad (480)$$

$$M_{AD}=+\frac{M}{2}\left\{1-\frac{n(n+2)}{(n+1)(n+3)}+\frac{1}{6n+2}\right\} \quad (481)$$

$$M_{BC}=+\frac{M}{2}\left\{\frac{n(n+2)}{(n+1)(n+3)}-\frac{6n+1}{6n+2}\right\} \quad (482)$$

$$M_{CD}=+\frac{M}{2}\left\{\frac{n}{(n+1)(n+3)}+\frac{1}{6n+2}\right\} \quad (483)$$

$$M_{DA} = +\frac{M}{2}\left\{\frac{n}{(n+1)(n+3)} - \frac{1}{6n+2}\right\} \quad (484)$$

If $n=s=p=1$, equations 480 to 484 take the form

$$M_{AB} = +\frac{5}{8}M \quad (485)$$

$$M_{AD} = +\frac{3}{8}M \quad (486)$$

$$M_{BC} = -\frac{M}{4} \quad (487)$$

$$M_{CD} = +\frac{M}{8} \quad (488)$$

$$M_{DA} = 0 \quad (489)$$

If there is a couple at B as well as at A and if the frame and loading are symmetrical about a vertical center line, that is, if the couples are equal in magnitude and opposite in sense, and if $n=s$

$$M_{AB} = -M_{BA} = \frac{Mn}{\alpha}(n+2p) \quad (490)$$

$$M_{AD} = -M_{BC} = \frac{M}{\alpha}(2n+3p) \quad (491)$$

$$M_{CD} = M_{DA} = \frac{Mn}{\alpha} \quad (492)$$

K. Moments in Frames Composed of a Large Number of Rectangles as a Skeleton-Construction Building Frame

52. *Effect of Restraint at One End of a Member upon Moment at Other End.*—Fig. 57 represents any member in flexure. The end A is

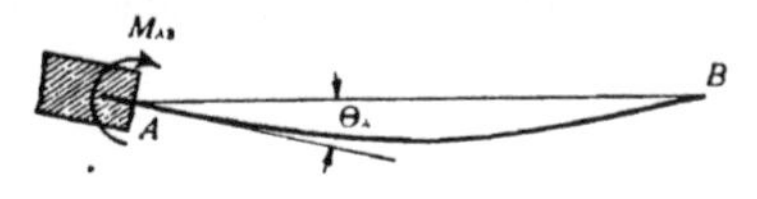

Figure 57

acted upon by a couple M_{AB} such that the tangent to the elastic curve at A makes an angle θ_A with AB. The magnitude of the moment M_{AB} depends not only upon the magnitude of θ_A, the moment of inertia of the section and the length of AB, but also upon the degree of restraint at B. This is illustrated by the following special problems.

Consider that AB is hinged at B. Applying equation (C) of Table 1 with R and H_{AB} equal to zero gives

$$M_{AB}=3EK\,\theta_A \qquad (493)$$

Consider that AB is fixed at B. Applying equation (A) of Table 1 with θ_B, R, and C_{AB} equal to zero gives

$$M_{AB}=4EK\,\theta_A \qquad (494)$$

Consider that $\theta_A=-\theta_B$. Applying equation (A) of Table 1 with R and C_{AB} equal to zero and with $\theta_A=-\theta_B$ gives

$$M_{AB}=2EK\,\theta_A \qquad (495)$$

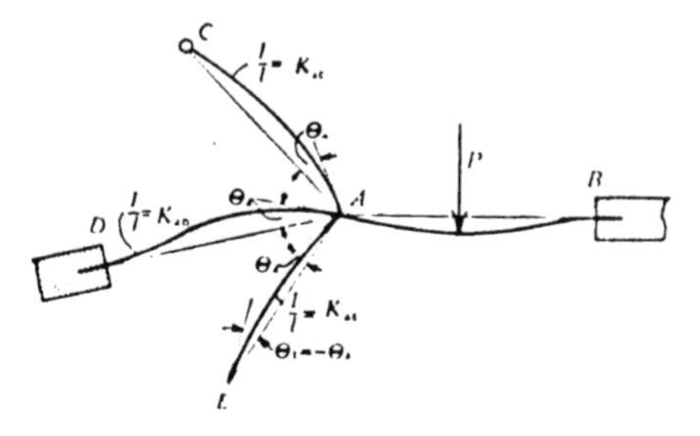

Figure 58

Fig. 58 represents a member AB having any degree of restraint at B and restrained at A by the members AC, AD, and AE. P represents any system of loads. AC is hinged at C, AD is fixed at D, and the restraint at E is such that $\theta_E = -\theta_A$. The moment at A in the member AB is taken as a measure of the restraint at A. Since A is in equilibrium, $M_{AB}+M_{AC}+M_{AD}+M_{AE}=0$. That is, the moment M_{AB} balances the three moments M_{AC}, M_{AD}, and M_{AE}. The moments M_{AC}, M_{AD}, and M_{AE} are therefore measures of the restraints which the members AC, AD, and AE exert on the member AB at A. From equations 493, 494, and 495

$$M_{AC}=3EK_{AC}\,\theta_A$$
$$M_{AD}=4EK_{AD}\,\theta_A$$
$$M_{AE}=2EK_{AE}\,\theta_A$$

These equations have the general form

$$M=EK\,\theta\,N \qquad (496)$$

in which N depends upon the restraints at C, D, and E, and might be termed a "restraint factor;" that is, the restraint which a member can exert upon a joint at one end equals $EK\,\theta$ times a factor N whose value depends upon the degree of restraint at the other end of the member. As derived, if the far end is hinged, $N=3$; if the far end is fixed, $N=4$; and if the angular rotation at the two ends is equal in magnitude but opposite in sense $N=2$. In general, N depends upon the restraints at C, D, and E.

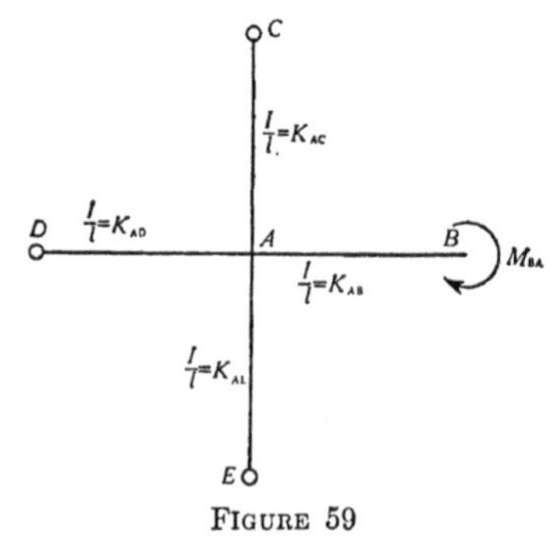

FIGURE 59

In Fig. 59, B is restrained by the couple M_{AB} and C, D, and E are hinged.

For equilibrium

$$M_{AB}+M_{AC}+M_{AD}+M_{AE}=0 \quad \text{(497)}$$

From equation (A), Table 1

$$M_{AB}=2EK_{AB}(2\theta_A+\theta_B) \quad \text{(498)}$$

$$M_{BA}=2EK_{AB}(2\theta_B+\theta_A) \quad \text{(499)}$$

Substituting the values of M_{AC}, M_{AD}, and M_{AE} from equation 493 and the value of M_{AB} from equation 498 in equation 497 gives

$$\theta_A=-\theta_B\,\frac{2K_{AB}}{4K_{AB}+3K_{AC}+3K_{AD}+3K_{AE}} \quad \text{(500)}$$

Substituting the value of θ_A from equation 500 in equation 499 gives

$$M_{BA}=EK_{AB}\ \theta_B\left[\frac{4(3K_{AB}+3K_{AC}+3K_{AD}+3K_{AE})}{4K_{AB}+3K_{AC}+3K_{AD}+3K_{AE}}\right] \quad \text{(501)}$$

If C, D, and E of Fig. 59 are fixed, the values of M_{AC}, M_{AD}, and M_{AE} of equation 497 are given by equation 494. Proceeding as before gives

$$M_{BA}=EK_{AB}\ \theta_B\left[\frac{4(3K_{AB}+4K_{AC}+4K_{AD}+4K_{AE})}{4K_{AB}+4K_{AC}+4K_{AD}+4K_{AE}}\right] \quad \text{(502)}$$

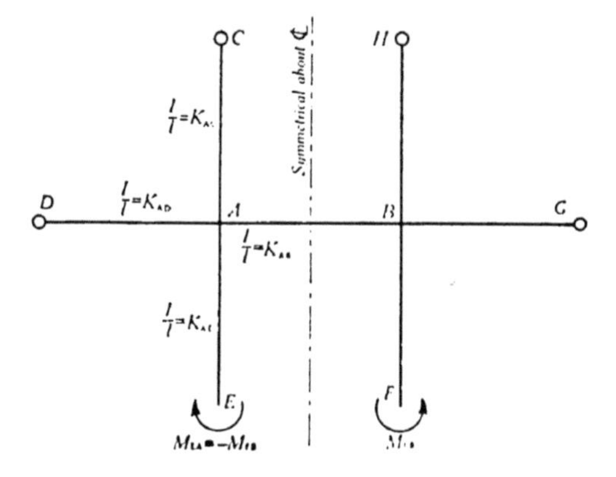

FIGURE 60

The frame represented by Fig. 60 is symmetrical about its vertical center line and is symmetrically loaded. θ_A therefore equals $-\theta_B$.

For the extremities of the members C, D, G, and H hinged, M_{AC} and M_{AD} are given by equation 493. M_{AB} is given by equation 495. From equation (A), Table 1

$$M_{AE} = 2EK_{AE}(2\theta_A + \theta_E) \quad \ldots \quad (503)$$

$$M_{EA} = 2EK_{AE}(2\theta_E + \theta_A) \quad \ldots \quad (504)$$

Substituting the values of M_{AB}, M_{AC}, M_{AD}, and M_{AE} in equation 497, solving for θ_A, and substituting the value of θ_A in equation 504 gives

$$M_{EA} = EK_{AE}\,\theta_E \left[\frac{4(2K_{AB}+3K_{AC}+3K_{AD}+3K_{AE})}{2K_{AB}+3K_{AC}+3K_{AD}+4K_{AE}} \right] \quad \ldots \quad (505)$$

If C and D are fixed

$$M_{EA} = EK_{AE}\,\theta_E \left[\frac{4(2K_{AB}+4K_{AC}+4K_{AD}+3K_{AE})}{2K_{AB}+4K_{AC}+4K_{AD}+4K_{AE}} \right] \quad \ldots \quad (506)$$

Equations 501, 502, 505, and 506 have the form $M = EK\,\theta\,N$ in which N corresponds to the quantity in the brackets.

It is to be noted that for the values of N in equations 501, 502, 505, and 506 the coefficient of K for the member in which the stress is to be determined is always 3 in the numerator and 4 in the denominator. For the members, furthermore, which restrain the member in which the moment is to be determined: if hinged at the far end the coefficient of K is 3; if fixed at the far end the coefficient of K is 4; and if the rotations of the two ends are equal in magnitude but opposite in sense the coefficient of the K is 2. These coefficients correspond to the coefficients of $EK\,\theta_A$ in the expressions for the moments M_{AC}, M_{AD}, and M_{AE} of Fig. 57.

53. *Moment in a Frame Composed of a Number of Rectangles Due to Vertical Loads.*—Fig. 61 represents a portion of a frame composed of a large number of rectangles. The portion considered is taken from the center of a frame symmetrical about a vertical line. The member AB carries any system of vertical loads symmetrical about the center line of AB. Under these conditions there is no horizontal deflection of the frame.

For equilibrium at A,

$$M_{AB} + M_{AD} + M_{AI} + M_{AH} = 0 \quad \ldots \quad (507)$$

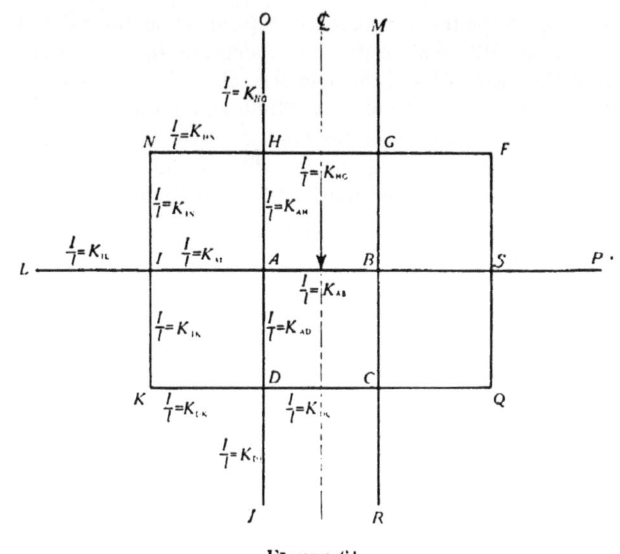

FIGURE 61

From equation 496, section 52

$$M_{AD}=EK_{AD}\,\theta_A\,N_{AD} \quad (508)$$

$$M_{AI}=EK_{AI}\,\theta_A\,N_{AI} \quad (509)$$

$$M_{AH}=EK_{AH}\,\theta_A\,N_{AH} \quad (510)$$

Substituting the values of M_{AD}, M_{AI}, and M_{AH} from equations 508, 509, and 510 in equation 507 gives

$$M_{AB}=-E\,\theta_A\left[K_{AD}\,N_{AD}+K_{AI}\,N_{AI}+K_{AH}\,N_{AH}\right] \quad (511)$$

Since the frame is symmetrical about a vertical center line, $\theta_A=-\theta_B$. From equation (C) of Table 1

$$M_{AB}=2EK_{AB}\,\theta_A-\frac{F}{l} \quad (512)$$

Eliminating θ_A from equations 511 and 512 gives

$$M_{AB}=-\frac{F}{l}\left[\frac{K_{AD}\,N_{AD}+K_{AI}\,N_{AI}+K_{AH}\,N_{AH}}{K_{AD}\,N_{AD}+K_{AI}\,N_{AI}+K_{AH}\,N_{AH}+2K_{AB}}\right] \quad (513)$$

The stress in the frame apparently depends upon the values of the N's in equation 513; that is, the stress depends upon the degrees of restraint of the extremities of the members, O, N, L, K, J, etc.

Although the degrees of restraint at these extremities are not known, it is known that the degree of restraint at each extremity is greater than if the extremity is hinged and less than if the extremity is fixed. If then the stresses are determined with the extremities hinged and again with the extremities fixed, although the true stresses will not be determined, they will be fixed between two limits.

54. *Extremities of Members Hinged.*—If the members are hinged at O, N, L, K, and J:

From equation 505, section 52,

$$N_{AD}=\frac{4(2K_{DC}+3K_{DJ}+3K_{DK}+3K_{AD})}{2K_{DC}+3K_{DJ}+3K_{DK}+4K_{AD}} \qquad (514)$$

$$N_{AH}=\frac{4(2K_{HG}+3K_{HN}+3K_{HO}+3K_{AH})}{2K_{HG}+3K_{HN}+3K_{HO}+4K_{AH}} \qquad (515)$$

From equation 501, section 52,

$$N_{AI}=\frac{4(3K_{AI}+3K_{IK}+3K_{IL}+3K_{IN})}{4K_{AI}+3K_{IK}+3K_{IL}+3K_{IN}} \qquad (516)$$

The restraint factor, N, in each of these equations has a value between 3 and 4.

Substituting the values of the N's from equations 514, 515, and 516 in equation 513 gives the value of M_{AB}.

The expression for M_{AB} in equation 513 is made up of three quantities, the three moments resisted by AD, AI, and AH. These moments are as follows:

$$M_{AD}=\frac{F}{l}\left[\frac{K_{AD}N_{AD}}{K_{AD}N_{AD}+K_{AI}N_{AI}+K_{AH}N_{AH}+2K_{AB}}\right] \qquad (517)$$

$$M_{AI}=\frac{F}{l}\left[\frac{K_{AI}N_{AI}}{K_{AD}N_{AD}+K_{AI}N_{AI}+K_{AH}N_{AH}+2K_{AB}}\right] \qquad (518)$$

$$M_{AH}=\frac{F}{l}\left[\frac{K_{AH}N_{AH}}{K_{AD}N_{AD}+K_{AI}N_{AI}+K_{AH}N_{AH}+2K_{AB}}\right] \qquad (519)$$

The values of the N's are given in equations 514, 515, and 516.

From the conditions for equilibrium at D and from the relation between M_{DA} and M_{AD} it can be proved that

$$M_{DA}=\frac{M_{AD}}{2}\left[\frac{3K_{DK}+3K_{DJ}+2K_{DC}}{3K_{DK}+3K_{DJ}+2K_{DC}+3K_{AD}}\right] \quad . \quad . \quad . \quad . \quad . \quad (520)$$

also

$$M_{IA}=\frac{M_{AI}}{2}\left[\frac{3K_{IK}+3K_{IN}+3K_{IL}}{3K_{IK}+3K_{IN}+3K_{IL}+3K_{AI}}\right] \quad . \quad . \quad . \quad . \quad . \quad (521)$$

and

$$M_{HA}=\frac{M_{AH}}{2}\left[\frac{3K_{HN}+3K_{HO}+2K_{HG}}{3K_{HN}+3K_{HO}+2K_{HG}+3K_{AH}}\right] \quad . \quad . \quad . \quad . \quad . \quad (522)$$

M_{DA} is made up of the three moments M_{DK}, M_{DJ}, and M_{DC}. These moments are as follows:

$$M_{DK}=-\frac{M_{AD}}{2}\left[\frac{3K_{DK}}{3K_{DK}+3K_{DJ}+2K_{DC}+3K_{AD}}\right] \quad . \quad . \quad . \quad . \quad (523)$$

$$M_{DJ}=-\frac{M_{AD}}{2}\left[\frac{3K_{DJ}}{3K_{DK}+3K_{DJ}+2K_{DC}+3K_{AD}}\right] \quad . \quad . \quad . \quad . \quad (524)$$

$$M_{DC}=-\frac{M_{AD}}{2}\left[\frac{2K_{DC}}{3K_{DK}+3K_{DJ}+2K_{DC}+3K_{AD}}\right] \quad . \quad . \quad . \quad . \quad (525)$$

In a similar manner M_{IA} can be divided into M_{IK}, M_{IN}, and M_{IL}, and M_{HA} can be divided into M_{HN}, M_{HO}, and M_{HG}.

55. *Extremities of Members Fixed.*—If the members are fixed at O, N, L, K, and J:

From equation 506, section 52,

$$N_{AD}=\frac{4(2K_{DC}+4K_{DJ}+4K_{DK}+3K_{AD})}{2K_{DC}+4K_{DJ}+4K_{DK}+4K_{AD}} \quad . \quad . \quad . \quad . \quad . \quad . \quad . \quad (526)$$

$$N_{AH}=\frac{4(2K_{HG}+4K_{HN}+4K_{HO}+3K_{AH})}{(2K_{HG}+4K_{HN}+4K_{HO}+4K_{AH})} \quad . \quad . \quad . \quad . \quad . \quad . \quad . \quad (527)$$

From equation 502, section 52,

$$N_{AI}=\frac{3K_{AI}+4K_{IK}+4K_{IL}+4K_{IN}}{K_{AI}+K_{IK}+K_{IL}+K_{IN}} \qquad (528)$$

Substituting the values of the N's in equation 513 gives the value of M_{AB}.

Equations 517, 518, and 519 are applicable. By substituting the values of the N's, given in equations 526, 527, and 528, M_{AD}, M_{AI}, and M_{AH} can be determined.

Proceeding as in the case where the extremities of the members are hinged it can be proved that

$$M_{DA}=\frac{M_{AD}}{2}\left[\frac{4K_{DK}+4K_{DJ}+2K_{DC}}{4K_{DK}+4K_{DJ}+2K_{DC}+3K_{AD}}\right] \qquad (529)$$

$$M_{IA}=\frac{M_{AI}}{2}\left[\frac{4K_{IK}+4K_{IN}+4K_{IL}}{4K_{IK}+4K_{IN}+4K_{IL}+3K_{AI}}\right] \qquad (530)$$

$$M_{HA}=\frac{M_{AH}}{2}\left[\frac{4K_{HN}+4K_{HO}+2K_{HG}}{4K_{HN}+4K_{HO}+2K_{HG}+3K_{AH}}\right] \qquad (531)$$

M_{DA} is made up of three parts, one part corresponding to each of the moments M_{DK}, M_{DJ}, and M_{DC}. These latter moments are proportional respectively to the parts of the numerator of equation 529: $4K_{DK}$, $4K_{DJ}$, and $2K_{DC}$. Similarly, M_{IK}, M_{IL}, and M_{IN} can be determined from M_{IA}; and M_{HN}, M_{HO}, and M_{HG} can be determined from M_{HA}.

To determine the effect of the degree of restraint of the extremities O, N, L, K, and J upon the moments in the frame, and also to determine the effect of the magnitude of the K's upon the moments in the frame, moments have been determined for frames having fixed and hinged extremities, for frames having all K's equal, and for frames for which the K's of the columns equal ten times the K's for the girders. The values of the M's are given in Table 15.

From Table 15, it is apparent that only members directly connected to the member carrying the load are subjected to moments sufficiently large to be considered in the design of the structure. Furthermore, the moments in the members adjacent to the member carrying the load are practically independent of the degree of restraint at the extremities O, N, L, K, J, etc. Therefore, the moments in the frame of Fig. 61, due to the load on AB, are given with sufficient accuracy for purposes of design by equations 513, 517, 518, 519, 520, 521, and 522, based upon the assumption that the extremities O, N, L, K, J, etc., are hinged. The equations based upon the assumption that the extremities are fixed give almost exactly the same results and could also be used.

TABLE 15

MOMENTS IN FRAME REPRESENTED BY FIG. 61

Moments are expressed in terms of $\frac{F}{l}$

Moment	Extremities of Members Hinged at O, N, L, K, and J		Extremities of Members Fixed at O, N, L, K, and J	
	All K's Equal	K's of Columns Equal 10 Times K's of Girders	All K's Equal	K's of Columns Equal 10 Times K's of Girders
M_{AB}	−.845	−.972	−.849	−.973
M_{AD}	+.281	+.462	+.282	+.462
M_{AH}	+.281	+.462	+.282	+.462
M_{AI}	+.283	+.048	+.285	+.049
M_{DA}	+.102	+.125	+.108	+.140
M_{HA}	+.102	+.125	+.108	+.140
M_{IA}	+.106	+.023	+.114	+.024
M_{DK}	−.038	−.011	−.043	−.012
M_{DJ}	−.038	−.107	−.043	−.122
M_{DC}	−.026	−.007	−.022	−.006
M_{HN}	−.038	−.011	−.043	−.012
M_{HO}	−.038	−.017	−.043	−.122
M_{HG}	−.026	−.007	−.022	−.006
M_{IK}	−.035	−.011	−.038	−.011
M_{IN}	−.035	−.011	−.038	−.011
M_{IL}	−.035	−.001	−.038	−.002

56. *Distribution of Loads for Maximum Moments in a Frame Composed of a Large Number of Rectangles.*—Referring to Fig. 61, a load on AB produces a moment M_{KD} having the same sign as M_{AB}. That being the case, a load on GF produces a moment M_{AB} of the same sign as the M_{AB} produced by the load on AB; therefore if AB and GF are loaded simultaneously the moment at A in AB is greater than if either AB or GF is loaded alone. Reasoning in a similar manner, the members can be selected which, if loaded, produce a moment at A in

the member AB having the same sign as the moment at the same point due to a load on AB. If all these members are loaded simultaneously, the moment at A in the member AB is a maximum.

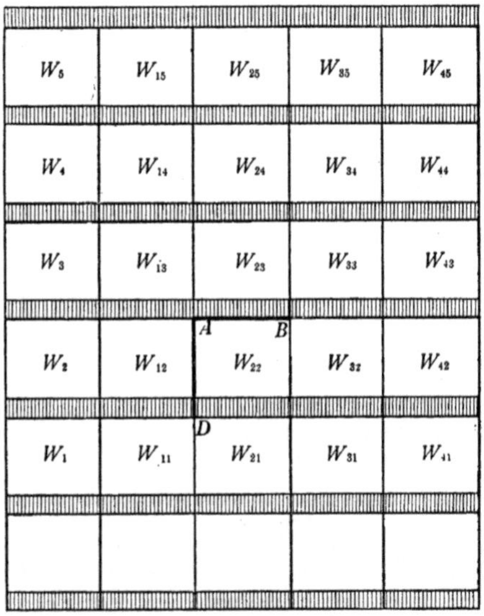

FIGURE 62

Fig. 62 represents a frame made up of similar rectangles. All girders are equally loaded. The moments M_{AB} and M_{AD}, due to these loads, as determined by the equations of section 53 are given in Table 16. For the frame of Table 16 the K's of all members are equal. The moments in similar frames for which the K's of all columns are equal and the K's of all girders are equal, but for which the K's of the columns do not equal the K's of the girders, have been determined. The relation between the maximum moments which it is possible to obtain in the girders, and the ratio of the K's of the columns to the K's of the girders, is presented graphically in Fig. 63. Similar data for the moment in the columns are given in Fig. 64.

Fig. 65 represents the loading which produces a maximum moment at A in the girder. Fig. 66 represents the loading which produces a maximum moment at A in the column.

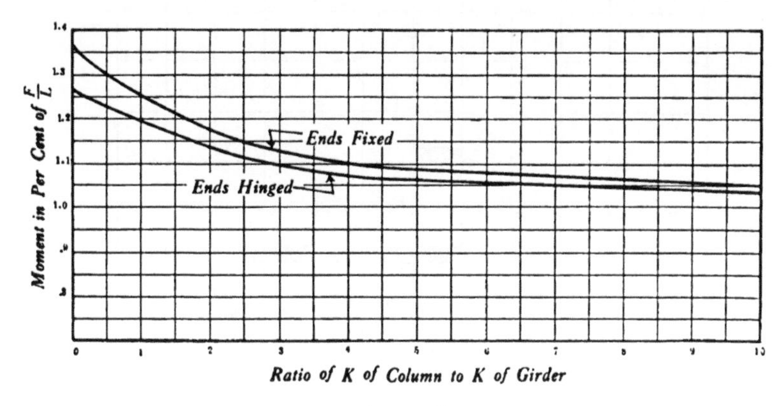

FIGURE 63

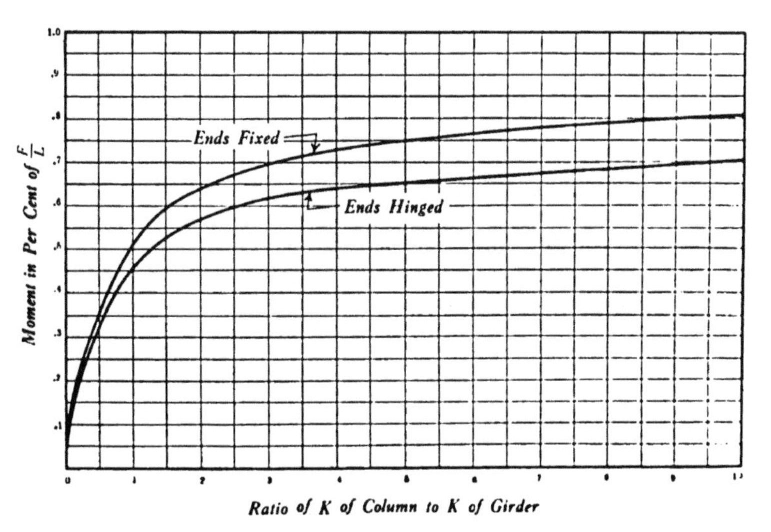

FIGURE 64

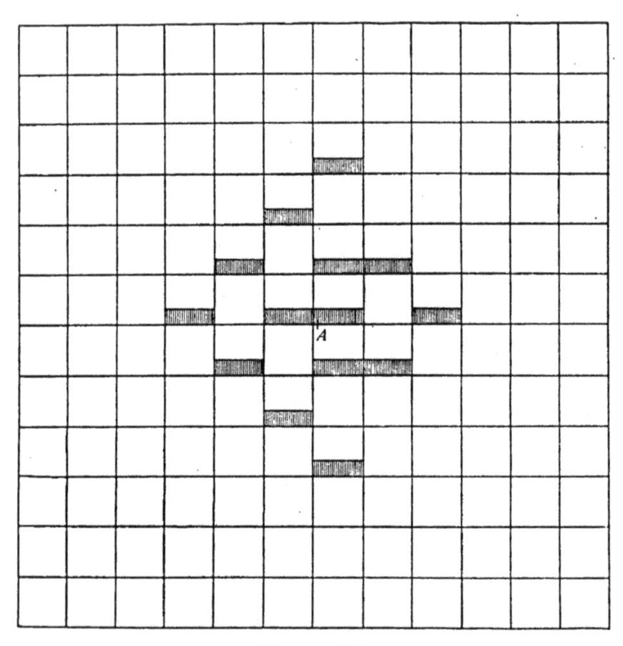

FIGURE 65

It is to be noted that for the frames of sections 52, 53, and 56 the horizontal deflection of one story of the frame relative to the other stories does not enter. If either the frame or the load is unsymmetrical there is a slight horizontal deflection. For the usual proportions of frames of engineering structures, the effect of this horizontal deflection is slightly to reduce the moments.

57. *Eccentric Load at Top of Exterior Column of a Frame. Connections of Girders to Columns Hinged.*—Fig. 67 represents a frame with eccentric loads at the tops of the exterior columns. The frame and the loading are symmetrical about the vertical center line of the frame. The connections of the girders to the columns are frictionless hinges The columns are continuous.

The moment in the column depends upon the restraint at 1. The degree of restraint at 1 is known to be between the restraint of a column hinged at 1 and a column fixed at 1. If, therefore, the moment is deter-

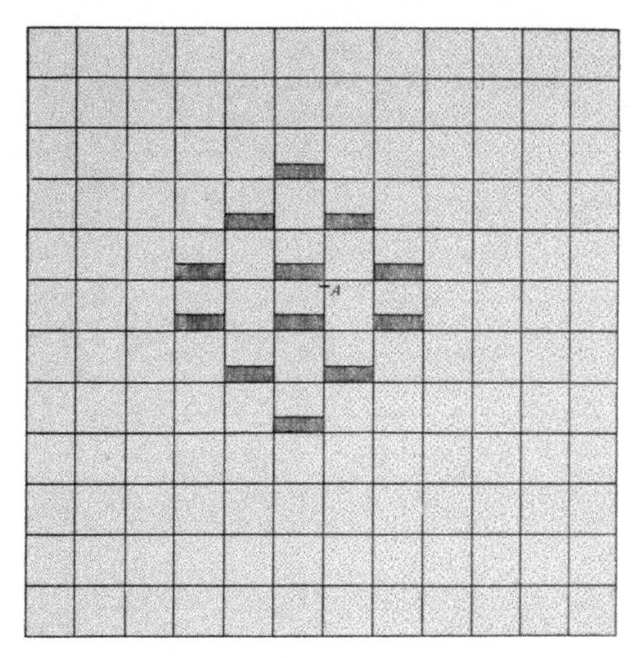

FIGURE 66

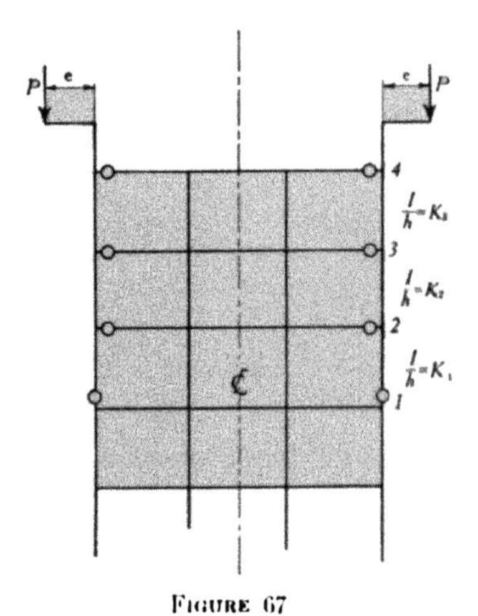

FIGURE 67

TABLE 16

MOMENTS IN FRAME REPRESENTED BY FIG. 62

K's of All Members Equal

Moment Produced by	M_{AB}		M_{AD}	
	Hinged	Fixed	Hinged	Fixed
$W2$			0	+.019
$W3$	+.035	+.038	+.035	+.038
$W11$			0	+.022
$W12$	+.038	+.043	−.102	−.108
$W13$	−.283	−.285	−.281	−.282
$W14$	+.038	+.043	+.038	+.043
$W21$			0	−.022
$W22$	−.026	−.022	+.102	+.108
$W23$	−.845	−.849	+.281	+.282
$W24$	−.026	−.022	−.038	−.043
$W32$	0	−.022	0	−.019
$W33$	+.106	+.114	−.035	−.038
$W34$	0	−.022		
$W43$	0	−.019		
Total	−.963	−1.003	0	0
Maximum	−1.180	−1.241	±.456	±.512

W's not given in table produce only very small moments.

mined for a column hinged at 1 and for a column fixed at 1 the true moment will be located between two limits.

Consider the column to be hinged at 1

$$M_{12}=0 \quad (532)$$

$$M_{21}=3EK_1\theta_2 \quad (533)$$

From equation 501, section 52,

$$M_{32}=4EK_2\theta_3\left(\frac{3K_2+3K_1}{4K_2+3K_1}\right) \quad (534)$$

$$M_{34}=-M_{32}=-4EK_2\theta_3\left(\frac{3K_2+3K_1}{4K_2+3K_1}\right)=2EK_3(2\theta_3+\theta_4) \quad (535)$$

$$M_{43}=2EK_3(2\theta_4+\theta_3) \quad (536)$$

Let $N_2=4\,\dfrac{3K_1+3K_2}{3K_1+4K_2}$

From equations 534 and 535

$$M_{43}=4EK_3\,\theta_4\left(\frac{N_2K_2+3K_3}{N_2K_2+4K_3}\right) \quad (537)$$

Also

$$M_{43}=Pe \quad (538)$$

From equations 537 and 538

$$\theta_4=\frac{Pe}{4EK_3}\,\frac{N_2K_2+4K_3}{N_2K_2+3K_3} \quad (539)$$

Eliminating θ_3 from equations 535 and 536 gives

$$M_{34}=-6EK_3\,\theta_4+2Pe \quad (540)$$

Substituting the value of θ_4 from equation 539 gives

$$M_{34}=\frac{Pe}{2}\,\frac{N_2K_2}{N_2K_2+3K_3} \quad (541)$$

$$M_{32}=-M_{34} \quad (542)$$

From the equations of Table 1

$$M_{32}=2EK_2(2\theta_3+\theta_2) \quad (543)$$

$$M_{23}=2EK_2(2\theta_2+\theta_3) \quad (544)$$

Eliminating θ_3 gives

$$2M_{23}-M_{32}=6EK_2\,\theta_2 \qquad (545)$$

From equations 533, 545, and 542, and since $M_{21}=-M_{23}$

$$M_{23}=-\frac{K_1}{2(K_1+K_2)}M_{34} \qquad (546)$$

If, therefore, the column is hinged at 1

$$M_{43}=Pe \qquad (538)$$

$$M_{34}=-M_{32}=\frac{Pe}{2}\left[\frac{N_2K_2}{N_2K_2+3K_3}\right] \qquad (541)$$

$$M_{21}=-M_{23}=\frac{K_1}{2(K_1+K_2)}M_{34} \qquad (546)$$

If the column is fixed at 1, letting N'_2 represent $4\,\frac{4K_1+3K_2}{4K_1+4K_2}$

$$M_{43}=Pe \qquad (547)$$

$$M_{34}=-M_{32}=\frac{Pe}{2}\;\frac{N'_2K_2}{N'_2K_2+3K_3} \qquad (548)$$

$$M_{21}=-M_{23}=\frac{2K_1}{4K_1+3K_2}M_{34} \qquad (549)$$

From a comparison of equations 538, 541, and 546 with 547, 548, and 549 it is apparent that the restraint at 1 does not materially affect the moments at 2 and 3. For purposes of design the moments as given by either equations 538, 541, and 546 or by equations 547, 548, and 549 are satisfactory. Moreover the average of the moments obtained by 538, 541, and 546, and 547, 548, and 549 approximate very closely the true moments.

If the K's are all equal, a condition often approximated in practice: For column hinged at 1

$$M_{43}=Pe \qquad (550)$$

$$M_{34} = -M_{32} = \frac{4}{15} Pe = 0.266\, Pe \quad \text{. . . . . (551)}$$

$$M_{21} = -M_{23} = \frac{1}{15} Pe = 0.067\, Pe \quad \text{. . . . . (552)}$$

For column fixed at 1

$$M_{43} = Pe \quad \text{. . . . . (553)}$$

$$M_{34} = -M_{32} = \frac{7}{26} Pe = 0.269\, Pe \quad \text{. . . . . (554)}$$

$$M_{21} = -M_{23} = \frac{2}{26} Pe = 0.077\, Pe \quad \text{. . . . . (555)}$$

It is to be noted that the frame considered is symmetrical about a vertical center line and is symmetrically loaded. If there is a load on the right-hand column only, the moments in that column will be slightly smaller than the moments given by the equation, and the other columns will be subjected to a small moment. The error in the moment in the loaded column and the neglected moment in the other columns increase as the ratio of the stiffness of the loaded column to the combined stiffness of the other columns increase. Although they have not been able to establish this statement mathematically, it is the opinion of the writers that if the equations of this section are applied to a frame that is either unsymmetrical or unsymmetrically loaded, the error due to the horizontal deflection of the frame is negligible for purposes of design.

58. *Eccentric Load at Top of Exterior Column of a Frame. Connections of Girders to Columns Rigid.*—Fig. 68 represents a frame with eccentric loads at the tops of the exterior columns. The frame and the loading are symmetrical about the vertical center line of the frame. The connections of the girders to the columns are rigid.

An exact determination of the moments in the frame is practically impossible. From previous similar work, however, it is known that of the moments produced by P on the right-hand side of the frame, only the moments at A, B, C, and F are large enough to be considered in the design of the frame. Furthermore, from previous work it is known that the moments at A, B, C, and F are practically independent of the degree of restraint at J, K, G, H, and D.

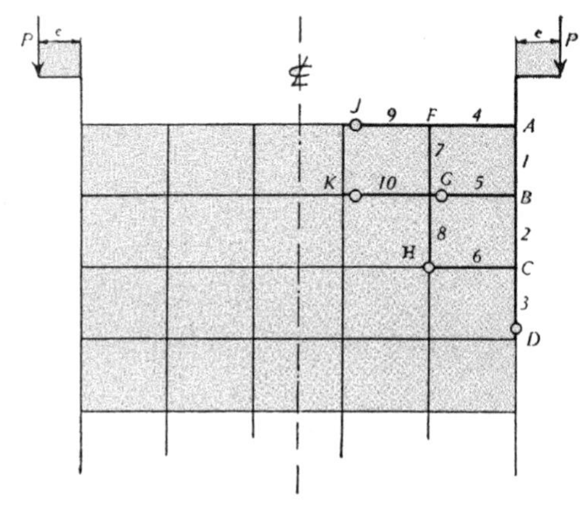

FIGURE 68

It is therefore considered that the girders are hinged at J, K, G, and H, and that the columns are hinged at H and D. Applying the same general method that was used in section 57, it can be proved that

$$M_{AB}=4EK_1\,\theta_A\left(\frac{3K_5+N_2K_2+3K_1}{3K_5+N_2K_2+4K_1}\right) \quad \text{. . . . . (556)}$$

$$M_{AF}=4EK_4\,\theta_A\left(\frac{3K_9+N_7K_7+3K_4}{3K_9+N_7K_7+4K_4}\right) \quad \text{. . . . . (557)}$$

in which

$$N_2=4\left(\frac{3K_3+3K_6+3K_2}{3K_3+3K_6+4K_2}\right)$$

$$N_7=4\left(\frac{3K_8+3K_{10}+3K_7}{3K_8+3K_{10}+4K_7}\right)$$

For A to be in equilibrium

$$M_{AB}+M_{AF}-Pe=0 \quad \text{. . . . . (558)}$$

From equations 556, 557, and 558 letting

$$N_1=4\left(\frac{3K_5+N_2K_2+3K_1}{3K_5+N_2K_2+4K_1}\right) \text{ and } N_4=4\left(\frac{3K_9+N_7K_7+3K_4}{3K_9+N_7K_7+4K_4}\right) \text{ gives}$$

$$M_{AB}=\frac{PeN_1K_1}{N_1K_1+N_4K_4} \qquad (559)$$

$$M_{AF}=\frac{PeN_4K_4}{N_1K_1+N_4K_4} \qquad (560)$$

Also

$$M_{BA}=\frac{M_{AB}}{2}\left(\frac{3K_5+N_2K_2}{3K_5+N_2K_2+3K_1}\right) \qquad (561)$$

$$M_{BG}=-\frac{M_{AB}}{2}\left(\frac{3K_5}{3K_5+N_2K_2+3K_1}\right) \qquad (562)$$

$$M_{BC}=-\frac{M_{AB}}{2}\left(\frac{N_2K_2}{3K_5+N_2K_2+3K_1}\right) \qquad (563)$$

$$M_{CB}=\frac{M_{BC}}{2}\left(\frac{K_3+K_6}{K_3+K_6+K_2}\right) \qquad (564)$$

$$M_{CH}=-\frac{M_{BC}}{2}\left(\frac{K_6}{K_3+K_6+K_2}\right) \qquad (565)$$

$$M_{CD}=-\frac{M_{BC}}{2}\left(\frac{K_3}{K_3+K_6+K_2}\right) \qquad (566)$$

$$M_{FA}=\frac{M_{AF}}{2}\left(\frac{3K_9+N_7K_7}{3K_9+N_7K_7+3K_4}\right) \qquad (567)$$

$$M_{FG}=-\frac{M_{AF}}{2}\left(\frac{N_7K_7}{3K_9+N_7K_7+3K_4}\right) \qquad (568)$$

$$M_{FJ}=-\frac{M_{AF}}{2}\left(\frac{3K_9}{3K_9+N_7K_7+3K_4}\right) \qquad (569)$$

If the $\frac{I}{l}$'s of the girders are all equal and are represented by K, if the $\frac{I}{l}$'s of the columns are all equal and are represented by $\frac{K}{n}$, and if the l's are all equal, $N_2=N_7=N=4\left(\frac{3n+6}{3n+7}\right)$, and equations 559 to 569 reduce to the form

$$M_{AB}=Pe\left[\frac{\left(\frac{3n+3+N}{3n+4+N}\right)}{\left(\frac{3n+3+N}{3n+4+N}\right)+n\left(\frac{6n+N}{7n+N}\right)}\right] \quad . \quad . \quad . \quad . \quad . \quad . \quad (570)$$

$$M_{AF}=Pe\left[\frac{n\left(\frac{6n+N}{7n+N}\right)}{\left(\frac{3n+3+N}{3n+4+N}\right)+n\left(\frac{6n+N}{7n+N}\right)}\right] \quad . \quad . \quad . \quad . \quad . \quad . \quad (571)$$

$$M_{BA}=\frac{M_{AB}}{2}\left(\frac{3n+N}{3n+3+N}\right) \quad . \quad . \quad . \quad . \quad . \quad . \quad . \quad . \quad . \quad (572)$$

$$M_{BG}=-\frac{M_{AB}}{2}\left(\frac{3n}{3n+3+N}\right) \quad . \quad . \quad . \quad . \quad . \quad . \quad . \quad . \quad . \quad (573)$$

$$M_{BC}=-\frac{M_{AB}}{2}\left(\frac{N}{3n+3+N}\right) \quad . \quad . \quad . \quad . \quad . \quad . \quad . \quad . \quad . \quad (574)$$

$$M_{CB}=\frac{M_{BC}}{2}\left(\frac{n+1}{n+2}\right) \quad . \quad . \quad . \quad . \quad . \quad . \quad . \quad . \quad . \quad . \quad (575)$$

$$M_{CH}=-\frac{M_{BC}}{2}\left(\frac{n}{n+2}\right) \quad . \quad . \quad . \quad . \quad . \quad . \quad . \quad . \quad . \quad . \quad (576)$$

$$M_{CD}=-\frac{M_{BC}}{2}\left(\frac{1}{n+2}\right) \quad . \quad . \quad . \quad . \quad . \quad . \quad . \quad . \quad . \quad . \quad (577)$$

$$M_{FA}=\frac{M_{AF}}{2}\left(\frac{3n+N}{6n+N}\right) \quad . \quad . \quad . \quad . \quad . \quad . \quad . \quad . \quad . \quad . \quad (578)$$

$$M_{FG}=-\frac{M_{AF}}{2}\left(\frac{N}{6n+N}\right) \quad . \quad . \quad . \quad . \quad . \quad . \quad . \quad . \quad . \quad . \quad (579)$$

$$M_{FJ}=-\frac{M_{AF}}{2}\left(\frac{3n}{6n+N}\right) \quad . \quad . \quad . \quad . \quad . \quad . \quad . \quad . \quad . \quad . \quad (580)$$

If all the K's are equal

$$M_{AB} = .500\ Pe \qquad (581)$$

$$M_{AF} = .500\ Pe \qquad (582)$$

$$M_{BA} = .172\ Pe \qquad (583)$$

$$M_{BG} = -.078\ Pe \qquad (584)$$

$$M_{BC} = -.094\ Pe \qquad (585)$$

$$M_{CB} = -.031\ Pe \qquad (586)$$

$$M_{CH} = .016\ Pe \qquad (587)$$

$$M_{CD} = .016\ Pe \qquad (588)$$

$$M_{FA} = .172\ Pe \qquad (589)$$

$$M_{FG} = -.094\ Pe \qquad (590)$$

$$M_{FJ} = -.078\ Pe \qquad (591)$$

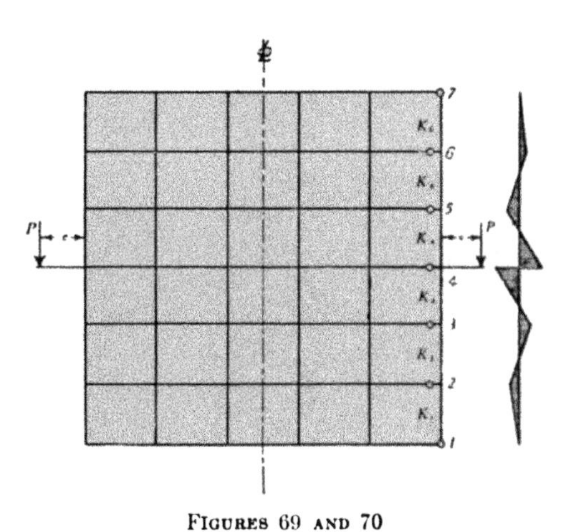

FIGURES 69 AND 70

59. *Eccentric Load at Middle Floor Level of Exterior Column of a Frame. Connections of Girders to Columns Hinged.*—Fig. 69 represents a frame with eccentric loads at the middle floor level of the exterior columns. The frame and the loading are symmetrical about the

vertical center line of the frame. The connections of the girders to the columns are hinged.

The moments are practically independent of the degree of restraint at 1 and 7. The column is therefore assumed to be hinged at 1 and 7.

$$-M_{43}-M_{45}+Pe=0 \quad \text{(592)}$$

$$M_{43}=Pe\left[\frac{K_3\left(\frac{N_2K_2+3K_3}{N_2K_2+4K_3}\right)}{K_3\left(\frac{N_2K_2+3K_3}{N_2K_2+4K_3}\right)+K_4\left(\frac{N_5+3K_4}{N_5+4K_4}\right)}\right] \quad \text{(593)}$$

$$M_{45}=Pe-M_{43} \quad \text{(594)}$$

$$M_{34}=-M_{32}=\frac{M_{43}}{2}\left[\frac{N_2K_2}{N_2K_2+3K_3}\right] \quad \text{(595)}$$

$$M_{54}=-M_{56}=\frac{M_{45}}{2}\left[\frac{N_5K_5}{N_5K_5+3K_4}\right] \quad \text{(596)}$$

$$M_{21}=-M_{23}=M_{34}\left[\frac{K_1}{2(K_1+K_2)}\right] \quad \text{(597)}$$

$$M_{67}=-M_{65}=M_{54}\left[\frac{K_6}{2(K_5+K_6)}\right] \quad \text{(598)}$$

In these equations

$$N_2=4\left(\frac{3K_1+3K_2}{3K_1+4K_2}\right)$$

$$N_5=4\left(\frac{3K_6+3K_5}{3K_6+4K_5}\right)$$

When all K's are equal

$$M_{43}=M_{45}=\frac{1}{2}Pe=0.5000\,Pe \quad \text{(599)}$$

$$M_{34}=M_{54}=\frac{2}{15}Pe=0.1333\,Pe \quad \text{(600)}$$

$$M_{21} = M_{67} = \frac{1}{30} Pe = 0.0333\, Pe \quad \text{. . . . .} \quad (601)$$

If the ends of the column are fixed at 1 and 7, with all K's equal

$$M_{43} = \frac{1}{2} Pe = 0.5000\, Pe \quad \text{. . . . .} \quad (602)$$

$$M_{34} = \frac{7}{52} Pe = 0.1346\, Pe \quad \text{. . . . .} \quad (603)$$

$$M_{21} = \frac{2}{52} Pe = 0.0385\, Pe \quad \text{. . . . .} \quad (604)$$

If there is a number of loads, the moment due to all the loads is the sum of the moments due to each load considered separately.

With a load at each floor, if all K's are equal and all values of Pe are equal, $M = \frac{1}{2} Pe$.

The moment diagram for the right-hand exterior column is represented by Fig. 70. The loading which produces a maximum stress in the column just below 4 is represented by Fig. 71. If all K's are equal

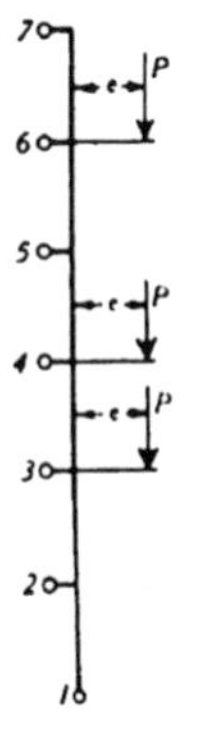

FIGURE 71

and all values of Pe are equal the maximum moment in a column that can be produced by eccentric loads is $\frac{Pe}{2} + \frac{2Pe}{15} + \frac{Pe}{30} = \frac{2Pe}{3}$

It is to be noted that the frame considered is symmetrical about a vertical center line and is symmetrically loaded. If there is a load on the right-hand column only, the moments in that column will be slightly smaller than the moments given by the equation, and the other columns will be subjected to a small moment. The error in the moment in the loaded column and the neglected moment in the other columns increase as the ratio of the stiffness of the loaded column to the combined stiffness of the other columns increases.

As in section 57, the error in the equations of this section due to the horizontal deflection of a frame under any vertical loading is undoubtedly negligible for purposes of design.

60. *Eccentric Load at Middle Floor Level of Exterior Column of a Frame. Connections of Girders to Columns Rigid.*—Fig. 72 represents

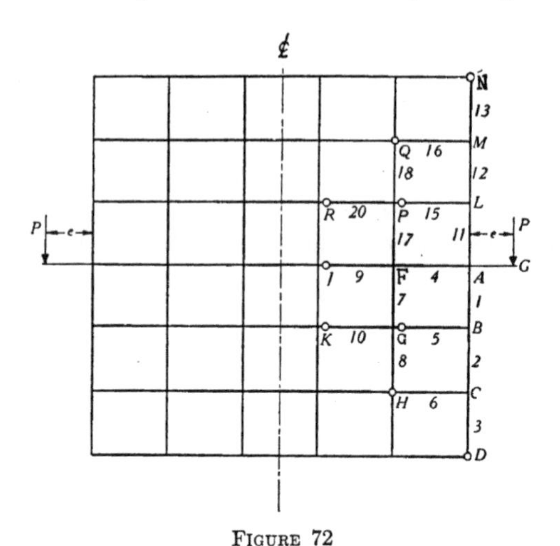

FIGURE 72

a frame with eccentric loads at the middle floor level of the exterior columns. The frame and the loading are symmetrical about the vertical center line of the frame. The connections of the girders to the columns are rigid.

From previous similar work it is known that of the moments produced by P on the right-hand side of the frame, only the moments at A, B, C, L, M, and F are large enough to be considered in the design of the frame. From previous work it is known, furthermore, that the moments at these points are practically independent of the degree of restraint at D, H, G, K, J, R, P, Q, and N. It will therefore be considered that the girders are hinged at H, G, K, J, R, P, and Q, and that the columns are hinged at D, H, Q, and N.

Applying the general method used in section 58, it can be proved

that letting $N_1=4\left[\frac{3K_5+N_2K_2+3K_1}{3K_5+N_2K_2+4K_1}\right]$

$$N_4=4\left(\frac{3K_9+N_7K_7+N_{17}K_{17}+3K_4}{3K_9+N_7K_7+N_{17}K_{17}+4K_4}\right)$$

and $N_{11}=4\left(\frac{3K_{15}+N_{12}K_{12}+3K_{11}}{3K_{15}+N_{12}K_{12}+4K_{11}}\right)$ gives

$$M_{AB}=\frac{PeN_1K_1}{N_1K_1+N_4K_4+N_{11}K_{11}} \quad \text{. . . . . (605)}$$

$$M_{AF}=\frac{PeN_4K_4}{N_1K_1+N_4K_4+N_{11}K_{11}} \quad \text{. . . . . (606)}$$

$$M_{AL}=\frac{PeN_{11}K_{11}}{N_1K_1+N_4K_4+N_{11}K_{11}} \quad \text{. . . . . (607)}$$

$$M_{LA}=\frac{M_{AL}}{2}\left(\frac{3K_{15}+N_{12}K_{12}}{3K_{15}+N_{12}K_{12}+3K_{11}}\right) \quad \text{. . . . . . . (608)}$$

$$M_{LP}=-\frac{M_{AL}}{2}\left(\frac{3K_{15}}{3K_{15}+N_{12}K_{12}+3K_{11}}\right) \quad \text{. . . . . . (609)}$$

$$M_{LM}=-\frac{M_{AL}}{2}\left(\frac{N_{12}K_{12}}{3K_{15}+N_{12}K_{12}+3K_{11}}\right) \quad \text{. . . . . . (610)}$$

$$M_{ML}=\frac{M_{LM}}{2}\left(\frac{K_{13}+K_{16}}{K_{13}+K_{16}+K_{12}}\right) \quad \text{. . . . . (611)}$$

$$M_{MQ}=-\frac{M_{LM}}{2}\left(\frac{K_{16}}{K_{13}+K_{16}+K_{12}}\right) \quad \text{. . . . . (612)}$$

$$M_{MN}=-\frac{M_{LM}}{2}\left(\frac{K_{13}}{K_{13}+K_{16}+K_{12}}\right) \quad \text{. . . . . (613)}$$

$$M_{BA}=\frac{M_{AB}}{2}\left(\frac{3K_5+N_2K_2}{3K_5+N_2K_2+3K_1}\right) \quad . \quad . \quad . \quad (614)$$

$$M_{BG}=-\frac{M_{AB}}{2}\left(\frac{3K_5}{3K_5+N_2K_2+3K_1}\right) \quad . \quad . \quad . \quad (615)$$

$$M_{BC}=-\frac{M_{AB}}{2}\left(\frac{N_2K_2}{3K_5+N_2K_2+3K_1}\right) \quad . \quad . \quad . \quad (616)$$

$$M_{CB}=\frac{M_{BC}}{2}\left(\frac{K_3+K_6}{K_3+K_6+K_2}\right) \quad . \quad . \quad . \quad (617)$$

$$M_{CH}=-\frac{M_{BC}}{2}\left(\frac{K_6}{K_3+K_6+K_2}\right) \quad . \quad . \quad . \quad (618)$$

$$M_{CD}=-\frac{M_{BC}}{2}\left(\frac{K_3}{K_3+K_6+K_2}\right) \quad . \quad . \quad . \quad (619)$$

$$M_{FA}=\frac{1}{2}M_{AF}\left(\frac{3K_9+N_7K_7+N_{17}K_{17}}{3K_9+N_7K_7+N_{17}K_{17}+3K_4}\right) \quad . \quad . \quad . \quad (620)$$

$$M_{FG}=-\frac{1}{2}M_{AF}\left(\frac{N_7K_7}{3K_9+N_7K_7+N_{17}K_{17}+3K_4}\right) \quad . \quad . \quad . \quad (621)$$

$$M_{FJ}=-\frac{1}{2}M_{AF}\left(\frac{3K_9}{3K_9+N_7K_7+N_{17}K_{17}+3K_4}\right) \quad . \quad . \quad . \quad (622)$$

$$M_{FP}=-\frac{1}{2}M_{AF}\left(\frac{N_{17}K_{17}}{3K_9+N_7K_7+N_{17}K_{17}+3K_4}\right) \quad . \quad . \quad . \quad (623)$$

in which

$$N_2=4\left(\frac{3K_3+3K_6+3K_2}{3K_3+3K_6+4K_2}\right)$$

$$N_7=4\left(\frac{3K_8+3K_{10}+3K_7}{3K_8+3K_{10}+4K_7}\right)$$

$$N_{12}=4\left(\frac{3K_{13}+3K_{16}+3K_{12}}{3K_{13}+3K_{16}+4K_{12}}\right)$$

$$N_{17}=4\left(\frac{3K_{18}+3K_{20}+3K_{17}}{3K_{18}+3K_{20}+4K_{17}}\right)$$

If the $\frac{I}{l}$'s of all girders are equal and are represented by K, and if the $\frac{I}{l}$'s of all columns are equal and are represented by $\frac{K}{n}$,

$$N_2 = N_7 = N_{12} = N_{17} = N = 4\left(\frac{3n+6}{3n+7}\right),$$

and equations 605 to 623 reduce to the form

$$M_{AB} = M_{AL} = \frac{Pe}{2}\left[\frac{\left(\frac{3n+3+N}{3n+4+N}\right)}{\left(\frac{3n+3+N}{3n+4+N}\right) + n\left(\frac{3n+N}{7n+2N}\right)}\right] \quad . \quad . \quad (624)$$

$$M_{AF} = \frac{nPe}{2}\left[\frac{\left(\frac{6n+2N}{7n+2N}\right)}{\left(\frac{3n+3+N}{3n+4+N}\right) + n\left(\frac{3n+N}{7n+2N}\right)}\right] \quad . \quad . \quad . \quad (625)$$

$$M_{BA} = M_{LA} = \frac{M_{AB}}{2}\left(\frac{3n+N}{3n+3+N}\right) \quad . \quad . \quad . \quad . \quad (626)$$

$$M_{BG} = M_{LP} = -\frac{M_{AB}}{2}\left(\frac{3n}{3n+3+N}\right) \quad . \quad . \quad . \quad . \quad (627)$$

$$M_{BC} = M_{LM} = -\frac{M_{AB}}{2}\left(\frac{N}{3n+3+N}\right) \quad . \quad . \quad . \quad . \quad (628)$$

$$M_{CB} = M_{ML} = \frac{M_{BC}}{2}\left(\frac{n+1}{n+2}\right) \quad . \quad . \quad . \quad . \quad (629)$$

$$M_{CH} = M_{MQ} = -\frac{M_{BC}}{2}\left(\frac{n}{n+2}\right) \quad . \quad . \quad . \quad . \quad (630)$$

$$M_{CD} = M_{MN} = -\frac{M_{BC}}{2}\left(\frac{1}{n+2}\right) \quad . \quad . \quad . \quad . \quad (631)$$

$$M_{FA} = \frac{M_{AF}}{2}\left(\frac{3n+2N}{6n+2N}\right) \quad . \quad . \quad . \quad . \quad (632)$$

$$M_{FJ} = -\frac{M_{AF}}{2}\left(\frac{3n}{6n+2N}\right) \quad \text{. . . . .} \quad (633)$$

$$M_{FG} = M_{FP} = -\frac{M_{AF}}{2}\,\frac{N}{6n+2N} \quad \text{. . . . .} \quad (634)$$

If all the K's are equal, equations 605 to 623 reduce to the form

$$M_{AB} = M_{AL} = 0.331\ Pe \quad \text{. . . . .} \quad (635)$$

$$M_{BA} = M_{LA} = 0.113\ Pe \quad \text{. . . . .} \quad (636)$$

$$M_{BG} = M_{LP} = -0.052\ Pe \quad \text{. . . . .} \quad (637)$$

$$M_{BC} = M_{LM} = -0.062\ Pe \quad \text{. . . . .} \quad (638)$$

$$M_{CB} = M_{ML} = -0.021\ Pe \quad \text{. . . . .} \quad (639)$$

$$M_{CH} = M_{MQ} = 0.010\ Pe \quad \text{. . . . .} \quad (640)$$

$$M_{CD} = M_{MN} = 0.010\ Pe \quad \text{. . . . .} \quad (641)$$

$$M_{AF} = 0.338\ Pe \quad \text{. . . . .} \quad (642)$$

$$M_{FA} = 0.131\ Pe \quad \text{. . . . .} \quad (643)$$

$$M_{FJ} = -0.038\ Pe \quad \text{. . . . .} \quad (644)$$

$$M_{FG} = M_{FP} = -0.046\ Pe \quad \text{. . . . .} \quad (645)$$

If there are loads at each of the points M, L, A, B, and C, the moment at any point, as A, can be determined by taking the sum of the moments due to each load separately. If the values of Pe are equal for all the points, if the $\frac{I}{l}$'s of all girders are equal and are represented by K, if the $\frac{I}{l}$'s of all columns are equal and are represented by $\frac{K}{n}$, and all values of $N = 4\left(\frac{3n+6}{3n+7}\right)$, the moments are

$$M_{AB} = M_{AL} = \frac{Pe}{2}\,\frac{\left[\dfrac{4.5n+3+N\left(\dfrac{3n+8}{4n+8}\right)}{3n+4+N}\right]}{\left(\dfrac{3n+3+N}{3n+4+N}\right)+n\left(\dfrac{3n+N}{7n+2N}\right)} \quad \text{. . . . .} \quad (646)$$

$$M_{AF}=\frac{\frac{nPe}{2}\left(\frac{6n+2N}{7n+2N}\right)+\left[\frac{\frac{N}{2n+4}-3}{3n+4+N}\right]}{\left(\frac{3n+3+N}{3n+4+N}\right)+n\left(\frac{3n+N}{7n+2N}\right)} \quad . \quad . \quad . \quad . \quad (647)$$

If the K's of the columns equal the K's of the girders

$$M_{AB}=M_{AL}=1\tfrac{1}{8}\quad(.331\ Pe)=0.372\ Pe \quad . \quad . \quad . \quad . \quad . \quad . \quad . \quad (648)$$

$$M_{AF}=0.254\ Pe \quad . \quad . \quad . \quad . \quad . \quad . \quad . \quad . \quad . \quad . \quad . \quad . \quad . \quad . \quad (649)$$

The moments M_{AL} and M_{AB} are a maximum when the frame is loaded as shown in Fig. 73. If the $\frac{I}{l}$'s of the girders all equal K, and if the $\frac{I}{l}$'s of the columns all equal $\frac{K}{n}$, and if all values of Pe are equal, the maximum moments are

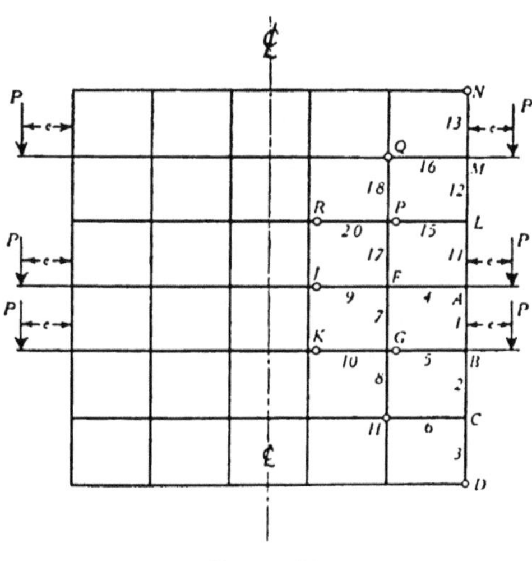

FIGURE 73

$$M_{AL}=M_{AB}=\frac{Pe}{2}\ \frac{\left[\dfrac{4.5n+3+N\left(\dfrac{6n+13}{4n+8}\right)}{3n+4+N}\right]}{\left(\dfrac{3n+3+N}{3n+4+N}\right)+n\left(\dfrac{3n+N}{7n+2N}\right)} \quad . \quad . \quad . \quad . \quad (650)$$

M_{AF} is maximum when alternate floors are loaded. For such oading

$$M_{AF}=n\ \frac{Pe}{2}\ \frac{\left(\dfrac{6n+2N}{7n+2N}\right)+\left[\dfrac{\left(\dfrac{N}{2n+4}\right)}{3n+4+N}\right]}{\left(\dfrac{3n+3+N}{3n+4+N}\right)+n\left(\dfrac{3n+N}{7n+2N}\right)} \quad . \quad . \quad . \quad . \quad (651)$$

If the K's are all equal and the values of Pe are all equal, the maximum moments are

$$M_{AB}=M_{AL}=0.454\ Pe \quad . \quad . \quad . \quad . \quad . \quad . \quad . \quad . \quad . \quad . \quad . \quad . \quad . \quad (652)$$

$$M_{AF}=0.358\ Pe \quad . \quad . \quad . \quad . \quad . \quad . \quad . \quad . \quad . \quad . \quad . \quad . \quad . \quad . \quad (653)$$

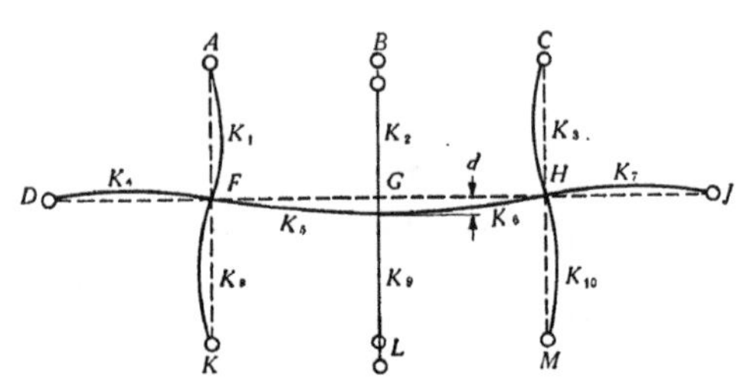

FIGURE 74

61. *Effect of Settlement of One Column of a Frame Composed of a Number of Rectangles.*—Fig. 74 represents a portion of a frame composed of a number of rectangles. The unstrained outline of the frame is represented by broken lines. The middle column BGL settles an amount represented by d. The strained outline of the frame is represented by the full lines. The points D, A, C, J, M, K, F, and H remain stationary.

For F to be in equilibrium

$$M_{FK}+M_{FD}+M_{FA}+M_{FG}=0 \quad . \quad . \quad . \quad . \quad . \quad . \quad . \quad . \quad . \quad . \quad (654)$$

Substituting the values of the M's as given by the equations of Table 1 gives

$$3E\theta_F\left(K_8+K_4+K_1\right)+2EK_5\left(2\theta_F+\theta_G-3\frac{d}{l_5}\right)=0 \quad (655)$$

or

$$\theta_F=\frac{6K_5\frac{d}{l_5}-2K_5\,\theta_G}{3K_4+3K_1+3K_8+4K_5} \quad (656)$$

$$M_{GF}=2EK_5(2\theta_G+\theta_F-3\frac{d}{l_5}) \quad (657)$$

Substituting the value of θ_F from equation 656 in equation 657 gives

$$M_{GF}=EK_5\left[4\theta_G\left(\frac{3K_4+3K_1+3K_8+3K_5}{3K_4+3K_1+3K_8+4K_5}\right)-\frac{6d}{l_5}\left(\frac{3K_4+3K_1+3K_8+2K_5}{3K_4+3K_1+3K_8+4K_5}\right)\right] \quad (658)$$

Similarly for the right-hand portion of the structure

$$M_{GH}=EK_6\left[4\theta_G\left(\frac{3K_7+3K_3+3K_{10}+3K_6}{3K_7+3K_3+3K_{10}+4K_6}\right)+\frac{6d}{l_6}\left(\frac{3K_7+3K_3+3K_{10}+2K_6}{3K_7+3K_3+3K_{10}+4K_6}\right)\right] \quad (659)$$

$$M_{GB}=3EK_2\theta_G \quad (660)$$

$$M_{GL}=3EK_9\theta_G \quad (661)$$

For C to be in equilibrium

$$M_{GF}+M_{GH}+M_{GB}+M_{GL}=0 \quad (662)$$

Substituting the value of the M's in equation 662 gives

$$\theta_G = \frac{6d\left[\frac{K_5}{l_5}\left(\frac{3K_4+3K_1+3K_8+2K_5}{3K_4+3K_1+3K_8+4K_5}\right)-\frac{K_6}{l_6}\left(\frac{3K_7+3K_3+3K_{10}+2K_6}{3K_7+3K_3+3K_{10}+4K_6}\right)\right]}{4K_5\left[\frac{3K_4+3K_1+3K_8+3K_5}{3K_4+3K_1+3K_8+4K_5}\right]+4K_6\left[\frac{3K_7+3K_3+3K_{10}+3K_6}{3K_7+3K_3+3K_{10}+4K_6}\right]+3K_2+3K_9} \quad (663)$$

From equation 656

$$\theta_F = \frac{6d\frac{K_5}{l_5}-2K_5\theta_G}{3K_4+3K_1+3K_8+4K_5} \quad (664)$$

Also

$$\theta_H = -\frac{6d\frac{K_6}{l_6}+2K_6\theta_G}{3K_7+3K_3+3K_{10}+4K_6} \quad (665)$$

If the portion of the frame DJ is symmetrical about G, $\theta_G=0$, and $\theta_F=-\theta_H$.

With θ_G, θ_F, and θ_H known the moments can be determined from the equations:

$$M_{FA}=3EK_1\theta_F \quad (666)$$

$$M_{FD}=3EK_4\theta_F \quad (667)$$

$$M_{FK}=3EK_8\theta_F \quad (668)$$

$$M_{FG}=2EK_5\left(2\theta_F+\theta_G-3\frac{d}{l_5}\right) \quad (669)$$

$$M_{GF}=2EK_5\left(2\theta_G+\theta_F-3\frac{d}{l_5}\right) \quad (670)$$

$$M_{GB}=3EK_2\theta_G \quad (671)$$

$$M_{GL}=3EK_9\theta_G \quad (672)$$

$$M_{GH}=2EK_6\left(2\theta_G+\theta_H+3\frac{d}{l_6}\right) \quad (673)$$

$$M_{HG}=2EK_6\left(2\theta_H+\theta_G+3\frac{d}{l_6}\right) \quad (674)$$

$$M_{HC}=3EK_3\theta_H \quad (675)$$

$$M_{HJ}=3EK_7\theta_H \quad (676)$$

$$M_{HM}=3EK_{10}\theta_H \quad (677)$$

If the members, instead of being hinged, are fixed at D, A, B, C, J, M, L, and K

$$\theta_G=\frac{6d\left[\frac{K_5}{l_5}\left(\frac{4K_4+4K_1+4K_8+2K_5}{4K_4+4K_1+4K_8+4K_5}\right)-\frac{K_6}{l_6}\left(\frac{4K_7+4K_3+4K_{10}+2K_6}{4K_7+4K_3+4K_{10}+4K_6}\right)\right]}{4K_5\left(\frac{4K_4+4K_1+4K_8+3K_5}{4K_4+4K_1+4K_8+4K_1}\right)+4K_6\left(\frac{4K_7+4K_3+4K_{10}+3K_6}{4K_7+4K_3+4K_{10}+4K_6}\right)+4K_2+4K_9} \quad (678)$$

$$\theta_F=\frac{6d\frac{K_5}{l_5}-2K_5\,\theta_G}{4K_4+4K_1+4K_8+4K_5} \quad (679)$$

$$\theta_H=-\frac{6d\frac{K_6}{l_6}+2K_6\,\theta_G}{4K_7+4K_3+4K_{10}+4K_6} \quad (680)$$

$$M_{FA}=4EK_1\theta_F \quad (681)$$

$$M_{FD}=4EK_4\theta_F \quad (682)$$

$$M_{FK}=4EK_8\theta_F \quad (683)$$

$$M_{FG}=2EK_5\left(2\theta_F+\theta_G-3\frac{d}{l_5}\right) \quad (684)$$

$$M_{GF}=2EK_5\left(2\theta_G+\theta_F-3\frac{d}{l_5}\right) \quad (685)$$

$$M_{GB}=4EK_2\theta_G \quad (686)$$

$$M_{GL} = 4EK_9\theta_G \quad (687)$$

$$M_{GH} = 2EK_6\left(2\,\theta_G + \theta_H + \frac{3d}{l_6}\right) \quad (688)$$

$$M_{HG} = 2EK_6\left(2\,\theta_H + \theta_G + \frac{3d}{l_6}\right) \quad (689)$$

$$M_{HC} = 4EK_3\theta_H \quad (690)$$

$$M_{HJ} = 4EK_7\theta_H \quad (691)$$

$$M_{HM} = 4EK_{10}\theta_H \quad (692)$$

If the portion of the frame DJ is symmetrical about G, $\theta_G = 0$, and $\theta_F = -\theta_H$.

The moments due to the settlement of one column for frames having the K's of all members equal and the lengths of all girders equal, and also for frames having the K's of all columns equal, the K's of all girders equal and the lengths of all girders equal are given in Table 17.

It is to be noted that the work in this section is based upon the assumption that there is no horizontal motion in the frame.

TABLE 17

MOMENTS IN A FRAME COMPOSED OF A NUMBER OF RECTANGLES, DUE TO SETTLEMENT OF ONE COLUMN

Moment	K's of All Members Equal Lengths of All Girders Equal		$\frac{I}{h}$ for All Columns Equals K; $\frac{I}{l}$ for All Girders Equals nK; Lengths of All Girders Equal	
	Ends Hinged	Ends Fixed	Ends Hinged	Ends Fixed
M_{GB} and M_{GL}	0	0	0	0
M_{FA}, M_{FK}, $-M_{HC}$, and $-M_{HM}$	$\frac{18}{13}\frac{KEd}{l}$	$\frac{3}{2}\frac{KEd}{l}$	$\frac{18n}{6+7n}\frac{KEd}{l}$	$\frac{3n}{1+n}\frac{KEd}{l}$
M_{FD} and $-M_{HJ}$	$\frac{18}{13}\frac{KEd}{l}$	$\frac{3}{2}\frac{KEd}{l}$	$\frac{18n^2}{6+7n}\frac{KEd}{l}$	$\frac{3n^2}{1+n}\frac{KEd}{l}$
M_{FG} and $-M_{HG}$	$-\frac{54}{13}\frac{KEd}{l}$	$-\frac{9}{2}\frac{KEd}{l}$	$\frac{-18n(2+n)}{6+7n}\frac{KEd}{l}$	$-3n\left(\frac{2+n}{1+n}\right)\frac{KEd}{l}$
M_{GF} and $-M_{GH}$	$-\frac{66}{13}\frac{KEd}{l}$	$-\frac{21}{4}\frac{KEd}{l}$	$\frac{-6n(6+5n)}{6+7n}\frac{KEd}{l}$	$\frac{-3n}{2}\left(\frac{4+3n}{1+n}\right)\frac{KEd}{l}$

XI. Three-legged Bent

62. *Three-legged Bent. Lengths of All Legs Different. Vertical Load on Left-hand Span—Legs Hinged at the Bases.*—Fig. 75 represents a three-legged bent. The lengths of all legs are different P represents the resultant of any system of loads on AB. The legs of the bent are hinged at D, C, and E.

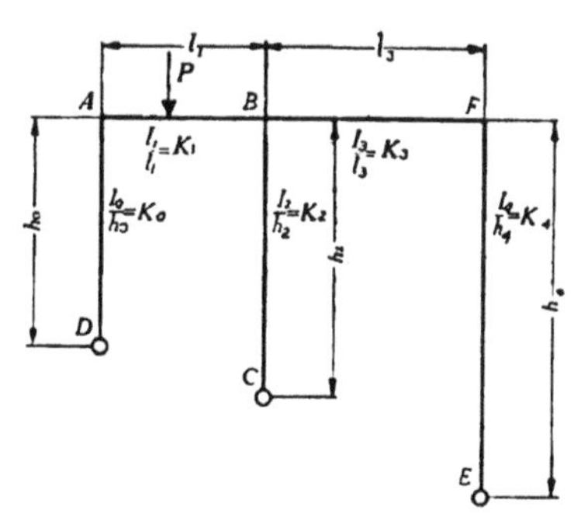

Figure 75

Since axial strains are neglected the vertical deflections of B relative to A and F are zero. Likewise the horizontal deflections of A B, and F are equal. This deflection is represented by d.

Applying the equations of Table 1 gives

$$M_{AD}=3EK_o\left(\theta_A-\frac{d}{h_o}\right) \qquad (693)$$

$$M_{AB}=2EK_1(2\theta_A+\theta_B)-C_{AB} \qquad (694)$$

$$M_{BA}=2EK_1(2\theta_B+\theta_A)+C_{BA} \qquad (695)$$

$$M_{BC}=3EK_2\left(\theta_B-\frac{d}{h_2}\right) \qquad (696)$$

$$M_{BF}=2EK_3(2\theta_B+\theta_F) \qquad (697)$$

$$M_{FB}=2EK_3(2\theta_F+\theta_B) \qquad (698)$$

$$M_{FE}=3EK_4\left(\theta_F-\frac{d}{h_4}\right) \qquad (699)$$

For equilibrium at A, B, and F

$$M_{AD}+M_{AB}=0 \quad (700)$$

$$M_{BA}+M_{BC}+M_{BF}=0 \quad (701)$$

$$M_{FB}+M_{FE}=0 \quad (702)$$

For the bent as a whole to be in equilibrium

$$\frac{M_{AD}}{h_o}+\frac{M_{BC}}{h_2}+\frac{M_{FE}}{h_4}=0 \quad (703)$$

Substituting the values of the moments from equations 693 to 699 inclusive in equations 700, 701, 702, and 703 gives

$$3EK_o\theta_A-3EK_o\frac{d}{h_o}+4EK_1\theta_A+2EK_1\theta_B=C_{AB} \quad (704)$$

$$4EK_1\theta_B+2EK_1\theta_A+3EK_2\theta_B-3EK_2\frac{d}{h_2}+4EK_3\theta_B+2EK_3\theta_F$$
$$=-C_{BA} \quad (705)$$

$$4EK_3\theta_F+2EK_3\theta_B+3EK_4\theta_F-3EK_4\frac{d}{h_4}=0 \quad (706)$$

$$\frac{3EK_o}{h_o}\theta_A-3EK_o\frac{d}{h_o^2}+\frac{3EK_2}{h_2}\theta_B-3EK_2\frac{d}{h_2^2}+\frac{3EK_4}{h_4}\theta_F-3EK_4\frac{d}{h_4^2}$$
$$=0 \quad (707)$$

These four equations contain only four unknowns, and it is therefore possible to combine the equations and solve for the unknowns. The resulting expressions, however, are so complex that it is more practicable to substitute numerical values for E, the K's and the h's and then solve for the unknowns, d and the θ's, by a process of elimination. Knowing d and the θ's, the moments can be determined from equations 693 to 699 inclusive.

For convenience in eliminating the unknowns, equations 704, 705, 706, and 707 are reproduced in Table 18. In this table the unknowns are at the tops of the columns and the coefficients of the unknowns are in the lines below. Equations A to D of Table 18 are identical with equations 704 to 707.

TABLE 18

EQUATIONS FOR THREE-LEGGED BENT

Lengths of All Legs Different. Vertical Load on Left-hand Span. Legs Hinged at the Bases.

No. of Equation	Left-hand Member of Equation				Right-hand Member of Equation
	θ_A	θ_B	θ_F	d	
A	$3K_o+4K_1$	$2K_1$	0	$-\frac{3K_o}{h_o}$	$+\frac{C_{AB}}{E}$
B	$2K_1$	$4K_1+3K_2+4K_3$	$2K_3$	$-\frac{3K_2}{h_2}$	$-\frac{C_{BA}}{E}$
C	0	$2K_3$	$4K_3+3K_4$	$-\frac{3K_4}{h_4}$	0
D	$-\frac{3K_o}{h_o}$	$-\frac{3K_2}{h_2}$	$-\frac{3K_4}{h_4}$	$+3\left[\frac{K_o}{h_o^2}+\frac{K_2}{h_2^2}+\frac{K_4}{h_4^2}\right]$	0

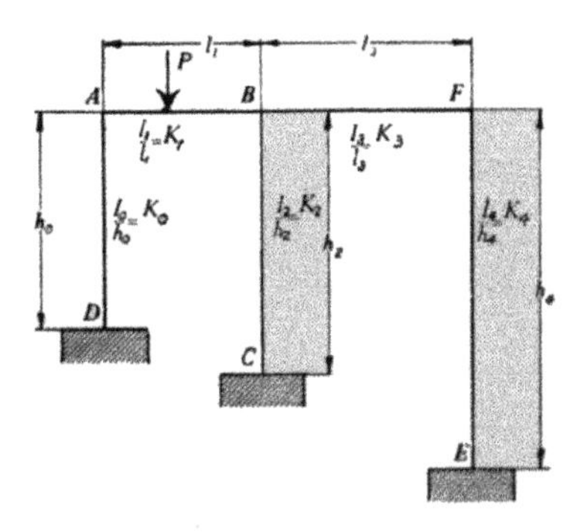

FIGURE 76

63. *Three-legged Bent. Lengths of All Legs Different. Vertical Load on Left-hand Span—Legs Fixed at the Bases.*—Fig. 76 represents a three-legged bent. The lengths of all legs are different. P represents the resultant of any system of loads on AB. The legs of the bent are fixed at D, C, and E.

From the equations of Table 1

$$M_{AD}=2EK_o\left(2\theta_A-3\frac{d}{h_o}\right) \quad (708)$$

$$M_{AB}=2EK_1(2\theta_A+\theta_B)-C_{AB} \quad (709)$$

$$M_{BA}=2EK_1(2\theta_B+\theta_A)+C_{BA} \quad (710)$$

$$M_{BC}=2EK_2\left(2\theta_B-3\frac{d}{h_2}\right) \quad (711)$$

$$M_{BF}=2EK_3(2\theta_B+\theta_F) \quad (712)$$

$$M_{FB}=2EK_3(2\theta_F+\theta_B) \quad (713)$$

$$M_{FE}=2EK_4\left(2\theta_F-3\frac{d}{h_4}\right) \quad (714)$$

$$M_{DA}=2EK_o\left(\theta_A-3\frac{d}{h_o}\right) \quad (715)$$

$$M_{CB}=2EK_2\left(\theta_B-3\frac{d}{h_2}\right) \quad (716)$$

$$M_{EF}=2EK_4\left(\theta_F-3\frac{d}{h_4}\right) \quad (717)$$

For equilibrium at A, B, and F, equations 700, 701, and 702 are applicable.

For the bent as a whole to be in equilibrium

$$\frac{M_{AD}+M_{DA}}{h_o}+\frac{M_{BC}+M_{CB}}{h_2}+\frac{M_{FE}+M_{EF}}{h_4}=0 \quad (718)$$

Substituting the values of the moments from equations 708 to 717 in equations 700, 701, 702, and 718 gives the equations of Table 19.

64. *Three-legged Bent. Lengths of All Legs Different. Any System of Loads.*—No matter what system of loads is applied to the three-legged bents of Figs. 75 and 76, equations similar to equations 693 to 699, 708 to 718 and 700 to 703 can be written. These equations will

contain no unknown quantities not found in the equations of section 62, and therefore they can be combined to obtain equations similar to the equations of Tables 18 and 19. The left-hand members of the equations for all bents with legs hinged at the bases will be identical with the left-hand members of the equations of Table 18, and the left-hand members of the equations for all bents with legs fixed at the bases will be identical with the left-hand members of the equations of Table 19.

The equations of Table 1 as applied to a three-legged bent having legs hinged at the bases and carrying a number of systems of loads are given in Table 20. Similar equations for a three-legged bent having legs fixed at the bases are given in Table 21. Four equations containing four unknowns, derived from the equations of Table 20 are given in Table 22, and four equations containing four unknowns based upon the equations of Table 21 are given in Table 23. The equations of Table 22 can be used to determine the stresses in a bent having legs hinged at the bases, and the equations of Table 23 can be used to determine the stresses in a bent having legs fixed at the bases. A numerical problem illustrating the use of the equation in Table 23 is presented in section 76.

TABLE 19

EQUATIONS FOR THREE-LEGGED BENT

Lengths of All Legs Different. Vertical Load on Left-hand Span. Legs Fixed at the Bases.

No. of Equation	Left-hand Member of Equation				Right-hand Member of Equation
	θ_A	θ_B	θ_F	d	
A	$4K_o+4K_1$	$2K_1$	0	$-6\frac{K_o}{h_o}$	$+\frac{C_{AB}}{E}$
B	$2K_1$	$4K_1+4K_2+4K_3$	$2K_3$	$-6\frac{K_2}{h_2}$	$-\frac{C_{BA}}{E}$
C	0	$2K_3$	$4K_3+4K_4$	$-6\frac{K_4}{h_4}$	0
D	$-6\frac{K_o}{h_o}$	$-6\frac{K_2}{h_2}$	$-6\frac{K_4}{h_4}$	$+12\left[\frac{K_o}{h_o^2}+\frac{K_2}{h_2^2}+\frac{K_4}{h_4^2}\right]$	0

The 's and d can be determined from the equations of Table 19 by a process of elimination. Knowing the 's and d, the moments can be determined from equations 708 to 717.

TABLE 20

FUNDAMENTAL EQUATIONS FOR THREE-LEGGED BENTS

Lengths of All Legs Different. Any System of Loads. Legs Hinged at the Bases.

No. of Equation	This part of the equation is independent of the loads	Additional term of the right-hand member of the equation, different for each system of loads, as follows:					
		Case I Vertical Load on Left-hand Span See Fig. 77	Case II Vertical Load on Right-hand Span See Fig. 78	Case III Horizontal Load to the Right on Left-hand Leg See Fig. 79	Case IV Horizontal Load to the Left on Right-hand Leg See Fig. 80	Case V Settlement and Sliding of Foundations See Fig. 81.	Case VI External Couple at Upper Left-hand Corner See Fig. 82
1	$M_{AD}=3EK_o\left(\theta_A-\frac{d}{h_o}\right)$	0	0	$+H_{AD}$	0	0	0
2	$M_{AB}=2EK_1(2\theta_A+\theta_B)$	$-C_{AB}$	0	0	0	$-6EK_1\frac{d_1}{l_1}$	0
3	$M_{BA}=2EK_1(2\theta_B+\theta_A)$	$+C_{BA}$	0	0	0	$-6EK_1\frac{d_1}{l_1}$	0
4	$M_{BC}=3EK_2\left(\theta_B-\frac{d}{h_2}\right)$	0	0	0	0	$+3EK_2\frac{d_2}{h_2}$	0

TABLE 20—CONTINUED

5	$M_{BF}=2EK_3(2\theta_B+\theta_F)$	0	$-C_{BF}$	0	0	$-6EK_3\frac{(d_3-d_1)}{l_3}$	0
6	$M_{FB}=2EK_3(2\theta_F+\theta_B)$	0	$+C_{FB}$	0	0	$-6EK_3\frac{(d_3-d_1)}{l_3}$	0
7	$M_{FE}=3EK_4\left(\theta_F-\frac{d}{h_4}\right)$	0	0	0	$-H_{FE}$	$+3EK_4\frac{d_4}{h_4}$	0
8	$0=M_{AD}+M_{AB}$	0	0	0	0	0	$-M$
9	$0=M_{BA}+M_{BF}+M_{BC}$	0	0	0	0	0	0
10	$0=M_{FB}+M_{FE}$	0	0	0	0	0	0
11	$0=\frac{M_{AD}}{h_o}+\frac{M_{BC}}{h_2}+\frac{M_{FE}}{h_4}$	0	0	$+\frac{M_D}{h_o}$	$+\frac{M_E}{h_4}$	0	0

M_D is the statical moment about point D of the horizontal loads on left-hand leg; M_E is the statical moment about point E of the horizontal loads on right-hand leg. The statical moment is positive when the indicated rotation is clockwise.

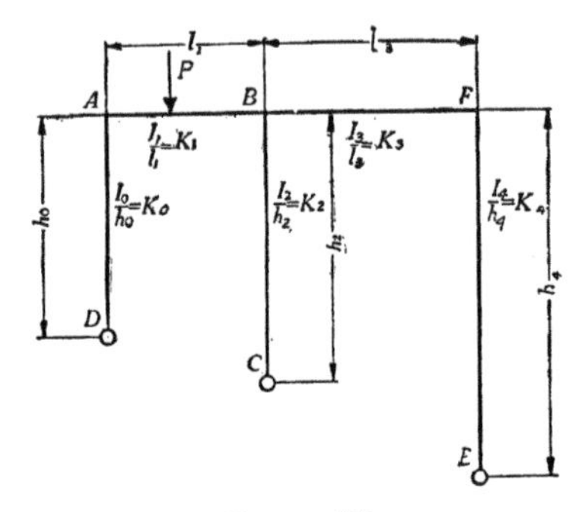

FIGURE 77

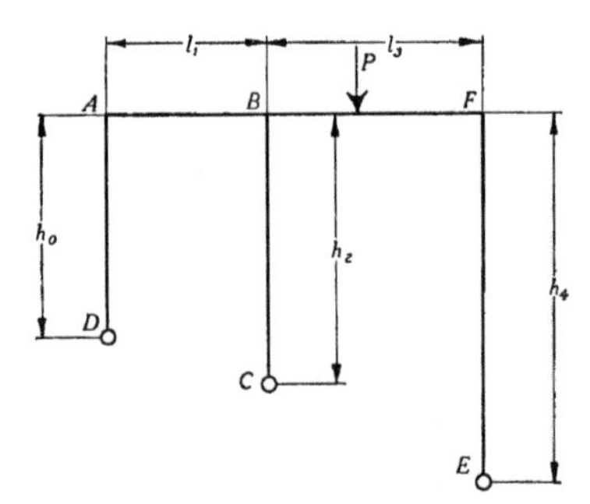

FIGURE 78

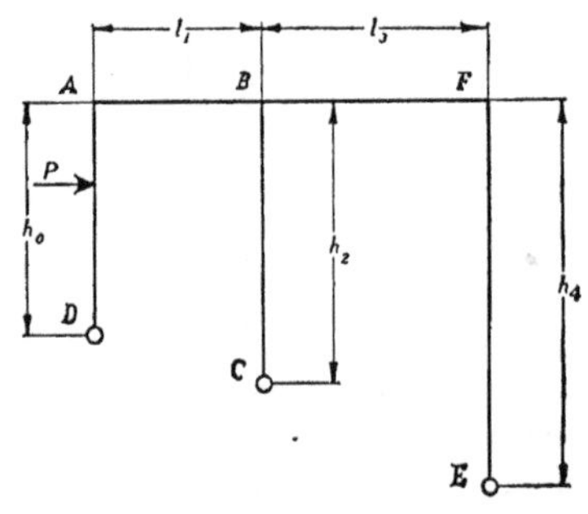

FIGURE 79

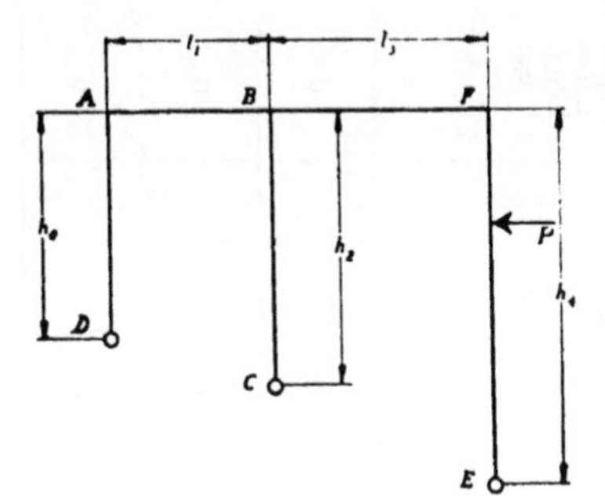

FIGURE 80

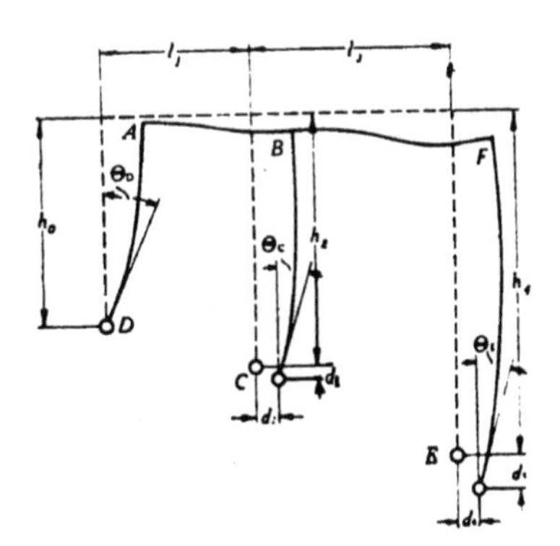

FIGURE 81

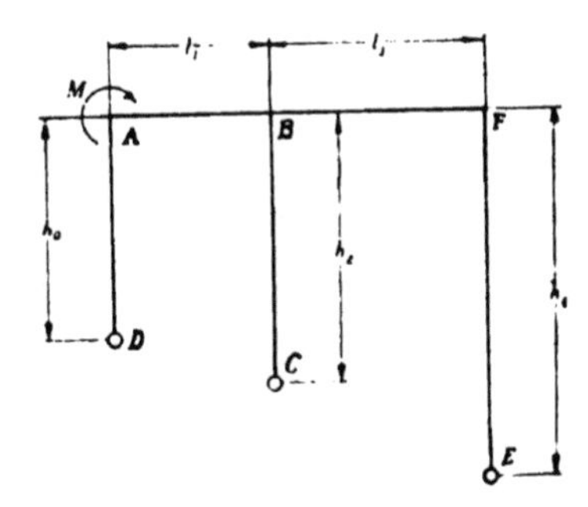

FIGURE 82

TABLE 21

FUNDAMENTAL EQUATIONS FOR THREE-LEGGED BENTS

Lengths of All Legs Different. Any System of Loads. Legs Fixed at the Bases.

No. of Equation	This part of the equation is independent of the loads	Additional term of the right-hand member of the equation, different for each system of loads, as follows:					
		Case VII Vertical Load on Left-hand Span See Fig. 83	Case VIII Vertical Load on Right-hand Span See Fig. 84	Case IX Horizontal Load to the Right on Left-hand Leg See Fig. 85	Case X Horizontal Load to the Left on Right-hand Leg See Fig. 86	Case XI Settlement and Rotation of Foundations See Fig. 87	Case XII External Couple at Upper Left-hand Corner See Fig. 88
1	$M_{AD}=2EK_o\left(2\theta_A-3\frac{d}{h_o}\right)$	0	0	$+C_{AD}$	0	$+2EK_o\theta_D$	0
2	$M_{AB}=2EK_1(2\theta_A+\theta_B)$	$-C_{AB}$	0	0	0	$-6EK_1\frac{d_1}{l_1}$	0
3	$M_{BA}=2EK_1(2\theta_B+\theta_A)$	$+C_{BA}$	0	0	0	$-6EK_1\frac{d_1}{l_1}$	0
4	$M_{BC}=2EK_2\left(2\theta_B-3\frac{d}{h_2}\right)$	0	0	0	0	$+6EK_2\frac{d_2}{h_2}+2EK_2\theta_C$	0

TABLE 21—CONTINUED

5	$M_{BF}=2EK_3(2\theta_B+\theta_F)$	0	$-C_{BF}$	0	0	$-6EK_3\frac{d_3-d_1}{l_3}$	0
6	$M_{FB}=2EK_3(2\theta_F+\theta_B)$	0	$+C_{FB}$	0	0	$-6EK_3\frac{d_3-d_1}{l_3}$	0
7	$M_{FE}=2EK_4\left(2\theta_F-3\frac{d}{h_4}\right)$	0	0	0	$-C_{FE}$	$+6EK_4\frac{d_4}{h_4}+2EK_4\theta_E$	0
8	$M_{DA}=2EK_o\left(\theta_A-3\frac{d}{h_o}\right)$	0	0	$-C_{DA}$	0	$+4EK_o\theta_D$	0
9	$M_{CB}=2EK_2\left(\theta_B-3\frac{d}{h_2}\right)$	0	0	0	0	$+6EK_2\frac{d_2}{h_2}+4EK_2\theta_C$	0
10	$M_{EF}=2EK_4\left(\theta_F-3\frac{d}{h_4}\right)$	0	0	0	$+C_{EF}$	$+6EK_4\frac{d_4}{h_4}+4EK_4\theta_E$	0
11	$0=M_{AD}+M_{AB}$	0	0	0	0	0	$-M$
12	$0=M_{BA}+M_{BF}+M_{BC}$	0	0	0	0	0	0
13	$0=M_{FB}+M_{FE}$	0	0	0	0	0	0
14	$0=\frac{M_{AD}+M_{DA}}{h_o}+\frac{M_{BC}+M_{CB}}{h_2}+\frac{M_{FE}+M_{EF}}{h_4}$	0	0	$+\frac{M_D}{h_o}$	$+\frac{M_E}{h_4}$	0	0

For definition of M_D and M_E see Table 20.

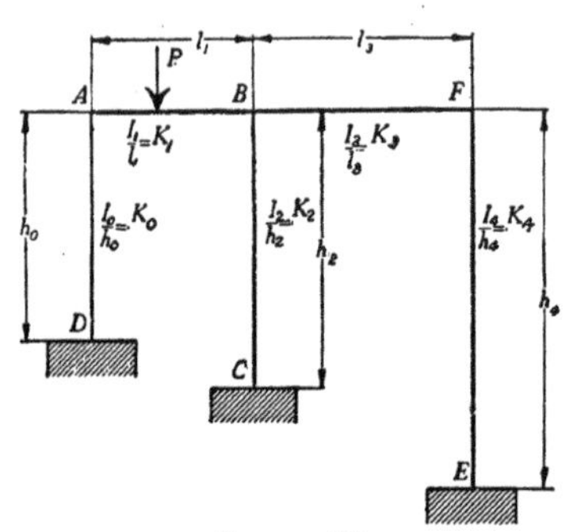

FIGURE 83

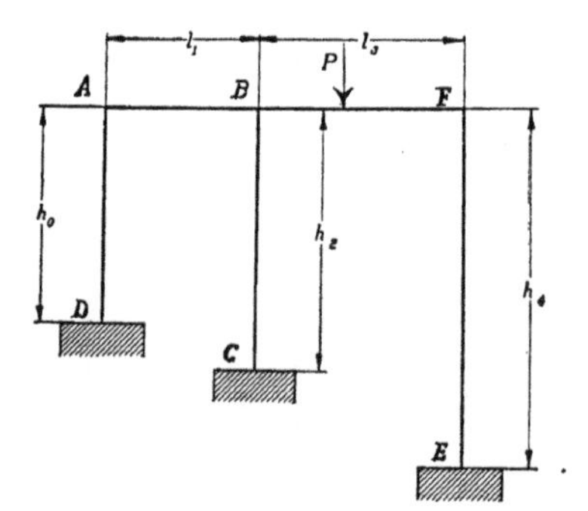

FIGURE 84

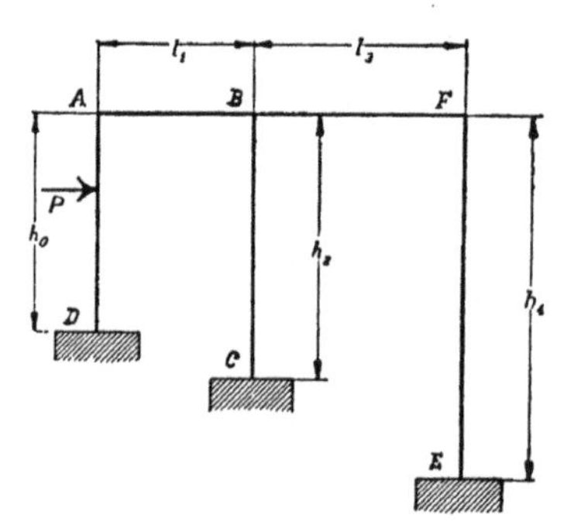

FIGURE 85

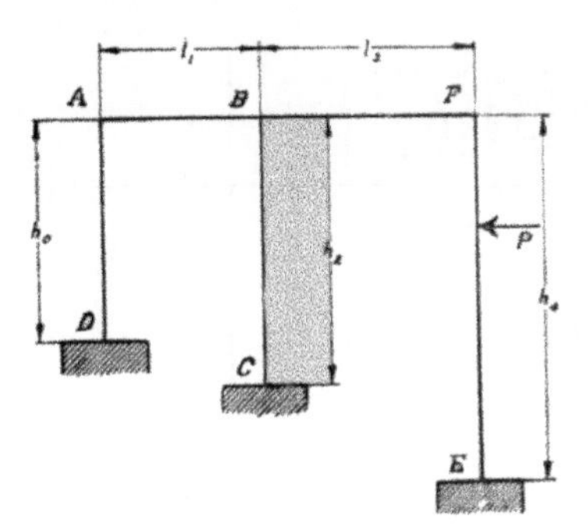

FIGURE 86

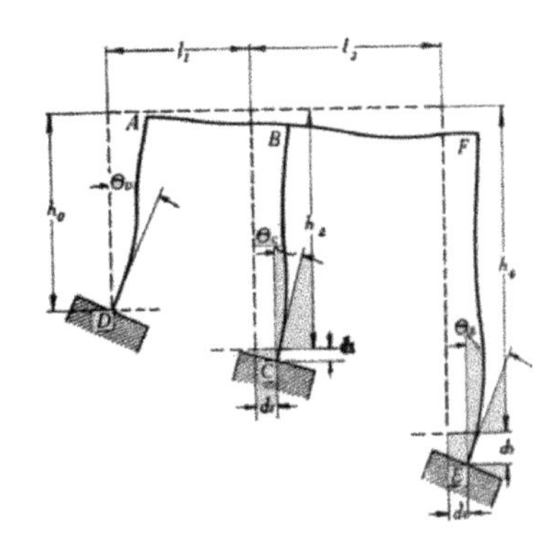

FIGURE 87

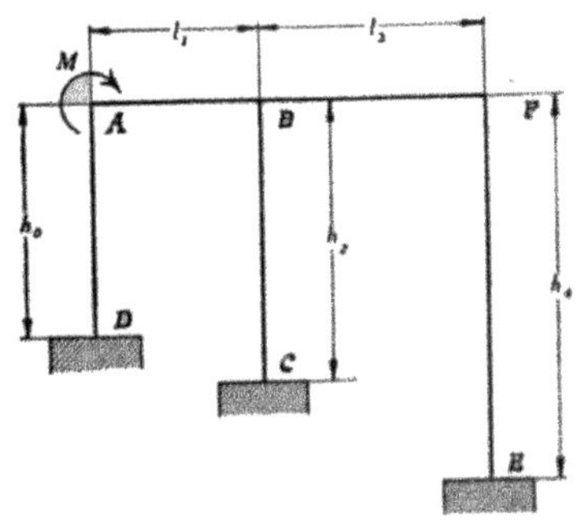

FIGURE 88

TABLE 22

EQUATIONS FOR THREE-LEGGED BENT

Lengths of All Legs Different. Any System of Loads. Legs Hinged at the Bases.

No. of Equation	Left-hand Member of Equation				Right-hand Member of Equation					
	θ_A	θ_B	θ_F	d	Case I Vertical Load on Left-hand Span See Fig. 77	Case II Vertical Load on Right-hand Span See Fig. 78	Case III Horizontal Load to the Right on Left-hand Leg See Fig. 79	Case IV Horizontal Load to the Left on Right-hand Leg See Fig. 80	Case V Settlement and Sliding of Foundations See Fig. 81	Case VI External Couple at Upper Left-hand Corner See Fig. 82
A	$3K_o+4K_1$	$2K_1$	0	$-3\frac{K_o}{h_o}$	$\frac{C_{AB}}{E}$	0	$-\frac{H_{AD}}{E}$	0	$6K_1\frac{d_1}{l_1}$	$\frac{M}{E}$
B	$2K_1$	$4K_1+3K_2+4K_3$	$2K_3$	$-3\frac{K_2}{h_2}$	$-\frac{C_{BA}}{E}$	$\frac{C_{BF}}{E}$	0	0	$6K_1\frac{d_1}{l_1}+6K_3\frac{d_3-d_1}{l_3}-3K_2\frac{d_2}{h_2}$	0
C	0	$2K_3$	$4K_3+3K_4$	$-3\frac{K_4}{h_4}$	0	$-\frac{C_{FB}}{E}$	0	$\frac{H_{FE}}{E}$	$6K_3\frac{(d_3-d_1)}{l_3}-3K_4\frac{d_4}{h_4}$	0
D	$\frac{K_o}{h_o}$	$\frac{K_2}{h_2}$	$\frac{K_4}{h_4}$	$-\left[\frac{K_o}{h_o^2}+\frac{K_2}{h_2^2}+\frac{K_4}{h_4^2}\right]$	0	0	$-\frac{1}{3Eh_o}\left[M_D+H_{AD}\right]$	$-\frac{1}{3Eh_4}\left[M_E-H_{FE}\right]$	$-K_2\frac{d_2}{h_2^2}-K_4\frac{d_4}{h_4^2}$	0

For definition of M_D and M_E see Table 20.

TABLE 23

EQUATIONS FOR THREE-LEGGED BENT

Lengths of All Legs Different. Any System of Loads. Legs Fixed at the Bases.

No. of Equation	Left-hand Member of Equation				Right-hand Member of Equation					
	θ_A	θ_B	θ_F		Case VII Vertical Load on Left-hand Span See Fig. 83	Case VIII Vertical Load on Right-hand Span See Fig. 84	Case IX Horizontal Load to the Right on Left-hand Leg See Fig. 85	Case X Horizontal Load to the Left on Right-hand Leg See Fig. 86	Case XI Settlement and Rotation of Foundations See Fig. 87	Case XII External Couple at Upper Left-hand Corner See Fig. 88
A	$4K_o+4K_1$	$2K_1$	0	$-6\frac{K_o}{h_o}$	$+\frac{C_{AB}}{E}$	0	$-\frac{C_{AD}}{E}$	0	$6K_1\frac{d_1}{l_1}-2K_o\theta_D$	$+\frac{E}{M}$
B	$2K_1$	$4K_1+4K_2+4K_3$	$2K_3$	$-6\frac{K_2}{h_2}$	$-\frac{C_{BA}}{E}$	$+\frac{C_{BF}}{E}$	0	0	$6K_1\frac{d_1}{l_1}-6K_2\frac{d_2}{h_2}+6K_3\frac{d_3-d_1}{l_3}-2K_2\theta_C$	0
C	0	$2K_3$	$4K_3+4K_4$	$-6\frac{K_4}{h_4}$	0	$-\frac{C_{FB}}{E}$	0	$+\frac{C_{FE}}{E}$	$6K_3\frac{d_3-d_1}{l_3}-6K_4\frac{d_4}{h_4}-2K_4\theta_E$	0
D	$\frac{K_o}{h_o}$	$\frac{K_2}{h_2}$	$\frac{K_4}{h_4}$	$-2\left[\frac{K_o}{h_o^2}+\frac{K_2}{h_2^2}+\frac{K_4}{h_4^2}\right]$	0	0	$-\frac{1}{6Eh_o}\left[M_D+C_{AD}-C_{DA}\right]$	$-\frac{1}{6Eh_4}\left[M_E+C_{EF}-C_{FE}\right]$	$-2K_2\frac{d_2}{h_2^2}-2K_4\frac{d_4}{h_4^2}-\frac{K_o\theta_D}{h_o}-\frac{K_2\theta_C}{h_2}-\frac{K_4\theta_E}{h_4}$	0

For definition of M_D and M_E see Table 20

65. *Three-legged Bent. Lengths and Sections of All Legs Equal. Lengths and Sections of Top Members Equal. Any System of Loads. Let* $\frac{I}{l}$ *of a Top Member be* n *times* $\frac{I}{h}$ *of a Leg. Legs Hinged at the Bases.*— Fig. 89 represents a three-legged bent. The lengths of the legs are equal, and the lengths of the top members are equal. The $\frac{I}{l}$ of a top member is designated as K, and the $\frac{I}{h}$ of a leg is designated as $\frac{K}{n}$. The legs of the bent are hinged at D, C, and E.

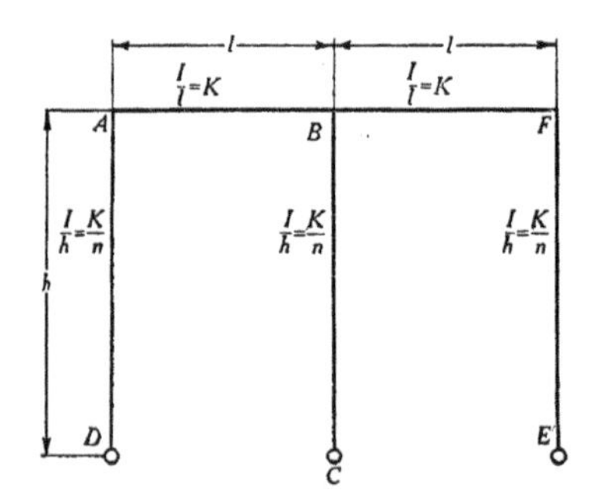

FIGURE 89

The equations of Table 1 as applied to the three-legged bent represented in Fig. 89 are given in Table 24 for a number of systems of loads. Four equations containing four unknowns, derived from the equations

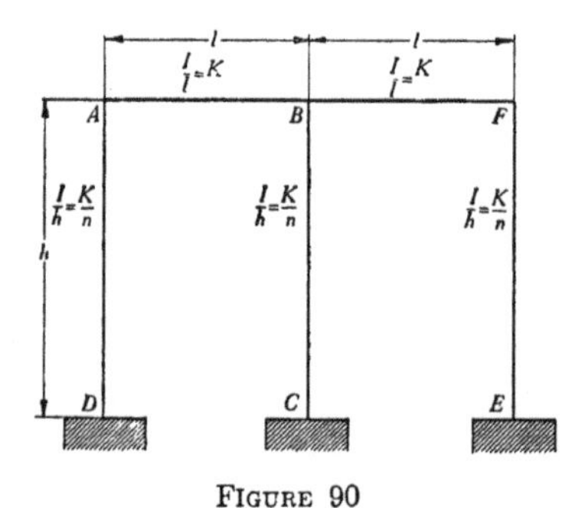

FIGURE 90

of Table 24, are given in Table 25, and the moments at the ends of the members as found from the equations of Tables 24 and 25 are given in Table 26.

66. *Three-legged Bent. Lengths and Sections of All Legs Equal. Lengths and Sections of Top Members Equal. Any System of Loads—Legs Fixed at the Bases.*—Fig. 90 represents a three-legged bent. The lengths of the legs are equal, and the lengths of the top members are equal. The $\frac{I}{l}$ of a top member is designated as K, and the $\frac{I}{h}$ is designated as $\frac{K}{n}$. The legs of the bent are fixed at D, C, and E.

The equations of Table 1 as applied to the three-legged bent represented in Fig. 90 are given in Table 27 for a number of systems of loads. Four equations containing four unknowns, derived from the equations of Table 27, are given in Table 28, and the moments at the ends of the members as found from the equations of Tables 27 and 28 are given in Table 29.

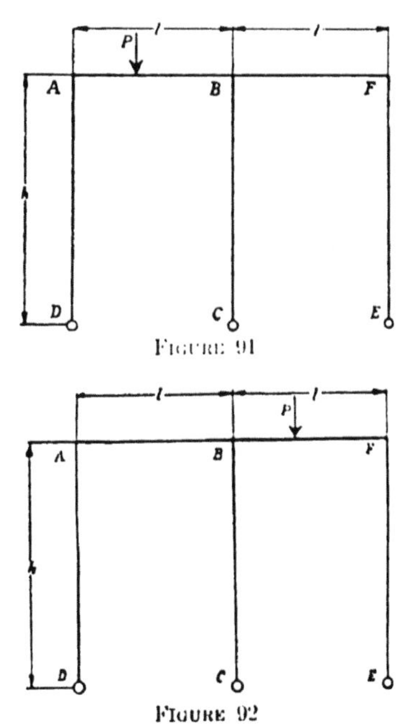

FIGURE 91

FIGURE 92

TABLE 24

FUNDAMENTAL EQUATIONS FOR THREE-LEGGED BENT

Lengths and Sections of Top Members Equal. $\left(\frac{I}{l} = K\right)$

Lengths and Sections of All Legs Equal. $\left(\frac{I}{h} = \frac{K}{n}\right)$

Any System of Loads. Legs Hinged at the Bases.

No. of Equation	This part of the equation is independent of the loads	Additional terms of the right-hand member of the equation, different for each system of loads, as follows:					
		Case I Vertical Load on Left-hand Span See Fig. 91	Case II Vertical Load on Right-hand Span See Fig. 92	Case III Horizontal Load to the Right on Left-hand Leg See Fig. 93	Case IV Horizontal Load to the Left on Right-hand Leg See Fig. 94	Case V Settlement and Sliding of Foundations See Fig. 95	Case VI External Couple at Upper Left-hand Corner See Fig. 96
1	$M_{AD} = 3E\frac{K}{n}\left(\theta_A - R\right)$	0	0	$+H_{AD}$	0	0	0
2	$M_{AB} = 2EK(2\theta_A + \theta_B)$	$-C_{AB}$	0	0	0	$-6EK\frac{d_1}{l}$	0
3	$M_{BA} = 2EK(2\theta_B + \theta_A)$	$+C_{BA}$	0	0	0	$-6EK\frac{d_1}{l}$	0

TABLE 24—CONTINUED

4	$M_{BC}=3E\ \frac{K}{n}\left(\theta_B-R\right)$	0	0	0	0	$+3E\ \frac{K}{n}\ \frac{d_2}{h}$	0
5	$M_{BF}=2EK(2\theta_B+\theta_F)$	0	$-C_{BF}$	0	0	$-6EK\ \frac{d_3-d_1}{l}$	0
6	$M_{FB}=2EK(2\theta_F+\theta_B)$	0	$+C_{FB}$	0	0	$-6EK\ \frac{d_3-d_1}{l}$	0
7	$M_{FE}=3E\ \frac{K}{n}\left(\theta_F-R\right)$	0	0	0	$-H_{FE}$	$+3E\ \frac{K}{n}\ \frac{d_4}{h}$	0
8	$0=M_{AD}+M_{AB}$	0	0	0	0	0	$-M$
9	$0=M_{BA}+M_{BF}+M_{BC}$	0	0	0	0	0	0
10	$0=M_{FB}+M_{FE}$	0	0	0	0	0	0
11	$0=M_{AD}+M_{BC}+M_{FE}$	0	0	$+M_D$	$+M_E$	0	0

M_D is the statical moment about point D of the horizontal loads on left-hand leg.
M_E is the statical moment about point E of the horizontal loads on right-hand leg.
The statical moment is positive when the indicated rotation is clockwise.

TABLE 25

EQUATIONS FOR THREE-LEGGED BENT

Lengths and Sections of Top Member Equal. $\left(\frac{I}{l}=K\right)$ Lengths and Sections of All Legs Equal. $\left(\frac{I}{h}=\frac{K}{n}\right)$

Any System of Loads. Legs Hinged at the Bases.

No. of Equation	Left-hand Member of Equation				Right-hand Member of Equation					
	θ_A	θ_B	θ_F	$3R$	Case I Vertical Load on Left-hand Span See Fig. 91	Case II Vertical Load on Right-hand Span See Fig. 92	Case III Horizontal Load to the Right on Left-hand Leg See Fig. 93	Case IV Horizontal Load to the Left on Right-hand Leg See Fig. 94	Case V Settlement and Sliding of Foundations See Fig. 95	Case VI External Couple at Upper Left-hand Leg See Fig. 96
A	$4n+3$	$2n$	0	-1	$\frac{n.C_{AB}}{EK}$	0	$\frac{-n.H_{AD}}{EK}$	0	$6n\frac{d_1}{l}$	$\frac{Mn}{EK}$
B	$2n$	$8n+3$	$2n$	-1	$\frac{-n.C_{BA}}{EK}$	$\frac{n.C_{BF}}{EK}$	0	0	$6n.\frac{d_3}{l}-\frac{3d_2}{h}$	0
C	0	$2n$	$4n+3$	-1	0	$\frac{-n.C_{FB}}{EK}$	0	$\frac{n.H_{FE}}{EK}$	$\frac{6n(d_3-d_1)}{l}-\frac{3d_4}{h}$	0
D	1	1	1	-1	0	0	$\frac{-n}{3EK}\left[M_D+H_{AD}\right]$	$\frac{-n}{3EK}\left[M_E-H_{FE}\right]$	$-\frac{d_2+d_4}{h}$	0

For definition of M_D and M_E see Table 24.

TABLE 26

IOMENTS DUE TO VARIOUS SYSTEMS OF LOADS ON A THREE-LEGGED BENT

engths and Sections of Top Members Equal. $\left(\frac{I}{l} = K\right)$

engths and Sections of All Legs Equal. $\left(\frac{I}{h} = \frac{K}{n}\right)$

egs Hinged at the Bases.

Moment	Case I Vertical Load on Left-hand Span See Fig. 91	Case II Vertical Load on Right-hand Span See Fig. 92	Case III Horizontal Load to the Right on Left-hand Leg See Fig. 93	Case IV Horizontal Load to the Left on Right-hand Leg See Fig. 94	Case V Settlement and Sliding of Foundation See Fig. 95	Case VI External Couple at Upper Left-hand Corner See Fig. 96
$M_{AD}=$	$+\frac{C_{AB}(10n+9)+C_{BA}(4n+3)}{4(n+1)(4n+3)}$	$-\frac{C_{BF}(4n+3)-C_{FB}(2n+3)}{4(n+1)(4n+3)}$	$+\frac{2nH_{AD}(16n+15)-M_D(4n+3)^2}{12(n+1)(4n+3)}$	$+\frac{2nH_{FE}(8n+9)-M_E(4n+3)^2}{12(n+1)(4n+3)}$	$+\frac{EK}{2hl}\ \frac{18h(n+1)(2d_1-d_3)-l[d_2(8n+6)+d_4(8n+9)]}{(n+1)(4n+3)}$	$+\frac{M(10n+9)}{4(n+1)(4n+3)}$
$M_{AB}=$	$-\frac{C_{AB}(10n+9)+C_{BA}(4n+3)}{4(n+1)(4n+3)}$	$+\frac{C_{BF}(4n+3)-C_{FB}(2n+3)}{4(n+1)(4n+3)}$	$-\frac{2nH_{AD}(16n+15)-M_D(4n+3)^2}{12(n+1)(4n+3)}$	$-\frac{2nH_{FE}(8n+9)-M_E(4n+3)^2}{12(n+1)(4n+3)}$	$-\frac{EK}{2hl}\cdot\frac{18h(n+1)(2d_1-d_3)-l[d_2(8n+6)+d_4(8n+9)]}{(n+1)(4n+3)}$	$+\frac{M(16n^2+18n+3)}{4(n+1)(4n+3)}$
$M_{BA}=$	$+\frac{(2n+3)[C_{AB}(2n+1)+C_{BA}(4n+3)]}{4(n+1)(4n+3)}$	$+\frac{C_{BF}(4n+3)(2n+1)+C_{FB}(4n^2-3)}{4(n+1)(4n+3)}$	$-\frac{(2n+3)[2nH_{AD}-M_D(4n+3)]}{12(n+1)(4n+3)}$	$-\frac{2nH_{FE}(10n+9)-M_E(4n+3)(2n+3)}{12(n+1)(4n+3)}$	$-\frac{EK}{2hl}\cdot\frac{6h(n+1)(2n+3)(2d_1-d_3)+l[d_2(8n+6)-d_4(10n+9)]}{(n+1)(4n+3)}$	$+\frac{M(2n+1)(2n+3)}{4(n+1)(4n+3)}$
$M_{BC}=$	$-\frac{C_{AB}+C_{BA}}{2(n+1)}$	$+\frac{C_{BF}+C_{FB}}{2(n+1)}$	$-\frac{2nH_{AD}+M_D(2n+3)}{6(n+1)}$	$+\frac{2nH_{FE}-M_E(2n+3)}{6(n+1)}$	$+\frac{EK}{h}\cdot\frac{2d_2-d_4}{n+1}$	$-\frac{M}{2(n+1)}$
$M_{BF}=$	$-\frac{C_{AB}(4n^2-3)+C_{BA}(4n+3)(2n+1)}{4(n+1)(4n+3)}$	$-\frac{(2n+3)[C_{BF}(4n+3)+C_{FB}(2n+1)]}{4(n+1)(4n+3)}$	$+\frac{2nH_{AD}(10n+9)+M_D(4n+3)(2n+3)}{12(n+1)(4n+3)}$	$+\frac{(2n+3)[2nH_{FE}+M_E(4n+3)]}{12(n+1)(4n+3)}$	$+\frac{EK}{2hl}\cdot\frac{6h(n+1)(2n+3)(2d_1-d_3)-l[d_2(8n+6)+d_4(2n+3)]}{(n+1)(4n+3)}$	$-\frac{M(4n^2-3)}{4(n+1)(4n+3)}$
$M_{FB}=$	$+\frac{C_{AB}(2n+3)-C_{BA}(4n+3)}{4(n+1)(4n+3)}$	$+\frac{C_{BF}(4n+3)+C_{FB}(10n+9)}{4(n+1)(4n+3)}$	$+\frac{2nH_{AD}(8n+9)+M_D(4n+3)^2}{12(n+1)(4n+3)}$	$+\frac{2nH_{FE}(16n+15)+M_E(4n+3)^2}{12(n+1)(4n+3)}$	$+\frac{EK}{2hl}\cdot\frac{18h(n+1)(2d_1-d_3)+l[d_2(8n+6)-d_4(16n+15)]}{(n+1)(4n+3)}$	$+\frac{M(2n+3)}{4(n+1)(4n+3)}$
$M_{FE}=$	$-\frac{C_{AB}(2n+3)-C_{BA}(4n+3)}{4(n+1)(4n+3)}$	$-\frac{C_{BF}(4n+3)+C_{FB}(10n+9)}{4(n+1)(4n+3)}$	$-\frac{2nH_{AD}(8n+9)+M_D(4n+3)^2}{12(n+1)(4n+3)}$	$-\frac{2nH_{FE}(16n+15)+M_E(4n+3)^2}{12(n+1)(4n+3)}$	$-\frac{EK}{2hl}\cdot\frac{18h(n+1)(2d_1-d_3)+l[d_2(8n+6)-d_4(16n+15)]}{(n+1)(4n+3)}$	$-\frac{M(2n+3)}{4(n+1)(4n+3)}$

For definition of M_D and M_E see Table 24.

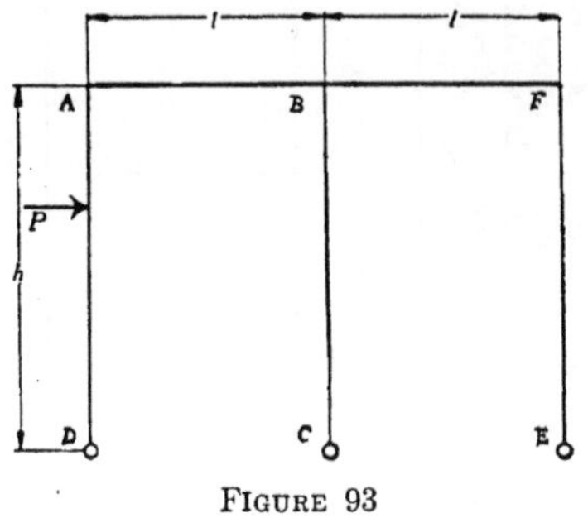

FIGURE 93

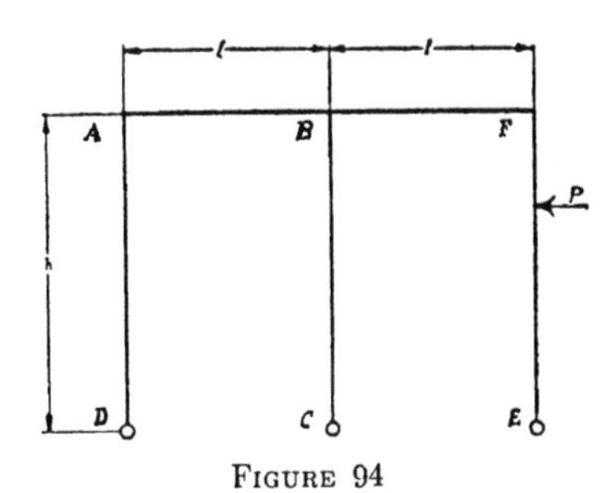

FIGURE 94

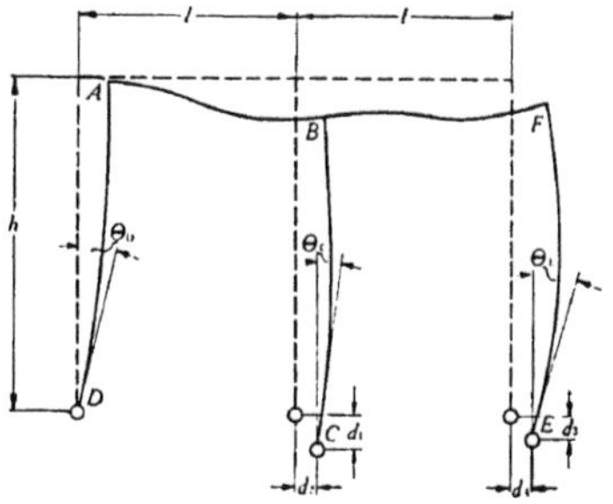

FIGURE 95

TABLE 27

FUNDAMENTAL EQUATIONS FOR THREE-LEGGED BENT

Lengths and Sections of Top Members Equal. $\left(\frac{I}{l} = K\right)$ Lengths and Sections of All Legs Equal. $\left(\frac{I}{h} = \frac{K}{n}\right)$

Any System of Loads. Legs Fixed at the Bases.

No. of Equation	This part of the equation is independent of the loads	Additional term of the right-hand member of the equation, different for each system of loads, as follows:				
		Case VII Vertical Load on Left-hand Span See Fig. 97	Case VIII Vertical Load on Right-hand Span See Fig. 98	Case IX Horizontal Load to the Right on Left-hand Leg See Fig. 99	Case X Horizontal Load to the Left on Right-hand Leg See Fig. 100	Case XII External Couple at Upper Left-hand Corner See Fig. 101
1	$M_{AD} = 2E\frac{K}{n}(2\theta_A - 3R)$	0	0	$+C_{AD}$	0	0
2	$M_{AB} = 2EK(2\theta_A + \theta_B)$	$-C_{AB}$	0	0	0	0
3	$M_{BA} = 2EK(2\theta_B + \theta_A)$	$+C_{BA}$	0	0	0	0
4	$M_{BC} = 2E\frac{K}{n}(2\theta_B - 3R)$	0	0	0	0	0
5	$M_{BF} = 2EK(2\theta_B + \theta_F)$	0	$-C_{BF}$	0	0	0

TABLE 27—CONTINUED

6	$M_{FB} = 2EK(2\theta_F + \theta_B)$	0	$+C_{FB}$	0	0	0
7	$M_{FE} = 2E\frac{K}{n}(2\theta_F - 3R)$	0	0	0	$-C_{FE}$	0
8	$M_{DA} = 2E\frac{K}{n}(\theta_A - 3R)$	0	0	$-C_{DA}$	0	0
9	$M_{CB} = 2E\frac{K}{n}(\theta_B - 3R)$	0	0	0	0	0
10	$M_{EF} \quad 2E\frac{K}{n}(\theta_F - 3R)$	0	0	0	$+C_{EF}$	0
11	$0 \quad M_{AD} + M_{AB}$	0	0	0	0	$-M$
12	$0 \quad M_{BA} + M_{BF} + M_{BC}$	0	0	0	0	0
13	$0 \quad M_{FB} + M_{FE}$	0	0	0	0	0
14	$0 \quad M_{AD} + M_{DA} + M_{BC} \quad M_{CB} \quad M_{FE} + M_{EF}$	0	0	$+M_D$	$+M_E$	0

M_D is the stat[illegible] moment about point D of the horizontal loads on left-hand leg.
M_E is the stat[illegible] moment about point E of the horizontal loads on right-hand leg.
[illegible] is positive when the [illegible] rotation is clockwise.

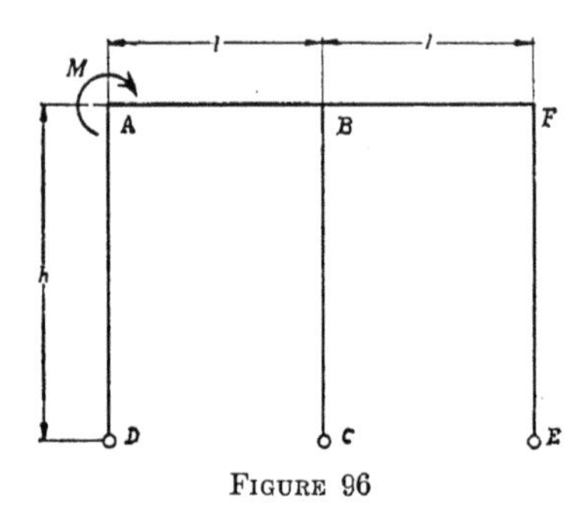

FIGURE 96

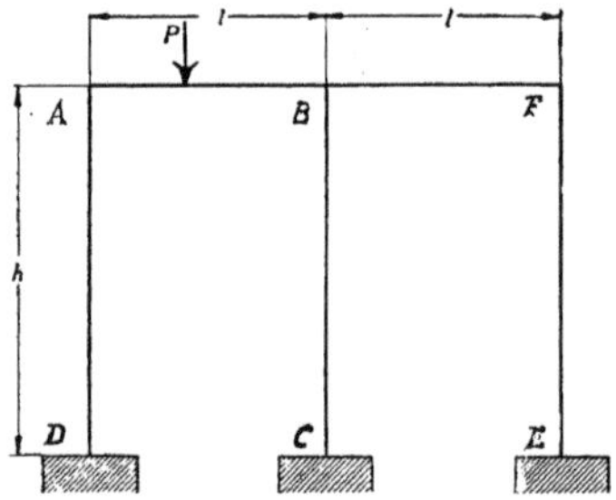

FIGURE 97

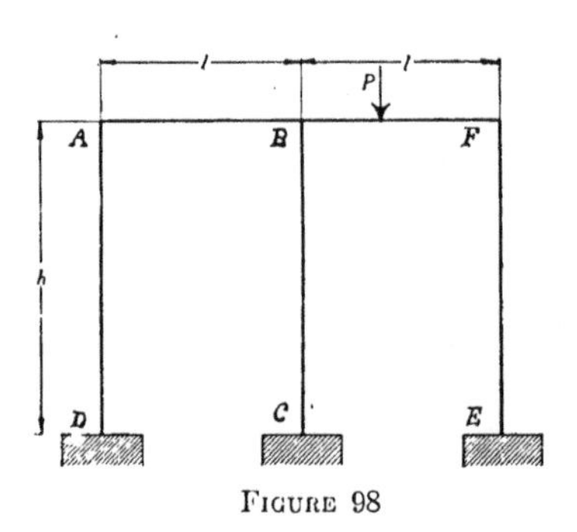

FIGURE 98

TABLE 28

EQUATIONS FOR THREE-LEGGED BENT

Lengths and Sections of Top Members Equal. $\left(\frac{I}{l} = K\right)$

Lengths and Sections of All Legs Equal. $\left(\frac{I}{h} = \frac{K}{n}\right)$

Any System of Loads. Legs Fixed at the Bases.

No. of Equation	Left-hand Member of Equation			Right-hand Member of Equation				
	θ_B	θ_F	$3R$	Case VII Vertical Load on Left-hand Span See Fig. 97	Case VIII Vertical Load on Right-hand Span See Fig. 98	Case IX Horizontal Load to the Right on Left-hand Leg See Fig. 99	Case X Horizontal Load to the Left on Right-hand Leg See Fig. 100	Case XII External Couple at Upper Left-hand Corner See Fig. 101
	$2n+2$	0	-1	$+\frac{nC_{AB}}{2EK}$	0	$\frac{nC_{AD}}{2EK}$	0	$+\frac{M_B}{2EK}$
	$4n\ 2$		-1	$-\frac{nC_{BA}}{2EK}$	$\frac{nC_{BF}}{2EK}$	0	0	0
		$2n+2$	-1	0	$-\frac{nC_{FB}}{2EK}$	0	$\frac{nC_{FE}}{2EK}$	0
				0	0	$\frac{n(M_D+C_{AD}-C_{DA})}{6EK}$	$\frac{n(M_E+C_{EF}-C_{FE})}{6EK}$	0

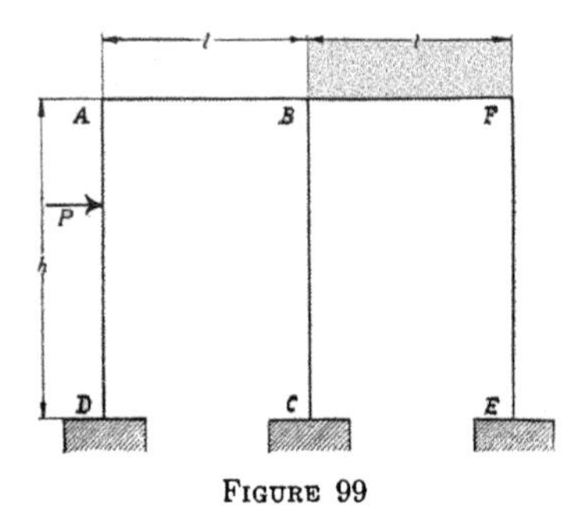

FIGURE 99

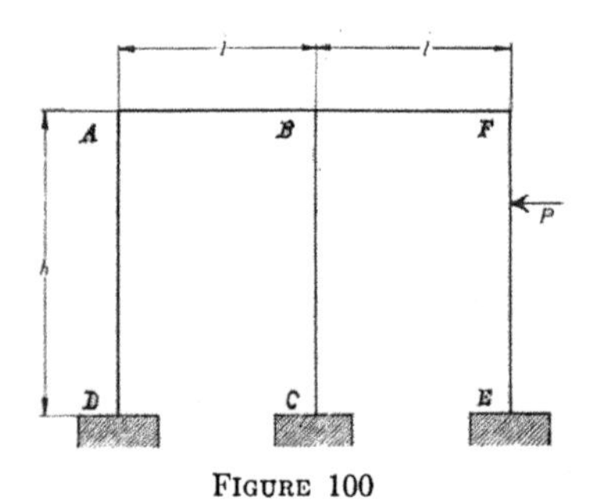

FIGURE 100

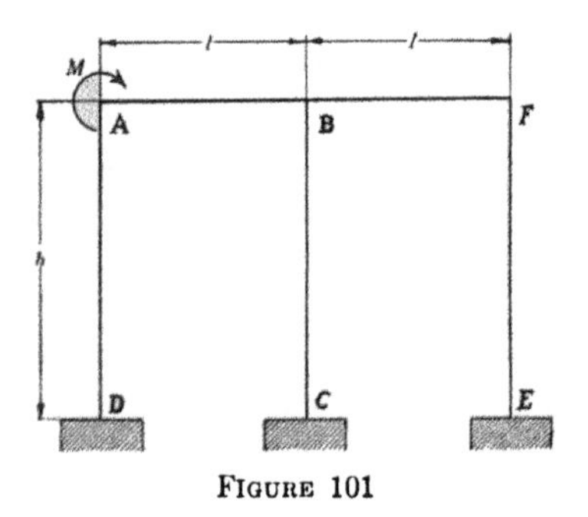

FIGURE 101

TABLE 29

MOMENTS DUE TO VARIOUS SYSTEMS OF LOADS ON A THREE-LEGGED BENT

Lengths and Sections of Top Members Equal. $\left(\frac{I}{l}=K\right)$

Lengths and Sections of All Legs Equal. $\left(\frac{I}{h}=\frac{K}{n}\right)$

Legs Fixed at the Bases.

No. of Equation	Moment	Case VII Vertical Load on Left-hand Span See Fig. 97	Case VIII Vertical Load on Right-hand Span See Fig. 98	Case IX Horizontal Load to the Right on Left-hand Leg See Fig. 99	Case X Horizontal Load to the Left on Right-hand Leg See Fig. 100	Case XII External Couple at Upper Left-hand Corner See Fig. 101
1	$M_{AD}=$	$+\frac{C_{AB}(11n^2+15n+2)+4nC_{BA}(n+1)}{2(n+1)(6n^2+9n+1)}$	$-\frac{4nC_{BF}(n+1)-nC_{FB}(n+3)}{2(n+1)(6n^2+9n+1)}$	$+\frac{nC_{AD}(10n^2+15n+3)-2n(M_D-C_{DA})(n+1)^2}{2(n+1)(6n^2+9n+1)}$	$+\frac{nC_{FE}(2n^2+3n-1)-2n(M_E+C_{EF})(n+1)^2}{2(n+1)(6n^2+9n+1)}$	$+\frac{M(11n^2+15n+2)}{2(n+1)(6n^2+9n+1)}$
2	$M_{AB}=$	$-\frac{C_{AB}(11n^2+15n+2)+4nC_{BA}(n+1)}{2(n+1)(6n^2+9n+1)}$	$+\frac{4nC_{BF}(n+1)-nC_{FB}(n+3)}{2(n+1)(6n^2+9n+1)}$	$-\frac{nC_{AD}(10n^2+15n+3)-2n(M_D-C_{DA})(n+1)^2}{2(n+1)(6n^2+9n+1)}$	$-\frac{nC_{FE}(2n^2+3n-1)-2n(M_E+C_{EF})(n+1)^2}{2(n+1)(6n^2+9n+1)}$	$+\frac{Mn(3n+1)(4n+5)}{2(n+1)(6n^2+9n+1)}$
3	$M_{BA}=$	$+\frac{nC_{AB}(3n^2+8n+4)+C_{BA}(n+1)(6n^2+13n+2)}{2(n+1)(6n^2+9n+1)}$	$+\frac{nC_{BF}(n+1)(6n+5)+nC_{FB}(3n^2+n-3)}{2(n+1)(6n^2+9n+1)}$	$-\frac{n(n+2)[C_{AD}(2n+1)-(M_D-C_{DA})(n+1)]}{2(n+1)(6n^2+9n+1)}$	$-\frac{nC_{FE}(4n^2+4n-1)-n(M_E+C_{EF})(n+1)(n+2)}{2(n+1)(6n^2+9n+1)}$	$+\frac{Mn(n+2)(3n+2)}{2(n+1)(6n^2+9n+1)}$
4	$M_{BC}=$	$-\frac{7nC_{AB}+2C_{BA}(4n+1)}{2(6n^2+9n+1)}$	$+\frac{2C_{BF}(4n+1)+7nC_{FB}}{2(6n^2+9n+1)}$	$-\frac{nC_{AD}(2n-3)+2n(M_D-C_{DA})(n+2)}{2(6n^2+9n+1)}$	$+\frac{nC_{FE}(2n-3)-2n(M_E+C_{EF})(n+2)}{2(6n^2+9n+1)}$	$-\frac{7Mn}{2(6n^2+9n+1)}$
5	$M_{BF}=$	$-\frac{nC_{AB}(3n^2+n-3)+nC_{BA}(n+1)(6n+5)}{2(n+1)(6n^2+9n+1)}$	$-\frac{C_{BF}(n+1)(6n^2+13n+2)+nC_{FB}(3n^2+8n+4)}{2(n+1)(6n^2+9n+1)}$	$+\frac{nC_{AD}(4n^2+4n-1)+n(M_D-C_{DA})(n+1)(n+2)}{2(n+1)(6n^2+9n+1)}$	$+\frac{n(n+2)[C_{FE}(2n+1)+(M_E+C_{EF})(n+1)]}{2(n+1)(6n^2+9n+1)}$	$-\frac{Mn(3n^2+n-3)}{2(n+1)(6n^2+9n+1)}$
6	$M_{FB}=$	$+\frac{nC_{AB}(n+3)-4nC_{BA}(n+1)}{2(n+1)(6n^2+9n+1)}$	$+\frac{4nC_{BF}(n+1)+C_{FB}(11n^2+15n+2)}{2(n+1)(6n^2+9n+1)}$	$+\frac{nC_{AD}(2n^2+3n-1)+2n(M_D-C_{DA})(n+1)^2}{2(n+1)(6n^2+9n+1)}$	$+\frac{nC_{FE}(10n^2+15n+3)+2n(M_E+C_{EF})(n+1)^2}{2(n+1)(6n^2+9n+1)}$	$+\frac{Mn(n+3)}{2(n+1)(6n^2+9n+1)}$
7	$M_{FE}=$	$-\frac{nC_{AB}(n+3)-4nC_{BA}(n+1)}{2(n+1)(6n^2+9n+1)}$	$-\frac{4nC_{BF}(n+1)+C_{FB}(11n^2+15n+2)}{2(n+1)(6n^2+9n+1)}$	$-\frac{nC_{AD}(2n^2+3n-1)+2n(M_D-C_{DA})(n+1)^2}{2(n+1)(6n^2+9n+1)}$	$-\frac{nC_{FE}(10n^2+15n+3)+2n(M_E+C_{EF})(n+1)^2}{2(n+1)(6n^2+9n+1)}$	$-\frac{Mn(n+3)}{2(n+1)(6n^2+9n+1)}$
8	$M_{DA}=$	$+\frac{nC_{AB}(4n+5)+C_{BA}(n+1)(2n+1)}{2(n+1)(6n^2+9n+1)}$	$-\frac{C_{BF}(n+1)(2n+1)-C_{FB}(2n^2+4n+1)}{2(n+1)(6n^2+9n+1)}$	$-\frac{C_{AD}(6n^3+27n^2+26n+2)}{6(n+1)(6n^2+9n+1)}-\frac{M_D(6n^2+9n+2)+C_{DA}(30n^2+45n+4)}{6(6n^2+9n+1)}$	$+\frac{C_{FE}(6n^3+9n^2-n-1)-(M_E+C_{EF})(n+1)(6n^2+9n+2)}{6(n+1)(6n^2+9n+1)}$	$+\frac{Mn(4n+5)}{2(n+1)(6n^2+9n+1)}$
9	$M_{CB}=$	$-\frac{C_{AB}(5n+1)+4nC_{BA}}{2(6n^2+9n+1)}$	$+\frac{4nC_{BF}+C_{FB}(5n+1)}{2(6n^2+9n+1)}$	$-\frac{C_{AD}(6n^2-3n-1)+2(M_D-C_{DA})(3n^2+6n+1)}{6(6n^2+9n+1)}$	$+\frac{C_{FE}(6n^2-3n-1)-2(M_E+C_{EF})(3n^2+6n+1)}{6(6n^2+9n+1)}$	$-\frac{M(5n+1)}{2(6n^2+9n+1)}$
10	$M_{EF}=$	$-\frac{C_{AB}(2n^2+4n+1)-C_{BA}(n+1)(2n+1)}{2(n+1)(6n^2+9n+1)}$	$-\frac{C_{BF}(n+1)(2n+1)+nC_{FB}(4n+5)}{2(n+1)(6n^2+9n+1)}$	$-\frac{C_{AD}(6n^3+9n^2-n-1)+(M_D-C_{DA})(n+1)(6n^2+9n+2)}{6(n+1)(6n^2+9n+1)}$	$+\frac{C_{FE}(6n^3+27n^2+26n+2)}{6(n+1)(6n^2+9n+1)}-\frac{M_E(6n^2+9n+2)-C_{EF}(30n^2+45n+4)}{6(6n^2+9n+1)}$	$-\frac{M(2n^2+4n+1)}{2(n+1)(6n^2+9n+1)}$

For definitions of M and M_E see Table 27.

XII. Effect of Error in Assumptions

67. *Error Due to Assumption that Axial Deformation is Zero.*—In determining the stresses in frames it has been assumed that the members of the frame do not change in length, but since the members are subjected to axial stress there must be a corresponding axial deformation. It remains to determine the effect of the axial deformation upon the stresses in the frames.

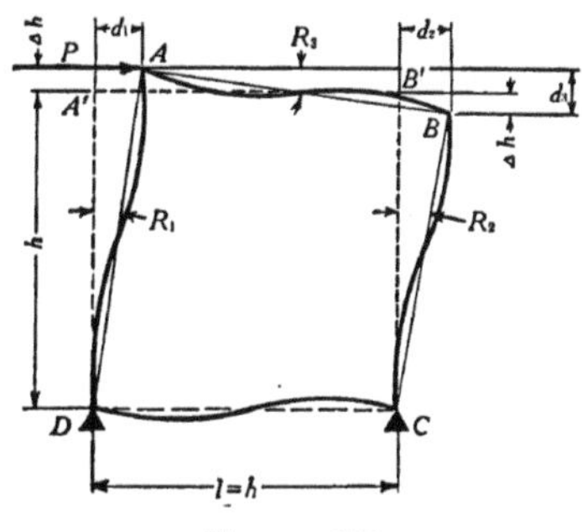

FIGURE 102

Fig. 102 represents a square frame with all sides identical in section. The frame is subjected to a single horizontal force P at the top, and is in equilibrium under the action of P and the reactions at D and C. Since there are no loads except at the ends of the members, H and C of Table 2 are zero for all members.

Case 1. Assume horizontal reactions at C and D equal. From equations 407 and 408 of section 48

$$M_{AB}=\frac{Ph}{4}$$

$$M_{BC}=-\frac{Ph}{4}$$

$$M_{CD}=\frac{Ph}{4}$$

$$M_{DA}=-\frac{Ph}{4}$$

$$\text{Shear in } AD = \frac{M_{AD} + M_{DA}}{h} = \frac{1}{2}P$$

Likewise

$$\text{Shear in } BC = \frac{1}{2}P$$

$$\text{Shear in } AB = \frac{1}{2}P$$

$$\text{Shear in } DC = \frac{1}{2}P$$

$$\text{Stress in } AB = \frac{1}{2}P \text{ compression}$$

$$\text{Stress in } CD = 0$$

$$\text{Stress in } AD = \frac{1}{2}P \text{ tension}$$

$$\text{Stress in } BC = \frac{1}{2}P \text{ compression}$$

Since *AB*, *AD*, and *BC* have the same sections, the same lengths, and are subjected to axial stresses of the same magnitudes, Δh, representing the change in length due to axial stress, is the same for all members.

Referring to Fig. 102

$$R_1 - R_2 = \frac{\Delta h}{h} = \frac{P}{2AE}$$

$$R_3 = \frac{2\Delta h}{l} = \frac{P}{AE}$$

Applying the equations of Table 1 gives

$$M_{AB} = 2EK(2\theta_A + \theta_B - 3R_3) \quad \ldots \ldots \ldots \quad (719)$$

$$M_{AB} = -2EK(2\theta_A + \theta_D - 3R_1) \quad \ldots \ldots \ldots \quad (720)$$

$$M_{BC} = 2EK(2\theta_B + \theta_C - 3R_2) \quad \ldots \ldots \ldots \quad (721)$$

$$M_{BC}=-2EK(2\theta_B+\theta_A-3R_3) \qquad (722)$$

$$M_{CD}=2EK(2\theta_C+\theta_D) \qquad (723)$$

$$M_{CD}=-2EK(2\theta_C+\theta_B-3R_2) \qquad (724)$$

$$M_{DA}=2EK(2\theta_D+\theta_A-3R_1) \qquad (725)$$

$$M_{DA}=-2EK(2\theta_D+\theta_C) \qquad (726)$$

For AD and BC to be in equilibrium

$$-M_{AB}+M_{BC}-M_{CD}+M_{DA}+Pl=0 \qquad (727)$$

Combining equations 719 to 727, substituting values for R_1, R_2, and R_3, and solving for the moments gives

$$M_{AB}=\frac{Ph}{4}\left[1-\frac{3K}{2Ah}\right] \qquad (728)$$

$$M_{BC}=-\frac{Ph}{4}\left[1-\frac{9K}{2Ah}\right] \qquad (729)$$

$$M_{CD}=\frac{Ph}{4}\left[1+\frac{3K}{2Ah}\right] \qquad (730)$$

$$M_{DA}=-\frac{Ph}{4}\left[1+\frac{9K}{2Ah}\right] \qquad (731)$$

In these equations $\frac{9K}{2Ah}$ represents the largest error due to neglecting the axial deformation in the members, and the ratio of $\frac{9K}{2Ah}$ to 1 represents the relative error.

Case 2. Assume horizontal reaction at C only. If the horizontal reaction is all taken at C

Stress in $AB=\frac{1}{2}P$ compression

Stress in $CD=\frac{1}{2}P$ compression

Stress in $AD = \frac{1}{2}P$ tension

Stress in $BC = \frac{1}{2}P$ compression

Since AB, CD, AD, and BC have the same sections, the same lengths, and are subjected to axial stresses of the same magnitude, Δh, representing the change in length due to axial stress, is the same for all members. The magnitude of Δh is given by the expression

$$\Delta h = \frac{Ph}{2AE}$$

As Δh for AB equals Δh for DC, $R_1 = R_2$

Since AD is in tension and BC in compression

$$R_3 = \frac{2\Delta h}{l} = \frac{P}{AE}$$

Equations 719 to 727 are applicable. Combining the equations, substituting the values of the R's and solving for the moments gives

$$M_{AB} = \frac{Ph}{4}\left(1 - \frac{6K}{2Ah}\right) \quad \text{. . . . . (732)}$$

$$M_{BC} = -\frac{Ph}{4}\left(1 - \frac{6K}{2Ah}\right) \quad \text{. . . . . (733)}$$

$$M_{CD} = \frac{Ph}{4}\left(1 + \frac{6K}{2Ah}\right) \quad \text{. . . . . (734)}$$

$$M_{DA} = -\frac{Ph}{4}\left(1 + \frac{6K}{2Ah}\right) \quad \text{. . . . . (735)}$$

In these equations $\frac{6K}{2Ah}$ represents the error due to neglecting the axial deformation in the members, and the ratio of $\frac{6K}{2Ah}$ to 1 represents the relative error.

Comparing Case 1 with Case 2, it is apparent that the error due to neglecting the axial deformations is smaller if the horizontal reaction is all at one corner than if half of it comes at each of the lower corners. The maximum relative error, for Case 1, is $\frac{9K}{2Ah}$. Substituting for K its value $\frac{I}{l}$, the expression $\frac{9K}{2Ah}$ becomes $\frac{9I}{2Ah^2}$. That is, the error due to neglecting the axial deformation in a square frame with all sides identical, subjected to a single horizontal force at the top, varies directly with the moments of inertia of the section of the members, inversely with the area of the members, and inversely with the square of the length.

If for Case 1 the frame, instead of being square as shown in Fig. 102, is a rectangle twice as wide as it is high, and if the sections of all members are identical, the moments in the frame due to a single horizontal force P at the top are

$$M_{AB}=\frac{Ph}{4}\left(1+\frac{9}{4}\frac{K}{Ah}\right) \quad \text{(736)}$$

$$M_{BC}=-\frac{Ph}{4}\left(1-\frac{15}{4}\frac{K}{Ah}\right) \quad \text{(737)}$$

$$M_{CD}=\frac{Ph}{4}\left(1-\frac{9}{4}\frac{K}{Ah}\right) \quad \text{(738)}$$

$$M_{DA}=-\frac{Ph}{4}\left(1+\frac{15}{4}\frac{K}{Ah}\right) \quad \text{(739)}$$

in which h and K are for the vertical members.

The maximum relative error in this case is $\frac{15}{4}\frac{K}{Ah}$ divided by 1.

It is to be noted that $\frac{15}{4}$, the coefficient of $\frac{K}{Ah}$ in equation 739, is less than $\frac{9}{2}$, the coefficient of $\frac{K}{Ah}$, in equation 731. Hence for two frames, one a rectangle twice as wide as it is high with sections of all members identical, and the other a square having all sides identical with the vertical sides of the rectangle, and both frames having a single horizontal force applied at the top, the error due to axial deformation is less for the rectangular frame than it is for the square frame.

If for Case 1 the frame, instead of being square as shown in Fig. 102, is a rectangle twice as high as it is wide, and if the sections of all members are identical, the moments in the frame due to a single horizontal force P at the top are

$$M_{AB}=\frac{Ph}{4}\left(1-\frac{51}{8}\frac{K}{Ah}\right) \quad (740)$$

$$M_{BC}=-\frac{Ph}{4}\left(1-\frac{51}{8}\frac{K}{Ah}\right) \quad (741)$$

$$M_{CD}=\frac{Ph}{4}\left(1+\frac{51}{8}\frac{K}{Ah}\right) \quad (742)$$

$$M_{DA}=-\frac{Ph}{4}\left(1+\frac{51}{8}\frac{K}{Ah}\right) \quad (743)$$

in which h and K are for the vertical members.

The maximum relative error in this case is $\frac{51}{8}\frac{K}{Ah}$ divided by 1.

It is to be noted that $\frac{51}{8}$, the coefficient of $\frac{K}{Ah}$ in equation 743, is more than $\frac{9}{2}$, the coefficient of $\frac{K}{Ah}$ in equation 731. Hence for two frames, one a rectangle twice as high as it is wide with sections of all members identical, and the other a square having all sides identical with the vertical sides of the rectangle, and both frames having a single horizontal force at the top, the error due to axial deformation is greater for the rectangular frame than it is for the square frame.

Table 30 gives the errors due to axial deformations in a number of rectangular frames. The largest error in the table is 3.70 per cent for a frame 10 feet square composed of 27 in.-83 lb.-I-beams. Although this frame is composed of members having sectional areas so large compared with their length that it would be impracticable to provide connections strong enough to develop the strength of the members, yet the error, 3.7 per cent, is well within the range of permissible error. For steel frames of the proportions common in engineering structures the error is well under 2 per cent. For concrete frames the ratio $\frac{I}{A}$ is much less than for steel frames, and the error due to neglecting the axial deformations is correspondingly less for concrete frames than for steel frames.

Table 30

Error in Stresses in Rectangular Frames Due to Neglecting Axial Deformation

Frame is subjected to a single horizontal force at the top. The horizontal reaction at the bottom is equally divided between the two lower corners. The sections of all members of a frame are identical.

Section	Square Frame		Frame Having Width Twice Height			Frame Having Height Twice Width		
	Height and Width feet	Error per cent	Height feet	Width feet	Error per cent	Height feet	Width feet	Error per cent
27"—I—83 lb.	20	.92	20	40	.77	40	20	.33
	15	1.64	15	30	1.37	30	15	.58
	10	3.70	10	20	3.08	20	10	1.31
24"—I—80 lb.	20	.70	20	40	.58	40	20	.25
	15	1.24	15	30	1.03	30	15	.43
	10	2.80	10	20	2.33	20	10	.99
20"—I—65 lb.	20	.48	20	40	.40	40	20	.17
	15	.85	15	30	.71	30	15	.30
	10	1.92	10	20	1.60	20	10	.68
15"—I—42 lb.	20	.28	20	40	.23	40	20	.10
	15	.49	15	30	.41	30	15	.17
	10	1.10	10	20	.92	20	10	.39
12"—I—31.5 lb.	20	.18	20	40	.15	40	20	.06
	15	.32	15	30	.27	30	15	.11
	10	.73	10	20	61	20	10	26

68. *Error Due to Assumption that Shearing Deformation is Zero.*—In the derivation of the fundamental propositions upon which the analyses are based, deflection due to shear was not considered. That being the case, R in the equations of Table 1 depends upon the bending deflection only. In the analyses of the rectangular frames the R's of the two vertical members are taken equal. This is true of the bending

deflections only when the shearing deformations in the two vertical members are equal. (Axial deformation is here neglected.) Likewise, in considering that the deflection due to bending for the top of the frame is zero, the shear deflection in the top of the frame is neglected.

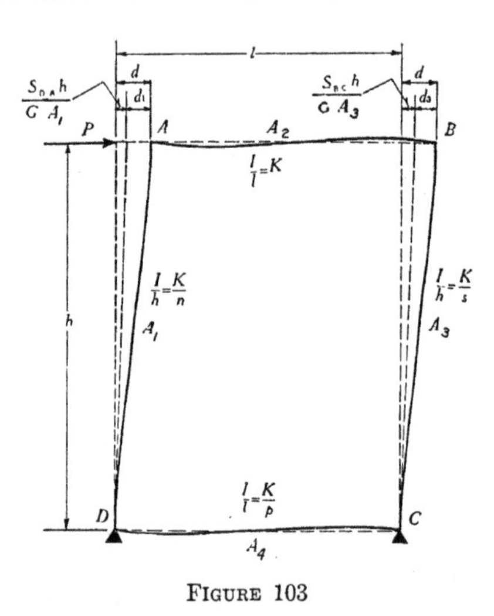

FIGURE 103

Fig. 103 represents a rectangular frame having a horizontal force applied at the top. The members of the frame are subjected to shear as follows:

$$\text{Shear in } DA,\ S_{DA} = \frac{M_{DA} - M_{AB}}{h} = -\frac{Ph}{h}\,\frac{\Delta_{DA} + \Delta_{AB}}{\Delta} \quad . \quad . \quad (744)$$

$$\text{Shear in } AB,\ S_{AB} = \frac{M_{AB} - M_{BC}}{l} = \frac{Ph}{l}\,\frac{\Delta_{AB} + \Delta_{BC}}{\Delta} \quad . \quad . \quad (745)$$

$$\text{Shear in } BC,\ S_{BC} = \frac{M_{BC} - M_{CD}}{h} = \frac{Ph}{h}\,\frac{\Delta_{BC} + \Delta_{CD}}{\Delta} \quad . \quad . \quad (746)$$

$$\text{Shear in } CD,\ S_{CD} = \frac{M_{CD} - M_{DA}}{l} = \frac{Ph}{l}\,\frac{\Delta_{CD} + \Delta_{DA}}{\Delta} \quad . \quad . \quad (747)$$

In these equations Δ_{AB}, Δ_{BC}, Δ_{CD}, and Δ_{DA} represent the quantities in the parentheses of equations 401, 402, 403, and 404, of section 48, respectively, and Δ represents the common denominator of the same equations.

The points D and C are considered to remain stationary, and axial strain is neglected.

The total horizontal deflection of A equals the total horizontal deflection of B. These deflections are represented in Fig. 103 by d. The shear deflection of DA is $\frac{S_{DA}h}{GA_1}$ and the shear deflection of BC is $\frac{S_{BC}h}{GA_3}$. The bending deflection of DA, d_1, is therefore given by the equation

$$d_1 = d - \frac{S_{DA}h}{GA_1} \quad (748)$$

and the bending deflection of BC, d_3, is given by the equation

$$d_3 = d - \frac{S_{BC}h}{GA_3} \quad (749)$$

The point C does not move vertically relatively to D, but there is a shear S_{CD} which produces a shear deflection in CD of $-\frac{S_{CD}l}{GA_4}$; therefore there must be an equal and opposite moment deflection of $\frac{S_{CD}l}{GA_4}$. Similarly, there is a bending deflection in AB of $\frac{S_{AB}l}{GA_2}$.

From the preceding equations

$$R_1 = \frac{d_1}{h} = \frac{d}{h} - \frac{S_{DA}}{GA_1} \quad (750)$$

$$R_2 = \frac{d_2}{l} = \frac{S_{AB}}{GA_2} \quad (751)$$

$$R_3 = \frac{d_3}{h} = \frac{d}{h} - \frac{S_{BC}}{GA_3} \quad (752)$$

$$R_4 = \frac{d_4}{l} = \frac{S_{CD}}{GA_4} \quad (753)$$

Applying the equations of Table 1 gives

$$M_{DA}=\frac{2EK}{n}(2\theta_D+\theta_A-3R_1) \quad (754)$$

$$M_{AB}=-\frac{2EK}{n}(2\theta_A+\theta_D-3R_1) \quad (755)$$

$$M_{AB}=2EK(2\theta_A+\theta_B-3R_2) \quad (756)$$

$$M_{BC}=-2EK(2\theta_B+\theta_A-3R_2) \quad (757)$$

$$M_{BC}=\frac{2EK}{s}(2\theta_B+\theta_C-3R_3) \quad (758)$$

$$M_{CD}=-\frac{2EK}{s}(2\theta_C+\theta_B-3R_3) \quad (759)$$

$$M_{CD}=\frac{2EK}{p}(2\theta_C+\theta_D-3R_4) \quad (760)$$

$$M_{DA}=-\frac{2EK}{p}(2\theta_D+\theta_C-3R_4) \quad (761)$$

For AD and BC to be in equilibrium

$$M_{DA}-M_{AB}+M_{BC}-M_{CD}+Ph=0 \quad (762)$$

Substituting the values of the shears from equations 744, 745, 746, and 747 in equations 750, 751, 752, and 753, substituting the resulting values of the R's in equations 754 to 761, combining the resulting equations and 762, and solving for the moments gives

$$M_{AB}=\begin{bmatrix}M_{AB}\text{ when shearing strain is}\\ \text{neglected, a positive quantity}\\ \text{(see equation 401, section 48).}\end{bmatrix}+\left[\frac{6PK}{\Delta^2}\frac{E}{G}\right]\left[B_{AB}D+O_{AB}Q\right] \quad (763)$$

$$M_{BC}=\begin{bmatrix}M_{BC}\text{ when shearing strain is}\\ \text{neglected, a negative quantity}\\ \text{(see equation 402, section 48).}\end{bmatrix}+\left[\frac{6PK}{\Delta^2}\frac{E}{G}\right]\left[B_{BC}D-O_{BC}Q\right] \quad (764)$$

$$M_{CD}=\begin{bmatrix}M_{CD}\text{ when shearing strain is}\\ \text{neglected, a positive quantity}\\ \text{(see equation 403, section 48).}\end{bmatrix}-\left[\frac{6PK}{\Delta^2}\frac{E}{G}\right]\left[B_{CD}D+O_{CD}Q\right] \quad (765)$$

$$M_{DA}=\begin{bmatrix}M_{DA}\text{ when shearing strain is}\\ \text{neglected, a negative quantity}\\ \text{(see equation 404, section 48).}\end{bmatrix}-\left[\frac{6PK}{\Delta^2}\frac{E}{G}\right]\left[B_{DA}D-O_{DA}Q\right] \quad (766)$$

in which

$$B_{AB}=6ns+5ps+2pn+p^2+2n+p-s$$

$$B_{BC}=6ns+5pn+2ps+p^2-n+2s+p$$

$$B_{CD}=6ns+5n+2ps+p-pn+2s+1$$

$$B_{DA}=6ns+5s-ps+p+2pn+2n+1$$

$$O_{AB}=\frac{h}{l}(ns-pn+s^2+5ps+2n+2s+6p)$$

$$O_{BC}=\frac{h}{l}(ns+5pn-ps+n^2+2n+6p+2s)$$

$$O_{CD}=\frac{h}{l}(ns+2ps+5n+6p+n^2+2pn-s)$$

$$O_{DA}=\frac{h}{l}(s^2+2ps+5s+6p+ns+2pn-n)$$

$$D=\frac{11ns+11nps+6ns^2+2s^2+10ps+12pn+2s+3p+2s^2p+2sp^2+3p^2}{A_1}$$

$$-\frac{6sn^2+2n^2+11nps+10pn+11sn+12ps+2n+3p+2n^2p+2np^2+3p^2}{A_3}$$

$$Q=\frac{10ns+12nps+3ns^2+2s^2+11ps+11pn+2s+6p+3sn^2+2n^2+2n}{A_4}$$

$$-\frac{12ns+10nps+3ns^2+11ps+11pn+2s^2p+2sp^2+6p^2+3sn^2+2n^2p+2np^2}{A_2}$$

$$\Delta=22(pns+ps+ns+np)+2(p^2s+ps^2+n^2p+p^2n+s^2+s+n^2+n)$$
$$+6(n^2s+ns^2+p^2+p)$$

If the members AB and CD are identical, equations 763, 764, 765, and 766 reduce to the form

$M_{AB} = M_{AB}$ when shearing deformation is neglected (positive quantity) $+k$ . . . . (767)

$M_{BC} = M_{BC}$ when shearing deformation is neglected (negative quantity) $+k$ . . . . (768)

$M_{CD} = M_{CD}$ when shearing deformation is neglected (positive quantity) $-k$ . . . . (769)

$M_{DA} = M_{DA}$ when shearing deformation is neglected (negative quantity) $-k$ . . . . (770)

in which

$$k = 6 \frac{E}{G} \frac{PK}{(n+s+6)^2} \left(\frac{s+3}{A_1} - \frac{n+3}{A_3} \right)$$

If the members AB and CD are identical, and if DA and BC are also identical but not necessarily identical with AB and CD, k of equations 767 to 770 is zero, showing that for a rectangular frame with opposite sides identical the shearing deformation does not affect the moment in the frame.

The error due to shearing deformation in a number of frames is given in Table 31.

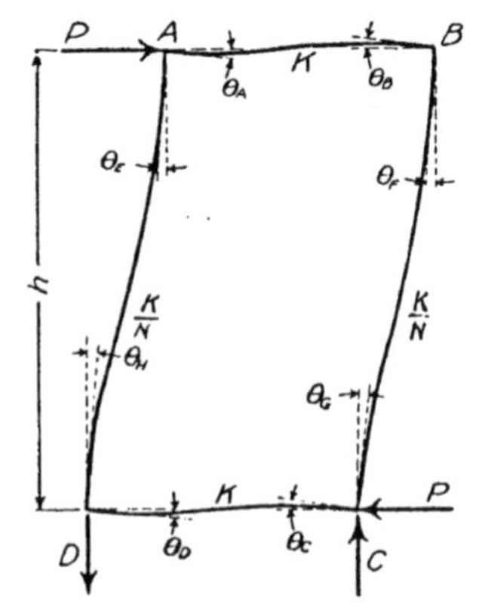

FIGURE 104

Table 31

Error in Stresses in Rectangular Frames Due to Neglecting Shearing Deformations

All frames are subjected to a single horizontal force at the top.

	Description of Frame						Error Due to Shearing Strain Per cent			
	Height in feet	Width in feet	Member	Area of Section in.^{2}	I in.^{4}	K in.^{3}	M_{AB}	M_{BC}	M_{CD}	M_{DA}
No Two Members Alike	30	50	AB BC CD DA	20 20 40 100	2500 300 5000 6000	4.16 .83 8.33 16.67	.07	.03	.09	.02
	24	40	AB BC CD DA	20 20 40 100	2500 300 5000 6000	5.21 1.04 10.41 20.81	.115	.05	.14	.03
	15	25	AB BC CD DA	20 20 40 100	2500 300 5000 6000	8.33 1.67 16.67 33.33	.30	.14	.35	.07
AB and CD identical	30	50	AB and CD BC DA	20 20 100	2500 300 6000	4.16 0.83 16.67	.032	.078	.078	.032
	24	40	AB and CD BC DA	20 20 100	2500 300 6000	5.21 1.04 20.81	.050	.122	.122	.056
	15	25	AB and CD BC DA	20 20 100	2500 300 6000	8.33 1.67 33.33	.127	.313	.313	.127
	50	30	AB and CD BC DA	20 20 100	2500 300 6000	6.95 0.50 10.00	.016	.072	.072	016
	40	24	AB and CD BC DA	20 20 100	2500 300 6000	8.69 0.62 12.50	025	.113	.113	025
	25	15	AB and CD BC DA	20 20 100	2500 300 6000	13.90 1.00 20.00	0.83	.290	.290	083

69. *Effect of Slip in the Connections upon the Moments in a Rectangular Frame Having a Single Horizontal Force at the Top.*—Fig. 104 represents a rectangular frame having a single horizontal force applied at the top. The sections of opposite sides of the frame are identical. The connections at the corners of the frame slip.

The changes in the slopes at the ends of the member AB are represented by θ_A at A and θ_B at B; likewise for the member CD the change in slope at C is represented by θ_C and at D by θ_D; for the member AD the change in slope at A is represented by θ_E and at D by θ_H; and for the member BC the change in slope at B is represented by θ_F and at C by θ_G. That is: slip at A equals $\theta_E-\theta_A$; slip at B equals $\theta_F-\theta_B$; slip at C equals $\theta_G-\theta_C$; and slip at D equals $\theta_H-\theta_D$.

Represent $\frac{I}{l}$ for AB and CD by K, and $\frac{I}{l}$ for AD and BC by $\frac{K}{n}$. Applying equation A, Table 1, and equating the sum of the moments at each of the points A, B, C, and D to zero and also equating the sums of the moments at the ends of the two members AD and BC to $-Ph$ gives

$$2EK(2\theta_A+\theta_B)+\frac{2EK}{n}(2\theta_E+\theta_H-3R)=0 \quad \text{(771)}$$

$$2EK(2\theta_B+\theta_A)+\frac{2EK}{n}(2\theta_F+\theta_G-3R)=0 \quad \text{(772)}$$

$$2EK(2\theta_C+\theta_D)+\frac{2EK}{n}(2\theta_G+\theta_F-3R)=0 \quad \text{(773)}$$

$$2EK(2\theta_D+\theta_C)+\frac{2EK}{n}(2\theta_H+\theta_E-3R)=0 \quad \text{(774)}$$

$$\frac{2EK}{n}(2\theta_E+\theta_H-3R+2\theta_H+\theta_E-3R+2\theta_F+\theta_G-3R+2\theta_G+\theta_F-3R)+Ph=0 \quad \text{(775)}$$

Letting A represent the quantity, slip at A divided by R, likewise letting B, C, and D represent the corresponding quantities at the points B, C, and D, respectively, gives

$$A=\frac{\theta_E}{R}-\frac{\theta_A}{R}$$

$$B=\frac{\theta_F}{R}-\frac{\theta_B}{R}$$

$$C=\frac{\theta_G}{R}-\frac{\theta_C}{R}$$

$$D=\frac{\theta_H}{R}-\frac{\theta_D}{R}$$

Substituting the values of θ_E, θ_F, θ_G, and θ_H from these equations in the preceding equations gives

$$\frac{2\theta_A}{nR}\left(1+n\right)+\frac{\theta_B}{R}+\frac{\theta_D}{nR}=\frac{1}{n}\left(3-2A-D\right) \quad \ldots\ldots (776)$$

$$\frac{2\theta_B}{nR}\left(1+n\right)+\frac{\theta_A}{R}+\frac{\theta_C}{nR}=\frac{1}{n}\left(3-2B-C\right) \quad \ldots\ldots (777)$$

$$\frac{2\theta_C}{nR}\left(1+n\right)+\frac{\theta_D}{R}+\frac{\theta_B}{nR}=\frac{1}{n}\left(3-2C-B\right) \quad \ldots\ldots (778)$$

$$\frac{2\theta_D}{nR}\left(1+n\right)+\frac{\theta_C}{R}+\frac{\theta_A}{nR}=\frac{1}{n}\left(3-2D-A\right) \quad \ldots\ldots (779)$$

$$2EKR=\frac{\frac{1}{3}Phn}{-\left[\frac{\theta_A}{R}+\frac{\theta_B}{R}+\frac{\theta_C}{R}+\frac{\theta_D}{R}\right]+4-(A+B+C+D)} \quad \ldots (780)$$

These five equations contain four unknown angles and one unknown deflection. Solving these equations and substituting the values for the θ's and R in the expressions for the moments gives

$$M_{AD}=\frac{1}{3}\;\frac{Ph}{4-(A+B+C+D)}\cdot\left[\frac{(6A+3B-9)+n(16A+5B+4C+5D-30)+n^2(6A+3D-9)}{(n+3)\,(3n+1)}\right] \quad \ldots\ldots (781)$$

$$M_{DA}=\frac{1}{3}\ \frac{Ph}{4-(A+B+C+D)}$$
$$\left[\frac{(6D+3C-9)+n\,(16D+5C+4B+5A-30)+n^2(6D+3A-9)}{(n+3)\ (3n+1)}\right] \quad . \quad . \quad . \quad (782)$$

$$M_{BC}=\frac{1}{3}\ \frac{Ph}{4-(A+B+C+D)}$$
$$\left[\frac{(6B+3A-9)+n(16B+5A+4D+5C-30)+n^2(6B+3C-9)}{(n+3)\ (3n+1)}\right] \quad . \quad . \quad . \quad (783)$$

$$M_{CB}=\frac{1}{3}\ \frac{Ph}{4-(A+B+C+D)}$$
$$\left[\frac{(6C+3D-9)+n(16C+5D+4A+5B-30)+n^2(6C+3B-9)}{(n+3)\ (3n+1)}\right] \quad . \quad . \quad . \quad (784)$$

The values of $\frac{\theta_A}{R}$, $\frac{\theta_B}{R}$, $\frac{\theta_C}{R}$, and $\frac{\theta_D}{R}$ determined from equations 776 to 779 substituted in equation 780 gives

$$R=\frac{1}{4}\left[(RA+RB+RC+RD)+\frac{1}{6}\frac{Ph(n+1)}{EK}\right] \quad . \quad . \quad . \quad (785)$$

in which RA, RB, RC, and RD represent the slips in the connections at A, B, C, and D respectively. If the slips are measured, R may be computed from equation 786. Knowing R and the slips, the values of A, B, C, and D may be computed. Substituting the values of A, B, C, and D in equations 781 to 784, the moments in a frame having connections which are not rigid may be determined.

If there is no slip in the connections the moment at each corner of the frame of Fig. 104 is $-\frac{1}{4}Ph$. The differences between $-\frac{1}{4}Ph$ and the moments given by equations 781, 782, 783, and 784 represent the effect of the slip in the connections.

If A, B, C, and D, of equations 781 to 784 inclusive, are equal to each other; that is, if the slips in all the connections of the rectangular frame represented by Fig. 104 are equal, equations 781 to 784 reduce to the form

$$M_{AD} = -\frac{1}{4}Ph \quad . \quad . \quad . \quad . \quad . \quad . \quad . \quad . \quad . \qquad (781a)$$

$$M_{DA} = -\frac{1}{4}Ph \quad . \quad . \quad . \quad . \quad . \quad . \quad . \quad . \quad . \quad . \quad . \qquad (782a)$$

$$M_{BC} = -\frac{1}{4}Ph \quad . \quad . \quad . \quad . \quad . \quad . \quad . \quad . \quad . \quad . \quad . \quad . \qquad (783a)$$

$$M_{CB} = -\frac{1}{4}Ph \quad . \quad . \quad . \quad . \quad . \quad . \quad . \quad . \quad . \quad . \quad . \qquad (784a)$$

That is, if the slips in all the connections of the rectangular frame shown in Fig. 104* are equal, the stresses in the frame are the same as they are in a similar frame having connections which are perfectly rigid.

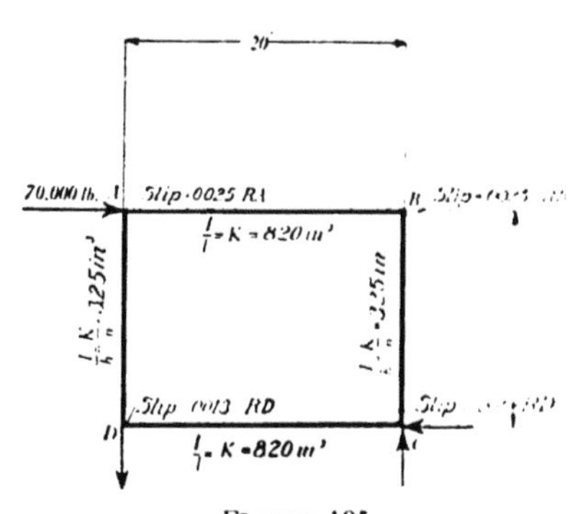

FIGURE 105

To illustrate the magnitude of the effect of slip in the connections upon the stresses in a rectangular frame consider the frame represented by Fig. 105. Equations 781, 782, 783, 784, and 785 are applicable. Substituting the values of the quantities given in Fig. 105 in equations 781 to 785 gives

*Wilson, W. M., and Moore, H. F. "Tests to Determine the R[illegible] ity of R[illegible] Steel Structures," Univ. of Ill. Eng. Exp. Sta., Bul 104, p 28, 1917

$M_{AD} = 3{,}030{,}000$ in. lb.
$M_{DA} = 3{,}240{,}000$ in. lb.
$M_{BC} = 3{,}050{,}000$ in. lb.
$M_{CB} = 3{,}300{,}000$ in. lb.

If there is no slip in the connections, each of these moments is 3,150,000 in. lb. The errors in the moments due to neglecting the slip in the connections are as follows:

M_{AD}, error $= -3.8$ per cent
M_{DA}, error $= +3.0$ per cent
M_{BC}, error $= -3.0$ per cent
M_{CB}, error $= +4.6$ per cent

For a given slip, the error due to slip is greater for a frame having short stiff members than it is for a frame having long flexible members.

XIII. Numerical Problems

The following numerical problems illustrate the use of the equations which have been derived.

70. *Girder Restrained at the Ends. Supports on Different Levels. Any System of Vertical Loads.*—Fig. 106 represents a 12 in.–31.5 lb. I-beam embedded in masonry at both ends. Because of upheaval by frost, or other causes, the beam which was originally horizontal now has one end higher than the other and the tangents to the elastic curve of the beam at the ends are inclined to the horizontal. The beam carries a concentrated load and a uniform load as shown. It is required to find the bending moments in the beam.

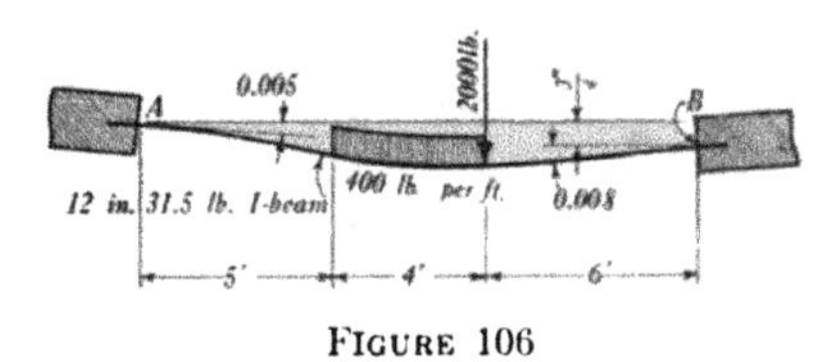

Figure 106

Equations 25 and 26, section 13, are applicable. E for steel is 30,000,000 lb. per sq. in.

$d_B = .75$ inch $\qquad \theta_A = +.005$

$I = 215.8$ in.^{4} $\qquad \theta_B = -.008$

$l = 180$ in. $\qquad R = \frac{.75}{180} = .00416$

$$K = \frac{215.8}{180} = 1.2 \text{ in.}^3$$

From Table 2, C_{AB} for a single concentrated load is

$\frac{Pab^2}{l^2}$ in which

$a = 108$ in.

$b = 72$ in.

$P = 2000$ lb.

$$\frac{Pab^2}{l^2} = 34560 \text{ in. lb.}$$

Also from Table 2, C_{AB} for a uniform load over a portion of the span is $\frac{w}{12l^2}\left[d^3(4l-3d)-b^3(4l-3b)\right]$ in which

$d=120$ in.

$b=\ 72$ in.

$w=400$ lb. per ft. $=\frac{400}{12}$ lb. per in.

$$\frac{w}{12l^2}\left[d^3(4l-3d)-b^3(4l-3b)\right]=37{,}200 \text{ in. lb.}$$

Total $C_{AB}=37{,}200+34{,}560=71{,}760$ in. lb.

Substituting the values of the constants in equation 25, section 13, gives

$M_{AB}=-826{,}300$ in. lb.

From Table 2, C_{BA} for the concentrated load is 52,000 in. lb., and C_{BA} for the uniform load is 32,700 in. lb.

Total $C_{BA}=52{,}000+32{,}700=84{,}700$ in. lb.

Substituting the values of the constants in equation 26, section 13, gives

$M_{BA}=-1{,}606{,}300$ in. lb.

It is to be noted in the solution of this problem, in solving for both M_{AB} and M_{BA}, that θ_A and R are both positive whereas θ_B is negative. The minus sign, moreover, is used before C_{AB} and the plus sign is used before C_{BA}. The signs are in accordance with the conventional method of fixing signs given at the bottom of Table 1.

If the supports had been on the same level and if the tangents to the elastic curves at the ends of the beams had been horizontal, the moments would have been given by the equations of section 11. Using the same values of C_{AB} and C_{BA} as before

$M_{AB}=-C_{AB}=-71{,}760$ in. lb.

$M_{BA}=+C_{BA}=+84{,}700$ in. lb.

71. *Girder Continuous over Four Supports. Supports on Different Levels. Any System of Vertical Loads.*—Fig. 107 represents a 20 in.-80 lb. I-beam supported on four supports and having hinged ends. The supports are all on different levels. It is required to find the moments in the girder.

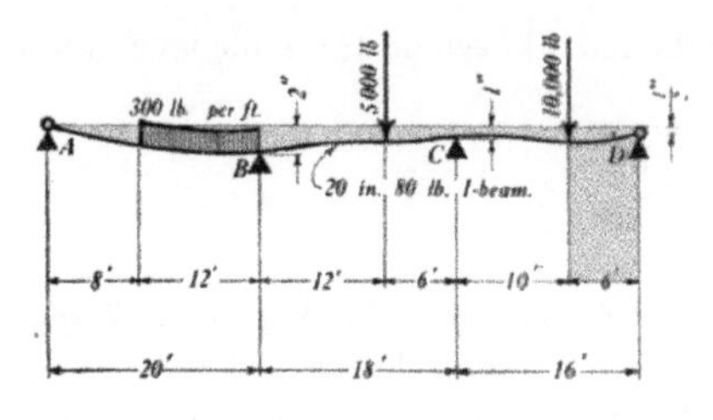

FIGURE 107

Equations (a), Table 6, are applicable.

$I_o = I_1 = I_2 = 1466$ in.^{4}

$l_o = 240$ in.

$l_1 = 216$ in.

$l_2 = 192$ in.

$$K_o = \frac{1466}{240} = 6.10 \text{ in.}^3$$

$$K_1 = \frac{1466}{216} = 6.79 \text{ in.}^3$$

$$K_2 = \frac{1466}{192} = 7.64 \text{ in.}^3$$

$$n_o = \frac{6.79}{6.10} = 1.11$$

$$n_1 = \frac{7.64}{6.79} = 1.125$$

$E = 30{,}000{,}000$ lb. per sq. in.

$d_A = 0$

$d_B = 2$ in.

$d_C = 1$ in.

$d_D = .5$ in.

From Table 2

$$H_{BA} = \frac{3600 \times 144}{8 \times 240 \times 240}\left(2 \times 240 - 144\right)^2 = 127{,}000 \text{ in. lb.}$$

$$H_{BC} = \frac{5000 \times 144 \times 72(216+72)}{2 \times 216 \times 216} = 160{,}000 \text{ in. lb.}$$

$$H_{CB} = \frac{5000 \times 144 \times 72(216+144)}{2 \times 216 \times 216} = 200{,}000 \text{ in. lb.}$$

$$H_{CD} = \frac{10000 \times 120 \times 72}{2 \times 192 \times 192}\left(192 + 72\right) = 310\,000 \text{ in. lb.}$$

Substituting these values in equations (a) of Table 6 gives

$M_{BC} = 4{,}071{,}500$ in. lb.

$M_{CD} = -1{,}987{,}500$ in. lb.

If the supports had all been on the same level, the moments would have been

$M_{BC} = -88{,}500$ in. lb.

$M_{CD} = -227{,}500$ in. lb.

72. *Girder Continuous over Five Supports. Supports on Different Levels. Any System of Vertical Loads. Ends of Girder Restrained.*—Fig. 108 represents a girder continuous over five supports. All the supports are on different levels and the girder is restrained at the ends. The slopes of the elastic curve at the ends are known. It is required to determine the moments in the girder in terms of I, the moment of inertia of the girder.

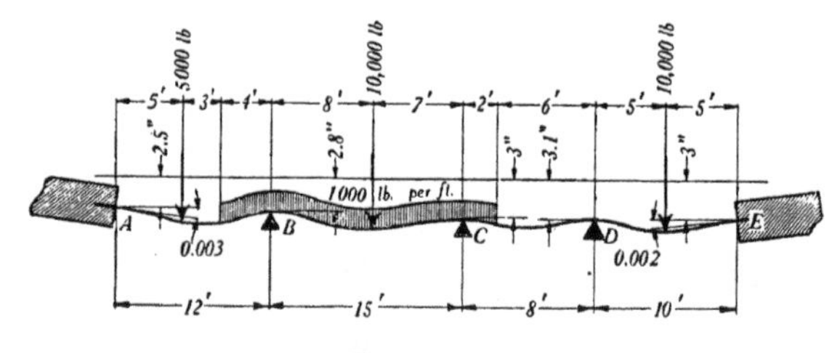

FIGURE 108

The solution of the problem involves writing the equations of three moments and solving these equations for the moments.

Cases (a), (d), and (e), Table 4, and the equations of Table 2 are applicable.

$$K_o = \frac{I}{12\times 12} = \frac{I}{144}\text{ in.}^3 \quad n_1 = \frac{K_1}{K_0} = \frac{4}{5} = .800 \quad l_o = 12\times 12 = 144\text{ in.}$$

$$K_1 = \frac{I}{12\times 15} = \frac{I}{180}\text{ in.}^3 \quad n_2 = \frac{K_2}{K_1} = \frac{15}{8} = 1.875 \quad l_1 = 12\times 15 = 180\text{ in.}$$

$$K_2 = \frac{I}{12\times 8} = \frac{I}{96}\text{ in.}^3 \quad n_3 = \frac{K_3}{K_2} = \frac{4}{5} = .800 \quad l_2 = 12\times 8 = 96\text{ in.}$$

$$K_3 = \frac{I}{12\times 10} = \frac{I}{120}\text{ in.}^3 \qquad l_3 = 12\times 10 = 120\text{ in.}$$

$d_A=2.5$ in. $\quad d_B-d_A=2.8\text{-}2.5=+0.3$ in. $\quad \theta_A=+.003$

$d_B=2.8$ in. $\quad d_C-d_B=3.0\text{-}2.8=+0.2$ in. $\quad \theta_E=-.002$

$d_C=3.0$ in. $\quad d_D-d_C=3.1\text{-}3.0=+0.1$ in. $\quad E=30{,}000{,}000$ lb. per sq. in

$d_D=3.1$ in. $\quad d_E-d_D=3.0\text{-}3.1=-0.1$ in.

$d_E=3.0$ in.

$$H_{AB}=\frac{5000\times5\times7\times19\times12}{2\times144}+\frac{12000(16)\times(2\times144-16)}{8\times144}$$
$$=-183{,}900 \text{ in. lb.}$$

$$H_{BA}=\frac{5000\times5\times7\times17\times12}{2\times144}+\frac{12000}{8\times144}\left(144-64\right)\left(2\times144-144-64\right)$$
$$=190{,}700 \text{ in. lb.}$$

$$H_{BC}=\frac{10000\times8\times7\times22\times12}{2\times225}+\frac{12000}{8}\times225=665{,}500 \text{ in. lb.}$$

$$H_{CB}=\frac{10000\times8\times7\times23\times12}{2\times225}+\frac{12000}{8}\times225=681{,}500 \text{ in. lb.}$$

$$H_{CD}=\frac{12000}{8\times64}\times\left(64-36\right)\left(2\times64-36-64\right)=18{,}400 \text{ in. lb.}$$

$$H_{DC}=\frac{12000}{8\times64}\left(4\right)\left(2\times64-4\right)=11{,}600 \text{ in. lb.}$$

$$H_{DE}=\frac{10000\times5\times5(15)\times12}{2\times100}=225{,}000 \text{ in. lb.}$$

$$H_{ED}=\frac{10000\times5\times5(15)\times12}{200}=225{,}000 \text{ in. lb.}$$

Substituting the values of these quantities in the equations of Table 4 gives

From equation (d)

$$6.4\ M_{BC}+2.0\ M_{CD}=1030\ I-2{,}978{,}000 \quad . \quad . \quad . \quad . \quad . \quad . \qquad (786)$$

From equation (a)

$$1.875\ M_{BC}+5.75\ M_{CD}+M_{DE}=130\ I-2{,}593{,}000 \quad . \quad . \quad . \quad (787)$$

From equation (e)

$$1.6\ M_{CD}+6.2\ M_{DE}=3900\ I-487{,}100 \quad . \quad . \quad . \quad . \quad . \quad . \quad . \quad . \quad (788)$$

From equations 786, 787, and 788

$$M_{DE}=(672\ I+9{,}500) \text{ in. lb.}$$

$$M_{CD}=-(166\ I+341{,}000) \text{ in. lb.}$$

$$M_{BC}=(213\ I-358{,}000) \text{ in. lb.}$$

Substituting the values of M_{BC} and M_{CD} in equation (a) of Table 4 gives

$$M_{AB}=(467\ I+18{,}750) \text{ in. lb.}$$

Also from equation (a) of Table 4

$$M_{ED}=-(530I-193{,}400) \text{ in. lb.}$$

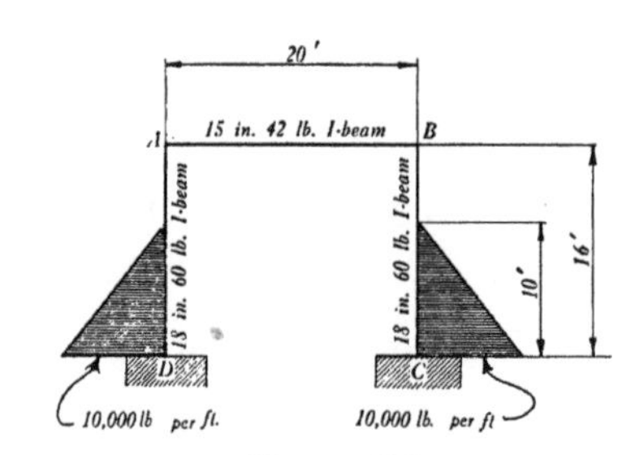

FIGURE 109

73. *Two-legged Rectangular Bent with Legs Fixed at the Bases. Bent and Loading Symmetrical about a Vertical Center Line.*—Fig. 109 represents a two-legged rectangular bent with legs fixed at the bases. Both legs are subjected to hydrostatic pressure. The bent and loading are symmetrical about a vertical center line. It is required to determine the moments in the frame.

Equations 126 and 127, section 25, are applicable.

$$n=\frac{K \text{ for } AB}{K \text{ for } AD}=\frac{\frac{441.8}{240}}{\frac{841.8}{192}}=.42$$

From Table 2

$$C_{AD}=\frac{Wa^2}{30l^2}(5l-3a) \text{ in which}$$

$$C_{DA}=\frac{Wa}{30l^2}(3a^2-10al+10l^2)$$

$$W=\frac{10{,}000\times 10}{2}=50{,}000 \text{ lb}$$

$a=120$ inches

$l=192$ inches

$C_{AD}=391{,}000$ in. lb.

$C_{DA}=984{,}400$ in. lb.

$$M_{AB}=M_{BC}=-\frac{0.42}{0.42+2}\times 391{,}000=-67{,}900 \text{ in. lb.}$$

$$M_{CB}=-M_{DA}=\frac{1}{0.42+2}\times 391{,}000+984{,}400=1{,}145{,}900 \text{ in. lb.}$$

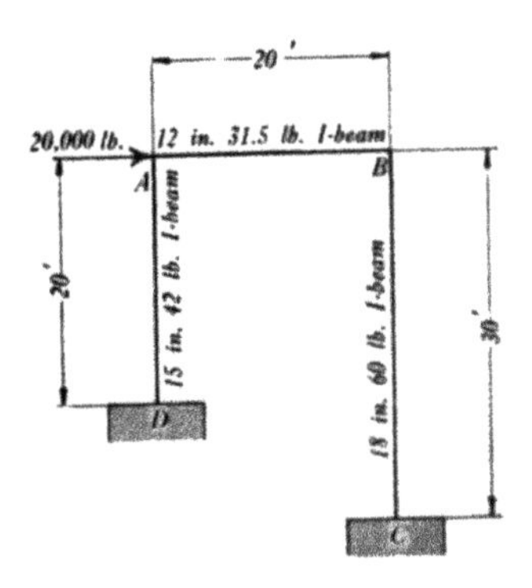

FIGURE 110

74. *Two-legged Rectangular Bent. One Leg Longer than the Other. Concentrated Horizontal Load at the Top. Legs Fixed at the Bases.*

Fig. 110 represents a rectangular bent having one leg longer than the other. There is a horizontal force of 20,000 lb. applied at the top. The bases of the legs are fixed. It is required to determine the moments in the bent.

Equations 252, 253, 254, and 255, section 33, are applicable.

$P=20{,}000$ lb.

$l=240$ inches

$h=240$ inches

$$q=\frac{240}{360}=0.667$$

$$K \text{ for } AB=\frac{215.8}{240}=0.90 \text{ in.}^3$$

$$K \text{ for } AD=\frac{441.8}{240}=1.84 \text{ in.}^3$$

$$K \text{ for } BC=\frac{841.8}{360}=2.34 \text{ in}^3$$

$$n=\frac{K \text{ for } AB}{K \text{ for } AD}=0.49$$

$$s=\frac{K \text{ for } AB}{K \text{ for } BC}=0.385$$

$$\Delta_o=2(3\times.49\times.385\times.385+4\times.49\times.385+.385\times.385+.385+3\times.49\times.385\times.667+3\times.49\times.49\times.385\times.667\times.667+.49\times.49\times.667\times.667+.49\times.667\times.667+4\times.49\times.385\times.667\times.667)=5.334$$

Substituting the values of the quantities in equations 252, 253, 254, and 255, of section 33, gives

$M_{AB}=1{,}100{,}000$ in. lb.

$M_{BA}=958{,}000$ in. lb.

$M_{DA}=-1{,}950{,}000$ in. lb.

$M_{CB}=-1{,}664{,}000$ in. lb.

75. *Two-legged Rectangular Bent. Settlement of Foundation.*—Fig. 111 represents a rectangular portal. Due to upheaval by frost or other causes the foundation originally at C' moves 2 inches to the right, settles 3 inches, and turns in a negative direction an angle of .01 radians. It is required to determine the moments in the portal.

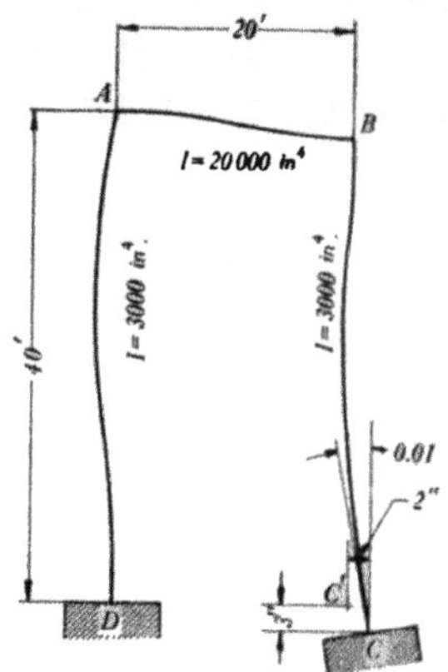

FIGURE 111

Equations 204, 205, 206, and 207, section 31, are applicable.

$$l = 240 \text{ inches}$$

$$h = 480 \text{ inches}$$

$$K = \frac{20{,}000}{240} = 83.4 \text{ in.}^3$$

$$n = \frac{K \text{ for } AB}{K \text{ for } AD} = \frac{83.4}{6.25} = 13.3$$

$$\theta_C = -.01 - \frac{3}{242} = -.0224$$

$$\theta_D = 0$$

$$d = 2 \text{ inches}$$

$$E = 30{,}000{,}000 \text{ lb. per sq. in}$$

$M_{AB} = 2{,}502{,}000{,}000 \quad [0.000816 - 0.001461 - 0.000831]$
$= -3{,}700{,}000$ in. lb.

$M_{BC} = 2{,}502{,}000{,}000 \quad [0.000816 - 0.001431 + 0.000831]$
$= +470{,}000$ in. lb.

$M_{CB} = 2{,}502{,}000{,}000 \quad [0.000877 - 0.003255 - 0.000831]$
$= -8{,}040{,}000$ in. lb.

$M_{DA} = -2{,}502{,}000{,}000 \quad [0.000877 - 0.003255 + 0.000831]$
$= +3{,}875{,}000$ in. lb.

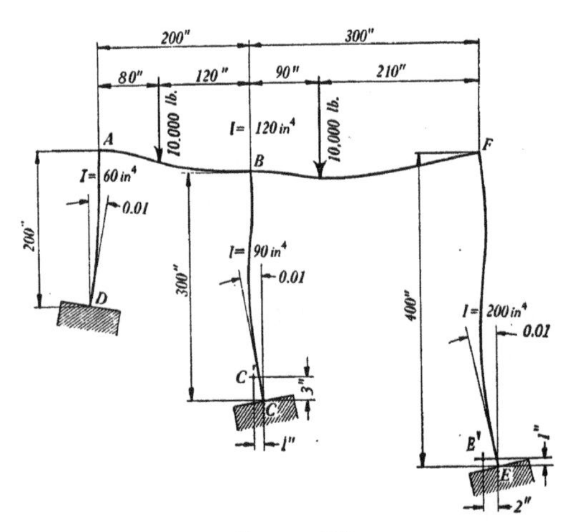

FIGURE 112

76. *Three-legged Bent. Lengths of All Legs Different. Loads on Top and Settlement of Foundations.*—Fig. 112 represents a three-legged bent having vertical loads on top. The legs are restrained at the bases. Due to upheaval by frost or other causes the foundation at D has rotated in a positive direction through an angle of 0.01 radians. Likewise the foundations at C and E have rotated in a negative direction through an angle 0.01. The foundation, originally at C', has moved, moreover, to the right 1 inch and has settled 3 inches. Likewise the foundation, originally at E', has moved to the right 2 inches and has settled 1 inch. It is required to determine the moments in the frame.

The equations of Table 21 are applicable. Since there are no horizontal loads on the legs nor external couples at the top, Cases IX, X, and XII will not enter.

The total right-hand members of the equations of Table 21 will therefore be the algebraic sum of the right-hand members due to Case VII, Case VIII, and Case XI.

$h_o = 200$ in. $\quad I_o = 60$ in.^{4}

$h_2 = 300$ in. $\quad I_1 = 120$ in.^{4}

$h_4 = 400$ in. $\quad I_2 = 90$ in.^{4}

$l_1 = 200$ in. $\quad I_3 = 120$ in.^{4}

$l_3 = 300$ in. $\quad I_4 = 200$ in.^{4}

$K_o = 0.30$ in.^{3} $\quad C_{AB} = 2{,}880{,}000$ in. lb.

$K_1 = 0.60$ in.^{3} $\quad C_{BA} = 1{,}920{,}000$ in. lb.

$K_2 = 0.30$ in.^{3} $\quad C_{BF} = 4{,}410{,}000$ in. lb.

$K_3 = 0.40$ in.^{3} $\quad C_{FB} = 1{,}890{,}000$ in. lb.

$K_4 = 0.50$ in.^{3} $\quad E = 30{,}000{,}000$ lb. per sq. in.

$$\frac{6d_1K_1}{l_1} = \frac{6\times3\times0.6}{200} = 0.054 \text{ in.}^3 \qquad K_o\theta_D = 0.003 \text{ in.}^3$$

$$K_2\theta_C = -0.003 \text{ in.}^3$$

$$\frac{6d_2K_2}{h_2} = \frac{6\times1\times0.3}{300} = 0.006 \text{ in.}^3 \qquad K_4\theta_K = -0.005 \text{ in.}^3$$

$$\frac{6(d_3-d_1)K_3}{l_3} = \frac{6\times(-2)\times0.4}{300}$$

$$= -0.016 \text{ in.}^3$$

$$\frac{6d_4K_4}{h_4} = \frac{6\times2\times0.5}{400} = 0.015 \text{ in.}^3$$

Substituting the values of these quantities in the equations of Table 23 gives equations 1, 2, 3, and 4 of Table 32. It is to be noted in the equations of Table 32 that the quantities under the headings Case VII and Case VIII are coefficients of $\frac{100}{E}$. Solving equations 1, 2, 3, and 4 by the method of elimination, as given in Table 32, gives

$$d = \frac{1}{E}\left[49{,}373{,}300 - 19{,}722{,}100\right] + 1.19314$$

$$\theta_F = \frac{1}{E}\left[238{,}155 - 797{,}034\right] - 0.0047062$$

$$\theta_B = \frac{1}{E}\left[-608{,}835 + 1{,}039{,}263\right] + 0.0061134$$

$$\theta_A = \frac{1}{E}\left[1{,}126{,}370 - 395{,}727\right] + 0.014278$$

Substituting these values of the θ's and d in the equations of Table 21 gives the moments in the frame. The moments are itemized and presented in Table 33.

TABLE 32

EQUATIONS FOR THE THREE-LEGGED BENT OF FIG. 112

No. of Equation	Left-hand Member of Equation				Right-hand Member of Equation		
	θ_A	θ_B	θ_F	d	Case VII Coefficients of $\frac{100}{E}$	Case VIII Coefficients of $\frac{100}{E}$	Case XI
1	3.6	1.2	0	−.0090	+28800	0	.04800
2	1.2	5.2	.80	−.0060	−19200	+44100	.03800
3	0	.8	3.60	−.0075	0	−18900	−.02100
4×1000	1.5	1.0	1.25	−.02792	0	0	−.01167
1	+1.0	+ .3333	0	−.002500	+ 8000	0	+.013333
2	+1.0	+4.3333	+ .6667	−.005000	−16000	+36750	+.031667
3	0						
4	+1.0	+ .6667	+ .8333	−.018611	0	0	−.007777
$(2-1)=a$	0	+4.0000	+ .6667	−.002500	−24000	+36750	+.018333
$(2-4)=b$	0	+3.6667	− .1667	+.013611	−16000	+36750	+.039444
a		+1.0	+.166667	−.000625	−6000	+ 9187.5	+.004583
b		+1.0	−.045454	+.003712	−4363.63	+10022.7	+.010757
3		+1.0	+4.500000	−.009375	0	−23625.0	−.026250
$(3-a)=c$		0	+4.333333	−.008750	+6000	−32812.5	−.030833
$(3-b)=d$		0	+4.545454	−.013087	+4363.63	−33647.7	−.037007
c			+1.0	−.0020192	+1384.61	−7572.11	−.0071154
d			+1.0	−.0028792	+ 960.00	−7402.50	−.0081415
$(c-d)=e$			0	+.0008600	+424.61	−169.61	+.0010261
e				+1.0	+493733	−197221	+1.19314

TABLE 33

MOMENTS IN THREE-LEGGED BENT OF FIG. 112

Moments are expressed in inch-pounds.

Moment	Due to Load on AB	Due to Load on BF	Due to Settlement of Foundations	Total
M_{DA}	+ 231500	− 59900	+294900	+ 466500
M_{AB}	− 907300	+ 297400	−371800	− 981700
M_{BA}	+ 1810400	+ 2019400	−665800	+ 3164000
M_{BC}	− 1026800	+ 1365400	+ 5300	+ 343900
M_{CB}	− 661500	+ 741900	−284700	− 204300
M_{BF}	− 783600	− 3384800	+660500	− 3507900
M_{FB}	− 106000	+ 1446200	+400800	+ 1741000
M_{EF}	− 132100	− 649100	−559600	− 1340800

XIV. Conclusions

Some outstanding features of the use of the slope-deflection equations, as brought out by the analysis in Part II, may well be emphasized.

Two general methods of using the equations have been illustrated. In one case, after the equations have been written for each member of a frame, by equating the sum of the moments at each joint to zero and employing one equation of statics, a number of equations is obtained which contain values of θ and R as the only unknowns. From these equations can be found values of θ and R, which, when substituted in the original slope-deflection equations, give values of the various moments. This method applies especially well to a frame in which a large number of members meet at each joint. Such a problem is generally best solved in numerical terms. Examples of this method are found in sections 23, 61, 64, and 65.

The procedure in the other case is more direct. The slope-deflection equations for each member may be combined to eliminate values of θ and R, leaving equations involving the unknown moments, the properties of the members, and the given loading of the frame. These equations may be solved directly for the moments. Examples of this method are found in sections 24, 35, and 49.

Special attention is called to the form of the equations, which are independent of the magnitude and location of the individual loads, except as the magnitude and location of such loads influence the numerical values of the quantities C and H of the equations. The quantities C and H are determinate and their numerical values may be readily found for any known system of loads.

It is well to note also that the treatment of continuous girders for which the supports are not on the same level is comparatively simple. This part of the work is of considerable value inasmuch as it permits the determination of the effect of the settlement of supports. The treatment of the settlement of foundations for two-legged bents is of equal importance.

Advantages of the slope-deflection equations which are worthy of appreciation are:

(1) The general form of the fundamental equation is easily memorized, and the equations may be written for all members of a structure with little effort. The value of the quantities C or H

for loaded members may be calculated by reference to Tables 2 and 3. It is frequently possible to simplify the equations through noting where values of θ and R must be equal to zero from the conditions of the problem.

(2) No integrations need be performed except possibly to find values of C or H, and there is little danger of the omission of the effect of a single indeterminate quantity, as there is in methods involving the work of internal forces or moments.

(3) The physical conception of a problem is easier than in the case where differentiation or integration is performed. When the slopes and deflections are determined, it is easy to visualize the approximate shape of the elastic curve of a member, whereas an expression involving the work of an indeterminate force or moment may have little physical meaning. Neither does the method of cutting a member and equating expressions for the linear and angular movement of the adjoining ends give so clear an idea of the actual deformation. To one unfamiliar with such a method, the determination of the sign of the movement of the ends of the member cut is also more or less difficult.

(4) It is shown in sections 67, 68, and 69 that the effect of axial and shearing deformations and of slip of joints may be calculated by the use of the slope-deflection equations. This makes possible a complete treatment of any problem, though it is shown that it is seldom necessary to make use of such refinements in an analysis.

(5) The use of the quantity K for $\frac{I}{l}$ of a member, and also of quantities n, s, and p, as ratios of K's for different members, is of great help in writing equations in a workable form. The restraint factor N is also useful in simplifying both analyses and final equations.

(6) Although the fact has not been brought out in this bulletin, these equations may be applied to many structures not composed of rectangular units. The determination of secondary stresses in bridge trusses is an example of such use which has been in print for some time. With trapezoidal and triangular frames, care must be taken in the use of the term R for adjoining members.

(7) While the method is readily applicable to all the problems solved in this bulletin, its advantage over other methods is seen when applied to structures which are statically indeterminate to a high degree and in which a number of members meets at each joint.

The use of statically indeterminate structures in recent years has grown rapidly and many new types of structures have been evolved. With the use of riveted connections in steel frames and the development of monolithic reinforced concrete structures of all sorts, it often happens that statically indeterminate stresses cannot be avoided. On the other hand, structures are frequently made of an indeterminate type for the purpose of securing economy of material. Rational methods of design will do much to inspire confidence in the reliability and economy of such structures, thus insuring their more widespread use.

It is felt that the treatment of statically indeterminate structures given in this bulletin will be helpful in giving information regarding such structures. The method has been explained in sufficient detail to enable the designing engineer to use it in the solution of his particular problems. It is believed that the fundamental principles can be quickly coördinated with the ordinary principles of mechanics so that the more complex problems and even the simpler ones may be studied from a new viewpoint.

LIST OF
PUBLICATIONS OF THE ENGINEERING EXPERIMENT STATION

Bulletin No. 1. Tests of Reinforced Concrete Beams, by Arthur N. Talbot. 1904. *None available.*

Circular No. 1. High-Speed Tool Steels, by L. P. Breckenridge. 1905. *None available.*

Bulletin No. 2. Tests of High-Speed Tool Steels on Cast Iron, by L. P. Breckenridge and Henry B. Dirks. 1905. *None available.*

Circular No. 2. Drainage of Earth Roads, by Ira O. Baker. 1906. *None available.*

Circular No. 3. Fuel Tests with Illinois Coal (Compiled from tests made by the Technological Branch of the U. S. G. S., at the St. Louis, Mo., Fuel Testing Plant, 1904–1907), by L. P. Breckenridge and Paul Diserens. 1908. *Thirty cents.*

Bulletin No. 3. The Engineering Experiment Station of the University of Illinois, by L. P. Breckenridge. 1906. *None available.*

Bulletin No. 4. Tests of Reinforced Concrete Beams, Series of 1905, by Arthur N. Talbot. 1906. *Forty-five cents.*

Bulletin No. 5. Resistance of Tubes to Collapse, by Albert P. Carman and M. L. Carr. 1906. *None available.*

Bulletin No. 6. Holding Power of Railroad Spikes, by Roy I. Webber. 1906. *None available.*

Bulletin No. 7. Fuel Tests with Illinois Coals, by L. P. Breckenridge, S. W. Parr, and Henry B. Dirks. 1906. *None available.*

Bulletin No. 8. Tests of Concrete: I, Shear; II, Bond, by Arthur N. Talbot. 1906. *None available.*

Bulletin No. 9. An Extension of the Dewey Decimal System of Classification Applied to the Engineering Industries, by L. P. Breckenridge and G. A. Goodenough. 1906. Revised Edition 1912. *Fifty cents.*

Bulletin No. 10. Tests of Concrete and Reinforced Concrete Columns, Series of 1906, by Arthur N. Talbot. 1907. *None available.*

Bulletin No. 11. The Effect of Scale on the Transmission of Heat through Locomotive Boiler Tubes, by Edward C. Schmidt and John M. Snodgrass. 1907. *None available.*

Bulletin No. 12. Tests of Reinforced Concrete T-Beams, Series of 1906, by Arthur N. Talbot 1907. *None available.*

Bulletin No. 13. An Extension of the Dewey Decimal System of Classification Applied to Architecture and Building, by N. Clifford Ricker. 1907. *None available.*

Bulletin No. 14. Tests of Reinforced Concrete Beams, Series of 1906, by Arthur N. Talbot. 1907. *None available.*

Bulletin No. 15. How to Burn Illinois Coal without Smoke, by L. P. Breckenridge. 1908. *None available.*

Bulletin No. 16. A Study of Roof Trusses, by N. Clifford Ricker. 1908. *None available.*

Bulletin No. 17. The Weathering of Coal, by S. W. Parr, N. D. Hamilton, and W. F. Wheeler. 1908. *None available.*

Bulletin No 18. The Strength of Chain Links, by G. A. Goodenough and L. E. Moore. 1908. *Forty cents.*

Bulletin No. 19. Comparative Tests of Carbon, Metallized Carbon and Tantalum Filament Lamps, by T. H. Amrine. 1908. *None available.*

Bulletin No. 20. Tests of Concrete and Reinforced Concrete Columns, Series of 1907, by Arthur N. Talbot. 1908. *None available.*

Bulletin No. 21. Tests of a Liquid Air Plant, by C. S. Hudson and C. M. Garland. 1908. *Fifteen cents.*

Bulletin No. 22. Tests of Cast-Iron and Reinforced Concrete Culvert Pipe, by Arthur N. Talbot. 1908. *None available.*

Bulletin No. 23. Voids, Settlement and Weight of Crushed Stone, by Ira O. Baker. 1908. *Fifteen cents.*

**Bulletin No. 24.* The Modification of Illinois Coal by Low Temperature Distillation, by S. W. Parr and C. K. Francis. 1908. *Thirty cents.*

Bulletin No. 25. Lighting Country Homes by Private Electric Plants, by T. H. Amrine. 1908. *Twenty cents.*

---

*A limited number of copies of bulletins starred is available for free distribution.

Bulletin No. 26. High Steam-Pressure in Locomotive Service. A Review of a Report to the Carnegie Institution of Washington, by W. F. M. Goss. 1908. *Twenty-five cents.*

Bulletin No. 27. Tests of Brick Columns and Terra Cotta Block Columns, by Arthur N. Talbot and Duff A. Abrams. 1909. *Twenty-five cents.*

Bulletin No. 28. A Test of Three Large Reinforced Concrete Beams, by Arthur N. Talbot. 1909. *Fifteen cents.*

Bulletin No. 29. Tests of Reinforced Concrete Beams: Resistance to Web Stresses, Series of 1907 and 1908, by Arthur N. Talbot. 1909. *Forty-five cents.*

**Bulletin No. 30.* On the Rate of Formation of Carbon Monoxide in Gas Producers, by J. K. Clement, L. H. Adams, and C. N. Haskins. 1909. *Twenty-five cents.*

**Bulletin No. 31.* Tests with House-Heating Boilers, by J. M. Snodgrass. 1909. *Fifty-five cents.*

Bulletin No. 32. The Occluded Gases in Coal, by S. W. Parr and Perry Barker. 1909. *Fifteen cents.*

Bulletin No. 33. Tests of Tungsten Lamps, by T. H. Amrine and A. Guell. 1909. *Twenty cents.*

**Bulletin No. 34.* Tests of Two Types of Tile-Roof Furnaces under a Water-Tube Boiler, by J. M. Snodgrass. 1909. *Fifteen cents.*

Bulletin No. 35. A Study of Base and Bearing Plates for Columns and Beams, by N. Clifford Ricker. 1909. *Twenty cents.*

Bulletin No. 36. The Thermal Conductivity of Fire-Clay at High Temperatures, by J. K. Clement and W. L. Egy. 1909. *Twenty cents.*

Bulletin No. 37. Unit Coal and the Composition of Coal Ash, by S. W. Parr and W. F. Wheeler. 1909. *None available.*

Bulletin No. 38. The Weathering of Coal, by S. W. Parr and W. F. Wheeler. 1909. *Twenty-five cents.*

**Bulletin No. 39.* Tests of Washed Grades of Illinois Coal, by C. S. McGovney. 1909. *Seventy-five cents.*

Bulletin No. 40. A Study in Heat Transmission, by J. K. Clement and C. M. Garland. 1910. *Ten cents.*

Bulletin No. 41. Tests of Timber Beams, by Arthur N. Talbot. 1910. *Thirty-five cents.*

**Bulletin No. 42.* The Effect of Keyways on the Strength of Shafts, by Herbert F. Moore. 1910. *Ten cents.*

Bulletin No. 43. Freight Train Resistance, by Edward C. Schmidt. 1910. *Seventy-five cents.*

Bulletin No. 44. An Investigation of Built-up Columns under Load, by Arthur N. Talbot and Herbert F. Moore. 1911. *Thirty-five cents.*

**Bulletin No. 45.* The Strength of Oxyacetylene Welds in Steel, by Herbert L. Whittemore. 1911. *Thirty-five cents.*

**Bulletin No. 46.* The Spontaneous Combustion of Coal, by S. W. Parr and F. W. Kressman. 1911. *Forty-five cents.*

**Bulletin No. 47.* Magnetic Properties of Heusler Alloys, by Edward B. Stephenson, 1911. *Twenty-five cents.*

**Bulletin No. 48.* Resistance to Flow through Locomotive Water Columns, by Arthur N. Talbot and Melvin L. Enger. 1911. *Forty cents.*

**Bulletin No. 49.* Tests of Nickel-Steel Riveted Joints, b Arthur N. Talbot and Herbert F. Moore. 1911. *Thirty cents.*

**Bulletin No. 50.* Tests of a Suction Gas Producer, by C. M. Garland and A. P. Kratz. 1912. *Fifty cents.*

Bulletin No. 51. Street Lighting, by J. M. Bryant and H. G. Hake. 1912. *Thirty-five cents.*

**Bulletin No. 52.* An Investigation of the Strength of Rolled Zinc, by Herbert F. Moore. 1912. *Fifteen cents.*

**Bulletin No. 53.* Inductance of Coils, by Morgan Brooks and H. M. Turner. 1912. *Forty cents.*

**Bulletin No. 54.* Mechanical Stresses in Transmission Lines, by A. Guell. 1912. *Twenty cents.*

**Bulletin No. 55.* Starting Currents of Transformers, with Special Reference to Transformers with Silicon Steel Cores, by Trygve D. Yensen. 1912. *Twenty cents.*

**Bulletin No. 56.* Tests of Columns: An Investigation of the Value of Concrete as Reinforcement for Structural Steel Columns, by Arthur N. Talbot and Arthur R. Lord. 1912. *Twenty-five cents.*

**Bulletin No. 57.* Superheated Steam in Locomotive Service. A Review of Publication No. 127 of the Carnegie Institution of Washington, by W. F. M. Goss. 1912. *Forty cents.*

---

*A limited number of copies of bulletins starred is available for free distribution.

**Bulletin No. 58.* A New Analysis of the Cylinder Performance of Reciprocating Engines, by J. Paul Clayton. 1912. *Sixty cents.*

**Bulletin No. 59.* The Effect of Cold Weather upon Train Resistance and Tonnage Rating, by Edward C. Schmidt and F. W. Marquis. 1912. *Twenty cents.*

Bulletin No. 60. The Coking of Coal at Low Temperatures, with a Preliminary Study of the By-Products, by S. W. Parr and H. L. Olin. 1912. *Twenty-five cents.*

**Bulletin No. 61.* Characteristics and Limitation of the Series Transformer, by A. R. Anderson and H. R. Woodrow. 1913. *Twenty-five cents.*

Bulletin No. 62. The Electron Theory of Magnetism, by Elmer H. Williams. 1913. *Thirty-five cents.*

Bulletin No. 63. Entropy-Temperature and Transmission Diagrams for Air, by C. R. Richards. 1913. *Twenty-five cents.*

**Bulletin No. 64.* Tests of Reinforced Concrete Buildings under Load, by Arthur N. Talbot and Willis A. Slater. 1913. *Fifty cents.*

**Bulletin No. 65.* The Steam Consumption of Locomotive Engines from the Indicator Diagrams, by J. Paul Clayton. 1913. *Forty cents.*

Bulletin No. 66. The Properties of Saturated and Superheated Ammonia Vapor, by G. A. Goodenough and William Earl Mosher. 1913. *Fifty cents.*

Bulletin No. 67. Reinforced Concrete Wall Footings and Column Footings, by Arthur N. Talbot. 1913. *Fifty cents.*

Bulletin No. 68. The Strength of I-Beams in Flexure, by Herbert F. Moore. 1913. *Twenty cents.*

Bulletin No. 69. Coal Washing in Illinois, by F. C. Lincoln. 1913. *Fifty cents.*

Bulletin No. 70. The Mortar-Making Qualities of Illinois Sands, by C. C. Wiley. 1913. *Twenty cents.*

Bulletin No. 71. Tests of Bond between Concrete and Steel, by Duff A. Abrams. 1914. *One dollar.*

**Bulletin No. 72.* Magnetic and Other Properties of Electrolytic Iron Melted in Vacuo, by Trygve D. Yensen. 1914. *Forty cents.*

Bulletin No. 73. Acoustics of Auditoriums, by F. R. Watson. 1914. *Twenty cents.*

**Bulletin No. 74.* The Tractive Resistance of a 28-Ton Electric Car, by Harold H. Dunn. 1914. *Twenty-five cents.*

Bulletin No. 75. Thermal Properties of Steam, by G. A. Goodenough. 1914. *Thirty-five cents.*

Bulletin No. 76. The Analysis of Coal with Phenol as a Solvent, by S. W. Parr and H. F. Hadley. 1914. *Twenty-five cents.*

**Bulletin No. 77.* The Effect of Boron upon the Magnetic and Other Properties of Electrolytic Iron Melted in Vacuo, by Trygve D. Yensen. 1915. *Ten cents.*

**Bulletin No. 78.* A Study of Boiler Losses, by A. P. Kratz. 1915. *Thirty-five cents.*

**Bulletin No. 79.* The Coking of Coal at Low Temperatures, with Special Reference to the Properties and Composition of the Products, by S. W. Parr and H. L. Olin. 1915. *Twenty-five cents.*

Bulletin No. 80. Wind Stresses in the Steel Frames of Office Buildings, by W. M. Wilson and G. A. Maney. 1915. *Fifty cents.*

**Bulletin No. 81.* Influence of Temperature on the Strength of Concrete, by A. B. McDaniel. 1915. *Fifteen cents.*

Bulletin No. 82. Laboratory Tests of a Consolidation Locomotive, by E. C. Schmidt, J. M. Snodgrass, and R. B. Keller. 1915. *Sixty-five cents.*

**Bulletin No. 83.* Magnetic and Other Properties of Iron Silicon Alloys. Melted in Vacuo, by Trygve D. Yensen. 1915. *Thirty-five cents.*

Bulletin No. 84. Tests of Reinforced Concrete Flat Slab Structures, by Arthur N. Talbot and W. A. Slater. 1916. *Sixty-five cents.*

**Bulletin No. 85.* The Strength and Stiffness of Steel under Biaxial Loading, by A. T. Becker 1916. *Thirty-five cents.*

**Bulletin No. 86.* The Strength of I-Beams and Girders, by Herbert F Moore and W. M. Wilson 1916. *Thirty cents.*

**Bulletin No. 87.* Correction of Echoes in the Auditorium, University of Illinois, by F. R. Watson and J. M. White. 1916. *Fifteen cents.*

Bulletin No. 88. Dry Preparation of Bituminous Coal at Illinois Mines, by E. A. Holbrook. 1916 *Seventy cents.*

---

*A limited number of copies of bulletins starred is available for free distribution.

**Bulletin No. 89.* Specific Gravity Studies of Illinois Coal, by Merle L. Nebel. 1916. *Thirty cents.*

**Bulletin No. 90.* Some Graphical Solutions of Electric Railway Problems, by A. M. Buck. 1916. *Twenty cents.*

Bulletin No. 91. Subsidence Resulting from Mining, by L. E. Young and H. H. Stoek. 1916. *None available.*

**Bulletin No. 92.* The Tractive Resistance on Curves of a 28-Ton Electric Car, by E. C. Schmidt and H. H. Dunn. 1916. *Twenty-five cents.*

**Bulletin No. 93.* A Preliminary Study of the Alloys of Chromium, Copper, and Nickel, by D. F. McFarland and O. E. Harder. 1916. *Thirty cents.*

**Bulletin No. 94.* The Embrittling Action of Sodium Hydroxide on Soft Steel, by S. W. Parr. 1917. *Thirty cents.*

**Bulletin No. 95.* Magnetic and Other Properties of Iron-Aluminum Alloys Melted in Vacuo, by T. D. Yensen and W. A. Gatward. 1917. *Twenty-five cents.*

**Bulletin No. 96.* The Effect of Mouthpieces on the Flow of Water through a Submerged Short Pipe, by Fred B Seely. 1917. *Twenty-five cents.*

Bulletin No. 97. Effects of Storage upon the Properties of Coal, by S. W. Parr. 1917. *Twenty cents.*

**Bulletin No. 98.* Tests of Oxyacetylene Welded Joints in Steel Plates, by Herbert F. Moore. 1917. *Ten cents.*

Circular No. 4. The Economical Purchase and Use of Coal for Heating Homes, with Special Reference to Conditions in Illinois. 1917. *Ten cents.*

**Bulletin No. 99.* The Collapse of Short Thin Tubes, by A. P. Carman. 1917. *Twenty cents.*

**Circular No. 5.* The Utilization of Pyrite Occurring in Illinois Bituminous Coal, by E. A. Holbrook. 1917. *Twenty cents.*

**Bulletin No. 100.* Percentage of Extraction of Bituminous Coal with Special Reference to Illinois Conditions, by C. M. Young. 1917.

**Bulletin No. 101.* Comparative Tests of Six Sizes of Illinois Coal on a Mikado Locomotive, by E. C. Schmidt, J. M. Snodgrass, and O. S. Beyer, Jr. 1917. *Fifty cents.*

**Bulletin No. 102.* A Study of the Heat Transmission of Building Materials, by A. C. Willard and L. C. Lichty. 1917. *Twenty-five cents.*

**Bulletin No. 103.* An Investigation of Twist Drills, by B. Benedict and W. P. Lukens. 1917. *Sixty cents.*

**Bulletin No. 104.* Tests to Determine the Rigidity of Riveted Joints of Steel Structures by W. M. Wilson and H. F. Moore. 1917. *Twenty-five cents.*

Circular No. 6. The Storage of Bituminous Coal, by H. H. Stoek. 1918. *Forty cents.*

Circular No. 7. Fuel Economy in the Operation of Hand Fired Power Plants. 1918. *Twenty cents.*

**Bulletin No. 105.* Hydraulic Experiments with Valves, Orifices, Hose, Nozzles, and Orifice Buckets, by Arthur N. Talbot, Fred B Seely, Virgil R. Fleming and Melvin L. Enger. 1918. *Thirty-five cents.*

**Bulletin No. 106.* Test of a Flat Slab Floor of the Western Newspaper Union Building, by Arthur N. Talbot and Harrison F. Gonnerman. 1918. *Twenty cents.*

Circular No. 8. The Economical Use of Coal in Railway Locomotives. 1918. *Twenty cents.*

**Bulletin No. 107.* Analysis and Tests of Rigidly Connected Reinforced Concrete Frames, by Mikishi Abe. 1918. *Fifty cents.*

**Bulletin No. 108.* Analysis of Statically Indeterminate Structures by the Slope Deflection Method, by W. M. Wilson, F. E. Richart and Camillo Weiss. 1918. *One dollar.*

---

*A limited number of copies of bulletins starred is available for free distribution.

THE UNIVERSITY OF ILLINOIS

THE STATE UNIVERSITY

Urbana

EDMUND J. JAMES, Ph.D., LL.D., President

THE UNIVERSITY INCLUDES THE FOLLOWING DEPARTMENTS:

The Graduate School

The College of Liberal Arts and Sciences (Ancient and Modern Languages and Literatures; History, Economics, Political Science, Sociology; Philosophy, Psychology, Education; Mathematics; Astronomy; Geology; Physics; Chemistry; Botany, Zoology, Entomology; Physiology; Art and Design)

The College of Commerce and Business Administration (General Business, Banking, Insurance, Accountancy, Railway Administration, Foreign Commerce; Courses for Commercial Teachers and Commercial and Civic Secretaries)

The College of Engineering (Architecture; Architectural, Ceramic, Civil, Electrical, Mechanical, Mining, Municipal and Sanitary, and Railway Engineering)

The College of Agriculture (Agronomy; Animal Husbandry; Dairy Husbandry; Horticulture and Landscape Gardening; Agricultural Extension; Teachers' Course; Household Science)

The College of Law (three years' course)

The School of Education

The Course in Journalism

The Courses in Chemistry and Chemical Engineering

The School of Railway Engineering and Administration

The School of Music (four years' course)

The School of Library Science (two years' course)

The College of Medicine (in Chicago)

The College of Dentistry (in Chicago)

The School of Pharmacy (in Chicago; Ph. G. and Ph. C. courses)

The Summer Session (eight weeks)

Experiment Stations and Scientific Bureaus: U. S. Agricultural Experiment Station; Engineering Experiment Station; State Laboratory of Natural History; State Entomologist's Office; Biological Experiment Station on Illinois River; State Water Survey; State Geological Survey; U. S. Bureau of Mines Experiment Station.

The library collections contain (November 1, 1918) 437,949 volumes and 108,289 pamphlets.

For catalogs and information address

THE REGISTRAR

URBANA, ILLINOIS

109

IVERSITY OF ILLINOIS BULLETIN
ISSUED WEEKLY
VI December 2, 1918 14
as second-class matter December 11, 1912, at the post office at Urbana, Illinois, under the Act of August 24, 1912. Acceptance for mailing at the special rate of postage provided for in section 1103, Act of October 3, 1917, authorized July 31, 1918]

THE ORIFICE AS A MEANS OF MEASURING FLOW OF WATER THROUGH A PIPE

RAYMOND E. DAVIS

HARVEY H. JORDAN

BULLETIN
ENGINEERING EXPERIMENT STATION
PUBLISHED BY THE UNIVERSITY OF ILLINOIS, URBANA

EUROPEAN AGENT
CHAPMAN & HALL, LTD., LONDO

THE Engineering Experiment Station was established by act of the Board of Trustees, December 8, 1903. It is the purpose of the Station to carry on investigations along various lines of engineering and to study problems of importance to professional engineers and to the manufacturing, railway, mining, constructional, and industrial interests of the State.

The control of the Engineering Experiment Station is vested in the heads of the several departments of the College of Engineering. These constitute the Station Staff and, with the Director, determine the character of the investigations to be undertaken. The work is carried on under the supervision of the Staff, sometimes by research fellows as graduate work, sometimes by members of the instructional staff of the College of Engineering, but more frequently by investigators belonging to the Station corps.

The results of these investigations are published in the form of bulletins, which record mostly the experiments of the Station's own staff of investigators. There will also be issued from time to time, in the form of circulars, compilations giving the results of the experiments of engineers, industrial works, technical institutions, and governmental testing departments.

The volume and number at the top of the front cover page are merely arbitrary numbers and refer to the general publications of the University of Illinois: *either above the title or below the seal is given the number of the Engineering Experiment Station bulletin or circular which should be used in referring to these publications.*

For copies of bulletins, circulars, or other information address the

ENGINEERING EXPERIMENT STATION,
URBANA, ILLINOIS.

UNIVERSITY OF ILLINOIS
ENGINEERING EXPERIMENT STATION

BULLETIN No. 109 DECEMBER, 1918

THE ORIFICE AS A MEANS OF MEASURING FLOW OF WATER THROUGH A PIPE

BY

RAYMOND E. DAVIS
ASSOCIATE IN CIVIL ENGINEERING

AND

HARVEY H. JORDAN
ASSISTANT PROFESSOR IN GENERAL ENGINEERING DRAWING

ENGINEERING EXPERIMENT STATION
PUBLISHED BY THE UNIVERSITY OF ILLINOIS, URBANA

CONTENTS

CONTENTS (CONTINUED)

LIST OF FIGURES

LIST OF TABLES

THE ORIFICE AS A MEANS OF MEASURING FLOW OF WATER THROUGH A PIPE

I. Introduction

1. *Scope of Bulletin.*—In this bulletin are given the results of tests made to determine the practicability of measuring the flow of water by means of the thin-plate circular orifice inserted in a pipe, to determine the experimental coefficients for calculating the velocity of the flow in the pipe and the discharge, and to determine the conditions most favorable to the use of such an orifice as a flow measuring device. An orifice thus inserted causes an abrupt change in the conditions of flow as the stream approaches and passes through the orifice and this change in the conditions of flow is accompanied by a considerable change in pressure head which may be readily measured and which varies with the velocity of flow in the pipe. The orifice inserted in a pipe line is looked upon more especially as a temporary or field device for measuring the flow of water through a pipe, being inexpensive, simple to construct, light in weight, and easy to install. There may be opportunities for its use in permanent installations where a continuous record is not necessary, loss in head is not an important factor, and the expense of any one of the common flow meters is not justified. In flanged pipe systems the thin-plate orifice may be inserted at a joint with little or no disturbance of existing piping; and in long pipe lines the loss in head caused by the orifice will be inconsiderable as compared with other losses. It would seem that this device might well be utilized, for example, in testing the efficiency of certain forms of pumps, in measuring the performance of individual wells of water works systems, in determining the water consumption for individual purposes in mills and factories, in measuring the discharge through city mains, and in distributing water for irrigation purposes. In planning and conducting the tests the conditions under which the orifice would ordinarily be found useful have been borne in mind, and no attempt has been made to refine apparatus or methods of observing beyond what might be expected in practice.

Tests were made to determine: (1) the positions of two cross-sections of the pipe, one section upstream from the orifice, the other downstream from the orifice, at which pressure head may be measured and the drop in pressure from one section to the other may be most favorably determined, (2) the relation between this drop in pressure head and the rate of discharge through the pipe, (3) the lost head occasioned by the orifice, (4) the effect of small deviations from what may be termed standard conditions, and (5) the proper size of orifice for given conditions. In the presentation of the results of these tests special attention has been given to the probable sources, magnitudes, and effects of the accidental and constant errors incidental to the observations. Attempt has been made through the use of tables and curves to render the results easily adaptable to all sizes of pipe from 4 inches to 20 inches in diameter and for all sizes of orifice up to five-sixths of the diameter of the pipe. In the belief that others than experienced hydraulicians may find use for the results, some attention has been given to matters connected with the construction, installation, and use of a form of apparatus adapted to normal practice.

2. *Other Experiments.*—The method of measuring water by inserting an orifice in the pipe line is not new. Partially closed valves have been calibrated as orifices for such use and experiments of a preliminary nature have been conducted recently with thin-plate orifices. In "Experiments on Water Flow through Pipe Orifices,"* Horace Judd reports the progress of somewhat similar experiments conducted at Ohio State University, the orifices being in plates of Monel metal 1/32 inch thick inserted in a 5-inch pipe.

In "Diaphragm Method of Measuring the Velocity of Fluid Flow in Pipes,"† Holbrook Gaskell, Jr., gives an account of a brief series of tests on 6-inch and 8-inch pipe.

3. *Acknowledgment.*—The tests were made in the Laboratory of Applied Mechanics of the University of Illinois in 1914-15 as graduate work in the Department of Theoretical and Applied Mechanics and in 1916-17 as an investigation of the University of Illinois Engineering Experiment Station. The investigations were under the general supervision of ARTHUR N. TALBOT, Professor of Municipal and Sanitary Engineering and in Charge of the Department of Theoretical and

*Jour. of the Am. Soc. of M. E., Sept. 1916.

†Minutes of Proc. of the Inst. of C. E., 1914.

Applied Mechanics. To Professor Talbot credit is due for valuable suggestions made during the preparation of this bulletin. To PROFESSORS M. L. ENGER and FRED B SEELY, of the Department of Theoretical and Applied Mechanics, credit is due for many helpful suggestions made during the progress of the work.

The tests treated in this bulletin are the outgrowth of brief experiments by C. S. MULVANEY, of the class of 1914, conducted under the direction of V. R. FLEMING, Assistant Professor of Applied Mechanics. The results of the experiments of Mr. Mulvaney were presented as an undergraduate thesis.

4. *Notation.*—Throughout the bulletin the following notation will be used:

V = mean velocity (ft. per sec.) in pipe at a section of normal uniform flow.

v = velocity (ft. per sec.) of the jet issuing from the orifice at the section of greatest contraction.

A = cross-sectional area of the pipe (sq. ft.).

Q = rate of discharge (cu. ft. per sec.).

D = diameter of the pipe (in.).

d = diameter of the orifice (in.).

h_a = lost head (ft. of water) caused by the orifice, including the loss due both to contraction and to sudden expansion of the jet.

h_b = drop in pressure head (ft. of water): or change in pressure head as registered by a differential gage one column of which is connected to the pipe at a section of normal uniform flow upstream from the orifice and the other column of which is connected to the pipe in the same plane as the section of greatest contraction of the jet downstream from the orifice.

C = coefficient of discharge.

C_c = coefficient of contraction.

C_v = coefficient of velocity.

g = acceleration of gravity (ft. per sec.^{2}).

C_p = factor depending on diameter of pipe.

K = velocity modulus; or velocity corresponding to a drop in pressure head (h_b) of 1 ft.

C_r = ratio factor, depending on $\frac{D}{d}$.

F = factor showing the dependence of h_a on diameter of pipe.

J = coefficient expressing the relation between h_a and h_b.

C' = coefficient of discharge for bevel-edged orifices.

II. Analytical Relation between Velocity, Drop in Pressure Head, and Lost Head

5. *Relation between Drop in Pressure Head and Velocity of Flow.*—In the use of the pipe orifice the velocity of flow in the pipe (or the rate of discharge) is calculated from the drop in the pressure head between a section upstream from the orifice and a section downstream from the orifice. It is desirable, therefore, to develop a simple rational expression for the velocity in terms of the drop in the pressure head. Furthermore, since the loss of head caused by the orifice in the pipe may be a determining factor in the selection of the proper orifice it is desirable that a rational expression be found for the lost head. Assumptions have been made concerning the behavior of the jet which have some experimental justification as will be discussed later.

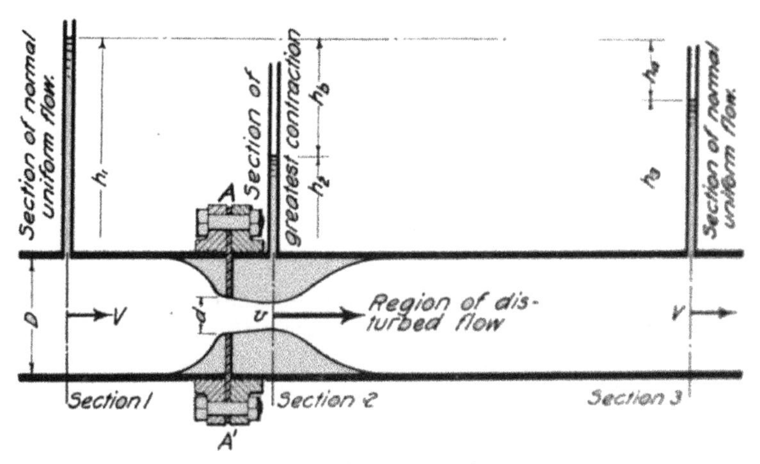

Fig. 1. Behavior of Jet

Fig. 1 represents a longitudinal section of a pipe with the water passing through an orifice at AA'. Let section 1 be a section of normal uniform flow on the upstream side of the orifice. As the stream approaches the orifice from this section it converges and, if it may be assumed to behave not unlike the jet issuing from the standard orifice in the open air, it continues to converge after leaving the orifice until the greatest contraction takes place at, let us say, section 2. The jet then expands until normal flow takes place at section 3.

Let h_1, h_2, and h_3 respectively be the pressure heads registered by piezometer tubes inserted at sections 1, 2, and 3, and let it be assumed that the pressure head h_2 is the same as it would be if the piezometer tube at section 2 were projecting through the wall of the pipe to the periphery of the jet at the contracted section.

Neglecting pipe friction, the drop in pressure head between the section of beginning of convergence and the section of greatest contraction (Fig. 1) is equal to the change in velocity head between the two sections plus the loss in head due to the contraction of the jet. Then by Bernoulli's Law

$$h_b = h_1 - h_2 = \frac{v^2}{2\,g\,C_v^2} - \frac{V^2}{2\,g}$$

Since $\frac{v}{V} = \frac{D^2}{C_c\,d^2}$ and $C = C_c\,C_v$

$$h_b = \left[\frac{\left(\frac{D}{d}\right)^4}{C^2} - 1\right]\frac{V^2}{2\,g} \quad . \quad . \quad . \quad . \quad . \quad . \quad . \quad . \quad (1)$$

$$V = \sqrt{\frac{2g\,h_b}{\frac{\left(\frac{D}{d}\right)^4}{C^2} - 1}} \quad . \quad . \quad . \quad . \quad . \quad . \quad . \quad . \quad (2)$$

It is evident that this drop in pressure head, h_b, is the maximum drop in head that may be found near the orifice and is somewhat greater than the sum of all losses occasioned by the orifice, since beyond the contracted section as the velocity decreases from v to V a portion of the velocity head is transformed into pressure head.

6. *Expression for Lost Head.*—Between two sections of uniform flow, one section on each side of the orifice, such as section 1 and section 3 (Fig. 1), a change in pressure head is accompanied by three losses—the loss due to contraction, the loss due to sudden expansion, and the loss due to pipe friction. If the latter factor is eliminated from consideration, then by methods similar to those used in the

derivation of the expression in the preceding article it may be shown that

$$h_a = h_1 - h_3 = \frac{V^2}{2g}\left[\left(\frac{D}{d}\right)^4\left(\frac{1}{C^2} - \frac{1}{C_c^2}\right) + \left(\left(\frac{D}{d}\right)^2\frac{1}{C_c} - 1\right)^2\right] \qquad (3)$$

This is the loss of head caused by the orifice. It is evidently always less than the drop in pressure head of equation (1) but approaches that quantity as the ratio $\frac{D}{d}$ increases.

In order properly to apply these equations, the sections previously referred to, particularly the section of beginning of convergence and the section of greatest contraction, must be studied experimentally. The method of modifying the rational equations to conform with the experimental results will be discussed later.

III. Tests and Results

7. *Scope of Experiments.*—The principal tests were made on 4-inch, 6-inch and 12-inch pipe with eight sizes of orifices for each size of pipe as recorded in Table 1. It will be noted that the diameters

Table 1

Diameters of Orifices and Pipes

4-inch Pipe $D=4.06$ in.		6-inch Pipe $D=6.12$ in.		12-inch Pipe $D=12.15$ in.	
d	$\frac{D}{d}$	d	$\frac{D}{d}$	d	$\frac{D}{d}$
0.497	8.17	0.75	8.16	1.50	8.10
0.667	6.09	1.00	6.12	2.00	6.08
1.00	4.06	1.49	4.10	3.00	4.05
1.34	3.04	2.00	3.06	4.00	3.04
1.67	2.43	2.50	2.44	5.00	2.43
2.00	2.03	3.01	2.03	6.00	2.03
2.68	1.52	4.02	1.52	8.00	1.52
3.33	1.22	5.01	1.22	10.00	1.22

of the orifices range from one-eight to five-sixths the diameter of the pipe, and that the eight ratios of diameter of pipe to diameter of orifice for the 4-inch series are approximately the same as for the 6-inch and 12-inch series.

The mean velocity in the pipe ranged from 0.01 ft. per sec. with the smallest orifice of each series to a maximum velocity of about 23 ft. per sec. with the largest orifice of the 4-inch series, 14 ft. per sec. with the 6-inch series, and 3¼ ft. per sec. with the 12-inch series.

Experiments were conducted to investigate phases of the problem as follows:

(1) Pressure Variations and Behavior of Stream

By observations of pressure variations along the pipe the position of the following sections was found: (a) the section at which the flow in the pipe changes from normal uniform flow and the stream begins to converge towards the orifice; (b) the section at which the jet issuing from the orifice becomes fully expanded and normal uniform flow is resumed; and (c) the section of greatest contraction of the jet issuing from the orifice. The change in pressure head was taken to indicate the change in area of stream flow, the section of greatest contraction being located where the pressure head is the least. The terms normal uniform flow, beginning of convergence, section of greatest contraction, and section where jet becomes fully expanded and normal uniform flow is resumed are used for convenience to describe the general phenomena at the several sections; they are not to be taken to define strictly the conditions of flow. The general behavior of the stream in the vicinity of the orifice was also studied.

(2) Relation between Drop in Pressure and Discharge

The observations of pressure variations along the pipe indicated that the two most favorable sections for gage connections, the sections giving the steadiest pressure and the most reliable drop in pressure head were a section at or slightly upstream from the section of beginning of convergence and one at the section of greatest contraction. For these sections a greater change in pressure head was found than for any other two sections in the vicinity of the orifice.

For measured rates of discharge the drop in pressure head as registered by a differential gage connected at approximately the section of beginning of convergence and at the section of greatest contraction was observed. From these tests coefficients of discharge will be derived.

(3) Lost Head

The lost head caused by the orifice, if pipe friction is disregarded, is represented by the change in pressure head between the section of beginning of convergence upstream from the orifice and the section of resumed normal uniform flow down stream from the orifice, or is the difference between the pressure heads at two sections of normal uniform flow, one section above and one below the orifice. Observations were made to determine the relation between lost head and discharge.

(4) Effect of Varying the Conditions

Tests were made to determine: (a) the effect of a small lateral displacement of the orifice, (b) the effect of throttling the gage on the precision of observations, (c) whether or not there is a measure able difference between the drop in pressure head with gage connections made at the top of each section and the drop in pressure head with the gage connections made at the bottom of each section, (d) the effect of number of connections at each section on the precision of observations, and (e) the effect of beveling the upstream edge of the orifice.

The drop in pressure was measured between a section at approximately the beginning of convergence and the section of greatest contraction. The tests were made only on 4-inch and 6-inch pipe.

8. *Apparatus.*—The pipe upon which tests were made was commercial steel pipe which had been in service some years. In general there was a small but measurable difference between the diameters of the adjoining sections of pipe between which the orifice plate was inserted, and, due to roughnesses and other variations, the error of measuring the diameter was perhaps 0.02 inch. The pipe was horizontal and straight for a sufficient distance on each side of the orifice to make the effect from bends negligible.

All orifices were circular in shape and were cut in 3/16-inch steel plates. In general the edges of the orifices were square. To find the effect of a deviation from this standard, experiments were also made with bevel-edged orifices, a bevel of 45 degrees being made on the upstream side of the orifice in such a way as to leave a thickness of metal of 1/32 inch at the throat.

Drop in pressure head was measured by the usual form of U-tube differential gage, the water gage being used for small differences, and the mercury gage for large differences.

The gage connections to the pipe at the sections under consideration were ¼-inch pipe nipples. Care was taken that no burr was left at the inner edge of the tapped holes and that the nipples did not protrude beyond the inner surface of the pipe. The pipe nipples were connected with the gages by ordinary rubber tubing.

9. *Water Supply and Measurement.*—The supply of water was obtained from the 60-foot standpipe of the Hydraulic Laboratory. For most of the tests the water level in the standpipe was within a few feet of the top and a nearly constant head was maintained by a 2200-gallon duplex pump, the pump being automatically regulated. For most work the head remained nearly constant during a given run,

FIG. 2. APPARATUS FOR OBSERVING PRESSURE CHANGES

but for large discharges surges occurred in the standpipe resulting in momentary fluctuations in pressure in the pipe.

When less than about 1/3 cu. ft. per sec., the discharge was measured by weight; when greater, by displacement.

Time of discharge was observed with a calibrated stop watch.

10. *Pressure Changes near Orifice.*—For these tests gage connections to the pipe were inserted at the sections indicated in Table 2,

TABLE 2

POSITION OF SECTIONS AT PRESSURE CONNECTIONS

Diameter of pipe and distances from the orifice to the various sections are given in inches.

Diam. of Pipe	Sec. A to Orif.	Sec. B to Orif.	Orif. to Sec. C	Orif. to Sec. D	Orif. to Sec. E	Orif. to Sec. F	Orif. to Sec. G	Orif. to Sec. H	Orif. to Sec. I	Sec A to Sec. I
4.06	3¼	2	½	2	3½	9½	15½	25½	49½	53
6.12	4½	2½	½	2⅛		8¼	20	38	71	77
12.15	5			4¼		16	40	76	127	132

section A being upstream from the orifice and the change in pressure head between adjacent sections was observed at from 20 to 40 rates of flow for each orifice. The maximum changes in pressure observed ranged from 0.01 ft. to 50 ft.

Fig. 2 shows the apparatus for observing changes in pressure head in the 12-inch pipe, the orifice plate being between the flanges near the left of the picture and the direction of flow being from left to right. At the sections adjacent to the orifice (section A and section D, Table 2), gage connections were inserted both in the top and in the bottom of the pipe, and the two nipples at each of these sections were brought to a union. Two connections were also used at section I. At other sections a single nipple was inserted in the bottom of the pipe.

The results of the observations are shown graphically in Fig. 3. For ease in making comparison, changes in pressure head along the pipe are shown in terms of the lost head caused by the orifice ratio of change in pressure head to lost head), an allowance for pipe friction having been made.

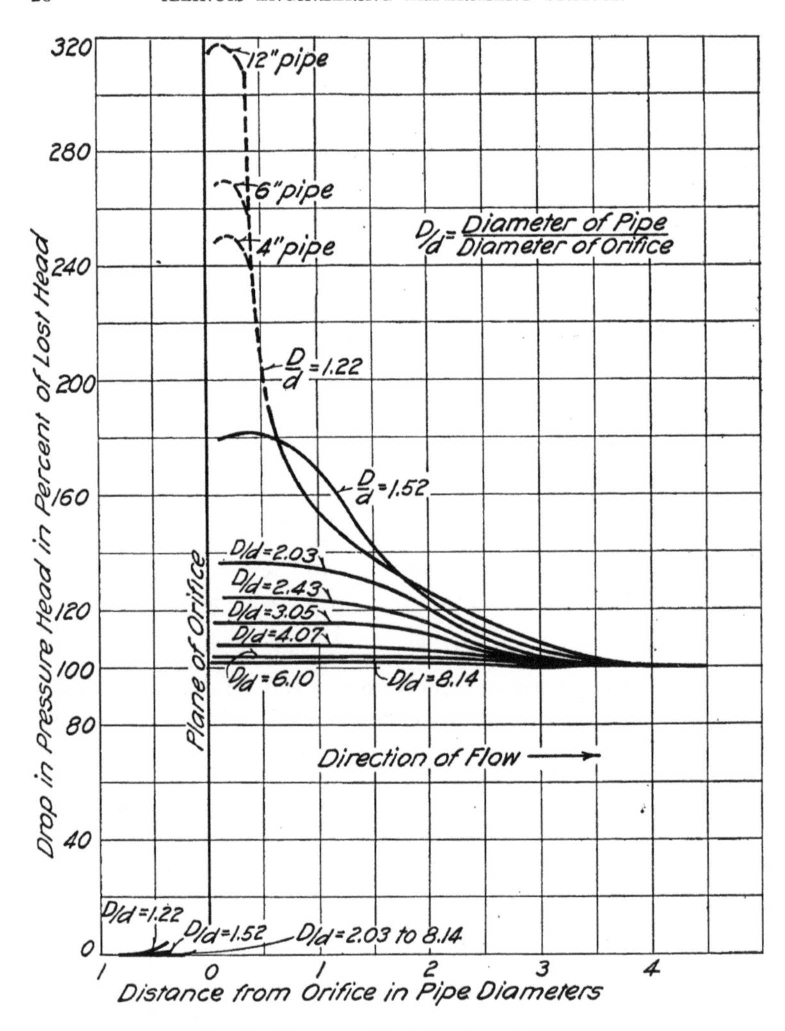

FIG. 3. PRESSURE VARIATIONS NEAR ORIFICE

For each $\frac{D}{d}$ the ordinate of the curve at any section was obtained by averaging the ratios for all rates of flow, there being nothing to indicate a consistent variation between these ratios at low rates of flow and at high rates of flow. That there are not three curves shown for each $\frac{D}{d}$ merely indicates that there is close agreement between the general behavior of the jet for one size of pipe and the general behavior for other sizes, and does not indicate that the size of pipe has no influence on the relation between velocity and change in pressure head. The dotted portion of the curve for $\frac{D}{d}=1.22$ is regarded as uncertain.

11. *Deductions on Behavior of the Jet.*—A study of the observations of the changes in pressure head (which are too numerous to be shown here) and of the curves in Fig. 3 warrants the following deductions:

(1) The location of the section of beginning of convergence as the stream approaches the orifice is approximately 0.8 of the diameter of the pipe upstream from the plane of the orifice, for $\frac{D}{d}=1.2$, and it gradually approaches the orifice as $\frac{D}{d}$ increases.

(2) The section of greatest contraction is at a fairly constant distance of 0.4 of the pipe diameter downstream from the plane of the orifice for values of $\frac{D}{d}$ of 1.5 or greater. As $\frac{D}{d}$ becomes less than 1.5, the section of greatest contraction gradually approaches the orifice.

(3) The distance from the orifice to the section at which the jet has fully expanded and normal uniform flow is resumed varies between 3 and 4 pipe diameters, increasing as $\frac{D}{d}$ decreases until $\frac{D}{d}$ becomes 1.5, and thereafter probably decreasing.

(4) Downstream from the contracted section there is a region of greatly disturbed pressure. As the diameter of the orifice approaches the diameter of the pipe, this region approaches the section of greatest contraction and the pressure fluctuations as shown by gage readings become more violent.

(5) The proportional accidental error of observing drop in pressure head between the section of beginning of convergence

upstream from the orifice and sections downstream from the orifice is in general least when the downstream section is at the point of greatest contraction. The error increases as the distance from this section is increased until the region of maximum pressure fluctuation is reached, and decreases thereafter as the downstream section approaches the section of resumed normal uniform flow.

(6) Except for $\frac{D}{d}=1.2$ (the largest orifice used with each size of pipe), there is no indication of a variation between the general behavior of the flow passing through the orifice used with one size of pipe and that passing the corresponding orifices used with other sizes of pipe. That is, for a particular value of $\frac{D}{d}$, except $\frac{D}{d}=1.2$, there is no indication that the curve showing the pressure variation for the 4-inch pipe should not coincide with that for the 6-inch pipe and for the 12-inch pipe. For $\frac{D}{d}=1.2$ there is evidence of considerable variation, though whether or not this is to any considerable extent due to size of pipe is problematical.

The preceding deductions indicate that, for the purpose of flow measurement, gage readings of drop in pressure should be taken with gage connections at the section of beginning of convergence or upstream therefrom and at the section of greatest contraction. The distances to these sections, as far as effect on measurement of drop in pressure head is concerned, may be considered as 0.8 D upstream from the orifice and 0.4 D downstream from the orifice. In the tests to determine the relation between drop in pressure and discharge these distances were used.

12. *Relation between Drop in Pressure Head and Discharge.*—Since the use of the pipe orifice as a flow measuring device depends upon the proper relations between the drop in pressure head and the velocity (or rate of discharge) a wide range of tests was made to determine their relation after the preliminary experiments had shown the proper arrangement of the apparatus as already discussed.

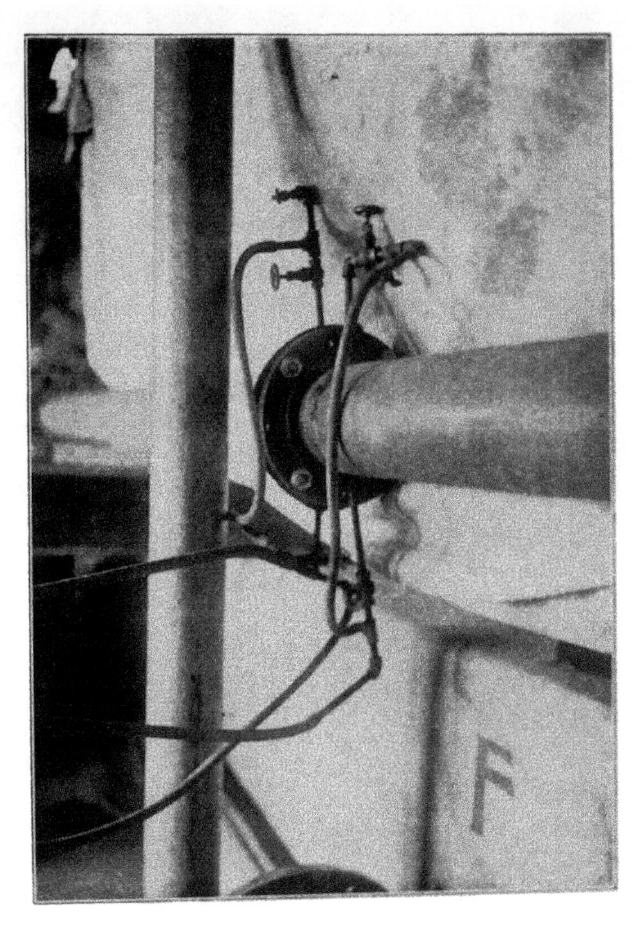

FIG. 4. PIPE CONNECTIONS NEAR ORIFICE

FIG. 5. DIFFERENTIAL GAGES

For various rates of flow the discharge was measured during an observed time interval and the drop in pressure head was read from the differential gage. The velocities in the pipe ranged from 0.01 ft. per sec. with the smallest orifice of each series to 24 ft. per sec. with the largest orifice of the 4-in. series; the drop in pressure ranged from 0.01 ft. to 56 ft. On the average 30 observations for as many rates of flow were made for each size of orifice.

Fig. 4 shows the gage connections with the pipe at the two sections adjacent to the orifice, the distances to these sections being 0.8

TABLE 3

SAMPLE DATA

Orifices in 4 in. pipe

Head in ft. of water. Time in sec. Discharge in cu. ft. per sec.

Obs. No.	Net Wt.	Cor. Time	Lost Head	Drop in Pressure Head	Discharge
		Orifice 1.67 in. diameter.			
183	345	240	0.073	0.088	0.0230
184	394	254½	0.086	0.104	0.0247
185	414	240	0.105	0.127	0.0276
186	332	180	0.123	0.151	0.0295
187	447	221½	0.144	0.174	0.0323
188	534	240	0.172	0.213	0.0356
189	1076	420	0.229	0.283	0.0410
190	525	180	0.293	0.360	0.0467
191	583	179½	0.363	0.454	0.0519
192	660	180	0.468	0.585	0.0587
193	751	182½	0.591	0.738	0.0658
194	839	180	0.760	0.950	0.0746
195	938	180	0.95	1.18	0.0834
196	977	165	1.22	1.52	0.0948
197	912	132¾	1.68	2.05	0.110
198	960	121¾	2.21	2.76	0.126
199	1515	160	3.17	3.95	0.152
200	1739	180	3.26	4.00	0.155
201	1884	179½	3.86	4.76	0.168
202	1857	164	4.45	5.48	0.181
203	1854	149½	5.36	6.68	0.199
204	1852	135	6.56	8.07	0.219
205	1825	120	7.98	9.97	0.244
206	1821	106	10.1	12.7	0.276
207	3256	174½	12.9	15.9	0.298
208	3373	155	16.7	20.6	0.348
209	3385	140	20.3	25.4	0.387
210	3129	120	23.7	29.5	0.417
		Orifice 2.00 in. diameter			
211	252	360	0.008	0.009	0.0112
212	301	360	0.011	0.013	0.0134
213	223	300	0.013	0.015	0.0144
214	284	300	0.013	0.018	0.0152
215	519	484	0.017	0.022	0.0172
216	246	200	0.020	0.026	0.0197
217	247	200	0.021	0.030	0.0197
218	683	490	0.030	0.039	0.0223
219	818	560	0.034	0.044	0.0234
220	578	360	0.040	0.052	0.0257
221	576	330	0.046	0.060	0.0279

TABLE 3 (Continued)

SAMPLE DATA

Orifices in 12-in. pipe

Head in ft. of water. Time in sec. Discharge in cu. ft. per sec.

Obs. No.	Net Wt.	Cor. Time	Lost Head	Drop in Pressure Head	Discharge
Orifice 1.50 in. diameter					
613a	224	600	0.011	0.011	0.00598
614a	210	500	0.013	0.013	0.00672
615a	198	400	0.017	0.018	0.00793
616a	205	360	0.021	0.022	0.00911
617	209	339	0.028	0.029	0.00987
618	216	300	0.036	0.036	0.0115
619	202	261	0.042	0.044	0.0124
620	203	241	0.050	0.052	0.0135
621	208	215	0.067	0.067	0.0155
622	258	240	0.079	0.080	0.0172
623	376	332	0.090	0.092	0.0181
624	342	240	0.112	0.113	0.0228
625	300	241	0.135	0.135	0.0199
626	372	240	0.176	0.176	0.0248
627	347	199	0.216	0.216	0.0279
628	392	199	0.274	0.277	0.0315
629	448	200	0.350	0.353	0.0358
630	497	200	0.428	0.437	0.0398
631	517	191	0.516	0.523	0.0433
632	585	191	0.663	0.675	0.0490
633	596	180	0.773	0.790	0.0530
634	650	180	0.915	0.940	0.0578
635	730	184	1.12	1.14	0.0635
636	830	179	1.50	1.53	0.0742
637	1040	202	1.86	1.91	0.0825
638	1512	260	2.39	2.44	0.0930
639	1542	240	2.97	3.04	0.103
640	1725	240	3.66	3.72	0.115
641	1644	200	4.79	4.89	0.132
642	1679	180	6.18	6.31	0.149
643	1902	180	7.88	8.04	0.169
644	1787	155	9.56	9.70	0.184
645	3286	259	11.5	11.7	0.203
646	3218	215	15.9	16.2	0.239
647	1800	103¼	21.2	21.6	0.279
648	3254	166½	27.2	27.8	0.312
649	3110	138	36.2	36.9	0.361
650	3398	145	41.4	42.1	0.375
Orifice 2.00 in. diameter					
651	224	360	0.009	0.010	0.00997
652	241	300	0.012	0.013	0.0129
653	206	240	0.016	0.018	0.0137
654	223	270	0.017	0.017	0.0132
655	258	240	0.026	0.028	0.0172
656	246	188½	0.037	0.040	0.0209
657	287	180	0.056	0.060	0.0255

D upstream and 0.4 *D* downstream from the orifice and there being two pressure openings at diametrically opposite points at each section. Fig. 5 shows the differential gages with connecting hose. The two gages on the left, one for mercury, the other for water, are connected to the sections adjacent to the orifice and register the drop in pressure head; the two gages on the right register the lost head due to the orifice plus

TABLE 3 (Concluded)

SAMPLE DATA

Orifices in 12-in. pipe

Head in ft. of water. Time in sec. Discharge in cu. ft. per sec.

Obs. No.	Cubic Feet			Cor. Time	Lost Head	Drop in Pressure Head	Cu. Ft Sec.
	Initial	Final	Net				
Orifice 6.00 in. diameter							
783	48.2	167.3	119.1	293½	0.136	0.179	0.407
784	54.7	170.3	115.6	233	0.197	0.265	0.497
785	57.7	168.3	110.6	271½	0.136	0.183	0.407
786	54.5	189.4	134.9	228½	0.276	0.370	0.589
787	53.3	200.9	147.6	217	0.371	0.497	0.680
788	57.2	209.6	152.4	194	0.490	0.66	0.786
789	45.1	193.7	148.6	168½	0.61	0.82	0.883
790	44.5	178.7	134.2	134	0.80	1.06	1.00
791	37.0	177.0	140.0	122½	1.04	1.40	1.14
792	13.6	190.9	177.3	133½	1.40	1.89	1.33
793	14.1	187.9	173.8	113½	1.84	2.47	1.53
794	1.4	204.1	205.5	119	2.38	3.18	1.72
795	15.5	184.3	168.8	78	3.76	5.07	2.16
796	4.4	209.9	205.5	85½	4.50	6.23	2.39
797	6.8	226.6	219.8	84½	5.27	7.12	2.60
798	11.0	204.5	193.5	72	6.46	8.94	2.69
799	2.3	236.6	234.3	79½	6.89	9.32	2.95
Orifice 8.00 in. diameter. Observed discharge given in pounds							
800			1443	122	0.005	0.010	0.189
801			1771	122½	0.008	0.0155	0.232
802			3245	235	0.007	0.014	0.221
803			3187	200	0.010	0.018	0.255
804			3197	176½	0.013	0.024	0.290
805			3148	162	0.015	0.0295	0.313
Discharge in Cubic Feet							
806	55.6	165.4	109.8	302½	0.023	0.040	0.363
807	57.8	169.1	111.3	264½	0.029	0.050	0.421
808	60.3	169.8	109.5	241½	0.035	0.063	0.454
809	11.3	167.4	156.1	293½	0.047	0.085	0.532
810	50.5	167.2	116.7	187½	0.070	0.110	0.622
811	48.2	170.6	122.4	164½	0.092	0.148	0.745
812	45.2	176.3	131.1	156½	0.125	0.200	0.838
813	52.7	182.1	129.4	136½	0.154	0.250	0.948
814	42.8	205.8	163.0	152½	0.206	0.325	1.07
815	51.1	202.4	141.3	125½	0.27	0.435	1.21
816	41.3	215.7	174.4	129½	0.30	0.53	1.35
817	36.3	225.2	188.9	121½	0.41	0.70	1.56
818	35.4	223.1	187.7	102	0.53	0.97	1.84
819	39.4	225.3	185.9	97	0.60	1.10	1.92
820	27.4	221.2	193.8	91	0.76	1.38	2.13
821	25.9	236.7	210.8	88	0.95	1.70	2.40
822	7.1	230.5	223.4	91½	1.07	1.78	2.44

the loss due to pipe friction within the length between gage connections. The lost head will be discussed in section 16.

A sample of the data thus obtained is shown in Table 3. The values of lost head there recorded have been corrected for pipe friction.

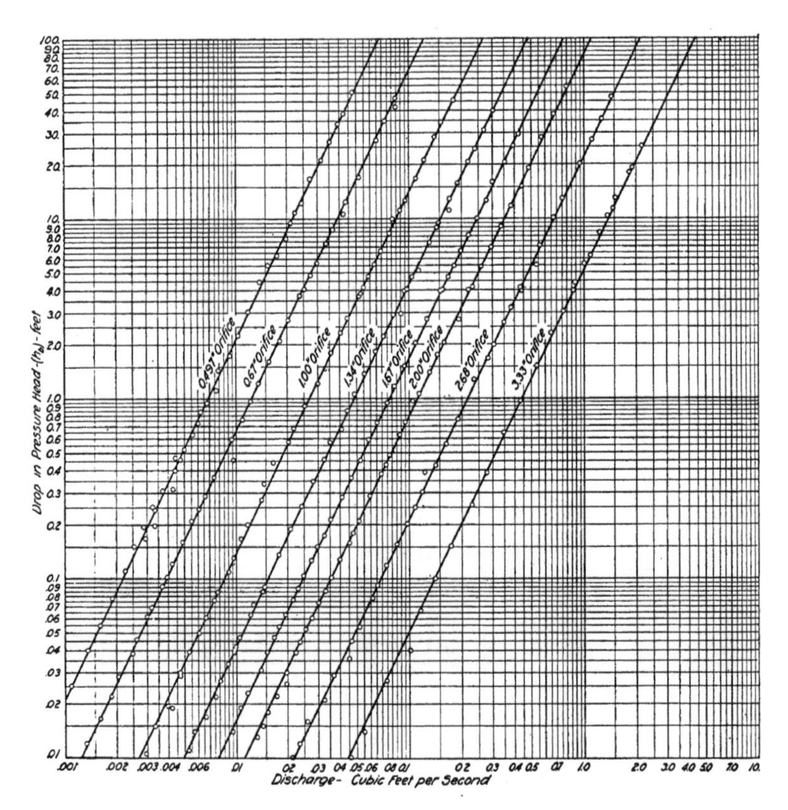

FIG. 6. EXPERIMENTAL DISCHARGE CURVES—4-INCH PIPE

The data of the tests are plotted logarithmically in Figs. 6, 7 and 8, the observed drop in pressure head in feet being plotted against measured discharge in cubic feet per second. The lines which have been drawn through the mean of the plotted points are seen to be straight and very nearly parallel. It will also be noted that the curves for one series bear the same general relation to one another as do the curves for the other two series.

The mean of the measured slopes of the 24 curves varies only slightly from 2.00. A study of the curves indicates that the variation of the slopes of the individual curves from the mean value is partly

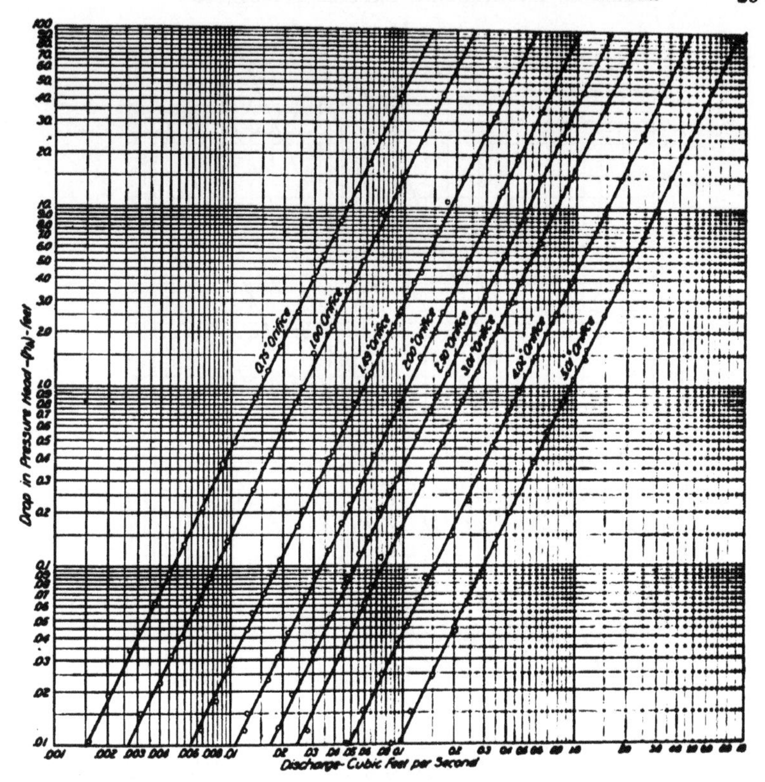

FIG. 7. EXPERIMENTAL DISCHARGE CURVES—6-INCH PIPE

accidental. Since the curves are plotted to logarithmic scale and have a slope of 2, h_b may be said to vary as V^2, which is in accord with the theoretical expression, equation (1) p. 12. To make the theoretical equation exactly fit the experimental curves it must be assumed that for a particular size of orifice and pipe the coefficient of discharge decreases slightly as the drop in pressure head increases, as may be seen in Table 4.

The values of the coefficients of discharge in Table 4 were obtained by substituting in equation (1) values of h_b and $V = \frac{Q}{A}$ taken from the experimental curves. It will be seen that for a particular value of $\frac{D}{d}$

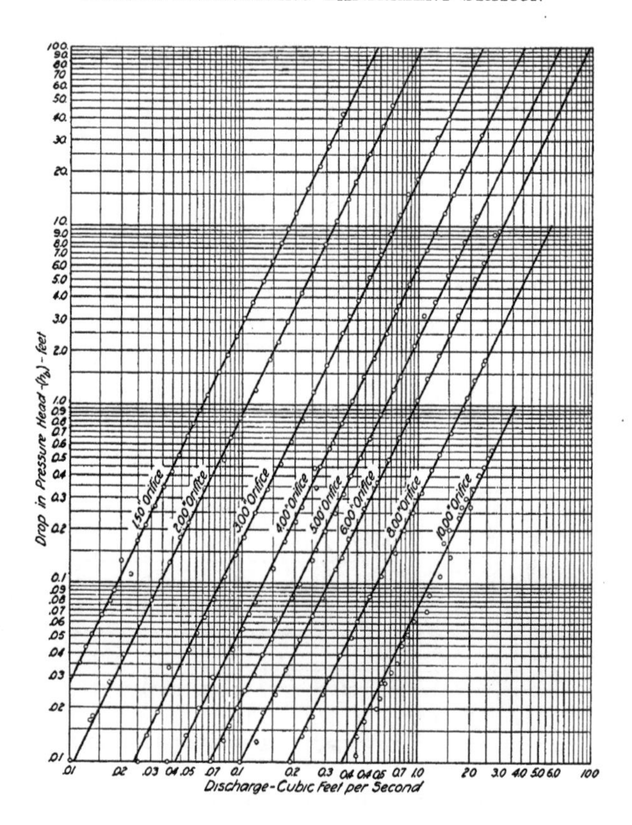

Fig. 8. Experimental Discharge Curves—12-inch Pipe

Table 4

Coefficient of Discharge—C

$\frac{D}{d}$	4-in. Pipe			6-in. Pipe			12-in. Pipe		
	h_b=0.1	h_b=1.00	h_b=50	h_b=0.1	h_b=1.00	h_b=50	h_b=0.1	h_b=1.00	h_b=10
1.22	.771	0.760	.....	0.753	0.746	.....	0.742	.....	.....
1.52	.670	0.663	0.651	0.659	0.651	.....	0.638	0.631	.....
2.03	.633	0.628	0.617	0.620	0.616	0.608	0.615	0.609	0.603
2.43	.622	0.619	0.609	0.609	0.606	0.600	0.608	0.602	0.596
3.05	0.617	0.614	0.606	0.604	0.602	0.597	0.605	0.600	0.594
4.07	.618	0.615	0.608	0.606	0.604	0.598	0.605	0.600	0.595
6.10	.626	0.624	0.618	0.613	0.611	0.601	0.610	0.606	0.602
8.14	.638	0.636	0.630	0.619	0.616	0.604	0.610	0.606	0.602

the coefficient of discharge not only decreases slightly as the drop in pressure head increases but also decreases as the diameter of the pipe increases.

For a better understanding of the general behavior of the coefficient of discharge, and also that the coefficient of discharge may be determined readily for other values of $\frac{D}{d}$ than those of the experiments, the data of Table 4 for the 4-inch pipe are shown graphically by the two upper curves of Fig. 9, values of the coefficients being plotted against $\frac{D}{d}$. The two curves—the full line for h_b=0.1 ft. and the dash line for h_b=50 ft.—are intended to show the extremes of the coefficients which may possibly be used. For $\frac{D}{d}$ greater than 3.0 the coefficient of discharge increases nearly as a straight line, a peculiarity which may perhaps be explained through the fact that as $\frac{D}{d}$ increases the distance from the wall of the pipe to the periphery of the jet at the section of greatest contraction also increases and there is consequently less likelihood of the pressure at the wall of the pipe being the same as that which exists at the periphery of the jet at the section of greatest contraction. As $\frac{D}{d}$ decreases from 3.00 the coefficient of discharge increases, and the rate at which the coefficient changes for small values of $\frac{D}{d}$ still further emphasizes the impracticability of attempting to make precise measurements of discharge when the diameter of the orifice is greater than two-thirds the diameter of the pipe.

The C curves for the 4-inch pipe are typical of those for the 6-inch and 12-inch for the lower values of h_b and $\frac{D}{d}$, but as h_b becomes larger the rate of increase of C diminishes for the higher values of $\frac{D}{d}$. The two lower curves of Fig. 9 illustrate this change in rate of increase of C.

13. *Simplified Equations for Velocity and Discharge.*—The theoretical equation for the determination of velocity in the pipe in terms

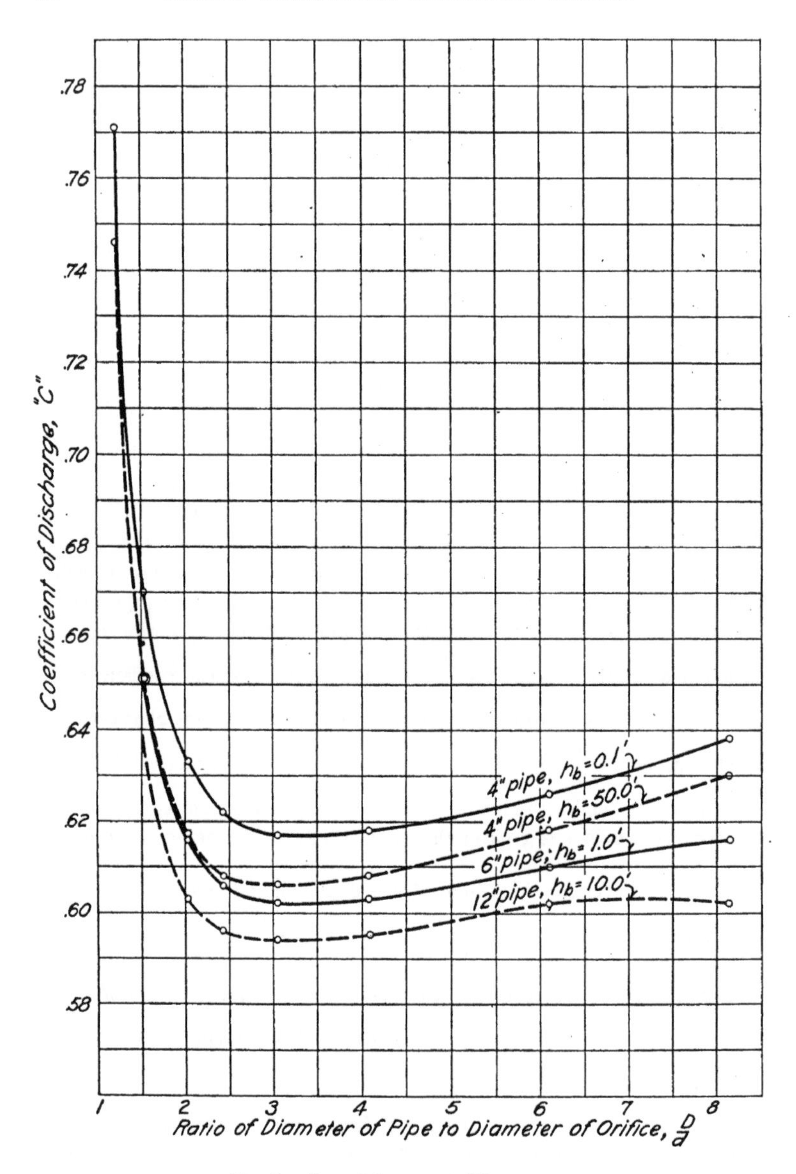

FIG. 9. COEFFICIENTS OF DISCHARGE

of the drop of pressure head (equation (2), p. 12) may be put in the form

$$V = K\sqrt{h_b} \qquad (4)$$

in which $K = \sqrt{\dfrac{2g}{\dfrac{\left(\frac{D}{d}\right)^4}{C^2} - 1}}$

The expression for the rate of discharge then is

$$Q = AK\sqrt{h_b} \qquad (5)$$

Within the range of the drop in pressure likely to be used in practice, K may with small error be regarded as a constant for a particular size of pipe and orifice. To simplify the work of computing velocities, values of K for various sizes of pipes are given in Table 5. For convenience, the term K will be called the *velocity modulus*. Some of the values of the velocity modulus have been carried to one more sig-

TABLE 5

VELOCITY MODULUS—K

Values of K in Equation $V = K\sqrt{h_b}$

$\frac{D}{d}$	Velocity Modulus					
	4-in. Pipe	5-in. Pipe	6-in. Pipe	8-in. Pipe	10-in. Pipe	12-in. Pipe
2.00	1.275	1.250	1.240	1.240	1.240	1.240
2.10	1.150	1.123	1.111	1.110	1.110	1.110
2.20	1.030	1.015	1.010	1.010	1.010	1.010
2.30	0.944	0.930	0.925	0.925	0.925	0.925
2.40	0.865	0.850	0.845	0.845	0.845	0.845
2.50	0.798	0.785	0.779	0.779	0.779	0.779
2.60	0.730	0.725	0.720	0.720	0.720	0.720
2.70	0.684	0.672	0.667	0.667	0.667	0.667
2.80	0.635	0.625	0.620	0.620	0.620	0.620
2.90	0.592	0.583	0.578	0.577	0.576	0.576
3.00	0.553	0.545	0.540	0.540	0.539	0.539
3.25	0.473	0.465	0.462	0.461	0.461	0.461
3.50	0.405	0.399	0.395	0.395	0.395	0.395
3.75	0.351	0.345	0.342	0.342	0.342	0.342
4.00	0.310	0.305	0.303	0.302	0.302	
4.50	0.244	0.240	0.238	0.237	0.237	0.237
5.00	0.197	0.194	0.192	0.192	0.192	0.192
5.50	0.165	0.161	0.160	0.160	0.160	0.160
6.00	0.138	0.136	0.135	0.135	0.135	0.135

nificant figure than the data may warrant in order to permit more nearly accurate interpolation. In interpolating for other values of $\frac{D}{d}$ or other sizes of pipe than those shown it may be assumed that the velocity modulus varies as a straight line.

The computations for the velocity modulus were based on the coefficients of discharge for a drop in pressure head of 1.0 ft. as shown in Table 4 and by the curves of Figs. 9 and 10. (Fig. 10 is explained in the following section.) Since K varies as C and since C decreases as h_b increases, the values of the velocity modulus are for a large drop in pressure head slightly too large and for a small drop in pressure head slightly too small; but within the limits h_b=0.1 ft. to h_b=10 ft., which seem to be about the limits which would be found practicable under ordinary conditions of flow, the error introduced by use of Table 5 is negligible.

14. *Application to Other Sizes of Pipes.*—In order to use equation (2) or (4) for sizes of pipe other than those used in the experiments herein recorded it is important to determine the effect of the diameter of the pipe on the coefficient of discharge. The proper coefficient of discharge is obtained by applying a *diameter factor, C_p,* which is the quantity by which the coefficient of discharge for a 6-inch pipe must be multiplied to produce the coefficient of discharge for a given size of pipe. The values of the diameter factor for various diameters of pipe are shown in Fig. 10. Since the curve is based on only

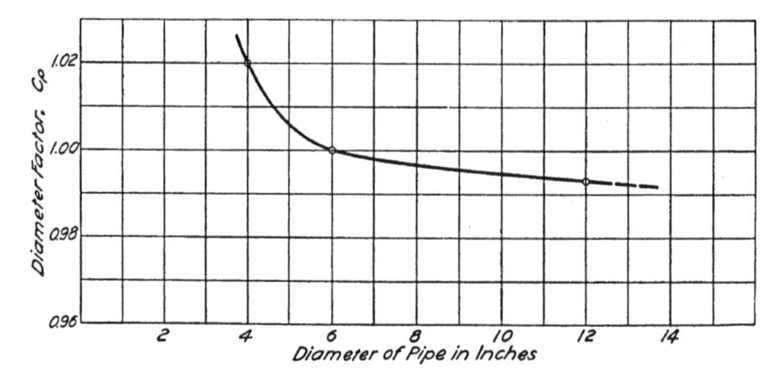

FIG. 10. DIAMETER FACTOR

three plotted points it is likely to be somewhat in error. Its use should be confined to values of $\frac{D}{d}$ between 2 and 6 and between 1 ft. and 10 ft. for h_b if accurate results are desired.

15. *Empirical Relation between Velocity and Drop in Pressure Head.*—From the discharge curves of Figs. 6, 7, and 8 the following empirical equation has been derived for the velocity in terms of the drop in pressure head,

$$V = 4.85 \frac{\sqrt{h_b}}{\left(\frac{D}{d}\right)^2 - \left(\frac{d}{D}\right)^3} \qquad (6)$$

This equation has an advantage over the rational expression, equation (2), in that the variable coefficient of discharge is eliminated. This empirical equation will give approximately the same results as equation (2) for those values of $\frac{D}{d}$ which are recommended for use in engineering practice, namely, for values of $\frac{D}{d}$ between 2 and 6.

16. *Empirical Formula for Lost Head.*—Since the amount of lost head caused by the orifice may be an important factor in the selection of the orifice to be used, or even in accepting the pipe orifice method of measuring the discharge, it is important to know the relation between the drop in pressure head and the lost head. In Fig. 5 the gages on the right registered the lost head due to the orifice plus the loss due to pipe friction within the length between gage connections.

The results of the experiments to determine this relation indicate that the rational expression, equation (3) p. 13, holds good when the ratio $\frac{D}{d}$ is large but gives results which are increasingly too large as $\frac{D}{d}$ decreases, until for $\frac{D}{d} = 1.2$ the lost head as computed by equation (3) is about 3 per cent less than the lost head as determined by experiment. In arriving at this conclusion, the coefficients of discharge as determined from the experiments on drop in pressure head and as shown in Table 4 were used in the rational expression and the coefficients of contraction employed were derived assuming a coefficient of velocity of 0.98.

Equation (3) is too complicated an expression to be used readily in computing. A simpler expression for the lost head and the discharge is given by the empirical equation

$$h_a = 0.0366\, F \left[\left(\frac{D}{d} \right)^2 - \frac{d}{D} \right]^{2.02} \cdot V^2 \quad . \quad . \quad . \quad . \quad . \quad (7)$$

in which F is a factor depending on the size of pipe and having the values 0.98 for 4-inch pipe, 1.02 for 6-inch pipe and 1.04 for 12-inch pipe. This expression gives lost heads correct within about 2 per cent except when $\frac{D}{d}$ is less than 1.5. For $\frac{D}{d} = 1.2$ the expression gives results about 3 per cent too small.

A convenient expression for determining the lost head, having given the drop in pressure head, is given by the equation

$$h_a = 0.84\, h_b \frac{\left[\left(\frac{D}{d} \right)^2 - \frac{d}{D} \right]^{2.02}}{\left[\left(\frac{D}{d} \right)^2 - \left(\frac{d}{D} \right)^3 \right]^{1.99}} = J h_b \quad . \quad . \quad . \quad (8)$$

The values of J in the foregoing equation for given values of h_b are shown by the curve in Fig. 11.

17. *Choice of Orifice.*—The choice of the size of orifice to be used under given conditions may depend upon four factors, the rate of flow in the pipe, the lost head that may be allowed, the desired precision of the discharge measurement, and the maximum drop in pressure head which the differential gage will register. Table 6 which shows approximate values of the drop in pressure head and the lost head for several velocities and values of $\frac{D}{d}$ is intended to be of assistance in estimating the size of orifice best adapted to particular requirements. In Fig. 12 the drop in pressure head for various ratios of $\frac{D}{d}$ is given in terms of the lost head.

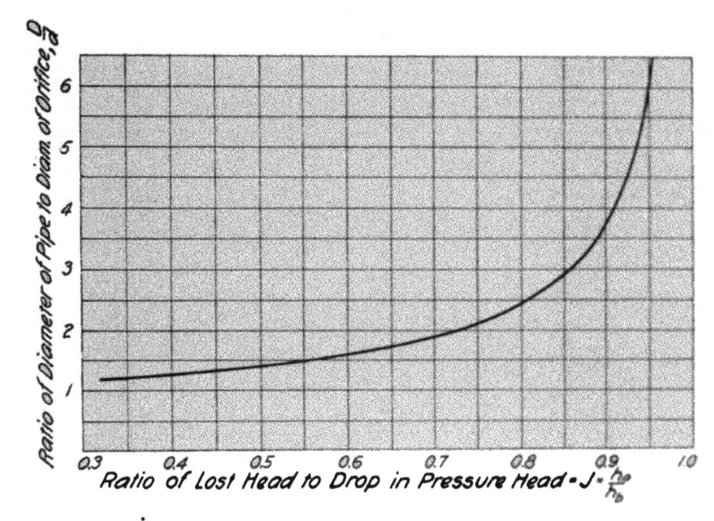

FIG. 11. RELATION BETWEEN LOST HEAD AND DROP IN PRESSURE HEAD

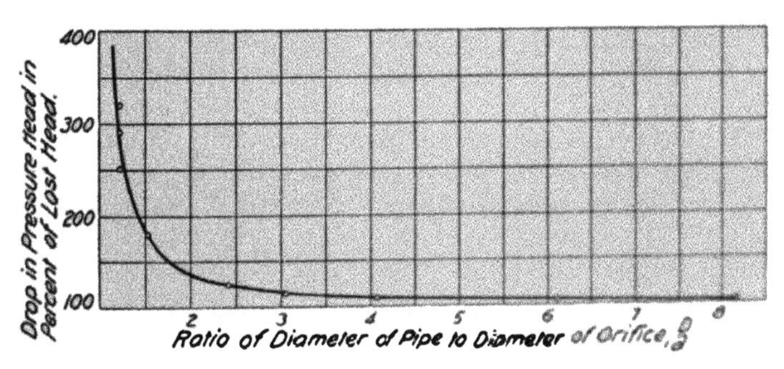

FIG. 12. RELATION BETWEEN MAXIMUM DROP IN PRESSURE AND SIZE OF ORIFICE

The relative error in the observation of the drop in pressure head varies inversely as $\frac{D}{d}$ but for $\frac{D}{d}$ equal to 2.0 or greater the error may be reduced to a negligible quantity. There are other factors which make it advisable to use a diameter of orifice not greater than one-half that of the pipe when conditions will allow. If the magnitude of

TABLE 6

APPROXIMATE VALUES OF DROP IN PRESSURE AND LOST HEAD

h_b=drop in pressure head in feet; h_a=lost head in feet

V	$\frac{D}{d}=1.2$		$\frac{D}{d}=1.5$		$\frac{D}{d}=2.0$		$\frac{D}{d}=2.5$		$\frac{D}{d}=3.0$		$\frac{D}{d}=4.0$	
	h_b	h_a	h_b	h_a	h_b	h_a	h_b	h_a	h_b	h_a	h_b	h_a
0.2											0.43	0.39
0.5					0.16	0.12	0.41	0.33	0.86	0.73	2.7	2.5
1			0.17	0.09	0.64	0.46	1.6	1.3	3.4	2.9	10.8	9.8
2			0.66	0.36	2.6	1.9	6.5	5.3	14	12	43	39
3	0.37	0.12	1.5	0.8	5.8	4.2	15	12	31	26		
4	0.65	0.21	2.6	1.4	10	7	26	21	55	47		
5	1.0	0.3	4.1	2.3	16	12	41	33				
6	1.5	0.5	5.9	3.2	23	17	59	48				
10	4.1	1.3	17	9	64	46						

the lost head is not an important factor, it is best that the drop in pressure head be greater than 1.0 ft. so that the relative error of measurement will be small, but the indications are that gage heights as small as 0.2 ft. may be measured with good results, provided the observer is experienced. Reliable observations may be taken with an orifice having a diameter two-thirds that of the pipe, but the probable error of reading the gage is likely to be several times what it is for the smaller orifices, even though the gage be carefully throttled. An orifice having a diameter much greater than two-thirds that of the pipe should not be used except for approximate discharge measurements.

As an illustrative example let it be required to measure the rate of discharge through an 8-inch pipe the velocity in which varies during the day through a large range, say from about 1 to 6 ft. per sec. Should the pipe-orifice method be used and, if so, what size of orifice should be installed? From Table 6 it will be noted that for $\frac{D}{d}=2$ the drop in pressure head is 0.64 ft. for a velocity of 1 ft. per sec. in the pipe, with a lost head of 0.46 ft. Likewise the drop in pressure head is 23 ft. for a velocity of 6 ft. per sec. with a lost head of 17 ft. Assuming that a lost head of 10 ft. is the greatest that should be allowed and that both a water gage and a mercury gage would be used for a range in drop of pressure head of 23 ft., it is seen that an orifice larger than 4 in. should be selected. One with $\frac{D}{d}=1.5$ gives a range in drop in pressure head from 0.17 to 5.9 ft. corresponding to a range in lost head of 0.09 to 3.2 ft. An orifice with $\frac{D}{d}=1.5$ would give somewhat less nearly accurate results than one with $\frac{D}{d}=2$ (as already discussed) but it would answer the purpose well, although perhaps it would be advisable to use a slightly larger orifice, one with $\frac{D}{d}=1.6$ or $\frac{D}{d}=1.75$.

18. *Computations for Discharge.*—For most precise computations use equation (2), p. 12, with coefficient of discharge taken from Table 4, p. 30. For sizes of pipe other than those for which coefficients are shown use the diameter factor as explained in section 14, p. 34. With $\frac{D}{d}$ from 1.5 to 6 and with care in making the observations, the error introduced may be kept lower than 2 per cent. With wider ranges of $\frac{D}{d}$ or with very high or low values in drop in pressure head this percentage will be increased, because of difficulties in reading the gage properly.

For less precise computations use equation (4) or equation (5), p. 33, with the proper velocity modulus taken from Table 5, p. 33. With pipe from 4 to 12 inches in diameter and with $\frac{D}{d}$ from 1.5 to 6, the maximum error introduced will be about 3 per cent. With wider ranges of $\frac{D}{d}$ or with very high or low values of drop in pressure head this percentage will be correspondingly increased.

The use of equation (6), p. 35, eliminates variable coefficients. For sizes of pipe from 4 to 12 inches, with $\frac{D}{d}$ from 2 to 4 the maximum error will not be greater than 3 per cent. With $\frac{D}{d}$ less than 2 or greater than 4 the error introduced may be 5 per cent or more.

Diagrams similar to Figs. 6, 7, and 8 may be constructed, from which the discharge for the more common sizes of pipe, with the proper range of $\frac{D}{d}$, may be obtained directly. The calculation, however, from equations (4) and (5) and Tables 6 and 7 is very simple.

To determine the lost head, having given $\frac{D}{d}$ and the drop in pressure head, use equation (8), p. 36.

19. *Errors and Precautions.*—In the use of a flow measuring device it is important to know the source of the errors likely to be met and to know the precautions and limitations which if observed will help to reduce the errors.

It is important that the edges of the orifices be sharp and square. The orifice plates in the present tests were 3/16 inch thick, but there is no reason for believing that the coëfficients derived from the experimental data would not apply equally well to orifices in plates of lesser thickness.

The orifice should be placed in a region where there is approximately uniform flow. This will necessitate the pipe upstream from the orifice being straight and free from abrupt changes in cross-sectional area for 10 diameters or more.

On account of the possibility of small openings becoming clogged it seems inadvisable to use gage connections having a diameter less than ¼ inch, particularly for permanent installation.

It is important that all burr be removed and also that the nipple does not protrude beyond the inner surface of the pipe; for small projections, by altering flow conditions, are likely to produce a systematic error of considerable magnitude in the gage readings.

For $\frac{D}{d}$ equal to 2.0 or greater a single nipple at each section will be sufficient. Although in the tests from which the coefficients of discharge were derived the sections at which nipples were placed were 0.8 D upstream from the orifice and 0.4 D downstream from the orifice, these distances may be altered somewhat without appreciably changing pressure conditions. (See Fig. 3, p. 20).

For $\frac{D}{d}$ less than 2.0 two opposite nipples at each section produce more reliable gage readings than do the single nipples, the pair at each section being joined together as illustrated in Fig. 4, p. 23. The indications are that the distance from the orifice to the section downstream from the orifice may not be appreciably altered from 0.4 D and that the distance from the orifice to the section upstream from the orifice should not be less than 0.8 D. On account of pipe friction this latter distance should not greatly exceed 0.8 D, particularly for high velocities.

A differential gage of the U-tube type, similar to those shown in Fig. 5, p. 24, answers all the requirements, is simple to construct, and may be used with mercury or water. The gage board should be graduated to 0.01 ft. and the graduations should extend under the gage tubes. For flushing the gage and the connecting tubes, and for regulating the height of the air columns when water is the differential, there should be a pet cock at the end of each gage tube. Stop cocks for throttling the gage should be placed near the ends of the gage tubes as shown in Fig. 5.

A gage that may be quickly constructed in an emergency and the essential parts of which are easily obtainable and readily portable is shown in Fig. 13. The glass tubes need not be more than a foot

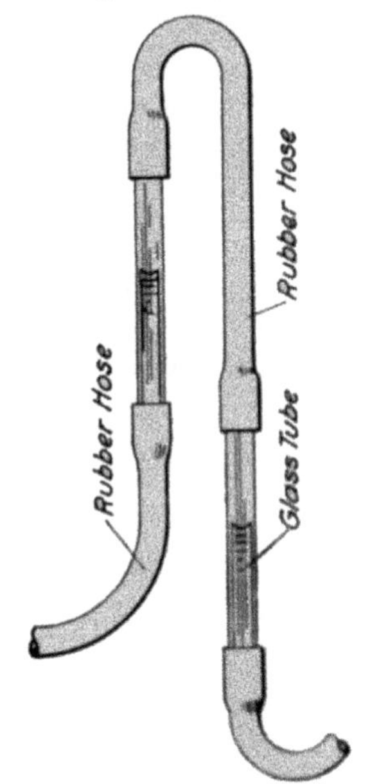

FIG. 13. EMERGENCY DIFFERENTIAL GAGE

long. The piece of rubber tubing connecting the glass tubes makes it possible to adjust the relative height of the tubes to suit the drop in pressure head. The gage may be fixed to an ordinary leveling rod or may be fixed to a plain board and the gage height measured with the ordinary pocket rule. This form of gage is not suitable for high pressures on account of the difficulty of making tight joints.

If the pressure openings are in the upper side of the pipe, it is important that provision be made for removing air, which is likely to become pocketed on either side of the orifice and in the gage connections, without forcing it through the hose or pipe connecting the pressure openings with the gage. If there are no pressure openings in the upper side of the pipe, pet cocks should be inserted. When opposite openings are inserted at each section provision should be made for ascertaining whether or not the tubes are free from obstruction and full of water. Fig. 4, p. 23, illustrates an arrangement of valves and pet cocks which makes this possible.

Preliminary to taking observations, with water flowing in the pipe, all cocks should be opened and the pipe and gage connections should be flushed. It is important that there should be no air in the pipe in the vicinity of the orifice, nor in the gage connections, for the presence of air is likely to change radically the gage reading.

If the foregoing precautions are observed under favorable conditions, the rate of discharge as measured by a pipe orifice should be accurate within 2 per cent. This means compares favorably with many other methods of measuring water. For unfavorable conditions the accuracy of the method will be as great as with most other methods under similarly unfavorable conditions.

IV. Effect of Deviation from Standard Conditions

20. *General Remarks.*—With the possibilities of utilizing the relation between drop in pressure and discharge for determining the rate of discharge through a pipe line in mind, it is important that something should be known concerning the effect of slight deviations from the standard conditions under which the tests were made, either in the arrangement or design of the apparatus or in the methods of handling the apparatus. To this end series of tests were made: (1) with gage connections fully opened and then with gage throttled, (2) with a single gage connection at each section, first with the nipple inserted in the upper side of the pipe and then in the under side, (3) with ¼-inch eccentricity between center of orifice and center of pipe, first with a single nipple at each section, and then with the standard connection of two opposite nipples at each section, (4) with 45 degree bevel-edged orifices, the bevel facing upstream.

21. *Throttling.*—The purpose of throttling is to reduce the effect of momentary pressure fluctuations, thereby causing, for a constant discharge, the fluid in the columns of the differential gage to remain in a nearly stationary position. The experiments indicate that for values of $\frac{D}{d}$ of 2 or greater, when there are no systematic surges along the pipe line (such as might be caused by pump action), there is no appreciable reduction in the magnitude of the accidental errors of observing brought about by throttling; but for $\frac{D}{d}$ less than 2, when the region of greatly disturbed flow (to which reference has previously been made) is at or near the section of greatest contraction, the accidental errors may be reduced to 20 or 25 per cent of what they are with gage connections fully open. The results of the tests also make it clear that systematic errors are likely to enter into the gage readings unless the throttling is done very carefully. If the gage is throttled too much and a change in the rate of flow takes place, it is likely to be several minutes before the columns become fully adjusted to the change in pressure. If the throttling is done

quickly, a gage reading taken immediately after is not likely to represent the mean difference in pressure head.

Briefly summed up, the experience of the observers warrants the following suggestions:

(1) Throttle only when necessary, which in general will be for values of $\frac{D}{d}$ less than 2, unless there are systematic surges along the pipe line.

(2) Throttling too much is often worse than not throttling at all, particularly if the rate of discharge is variable. Accurate observations may be taken when there is a considerable fluctuation in the heights of the gage columns.

(3) When throttling, best results are to be obtained if the throttle cocks are opened prior to each observation and simultaneously and slowly closed the desired amount. It is important that there should be no leaks in the gage connections. A leakage of a few drops per minute from the throttle cock is likely to produce a large error in gage reading.

(4) When using the water gage, observations may be most nearly accurately taken when the gage is so throttled that the two water columns fluctuate the same amount and in unison. In general this arrangement will require that the gage column connection for the section of greatest contraction be throttled more than that for the section of beginning of convergence.

22. *Position and Number of Gage Connections.*—The tests to determine the effect of the position and number of gage connections warrant the following statements:

(1) If the orifice is concentric with the pipe, a change in the circumferential position of the pipe nipples inserted at the section of beginning of convergence and the section of greatest contraction will cause no variation in gage readings. If only one nipple is to be inserted at each section, there is an advantage in placing it in the under side of the pipe if air is present in the discharge, and in the upper side of the pipe if sediment is being carried.

(2) For $\frac{D}{d}$ equal to 2 or more there is no material advantage

in having more than one pressure opening at each section. For $\frac{D}{d}$ less than 2 the momentary fluctuations in pressure, as registered by the gage, may be appreciably reduced by having two opposite openings at each section, these being connected as shown in Fig. 4, p. 23. It seems probable that additional openings would still further reduce the gage fluctuations.

23. *Eccentricity.*—The tests to determine the effect of eccentricity were made on only the 4-inch pipe. The eccentricity was ¼ inch and the displacement of the orifice was vertical. The apparatus was so arranged that the gage could be made to register the difference in pressure obtained for the two sections by connecting to the differential gage (a) the openings in the top of the pipe, (b) the openings in the bottom of the pipe, and (c) all four openings.

The results indicate that regardless of the size of orifice an eccentricity of as much as $D/16$ will produce no effect on gage readings, provided the gage is connected to two opposite pressure openings at each section. But for $\frac{D}{d}$ less than 2 the gage readings will be changed considerably when the gage is connected to a single pressure opening at each section. For $\frac{D}{d}=1.5$ this change is about 4 per cent, and for $\frac{D}{d}=1.2$, about 18 per cent. It seems probable that a corresponding eccentricity with other sizes of pipe would produce a similar effect. Considering the difficulties of ascertaining whether or not the orifice is properly centered under the ordinary conditions of practice, the necessity for gage connections on opposite sides of the pipe at each section, when $\frac{D}{d}$ is less than 2, will be readily appreciated.

The effect of small longitudinal displacements of the gage connections is well illustrated by the curves of Fig. 3, p. 20. It is evident that for $\frac{D}{d}$ less than 2, in order that no measurable change in the gage readings be produced, no considerable change in the longitudinal position of the prssure openings for the section of beginning of convergence may be made, nor may the pressure openings for the section of greatest contraction be placed nearer the orifice than four-tenths the diameter of the pipe.

24. *Bevel-edged Orifices.*—The method of conducting tests with bevel-edged orifices was identical with that used in determining the relation between drop in pressure and discharge for thin square-edged orifices. Tests were made on 4-inch and 6-inch pipe with five sizes of orifice for each size of pipe. For each of the two series of tests the ratios of diameter of pipe to diameter of orifice were 1.22, 1.52, 2.03, 3.04, and 6.08.

The orifices were cut in 3/16-inch plates with the upstream edge on a 45-degree bevel. The thickness of metal at the small diameter of each orifice was 1/32 inch, and the large diameter was 5/16 inch greater than the small diameter.

In instituting the tests the thought was that the effect of the beveled edge might go to produce a coefficient of discharge less affected by small changes in $\frac{D}{d}$ as the diameter of the orifice approaches that of the pipe, and thus make it practicable, with high velocities, to use a larger sized orifice than the experiments with thin-plate orifices indicate may be used with precision.

TABLE 7

COEFFICIENTS OF DISCHARGE FOR BEVEL-EDGED ORIFICES—C'

$\frac{D}{d}$	Pipe Diam.	Coef. of Discharge $h_b=0.01$	$h_b=0.1$	$h_b=1.0$	$h_b=10$	$h_b=50$	Pipe Diam.	Coef. of Discharge $h_b=0.01$	$h_b=0.1$	$h_b=1.0$	$h_b=10$	$h_b=50$
1.22	4-in.	0.870	0.852	0.832	0.811	0.800	6-in.	0.830	0.813	0.795	0.775	0.764
1.52		0.802	0.784	0.765	0.747	0.738		0.762	0.744	0.727	0.709	0.701
2.03		0.773	0.759	0.745	0.731	0.724		0.735	0.722	0.709	0.696	0.690
3.04		0.753	0.741	0.730	0.718	0.713		0.735	0.724	0.713	0.702	0.697
6.08		0.756	0.750	0.745	0.739	0.737		0.746	0.740	0.735	0.729	0.727

The coefficients of discharge, deduced from discharge curves similar to those of Figs. 6 and 7, pp. 28 and 29, are shown in Table 7. A comparison of these coefficients with corresponding values in Table 4, p. 30, will show that the effect of beveling has been materially to increase the coefficients of discharge and also to produce a coefficient of discharge that, as the drop in pressure head increases and as the size of pipe increases, decreases much more rapidly. In Fig. 14 the coeffi-

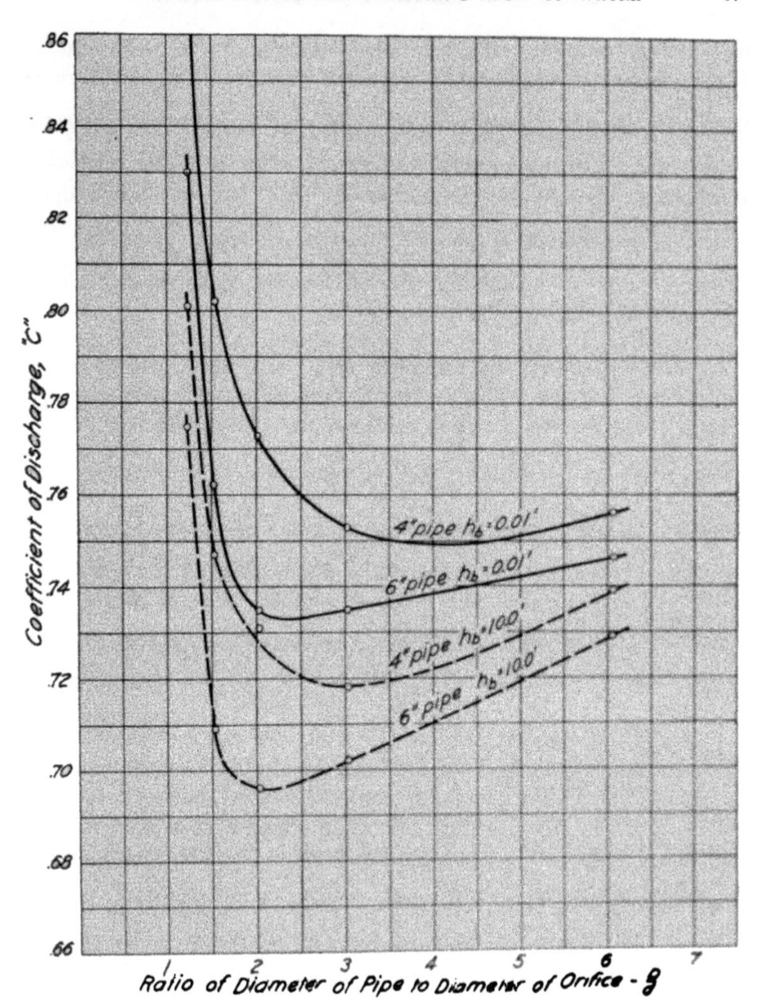

FIG. 14. COEFFICIENTS OF DISCHARGE FOR BEVEL-EDGED ORIFICES

cients of Table 7 for h_b=0.01 ft. and h_b 10 ft. are plotted against values of $\frac{D}{d}$. A comparison of the curves drawn through these points with corresponding curves for square-edged orifices, Fig. 9, p. 32, will still further emphasize the extreme variability of the coefficients of discharge under consideration. As an offset to these undesirable characteristics, for a given change in the value of $\frac{D}{d}$ when that ratio is less than 2, there is considerably less variation in the coefficient of discharge for the bevel-edged orifice than for the square-edged orifice. This statement is especially true for values of $\frac{D}{d}$ less than 1.5, where the rate of change for the bevel-edged orifice is about two-thirds that for the square-edged orifice.

It is worth noting that the coefficient of discharge for each orifice approaches that for the corresponding square-edged orifice as the drop in pressure head increases, and that the rate of this increase becomes greater as $\frac{D}{d}$ decreases and becomes greater as the diameter of the pipe increases. For example, for the 4-inch pipe and $\frac{D}{d}=2.03$ the variation between the two coefficients of discharge is 17 per cent when $h_b=0.1$ ft. and 15 per cent when $h_b=50$ ft.; for the 6-inch and $\frac{D}{d}=2.03$ the variation is 14 per cent when $h_b=0.1$ ft. and 12 per cent when $h_b=50$ ft.

A study of the observations indicates that the accidental errors of reading the gage are about the same for the bevel-edged orifice as for the square-edged orifice. The systematic errors, however, or those which will not be eliminated by increasing the number of observations, are those to which most attention must be given; and among the sources of systematic error when $\frac{D}{d}$ is small, that source most likely to be productive of the greatest error in the computed discharge under the conditions of ordinary practice seems to be in the determination of the ratio $\frac{D}{d}$. Since for small values of $\frac{D}{d}$ the coefficient of discharge for bevel-edged orifices varies less for a given change in $\frac{D}{d}$ than does the coefficient for square-edged orifices, it seems reasonable to believe that for values of $\frac{D}{d}$ less than 1.5 there may be a slight advantage in using the bevel-edged orifice, although the present experiments are not sufficiently comprehensive to confirm fully this belief.

The curves in Figs. 9 and 14 are of value as an indication of the importance of having the edges of the square-edged orifice sharp and square if the coefficients of discharge for square-edged orifices are expected to hold good. In permanent installations this will make it important that the orifice plate be of some material which does not rust or corrode easily.

V. Summary

25. *General Applicability of Pipe Orifice Method.*—The foregoing discussion has shown that the thin-plate orifice inserted in a pipe may be used with confidence for measuring the discharge of water through pipes. Like nearly all methods of measuring water it is subject to some limitations although it helps to fill a growing need which has been partly filled by the pitometer and by the injection of chemicals. The pipe orifice is in effect a portable Venturi meter, the disadvantage of the pipe orifice being the relatively large lost head caused by the obstruction of the orifice plate; however since the pipe orifice method is probably best adapted to temporary use the lost head may in general be unimportant. In a long pipe line also the lost head caused by the orifice would be relatively small. Cases in which the pipe orifice should be of particular value have already been suggested in the introduction.

Although all the deductions and conclusions given in this summary apply to the measurement of water, attention should be called to the fact that the pipe orifice is adapted to measuring the discharge of air, gas, and steam through pipes.

26. *Conclusions.*—The following points are important as a guide to the proper use of the pipe orifice method of measuring the discharge of water through a pipe:

(1) The two sections of the pipe between which change in pressure head may be most reliably determined are the section at which normal flow is discontinued and the stream begins to converge as it approaches the orifice and the section of greatest contraction of the jet after it leaves the orifice. Regardless of the size of pipe, for all sizes of orifice which it is feasible to use, the distance from the plane of the orifice to the section of beginning of convergence may be taken as eight-tenths the pipe diameter, and the distance to the section of greatest contraction as four-tenths the pipe diameter (section 11, p. 21).

(2) The drop in pressure head between these two sections is greater than that to be found for any other two sections near the orifice.

(3) Having given the measured difference between the pressure head at the section of beginning of convergence and the pressure head at the section of greatest contraction the discharge may be determined through the use of equation (2), p. 12, or through the use of equation (6), p. 35.

(4) The coefficient of discharge to be used in equation (2) is a variable quantity (Table 4, p. 30). It decreases as the size of pipe increases; it decreases slightly as the drop in pressure head increases; it has a minimum value for orifices having a diameter of one-third that of the pipe and increases as the diameter of the orifice becomes greater or becomes less than one-third the diameter of the pipe.

(5) The lost head caused by any given orifice in the pipe in terms of the velocity in the pipe may be determined by equation (3), p. 13.

(6) The lost head is always less than the drop in pressure head but approaches it in value as the ratio of the diameter of the pipe to the diameter of the orifice increases (section 6, p. 13).

(7) Due to the fluctuations of the liquid in the gage tubes the systematic error of reading the gage increases as the ratio of the diameter of the pipe to that of the orifice decreases, but when that ratio $\frac{D}{d}$ is 2 or greater the error may under normal conditions of flow be reduced to a negligible quantity by a proper manipulation of apparatus. As $\frac{D}{d}$ becomes less than 2 the accidental error of reading the gage increases very rapidly (section 19, p. 40) and also small errors in the measurement of the diameter of the pipe or the diameter of the orifice are likely to be the constant sources of an error of increasing magnitude in the computed discharge (section 14, p. 34). The indications are that, for favorable conditions of flow and with care in installing the apparatus and in observing, discharge may be determined generally within 2 per cent when the diameter of the orifice is not in excess of two-thirds that of the pipe, but this size of orifice seems to be about the maximum that can be used except for approximate determinations of discharge. When the magnitude of the lost head is not the controlling factor in the choice of size of orifice, best results

are likely to be obtained if the diameter of the orifice is not greater than one-half that of the pipe.

(8) For orifices having a diameter greater than one-half that of the pipe the use of two opposite pressure openings at each section is important because of the probability of the orifice being somewhat eccentric with the pipe, unless greater care is taken in placing the orifice than will usually be found practicable (sections 22, p. 44, and 23, p. 45). Systematic errors of observing may be greatly reduced by proper throttling (section 21, p. 43).

(9) The coefficient of discharge for bevel-edged orifices is a much more variable quantity and is materially greater than the coefficient of discharge for thin square-edged orifices. The use of the bevel-edged orifice seems not to be practicable, except for approximate measurements when the orifice diameter is greater than two-thirds of the pipe diameter (section 24, p. 46).

LIST OF
PUBLICATIONS OF THE ENGINEERING EXPERIMENT STATION

Bulletin No. 1. Tests of Reinforced Concrete Beams, by Arthur N. Talbot. 1904. *None available.*

Circular No. 1. High-Speed Tool Steels, by L. P. Breckenridge. 1905. *None available.*

Bulletin No. 2. Tests of High-Speed Tool Steels on Cast Iron, by L. P. Breckenridge and Henry B. Dirks. 1905. *None available.*

Circular No. 2. Drainage of Earth Roads, by Ira O. Baker. 1906. *None available.*

Circular No. 3. Fuel Tests with Illinois Coal (Compiled from tests made by the Technological Branch of the U. S. G. S., at the St. Louis, Mo., Fuel Testing Plant, 1904–1907), by L. P. Breckenridge and Paul Diserens. 1908. *Thirty cents.*

Bulletin No. 3. The Engineering Experiment Station of the University of Illinois, by L. P. Breckenridge. 1906. *None available.*

Bulletin No. 4. Tests of Reinforced Concrete Beams, Series of 1905, by Arthur N. Talbot 1906. *Forty-five cents.*

Bulletin No. 5. Resistance of Tubes to Collapse, by Albert P. Carman and M. L. Carr. 1906 *None available.*

Bulletin No. 6. Holding Power of Railroad Spikes, by Roy I. Webber. 1906. *None available.*

Bulletin No. 7. Fuel Tests with Illinois Coals, by L. P. Breckenridge, S. W. Parr, and Henry B Dirks. 1906. *None available.*

Bulletin No. 8. Tests of Concrete: I, Shear; II, Bond, by Arthur N. Talbot. 1906. *None available.*

Bulletin No. 9. An Extension of the Dewey Decimal System of Classification Applied to the Engineering Industries, by L. P. Breckenridge and G. A. Goodenough. 1906 Revised Edition 1912. *Fifty cents.*

Bulletin No. 10. Tests of Concrete and Reinforced Concret Columns, Series of 1906, by Arthur N. Talbot. 1907. *None available.*

Bulletin No. 11. The Effect of Scale on the Transmission of Heat through Locomotive Boiler Tubes, by Edward C. Schmidt and John M. Snodgrass. 1907. *None available.*

Bulletin No. 12. Tests of Reinforced Concrete T-Beams, Series of 1906, by Arthur N Talbot 1907. *None available.*

Bulletin No. 13. An Extension of the Dewey Decimal System of Classification Applied to Architecture and Building, by N. Clifford Ricker. 1907. *None available.*

Bulletin No. 14. Tests of Reinforced Concrete Beams, Series of 1906, by Arthur N Talbot 1907. *None available.*

Bulletin No. 15. How to Burn Illinois Coal without Smoke, by L. P. Breckenridge. 1908 *None available.*

Bulletin No. 16. A Study of Roof Trusses, by N. Clifford Ricker. 1908. *None available.*

Bulletin No. 17. The Weathering of Coal, by S. W. Parr, N. D. Hamilton, and W. F. Wheeler 1908. *None available.*

Bulletin No. 18. The Strength of Chain Links, by G. A. Goodenough and L. E. Moore 1908. *Forty cents.*

Bulletin No. 19. Comparative Tests of Carbon, Metallized Carbon and Tantalum Filament Lamps, by T. H. Amrine. 1908. *None available.*

Bulletin No. 20. Tests of Concrete and Reinforced Concrete Columns, Series of 1907, by Arthur N. Talbot. 1908. *None available.*

Bulletin No. 21. Tests of a Liquid Air Plant, by C. S. Hudson and C. M. Garland. 1908 *Fifteen cents.*

Bulletin No. 22. Tests of Cast-Iron and Reinforced Concrete Culvert Pipe, by Arthur N. Talbot 1908. *None available.*

Bulletin No. 23. Voids, Settlement and Weight of Crushed Stone, by Ira O. Baker. 1908 *Fifteen cents.*

**Bulletin No. 24.* The Modification of Illinois Coal by Low Temperature Distillation, by S. W. Parr and C. K. Francis. 1908. *Thirty cents.*

Bulletin No. 25. Lighting Country Homes by Private Electric Plants by T. H. Amrine 1908. *Twenty cents.*

---

*A limited number of copies of bulletins starred are available for free distribution.

Bulletin No. 26. High Steam-Pressure in Locomotive Service. A Review of a Report to the Carnegie Institution of Washington, by W. F. M. Goss. 1908. *Twenty-five cents.*

Bulletin No. 27. Tests of Brick Columns and Terra Cotta Block Columns, by Arthur N. Talbot and Duff A. Abrams. 1909. *Twenty-five cents.*

Bulletin No. 28. A Test of Three Large Reinforced Concrete Beams, by Arthur N. Talbot. 1909. *Fifteen cents.*

Bulletin No. 29. Tests of Reinforced Concrete Beams: Resistance to Web Stresses, Series of 1907 and 1908, by Arthur N. Talbot. 1909. *Forty-five cents.*

**Bulletin No. 30.* On the Rate of Formation of Carbon Monoxide in Gas Producers, by J. K. Clement, L. H. Adams, and C. N. Haskins. 1909. *Twenty-five cents.*

**Bulletin No. 31.* Tests with House-Heating Boilers, by J. M. Snodgrass. 1909. *Fifty-five cents.*

Bulletin No. 32. The Occluded Gases in Coal, by S. W. Parr and Perry Barker. 1909. *Fifteen cents.*

Bulletin No. 33. Tests of Tungsten Lamps, by T. H. Amrine and A. Guell. 1909. *Twenty cents.*

**Bulletin No. 34.* Tests of Two Types of Tile-Roof Furnaces under a Water-Tube Boiler, by J. M. Snodgrass. 1909. *Fifteen cents.*

Bulletin No. 35. A Study of Base and Bearing Plates for Columns and Beams, by N. Clifford Ricker. 1909. *Twenty cents.*

Bulletin No. 36. The Thermal Conductivity of Fire-Clay at High Temperatures, by J. K. Clement and W. L. Egy. 1909. *Twenty cents.*

Bulletin No. 37. Unit Coal and the Composition of Coal Ash, by S. W. Parr and W. F. Wheeler. 1909. *None available.*

Bulletin No. 38. The Weathering of Coal, by S. W. Parr and W. F. Wheeler. 1909. *Twenty-five cents.*

**Bulletin No. 39.* Tests of Washed Grades of Illinois Coal, by C. S. McGovney. 1909. *Seventy-five cents.*

Bulletin No. 40. A Study in Heat Transmission, by J. K. Clement and C. M. Garland. 1910. *Ten cents.*

Bulletin No. 41. Tests of Timber Beams, by Arthur N. Talbot. 1910. *Thirty-five cents.*

**Bulletin No. 42.* The Effect of Keyways on the Strength of Shafts, by Herbert F. Moore. 1910. *Ten cents.*

Bulletin No. 43. Freight Train Resistance, by Edward C. Schmidt. 1910. *Seventy-five cents.*

Bulletin No. 44. An Investigation of Built-up Columns under Load, by Arthur N. Talbot and Herbert F. Moore. 1911. *Thirty-five cents.*

**Bulletin No. 45.* The Strength of Oxyacetylene Welds in Steel, by Herbert L. Whittemore. 1911. *Thirty-five cents.*

**Bulletin No. 46.* The Spontaneous Combustion of Coal, by S. W. Parr and F. W. Kressman. 1911. *Forty-five cents.*

**Bulletin No. 47.* Magnetic Properties of Heusler Alloys, by Edward B. Stephenson. 1911. *Twenty-five cents.*

**Bulletin No. 48.* Resistance to Flow through Locomotive Water Columns, by Arthur N. Talbot and Melvin L. Enger. 1911. *Forty cents.*

**Bulletin No. 49.* Tests of Nickel-Steel Riveted Joints, by Arthur N. Talbot and Herbert F. Moore. 1911. *Thirty cents.*

**Bulletin No. 50.* Tests of a Suction Gas Producer, by C. M. Garland and A. P. Kratz. 1912. *Fifty cents.*

Bulletin No. 51. Street Lighting, by J. M. Bryant and H. G. Hake. 1912. *Thirty-five cents.*

**Bulletin No. 52.* An Investigation of the Strength of Rolled Zinc, by Herbert F. Moore. 1912. *Fifteen cents.*

**Bulletin No. 53.* Inductance of Coils, by Morgan Brooks and H. M. Turner. 1912. *Forty cents.*

**Bulletin No. 54.* Mechanical Stresses in Transmission Lines, by A. Guell. 1912. *Twenty cents.*

**Bulletin No. 55.* Starting Currents of Transformers, with Special Reference to Transformers with Silicon Steel Cores, by Trygve D. Yensen. 1912. *Twenty cents.*

**Bulletin No. 56.* Tests of Columns: An Investigation of the Value of Concrete as Reinforcement for Structural Steel Columns, by Arthur N. Talbot and Arthur R. Lord. 1912. *Twenty-five cents.*

**Bulletin No. 57.* Superheated Steam in Locomotive Service. A Review of Publication No. 127 of the Carnegie Institution of Washington, by W. F. M. Goss. 1912. *Forty cents.*

---

*A limited number of copies of bulletins starred is available for free distribution.

**Bulletin No. 58.* A New Analysis of the Cylinder Performance of Reciprocating Engines, by J. Paul Clayton. 1912. *Sixty cents.*

**Bulletin No. 59.* The Effect of Cold Weather upon Train Resistance and Tonnage Rating, by Edward C. Schmidt and F. W. Marquis. 1912. *Twenty cents.*

Bulletin No. 60. The Coking of Coal at Low Temperatures, with a Preliminary Study of the By-Products, by S. W. Parr and H. L. Olin. 1912. *Twenty-five cents.*

**Bulletin No. 61.* Characteristics and Limitation of the Series Transformer, by A. R. Anderson and H. R. Woodrow. 1913. *Twenty-five cents.*

Bulletin No. 62. The Electron Theory of Magnetism, by Elmer H. Williams. 1913. *Thirty-five cents.*

Bulletin No. 63. Entropy-Temperature and Transmission Diagrams for Air, by C. R. Richards. 1913. *Twenty-five cents.*

**Bulletin No. 64.* Tests of Reinforced Concrete Buildings under Load, by Arthur N. Talbot and Willis A. Slater. 1913. *Fifty cents.*

**Bulletin No. 65.* The Steam Consumption of Locomotive Engines from the Indicator Diagrams, by J. Paul Clayton. 1913. *Forty cents.*

Bulletin No. 66. The Properties of Saturated and Superheated Ammonia Vapor, by G. A. Goodenough and William Earl Mosher. 1913. *Fifty cents.*

Bulletin No. 67. Reinforced Concrete Wall Footings and Column Footings, by Arthur N. Talbot. 1913. *Fifty cents.*

Bulletin No. 68. The Strength of I-Beams in Flexure, by Herbert F. Moore. 1913. *Twenty cents.*

Bulletin No. 69. Coal Washing in Illinois, by F. C. Lincoln. 1913. *Fifty cents.*

Bulletin No. 70. The Mortar-Making Qualities of Illinois Sands, by C. C. Wiley. 1913. *Twenty cents.*

Bulletin No. 71. Tests of Bond between Concrete and Steel, by Duff A. Abrams. 1914. *One dollar.*

**Bulletin No. 72.* Magnetic and Other Properties of Electrolytic Iron Melted in Vacuo, by Trygve D. Yensen. 1914. *Forty cents.*

Bulletin No. 73. Acoustics of Auditoriums, by F. R. Watson. 1914. *Twenty cents.*

**Bulletin No. 74.* The Tractive Resistance of a 28-Ton Electric Car, by Harold H. Dunn. 1914. *Twenty-five cents.*

Bulletin No. 75. Thermal Properties of Steam, by G. A. Goodenough. 1914. *Thirty-five cents.*

Bulletin No. 76. The Analysis of Coal with Phenol as a Solvent, by S. W. Parr and H. F. Hadley. 1914. *Twenty-five cents.*

**Bulletin No. 77.* The Effect of Boron upon the Magnetic and Other Properties of Electrolytic Iron Melted in Vacuo, by Trygve D. Yensen. 1915. *Ten cents.*

**Bulletin No. 78.* A Study of Boiler Losses, by A. P. Kratz. 1915. *Thirty-five cents.*

**Bulletin No. 79.* The Coking of Coal at Low Temperatures, with Special Reference to the Properties and Composition of the Products, by S. W. Parr and H. L. Olin. 1915. *Twenty-five cents.*

Bulletin No. 80. Wind Stresses in the Steel Frames of Office Buildings, by W. M. Wilson and G. A. Maney. 1915. *Fifty cents.*

Bulletin No. 81. Influence of Temperature on the Strength of Concrete, by A. B. McDaniel. 1915. *Fifteen cents.*

Bulletin No. 82. Laboratory Tests of a Consolidation Locomotive, by E. C. Schmidt, J. M Snodgrass, and R. B. Keller. 1915. *Sixty-five cents.*

**Bulletin No. 83.* Magnetic and Other Properties of Iron Silicon Alloys, Melted in Vacuo, by Trygve D. Yensen. 1915. *Thirty-five cents.*

Bulletin No. 84. Tests of Reinforced Concrete Flat Slab Structures, by Arthur N. Talbot and W. A. Slater. 1916. *Sixty-five cents.*

**Bulletin No. 85.* The Strength and Stiffness of Steel under Biaxial Loading, by A. T. Becker 1916. *Thirty-five cents.*

**Bulletin No. 86.* The Strength of I-Beams and Girders, by Herbert F. Moore and W. M. Wilson 1916. *Thirty cents.*

**Bulletin No. 87.* Correction of Echoes in the Auditorium, University of Illinois, by P. R. Watson and J. M. White. 1916. *Fifteen cents.*

Bulletin No. 88. Dry Preparation of Bituminous Coal at Illinois Mines, by E. A Holbrook 1916 *Seventy cents.*

*A limited number of copies of bulletins starred is available for free distribution.

**Bulletin No. 89.* Specific Gravity Studies of Illinois Coal, by Merle L. Nebel. 1916. *Thirty cents.*

**Bulletin No. 90.* Some Graphical Solutions of Electric Railway Problems, by A. M. Buck. 1916. *Twenty cents.*

Bulletin No. 91. Subsidence Resulting from Mining, by L. E. Young and H. H. Stoek. 1916. *None available.*

**Bulletin No. 92.* The Tractive Resistance on Curves of a 28-Ton Electric Car, by E. C. Schmidt and H. H. Dunn. 1916. *Twenty-five cents.*

**Bulletin No. 93.* A Preliminary Study of the Alloys of Chromium, Copper, and Nickel, by D. F. McFarland and O. E Harder. 1916. *Thirty cents.*

**Bulletin No. 94.* The Embrittling Action of Sodium Hydroxide on Soft Steel, by S. W. Parr 1917. *Thirty cents.*

**Bulletin No. 95.* Magnetic and Other Properties o f Iron-Aluminum Alloys Melted in Vacuo, by T. D. Yensen and W. A. Gatward. 1917. *Twenty-five cents.*

**Bulletin No. 96.* The Effect of Mouthpieces on the Flow of Water through a Submerged Short Pipe, by Fred B Seely. 1917. *Twenty-five cents.*

Bulletin No. 97. Effects of Storage upon the Properties of Coal, by S. W. Parr. 1917. *Twenty cents.*

**Bulletin No. 98.* Tests of Oxyacetylene Welded Joints in Steel Plates, by Herbert F. Moore 1917. *Ten cents.*

Circular No. 4. The Economical Purchase and Use of Coal for Heating Homes, with Special Reference to Conditions in Illinois. 1917. *Ten cents.*

**Bulletin No. 99.* The Collapse of Short Thin Tubes, by A. P. Carman. 1917. *Twenty cents.*

**Circular No. 5.* The Utilization of Pyrite Occurring in Illinois Bituminous Coal, by E. A. Holbrook. 1917. *Twenty cents.*

**Bulletin No. 100.* Percentage of Extraction of Bituminous Coal with Special Reference to Illinois Conditions, by C. M. Young. 1917.

**Bulletin No. 101.* Comparative Tests of Six Sizes of Illinois Coal on a Mikado Locomotive, by E. C Schmidt, J. M. Snodgrass, and O. S. Beyer, Jr. 1917. *Fifty cents.*

**Bulletin No. 102.* A Study of the Heat Transmission of Building Materials, by A. C. Willard and L. C. Lichty. 1917. *Twenty-five cents.*

**Bulletin No. 103.* An Investigation of Twist Drills, by B. Benedict and W. P. Lukens. 1917. *Sixty cents.*

**Bulletin No. 104.* Tests to Determine the Rigidity of Riveted Joints of Steel Structures, by W M. Wilson and H. F. Moore. 1917. *Twenty-five cents.*

Circular No. 6. The Storage of Bituminous Coal, by H. H. Stoek. 1918. *Forty cents.*

Circular No. 7. Fuel Economy in the Operation of Hand Fired Power Plants. 1918. *Twenty cents.*

**Bulletin No. 105.* Hydraulic Experiments with Valves, Orifices, Hose, Nozzles, and Orifice Buckets, by Arthur N. Talbot, Fred B Seely, Virgil R. Fleming, and Melvin L. Enger. 1918. *Thirty five cents.*

**Bulletin No. 106.* Test of a Flat Slab Floor of the Western Newspaper Union Building, by Arthur N. Talbot and Harrison F. Gonnerman. 1918. *Twenty cents.*

Circular No. 8. The Economical Use of Coal in Railway Locomotives. 1918. *Twenty cents*

**Bulletin No. 107.* Analysis and Tests of Rigidly Connected Reinforced Concrete Frames, by Mikishi Abe. 1918. *Fifty cents.*

**Bulletin No. 108.* Analysis of Statically Indeterminate Structures by the Slope Deflection Method, by W. M. Wilson, F. E. Richart, and Camillo Weiss. 1918. *One dollar.*

**Bulletin No. 109.* The Orifice as a Means of Measuring Flow of Water through a Pipe, by R. E. Davis and H. H. Jordan. 1918. *Twenty-five cents.*

---

*A limited number of copies of bulletins starred are available for free distribution.

THE UNIVERSITY OF ILLINOIS

THE STATE UNIVERSITY

Urbana

EDMUND J. JAMES, Ph. D., LL. D., President

THE UNIVERSITY INCLUDES THE FOLLOWING DEPARTMENTS:

The Graduate School

The College of Liberal Arts and Sciences (Ancient and Modern Languages and Literatures; History, Economics, Political Science, Sociology; Philosophy; Psychology, Education; Mathematics; Astronomy; Geology; Physics; Chemistry, Botany, Zoology, Entomology; Physiology; Art and Design)

The College of Commerce and Business Administration (General Business, Banking, Insurance, Accountancy, Railway Administration, Foreign Commerce; Courses for Commercial Teachers and Commercial and Civic Secretaries)

The College of Engineering (Architecture; Architectural, Ceramic, Civil, Electrical, Mechanical, Mining, Municipal and Sanitary, and Railway Engineering)

The College of Agriculture (Agronomy; Animal Husbandry; Dairy Husbandry; Horticulture and Landscape Gardening; Agricultural Extension; Teachers' Course; Household Science)

The College of Law (three-year and four-year curriculums based on two years and one year of college work respectively)

The College of Education

The Curriculum in Journalism

The Curriculums in Chemistry and Chemical Engineering

The School of Railway Engineering and Administration

The School of Music (four-year curriculum)

The School of Library Science (two-year curriculum for college graduates)

The College of Medicine (in Chicago)

The College of Dentistry (in Chicago)

The School of Pharmacy (in Chicago; Ph. G. and Ph. C. curriculums)

The Summer Session (eight weeks)

Experiment Stations and Scientific Bureaus; U. S. Agricultural Experiment Station; Engineering Experiment Station; State Laboratory of Natural History; State Entomologist's Office; Biological Experiment Station on Illinois River; State Water Survey; State Geological Survey; U. S. Bureau of Mines Experiment Station.

The library collections contain (March 1, 1919) 431,654 volumes and 53,858 pamphlets.

For catalogs and information address

THE REGISTRAR

URBANA, ILLINOIS

Lightning Source UK Ltd.
Milton Keynes UK
UKHW020431210219
337573UK00007B/1618/P